Industrial Plastics
Theory and Application

Delmar Publishers' Online Services
To access Delmar on the World Wide Web, point your browser to:
http://www.delmar.com/delmar.html
To access through Gopher: gopher://gopher.delmar.com
(Delmar Online is part of the "thomson.com", an Internet site with information on more than 30 publishers of the International Thomson Publishing organization.)
For information on our products and services:
email:info@delmar.com
or call 800-347-7707

Industrial Plastics

Theory and Application

THIRD EDITION

Terry L. Richardson, Ph.D.
Northern State University
Aberdeen, South Dakota

Erik Lokensgard
Eastern Michigan University
Ypsilanti, Michigan

Albany • Bonn • Boston • Cincinnati • Detroit • London • Madrid
Melbourne • Mexico City • New York • Pacific Grove • Paris
San Francisco • Singapore • Tokyo • Toronto • Washington

NOTICE TO THE READER

Publisher does not warrant or guarantee any of the products described herein or perform any independent analysis in connection with any of the product information contained herein. Publisher does not assume, and expressly disclaims, any obligation to obtain and include information other than that provided to it by the manufacturer.

The reader is expressly warned to consider and adopt all safety precautions that might be indicated bv the activities herein and to avoid all potential hazards. By following the instructions contained herein, the reader willingly assumes all risks in connections with such instructions.

The publisher makes no representation or warranties of any kind, including but not limited to, the warranties of fitness for particular purpose of merchantability, nor are any such representations implied with respect to the material set fourth herein, and the publisher takes no responsibility with respect to such material. The publisher shall not be liable for any special, consequential, or exemplary damages resulting, in whole or part, from the readers' use of, or reliance upon, this material.

Cover photo courtesy of The Nalge Company
Cover Design by The Color Shop

Delmar Staff:
Publisher: Robert Lynch
Senior Administrative Editor: John Anderson
Senior Project Editor: Christopher Chien
Production Manager: Larry Main
Art and Design Coordinator: Nicole Reamer
Editorial Assistant: John Fisher

Printed in the United States of America

For more information, contact:

Delmar Publishers
3 Columbia Circle, Box 15015
Albany, New York 12212-5015

International Thomson Publishing Europe
Berkshire House 168-173
High Holborn
London, WC1V 7AA
England

Thomas Nelson Australia
102 Dodds Street
South Melbourne, 3205
Victoria, Australia

Nelson Canada
1120 Birchmount Road
Scarborough, Ontario
Canada MIK 5G4

International Thomson Editores
Campos Eliseos 385, Piso 7
Cot Polanco
11560 Mexico D F Mexico

International Thomson Publishing GmbH
Königswinterer Strasse 418
53227 Bonn
Germany

International Thomson Publishing Asia
221 Henderson Road
#05-10 Henderson Building
Singapore 0315

International Thomson Publishing-Japan
Hirakawacho Kyowa Building, 3F
2-2-1 Hirakawacho
Chiyoda-ku, Tokyo 102
Japan

2 3 4 5 6 7 8 9 10 XXX 02 01 00 99 98 97

Library of Congress Cataloging-in-Publication Data

Richardson, Terry L.
Industrial plastics: theory and application / Terry L. Richardson, Erik Lokensgard. -- 3rd ed.
p. cm.
Includes bibliographical references and index.
ISBN 0-8273-6558-6
1. Plastics. I. Lokensgard, Erik. II. Title.
TP1120.R49 1996 96-6775
668.4--dc20 CIP

Contents

Preface

This third edition of *Industrial Plastics: Theory and Applications* combines new material, updated revisions of chapters in the second edition, and some unchanged chapters. The intent of this edition is to expand areas of growing importance and streamline other topics. A major contrast with the second edition is the addition of laboratory activities for most chapters. Some of the activities were reclaimed from the first edition, but most are new presentations.

The philosophy imbedded in the laboratory activities is that practical applications are essential for thorough understanding of many theoretical concepts. The activities contain tried approaches, but also include suggestions for further practical investigations. It is hoped that students and instructors will build on the laboratory activities and customize them for available equipment and materials.

All chapters have vocabulary lists, and review questions. The confidential and proprietary details outlined or other information provided is intended only as a guide. It is not to be taken as a license under which to operate or as a recommendation to infringe on any patents.

Chapter 1, provides an historical introduction to plastics and includes expanded details about natural plastics.

Chapter 2, on the current status of the plastics industry, features U.S. consumption of major materials, recycling, disposal, and significant organizations within the industry. The sections on recycling and disposal contain an abundance of current information.

Chapter 3, on elementary polymer chemistry, has reduced the content on general organic chemistry. This chapter attempts to present basics about plastics in a very practical context.

Chapter 4, on health and safety, takes its organization from required sections in Material Safety Data Sheets. The intent of this organization is that students should be adept at reading and understanding MSDS for plastics materials.

Chapter 5, on elementary statistics, is a new chapter. It provides an introduction to statistical principles without presentation on hypothesis testing. It relies on graphical techniques instead of numerical procedures.

Chapter 6, on properties and tests, has been updated to include ISO tests as well as ASTM procedures.

Chapter 7, on ingredients of plastics, includes updated material on heavy metals in colorants and blowing agents.

Chapter 8, on selection of commercial plastics, focuses on polymerization types, melt index values, and understanding of various grades of plastics.

Chapter 10, on molding processes, contains a major new section on injection-molding safety.

Chapters 9, 11, 12, 14, 15, 16, 17, 18, and 19, now contain laboratory activities. Chapters 20, 21, 22 and 23 are identical with the second edition. Appendices A, B, and C have been updated. Appendix D, on material identification, has been moved into the appendix format and removed from a chapter.

Appendix E on themoplastics, and Appendix F on thermosets, have been made appendices instead of chapters. The rationale for this alteration is that many students were overwhelmed with the encyclopedic nature of the content. Placing this material in an appendix implies that students will use this appendix to seek information on a selected number of materials, rather than try to read and understand the entire alphabetical group of thermosets and thermoplastics.

It is hoped that the reorganization of this book and the added materials will enhance the ease of use and depth of content.

Chapter 1

Historical Introduction to Plastics

Introduction

Life without plastics is rather hard to imagine. Daily activities rely on plastic items such as milk jugs, eyeglasses, telephones, nylons, automobiles, and videotapes. Yet, not much more than one hundred years ago, the plastics taken for granted today did not exist. Long before the development of commercial plastics, some existing materials displayed unique features. They were strong, translucent, lightweight, and moldable, but only a few substances combined these qualities. Today, these materials have the name *natural plastics*. They provide the starting point for a brief history of plastic materials.

This chapter will provide information about advantages of early plastics, and the difficulties that attended their manufacturing. It sets modern materials and processes in an historical context and indicates the powerful influence of pioneers in the plastics industry. The topics included are:

- I. Natural plastics
  - A. Horn
  - B. Shellac
  - C. Gutta percha
- II. Early modified natural materials
  - A. Rubber
  - B. Celluloid
- III. Early synthetic polymers
- IV. Commercial synthetic plastics

Natural Plastics

Medieval England provides a starting point for this section. In medieval times, English surnames (last names) indicated professions. Some of these professions are easily recognized today. The occupational reference for names such as Smith, Baker, Carpenter, Weaver, Taylor, Cartwright, Barber, Farmer, and Hunter is obvious. The occupational origin of other names, such as Fuller, Tucker, Cooper, and Horner, is less familiar.

> Little Jack Horner
> Sat in a corner,
> Eating his Christmas pie;
> He put in his thumb
> And pulled out a plum,
> And said, "What a good boy am I."

This rhyme reveals that Jack was not hungry or poor, and did not have to share his Christmas treat with other family members. He enjoyed a special treat alone. Apparently, Jack's father had a comfortable income. What did Jack's father, or perhaps grandfather, do? He was a horner, a man who made small items from horns, hooves, and occasionally from tortoise shells.

A typical response to horn working is to dismiss it as quaint, irrelevant, or disgusting. The horner's craft was smelly and often unpleasant. Today, horners can only be found in rare historically oriented craft museums. However, horn working is not the least bit irrelevant to the plastics industry. The unique properties of horn inspired a search for substitutes. The quest for synthetic horn led to early plastics and the beginnings of the modern plastics industry.

Horn

Spoons, combs, and lantern windows were common products made by horners in England and Europe during the Middle Ages. Horn spoons were strong and lightweight. They did not rust, corrode, or give an undesirable taste to food. Horn combs were flexible, smooth, glossy, and often decorative. Lantern windows exploited the translucent quality of horn, as seen in Figure 1-1. They also flexed without shattering, and withstood some impact. No other material provided this combination of properties.

Making utilitarian objects from natural polymers did not begin in the Middle Ages. One of the oldest known uses of horn dates from the times of the Pharaohs of Egypt. Approximately 2000 B.C. ancient Egyptian craftsmen formed ornaments and food utensils by softening tortoise shells in hot oils. When the shell was sufficiently pliable, they pressed it into the desired shape. They trimmed the rough shapes, scraped and sanded them, and finally polished them to a high luster with fine powders.

Little Jack Horner's forefather worked in a manner similar to the ancient Egyptians. He softened pieces of cow horn by boiling in water or soaking in alkaline solutions, and then pressed the pieces flat. Some horns were delaminated along growth lines, yielding thin sheets. If thicker pieces were needed, several thinner sheets were welded together. After creating the desired thickness, horn pieces were squeezed into molds to create a useful shape. Sometimes, horners dyed the pieces to make them look like expensive tortoise shell.

Two items are particularly important for this history, because they involved differing techniques: combs, and buttons.

Combs. Some English horners emigrated to the American colonies and established small businesses. By 1760, horn workers were well established in Massachusetts. Leominster, Massachusetts became a center for the comb business and earned the name, comb city.

In comb factories, craftsmen sawed flattened horn pieces to size, cut in teeth with fine saws, smoothed rough edges, colored and polished the combs. The final operation was bending. A contoured wooden form imparted a curve to a softened comb and maintained the shape as the comb cooled.

Figure 1-2 shows a photograph of a comb made from tortoise shell. Notice that several teeth are slightly warped. Even in the most carefully made combs, the thin teeth were easily broken. Notice also that the comb is generally uniform in cross section. Combs were generally not embossed with raised artistic motifs, because shell and horn do not flow easily.

Although comb makers in Massachusetts developed machines to mechanize manufacturing, they could not establish stable production. The fault lay not in the machines, but in the material. Clamping fixtures and cutter movements required uniform, flat workpieces. Horn was not flat or uniform in size and flexability.

Fig. 1-1. This candle lantern displays horn windows. (From the Collections of the Henry Ford Museum and Greenfield Village)

Fig. 1-2. This well preserved tortoise shell comb shows only one broken tooth. (From the Collections of the Henry Ford Museum and Greenfield Village)

The lack of dimensional consistency and the low "flowability" of horn encouraged comb manufacturers to search for substitutes. The inherent waste caused by the shape of horns also promoted interest in alternatives.

Buttons. Horn button makers faced a different set of problems. Flat, utilitarian buttons were molded from horn pieces, cut out in pre-sized blanks, and then pressed in heated molds. Customers, however, also wanted decorative buttons to compliment fine garments. Hand carved buttons of ivory had been available for centuries, but they were expensive and one-of-a-kind. In order to make embossed and raised motifs, the molding material needed to flow easily in the mold. To achieve this flow, the horners developed molding powders of ground horn. Horn buttons were often ground cow hoofs, colored with a water solution. The horn powder was poured into molds and then compressed or rolled out into sheets. The sheets were cut into small blanks with tools similar to small cookie cutters. The blanks were then compressed in a mold to achieve the three dimensional surfaces. Figure 1-3 shows two horn-buttons, one with a prominant relief.

Buttons required rather undemanding physical properties. They were thick enough to be strong, and devoid of fragile teeth. The impetus to seek alternatives came from the actual working of the horn. Removing the tissue mass and cleaning the slimy membrane from the inside of the horn was dirty work, accompanied by strong odors from boiled horns. When shellac became readily available, horners carefully evaluated its qualities.

Shellac

When Marco Polo returned to Europe from his travels in Asia in about 1290, he brought back shellac. He had found shellac in India, where people had been using it for centuries. They had discovered the unique properties of a natural polymer that came from insects instead of cow horn.

Fig. 1-3. These black horn buttons show the three dimensional relief possible with horn molding compounds. (From the collection of Evelyn Gibbons)

The insect that produced a polymer was a small bug called the *lac*, native to India and southeast Asia. A female lac inserts a stinger like probiscus into a twig or small branch of a tree. She lives off the sap drawn from the host plant and exudes a thick liquid, which dries slowly. As the deposit of hardened liquid grows, the insect becomes immobilized. After the male lac fertilizes the female, she increases the juice excretions, and finally is totally covered. Inside this deposit, she then lays hundreds of eggs and eventually dies. When the eggs hatch, the young insects eat their way out of the covering and go off to repeat the cycle.

The hardened excretion has unique properties. When cleaned, dissolved in alcohol, and applied to a surface, it makes a shinny, almost transparent coating. The name *shellac* was descriptive, since it came from the *shell* of the *lac*. In addition to use as a protective coating for furniture and floors, solid shellac was moldable.

Under heat and pressure, shellac will flow into the recesses of intricate and detailed molds. Since pure shellac is brittle and weak, compounds containing various fibers were developed to give the moldings some strength. An early product made from molded shellac was the daguerreotype case, like the one seen in Figure 1-4. Their manufacture in the United States began about 1852.

In addition to such cases, shellac was molded into buttons, knobs, and electrical insulators. The shellac molding business was well established by 1870. The business got a big boost when phonograph records were made from shellac. Shellac molding materials could accurately reproduce the intricate detail needed for sound. Molded shellac parts maintained a nitch in

Fig. 1-4. This daguerreotype case, molded in about 1855, contains shellac and woodflour. The detail is remarkable.

the growing plastics industry until the 1930s, when synthetic plastics finally surpassed their qualities.

The desirable characteristics of this material were offset by several undesirable traits. The amount and quality of the lac harvest were effected by predator insects, insufficient rain, wide temperature varaitions, hot winds, and the geographic region of India. In a drought, farmers harvested twigs hosting live lac and eggs. They stored the lac brood in pits and kept the sticks and twigs wet with cool water. The alternative to this burdensome task was the death of the lac brood stock.

Under normal conditions, the farmers collected the encrusted twigs after the larva left the sheltering deposit. They then scraped the hardened residue off and cleaned it. Cleaning was not a simple process, due to sand, dirt, dead lac bodies, leaves, and wood fibers.

Once the shellac was ready for use as a coating or a molding powder, problems persisted. The largest problem was moisture absorption. If a shellac molding or coating gets wet, it absorbs water. If soaked for 48 hours, it will absorb up to 20 percent water and change to a whitish color. Antique furniture suffered from *water rings* caused by condensation on containers of ice water. Shellac also takes on moisture from the atmosphere. In high humidity environments, it will absorb enough water to whiten shellac finishes. In moldings, moisture absorption could lead to cracking. Even such stable forms as buttons cracked due to moisture absorption.

The color of shellac was not consistent. The most common colors, yellow and orange, depended on the type of tree the lac infested. To create white shellac, chlorine bleaches were used to lighten the natural color. However, the bleaching process, also affected its solubility in alcohol. Bleached shellac oftened coalesced into a gummy, worthless lump.

Another problem involved aging. Shellac finishes and moldings darkened with age. Old shellac became insoluble in alcohol. Shellac finishes stored in steel cans also absorbed iron, which caused the finish to turn grey or black.

These problems caused manufacturers to seek alternatives. During the 1920s and 30s, new plastics began replacing shellac. In response, shellac producers tried to improve its qualities. Since shellac contained several polymers, they hoped to separate out the most desirable portion by fractional distillation. This effort did not result in a material that could withstand the competition from the synthetic plastics.

Gutta Percha

Gutta percha is a natural polymer with remarkable properties. It is produced by the Palaquium gutta trees which are indigenous to the Malay peninsula. In 1843, William Montgomerie reported that in Malaya, gutta percha was used to make knife handles. The material was softened in hot water, and then pressed by hand into a desired shape. His report stirred interest in the material and led to the formation of the Gutta Percha Company, which remained active until 1930. This company manufactured molded items.

The characteristics of gutta percha are unusual. At room temperature, it is a solid. It can be dented, but does not break easily. When heated, it can be drawn out into long strips, which will not rebound like rubber. Gutta percha is highly inert and resists vulcanization. Its resistance to chemical attack made it an excellent insulator for electric wires and cable. When long strips of extended gutta percha were wound tightly around a wire, the resulting cable was flexible, water proof, and impervious to chemical attack.

The first successful underwater telegraph cable ran across the English Channel from Dover to Calais. Its success depended on gutta percha insulation. In the United States, the Morse Telegraph Company laid a cable insulated with gutta percha across the Hudson River in 1849. Gutta percha also protected the first transatlantic cable, laid in 1866. Figure 1-5 shows the use of gutta percha in the first transatlantic cable.

Like other natural materials, gutta percha was inconsistent. Contamination created regions in the insulation that were low in resistance to electricity. These areas eventually lost the ability to insulate, leading to the shorting out of the electric circuit. Despite these problems, it remained unsurpassed as an insulator until the development of synthetic plastics in the 1920s and 30s. Only then did gutta percha decline in importance in electrical applications.

Early Modified Natural Materials

It was difficult to harvest, gather, or purify the natural plastics. Using these materials in manufacturing processes was arduous. Virtually any material that held potential as a substitute for horn and shellac received attention. Many materials were complete failures. Others failed in their natural condition, and became useful only when chemically altered.

Casein, a material made from milk curd, appeared to have some value as artificial horn. Dried milk curds were ground into powder, plasticated with water, and the resulting dough was molded into various shapes. This attempt met failure, because the molded items dissolved when wet. Casein held no significance as a rival to horn until 1897. In that year, a German printer, Adolf Spitteler learned how to harden the casein dough with formaldehyde. The hardened casein was called *Galalith*, which means milkstone. It was a moldable plastic used for buttons, umbrella handles, and other small items.

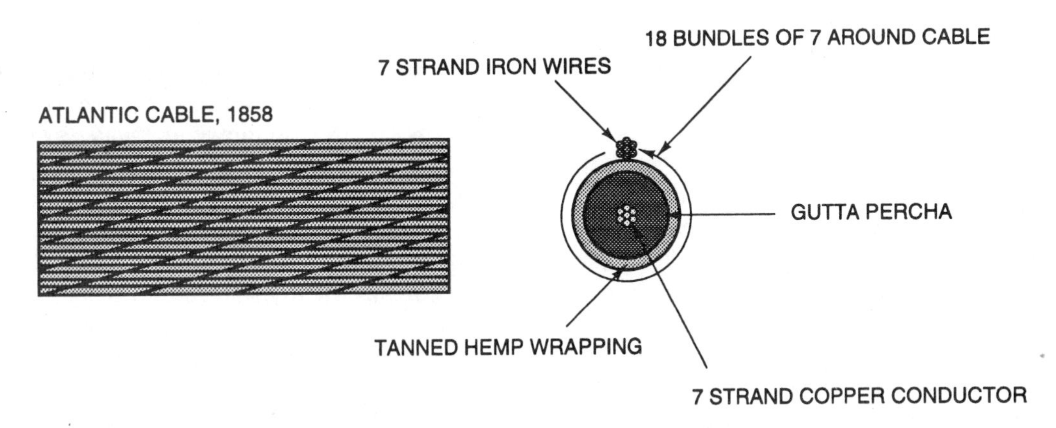

Fig. 1-5. The first transatlantic cable had an overall diameter of 0.62 inches and contained 1 pound of gutta percha in every 23 feet of cable. The amount of gutta percha used for the entire cable was over 260 tons.

The importance of Galalith is that it exemplifies a group of materials which originate in nature and become useful for manufacturing only after chemical modification. One of the earliest and most important materials in this category is rubber.

Rubber

Natural rubber, also called *gum rubber*, is natural latex. It is found in the sap or juice of many plants and trees. The white, sticky juice of the milkweed plant is rich in latex. Several trees also produce natural latex in great quantities. To simplify rubber production, the Hevea brasiliensis, a prolific producer of latex, was cultivated in large plantations in India.

Compared to gutta percha, natural rubber had little industrial significance. Natural rubber is extremely sensitive to temperature. When the weather is hot, it becomes very soft. When the ambient temperature is cool or cold, it becomes stiff. One of the first uses of gum rubber was to make cloth waterproof.

Charles Macintosh got a patent for waterproof cloth in 1823. He pressed a layer of rubber between two pieces of cloth. This solved one problem. In comfortable temperatures, the gum rubber becomes tacky. By putting the rubber between two pieces of fabric, he avoided any tacky feeling. He manufactured some waterproof jackets, Mackintoshs, but they had all the problems of gum rubber. In cold weather, the jackets were stiff, and frequently cracked. When it was hot, the jackets melted. In addition to getting sticky in warm weather, gum rubber decomposed easily, creating strong and foul odors.

In 1839, Charles Goodyear discovered that kneading powdered sulfur into rubber tremendously improved its characteristics. That finding did not happen easily. Goodyear spent years trying to alter gum rubber. He tried blending it with ink, castor oil, soup, and even cream cheese. He finally mixed gum rubber with powdered sulfur and heated the blend. The resulting rubber was stronger, tougher, less sensitive to temperature, and more resilient than before. He had learned how to *vulcanize* the gum rubber. Kneading in small amounts of sulfur produced flexible rubber. Large amounts of sulfur, up to 50 percent, yielded *ebonite*, a rubber so hard it could shatter like glass.

In 1844, Goodyear received an American patent for his discovery. He hoped that a large display at the London Exhibition in 1851 would set him on the road to riches. Goodyear put considerable effort into his display, which was called the Vulcanite Court. Its walls, roof and furniture were made of rubber. His display included combs, buttons, canes, and knife handles molded of hard rubber. His products of flexible rubber featured large rubber balloons and a rubber raft. Goodyear also set up an exhibit in 1855 at the Paris Exposition. In Paris, he featured electric wires insulated with hard rubber, toys, sporting equipment, dental plates, telegraph equipment, and fountain pens.

These displays convinced many people that vulcanized rubber held immense commercial potential. However, before Goodyear could win personal wealth from his idea, he died in 1860. He did not live to see the rise of the rubber industry, which became significant during the Civil War. During that period, the Union Army purchased rubber products valuing $27 million. The Goodyear Company moved into the forefront of the new rubber industry.

Horners were particularly interested in hard rubber as a substitute for horn. In England, comb makers purchased tons of hard rubber. They preferred it to horn or tortoise shell because it reduced waste.

While rubber was less wasteful, it did not have an advantage in appearance. The highly sulfur loaded material was usually black or dark brown. It could not replace the many horn products which imitated tortoise shell or ivory. This appearance limitation prevented ebonite from sweeping other materials aside.

Vulcanized rubber was one of the first modified natural polymers. Without vulcanization, gum rubber was of limited utility. Vulcanized rubber, both flexible and hard, became a very significant industrial material.

Celluloid

To make celluloid, cellulose in the form of cotton linters, underwent a series of chemical modifications. One alteration was the conversion of cotton into nitrocellulose. In 1846, a Swiss chemist C. F. Schönbein discovered that a combination of nitric acid and sulfuric acid transformed the cotton into a high explosive. Explosive nitrocellulose is highly nitrated. Moderately nitrated cellulose is not explosive, but is useful in other ways.

Moderately nitrated cellulose is called *pyroxylin*, a material that dissolves in several organic solvents. When applied to a surface, the solvents evaporate and leave behind a thin, transparent film. This film was named *collodion*. Collodion found widespread use as a carrier for photosensitive materials. Anyone familiar with the photographic processes common in the 1850s and 60s observed dried collodion. When a thick layer of collodion dried, the resulting material was hard, water resistant, somewhat elastic, and very similar to horn.

Alexander Parkes, a British business man, decided to focus his efforts on developing collodion into an industrial material. Parkes lived in Birmingham, England, and had considerable experience in working with natural polymers. He had worked with gum rubber, gutta percha, and chemically treated gum rubber. He understood the qualities of natural plastics and their limitations. In 1862 he announced a new material, which he called *Parkesine*.

He claimed that Parkesine was a substance "partaking in a large degree of the properties of ivory, tortoise-shell, horn, hard wood, india rubber, gutta percha, etc., and which will,... to a considerable extent, replace such materials..." He founded a company in 1866 to sell his new material, but his expectations did not match the reality. When mixing pyroxylin with various stiff oils, he used several solvents. When the solvents evaporated, the new plastic shrunk excessively. Combs became so warped and twisted that they were useless. Parkes did not find buyers flocking to his door to purchase his material. On the contrary, his company failed in two years.

Parkes' failure did not turn others away from the effort to convert hardened collidion into an industrial material. An American, John W. Hyatt, also turned his attention to the problem. In 1863, he decided to try for a $10,000 reward, promised to anyone who could find a substitute for ivory billiard balls. Hyatt made a few billiard balls out of shellac and wood pulp, similar to the material used for daguerreotype cases. These were poor substitutes because they lacked the elasticity of ivory.

Hyatt then set out to make a solid material from pyroxylin. In 1870, he received a patent on the process for making a new material, which he called *celluloid*. He mixed powdered pyroxylin with pulverized gum camphor. To evenly disperse the powders, Hyatt wet the mixture. He then removed the water by pressing it with blotting paper. The material, by then a fragile block, was placed in a mold, heated and pressed. The result was a block of material which was uniform throughout. This block could be used as molding compound but was usually shaved into sheets, which then needed seasoning to remove residual water. Figure 1-6 shows the shaving of a large block of celluloid into sheets.

John and his brother, Isiah S. Hyatt, established a few companies to use their new material. The first was the Albany Dental Plate Company, established in 1870. The Albany Billiard Ball Company was their second try. In both cases, the applications they selected for celluloid were failures. Dental plates were a very poor choice, because they tasted of camphor. Some dental plates softened, warped, or flaked. They were not nearly as good as dental plates made with hard rubber, and never seriously competed with them for markets. The celluloid billiard balls, pictured in Figure 1-7, had the same problems as the shellac balls. Hyatt's company abandoned celluloid in favor of gutta percha.

Celluloid was a very good substitute for horn. It could easily imitate ivory, tortoise shell, and horn. Celluloid became a commercial success, and the Celluloid Manufacturing Company brought the Hyatts substantial profits. By 1874, celluloid combs and mirrors were readily available. Between 1890 and 1910, comb manufacturers in Leominster, Massachussets switched almost completely to celluloid. Figure 1-8 shows these products.

Instead of trying to monopolize celluloid manufacturing, the Hyatts licensed a number of companies to use their material. Between 1873 and 1880, they began affiliations with the Celluloid Harness Trimming Company, the Celluloid Novelty Company, the Celluloid Waterproof Cuff and Collar Company, the Celluloid Fancy Goods Company, the Celluloid Piano Key Company, and the Celluloid Surgical Instrument Company. This list indicates that the celluloid products were generally small items related to clothing or novelties.

Celluloid was not adequate for most industrial applications. One example of its failure in the engineering materials market was safety glass. Layers of celluloid were placed between two pieces of glass to make safety glass for automobiles. The problem was that exposure to sunlight caused yellowing and deterioration. Celluloid did meet the needs of one major application that could never have been fulfilled by ivory, tortoise shell, horn, or hard rubber—it was used for photographic film.

By 1895, motion pictures based on celluloid roll film were available to some audiences. Celluloid made

Fig. 1-6. This machine shaved sheets of celluloid from large blocks. The block shown contains spots of various colors to make the celluloid appear like tortoise shell. (Monsanto Chemical Co.)

possible the early silent films and the famous celluloid personalities. The biggest problem with celluloid film was flammability. Carbon arcs provided light for projection, but when films jammed in the projector, the intense heat caused the ignition of film. In disastrous theatre fires, hundred of people lost their lives. However, those deaths did not limit the use of celluloid for motion pictures. No other material could perform like celluloid. Not until safety film was invented in the 1930s was a photographic base available that eliminated the fire hazard.

The consumption of celluloid rose until the mid-1920s. It did not mold easily, and was used mostly as a fabricating material. Because of this, moldings of plastics continued to be dominated by shellac. Starting in the 1920s, more robust synthetic polymers took over

Fig. 1-7. The Hyatt billiard ball, an early celluloid product. (Celanese Plastic Materials Co.)

Fig. 1-8. A comb and brush produced from celluloid in 1880. (Celanese Plastic Materials Co.)

celluloid applications. Ping-pong balls are one of the few remaining products made of celluloid.

Early Synthetic Plastics

Dr. Leo H. Baekeland was a research chemist who searched for a substitute for shellac and varnish. In June 1907, when working with the chemical reaction of phenol and formaldehyde, he discovered a plastic material that he named *Bakelite*. Phenol and formaldehyde came from chemical companies, not from nature. This marked a major difference between Bakelite and the modified natural plastics.

In his notebook, Baekeland wrote that with some improvements, his material might be "a substitute for celluloid and hard rubber." He reported his finding to the New York section of the American Chemical Society in 1909, and claimed that Bakelite made excellent billiard balls because its elasticity was very similar to that of ivory.

His company, the General Bakelite Company, was established in 1911. Bakelite use grew rapidly, and in contrast to celluloid, it found applications beyond the area of novelties and fashion apparel. Other companies began production of *phenolics*, which are plastics very similar to Bakelite. By 1912, the Albany Billiard Ball Company, a company founded by J. W. Hyatt, adopted Bakelite for billiard balls. In 1914, Western Electric began to use phenolic resins for a telephone earpiece. That same year, Kodak cameras used phenolic for end panels.

In 1916, Delco began to use phenolics for molded insulation in automotive electrical systems. By 1918, molded phenolic resins appeared in dozens of automotive parts. In World War I, aircraft and communication systems increasingly used molded phenolics parts. Phenolics have not yet been made obsolete. In 1991, the U.S. plastics industry used 165 million pounds of phenolics.

With Bakelite, a new era began in plastics. Previously, plastics were natural—or chemical modifications of natural materials. Bakelite proved that it was possible to do in a laboratory or factory what the lac insects and rubber trees did in nature. In fact, the controlled conditions in a factory allowed the production of purer and more uniform materials than anything produced from trees, insects, or horns.

Commercial Synthetic Plastics

Bakelite was the first in a long, still continuing stream of new plastics. The following list (Table 1-1) provides a partial chronology of the growth of plastics. Details about many of the materials in this list are available in Appendices E and F.

Summary

For centuries, natural plastics combined light weight, strength, water resistance, translucency and moldability. Their potential was obvious, but the materials were difficult to gather, or they were only available in limited volumes or sizes. All over the world, people tried to improve natural plastics or find substitutes.

The manufacture of modified natural plastics converted natural raw materials, such as cotton linters or gum rubber, into new and better forms. Celluloid did surpass horn in many qualities. However, the modified materials still relied on natural sources for the prime ingredient. Not until the development of Bakelite was it possible to create in a factory a material that rivaled nature. Bakelite opened the door to the development of a host of synthetic polymers, many tailored to meet specific requirements.

The search for improved materials continues today. Many modern fibers are a result of attempts to create artificial silk. Composite materials are now taking over applications previously reserved for metals. The possibilities for new substitutes seem endless. Leo Baekeland saw boundless potential in phenolic plastics, and used the infinity symbol to represent its uses. That symbol applies today to the unlimited future facing those who strive to find and use new polymers.

Vocabulary

The following vocabulary words are found in this chapter. Look up the definition of any of these words you do not understand as they apply to plastics in the glossary, Appendix A.

Bakelite
Casein
Celluloid
Collodion
Galalith
Gutta Percha
Lac
Nitrocellulose
Parkesine
Pyroxylin
Shellac
Vulcanize

Table 1-1. Chronology of Plastics

Date	Material	Example
1868	Cellulose nirate	Eyeglass frames
1909	Phenol-formaldehyde	Telephone handset
1909	Cold molded	Knobs and handles
1919	Casein	Knitting needles
1926	Alkyd	Electrical bases
1926	Analine-formaldehyde	Terminal boards
1927	Cellulose acetate	Toothbrushes, packaging
1927	Polyvinyl chloride	Raincoats
1929	Urea-formaldehyde	Lighting fixtures
1935	Ethyl cellulose	Flashlight cases
1936	Acrylic	Brush backs, displays
1936	Polyvinyl acetate	Flash bulb lining
1938	Cellulose acetate butyrate	Irrigation pipe
1938	Poystyrene or styrene	Kitchen housewares
1938	Nylon (polyamide)	Gears
1938	Polyvinyl acetal	Safety glass interlayer
1939	Polyvinylidene chloride	Auto seat covers
1939	Melamine formaldehyde	Tableware
1942	Polyester	Boat hulls
1942	Polyethylene	Squeezable bottles
1943	Flourocarbon	Industrial gaskets
1943	Silicone	Motor insulation
1945	Cellulose propionate	Automatic pens and pencils
1947	Epoxy	Tools and jigs
1948	Acrylonitrile-butadiene styrene	Luggage
1949	Allylic	Electrical connectors
1954	Polyurethane or urethane	Foam cushions
1956	Acetal	Automotive parts
1957	Polypropylene	Safety helmets
1957	Polycarbonate	Appliance parts
1959	Chlorinated polyether	Valves and fittings
1962	Phenoxy	Bottles
1962	Polyallomer	Typewriter cases
1964	Ionomer	Skin packages
1964	Polyphenylene oxide	Battery cases
1964	Polyimide	Bearings
1964	Ethylene-vinyl acetate	Heavy-gauge flexible sheeting
1965	Parylene	Insulating coatings
1965	Polysulfone	Electrical and electronic parts
1965	Polymethylpentene	Food bags
1970	Poly(amide-imide)	Films
1970	Thermoplastic polyester	Electrical and electronic parts
1972	Thermoplastic polyimides	Valve seats
1972	Perfluoroalkoxy	Coatings
1972	Polyaryl ether	Recreation helmets
1973	Polyethersulfone	Oven windows
1974	Aromatic polyesters	Circuit boards
1974	Polybutylene	Pipes
1975	Nitrile barrier resins	Packaging
1976	Polyphenylsulfone	Aerospace components
1978	Bismaleimide	Circuit boards
1982	Polyetherimide	Ovenable containers
1983	Polyetheretherketone	Wire coating
1983	Interpenetrating Networks (IPN)	Shower stalls
1983	Polyarylsulfone	Lamp housings
1984	Polyimidesulfone	Convey links
1985	Polyketone	Automotive engine parts
1985	Polyether sulfonamide	Cams
1985	Liquid-crystal polymers	Electronic Components

Questions

1-1. Why was machine manufacture of horn combs often unsuccessful?

1-2. Why weren't daguerreotype cases made from horn or horn powders?

1-3. Casein comes from __?__.

1-4. Moderately nitrated nitrocellulose is called __?__.

1-5. What is the difference between collodion and moderately nitrated nitrocellulose? __?__.

1-6. What is the difference between parkesine and celluloid?

1-7. Bakelite comes from a chemical reaction between __?__ and __?__.

Activities

1-1. Write a report tracing the development of one piece of sporting equipment that has a long history. Examples include golf clubs, tennis rackets, tennis balls, snow skis, tennis shoes, billiard balls, and fishing rods. Try to gather information on the manufacturing processes involved.

As an example of the changes in sports equipment, the following section begins to treat golf balls:

> Golf balls began as ovals or spheres of wood, ivory or iron. Featheries, which were leather pouches tightly stuffed with feathers, replaced the solid balls. Featheries could not withstand moisture, and prevented any play on wet grass or in rain.
>
> Between 1846 and 1848, balls of solid gutta percha began to replace featheries. They were fine in the rain, but fractured on cold days. However, the pieces could be remolded into a useable ball. Figure 1-9 shows a mold for balls of gutta percha.
>
> In about 1899, a new ball made balls of gutta percha obsolete. It consisted of stretched elastic thread wound into a sphere and covered with gutta percha or *balata*, a natural rubber which could be vulcanized. These balls required high quality elastic thread, and the equipment to tightly and uniformly wind the balls. The new balls were much more elastic and hit further.
>
> In 1966, the balls of wound elastic thread became outdated when a ball molded of a solid synthetic type of rubber came to the market. It was much tougher than the previous balls. It depended on new materials, which were just being developed by chemical companies. Since then, new outer coverings have been developed that are almost impossible to cut with a golf club.*

1-2. Trace the development of a plastics manufacturing company that has a long history. Companies like DuPont, Celeanse, and Monsanto have ties to companies that used or manufactured celluloid. Companies that had factories in Leominster, Massachusetts often have ties back to horn comb making. For example, the Foster Grant company, famous for its sunglasses, began operations in Leominster to make plastic combs. Many of its early employees likely had been horn workers.

1-3. Investigate natural plastics

Equipment. Heated platen press or unheated press, saws, polished plates or stainless steel mirror stock.

Fig. 1-9. This mold shaped gutta percha into golf balls. (U.S. Golf Association Museum)

*Source: Martin, John S. *The Curious History of the Golf Ball: Mankinds Most Fascinating Sphere.* New York: Horizon Press, 1968.

CAUTION: Perform activities in a laboratory environment under supervision.

Solid Horn

Note: *This activity does not reflect current practices in the plastics industry. It does demonstrate the qualities of molded horn.*

a. Acquire cow horns. If fresh, boil them for about 30 minutes to prevent decay. Saw the horns and remove the tissue mass. Scrape the membrane from the inside of the horn pieces. When finished, the half horn will appear as shown in Figure 1-10.

b. Cut a small piece, about 1 inch square. Measure length, width, and thickness. Soak in boiling water for about 15 minutes. Place between polished plates and squeeze in a platen press. If heated, keep heat about 250 °F. Keep pressed and let cool. Examine resulting horn piece. If original piece appeared off-white in color, the flattened and thinned piece should be quite transparent. Figure 1-11 here shows some rather transparent pieces of flattened horn. The relative smoothness of the press plates will effect the transparency of the horn.

c. Measure flattened horn

d. Resoak in boiling water.

e. Remeasure. Did horn return to original dimensions?

f. Put food color in water to see how readily the horn takes dye.

g. Examine egg cup spoons made from horn, available in some kitchen utensil retailers. Make a wooden form for a similar spoon. You may have seen a shoe horn, like the one shown in Fig. 1-12.

Powdered horn or hoof. Cow hooves are available in pet stores for dog chews.

a. Make powder by filing with coarse file.

b. Compact with a plunger in a cylinder, as seen in Figure 1-13. The rectangular piece of material is a rubber *bumper*.

c. Explore surface detail by pressing against a coin—see how well the detail transfers.

d. Experiment with colors.

Shellac. It is available in stick form for furniture scratch repair. Also available as flake from fine woodworking supply stores. (See Figure 1-14.)

Fig. 1-10. This half horn is ready for cutting and flattening.

Fig. 1-11. This photograph demonstrates the transparent qualities of horn.

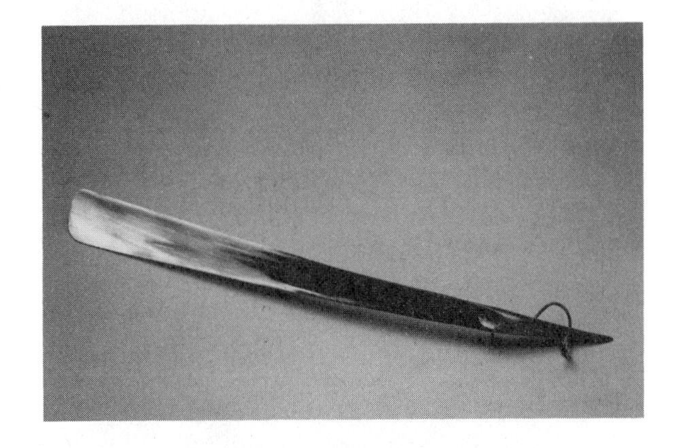

Fig. 1-12. Molding altered the shape and surface finish of this shoe horn.

Fig. 1-13. This type of simple cylinder and plunger can mold shellac and horn powders.

Fig. 1-14. Shellac in flake form is rather rare, but still available.

Fig. 1-15. This shellac molding, made with a quarter as the impression piece, indicates the fine detail possible with molded shellac.

a. Do simple moldings using cylinder, plunger, and a coin. Figure 1-15 shows a shellac molding using a quarter as the impression. Note the reproduction of details. Shellac will flow at a temperature just above 100 °C.

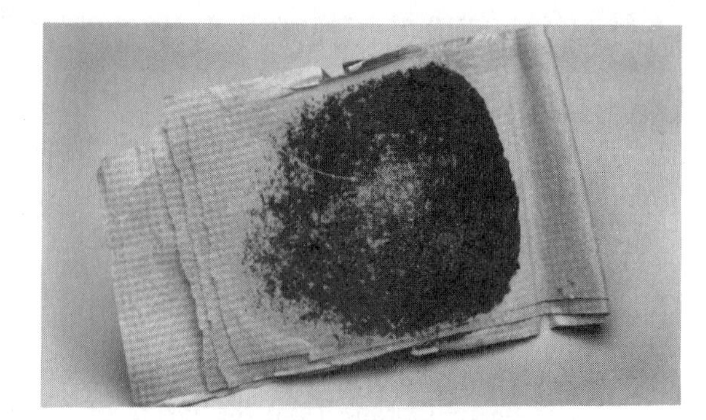

Fig. 1-16. This thin laminate is significantly more flexible than a similar sheet of unreinforced shellac.

Without a device to apply pressure to the cylinder, the shellac will raise the cylinder and flash excessively. The rubber "bumper" keeps pressure on the cylinder during compression.

b. Compare powdered hoof moldings to shellac moldings.

c. Investigate effect of fibers

(1) Create sandwich of paper towels and shellac.

(2) Press sandwich in heated platen press.

(3) Does shellac flow through paper? How much does paper improve the strength of shellac? Figure 1-16 shows a thin laminate, containing two layers of paper toweling and a small amount of shellac.

(4) Create a molding compound using powdered shellac and glass or cotton fibers. How greatly do the fibers improve the physical properties of the moldings?

References

Borglund, Erland, and, Jacob Flavensgaard. *Working in Plastic, Bone, Amber, and Horn*. New York: Reinhold Book Corp., 1968.

Friedal, Robert. *Pioneer Plastic; The Making and Selling of Celluloid*. University of Wisconsin Press, 1983.

Luscomb, Sally C. *The Collector's Encyclopedia of Buttons*. New York: Bonanza Books, 1967.

Mark, Herman F. "Polymer chemistry: the past 100 years," *Chemical and Engineering News*, April 6, 1976, pages 176–189.

Chapter 2

Current Status of the Plastics Industry

Introduction

Chapter one used the words *polymers*, *rubber* and *plastics* without providing thorough definitions. *Polymers* are natural or synthetic organic compounds. Natural polymers include horn, shellac, gutta percha, and gum rubber. Synthetic polymers appear in thousands of plastics products, clothing, automobile parts, finishes, and cosmetics. Whether natural or synthetic, polymers have chemical structures characterized by repeating small units called *mers*. In order for a compound to be a polymer, it should have at least 100 mers. Many polymers found in plastics products have 600 to 1000 mers.

The word *plastics* comes from the Greek word *plastikos*, which means "to form or fit for molding." A more definitive explanation comes from The Society of the Plastics Industry. It identifies plastics as:

> Any one of a large and varied group of materials consisting wholly or in part of combinations of carbon with oxygen, nitrogen, hydrogen, and other organic or inorganic elements which, while solid in the finished state, at some stage in its manufacture is made liquid, and thus capable of being formed into various shapes, most usually through the application, either singly or together, of heat and pressure.

In this book, the word plastics will always end with an "s" when it refers to a material. The word plastic, without an "s," will be considered an adjective meaning formable. Because plastics are closely related to resins, the two are often confused. *Resins* are gum-like solid or semisolid substances used in making such products as paints, varnishes, and plastics. A resin is not a plastics unless it has become a "solid in the finished state."

The English chemist Joseph Priestley coined the word *rubber* after he noticed that a piece of natural latex was good to rub out pencil marks. Natural rubber is one material in a group called elastomers. *Elastomers* are natural or synthetic polymeric materials that can be stretched to at least 200 percent of their original length and, at room temperature, return quickly to approximately their original length.

Although elastomers and plastics have been considered as separate categories of materials, the distinction between them has eroded significantly. In the early development of plastics and rubber materials, the plastics tended to be stiff, while the rubbers tended to be flexible. Now many plastics exhibit characteristics traditionally available only in rubber. While rubbers still have unique characteristics—especially the ability to retract rapidly—the categories now partially overlap.

In recent years, a family of materials called thermoplastic elastomers (TPEs) have partially bridged the gap between traditional rubbers and plastics. Within the TPEs, subsets include: thermoplastic elastomers based on urethanes, polyesters, styrenics, and olefins. By far the most dominant are the TPOs, which stands for thermoplastic olefin elastomers. They have replaced many grades of rubber, particularly in automotive parts. The TPEs have supplanted many traditional rubber products because of their ease of processing.

Plastics and elastomer products do not contain 100 percent polymer. They generally consist of one or more polymers plus various additives. (See Chapter 7 for discussion of additives and their effects.) The rela-

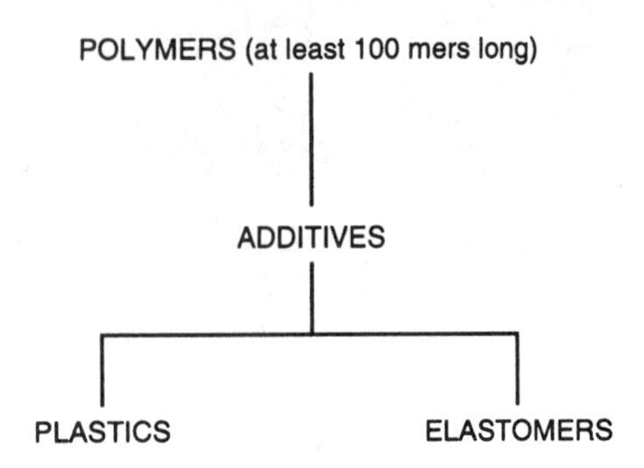

Fig. 2-1. Plastics and elastomers contain additives to make them reliable in a wide variety of environments. Even natural color plastics contain additives.

tionship between these basic terms appears in Figure 2-1. Note that polymers is the umbrella category. With the presence of additives, some polymers become plastics or elastomers.

This chapter will present the current status of the plastics industry, with emphasis on the United States. The content outline is:

I. Major plastics materials
II. Recycling of plastics
  A. Bottle deposit laws and their effect
  B. Curbside recycling
  C. Automotive recycling
  D. Chemical recycling
  E. Recycling in Germany
III. Disposal by incineration or degradation
  A. History in incineration in the United States
  B. Advantages of incineration
  C. Disadvantages of incineration
  D. Degadable plastics
IV. Organizations in the plastics industry
  A. Publications for the plastics industry
  B. Trade newspapers

Major Plastics Materials

During the 10 year period from 1984 to 1994, the sales of plastics in the United States grew at a average annual rate of about 5 percent. Figure 2-2 graphs these sales figures and displays a rather steady upward trend. According to a 1992 report of the Society of the Plastics Industry, in 1991, the plastics industry, including plastics manufacturing within other categories of industry, generated shipments of $271 billion, almost 11 percent of all manufacturing shipments. Plastics manufacturing also accounted for nearly 3 percent of the U.S. workforce.

To get a clearer picture of the plastics consumption, the sales data from 1994 were selected. Although the numbers change each year, the relative consumption and application for various materials have been rather consistent. Table 2-1 contains the sales volumes.

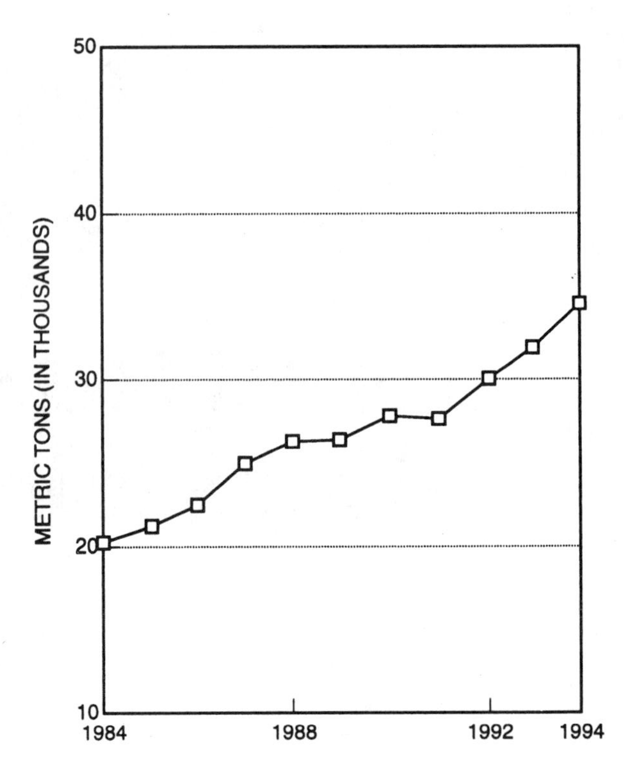

Fig. 2-2. U.S. sales of plastics grew at about 5 percent annually between 1984 and 1994. (adapted from *Modern Plastics*)

The eight plastics with the largest sales accounted for about 82 percent of all sales. Those eight were:

Low Density Polyethylene (LDPE)	6394
High Density Polyethylene (HDPE)	5280
Polyvinyl chloride (PVC)	5056

Table 2-1. U.S. Sales of Plastics

Acrylonitrile-butadiene-styrene (ABS)	677
Epoxy	274
Nylon	419
Phenolic	1465
Polyacetal	97
Polycarbonate	316
Polyester, thermoplastic	1574
Polyester unsaturated	1333
High Density Polyethylene	5280
Low Density Polyethylene	6394
Polyphenylene alloys	109
Polypropylene	4433
Polystyrene	2671
Polyurethane	1707
Polyvinyl chloride	5056
Styrene acrylonitrile	59
Thermoplastic elastomers	394
Urea and melamine	993
Others	1495
Total	34 736
Units: 1000 metric tons	

Adapted from *Modern Plastics*, January 1995.

Polypropylene (PP)	4433
Polystyrene (PS)	2671
Polyurethane (PU)	1707
Polyester thermoplastic (PET)	1564
Phenolic	1465
Total	28570

Units: 1000 metric tons

To achieve an expectation for the utilization of these eight plastics, projections were determined based on the 1994 sales data and an annual growth rate of 5 percent. Figures 2-3 through 2-10 provide graphic representation of expected usages in the year 2000.

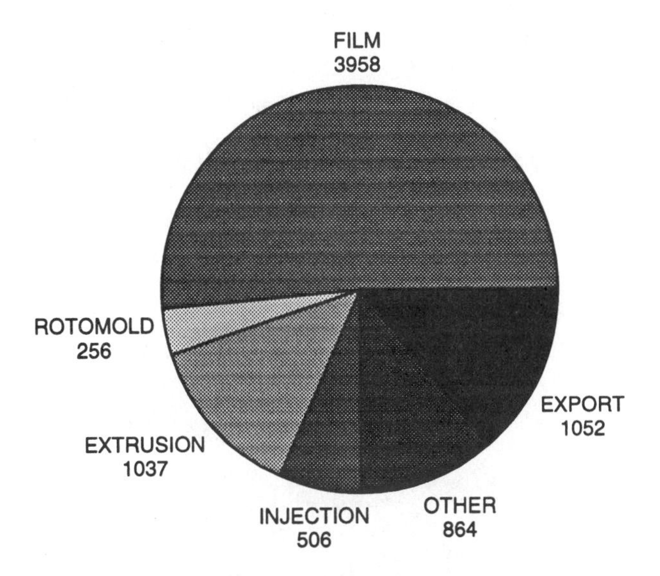

Fig. 2-3. LDPE expected utilization: Projected for year 2000. (adapted from *Modern Plastics*)

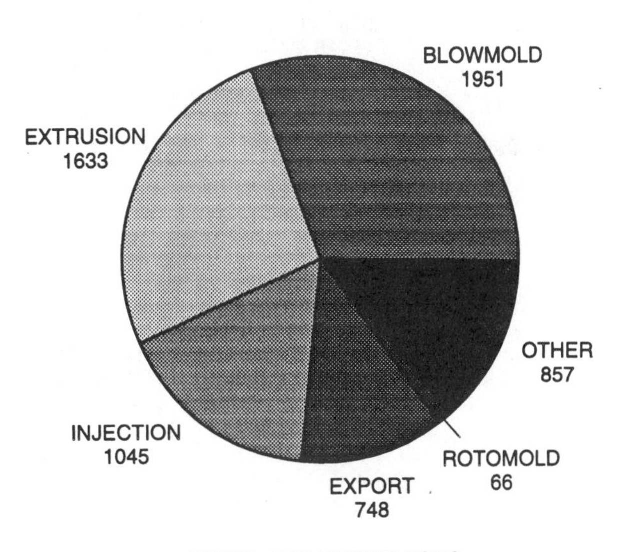

Fig. 2-4. HDPE expected utilization: Projected for year 2000. (adapted from *Modern Plastics*)

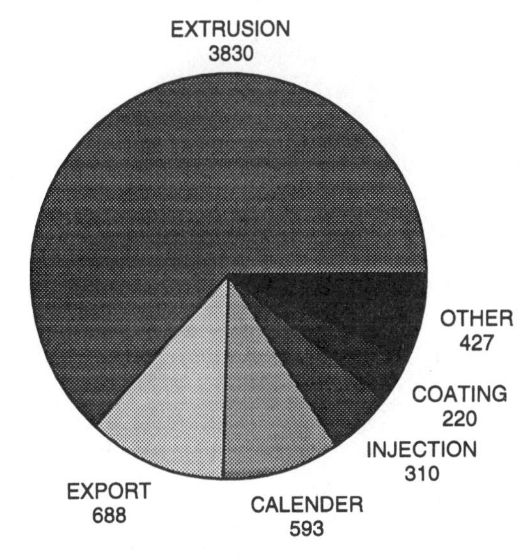

Fig. 2-5. PVC expected utilization: Projected for year 2000. (adapted from *Modern Plastics*)

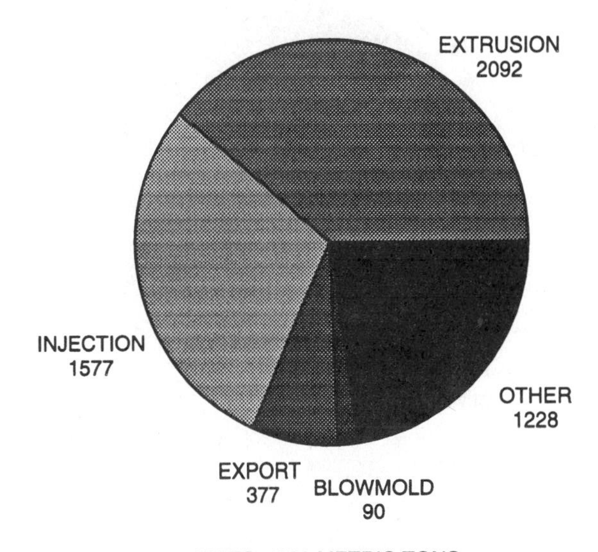

Fig. 2-6. PP expected utilization: Projected for year 2000. (adapted from *Modern Plastics*)

The single largest expected use for LDPE was film, for HDPE was bottles for liquid food, for PVC was pipe, for PP was fibers, for PS was cassettes, and PU was foam for furniture, for PET was soft drink bottles, and for phenolic was plywood adhesive. The steady growth of plastics sales reflects the ability of plastics products to fulfill an increasing number of consumer demands. The increasing utilization of plastics has also caused concern for the role of plastics in environmental pollution.

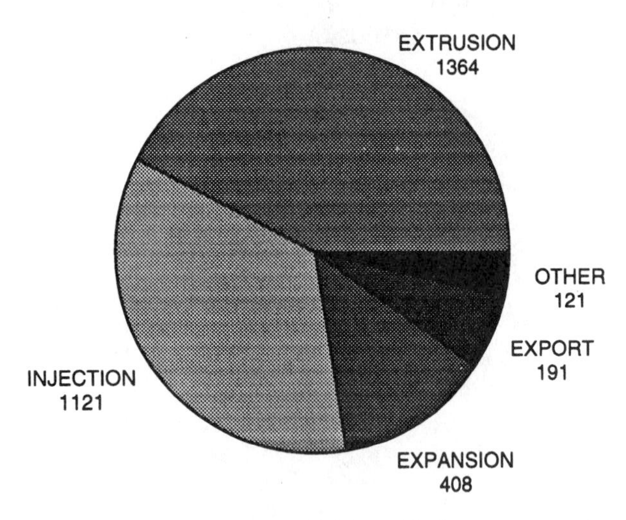

Fig. 2-7. PS expected utilization: Projected for year 2000. (adapted from *Modern Plastics*)

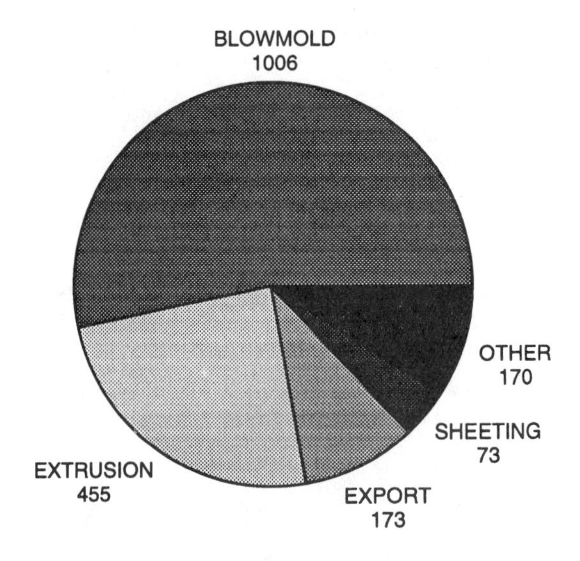

Fig. 2-9. PET expected utilization: Projected for year 2000. (adapted from *Modern Plastics*)

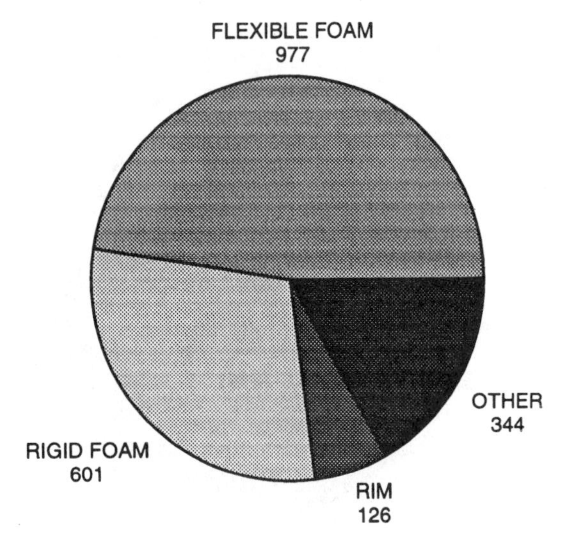

Fig. 2-8. LDPE expected utilization: Projected for year 2000. (adapted from *Modern Plastics*)

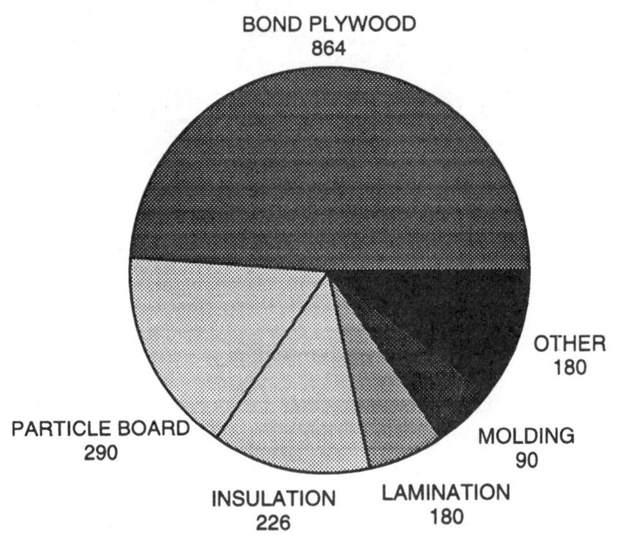

Fig. 2-10. Phenolic expected utilization: Projected for year 2000. (adapted from *Modern Plastics*)

Recycling of Plastics

The first Earth Day, which occurred in 1970, signaled the development of a new level of awareness and concern about the environment. During the 1970s, a number of anti-litter campaigns sprang up. In 1976, the federal government passed the Resource Conservation and Recovery Act (RCRA). It promoted reuse, reduction, incineration, and recycling of materials. The combined effects of public concern and legislation created major changes in two arenas: hazardous waste management, and recycling of non-hazardous materials. Treatment of hazardous materials in the plastics industry appears in Chapter 4.

Recycling is a term generally reserved for post-consumer waste materials. In contrast, reuse or reprocessing usually handles waste materials generated during manufacturing. Plastics industries have for decades reused the material in defective parts, trimmings and other manufacturing scrap. This usage ranges from very small-scale reprocessing in small companies, to huge programs that generate thousands of tons of reprocessed materials. Systematic recycling of post-consumer plastics occurs through two main channels: returns for bottle deposits, and curbside pick-up programs for recyclable materials.

Bottle Deposit Laws and Their Effect

Efforts to encourage or force recycling of bottles have involved state laws, federal proposals, and state mandates. The recycling of PET containers reflects the effectiveness of these efforts.

State legislation. During the 1970s, five states enacted bottle deposit laws. In chronological sequence, they were: Oregon, Vermont, Maine, Michigan, and Iowa. In the 1980s, five more states required deposits, they were: Connecticut, Delaware, Massachusetts, New York, and California. The first deposit law took effect in 1972 in Oregon, and the latest occurred in 1987 in California. The deposit law states require 5 cents deposit, except Michigan which requires 10 cents.

Because Michigan was the only state to charge more than 5 cents, it has held a rather unique position in bottle recycle efforts. The Michigan Department of Natural Resources indicated that the deposit law caused a 90 percent reduction in litter along highways and parks. In addition, it diverts approximately 700 000 tons of refuse from landfills annually. The deposit law brings about 95 percent of 4.5 billion containers manufactured annually back into recycling facilities.

Federal legislation. Concurrent with the initiation of state laws, federal legislators and officials investigated deposit laws. In 1976, a Federal Energy Administration study recommended a nationwide deposit law. In 1977, the General Accounting Office favored deposits. In 1978, the Office of Technological Assessment issued a report favoring deposits. In 1981, Senate Bill 709 for deposits received hearings, but no action. In 1983, Senate Bill 1247 and House of Representatives Bill 2960 proposed 5 cent deposit on carbonated beverages. No action was taken. Since then, several variations on deposit laws have received attention every year, but none have passed into law. Consequently, bottle deposit laws have remained at the state level.

State recycled content mandates for bottles. Due to pressure from the federal government and concern at the state level, by the end of 1994, 40 states had established legislated goals for litter control or recycling. In an effort to force more thorough recycling, some states also passed recycling content mandate laws. These laws often replaced laws enacted in the early 1990s which subjected various plastics to bans. For example, some states placed polystyrene foam packaging under a ban. In many cases, these laws were not enforced. In contrast to the ban rulings, the recycled content mandates may find better compliance.

These laws usually specify the percentage of post-consumer recycled (PCR) content in various types of containers. Starting in 1994 in Florida, bottles and jars must have 25 percent PCR. In California, non-food rigid containers are required to contain at least 25 percent post-consumer recycled content in 1995. Containers which can be reused or refilled five times are exempt. California also required trash bags to have 10 percent recycled content in 1994, and 30 percent in 1995. In Oregon, any rigid container must also contain 25 percent PCR. Florida requires a one-cent advance disposal fee paid by wholesale distributors unless the containers have 25 percent recycled content.

Table 2-2. Recycling of PET

1982	1989	1993
18	88.5	203.5

Units: 1000 metric tons

PET bottle recycling. The general sequence of events is that, to redeem the deposit on PET (polyester thermoplastic) soft-drink bottles, consumers return them to stores. The stores sort the containers according to manufacturers. The drivers of beverage delivery trucks often have to pick-up large plastics bags full of empty plastics bottles. They deliver the bottles to a processing facility, where the containers are sorted. To make shipping more cost effective, densification equipment squeezes the bottles into bales, which are sold to recycling companies. The recyclers unbale the containers, chop them into flakes, clean, wash, dry, and in some cases, reprocess the materials. Table 2-2 shows the growth of PET recycling.

Finding uses for the recycled PET has not always been an easy task. For years, the Federal Food and Drug Administration (FDA) did not allow recycled materials to be used in food contact applications. Consequently, PET had to be used in non-food applications. A major use for recycled PET was as fiber. For

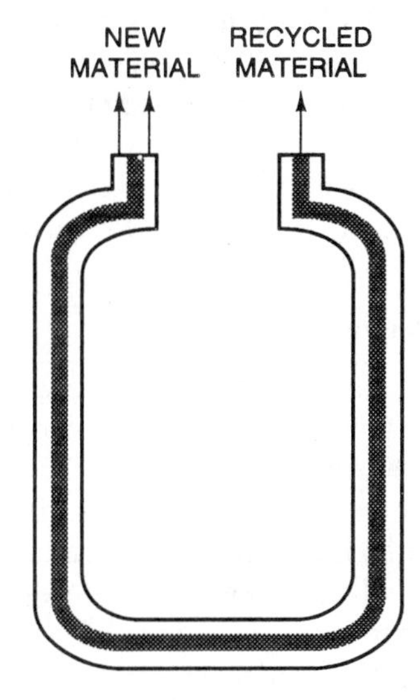

Fig. 2-11. An inner layer of recycled material is sandwiched between two layers of prime material.

example, 35 soft-drink bottles provide enough material for the fiberfill used in one sleeping bag. Other products are polyester fabrics for T-shirts and blankets.

Another possibility was to make multilayer containers, with an inner of virgin material, and middle layer of recycled material, and an outer layer of virgin material. Figure 2-11 shows a sketch of this style container. This allows use of recycled material, yet provides control of the food contact layer and the outer or show surface.

Yet another possibility was to chemically depolymerize the PET, and then use the resulting materials for polymerization into "new" PET. This option has found limited success, because it produces PET that is more expensive than newly manufactured PET. According to one estimate in 1994, recycled PET made by depolymerization cost 20 to 30 cents per pound more than virgin material.

In an important decision was made in August 1994. The U.S. Food and Drug Administration approved the use of 100 percent recycled PET for food-contact packaging. This was the first time the FDA approved 100 percent recycled content in food and beverage packaging. That means that PET bottles for soft-drinks can be reprocessed into new food-use bottles.

To win this approval, a recycling facility in Michigan had to develop new ways to thoroughly clean the recycled material. The new process features high-intensity washing, temperatures of about 500 °F, and other cleaning techniques. It is not yet known if curbside materials will be clean enough to be economically feasible into the same process.

In 1993, the recycling rate for all PET packaging was approximately 30 percent. In states with bottle deposit laws, the containers covered by deposits are returned at a rate of about 95 percent. Since deposit laws cover 18 percent of the American population, that means that the one-fifth of the population under deposit laws accounts for about 60 percent of the total recycled PET bottles. Although these figures might indicate support for bottle deposits, opponents of bottle deposits assert that comprehensive waste-management programs are far more effective than forced deposits.

Curbside Recycling

About 7000 communities across the United States have curbside pick up of recyclable materials. These programs serve over 15 million households and thousands of businesses. In small communities, the programs tend to be managed by companies or agencies established to fill that need. Many medium to large communities hire a nationwide solid waste management company to organize the collection, to establish and maintain facilities, and to locate buyers for recycled materials.

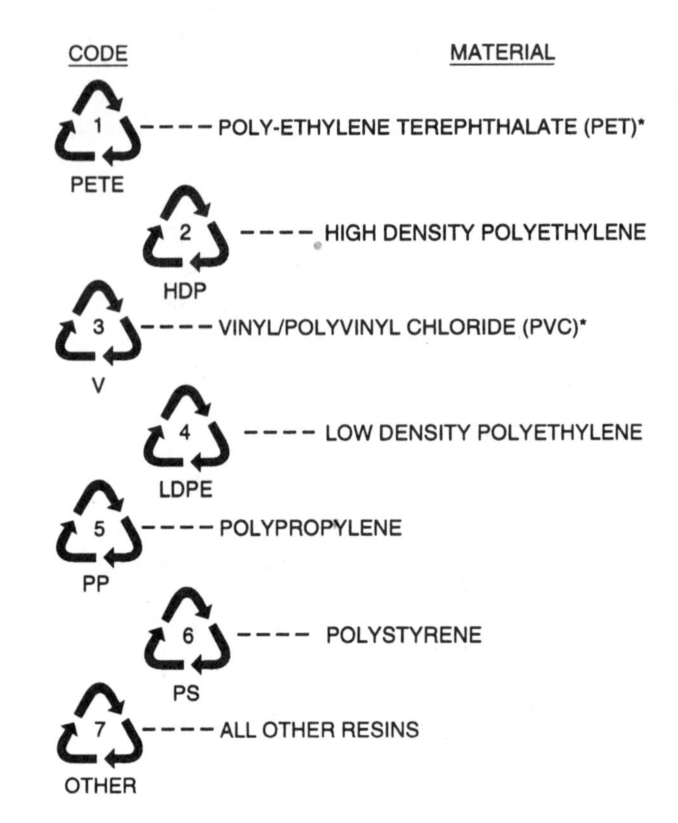

Fig. 2-12. These codes were recommended by the Plastic Bottle Institute.

Identification coding. Unlike bottle deposit materials, in which the type of plastics were clearly known, it was difficult to distinguish the types of plastics obtained in the curbside recycling programs. In some cases, mistaken identity of a few containers could ruin a large amount of otherwise useful material. For example, PET bottles and PVC bottles are often impossible to distinguish by appearance. If a small amount of PVC is mixed into a large batch of PET, the PET will be ruined.

To avoid this and other similar problems, in 1988, the Plastic Bottle Institute of the Society of the Plastics Industry established a system for identifying plastics containers. Each code has a number in the triangular symbol and an abbreviation below it, as shown in Figure 2-12.

The "chasing arrows" symbol has come to imply recycling. Some individuals and organizations feel that the recycling symbol is deceptive, for when they tried to recycle containers marked 3 or 6, they frequently found no one who would accept those materials. They thought the symbol implied recycling, not just the capability of the material to be recycled. However, systematic and widespread recycling of plastics will require some identification system, with or without the chasing arrows symbol.

Collection. Most recycling programs accept metals, plastics, and paper/cardboard. Some programs require residents to extensively separate the materials. For the plastics, the residents look at the recycling codes and sort plastics into selected groups. During pickup, the materials go into bins or containers to preserve the sorting. Most programs accept plastics number 1 (PET), and 2 (HDPE). Requiring residents to sort the plastics containers has two major drawbacks. First, they often make mistakes in sorting. Second, and often more important, the pickup drivers need more time to place the sorted materials into appropriate containers. Some cities have found that a central sortation line is much more efficient. One very basic manual sorting line requires six employees working only 3.5 hours to process the containers from 5000 houses.

Because of the efficiency of centralized sorting, many large communities accept commingled materials. Residents in many cities throw plastics 1, 2, and 6, plus aluminum and steel cans, newsprint, cardboard, scrap metal, and other paper into one large container. The importance of compaction effects the decision to accept commingled materials. Uncompacted materials can occupy 10 cubic yards, but if the items were compacted, they would need 3 cubic yards. Compaction in the collection truck can result in savings, primarily in transportation costs. Some trucks can compact paper and cardboard—leaving glass, metal and plastics loose.

Figure 2-13 shows a collection truck. This style features two large bins, one for paper and cardboard, the other for commingled containers. Figure 2-14 shows the dumping of the commingled containers. Since the plastics containers account for the most volume of containers, some trucks are outfitted with special compacting equipment. Figure 2-15 shows a small compaction bin for crushing plastics containers, particularly milk jugs. Without the ability to compact milk jugs, the drivers would have to make more trips to unload.

Sortation. The collection trucks deliver materials to a materials recovery facility (MRF). In 1995, about 750 MRFs were operating in the United States. Only about 60 had automatic sortation lines, leaving almost 700 with mostly manual procedures. Although there is interest in automatic sortation, many new facilities have manual techniques. Even low level automatic systems, which can handle only a limited stream of waste plastics, cost about $100,000.

Trucks tip the collected materials into initial sorting equipment. Figure 2-16 shows a conveyor that delivers the commingled containers into the sorting equipment. Many of the manual MRFs can magnetically separate all ferrous containers. Manual sortation can be a very simple process that involves depositing various materials in bins or tubs. Manual sortation can also occur on a picking line. Picking lines feature conveyor belts that move recycled materials past employees who sort them into various categories.

Some recycling facilities accept foamed polystyrene. Foamed PS presents several problems. Because its bulk density is so low, the initial storage point must be rather large. Figure 2-17 shows an 8 ft tall metal storage

Fig. 2-14. The plastics containers consume far greater volume than steel or glass containers. (Courtesy of the City of Ann Arbor, Solid Waste Dept.)

Fig. 2-13. The bins in this style truck allow separation of paper from mixed containers. (Courtesy of the City of Ann Arbor, Solid Waste Dept.)

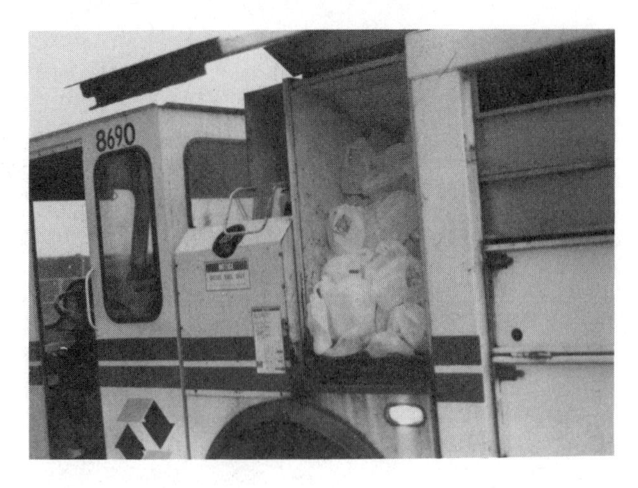

Fig. 2-15. The compaction bin reduces the volume of collected plastics bottles. (Courtesy of the City of Ann Arbor, Solid Waste Dept.)

Fig. 2-16. To remove the steel from the mixed container stream, a magnetic drum pulls cans out of the main flow. (Courtesy of the City of Ann Arbor, Solid Waste Dept.)

Fig. 2-17. This enclosed bin protects scrap PS foam from the elements. When full, a truck delivers the foam to a MRF for baling.

Fig. 2-18. This tightly packed bale of PS foam came from a vertical style baler.

container. A full semi-truck load of PS foam for recycling will weigh about 1500 lbs. Even bales of foamed PS are very light. Figure 2-18 shows a bale of PS foam. These bales weigh from 80 to 90 lbs. In contrast, a similarly sized bale of HDPE weighs about 450 lbs.

Recycling of PCR HDPE

Because HDPE does not fall under most bottle deposit laws, the HDPE recycling program is the most effective in reclaiming plastics. Table 2-3 shows the growth of HDPE recycling.

The recovery rate on natural HDPE bottles, (predominantly one-gallon and one-half-gallon milk jugs) was slightly less than 25 percent in 1993. In contrast, the rate for all HDPE packaging was about 10 percent. The total packaging sales in 1993 was 1929000 metric tons. The total sales of HDPE for the same year was 4820000 metric tons, which indicated that packaging of the type that appears in the post-consumer waste stream is only 25 percent of the total sales. The other products eventually find their way to landfills or incinerators.

In 1995, the demand for recycled materials was up and the recycling companies were making profits. The recycled HDPE, number 2, had demand of 500 million lbs in 1994, which was about one-third more than what came through recycle streams. Because of this high demand, a number of major manufacturers began to market materials containing recycled post-consumer plastics. Among these companies were Dow, Eastman Chemical, and Hoechst Celanese.

While only a limited number of MRFs handle polystyrene foams, almost all accept PET, natural HDPE, and mixed colored HDPE. Picking lines and automatic sorting equipment separate natural from colored HDPE. When the volume of either type is great enough, the stored containers will go to a baling machine. Figure 2-19 shows a view of a large bin of mixed color HDPE.

Balers are of two major types, horizontal and vertical. The distinction refers to the direction of movement of the ram, which compacts the containers. Vertical balers may be the most simple type, which require hand loading. Because vertical balers force the containers into a closed rectangular space, the bales so produced are often very tight.

Horizontal balers raise the containers up a long conveyor and drop them into a chamber. Figure 2-20 shows the feed conveyor of a horizontal baler. The containers fall into a chamber and a ram presses them

Table 2-3. Recycling of HDPE

	1982	1989	1993
HDPE	-	59	216.4
Units: 1000 metric tons			

Fig. 2-19. When the bin is full, this pile of mixed-color HDPE will go to the baler. (Courtesy of the City of Ann Arbor, Solid Waste Dept.)

into a bale. Wire tying equipment finishes the bales. In contrast to many vertical balers, which often make tight bales, some horizontal balers produce bales which are not tightly packed. This occurs because the ram on some horizontal balers pushes against the resistance provided by previously generated bales. Figure 2-21 shows bales exiting the machine. The difficulty

Fig. 2-20. The conveyor lifts the containers to the top of the baling chamber. (Courtesy of Resource Recovery Systems, Inc.)

with loose bales is that they may break up during handling with fork lifts. Figures 2-22 shows two bales of HDPE on a fork lift.

Because most MRFs do not have equipment to reprocess the plastics, they sell the bales to a reprocess-

Fig. 2-21. This photo shows a bale of PET exiting the baling machine.(Courtesy of Resource Recovery Systems, Inc.)

Fig. 2-22. A fork lift carries bales to a waiting semi- trailer. (Courtesy of the City of Ann Arbor, Solid Waste Dept.)

ing company. Transportation from the MRF to the reprocessing company generally occurs in semi-trucks.

Because a reprocessing facility acquires baled HDPE from several sources, the quality differences in various types of baling equipment is apparent. Figure 2-23 shows tightly-packed bales. In contrast, Figure 2-24 shows a loose bale—one in danger of breaking up before reaching the reprocessing equipment.

The first step in reprocessing is to unbale the milk-jugs and feed them into a chopper. This provides an opportunity for some control on the materials going into the reprocessing system. The flakes coming from the chopper contain irregular shapes and considerable contamination. Figure 2-25 shows unwashed chopped flakes. An blower system then separates the fine pieces out, since they are too light to fall through a controlled updraft. This process is called *elutriation*. Elutriation refers to a purification by straining, washing, or decanting. In this context, the purification is done with air. Figure 2-26 shows the types of fines, paper, and dirt removed during the first elutriation process.

Storage bins hold the flake until it enters a washer. Figure 2-27 shows a washer for HDPE flake. After a pre-weighed charge of flakes drops into the washer, a determined amount of water enters, and a vigorous washing cycle begins. Some systems utilize detergents to assist in cleaning. Others rely on the abrasive characteristics of the flakes to scrub each other.

The washer dumps the load of water and flakes out an outlet pipe, as seen in Figure 2-28. The charge of flakes then enters a flotation tank. Slow-moving paddles move the flake, which floats in water, through the tank. The paddles also agitate the flakes, and cause heavy particles to fall to the bottom of the tank. In the tank, dirt, sand, and most plastics other than polyethylene and polypropylene, settle out.

The paddles of the flotation tank lift the flakes to an exit chute, which delivers them to a centrifugal dewatering device. It spins the water out of the flakes and then delivers the flakes to a rapid drying treatment.

Fig. 2-23. These large, tight bales promote easy handling (Courtesy of Resource Recovery Systems, Inc.)

Fig. 2-24. This loose bale may break when moved with a fork lift. (Courtesy of Michigan Polymer Reclaim)

Fig. 2-25. Unwashed flakes contain many types of contamination, including paper, dirt, pebbles, and undesired plastics. (Courtesy of Michigan Polymer Reclaim)

Fig. 2-26. This photo reveals dust, lint, paper and plastic films. The particles were light enough for separation by an elutriation system. (Courtesy of Michigan Polymer Reclaim)

Fig. 2-27. The cylindrical machine is a high intensity washer. (Courtesy of Michigan Polymer Reclaim)

Fig. 2-28. The washer dumps a load of scrubbed flakes and water into the flotation tank. (Courtesy of Michigan Polymer Reclaim)

After drying, a second elutriation system creates a controlled updraft, which separates fines from the washed flakes. Figure 2-29 shows a blower used for this separation. This elutriation yields mostly label films. Figure 2-30 shows the type of fines collected by the second air sorting. Blowers deliver the clean flakes to storage bins, such as the one seen in Figure 2-31. If the storage bins are full, flakes can be stored in *gaylords*, which are large boxes containing approximately one cubic yard of material.

The clean flakes (as seen in Figure 2-32) then enter the feed throat of an extruder. The extruder melts the flakes, and forces the melted material through a die. Figure 2-33 shows an extruder prepared for this operation. This extruder has a water-wall type pelletizer, which chops strands of extruded material just after it exits the die. The particles are thrown into a cylindrically shaped wall of water. Figure 2-34 shows this type of pelletizer head. The pellets cool rapidly. They are then dewatered, and dumped into a gaylord for shipment to a processing facility, which will make new products. Since the HDPE found in milk jugs is blow-molding grade material, it frequently goes into additional blow-molded products.

Fig. 2-29. This photo shows the blower unit that separates out the fines and lightweight contaminants. (Courtesy of Michigan Polymer Reclaim)

Fig. 2-30. This group of fines consists primarily of films used in labels on HDPE bottles. (Courtesy of Michigan Polymer Reclaim)

Automatic sortation. Manual sortation has two serious limitations: It is not effective in distinguishing PVC from PET, and it becomes unworkable when the

Fig. 2-31. Below the storage bin is a gaylord, which can store flakes if the bin fills up. Notice the stockings hanging around the bin. The air used to deliver the flakes into the bin must escape. The stockings catch any fines in this exhausting air stream. (Courtesy of Michigan Polymer Reclaim)

Fig. 2-33. Clean flakes drop into the feed hopper at the rear of this extruder. (Courtesy of Michigan Polymer Reclaim)

Fig. 2-34. This photo shows a water-wall type pelletizer. It has the advantage of keeping the die hot and dry, yet cooling hot pieces of plastics in a water bath. (Courtesy of Michigan Polymer Reclaim)

Fig. 2-32. These are clean flakes. Notice the one dark piece. It is a portion of a milk bottle top and consists of blue polypropylene. It is a contaminant, but since PP and HDPE have similar densities, it cannot be excluded by a flotation system. (Courtesy of Michigan Polymer Reclaim)

total volume of materials is very high. To develop faster systems, companies have developed various types of automatic sorting devices. Many automatic systems include similar equipment to prepare the containers for identification. They all use bale loaders, unbaling machines, and screens to remove rocks, dirt, and extremely large or small containers. Great diversity occurs in the technical method used to identify various plastics. Identification systems have two major components: a method to separate and convey the containers, and a method to identify the plastics.

The separation and transportation usually involves high speed conveyors. If there is a single detector in the system, the conveyors must singulate the containers and bring them past the detector one at a time. Frequently, an air blast is used to separate one container from the next. In addition to air blasts, vibratory conveyors also aid in singulation. If a single detector fails to recognize the container, or if the container is improperly positioned for optimal recognition, it may be unidentified. To increase identification rates, some systems have multiple detectors with a singulating conveying system. Some systems are capable of multiple containers, and require multiple detectors for this application. Figure 2-35 shows several possible configurations for automatic identification.

The general purpose of the detector is to determine the chemical make-up of a container. Once that is complete, computer-based systems must track the location of the known container and then activate the air jet to eject to the appropriate takeoff conveyor. The four major types of detectors are: optical, x-ray, single wavelength infrared (IR), and multiple wavelength IR.

Optical systems rely on vision systems that determine the color of a container. X-ray sensor can distinguish PET from PVC by sensing the presence of chlorine atoms in the PVC. Chemical detection can occur rapidly—with some systems requiring less than 20 milliseconds. Single wave length IR systems can determine opacity and, based on the results, sort containers into clear, translucent, and opaque streams. Multiple wavelength IR can determine the chemical constitution of a container by comparing its results to a known standard. Compared to the single wavelength IR systems, the multiple wavelength systems require more time for identification.

Automated sorting systems can be modest, with equipment to recognize only a few materials. A basic system can distinguish three main classes: natural HDPE AND PP, PET and PVC, and mixed-color HDPE. A slightly more powerful system would distinguish PET from PVC.

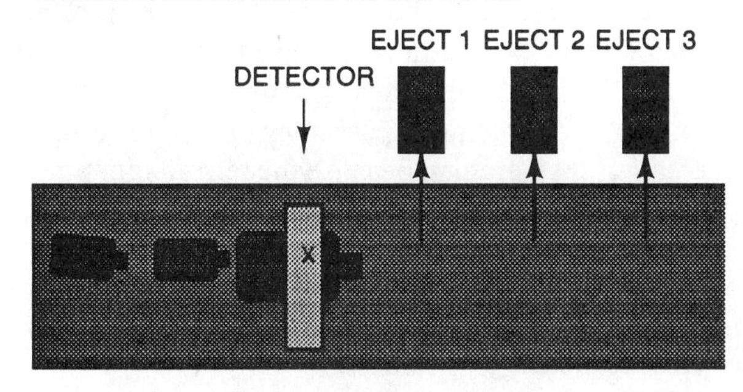

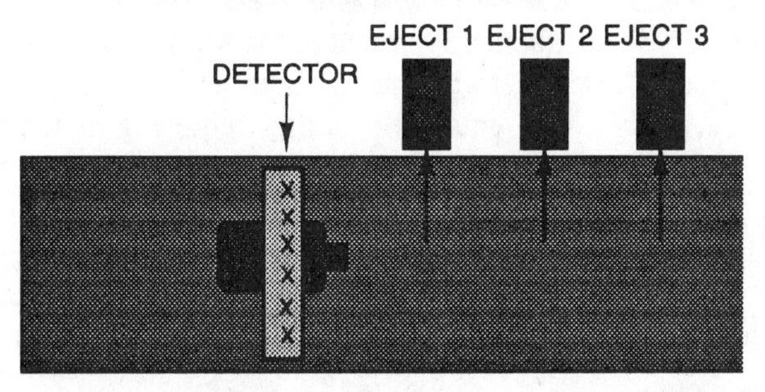

Fig. 2-35. Singulating conveying systems present containers to the detection equipment one-at-a-time.

Some color systems claim the ability to distinguish millions of shades of color. IR systems extend the capabilities further. After identification is complete, the selected containers are blown by air jets onto appropriate conveyors or hoppers. These automated systems can process containers at a rate of two to three containers per second or about 1500 lbs per hour. To improve on that rate, multiple lines are needed. However, the costs of such systems can reach $1 million dollars.

Chopped commingled materials. Elimination of all sorting can simplify the systems in MRFs. However, the chopping of commingled plastics results in mixed chopped flakes. Sorting the flakes into appropriate material streams can be done in various ways.

Flotation systems can distinguish materials based on differences in density. Froth flotation separates plastics by differing surface wetting potentials.

Systems based on an optical sorting technology bring a stream of chopped flakes past detectors. If the detectors indicate the presence of an unwanted flake, air blasts remove it from the stream. The efficiency and speed of these systems is still under development. Currently, one pass through the system can remove about 98 percent of contaminates. However, to clean the stream to the level of 10 ppm, several additional passes are required. Similar systems based on magnetic separation can distinguish PVC from PET.

Automotive Recycling

Approximately 10 million automobiles are discarded every year in the U.S. They arrive first at automotive dismantlers. After removing parts useful for resale, the dismantlers hand the remains over to shredders. There are approximately 180 shredders in the United States. After turning a car into pieces, the facilities sort ferrous from non-ferrous metals, and feed these materials to foundries, and steel mills for reprocessing. About 75 percent of the materials in automobiles are recycled. The shredder residue, sometimes called shredder fluff, contains plastics, glass, fabrics, adhesives, paint, and rubber. In 1993, the shredder fluff amounted to 3 million tons, all of which had to be landfilled.

In an effort to reduce the shredder residue, the major automobile companies have established procedures and guidelines for recycling. To make recycling work throughout the automotive industry, several major automobile companies have created the Vehicle Recycling Partnership. As part of the efforts, they established a code—SAE code J1344—for marking plastics to aid in identification during disassembly. This code utilizes the ISO (International Organization for Standardization) designations for plastics. (For information on ISO tests, see Chapter 6.)

In addition to efforts aimed at the entire automobile industry, major companies have established internal guidelines for recycling. For example, Ford Motor Company established recycling guidelines in 1993. These guidelines promote the use of recycled materials and encourage the reduction of painted plastics. Molded-in color reduces volatile emissions from painting operations and makes recycling easier. The guidelines also recommend the use of a limited number of plastics. It suggests the use of PP, ABS, PE, PA, PMMA, and PC. PVC should be used only where separation and recycling techniques are established. In addition, suppliers should "be encouraged to...take materials back for recycling at the end of the vehicle's useful life."

Chemical Recycling

Chemical recycling involves two levels of depolymerization. The original polymerization of some plastics is reversible. Depolymerization yields monomers, which can be used to make new polymers. Other plastics do not depolymerize to immediately useful monomers.

The type of depolymerization that produces useful monomers is called *hydrolysis*. It is beyond the scope of this discussion to throuoghly explain hydrolysis and the chemical reactions it utilizes. It is, however, a technique that is feasible for polyesters, polyamides and polyurethanes. The chemical techniques vary depending on the type of plastics. For years, companies making PET have used a form of hydrolysis to convert manufacturing waste PET to monomer.

Some plastics can be converted to monomers with thermal depolymerization. This process involves heating the plastics in the absence of oxygen. This process is also labeled pyrolysis. Acrylic, polystyrene, and some grades of acetal, yield monomers under these conditions.

Other plastics arose through irreversible reactions, and consequently can not be depolymerized into monomers. They can be depolymerized into useful petrochemical materials by pyrolytic liquefaction. Materials appropriate for this treatment are HDPE, PP, and PVC.

In simple pyrolytic liquefaction, pieces of plastics enter a tube that is heated to about 1 000 °F. The plastics melt, decompose into vapors, and the vapors condense into liquids. Liquids are used as feed stocks in petrochemical plants. Some facilities hope to make gasoline from the plastics.

Recycling in Germany

In the United States, some packaging and container manufacturers now face various recycling laws and mandates. Such laws do not currently extend to producers of appliances, automobiles, and electronics. In contrast, Germany has placed full responsibility for recovering, reprocessing or disposal of packaging on the manufacturers. Laws requiring takebacks for appliances, electronics, disposable, and automobiles are under consideration.

To meet this challenge, over 600 manufacturers and distributors combined their forces and created a system called Duales System Deutschland (DSD). DSD is a non-profit company which collects, sorts and arranges for reprocessing. To pay for DSD, each manufacturer pays a fee based on the weight and content of packaging material. During 1993, DSD collected 360000 tons of materials, but the capacity of the recycling equipment was about 115000 tons. A major problem was that much of the material collected by DSD was dirty and not recyclable. Consequently, the cost of sorting the collected materials have reached as much at $2000 per ton. As a result of the excessive volume and cost, members of the DSD had to store huge amounts of material and search for opportunities to sell the collected material. Incineration and landfilling have been investigated as temporary solutions.

Disposal by Incineration or Degradation

History of Incineration in the United States

Burning of solid wastes in open dumping areas was common for centuries; however, the concerns for environmental protection closed open dumping and burning areas, and replaced them with landfills. Some communities constructed special incinerators to dispose of the solid wastes. By 1960, approximately 30 percent of municipal solid waste (MSW) was handled in incinerators without any attempt to recover or use the heat. This method grew until the early 1970s. The Clean Air Act, which was passed in 1970, closed about one-half of incinerators, because the retrofits for pollution control were considered too expensive. To make incineration viable, some method to pay for the cost of equipping a facility was required. One possible solution was to use the solid waste as fuel in the generation of electrical energy or steam.

During the late 1970s, through 1980, there was a tremendous increase in waste-to-energy (WTE) facilities. By 1990, about 15 percent of the MSW was incinerated. These waste-to-energy facilities were expensive, with a well-equipped facility requiring at least $50 million to construct. In 1991, there were 168 active incinerators in the U.S. In comparison, Japan had about 1 900, and Western Europe over 500 incinerators.

Advantages of Incineration

One of the advantages of incineration is that it does not require sorting of the solid wastes. The entire collection of paper, plastics, and other materials can go into the incinerator. It provides a 80 to 90 percent reduction in the volume of the solid waste, turning many cubic yards of material into a few pounds of ash. WTE plants incur an operating cost of about $10 to $20 per ton. To offset the cost, the WTEs have two sources of revenue. They charge city solid waste organizations to deliver truck loads of compacted MSW. Incinerator facilities receive approximately $50 per ton for accepting the MSW. In addition, the incinerators sell electricity. The cost of electricity from these facilities is comparable to other electric generating plants—generally not significantly lower.

For many cities, the alternative to incineration is landfilling. However, landfill costs are high in highly populated regions of the country. In New York and Massachusetts, landfill costs can exceed $50 to $60 per ton.

Disadvantages of Incineration

When new incinerators are proposed, some citizens groups vehemently oppose them. The grounds for trying to stop new incinerators involve two factors: the incinerator ash, and the emissions from the incineration process.

Incinerators generate two types of ash, bottom ash and fly ash. *Bottom ash* comes from the bottom of the incineration chamber and contains noncombustible materials. Mass burn incinerators deliver the full contents of the compacted load of waste into the burning chamber. Bricks, rocks, steel, iron, and glass go into the bottom ash, along with the residues of combustion. The *fly ash* is material collected from the smoke stack gases by pollution control equipment.

Fly ash often contains relatively high concentrations of heavy metals, and some hazardous chemicals. In contrast, the bottom ash usually contains less toxic materials. Some incinerator facilities mix the fly ash and bottom ash together.

A point of conflict concerns the danger to citizens from the ashes, and the proper disposal method for the ashes. If the ash is considered a toxic material, then it must be landfilled in a special landfill designed for toxic materials. That raises the cost tremendously, compared to regular landfills.

The emissions from incinerators often contain various levels of furans, dioxins, arsenic, cadmium, and chromium. These are all very toxic materials and many people fear potential adverse health effects. Proponents of incineration argue for comparisons between incinerators and power plants that burn pulverized coal. Environmentalists argue that the emissions are potentially carcinogenic and should be banned immediately.

A particular concern for the plastics industry is dioxin. When compounds that contain chlorine, such as PVC and paper whitened by chlorine bleaching, are incinerated at high temperatures, various chlorinated chemicals are produced. When the gases containing these chemicals cool to about 300 °C., dioxin forms. The Environmental Protection Agency (EPA) views dioxin as a probable human carcinogen and a dangerous non-carcinogenic health threat.

An EPA report, published in 1995, cites medical waste incinerators as the biggest source of dioxin in the U.S., with the second largest source being municipal solid waste incinerators. Although the medical waste incinerators burn small volumes compared to MSW incinerators, the medical waste stream has a high PVC content. In addition, there are many more medical waste incinerators than MSW incineration facilities. In 1994, over 6700 medical waste incinerators were operational.

Two approaches have been proposed: first, the elimination of all chlorinated wastes from incinerators; and second, the improvement of emissions controls for incinerators. Reducing dioxin emissions will require significant investments in pollution-control equipment, perhaps involving at least 60 percent of existing incinerators. If the new air-quality regulations are enforced, about 80 percent of the existing medical incinerators will probably cease operation. To handle medical waste, a small number of extremely well-controlled incinerators may be able to meet the standards. Hauling the waste to a few incinerators will also add to the cost of handling medical waste.

Another possibility is to deliver medical waste to huge autoclaves, which heat the materials to a temperature that kills biohazards. After autoclaving, the waste materials then go into a traditional landfill.

Degradable Plastics

Considerable controversy surrounds biodegradable plastics. A focus for the controversy was rings or yokes used to hold six-packs and eight-packs of beer or soft-drink cans or bottles. It was proven by various environmental groups that some animals, particularly sea birds, were getting stuck or trapped in the rings. The response to this was legislation requiring that such rings, yokes, or other devices be degradable.

Raw material suppliers introduced photo- and biodegradable grades of material for these applications. The photodegradable materials contain chemicals that are sensitive to sunlight, and cause the rings to disintegrate. Biodegradable grades often contain corn starch or other starches, which are attacked by microorganisms in water or soil. The rings also disintegrate over time. In all cases, the thinner the packaging material, the quicker the physical deterioration.

Currently, 28 states have laws requiring degradable connecting devices. Michigan requires degradation within 360 days, Florida allows only 120 days. Other states do not specify the time lapse permitted.

This did not end the controversy—environmental groups attacked the degradable materials. They argued that bio- or photodegradable plastics can contaminate otherwise useful recycled plastics. They also view the degraded materials as a potential threat to water purity. After sufficient bio- or photodegradation, the rings break down into small pieces. However, these small chips are often rather chemically stable. For example, a grocery sack made of polyethylene with corn starch as an additive will disintegrate into tiny particles of polyethylene. It will appear to be gone from the road side, but the tiny pieces are still there. They are likely to remain intact for long periods of time.

A few companies have developed plastics that truly degrade and after a period of time, these materials turn into carbon dioxide and water. Even with these materials, the rate of degradation is effected by the thickness of the material. Films degrade rather rapidly, but thicker products take months or years to degrade.

Organizations in the Plastics Industry

The plastics industry supports a large number of organizations, ranging from international, all-encompassing societies to small, highly specific groups. Some of the larger groups will receive attention: Many of the smaller organizations will receive mention only.

The Society of the Plastics Industry (SPI)

The SPI was established in 1937 to serve as "the voice of the plastics industry." According to its mission statement, the SPI seeks to "promote the development of the plastics industry and enhance public understanding of its contributions while meeting the needs of society." The Society has a complex structure, involving 26 divisions, 4 regional offices, 3 special purpose groups, and 17 special services.

One of its most familiar activities is organizing the National Plastics Exposition (NPE), and International Plastics Exposition, every three years. This exposition began in 1946 and now draws visitors from 75 countries. The exposition brings together equipment manufacturers, raw materials suppliers, laminators, fabricators, mold makers, and manufacturers of plastics.

Because of the success and growth of NPE, the SPI has recently planned a second exposition, Plastics USA. The first Plastics USA occurred in October 1992. This three-day event, held 15 to 16 months after NPE, is similar to NPE, but smaller in scale.

NPE and Plastics USA are major shows. Specialty shows organized by SPI include:

- Composites Institute Annual Conference and EXPO
- PLASTICOS, held in Mexico City
- The Polyurethanes World Congress
- Annual Conference and New Product Design
- Competition of the Structural Plastics Division.

The 29 divisions cover all facets of the plastics industry. The larger divisions have standing committees and annual awards for scholarly papers or design innovations. For details about these divisions, contact the SPI for a Member Services Guide.

The special purpose groups include the Degradable Plastics Council, the Polyolefins Fire Performance Council, and the Styrene Information and Research Center.

The special services committees focus on managerial and public relations issues. These committees form the Societies formal link with governmental agencies. One example is the work of the Food, Drug and Cosmetic Packaging Materials Committee, which reviews Food and Drug Administration (FDA) policies. It also provides an avenue for discussion of common interests.

The SPI maintains a publications service. The free SPI Literature Catalog is available from the Literature Sales Department. To contact the Society, write:

SPI
1275 K Street, NW
Suite 400
Washington, DC 20005

Society of Plastics Engineers (SPE)

In 1942, 60 salesmen and engineers met near Detroit, Michigan and began the SPE "to promote scientific and engineering knowledge related to plastics." In 1993, its members numbered over 37800—divided among 91 sections in 19 countries. Large local sections hold monthly meetings, which provide social as well as technical benefit to members. The SPE maintains its world headquarters in Brookfield, Connecticut, and recently opened a European office in Brussels, Belgium.

The SPE supports and promotes formal education related to plastics. It encourages students to engage in research projects on plastics materials and processes and provides scholarships for both undergraduate and graduate study. Students can join SPE student chapters, which number 89 throughout the world.

The SPE provides continuing education for its members through a wide range of seminars and conferences. Regional and national technical conferences provide members with access to the most current research in the field. The largest conference is the Annual Technical Conference and Exhibition (ANTEC). In 1943, the SPE held its first Annual Technical Conference, which had 59 exhibitors and less than 2000 visitors. In 1993, ANTEC drew about 5000 members and hosted over 650 technical presentations. Specialized seminars deal with various aspects of the plastics industry in great depth.

Every month, the SPE publishes *Plastics Engineering*, a magazine containing articles on current development in plastics. In addition to *Plastics Engineering,* the Society also publishes three technical journals that include articles and papers of scientific and scholarly merit. The journals are: *Journal of Vinyl Technology*, published 4 times a year, *Polymer Engineering & Science*, published in 24 issues per year, and *Polymer Composites,* six issues per year.

The SPE publishes a catalog annually, containing a variety of books from world-renowned publishers. The books cover two broad areas: one concerning engineering and processing of plastics materials, and the other on polymer science. The catalog also offers the proceedings of regional technical conferences and the ANTEC proceedings and subscriptions to the SPE journals.

The SPE holds the position as the world's largest technical organization for scientific and engineering knowledge relating to plastics. For further information or membership applications, write to:

Society of Plastics Engineers
14 Fairfield Drive
Brookfield CT 06804

The Plastics Institute of America (PIA)

The PIA began in 1961 as a non-profit corporation. It sponsors graduate research at universities. The recipients are graduate students studying polymer science and engineering at U.S. and Canadian universities. It also sponsors technical instruction for industry personnel. This education takes the form of technical courses offered each spring and fall in cities across the United States. Topics include injection molding, plastics recycling, extrusion techniques, and polymer testing. The PIA also holds three annual industrial conferences: RECYCLINGPLAS, on opportunities in recycling; FOODPLAS, on plastics in food packaging; and CONSTRUCTIONPLAS, concerning plastics in construction.

For information about the PIA and membership applications, contact:

The Plastics Institute of America, Inc.
277 Fairfield Rd., Suite 100
Fairfield, NJ 07004-1932

Society for the Advancement of Material and Process Engineering (SAMPE)

This society has members who are involved in the development of materials and process—chiefly materials and process engineers. It publishes the *SAMPE Journal* bimonthly and the *SAMPE* quarterly.

For information about SAMPE contact:

SAMPE
P.O. Box 2459
Covina, CA 91722

American Society of Electroplated Plastics, Inc. (ASEP)

This society serves individuals and companies with information about electroplated plastics. It also develops standards for the quality of plating and manufacturing of plated plastic products. It published *ASEP Standards and Guidelines for Electroplated Plastics.* For more information, contact:

ASEP
1101 14th St. N.W., Suite 1100
Washington D.C. 20005-5601

American Society for Plasticulture (ASP)

This society focuses on the use of plastics in agriculture. It publishes *Agri-Plastics Report*, which contains technical information about plastics in agriculture. It sponsors a special committee on plastics disposal. For more information, contact:

ASP
P.O. Box 860238
St. Augustine, FL 32086

Other Organizations

American Polyolefin Association, Inc.
2312 East Mall
Ardentown, DL 19810

International Association of Plastics Distributors (IAPD)
6333 Long St., Ste, 340
Shawnee, KS 66216

Association of Post-consumer Plastics Recyclers (APR)
c/o Wellman Inc.
1040 Broad St.
Suite 302
Shrewsbury, NJ 07702

Association of Rotational Molders (ARM)
2000 Spring Road
Suite 511
Oak Brook, IL 60521

Center for Plastics Recycling Research
Rutgers University
Building 4109, Livingston Campus
New Brunswick, NJ 08903

National Association for Plastic Container Recovery (NAPCOR)
3770 Nations Bank Corporate Center
100 N. Tryon St.
Charlotte, NC 28202

National Plastics Center and Museum
P.O. Box 639
Leominster, MA 01453

National Recycling Coalition (NRC)
1101 30th St. N.W. Ste 305.
Washington, DC 20007

Plastic Lumber Trade Association (PLTA)
c/o Plastic Lumber Company
540 S. Main St.
Building 7
Akron, OH 44311-1010

Plastics Institute of America Inc. (PIA)
277 Fairfield Road
Suite 307
Fairfield, NJ 07004-1931

Polymer Processors Association
4040 Embassy Parkway
Akron, OH 44333

Polystyrene Packaging Council (PSPC)
1275 K St. N.W.
Suite 400
Washington DC 20005

Polyurethane Foam Association
P.O. Box 1459
Wayne, NJ 07474-1459

Polyurethane Manufacturers Association (PMA)
800 Roosevelt Road
Building C, Suite 20
Glen Ellyn, IL 60137

SMC Automotive Alliance
26677 W. Twelve Mile Road
Southfield, MI 48034

Publications for the Plastics Industry

Modern Plastics is a monthly magazine published by McGraw-Hill, Inc. It includes technical papers, market and business reports, and plastics-related advertising. For subscription information, contact:
Modern Plastics
P.O. Box 602
Hightstown, NJ 08520

Plastics World is published monthly by The Cahners Publishing Company. It includes business, technical articles. For subscription information, contact:
Plastics World
P.O. Box 5391
Denver, CO 80217-5391.

Plastics Technology is published monthly by Bill Communications, Inc. It features articles for plastics processors and a pricing update. Plastics Technology also publishes PLASPEC, which provides information on various grades of materials. For more information:
Plastics Technology
633 Third Ave.
New York, NY 10017-6743

Trade Newspapers

Plastics News is a weekly publication from Crain Communications Inc. It includes short articles on business topics. It reports on sales and acquisitions of processing plants, new materials and designs, seminars, and resin prices. For subscription information:
Subscription Department, Plastics News.
965 E. Jefferson
Detroit, MI 48207-3185

Plastics Machinery and Equipment is a monthly publication of Advanstar Communications, Inc. It reports on advances in plastics processing equipment. The articles include considerable technical data on processes and materials. It also features advertising from equipment makers. For subscription information, contact:
Plastics Machinery & Equipment Magazine
131 W. First St.
Duluth, MN 55802

Vocabulary

The following vocabulary words are found in this chapter. Look up the definition of any of these words you do not understand as they apply to plastics in the glossary, Appendix A.

Biodegradable
Bottom ash
Clean Air Act
Elastomers
Elutriation
Flotation tank
Fly ash
Gaylord
Hydrolysis
IR
MRF
MSW
PCR
Photodegradable
Picking line
Plastics
Polymers
Pyrolysis
RCRA
SAE code J1344
TPE
WTE

Questions

1-1. Explain the difference between recycling and reprocessing.

1-2. What is pyrolysis?

1-3. How large is a gaylord?

1-4. How do plastics and elastomers differ?

1-5. How are resins different from plastics?

1-6. Approximately what percentage of the U.S. workforce is directly involved in plastics manufacturing?

1-7. How effective are bottle deposit laws?

1-8. Containers marked with the number 5 contain what type of plastics?

1-9. What does the acronym MRF stand for?

1-10. What are the four major types of detectors used in automatic sorting systems for plastics?

1-11. Explain hydrolysis.

1-12. What conditions are required for the formation of dioxins during incineration?

Activities

Recycling HDPE

Introduction. Recycling of post-consumer HDPE is one of the biggest and most successful plastics recycling efforts. The bulk of the recycled HDPE comes through curbside pickup of HDPE bottles and containers—particularly milk jugs. Some companies advertize their use of recycled materials to demonstrate their environmental concern. Figure 2-36 shows a tool for removing bicycle tires from rims.

Procedure

2-1. If a curbside recycling program exists nearby, determine which plastics are acceptable. If the program accepts both PET and PVC, how do they guarantee that the PVC does not contaminate the PET?

2-2. Does the program expect the participants to sort the materials? Is the collection put into a truck with specific compartments, or is the material commingled?

2-3. Is there a MRF nearby? If so, how does it sort the plastics? Does it have any automated sorting equipment?

2-4. Does the program distinguish between natural HDPE and colored HDPE? If yes, how does this sorting occur?

2-5. How are the sorted HDPE containers handled? Is a baler used? Is the material chopped into flakes?

2-6. Who buys the HDPE from the recycling organization? What is the current price for baled material, dirty chopped flakes, clean chopped flakes, and reprocessed pellets?

2-7. How much difference is there between the price of recycled materials and virgin materials?

2-8. Is there a mandate on the percentage of post-consumer recycled materials in packaging? If yes, what percentage must be PCR?

2-9. What technologies are required for a company to turn bales for crushed HDPE containers into clean pellets or flakes?

2-10. Write a report summarizing your findings.

Additional recycling activity. This activity is not feasible without adequate equipment. If processing equipment and choppers are available, try to determine the difference between recycled material and virgin material.

CAUTION: Use processing equipment only under the guidance of trained supervisors.

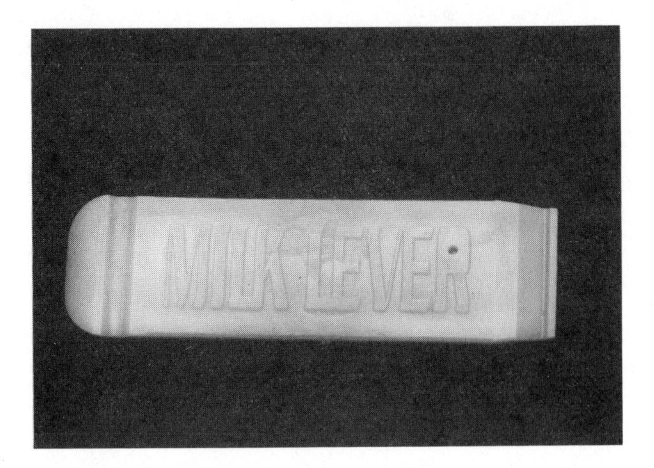

Fig. 2-36. The marketing of this bike tire removal tool appeals to the environmental concern of customers.

Procedure

2-1. Acquire milkjugs from a local recycling company. Generally, such companies are pleased to donate some HDPE milkjugs for students learning about recycling.

2-2. List the decisions needed. For example, how should labels be treated? How should cleaning occur? If a chopper is small and cannot handle complete bottles, how should they be reduced in size? Figure 2-37 shows pieces of milk jugs cut into a size that fits into a small chopper, along with chopped flakes. Should contaminated bottles be rejected? What determines the difference between contaminated and not contaminated? What should be done with screw-on or pop-on closures?

Fig. 2-37. These milk jug pieces were an appropriate size for a small chopper. The fines were not removed from the flakes.

2-3. Determine a procedure to follow, and process enough jugs to yield 3 to 5 lbs of chopped flakes. Should the flakes be washed? If so, how will this occur. Figure 2-37 also shows flakes produced from the milk jug pieces.

2-4. Extrude the clean flakes to produce pellets. Is there any odor of sour milk attached to the pellets?

2-5. Is there a color difference between virgin pellets and recycled pellets?

2-6. One way to examine the recycled material for contaminants is to make it into film. If blown film equipment is available, try to process the recycled material. If not, squeeze out the thinnest film possible using a heated platen press. Are impurities visible in the recycled materials? A microscope or magnifying glass will help answer this question.

2-7. Stretch, tear, and bend film samples of virgin and recycled materials. Do they appear to possess differing physical properties?

2-8. Write a report summarizing the findings.

Chapter 3

Elementary Polymer Chemistry

Introduction

Many people have no difficulty recognizing such common metals as copper, aluminum, lead, iron, and steel. They also know the difference between oak, pine, walnut, and cherry. In addition to identifying several metals and woods, many people also understand some physical qualities of these materials. For example, they know that steel is harder and stronger than copper.

In contrast, the significant characteristics, and even the names of major types of plastics are often unknown. This chapter will discuss several commercial plastics. It will acquaint the reader with polymer names and chemical structures. In order to do this, some knowledge of basic chemistry is required. This chapter assumes the reader has an understanding of basic chemistry, elements, the periodic table, and some chemical structures. In this and following chapters, every reference to an atom will also include its chemical symbol. The outline for this chapter is:

I. Review of basic chemistry
II. Hydrocarbon molecules
III. Macromolecules
   A. Carbon chain polymers
   B. Carbon and other elements in the backbone
IV. Molecular organization
   A. Amorphous and crystalline polymers
V. Intermolecular forces
VI. Molecular orientation
   A. Uniaxial
   B. Biaxial
VII. Thermosets

Review of Basic Chemistry

Understanding basic chemistry is crucial to learning about plastics. Chemical terms are used to explain the names and properties of polymers. Chemical structures determine the unique characteristics of polymers, as well as their limitations. It is convenient to begin this section with a review of molecules and chemical bonds.

Molecules

Molecules occur when two or more atoms combine. The properties of molecules stem from three major factors: the elements involved, the number of atoms joined together, and the type of chemical bonds present. The number of atoms joined determines the size of the molecule, and the bonds determine its strength.

For example, water (H_2O) is made up of hydrogen atoms chemically combined with oxygen. Water consists of two hydrogen atoms and one oxygen atom—three atoms in total. Since only three atoms are involved, water is a very small molecule. The molecular mass of water is 18 because the number of atomic mass units (amu's) is 16 for oxygen and 1 for each hydrogen.

Chemical bonding concerns the ways atoms can attach to each other. There are three basic categories of primary chemical bonds: 1) *metallic bonds*, 2) *ionic bonds*, and 3) *covalent bonds*. Of these three, covalent bonding is the most important for plastics. Covalent chemical bonding involves the sharing of electrons between two atoms. The exact mechanism of this sharing is beyond the scope of this chapter; however, it is important to know that covalent bonds have known

C:C or C—C	C::C or C═C	C⋮⋮C or C≡C
Single bonds	*Double bonds*	*Triple bonds*

Fig. 3-1. Types of covalent bonds

strengths and lengths, which depend on the atoms that are combined. Covalent bonds can involve varying numbers of electrons, as shown in Figure 3-1. The bond that consists of the least number of electrons (2) is called a *single covalent bond.* If more electrons are involved (4 or 6), the bonds are called *double* or *triple covalent bonds*, respectively.

Unless otherwise specified, comments about covalent bonding will refer only to single covalent bonds.

Hydrocarbon Molecules

Hydrocarbons are materials that consist mainly of carbon and hydrogen. Pure hydrocarbons contain only carbon and hydrogen. When hydrocarbon molecules have only single covalent bonds, they are considered *saturated.* The word saturated implies that the bonding cites are fully "loaded-up." In contrast, *unsaturated molecules* contain some double bonds. Since double bonds are more chemically reactive than single bonds, the saturated molecules tend to be more stable than unsaturated molecules. The following table (Table 3-1) lists saturated hydrocarbon molecules.

As indicated in Table 3-1, molecules with 1, 2, 3 or 4 carbon atoms have boiling points below 0°C. This

Table 3-1 Saturated Hydrocarbon Molecules with Melting and Boiling Points.

Formula	Name	Melting Point, °C	Boiling Point, °C
CH_4	Methane	−182.5	−161.5
C_2H_6	Ethane	−183.3	−88.6
C_3H_8	Propane	−187.7	−42.1
C_4H_{10}	Butane	−138.4	−0.5
C_5H_{12}	Pentane	−129.7	+36.1
C_6H_{14}	Hexane	−95.3	68.7
C_7H_{16}	Heptane	−90.6	98.4
C_8H_{18}	Octane	−56.8	125.7
C_9H_{20}	Nonane	−53.5	150.8
$C_{10}H_{22}$	Decane	−30	174
$C_{11}H_{24}$	Undecane	−26	196
$C_{12}H_{26}$	Undecane	−10	216
$C_{15}H_{32}$	Pentadecane	+10	270
$C_{20}H_{42}$	Eicosane	36	345
$C_{30}H_{62}$	Triacontane	66	distilled at
$C_{40}H_{82}$	Tetracontane	81	reduced pres-
$C_{50}H_{102}$	Pentacontane	92	sure to avoid
$C_{60}H_{122}$	Hexacontane	99	decomposition
$C_{70}H_{142}$	Heptacontane	105	

means that they are gasses at room temperature. The molecules from 5 to 10 carbons are rather volatile liquids at room temperature. When the number of carbons increases beyond 20, the materials become solid.

The hydrocarbon molecule with 8 carbons is octane, which is familiar because of its use as automotive fuel. The molecular mass of octane is 114 amu's, and its chemical structure is:

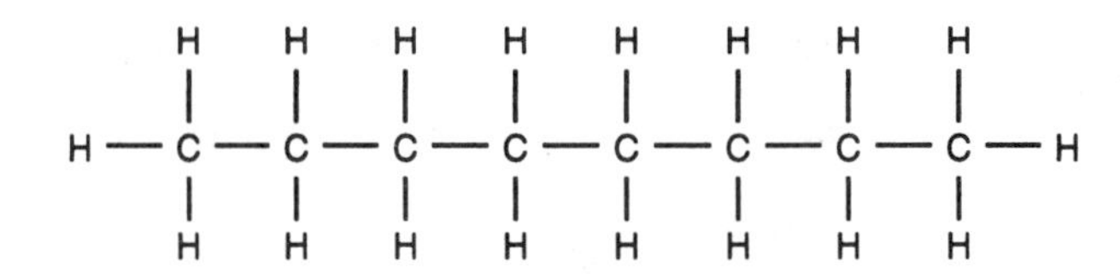

Fig. 3-2. Chemical structure of octane.

Note: Carbon has 12 amu's and hydrogen 1.

$$(8 \times 12) + (18 \times 1) = 114.$$

Notice that the hydrogens are connected (with single covalent bonds) only to carbons. There are no bonds between the hydrogens. This means that the structural integrity of the molecule is provided by the bonds between the carbons. If we imagine the hydrogens removed, what would remain is a row of 8 carbon atoms. This line is called the *backbone* of the molecule. The strength of the molecule is substantially determined by the strength of the carbon-to-carbon single bonds.

As the number of carbons in the backbone increases, the molecules get longer and longer. Light liquids give way to viscous liquids or oils. Oils become greases. Greases turn to waxes, which are weak solids. Weak solids become flexible solids. Flexible solids turn to rigid solids. Eventually, the molecules get so long

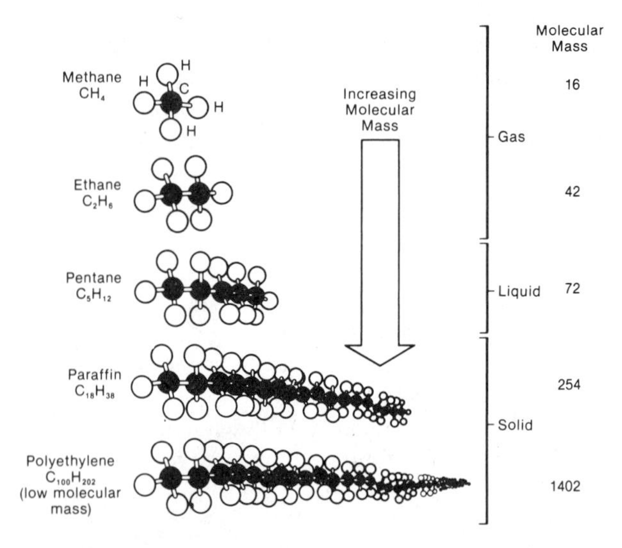

Fig. 3-3. Growing chain length (molecular mass) from a molecule of gas to a solid plastics.

they become rigid and strong at room temperature. Figure 3-3 graphically portrays this sequence of changes.

Macromolecules

The simplest plastic structure is *polyethylene*, a saturated hydrocarbon with the abbreviation PE. A common PE molecule contains approximately 1 000 carbon atoms in its backbone, also called a carbon chain. Because of their large size, molecules of plastics materials are often called *macromolecules.*

Although the molecules are very large, they cannot easily be seen. A common PE molecule is approximately 0.0025 mm, stretched out. The thickness of a sheet of typing paper is about 0.076 mm. It would require 30 stretched-out molecules, end to end, to reach from the top side to the bottom side of the paper sheet.

Because this kind of molecule—a macromolecule—is so long, chemists don't show its extended chemical structure. An abbreviated method is used. The smallest repeating structure is named the *mer*. It is drawn in brackets as follows:

$$\left[\begin{array}{ccc} & H & H \\ & | & | \\ - & C - & C - \\ & | & | \\ & H & H \end{array}\right]_n$$

The little *n* is called the subscript. It stands for the degree of polymerization. *Polymerization* means joining many mers together. (Chapter 8 will discuss various polymerization techniques.) The *degree of polymerization* (DP) represents the number of mers joined in one molecule. If the DP is 500, that means that 500 of the repeating units are linked together. The molecule then would be PE, 1000 carbons long, with a molecular mass of 14000 amu's. The molecular mass is calculated by multiplying the DP (in this case 500) times the molecular mass of the mer (in this case 28). If the DP was 9, the material would be paraffin, a hydrocarbon wax with a backbone 18 carbons long.

Carbon Chain Polymers

There are literally thousands of plastics, and polymer chemists are constantly developing new polymers. However, most industrial plastics contain a rather limited number of elements. A large number of plastics consist of carbon (C), hydrogen (H) and the following atoms:

oxygen (O)
nitrogen (N)
chlorine (Cl)
fluorine (F)
sulfur (S)

Table 3-2. Plastics Involving Single Substitutions

X Position	Material Name	Abbreviation
H	Polyethylene	PE
Cl	Polyvinyl chloride	PVC
Methyl group	Polypropylene	PP
Benzene ring	Polystyrene	PS
CN	Polyacrylonitrile	PAN
$OOCCH_3$	Polyvinyl acetate	PvaC
OH	Polyvinyl alcohol	PVA
$COOCH_3$	Polymethyl acrylate	PMA
F	Polyvinyl fluoride	PVF

(Methyl group is:
$$\begin{array}{ccc} & | & \\ H - & C & - H \\ & | & \\ & H & \end{array}$$

(Benzene ring is:

Homopolymers. The chemically simplest plastics are called *homopolymers,* because they contain only one basic structure. A convenient approach to understanding several plastics is to rewrite the structural formula of PE as follows:

$$\left[\begin{array}{ccc} & H & H \\ & | & | \\ - & C - & C - \\ & | & | \\ & H & X \end{array}\right]_n$$

If a hydrogen (H) goes in the *X* position, the material is PE. If a chlorine (Cl) goes in the *X* position, the material is polyvinyl chloride, PVC. Table 3-2 lists several other plastics which have this form.

In some cases, two hydrogen (H) atoms are replaced. Again, it is convenient to rewrite the PE structure as:

$$\left[\begin{array}{ccc} & H & Y \\ & | & | \\ - & C - & C - \\ & | & | \\ & H & X \end{array}\right]_n$$

See Table 3-3.

Table 3-3. Plastics Involving Two Substitutions

X Position	Y Position	Material Name	Abbrev.
F	F	Polyvinylidene fluoride	PVDF
Cl	Cl	Polyvinyl dichloride	PVDC
$COOCH_3$	CH_3	Polymethyl methacrylate	PMMA
CH_3	CH_3	Polyisobutylene	

If 3 or more hydrogens are replaced, the new atoms are often fluorine, and the resulting plastics are fluoroplastics. If all four hydrogens are replaced with fluorine, the material is PTFE, polytetrafluoroethylene, which is sold under the trade name Teflon.

Copolymers. Up to this point, all the plastics mentioned contain only one type of functional group. The structural formulas require only one bracket of the H H type.

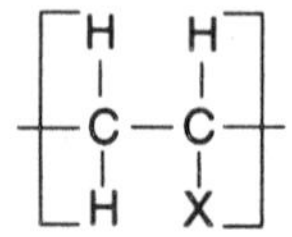

```
 ┌H   H┐
  |   |
 ─C ─ C─
  |   |
 └H   X┘
```

These materials are *homopolymers*. However, some plastics combine two or more different functional groups—two or more different mers. If only two different mers are involved, the material is called a *copolymer*.

An example of a copolymer is styrene-acrylonitrile (SAN). Using the information in Table 3-1, we can draw its chemical structure. First, the styrene structure is

```
  H   H
  |   |
 −C − C−
  |   |
  H   ⌬
```

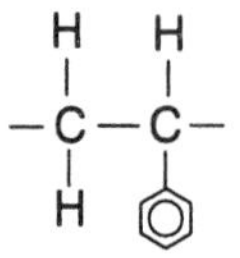

Next, the acrylonitrile structure is

```
  H   H
  |   |
 −C − C−
  |   |
  H   C
      |||
      N
```

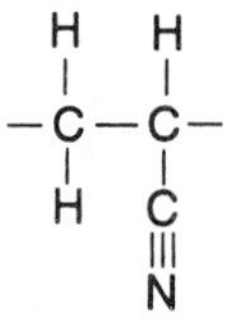

To complete the chemical structure of the copolymer, place the two structures side by side. The resulting structure represents SAN, styrene-acrylonitrile.

```
 ┌H   H┐ ┌H   H┐
  |   |   |   |
 ─C ─ C───C ─ C─
  |   |   |   |
 └H   ⌬┘n H   C
              |||
         └    N┘m
```

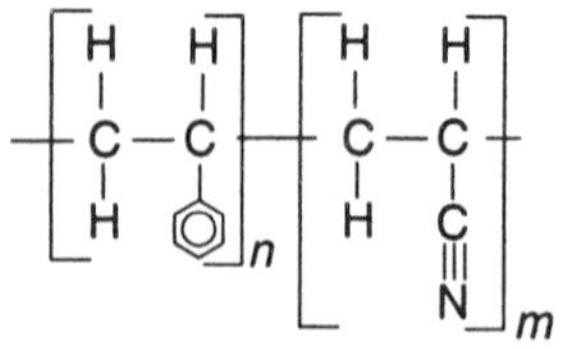

When two different mers are involved, four possible combinations arise: 1) if the mers alternate, ABABABABAB, the material is an *alternating copolymer,* 2) if the mers join up in a haphazard or random fashion, ABBAAAABAABBBABBABBBBAA, the material is a *random copolymer,* 3) if the mers join up in chunks of like mers, AAAABBBAAAABBBAAAABBBAAAABBB, the material is a *block copolymer*, and 4) if the backbone remains made up of one mer, and joining side groups of the second mer, the material is called a *graft copolymer.* Its structure is

```
AAAAAAAAAAAAAAAAAAAAAA
    B              B
    B              B
    B              B
```

Printed chemical structures do not clearly indicate whether a copolymer is alternating, random, or block. Printed structures identify the mers, but are too short to show the larger arrangement. Don't assume a material is an alternating copolymer because the structure shows two mers, side by side.

All the information needed for the chemical structure of SAN is provided in Table 3-2. This table also specifies other copolymers. For example, polyethylene-vinyl acetate (EVA), and polyethylene methyl acrylate (EMAC), are two more copolymers based on the materials listed in Table 3-2.

Terpolymers. If three separate mers combine to form a material, it is identified as a *terpolymer*. Terpolymers can also exhibit alternating, random, block, or branch structures. An example of a terpolymer, built from the materials in Table 3-1, is acrylic-styrene-acrylonitrile (ASA).

The acrylic structure is:

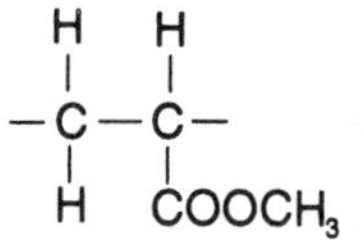

```
  H   H
  |   |
 −C − C−
  |   |
  H   COOCH3
```

The styrene structure is

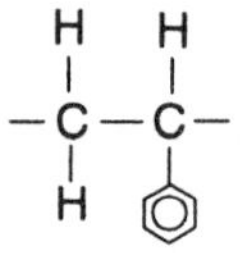

```
  H   H
  |   |
 −C − C−
  |   |
  H   ⌬
```

The acrylonitrile structure is:

```
  H   H
  |   |
 −C − C−
  |   |
  H   C
      |||
      N
```

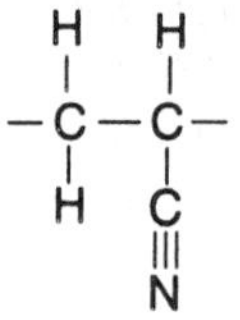

Putting these three pieces together results in the chemical structure for ASA.

Carbon and Other Elements in the Backbone

The types of macromolecules discussed so far have all carbon backbones. This is not true of all common macromolecules. Many include oxygen, nitrogen, sulfur, or benzene rings in the main chain. Most of these materials are homopolymers, with decreasing counts of co- and terpolymers. These materials tend to be unique in chemical structure and are not easily categorized. Specific details about the structures of these materials will appear in Chapter 8.

Molecular Organization

Molecular organization deals with the arrangement of molecules rather than the details about elements and chemical bonding (which is termed molecular structure). This section treats the major categories of molecular arrangement and the effects of arrangement on selected properties.

Amorphous and Crystalline Polymers

Plastics exhibit two basic types of molecular arrangement, *amorphous* and *crystalline*. In amorphous plastics, the molecular chains have no order. They are randomly twisted, kinked, and coiled, as shown in Figure 3-4.

Amorphous plastics can be rather easily identified because they are transparent if no fillers or color pigments are present. Department store display cases are often acrylic, because of the high clarity of this amorphous polymer.

Some plastics develop crystalline regions. In those regions the molecules take on a highly ordered structure. How this occurs is beyond the scope of this chapter, however it is generally accepted that the polymer chains fold back and forth, producing highly ordered crystalline regions, as seen in Figure 3-5.

Plastics do not fully crystallize like metals do. Crystalline plastics are more accurately termed semicrystalline materials. This means that they consist of crystalline regions surrounded by noncrystalline, amorphous areas. (See Figure 3-6.)

Fig. 3-4. Amorphous arrangement.

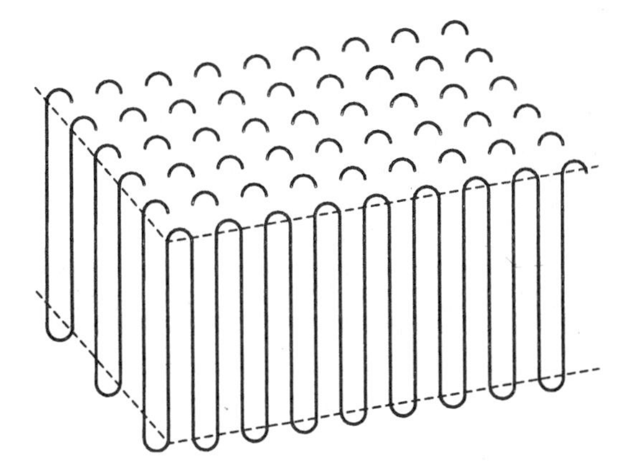

Fig. 3-5. A crystalline region.

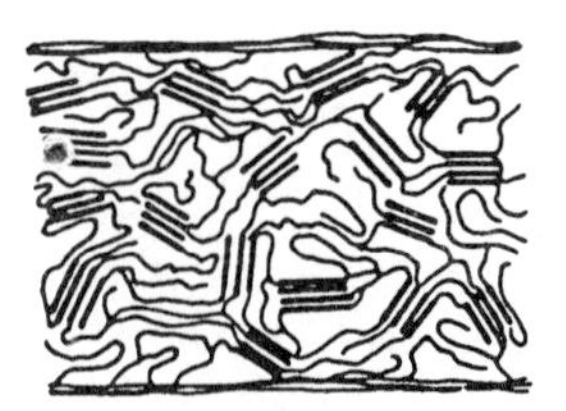

Fig. 3-6. Mixture of amorphous and crystalline regions.

Although plastics have carbon backbones, some of them crystallize, and others remain amorphous. A major factor in this difference is the regularity and flexibility of the polymer chain.

Since hydrogen (H) is the smallest atom, any atom that replaces hydrogen will be larger than it. Single atom replacements create small "bumps" on a polymer chain. Small groups, such as methyl (1 carbon, 3 hydrogens) or ethyl (2 carbons, 5 hydrogens) represent medium-sized bumps on the chain. Groups containing roughly 10 or more atoms, such as a benzene ring (6 carbons, 6 hydrogens) are large bumps on a molecule.

A group of atoms attached to a backbone is often called *side group*, especially if it is chemically different from the main chain. If the chemical structures attached to a backbone are identical to the backbone, the molecule is often considered a *branched* structure. Chain branching concerns both the size (usually the length) of the branch and the frequency of branching. When the branches are long and/or numerous, they prevent molecules from getting very close. When the branches are small or infrequent, the molecules can "snuggle up" and form denser solids.

Drawings that try to depict the shape and size of molecules come in a variety of types. The graphically simplest model is a line formula, which shows only the bonds, not the atoms. A line formula for PE is shown Figure 3-7. Notice that the carbon atoms create a zigzag pattern, because the bonds are slightly angular. The bond angles between two carbon atoms in a carbon backbone polymer is 109.5°.

When chains fold and coil, they twist or rotate between carbons. Figure 3-8 shows how coiled PE molecule in a line formula might appear.

Keep in mind that a coiled molecule is a three-dimensional structure, not well represented by a two-dimensional drawing. Figure 3-9 is an attempt to draw a kinked backbone. For clarity, all atoms connected to the carbon backbone were omitted.

Fig. 3-7. Line formula for polyethylene.

Fig. 3-8. Line formula of a coiled molecular chain.

Fig. 3-9. Graphic representation of a kinked backbone.

A space-filling model is one that tries to show a three-dimensional structure, but straightens and flattens out the backbone. Figure 3-10 shows a space-filling drawing of a polyethylene molecule.

This image will be successful if it conveys the regularity or "smoothness" of the PE molecule. Since it is so even, it crystallizes readily.

In PVC, a chlorine replaces one hydrogen per mer. A model of this molecule is shown in Figure 3-11. The bumps caused by the chlorine atoms affect the ability of the molecules to crystallize. It is partially crystalline, but less than PE.

In PS, a benzene ring replaces one hydrogen per mer. The resulting structure is very bumpy. A drawing of a benzene ring appears in Figure 3-12. To reduce confusion, the zigzag line in the following drawing of a section of a PS molecule (Fig. 3-13) indicates the backbone. The benzene rings prevent crystallization and cause PS to be amorphous throughout.

Optical effects of crystallinity. The amorphous materials are transparent because the haphazard arrangement of the chains does not uniformly disrupt light. In contrast, semicrystalline polymers have highly ordered crystalline regions. These crystalline regions significantly deflect light. The result is that semicrystalline materials are usually translucent or opaque.

A short demonstration can make this difference apparent. When sufficiently heated, semicrystalline polymers loose their crystalline regions and become completely amorphous. A piece of a plastic milk jug (high density polyethylene, HDPE) is readily available and contains only PE, with no fibers or colorants.

At room temperature, the piece of HDPE is translucent. If heated on a hot plate or in a flame, the regions that get hot enough will become transparent.

CAUTION: Do not ignite the material!! Upon cooling, the transparent regions will return to translucent.

The crystalline regions are disrupted by the heat, causing the material to become amorphous throughout. On cooling, some regions become highly ordered, other regions remain disordered. The ordered regions cause light beams to be difracted, rather than pass through with little disturbance.

Dimensional effects of crystallinity. Besides the optical differences, when melted plastics cool to a solid state, the crystalline materials shrink more than amorphous ones. That is because when the crystal regions form, those regions require less volume than they did when amorphous, due to the closeness of the folded chains. That results in a greater shrinkage. The sim-

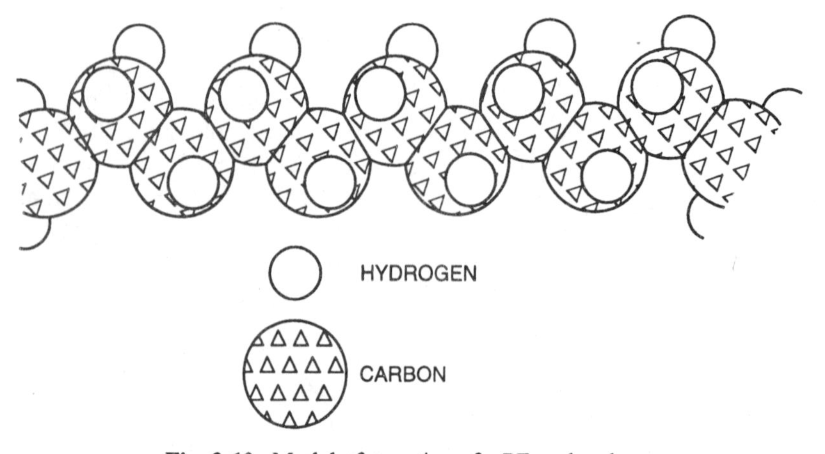

Fig. 3-10. Model of a section of a PE molecule.

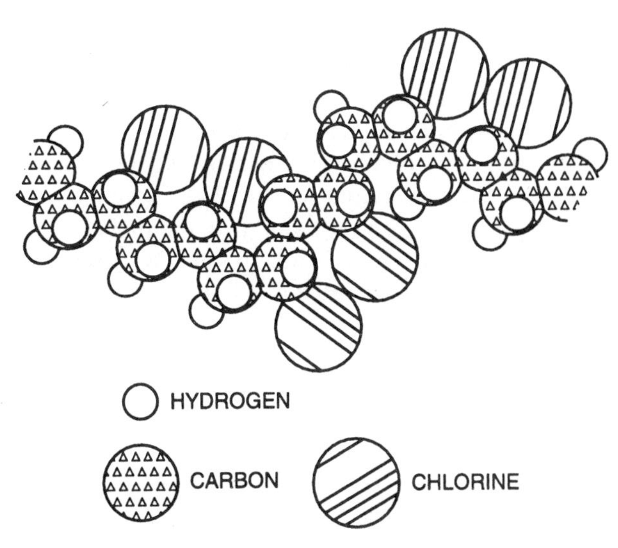

Fig. 3-11. Model of a section of a PVC molecule.

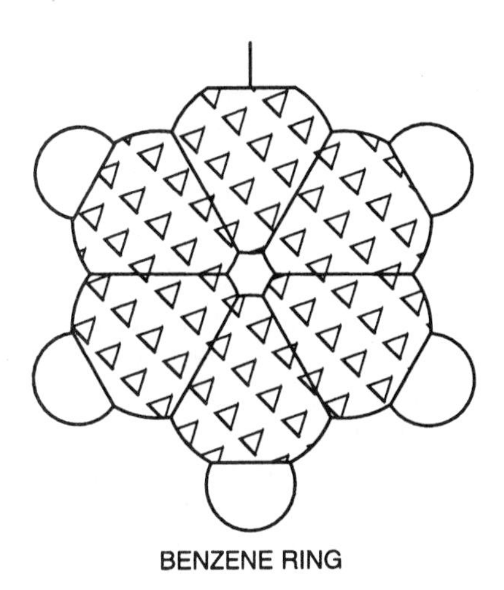

Fig. 3-12. Model of a benzene ring.

plest, hand operated injection molding machine can demonstrate this difference. Using the same mold, inject a natural crystalline polymer, such as HDPE. After purging out the HDPE, inject a natural amorphous polymer, like PS. The difference in length is readily noticeable, especially if the part is at least two inches long.

Intermolecular Forces

The size of atoms and side groups affects the crystallinity of plastics. The amount of crystallinity affects optical characteristics and shrinkage. However, there are major differences in some physical properties that are not explained by crystallinity. Two such properties are melting point, and tensile strength. A good example is provided by two types of polyamide, (nylon). Nylon 6 has a melting point of 220 °C (428 °F) and a tensile strength of 78 MPa (11 000 psi). In contrast, nylon 12 has a melting point of 175 °C (347 °F) and a tensile strength of 50 MPa (7 100 psi).

Intermolecular interactions are a major factor in these differences. *Intermolecular interactions* are attractions between molecules or between atoms in different molecules. These forces are very different from chemical bonds, which are much stronger. However, these intermolecular forces influence the amount of energy needed to disrupt (melt, break, stretch, dissolve) the materials. Three types of interactions are important

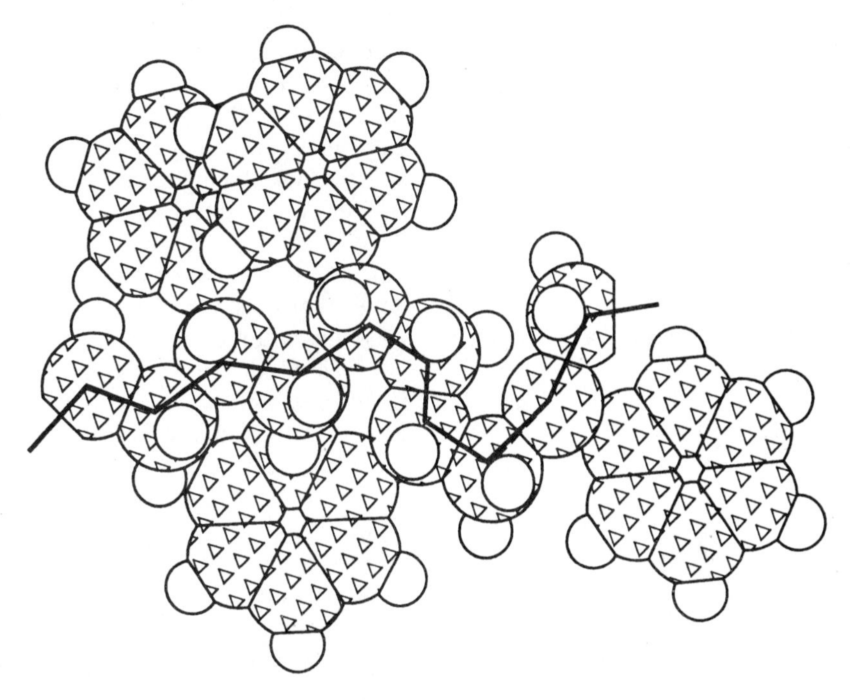

Fig. 3-13. Model of a section of a PS molecule.

in plastics: 1) Van der Waal's forces, 2) *dipole interactions*, and 3) *hydrogen bonds*. Van der Waal's forces occur between all molecules. They contribute only slightly to differences between various polymers. Dipole interactions occur when molecules or portions of molecules exhibit *polarity*, or unbalanced electrical charges. Hydrogen bonding is a special case of dipole interaction and requires a bond between hydrogen and oxygen or hydrogen and nitrogen. Hydrogen bonds are the strongest of the intermolecular forces.

In plastics, hydrogen bonds are very important. To better understand the effect of hydrogen bonds, think about the physical properties of water and methane.

	Molecular Weight	**Melting Point**	**Boiling Point**
H_2O	18	0 °C.	100 °C.
CH_4	16	–183 °C.	–162 °C.

These two molecules are of similar size. Why are these properties so different? The answer lies in hydrogen bonds. Water abounds in hydrogen bonds, and methane has no hydrogen bonds. In the following figure (Fig. 3-14) the dotted line represents the attraction between an oxygen in one molecule and a hydrogen in a neighboring molecule.

Nylon, a plastic that contains nitrogen in the backbone, is an excellent example of hydrogen bonding. Nylon fibers are elastic because the hydrogen bonds act like springs. Fibers spring back after stretching because of the force of the hydrogen bonds. If it were not for hydrogen bonds, nylon stockings would sag.

To know whether a plastic has dipoles or hydrogen bonds, it is not necessary to memorize the bonding characteristics of various plastics. A simple set of conditions can signal the presence of dipoles and/or hydrogen bonds.

The following combinations of atoms signal a permanent dipole. Dipoles occur if we find a:

a. carbon-chlorine single bond
b. carbon-fluorine single bond
c. carbon=oxygen double bond

The following combinations signal a hydrogen bond:

a. a carbon-OH single bond
b. a nitrogen-hydrogen single bond

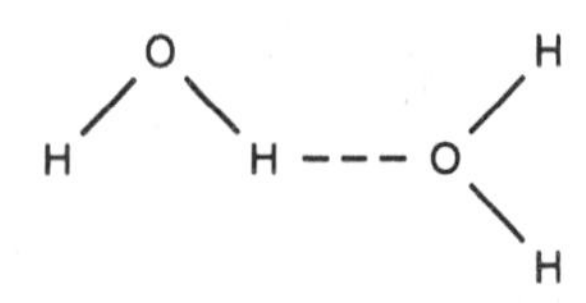

Fig. 3-14. Hydrogen bonding in water.

When hydrogen bonds or permanent dipoles are present, the physical properties change. An explanation of the differences between nylon 6 and nylon 12 can be derived when analyzing the secondary bonding forces.

Nylon 6 has 1 dipole and 1 hydrogen bond every 6 carbons in the main chain. Nylon 12 has 1 dipole and 1 hydrogen bond every 12 carbons in the main chain. The additional secondary bonds in nylon 6 cause it to be stronger and have a higher melting point.

Molecular Orientation

Under normal conditions, amorphous macromolecules are not straight, but are kinked, coiled, twisted, and wrapped around one another. When they melt, the molecules retain much of their intertwining. When crystalline plastics melt, the crystalline regions unfold and the entire structure becomes amorphous. When a melted plastics moves or flows, some of the molecules stretch out. If the flow rate is high, the molecules will stretch out to almost straight. This is called *orientation*.

When highly oriented amorphous plastics cool, the molecules will return to the coiled, kinked state if they have the chance. This is dependant on the cooling rate. If cooling is slow, they will have the time to rearrange and will coil. If the cooling time is very short, then the stretched out molecules will freeze before coiling.

If oriented semi-crystalline plastics cool rapidly, some of the orientation will be locked in. In addition, the cooling rate will also influence the degree of crystallinity. Slow cooling allows a greater degree of crystallinity. Rapid cooling inhibits some crystal formation. Since changes in the degree of crystallinity can alter the dimensions of parts, control of cooling rate is of great practical importance.

When molecules freeze in a stretched out condition, they "don't like it." The molecules are stressed and "would like to" get into a less stressed state. The stresses locked in by rapid freezing are called *residual stresses*. Given the chance, the molecules will coil. In many materials, that will happen very slowly over time, but it will also occur rapidly if the material gets hot enough. Then the material will change shape, and generally become unfit for use.

If a plastic part cannot be allowed to change slowly over time or must withstand spikes of high heat, then the stress relief must occur in the manufacturing process. This is called *annealing*, and usually involves controlled heating of parts. After the parts change shape, they are sometimes machined or pressed to achieve desired dimensions.

Frequently, relatively thick plastic objects will cool slowly enough that the center portions are not oriented. However, very thin objects may exhibit high orientation throughout. A good example is a thin-wall

Fig. 3-15. Directional fractures in molded PS container.

drinking glass, injection molded of PS. When these containers break, the fracture lines show directionality, as indicated in the following sketch.

These glasses will break easily only in one direction. That is because the molecules are oriented in one direction. The technical term for this is *uniaxial orientation.*

Uniaxial Orientation

Since highly oriented materials often appear identical to materials with low residual stresses, simple tests can help identify orientation. These tests involve fracture, stretch, and tear characteristics.

Some thin sheets used in thermoforming processes are highly oriented. They will crack easily in one direction, but not in the perpendicular direction. Pieces of these sheets will curl up when heated. The direction of curl indicates the direction of orientation.

Teflon tape, sold in rolls for sealing pipe fittings, shows orientation in its stretch differences. A piece of this tape will stretch easily in the crosswise direction. It will not stretch so readily lengthwise.

Some labels on two-liter pop bottles are a thin film of plastic stretched around the bottle. This film may tear easily in one direction, and not tear perpendicularly.

A dramatic demonstration of orientation occurs when test bars of HDPE are pulled at a low rate—less than 1 in. per minute. They will stretch out several-hundred percent and become fibrous. In this condition, the fibers can be separated by hand.

Biaxial Orientation

Biaxial orientation means that a plastic object or sheet contains molecules that are stretched in two directions, usually perpendicular to each other. When heated, biaxially oriented materials shrink in two directions. In contrast, uniaxially oriented materials shrink dramatically in one direction, and many even increase in length in the other direction.

Some materials are intentionally biaxially oriented, such as shrink-wrap and window-covering materials that shrink when heated with a hair dryer. A similar product is *Shrinky-Dinks*, for handicrafts. These materials are heated, then stretched in two directions prior to rapid cooling. Later, heating produces high shrinkage in two directions.

A familiar product, two-liter pop bottles, also have biaxial orientation. They are blown up like balloons, and that blowing process stretches the molecules in two directions. Pouring very hot water into a two-liter bottle causes vertical and circumferential shrinkage. When trying this, be sure to place the bottle in a sink, so that when it shrinks, excess water is safely contained.

Thermosets

So far, all the materials discussed in this chapter have been long-chain hydrocarbons. Even though the molecules are very long, they do have ends. The long chains are not connected to each other. Because the molecules are not chemically tied together, they can slide past each other when pulled. If the chains are smooth, the molecules can slide significantly. When heated, the molecules can move, and the material will soften or melt if heated sufficiently.

Plastics that consist of disconnected chains are called *thermoplastics.* As that word implies, when they are heated (thermo), they become soft and formable (plastic). In contrast, some plastics are *thermoset.* When heated, they do not soften or become pliable. Most thermosets cure into non-melting, insoluble solids. The chemical basis for the characteristic is that the molecules in thermosets are chemically linked to each other. The chemical bonds between molecules are called *crosslinks.* (See Figure 3-16.) It is theoretically possible that large thermoset objects, such as the hoods of diesel trucks, are actually one immense molecule.

The image of a bowl of spaghetti can also help explain the difference between thermosets and thermoplastics. Tying one strand of spaghetti to another is like creating one crosslink. If many knots tied many strands

Fig. 3-16. Cross-linked molecules.

to many neighboring strands, then the bowl would have *cured*.

Thermosets fall into two large categories: 1) rigid, and 2) flexible. The rigid thermosets often find applications in high heat environments. They don't soften under heat, and will char at high temperatures. Because thermosets are tightly chemically bonded, they also tend to resist attack by solvents.

Flexible thermosets have a long history. In Chapter 1, Charles Goodyear and the vulcanization of rubber received attention. Chemically speaking, Goodyear found a way to cause the formation of crosslinks between rubber molecules. Another group of flexible thermosets is based on urethane. Foams for car seats, sofas, furniture, and beds come from polyurethane.

Both rigid and flexible thermosets cannot be recycled or reprocessed like the thermoplastics. Rubber tires can't be chopped up and used again to make new tires. Current recycling programs for consumer items focus on thermoplastic materials, which are reprocessable. For more information on efforts to recycle thermosets, see Chapter 2.

Vocabulary

The following vocabulary words are found in this chapter. Look up the definition of any of these words you do not understand as they apply to plastics in the glossary, Appendix A.

Alternating copolymer
Amorphous
Annealing
Atomic mass units
Backbone
Benzene
Block copolymer
Branching
Copolymer
Covalent bonding
Cross-linking
Crystallization
Degree of polymerization
Electric dipole
Graft copolymer
Hydrocarbon
Intermolecular forces
Macromolecules
Mer
Molecule
Molecular mass
Orientation
  uniaxial
  biaxial
Polymerization
Primary bonds
Random copolymer
Saturated/unsaturated
Single covalent bonds
Van der Waals' forces

Questions

3-1. A bond between two carbon atoms is a __?__ bond.

3-2. A CH_3 group is called a __?__ group.

3-3. A dash between atoms indicates a __?__ bond.

3-4. The small repeating units that make up a plastic molecule are called __?__.

3-5. What does poly mean?

3-6. Three types of intermolecular forces found in plastics are: __?__, __?__, and __?__.

3-7. Crystalline plastics are often more rigid and not as transparent as __?__ plastics.

3-8. A __?__ material may be softened repeatedly when heated and hardened when cooled.

3-9. The term used to describe the tying together of adjacent polymer chains is __?__.

3-10. If two different mers make up the composition of a polymer, it is called a __?__.

3-11. A hydrocarbon molecule which contains some double bonds is called __?__.

3-12. Is PVC a copolymer? Explain.

3-13. What is the general structure of an alternating copolymer?

3-14. If a carbon chain has methyl side groups, is it branched?

3-15. If a material is transparent, is it crystalline?

3-16. What effect do dipoles and hydrogen bonds have on the melting point of polymers?

3-17. Is residual stress synonymous with orientation?

Activities

Oriented Thermoforming Sheet

Equipment. Toaster oven or heat lamp, tongs, calipers, oriented thermoforming sheet.

Procedure

3-1. Locate some highly oriented thermoforming sheets. To tell if the material is highly oriented, bend it to see if it fractures easily in one direction, but bends without fracture in the other direction. This phenomenon appears most clearly in rather thin sheets.

Fig. 3-17. A toaster oven for stress relief of small samples.

3-2. Cut out squares, approximately 75 mm by 75 mm. Carefully measure length, width, and thickness. Mark the piece for identification of length and width.

3-3. Heat until the stress relief occurs. A toaster oven fitted with hardware mesh can provide a convenient source of heat.(Fig.3-17)

Without mesh, samples may fall through the toaster tray. When the sample is sufficiently hot, the material may curl up. Quickly remove it before it melts and flatten immediately.

CAUTION: Use proper protective clothing and equipment to avoid burns. Conduct this activity in a well ventilated area.

3-4. When cool, measure length, width, and thickness. Record the change in dimensions as a percent.

3-5. If the molecules disoriented completely, the samples should bend with equal toughness in both directions.

3-6. Repeat the test on other samples. Injection molded PS drinking glasses/cups also show huge change when heated.

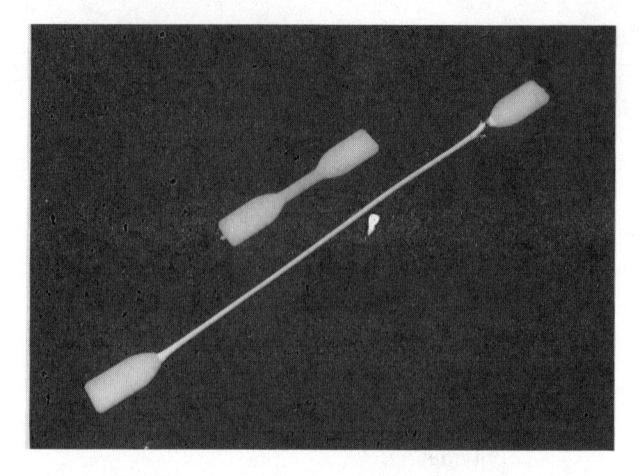

Fig. 3-18. The extreme elongation of HDPE, when pulled slowly, causes a high degree of molecular orientation.

Orientation of HDPE

Equipment. Tensile tester, HDPE tensile bars.

Procedure

3-1. Slowly pull a tensile test bar of HDPE. Do not exceed 1 in. per minute of cross-head speed. The sample should extend several hundred percent. Figure 3-18 shows a sample before and after slow extension.

3-2. Calculate the ultimate strength of the sample.

3-3. Cut a section of the thinned-out region. Mount this piece in a tensile tester and pull it to failure. Calculate the ultimate strength of the oriented strip. Compare the strength resulting from step 2 with the strength resulting from step 3.

3-4. Pull off "threads" of oriented polyethylene using only finger strength. Figure 3-19 shows the threads that formed when the sample failed. These threads can be further separated with fingers.

3-5. Stress relieve a piece of the thinned-out region. After heating, was it thicker than before?

Storm Window Covering

Obtain a piece of shrink film for window covering. Measure it, then stress relieve it with heat. Measure the annealed piece. Calculate the percentage reduction in length and width. Was the amount of stress in the original film equal in each direction?

Fig. 3-19. The tendency of this sample to "fray" or "thread" demonstrates an effect of molecular orientation.

Chapter 4

Health and Safety

Introduction

Most industries present their workers with a wide range and variety of potential health and safety hazards. One method of categorizing these hazards is to divide them into three types: physical, biomechanical, and chemical hazards. The outline for this chapter is:

I. Physical hazards
II. Biomechanical hazards
III. Chemical hazards
IV. Sources of chemical hazards
V. Reading and understanding MSDS
   A. Section I: General information
   B. Section II: Composition
   C. Section III: Physical Properties
   D. Section IV: Fire and explosion hazard data
   E. Section V: Health hazard data
   F. Section VI: Reactivity data
   G. Section VII: Spill or leak procedures
   H. Section VIII: Occupational protective measures
   I. Section IX: Special precautions
   J. Section X: Transportation

Physical Hazards

Physical hazards include: machine motions, electrical systems, hydraulic and pneumatic pressure systems, noise, heat, vibrations and other potential dangers. Physical hazards also encompass ionizing, ultraviolet, microwave, or thermal radiation.

To guard against physical hazards, industrial safety personnel must guarantee that machines have proper safety devices, guards, and warning systems. Protection against noise, vibrations, and radiation may involve protective devices for the personnel, including safety glasses, ear plugs, and various shields. Two-hand controls may be installed to ensure that the operator's hands are not near cutting blades, moving shafts, shearing blades, or hot surfaces.

Biomechanical Hazards

Biomechanical hazards are often related to repetitive motions. Ergonomics deals with those types of actions that are not of immediate danger, but may lead to injury if repeated over days, weeks and months. Reducing or eliminating these hazards requires proper hand-tool and machine design, good visual conditions and air quality. Fatigue generated by poor conditions may lead to accidents. In addition to physical injuries, psychological and mental problems can arise. Some problems possibly linked with conditions are anxiety and irritability, substance abuse, sleeping difficulties, neuroses, and headaches and gastrointestinal symptoms.

Chemical Hazards

While the plastics industry has its share of physical and biomechanical hazards, the largest hazard is chemical. Many of the compounds and processes used in the plastics industry are potentially dangerous.

The inhalation of toxic substances and absorption through the lungs account for nearly 90 percent of toxicity cases in the plastics industry. In some cases, workers expose themselves to danger because they are unaware of the hazards. To increase employee awareness, many reputable companies carry on safety education programs to protect and inform workers. This chapter discusses the chemical health and safety hazards and their correction and prevention.

Sources of Chemical Hazards

Between 1995 and the year 2000, the overall consumption of plastics in the United States is likely to rise at a rather steady rate—if trends continue without major upsets. During those years, the following plastics, listed here in order of sales volumes, should dominate sales.

Polyethylene, LD
Polyethylene, HD
Polyvinyl chloride
Polyproplyene
Polystyrene
Polyurethane
Phenolic
Polyester, thermoplastic

In order of volume processed, the major techniques for converting these materials into products will likely be:

Extrusion
Injection molding
Blow molding
Polyurethane foaming
Phenolic adhesive application
Polystyrene expanding.

The dominating materials and processes indicate that solid pelletized plastics is the major form. Consequently, health and safety issues for pellets receive primary attention. Powders and liquids will receive comment when they are specially hazardous. The polyolefins on the list above, namely high and low density polyethylene and polypropylene, pose low hazard during processing. Thermoplastic polyester also raises minor hazards. Concerns about these materials focus on additives, and the possible toxic effects of these additives. For more information about additives, see Chapter 7.

The two thermoset materials, polyurethane and phenolic, potentially expose humans to hazardous byproducts of polymerization. Discussion of these hazards appears later in this chapter.

Reading and Understanding MSDS

Hazardous methods and materials are common in the plastics industry. In the United States, a Material Safety Data Sheets (MSDS) accompany any purchase of hazardous industrial raw materials. Federal Standard 313B, which provides guidelines for the preparation of MSDS, defines what the term hazardous means. This definition is broad and covers plastics because in the course of normal use, plastics "may produce dusts, gases, fumes, vapors, mists, or smokes" which are dangerous.

A similar definition of the word hazardous appears in the Code of Federal Regulations (CFR). They are substances that have been found capable of posing an unreasonable risk to health, safety, and property.

When a customer makes repeated purchases of the same material, an MSDS is sent with the first order every year. Although each producer of raw materials is responsible for creating the MSDS, the guidelines require certain categories of information. A thorough understanding of MSDS as they relate to the plastics industry can promote safety for all personnel.

To organize the following discussion, the sections recommended in Federal Standard 313B for all MSDS will serve as an outline.

Section I: General Information

Section I contains information about the product name and the manufacturer's identity. This section usually includes emergency telephone numbers, and the trade and chemical family name of the material. A product identification section indicates the chemical name of the product.

For example, LEXAN, manufactured by GE Plastics, is a type of polycarbonate appropriate for injection molding. Its chemical name is Poly(Bisphenol-A carbonate). Every chemical has a CAS number. CAS stands for Chemical Abstracts Services Registry Number. In addition to cataloging chemical substances, the CAS Registry provides an unambiguous identification of materials. Chemical companies promote their materials with trade names, such as LEXAN. To know if LEXAN is chemically identical to MERLON—a polycarbonate made by Miles Chemical Company—compare the CAS numbers. The number for both brands is 25971-65-5.

Section II: Composition

Section II contains information on hazardous ingredients. Since polycarbonate is not a controlled product, the MSDS provides no additional data in this section. However, in addition to the major constituents of a material, all hazardous additives, fillers or colorants must appear in this section. A MSDS for a grade of ABS lists the following ingredients.

CAS #	Chemical Name	OSHA PEL Units	ACGIH TLV Units
7631-86-9	Silica	0.05 mg/m^3	0.05 mg/m^3
100-42-5	Styrene	50.0 ppm	50.0 ppm
1333-86-4	Carbon black	3.5 mg/m^3	3.5 mg/m^3

It is very important to understand in detail what this information means. The acronym OSHA stands for the Occupational Safety and Health Administration. ACGIH stands for The American Conference of Governmental Industrial Hygienists. Both these organizations publish standards for exposure to various industrial materials.

OSHA utilizes a measure called the PEL, or Permissible Exposure Limits. A PEL indicates not only the amount of exposure permissible, but also the time of exposure. Thus the PEL is a time-weighted average, or TWA. A TWA represents the exposure level considered acceptable for an 8 hour day as part of a 40 hour week. In addition to the PEL values, OSHA reports a REL, or Recommended Exposure Limits, and a STEL, or short-term exposure limit.

In this example, the Permissible Exposure Limit for silica is 0.05 mg/m^3. The unit, mg/m^3, is appropriate for dusts, powders, or fibers.

The abbreviation TLV means Threshold Limit Value. This is the value recommended by the American Congress of Governmental Industrial Hygienists. The TLV is also a time-weighted-average, acceptable for 8 hours a day as part of a 40 hour week.

The ACGIH also has two additional categories of TLVs. TLV-STEL is the first category. STEL stands for Short-Term Exposure Limit. It indicates acceptable exposure for 15 minutes, and should not be exceeded at any time during a 8 hour day, even if the TWA for the day is within the limits.

The TWA is generally lower than the STEL. The procedure for exposures, which are above the TWA yet below the STEL, is clearly specified. Such exposures should be:

- no longer than 15 minutes
- no more than 4 times per day

The second additional category of TWA is a ceiling value. It is a value that should not be exceeded during any part of a workday.

In this example, silica and carbon black are powders or dusts. Styrene is as hazardous as a gas, and consequently, the PEL or TLV units are ppm, parts per million.

It is important to clearly distinguish between PEL or TLV levels and the percentage of an ingredient by weight. The acrylonitrile-butadiene-styrene (ABS) used as an example contained 3 percent of carbon black, and 0.2 percent of residual styrene monomer. Residual styrene monomer is material that did not combine to produce polymer molecules, but remained trapped in the polymer. The silica was present at 5 percent by weight. Since carbon black and silica are powdered solid materials, they are encapsulated by the major plastics and are very unlikely to exist in a *free* form. Consequently, the TLV and PEL values have little practical application. However, the monomeric styrene can escape as a gas when the material is at processing temperatures. Its TLV and PEL limits are of practical value.

Styrene as a health hazard. Styrene is so important to the plastics industry that it warrants special attention. Styrene holds a crucial position in the plastics industry, because it is a building-block for thermoplastic styrenics, which include polystyrene, impact polystyrene, SAN, ABS, and others. In addition, styrene appears in polyester casting resins.

In thermoplastic styrenic resins, monomeric styrene is a very minor ingredient. Some ABS plastics contain less than 0.2 percent of monomeric styrene. In addition to the residual monomer, commercial styrenic plastics yield styrene during thermo-oxidative degradation. These sources can combine to supply styrene into the air of thermoplastic molding and forming processes. One study found a range of 1 to 7 ppm of styrene in the atmosphere in a plant injection molding polystyrene.

The potential hazards of styrene from thermoplastics is tiny compared to the danger from polyester resins. In particular, open mold operations are common for the production of boats, yacht hulls, large tanks or pipes, tub and shower stalls, and truck or tractor tops. Polyester resins provide a matrix for reinforcing glass fibers, either as cloth, matt, or chopped strand. Common processes are *chop-and-spray* and *hand-layup*. In chop-and-spray, catalyzed resin and glass strands meet in a chop/mix head. Operators direct the coated glass fibers into a form. In hand-layup, operators position reinforcing glass layers in molds and apply resin, sometimes with airless spray guns.

What ever the process, open-mold work exposes operators to styrene vapors. Polyester resins contain about 35 percent styrene by weight. Although MSDS recommends that personnel "avoid breathing vapors," this is attainable only if operators wear self-contained breathing apparatus. Some companies making large yacht hulls do fit their operators with such equipment. However, many manufacturers rely on ventilation systems.

Ventilation systems have varying efficiencies. If the ventilation systems are not adequate, the concentrations may exceed the current TLV of 50 ppm. One study found styrene concentrations of 109 ppm when an open mold had a surface area of 1.3 m^3. An 8.3 m^3 surface yielded 123 ppm. Another study found 120 ppm for a chop-and-spray operation, and 86 ppm for a roll-out station.

However, if the ventilation system is well placed and powerful, the concentrations drop. One yacht company consumes 8000 lbs of polyester resin per week. Its plant has local ventilation at each hull work station capable of exhausting 17000 cubic feet per minute (CFM). In addition, the system for the entire building guarantees 10–15 air changes per hour. Under these circumstances, the hull lamination workers had exposure rates of 17–25 ppm. According to another study, the average exposure for yacht companies is 37 ppm. Small boat companies showed an average of 82 ppm. This was probably due to less efficient ventilation systems.

Even lower exposures occur in closed-mold, or press-mold operations. Press-mold companies had exposures between 11 and 26 ppm. In order to achieve lower exposures, some companies are converting some or all operations to closed molds. If the ACGIH lowers the TLV for styrene, the incentive to eliminate open molds will be even greater.

Table 4-1 lists the limits currently recommended by The American Conference of Governmental Industrial Hygienists (ACGIH). These recommendations are available in *1994-1995 Documentation of the Threshold Limit Values for Chemical Substances and Physical Agents and Biological Exposure Indices.* The ACGIH regularly publishes updates to the values, to keep the documentation up-to-date.

The ACGIH also rates materials as *carcinogens*, or cancer causing. A1 is the rating for confirmed human carcinogens. A2 is for suspected human carcinogens, and A3 for animal carcinogens.

Table 4-1. Threshold Limit Values for Selected Chemicals

Material	Flash Point (°C)	TLV (ppm)	TLV (mg/m^3)	Health Hazard
Acetaldehyde A3		STEL25	180	Animal carcinogen
Acetone (dimethyl ketone)	−18	750	1780	Skin irritation, moderate narcosis from inhalation
Acrylonitrile (vinyl cyanide) A2	5	2	4.3	Absorbed through skin, inhalation carcinogenic
Ammonia		25	17	
Asbestos Al (as amosite)			0.5 fibers/cc	Respiratory (inhalation) diseases, carcinogenic
Benzene (benzol) A2	11	10	32	Poisoning by inhalation, carcinogenic skin irritation and burns
Bisphenol A				Skin and nasal irritation
Boron fibers				Irritant, respiratory discomfort
Carbon dioxide	gas	5000	9000	Asphyxiation possible, chronic (inhalation) poisoning in small amounts
Carbon monoxide	gas	25	29	Asphyxiation
Carbon tetrachloride		5	31	Inhaled and absorbed, chronic poisoning in small amounts, carcinogenic A3
Chlorine		0.5	1.5	Bronchial distress, poisoning and chronic effects
Chlorobenzene (phenyl chloride)	29	10	46	Absorbed and inhaled, paralyzing in acute poisoning
Cobalt A3 (as metal dust and fume)			0.02	Possible (inhalation) pneumoconiosis and dermatitis
Cyclohexane	−20	300	1030	Liver and kidney damage, inhalation
Cyclohexanol	66	50	206	Inhaled and absorbed, possible organ damage
0-Dichlorobenzene	66	25	150	Possible liver damage, inhalation, percutaneous
1,2-Dichloroethane (ethylene dichloride)	13	10	40	Anesthetic and narcotic, inhalation, possible nerve damage

Table 4-1. Threshold Limit Values for Selected Chemicals (Continued)

Material	Flash Point (°C)	TLV (ppm)	TLV (mg/m^3)	Health Hazard
Epichlorchydrin	95	2	7.6	Highly irritating to eyes, percutaneous and respiratory tract, carcinogenic
Ethanol (ethyl alcohol)	13	1000	1880	Possible liver damage, narcotic effects
Fluorine		1	1.6	Respiratory distress, acute in high concentrations
Formaldehyde A2		0.3 ceiling		Skin and bronchial irritations, inhalation, carcinogenic
Glass fiber		10		
Hydrogen chloride		5 ceiling	7.5 ceiling	Eye, skin, and mucous membrane irritant
Hydrogen fluoride		3 ceiling	2.6 ceiling	
Isopropyl alcohol		400	983	Narcotic, irritation to respiratory tract, dermatitis
Methanol (methyl alcohol)	11	200	262	Chronic if inhaled, poisoning may cause blindness
Methyl acrylates	3	10	35	Inhaled and absorbed, liver, kidney, and intestinal damage
Methylene chloride A2	gas	50	174	Possible liver damage, narcotic, severe skin irritation, moderate narcosis through inhalation, ingestion, carcinogenic
Methyl ethyl ketone	-6	200	590	Slightly toxic, effects disappearing after 48h
Mica			3	Pneumoconiosis, respiratory discomfort
Nickel			1	Chronic eczema, carcinogenic
Phenol (carbolic acid)	80	5	19	Inhaled, absorbed through skin, narcotic, tissue damage, skin irritation
Phosgene		0.1	0.4	Lung damage
Pyridine	20	5	16	Liver and kidney damage
Silane				Inhalation, organ damage
Silica (fused)			0.1	Respiratory discomfort, silicosis, potential carcinogen
Styrene		50	213	Eye and mucous membrane irritant
Toluene	4	50	188	Similar to benzene, possible liver damage
Vinyl acetate A3	8	10	35	Inhalation, irritant
Vinyl chloride A1		5	13	Carcinogenic

Source: 1994–1995 *Threshold Limit Values for Chemical Substances and Physical Agents and Biological Exposure Indices*, American Conference of Governmental Industrial Hygienists, Inc., Cincinnati, Ohio.

Section III: Physical Properties

This section does not treat the ingredients separately, but considers the material one substance. Examples of the data are evaporation rate, melting point, boiling point, specific gravity, solubility in water, and form. For pelletized plastics, some of these characteristics are NE, which stands for Not Established.

Section IV: Fire and Explosion Hazard Data

Since most pelletized plastics are not explosive, this section generally concentrates on fire fighting. In the presence of sufficient heat and oxygen, most plastics will burn and yield carbon dioxide and water vapor. Many plastics can be made self-extinguishing or fire retardant. All thermosets are self-extinguishing. Glass and other inorganic reinforcement may reduce flammability. Most flame-retardant additives act by interfering chemically with the flame reactions.

Many MSDSs recommend water as the best medium for extinguishing fires. They also warn of hazardous chemicals produced during combustion, such as dense black smoke, carbon monoxide, hydrogen cyanide, and ammonia.

Deaths due to fire have several causes, including direct burns, oxygen deficiency, and exposure to toxic chemicals. Carbon monoxide is one of the greatest dangers in a fire. It is a colorless, odorless gas that can produce unconsciousness in less than three minutes.

Combustion often produces toxic by-products. Table 4-2 gives the relative fire toxicity of selected polymers and fibers. The values in Table 4-2 come from experiments using test animals exposed to gases generated by pyrolysis of selected materials. Such data do not directly predict fire toxicity in humans. Note that wool and silk are the most toxic materials, with wool rated as more toxic than many polymers.

Many commercial plastics have flash points so high that MSDSs report the flash point as "Not Applicable." On a grade of nylon, one MSDS reported a flash point of 400 °C (752 °F), determined by the American Society for Testing and Materials, open cup method—ASTM D-56. In the open cup method, a material is heated in an open container. The flash point is the lowest temperature at which enough vapors are given off to form an ignitable mixture of vapor and air immediately above the surface of the melt or liquid.

In contrast to the flash points of commercial plastics, the flash points of many liquids have practical implications. According to the Occupational Safety and Health Administration (OSHA) and the National Fire Prevention Association (NFPA), a *flammable* liquid is any material having a flash point below 38 °C [100 °F]. *Combustible* liquids are those with flash points at or above 38 °C. Table 4-1 lists the flash points of selected materials. Note that many hydrocarbons have low flash points, often below 0 °C. These flammable materials pose a severe fire hazard if not properly handled.

Table 4-2. Relative Fire Toxicity of Selected Polymers and Fibers

Material	Approximate Time to Death, min	Approximate Time to Incapacitate, min
Acrylonitrile-butadiene-styrene	12	11
Bisphenol A polycarbonate	20	15
Chlorinated polyethylene	26	9
Cotton fiber, 100%	13	8
Polyamide	14	12
Polyaryl sulfone	13	10
Polyester fiber, 100%	11	8
Polyether sulfone	12	11
Polyethylene	17	11
Polyisocyanurate rigid foam	22	19
Polymethyl methacrylate	16	13
Polyphenyl sulfone	15	13
Polyphenylene oxide	20	9
Polyphenylene sulfide	13	11
Polystyrene	23	17
Polyurethane flexible foam	14	10
Polyurethane rigid foam	15	12
Polyvinyl chloride	17	9
Polyvinyl fluoride	21	17
Polyvinylidene fluoride	16	7
Silk fiber, 100%	9	7
Wood	14	10
Wool fiber, 100%	8	5

Note: Information from Fire Safety Center of the University of San Francisco with the support of the National Aeronautics and Space Administration. All data modified to show approximate values.

Section V: Health Hazard Data

This section deals with the possible routes of entry of toxic substances into humans. The most common routes are ingestion, inhalation, skin, and eyes. In addition to toxicity, this section also reports on chronic effects and carcinogenicity.

Ingestion. Ingestion of pellets is rather unlikely, and some companies state "Not a probable route of exposure." Other companies are more cautious and provide statements such as: "The oral LD-50 in rats is in excess of 1000 milligrams per kilogram of body weight. Two-week feeding tests with dogs and rats showed no evidence for gross pathological change."

Toxicity ratings utilize the term LD_{50} or LD-50. LD stands for Lethal Dose, and the subscript 50 means that this dose is capable of killing 50 percent of a population of experimental animals. Many pelletized plastics are rather inert and experimental animals can eat large quantities with little effect.

Discussions of toxicity levels often rely on a few general categories. When lethal doses are less than 1 mg or 10 ppm, the material is *extremely toxic.* If the lethal range is less than 100 ppm or 50 mg, the material is *highly toxic. Moderate toxicity* refers to doses less than 1000 ppm or 500 mg. Toxicity ratings over 1000 ppm or 500 mg should be considered *slightly toxic.*

Toxicity ratings refer to the total weight of the subject. A MSDS might contain a rating similar to the following:

ORAL LD_{50} 264 mg/Kg

This means that this moderately toxic substance kills 50 percent of experimental guinea pig populations. If the toxicity to humans is identical to the response shown by guinea pigs, then the lethal oral dose for a human with a body weight of 70 kg would be 264 g × 70, which yields 18480 mg or 18.48 g.

Inhalation. Some MSDSs view inhalation of pelletized plastics as unlikely because of the physical form. Others provide the following data:

INHALATION LC_{50} NO DATA AVAILABLE

The abbreviation LC stands for lethal concentration. It normally refers to a vapor or gas. The units for numeric values for vapors and gases are ppm (parts per million) at standard temperature and pressure.

While inhalation of pellets is unlikely, inhalation of gases and vapors is a major concern. Many studies have measured the odor response of humans, to determine how effectively odor warns humans of possible dangers. Many gases are easily detected. An example is acetaldehyde, a chemical used in the production of some phenolic resins. Overheated PET also releases small amounts of acetaldehyde. Acetaldehyde has a STEL of 25 ppm, but its air odor threshold is 0.050 ppm. Since it is easily detected at concentrations well below the TLV, odor can provide warning about exposure to this material.

However, some gases provide no odor warning. Vinyl chloride is the primary monomer used in the polymerization of PVC. It has a TLV of 5 ppm and an odor threshold of 3000 ppm. It is also classed Al—known human carcinogen. Odor can provide no warning of exposure to this dangerous material.

Table 4-3 lists chemicals selected used in the production of plastics. Some of them are also decomposition products, formed when selected plastics are overheated.

The inhalation of isocyanates is a hazard associated with the manufacture of polyurethane products. Polyurethane involves the polymerization of TDI (toluene diisocyanate) or MDI (methylene diisocyanate). This polymerization routinely occurs during the manufacture of polyurethane foams and reaction-injection molding of polyurethane products. In some construction projects, foamed polyurethane is sprayed on the inside of walls and roofs. According to the ACGIH, the TLV for both isocyanates is 0.005 ppm, with the additional restriction that TDI has a STEL of 0.02 ppm. These levels are very low and require significant effort to achieve and maintain.

Dermal. A *dermal* exposure is one in which the substance comes in contact with the skin. Skin is an effective barrier against some chemicals. They pose no hazard through dermal exposure. Other chemicals and materials may irritate the surface of the skin. Their danger is limited. Other materials can penetrate the skin and cause sensitization. Epoxies may cause sensitization after repeated exposures. The most severe hazard is when a chemical penetrates the skin, enters the blood stream, and acts directly on bodily systems. This is called a *systemic hazard.*

Table 4-3. Comparison of TLVs to Odor Thresholds for Selected Chemicals

Name	TLV (ppm)	Air Odor Threshold (ppm)
Acetaldehyde	STEL 25	0.05
Acrylonitrile	2	17
Ammonia	25	5.2
1,3 Butadiene A2	2	1.6
Formaldehyde A2	0.3 ceiling	0.83
Hydrogen chloride	5	0.77
Hydrogen cyanide	10	0.58
Hydrogen fluoride	3	0.042
Methyl methacryiate	100	0.083
Phenol	5	0.040
Styrene	50	0.32
Tetrahydrofuran	200	2.0
Vinyl acetate A3	10	0.5
Vinyl chloride Al	5	3000

Pelletized plastics can not penetrate the skin, but may cause skin irritation or dermatitis—particularly if the material contains abrasive fibers, such as glass. Skin contact with molten plastics can cause severe burns.

Liquid materials may pose a severe dermal hazard. A catalyst used to harden an epoxy resin has the following data in its MSDS.

Diethylene Triamine **CAS # 111-40-0**

ACGIH

TLV	STEL
1 ppm (skin)	NE

OSHA

PEL	STEL
1 ppm (skin)	NE

This substance will enter through the skin and has a low TLV and PEL. This indicates that it is very dangerous and requires special precautions. If this substance received laboratory testing on animals, it may have a lethal dose (LC) value.

Eyes. Eyes may be injured by mechanical irritation in the event that pellets get into a person's eyes. Liquids and gases may cause severe eye damage. For example, methylenedianiline is an ingredient in a catalyst for hardening an epoxy resin. One MSDS indicates that methylenedianiline causes irreversible blindness in cats, and visual impairment in cattle.

Carcinogenicity. Solid plastics in pelletized form are often not regulated as carcinogenic. However, residual monomers may have links to cancer. For example, a MSDS of ethylene-vinyl acetate (EVA) lists vinyl acetate as a hazardous ingredient, which is present at a maximum of 0.3 percent. When EVA is polymerized, a small amount of the vinyl acetate monomer remains. Extensive exposure of test animals to vinyl acetate monomer at 600 ppm caused some carcinomas in nose and airpassages of some animals. Consequently it bears the A3 notation.

Section VI: Reactivity Data

Since most pelletized plastics are very stable, they are not reactive under normal conditions. Some materials will react with strong acids and oxidizing agents. Many remain inert.

However, most plastics degrade when sufficiently hot. The degradation of plastics is considered thermo-oxidative degradation, which is thermal degradation in the presence of oxygen. A few plastics begin to degrade at normal processing temperatures, and others begin to degrade when heated beyond normal processing ranges. In either case, hazardous gases and vapors enter the air and might enter humans by inhalation.

The following list specifies such decomposition products.

- At 230 °C, POM releases formaldehyde.
- At 100 °C, PVC releases HCl.
- At 300 °C, PET releases acetaldehyde.
- At 300 °C, nylons release carbon monoxide and ammonia.
- At 340 °C, nylon 6 releases e-caprolactam.
- At 250 °C, fluoroplastics release HF. Inhaling fumes containing decomposition products of fluoroplastics may cause influenza-like symptoms. These are sometimes called "polymer fume fever," and include fever, cough, and malaise.
- At 100° C, PMMA releases MMA.

Thermal degradation of PVC. The potential of PVC to decompose is a serious problem. In both extrusion and injection-molding processes, PVC can decompose catastrophically if over heated and held too long at processing temperatures in the barrel of a machine.

Figure 4-1 shows decomposed polymer which was removed from the barrel and nozzle of an injection molding machine. Please notice that what remains after decomposition is tightly packed carbon. When packed into the nozzle and barrel of an injection molding machine, this carbon prevents purging of the PVC, which remains in the machine.

The sequence of PVC degradation often involves initial discoloration of the PVC, and perhaps the beginnings of small black specks in the parts. If degradation continues, dust and fumes will "spit" out of the nozzle. The fumes will contain high concentrations of hydrogen chloride, which is highly toxic. In such an event, all personnel should be immediately evacuated and anyone near the molding machine should wear a respirator appropriate for organic vapors and acids.

Attempting to purge the remaining PVC may prove futile, if the nozzle and barrel contain carbon powder. This material will not purge. It may be best to

Fig. 4-1. Decomposed PVC.

shut down the machine and return after the machine is cool. The nozzle and end cap will need to come off in order to remove the carbon powder.

The potential of PVC to degrade in this manner may be increased by the addition of flame retardants, especially flame retardants that contain zinc. Manufacturers of zinc containing flame retardants urge customers to use caution to avoid "catastrophic zinc failure."

Thermal degradation of POM. *Polyacetal,* also called POM or *polyoxymethylene,* is an engineering thermoplastic. It thermally degrades by depolymerization and releases formaldehyde into the air. Although antioxidants can retard the depolymerization, they do not prevent the chemical breakdown. Ventilation systems also fail to remove all formaldehyde; consequently, the air near injection molding machines and extruders may contain excess amounts of formaldehyde. One MSDS indicates that heating above 230 °C causes the formation of formaldehyde.

Molding technicians may be overexposed to formaldehyde, particularly when purging molding machines. Some technicians have said that the fumes from acetal purgings can "knock you down." Such statements indicate exposures to formaldehyde well over the TLV of 0.3 ppm ceiling.

Studies in 4 injection-molding plants found formaldehyde concentrations of 0.05 ppm to 0.19 ppm. In these plants, the molding machines had local exhaust systems. The study did not measure the effect of purgings.

Thermal degradation of phenolics. Phenolic resins find major use in adhesive applications, particuluary in the manufacture of plywood and particleboard. Phenolic molding compounds are common in compression and transfer moldings. One MSDS warns that processing phenolics may release small amounts of ammonia, formaldehyde and phenol. Phenol has a TLV of 5 ppm, a LD_{50} (rat) of 414 mg/kg, and a LC_{50} (inhalation rat) of 821 ppm. Formaldehyde has a ceiling of 0.3 ppm. One MSDS asserts that the symptoms of exposure to formaldehyde include eye, nose, throat and upper respiratory irritation, tearing, and nose stuffiness. These symptoms often occur at concentrations in the range of 0.2 to 1.0 ppm, and become more severe above 1 ppm.

Thermal degradation of nylon 6. Nylon 6 thermally degrades into the monomer from which it is formed, namely e-caprolactam. In addition, many grades of nylon contain residual caprolactam, often less than 1 percent by weight. The ACGIH has established 5 ppm as the TLV for caprolactam vapor. Caprolactam is a toxic material. The LD_{50} (rat) is 2.14 mg/kg.

Normal injection-molding operations yield some caprolactam vapor; purgings release greater amounts; and extrusion operations add a steady amount into the workplace atmosphere.

One study, involving two injection-molding plants and one extrusion operation, found concentrations of 0.01 to 0.03 ppm for e-caprolactam. This is well below the TLV of 5 ppm. However, the study did not consider purging of molding machines.

Thermal Degradation of PMMS. Polymethylmethacrylate (PMMA) is often called acrylic, and is familiar in sheet form, sold under the trade name *Plexiglas.* It is molded, extruded, and thermoformed into numerous products. During processing, PMMA thermally degrades into MMA—methylmethacrylate. Besides the degradation products, most PMMA contains a small amount of residual monomer. One MSDS reports that acrylic molding pellets contain less that 0.5% by weight of methyl methacrylate.

A study involving one injection-molding plant, two thermoforming operations, and one extrusion location indicated concentrations ranging from a low of 0.06 mg/m^3 for injection molding, to a high of 4.6 mg/m^3 for thermoforming at 160 °C. These are also well below the TLV for MMA is 410 mg/m^3 [100 ppm]. If the processing temperature rose dramatically, due to a uncontrolled or "run-away" heater, the release of MMA could also rise considerably.

Section VII: Spill or Leak Procedures

Section VII gives spill or leak procedures. In the case of pelletized materials, this amounts to sweeping up the spilled pellets. For liquids, the procedures are much more elaborate.

Section VIII: Occupational Protective Measures

This section lists protective measures for the working space and for the individuals.

Workplace protection. Adequate ventilation is needed for processing areas. The ACGIH has prepared guidelines for Industrial Ventilation, which are available from the ACGIH Committee on Industrial Ventilation, P.O. Box 116153, Lansing, MI 48901.

Personal protective devices include safety glasses or goggles for eye and face protection; skin protection with gloves, long sleeves, and face shields; ear plugs for hearing protection; and respiratory protection through the use of a respirator whenever processing fumes are out of control. Respirators should also be worn when secondary operations such as grinding, sanding or sawing create excessive dusts.

The ACGIH has published *Guidelines for the Selection of Chemical Protective Clothing*, and the American National Standards Institute (ANSI) and OSHA standards address eye and face protection. Protective barrier cream systems may be used to protect workers from minor skin irritants and can curb the incidence of dermatitis.

Allergic-reactive people should be warned of possible bronchial or skin irritants. Any irritation to the skin, eyes, nose, or throat should be dealt with promptly.

Section IX: Special Precautions

Many MSDSs provide a separate section on handling and storage. A hazard in handling concerns typical secondary operations of grinding, sanding and sawing. These operations produce dusts. Dusts are potentially explosive. Table 4-4 lists the explosion characteristic of dusts found in the plastics industry.

Section X: Transportation

Section X concerns transportation. Most pelletized plastics have no special restrictions for transportation, and consequently are not regulated.

Vocabulary

The following vocabulary words are found in this chapter. Look up the definition of any of these words that you do not understand as they apply to plastics, in the glossary, Appendix A.

ACGIH
Carcinogen
CAS number
Ceiling
Combustable
Depolymerization
Flammable
Flash point
Inhalation
LD-50
MSDS
PEL
REL
STEL
TLV

Table 4-4. Explosion Characteristics of Selected Dusts Used in the Plastics Industry

Type of Dust	Ignition Temperature, °C [°F]	Explosibility	Ignition Sensitivity
Cornstarch	400 [752]	Severe	Strong
Wood flour, white pine	470 [878]	Strong	Strong
Acetal, linear	440 [824]	Severe	Severe
Methyl methacrylate polymer	480 [896]	Strong	Severe
Methyl methacrylate-ethyl acrylate-styrene copolymer	440 [824]	Severe	Severe
Methyl methacrylate-styrene-butadiene-acrylonitrile copolymer	480 [896]	Severe	Severe
Acrylonitrile polymer	500 [932]	Severe	Severe
Acrylonitrile-vinyl pyridine copolymer	510 [950]	Severe	Severe
Cellulose acetate	420 [788]	Severe	Severe
Cellulose triacetate	430 [806]	Strong	Strong
Cellulose acetate butyrate	410 [770]	Strong	Strong
Cellulose propionate	460 [860]	Strong	Strong
Chlorinated polyether alcohol	460 [860]	Moderate	Moderate
Tetrafluoroethylene polymer	670 [1238]	Moderate	Weak
Nylon polymer	500 [932]	Severe	Severe
Polycarbonate	710 [1310]	Strong	Strong
Polyethylene, high-pressure process	450 [842]	Severe	Severe
Carboxy polymethylene	520 [968]	Weak	Weak
Polypropylene	420 [788]	Severe	Severe
Polystyrene molding compound	560 [1040]	Severe	Severe
Styrene-acrylonitrile copolymer	500 [932]	Strong	Strong
Polyvinyl acetate	550 [1022]	Moderate	Moderate
Polyvinyl butyral	390 [734]	Severe	Severe
Polyvinyl chloride, fine	660 [1220]	Moderate	Weak
Vinylidene chloride polymer, molding compound	900 [1652]	Moderate	Weak
Alkyd molding compound	500 [932]	Weak	Moderate
Melamine-formaldehyde	810 [1490]	Weak	Weak
Urea-formaldehyde molding compound	460 [860]	Moderate	Moderate
Epoxy, no catalyst	540 [1004]	Severe	Severe
Phenol formaldehyde	580 [1076]	Severe	Severe
Polyethylene terephthalate	500 [932]	Strong	Strong
Styrene-modified polyester-glass-fiber mix	440 [824]	Strong	Strong
Polyurethane foam	510 [950]	Severe	Severe
Coumarone-indene, hard	550 [1022]	Severe	Severe
Shellac	400 [752]	Severe	Severe
Rubber, crude	350 [662]	Strong	Strong
Rubber, synthetic, hard	320 [608]	Severe	Severe
Rubber, chlorinated	940 [1724]	Moderate	Weak

Source: Compiled in part from *The Explosibility of Agricultural Dusts*, R1 5753, and *Explosibility of Dusts Used in the Plastics Industry*, R1 5971, U. S. Department of Interior.

Questions

4-1. A highly toxic, colorless, odorless gas is named ___?___ .

4-2. Combustible liquids are those with flash points at or above ___?___ °C.

4-3. Name three natural materials that may be more toxic than plastics.

4-4. Name three broad categories of hazards in working with and processing chemicals.

4-5. The overheating of ___?___ plastics may cause polymer-fume fever with flulike symptoms.

4-6. Liquids that have a flash point below 38 °C are called ___?___ .

4-7. What is the ignition temperature of cellulose acetate plastics dust? (See Appendix D.)

4-8. Many plastics degrade when overheated. Name one that releases gaseous hydrochloric acid when heated.

4-9. What time period is attached to a STEL?

4-10. Approximately what percent of polyester resin is styrene?

4-11. Approximately what percent of polystyrene pellets is styrene?

4-12. What is a systemic hazard?

Activities

4-1. Investigate CAS numbers by acquiring MSDS on several grades of the same basic material. For example, find MSDS for various melt index grades and various colors of polyethylene.

- Does a high melt index polyethylene have a different CAS number than a low melt index polyethylene?
- Does red polyethylene have a different CAS number than green polyethylene?

4-2. Investigate the percentage of styrene in various brands of polyester resin. Acquire MSDSs on polyester resins from several manufacturers. Do some companies have reduced amounts of monomeric styrene in the resin?

4-3. Investigate the percentage of residual styrene in polystyrene pellets. Acquire MSDSs from several manufacturers of polystyrene. Do some companies offer polystyrene with reduced residual monomer?

Chapter 5

Elementary Statistics

Introduction

Global competition is a fact of life for most industries. In the plastics industry, world-wide competition effects small companies as well as large ones. Many companies have focused their efforts on improving the quality of their products. They hope to withstand the competition by offering consumers reliable items of high quality. However, in order to make high quality products, the raw materials and the processes must also be high quality.

Buying quality materials is in itself a complex task. Frequently, a major problem in acquiring materials is knowing how uniform or consistent they are. Sales agents describe the degree of uniformity in statistical terms. Purchasing agents and others involved in buying raw materials need to understand basic statistical concepts.

Similarly, manufacturing quality products involves controlling variations in the manufacturing process so that the products are uniform and consistent. To reduce unwanted variations in products, manufacturing personnel seek to minimize random changes in processing. This effort also relies on statistics, because statistical procedures can accurately document the repeatability of production equipment.

Understanding and using statistics is essential for companies as they fight to withstand the global competition. This chapter will introduce a few basic statistical techniques. It assumes the reader has basic math computational skills. It also assumes that the reader has no previous exposure to statistics. The content outline for this chapter is:

I. Calculating the mean
II. The normal distribution
III. Calculating the standard deviation
IV. The standard normal distribution
V. Graphic representation of hardness test results
VI. Sketching graphs
VII. Graphic comparison of two group
VIII. Summary

Calculating the Mean

Comparisons of size or shape abound in everyday life. When someone says, "Look at that big house," an average house is the point of comparison. When a tall man walks by, he stands out because of his difference from an average man.

Most women are rather close to the average height. Many women are somewhat taller or shorter than the average. Fewer women are much taller or much shorter than average. Only a rare few are very much taller or very much shorter than average. When an extremely tall woman walks by, a mental comparison to the average identifies her as tall. A different comparison, not to the average, but to the range of height possibilities, places her as rare or unique. Quantifying the average and the range of possibilities is the topic of the next few sections.

When calculating an average, the first step is to define the type of average. The *mean*, the *median*, and the *mode* are averages, but this book will treat only the mean.* To calculate a mean, add the values in a group to create a sum. Then divide the sum by the number of values in the group.

The following example illustrates this procedure.

12
11
10
9
8

The sum of the values (12 + 11 + 10 + 9 + 8) is 50. The number of values in this set is 5. Instead of writing out the phrase, "the number of values in a set," statisticians use n as an abbreviation. In this case, $n = 5$. Dividing the sum (50) by n (5) results in 10.

$$50/5 = 10$$

The mean is 10. It is often abbreviated $\bar{x}$, which is pronounced x-bar.

A *distribution* is a collection of values. The set of numbers used above for calculation of the mean is a distribution. Virtually any collection of numbers is a distribution. However, statistical analysis relies on only a few standard distributions to explain the countless groups of collected data. In this book, the only standard distribution treated is the *normal distribution*.

The Normal Distribution

For a distribution to be normal, it must exhibit two characteristics. First, it must show *central tendency*. That means that the values must cluster around one central point. Second, it must be rather well centered around the mean. In other words, it must be *symmetrical*. An example will help clarify the ideas of central tendency and symmetry.

Imagine 1000 randomly selected women standing on a football field in groups according to their height. The height groups change in one inch increments. A person looking down from an overhead blimp would see the shape illustrated in Figure 5-1. Figure 5-1 contains 1000 small circles, one for each person.

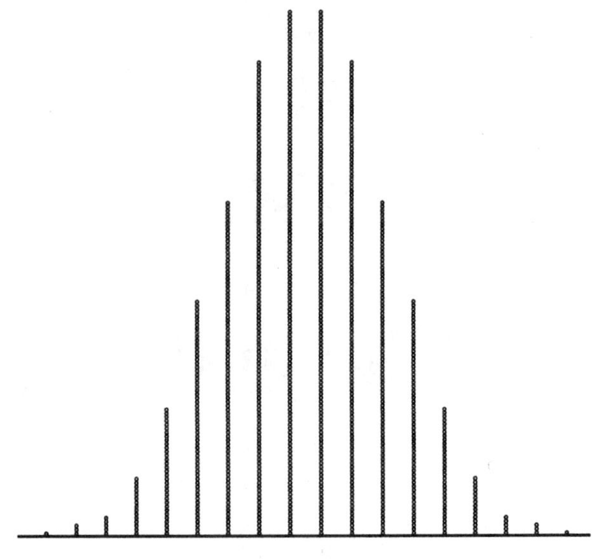

Fig. 5-1. 1000 women grouped according to height. (Data adapted from: National Center for Health Statistics, Height and Weight of Adults, Ages 18–74 Years, by Socioeconomic and Geographic Variables, United States, 1971-74.)

If each woman held up a large card, as stadium crowds sometimes do to create words or images, it would look like Figure 5-2.

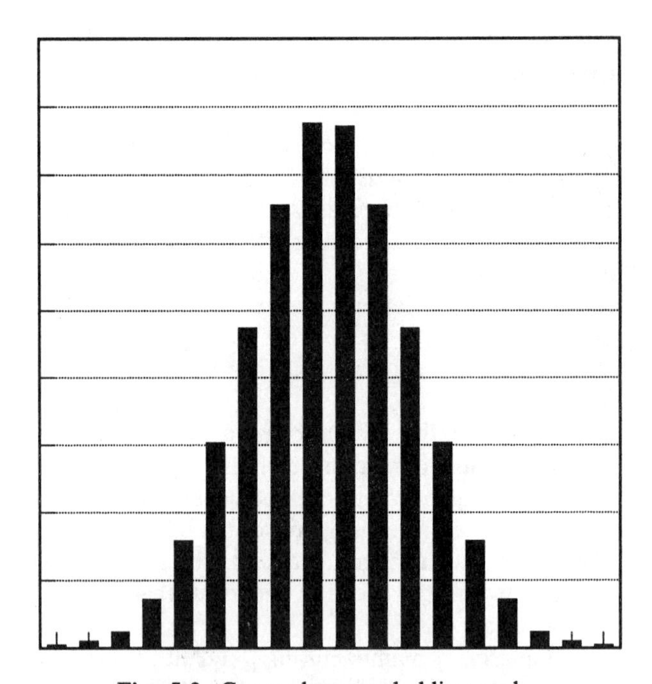

Fig. 5-2. Grouped women holding cards.

Figure 5-2 is similar to a *histogram*, which is a vertical bar graph of the frequency in each height group. Converting Figure 5-2 into a histogram requires labels for the height groups along the x-axis and the frequency count along the y-axis, as seen in Figure 5-3.

If the person at the top of each row grabbed onto a long rope, and all the remaining women left the field, the following shape would appear.

*The mode is the most numerous value in a distribution. The median is the middle value, which has equal numbers of values above and below it.

In Figure 5-4, the rope runs as a straight line from one person to the next. Moving the rope into a smooth line results in Figure 5-5.

This shape is called a curve. This particular curve is the *bell curve*, because it resembles the shape of a bell. There are many shapes for bell curves. Figure 5-6 shows variations of bell curves.

In order to distinguish these curves, a measure of the *spread* of the bell is necessary. The next section presents a method for calculating a numerical measure of this spread.

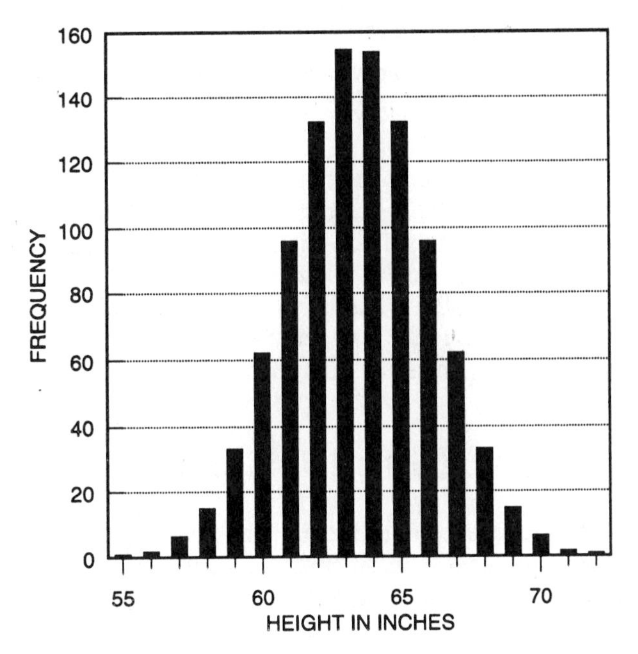

Fig. 5-3. Histogram: Women by height.

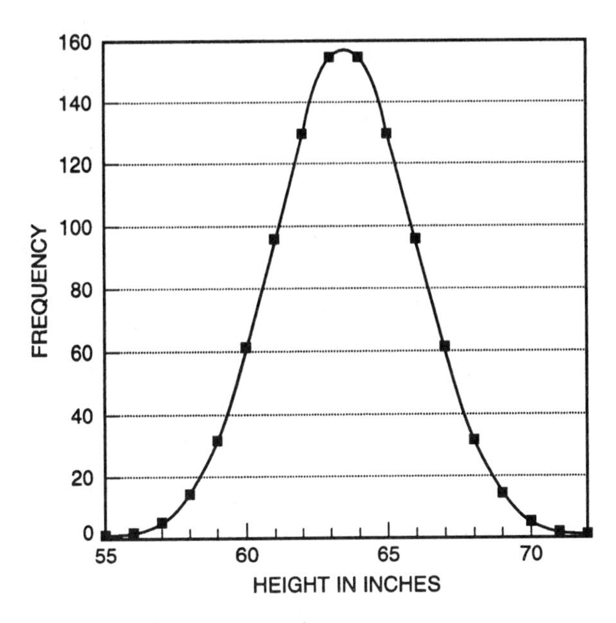

Fig. 5-5. Bell curve: Women by height.

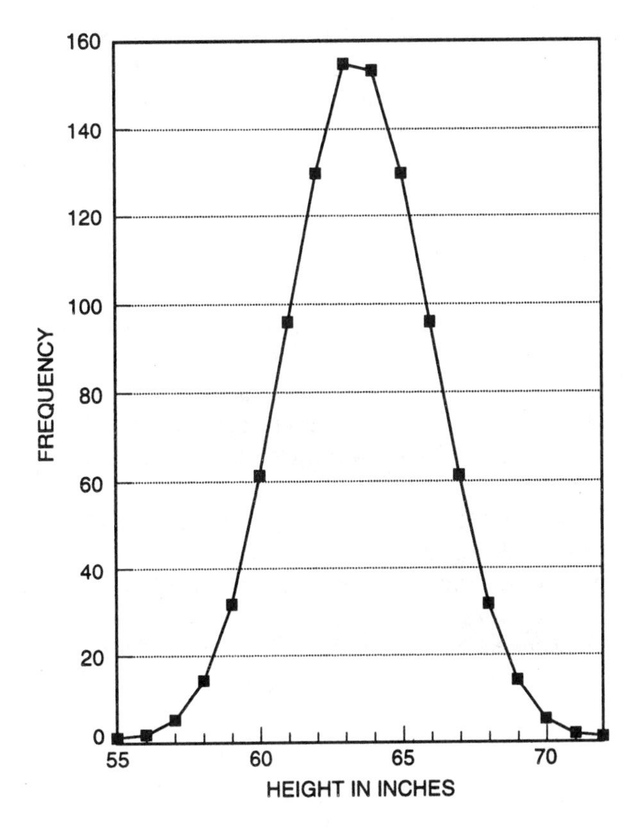

Fig. 5-4. Curve composed of straight-line segments: Women by height.

Calculation of the Standard Deviation

Carefully examine the following distributions. Notice that these distributions are made up. They are not normal distributions, but they do illustrate an important point.

A	B	C	D	E
12	14	18	11	10
11	12	14	10	10
10	10	10	10	10
9	8	6	10	10
8	6	2	9	10
$\bar{x} = 10$	$\bar{x} = 10$	$\bar{x} = 10$	$\bar{x} = 10$	$\bar{x} = 10$

The means of distributions A, B, C, D and E are all 10. However, these 5 distributions are not equivalent. E has values that are the closest together. C has values that are most spread out. How is it possible to describe the spread numerically?

One approach is to determine the difference between each value and the mean. For distribution A, the largest value is 12. After subtracting the mean (10), the resultant value is 2. Repeating the same procedure for each value yields the results summarized below. The column marked *d* stands for the deviation between each value and the mean.

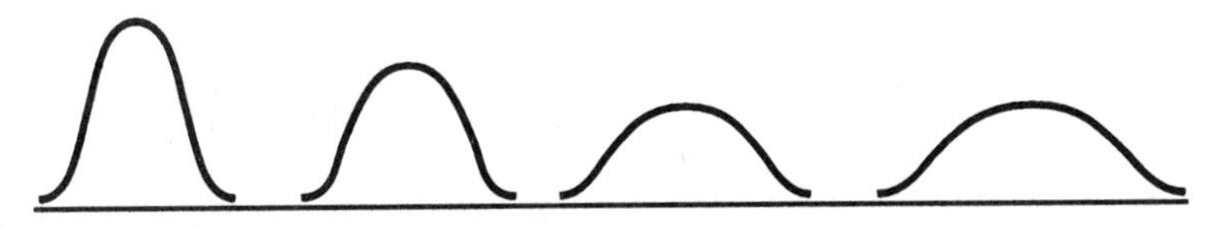

Fig. 5-6. Variations of bell curves.

A	*d*
12	2
11	1
10	0
9	–1
8	–2

There is a problem with the column of deviations, because summing them results in zero. This prevents further calculations to arrive at a numerical measure of the spread. To overcome the difficulties caused by that zero, square all the deviations. Squaring will eliminate the negative values.

	$(d)^2$
	4
	1
	0
	1
	4
sum =	10

The 10 is *the sum of the squared deviations from the mean.* This sum is not zero, so work can continue. Next calculate the average deviation by dividing the sum by n-1. In other words, 10 divided by 4.

When calculating the mean, the divisor was n. Here, the divisor is 4, $(n - 1)$, instead of 5, (n). The reason for altering the divisor is beyond the scope of this chapter. When n is 30 or greater, the difference between n, 30 and $n - 1$, 29 becomes insignificant. However, when the total numberms of values collected is small, the difference between dividing by n and dividing by n-1 can be important. To be on the safe side, use $n - 1$.

Squaring the values did eliminate the zero sum, but it also introduced squared units. If the original units were pounds, the squared values are pounds squared. To get back to the original units, take the square root of the average squared deviations.

$$10/4 = 2.5$$

$$\sqrt{2.5} = 1.58$$

The final number, 1.58, is the *standard deviation.* The calculation steps for distribution B are identical.

B	*d*	$(d)^2$
14	4	16
12	2	4
10	0	0
8	–2	4
6	–4	16
x = 10		40

$$40/n = 40/4 = 10$$

$$\sqrt{10} = 3.16$$

To practice this procedure, calculate the standard deviations for distributions C, D, and E. The results are:

Standard deviation for C = 6.32

Standard deviation for D = 0.71

Standard deviation for E = 0

If a distribution is normal, the mean and standard deviation describe it completely. By convention, the total area under a bell curve is constant. Changing the standard deviation causes the height and width to change. Narrow distributions are tall; broad distributions are short. Refer to Figure 5-6 to see the changes in the curve as the standard deviation increases.

The Standard Normal Distribution

Mathematicians noticed that bell curves described many physical measurements, including height, weight, length, temperature, and density. They worked to identify a *generic* bell curve that could be used to explain details about unique curves. This generic curve is the *standard normal distribution.* A standard normal distribution has several important characteristics. It has a mean of zero and and a standard deviation equal to one. Figure (5-7) shows the percentages of area included in each major section.

Graphical Representation of Hardness Test Results

To put this statistical information to use, consider an experiment on a piece of clear acrylic material,—often called *Plexiglas*, after a popular trade name. The experiment has two goals—first, to measure the hardness of the piece, and second to determine the uniformity of hardness.

Before beginning any testing activities, state the expected results. In this case, all testing occurrs on one piece of material. If the material is not badly flawed, the hardness results should be quite uniform.

To check these expectations against reality, a Rockwell hardness tester was used; 100 spots were

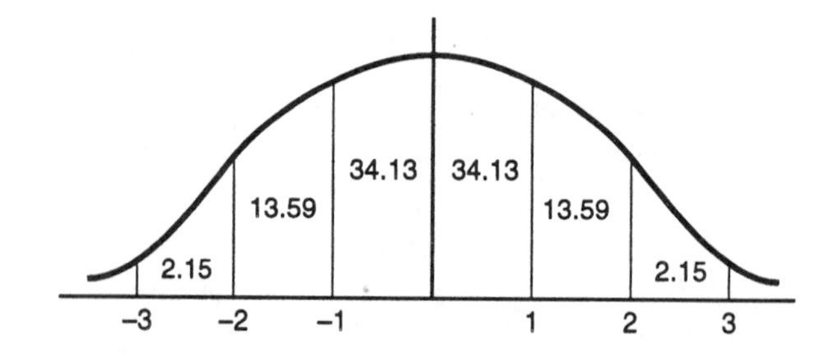

Fig. 5-7. Standard normal distribution with area regions identified.

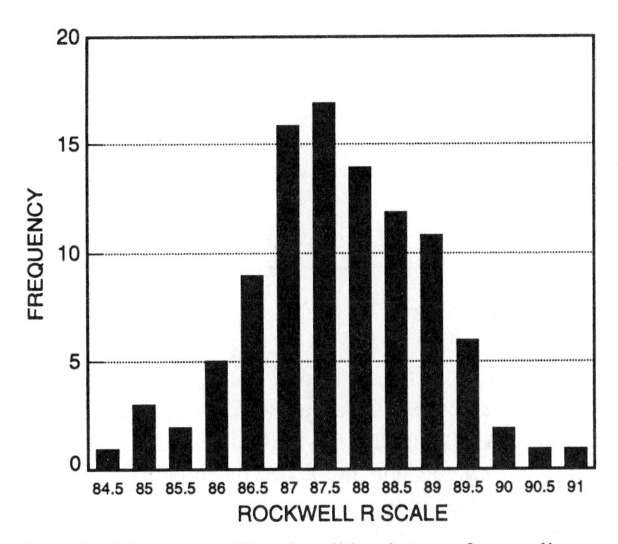

Fig. 5-8. Histogram of Rockwell hardnesses for acrylic.

tested. (For further information on Rockwell hardness tests, see Chapter 6.) A histogram graphically represents the frequency of readings at each hardness.

This histogram (Fig. 5-8) shows two important characteristics. First, it shows central tendency, since the values cluster around one point. Second, it shows symmetry. Although it is not perfectly symmetrical, it does match the theoretical normal distribution well. A number of statistical techniques can evaluate this match, but they are beyond the scope of this chapter. These graphic techniques rely on informed judgment. If a distribution of interest shows central tendency and considerable symmetry, consider it normal. If the empirical data are clearly not normal, then halt further analysis. The sketches in Figure 5-9 should help when making a decision.

Since the distribution of the hardness test results is normal, the next step is to calculate the mean and standard deviation.

Mean: 87.7 (Rockwell R scale)

Standard Deviation: 1.24

Usually, testing is far too expensive to gather this many results. A common industrial practice is to stop with 5 to 10 tests. However, this raises a problem. If a technician gathers only 10 data points, is it fair to assume the data follow the normal distribution?

The answer usually relies on past experience. If previous tests indicated normalcy, then it is assumed that future tests of the same type will also be normal. In this case, another 10 points were tested on the same piece of acrylic. The results were:

Mean: 87.4

Standard deviation: 1.10

These results clearly go along with the previous larger sample.

An example of the practical use of these statistical procedures concerns the effect of heat on the hardness of acrylic. Suppose acrylic was selected for use in a department store display case. Making the display involves bending acrylic sheets. To bend the material, it must be hot.

Because employees in department stores routinely clean the displays by washing with window cleaner and drying with paper towels, the management wanted to know if the corners would scratch easily, making the display appear old and worn in a short time.

To answer the question, a test piece of acrylic (of the same type as previously tested) was placed on a strip heater. A band approximately 20 mm wide was heated to 150 °C (300 °F). After the plastic cooled to room temperature, another 10 hardness readings were taken. The results were:

Mean: 80.3

Standard deviation: 1.68

What happened to the hardness of the acrylic? Observation of the readings indicates that it decreased, from 87.4 to 80.3. Is that a big difference? Does that mean that after heating, it is a little softer, a lot softer, or extremely soft?

To get a handle on the amount of change in hardness, comparisons must take the standard deviation into account. The standard deviation for the group of 10 readings before heat was 1.10, and the standard deviation for the group of 10 after heating was 1.68. It is possible to sketch bell curves for each group, but the two curves would be different in shape, because the deviations are different. Figure 5-10 shows two very different curves.

To eliminate such differing shapes, pool the standard deviations. *Pooling* is simply calculating the mean of the two values. In this case:

$$1.10 + 1.68 = 2.78$$

$$2.78/2 = 1.39$$

Fig. 5-9. Normal and non-normal distributions.

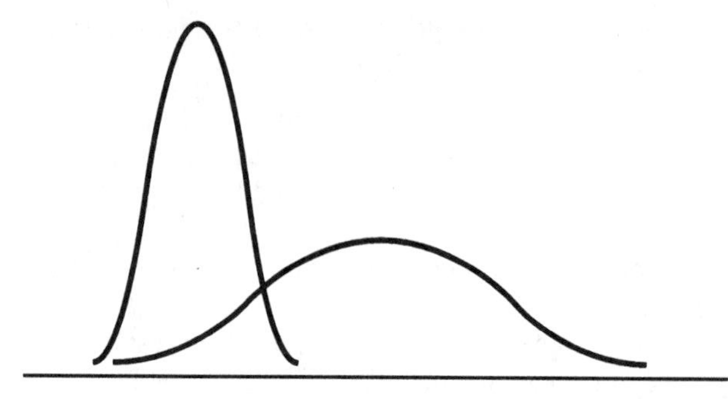

Fig. 5-10. Bell curves showing large differences in standard deviations.

Since the raw data for the hardness test were accurate to the nearest one half unit, round the standard deviation and mean to the nearest half unit also. Do the rounding on the standard deviation after completing the pooling calculation. The pooled standard deviation becomes 1.5; the mean unheated becomes 87.5; and the mean heated becomes 80.5.

A problem arises when the deviations of two groups are not similar. Because the tests involved the same material, the same test equipment, the same time period, and the same environmental conditions, it is reasonable to expect the random variations in the two groups to be similar. Pooling assumes the differences between the two standard deviations come from random causes. However, sometimes the differences are not the same. Statistical techniques are available to determine if pooling is permitted. However, in the absence of advanced statistical knowledge, use the following rules to decide if pooling is allowed.

1. Create a ratio of the two standard deviations with the largest as the numerator and the smallest as the denominator. If the value of that ratio equals 1.5 or less, then pool.
2. If the ratio is larger than 1.5, do not pool. Review the test procedure, looking for a difference in the treatment of the groups. If a difference appears, run the test again to see if the deviations of the two groups come closer to each other.

Sketching Graphs

To create an accurately scaled graphic, follow these steps. First, after pooling the deviations, draw a base line (Fig. 5-11).

Second, arbitrarily locate and label the mean of one group, and make three equally spaced marks on each side of the mean as shown in Figure 5-12.

Third, label these marks as ±1, ±2, and ±3 standard deviations; also attach the calculated deviations to the appropriate marks (Fig. 5-13).

Fig. 5-11. Base line.

Fourth, sketch in one bell curve, making it symmetrical, centered at the mean, and almost touching the base line at ±3 standard deviations (Fig. 5-14).

Fifth, extend and label the increments along the base line to reach the mean of the second group. In this example, more increments are needed to the left of −3 (Fig. 5-15).

Sixth, locate the mean of the second group on the base line. Notice that the mean of the second group did not line up exactly with the increments on the base line. Only in rare cases will it align exactly. Sketch in a bell curve identical in height and width to the one already drawn (Fig. 5-16).

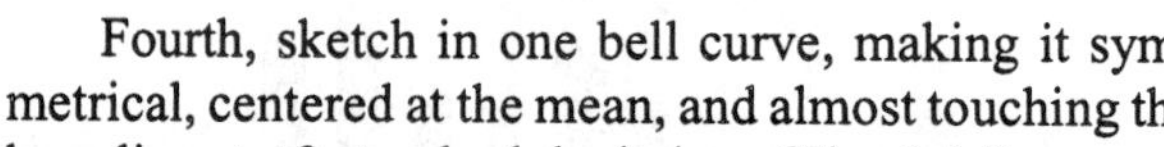

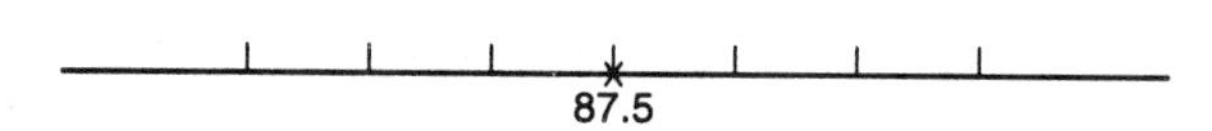

Fig. 5-12. Base line with mean and increments.

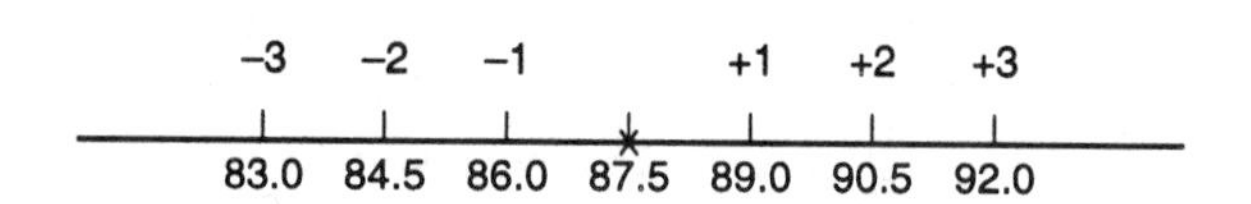

Fig. 5-13. Base line with labeled increments.

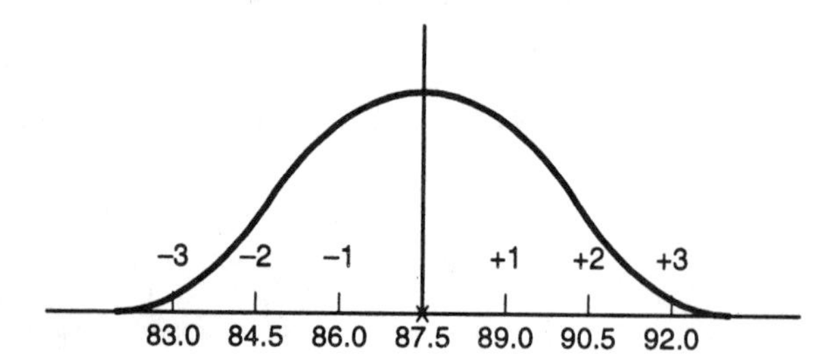

Fig. 5-14. Sketch of one bell curve.

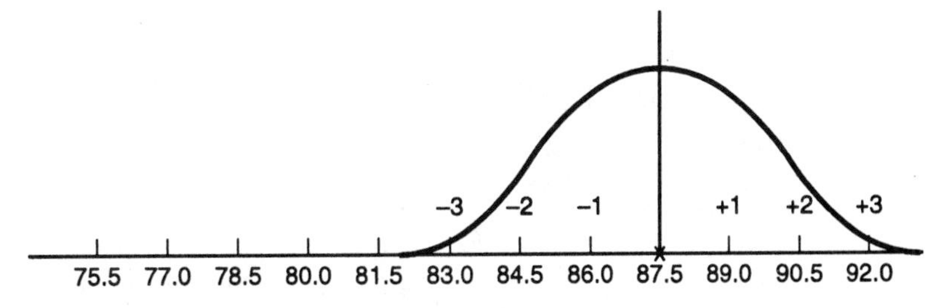

Fig. 5-15. Base line extended and labeled.

Graphic Comparison of Two Groups

Look at the curves. Do two distinct bells occur? Do the two bells overlap almost completely? Do they overlap partially? For the purposes of this chapter, evaluations rely on interpreting the graphics and on a few simple decision criteria. If the two bells don't touch, the answer is clear: The difference is very significant. If the two bells overlap completely, then there is no difference. Assuming the use of 10 pieces in each sample, the following criteria are appropriate.

1. If the difference between the two means is smaller in size than one standard deviation, there is no difference (Fig. 5-17).
2. If the difference between the two means is four standard deviations or more, the difference is significant. In Figure 5-16, the difference between the means is slightly less than five standard deviations.
3. If the difference is between one and four standard deviations, graphic analysis is insufficient for a definitive answer (Figure 5-18). A host of statistical techniques are available to determine the significance in cases when the graphics are inconclusive. Without those procedures, report the results as inconclusive or repeat the test to see if new results will be conclusive.

Summary

This chapter introduced the normal distribution, calculation of the mean and standard deviation, the standard normal distribution, and graphic techniques to compare two samples. Several criteria for decisions help determine if data are normal; if it is appropriate to pool the standard deviations; and how to accurately sketch bell curves. These procedures reappear in other chapters of this book.

For the most reliable conclusions, be sure to observe the necessary conditions. They are:

1. A normal distribution in each group.
2. The standard deviation for the first group should be equal or almost equal to the deviation in the second group.
3. A sample size of 10.

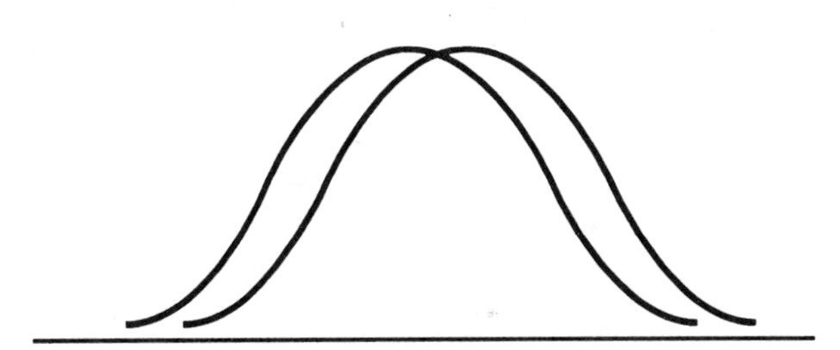

Fig. 5-17. Overlapping curves.

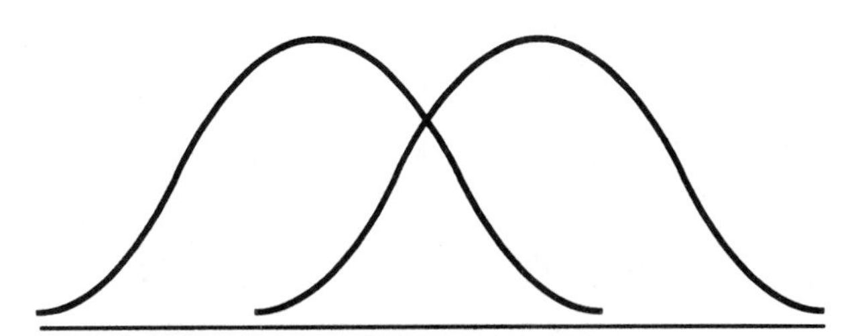

Fig. 5-18. Inconclusive curves.

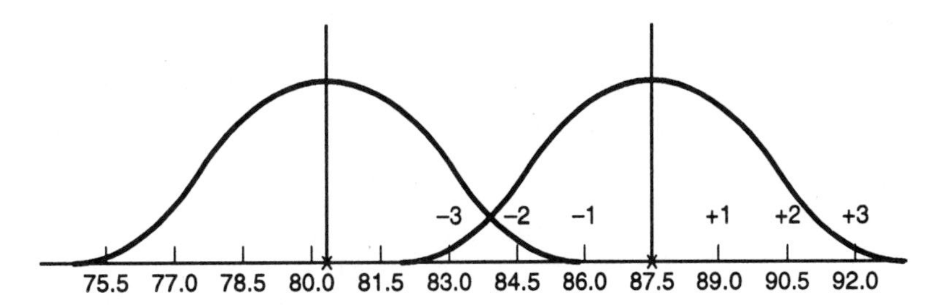

Fig. 5-16. Sketch of second bell curve.

Vocabulary

The following vocabulary words are found in this chapter. Look up the definition of any of these words you do not understand as they apply to plastics in the glossary, Appendix A.

Bell curve
Central tendency
Distribution
Histogram
Mean
n
Normal distribution
Pooled standard deviation
Standard deviation
Standard normal distribution
Symmetrical distribution
x

Questions

5-1. X is an abbreviation that stands for ___?___ .

5-2. A normal distribution must exhibit ___?___ and ___?___ .

5-3. A vertical bar graph of frequency and groups is a ___?___ .

5-4. Squaring the deviations from the mean has the goal of ___?___ .

5-5. The difference between n and n-1 becomes insignificant if n is ___?___ or larger.

5-6. The mean of the standard normal distribution is ___?___ .

5-7. The percentage of area of the standard normal distribution between the mean and +1 standard deviation is ___?___ .

5-8. The percent of area of the standard normal distribution beyond +3 is ___?___ .

5-9. What is the basis for assuming a small sample (10 pieces or less) is normal?

5-10. What does pooling assume?

5-11. If doubling the smaller standard deviation results in a number less than the larger deviation, is pooling appropriate?

5-12. Find distributions A, B, C, D, and E on page 000. Do these distributions appear normal?

Activities

Measuring Standard Deviation

Equipment. Ruler, printed on page 66, (Fig. 5-19) pencil, and calculator. (When using a calculator to determine standard deviation, check to see if it uses n or n-1 in the calculation.)

Procedure

5-1. Measure hand reach, making sure that only fingers are included, not fingernails. If the number of people measured is at least 30 and the data normal, the percentages listed in Figure 5-12 will apply directly. Record your findings to the nearest unit of the scale. Measure both right and left hands, and be sure to record if the subjects are male or female and at least 18 years old.

5-2. Make a histogram of all results.

5-3. Make two histograms, one for females, and another for males.

5-4. Make four histogram, one for right hands of females, a second for left hands of females,

a. a third and fourth for right hands of males and left hands of males.

5-5. Do the plots indicate normal data?

5-6. Calculate means and standard deviations for normal data. It is pointless to complete these calculations if the data are not normal.

Weighing Standard Deviation

Equipment. Scale.

Pressure. Exercises based on weights are easy to create. Find some manufactured items intended to be identical. Carefully weigh them, calculate mean and standard deviation. If virtually identical items are available from competing manufacturers, comparisons are appropriate.

One exercise concerns the weight of plastic pellets. This will require a scale accurate to 0.001 gram. Try to guarantee that the pellets are representative of a bag or box of pellets. In industrial sampling of gaylords—1000 lb boxes—a grain sampler guarantees that the sample includes pellets from the top, middle, and bottom of the box.

5-1. Find chopped pellets, identifiable by a cylindrical shape and cut ends.

a. Weigh 30 chopped pellets. Calculate mean and standard deviation.

b. Weigh 30 chopped pellets from another manufacturer.

c. Are the means identical?

d. Does one manufacturer have a more consistent chop than the other?

5-2. Find pellets that are oval or spherical in shape. These pellets will show no marks from chopper blades.

a. Weigh 30 oval/spherical pellets, and calculate mean and standard deviation.

b. Are the spherical pellets heavier than the chopped ones?

c. Is the process for making spherical pellets more consistent than the chopping process?

Representative results:

Chopped PP: mean 0.0174 g sd. 0.0015 g

Chapped PS: mean 0.0164 g sd. 0.0024 g

HIPS (spherical): mean 0.0352 g sd. 0.0050 g

5-3. Weigh individual pellets of the chopped type and the spherical type. Do the data exhibit normalcy?

5-4. Are chopped pellets more uniform or less uniform in weight than spherical type?

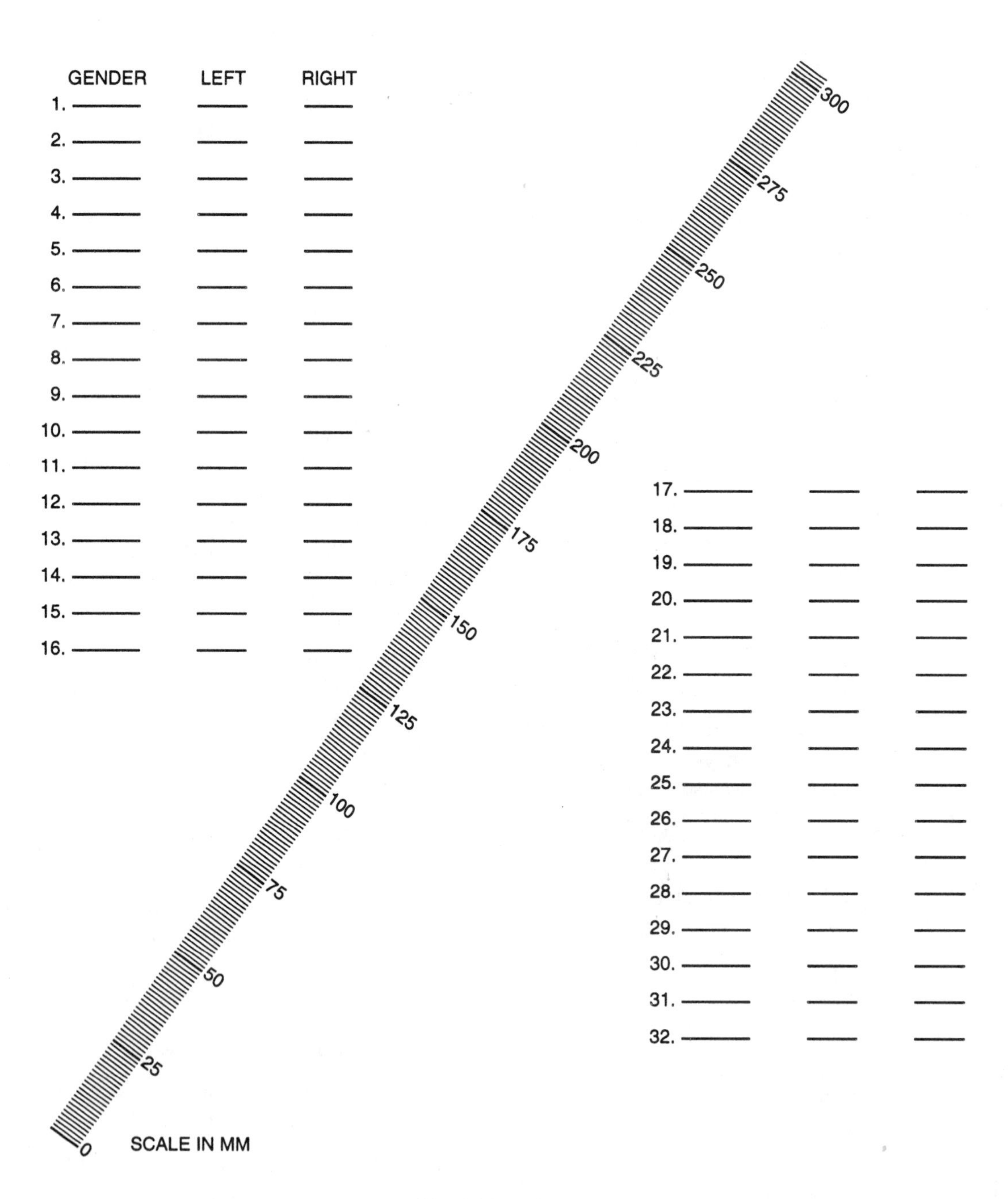

Fig. 5-19. Ruler.

Chapter 6

Properties and Tests of Selected Plastics

Introduction

Virtually every segment of the plastics industry relies on test data to direct its activities. Raw-materials manufacturers use testing to maintain control of their processes and to characterize their products. Designers base their selection of plastics for new products on the results of standard tests. Mold and tool makers depend on shrinkage factors to build molds which will produce finished parts that meet dimensional requirements. Plastics manufacturers use test results to help establish process parameters. Quality-control personnel check to see if products meet customers' requirements, which often cite standard tests. A thorough understanding of testing is essential to many positions within the plastics industry.

This chapter will discuss the most common tests for plastics, which have been grouped into categories. The outline for this chapter is:

I. Testing Agencies
   A. ASTM
   B. ISO
   C. SI units
II. Mechanical properties
   A. Tensile strength (ISO 527, ASTM D-638)
   B. Compressive strength (ISO 604, ASTM D-695)
   C. Shear strength (ASTM 732)
   D. Impact strength
   E. Flexural strength (ISO 178, ASTM D-790 and D-747)
   F. Fatigue and flexing (ISO 3385, ASTM D-430 and D-813)
   G. Damping
   H. Hardness
   I. Abrasion resistance (ASTM D-1044)
III. Physical properties
   A. Density and relative density (ISO 1183, ASTM D-792 and D-1505)
   B. Mold Shrinkage (ISO 2577, ASTM D-955)
   C. Tensile creep (ISO 899, ASTM D-2990)
   D. Viscosity
IV. Thermal properties
   A. Thermal Conductivity (ASTM C-177)
   B. Specific heat (heat capacity)
   C. Thermal expansion
   D. Deflection temperature (ISO 75, ASTM D-648)
   E. Ablative plastics
   F. Resistance to cold
   G. Flammability (ISO 181, 871, 1210, ASTM D-635, D-568, and E-84)
   H. Melt index (ISO 1133, ASTM D-1238)
   I. Glass transition temperature
   J. Softening point (ISO 306, ASTM D-1525)
V. Environmental properties
   A. Chemical properties
   B. Weathering (ISO 45, 85, 4582, 4607, ASTM D-1435 and G-23)
   C. Ultraviolet resistance (ASTM G-23 and D-2565)
   D. Permeability (ISO 2556, ASTM D-1434, and E-96)

E. Water absorption (ISO 62, 585, 960, ASTM D-570)
F. Biochemical resistance (ASTM G-21 and G-22)
G. Stress cracking (ISO 4600, 6252, ASTM D-1693)

VI. Optical properties
A. Specular gloss
B. Luminous transmittance (ASTM D-1003)
C. Color
D. Index of refraction (ISO 489, ASTM D-542)

VII. Electrical properties
A. Arc resistance (ISO 1325, ASTM D-495)
B. Resistivity (ISO 3915, ASTM D-257)
C. Dielectric strength (ISO 1325, 3915, ASTM D-149)
D. Dielectric constant (ISO 1325, ASTM D-150)
E. Dissipation factor (ASTM D-150)

Testing Agencies

Several national and international agencies establish and publish testing specifications for industrial materials. In the United States, the standards generally come from the American National Standards Institute, the United States military services, and the American Society for Testing and Materials (ASTM). A major international organization similar to the ASTM is the International Organization for Standardization (ISO).

ASTM

The ASTM is an international, nonprofit technical society devoted to "...the promotion of knowledge of the materials of engineering, and the standardization of specifications and methods of testing." The ASTM publishes testing for specifications for most industrial materials. Plastics testing comes under the jurisdiction of the ASTM Committee D on plastics. The ASTM annually publishes the *Book of ASTM Standards*, which includes approximately 15 volumes. Most volumes consist of several sections, each one bound separately. A full set of ASTM standards amounts to about 70 sections. The three sections comprising volume 8 deal with plastics.

ISO

The International Organization for Standardization (ISO) contains national standards organizations from over 90 countries. "The object of ISO is to promote the development of standards in the world with a view to facilitating international exchange of goods and services, and to developing co-operation in the sphere of intellectual, scientific, technological and economic activity." The *ISO Standards Handbook 21* consists of two volumes and includes tests on plastics materials and products.

Several U.S. companies that manufacture plastics are adding ISO methods to their testing capabilities. Manufacturers that hope to initiate material sales in Europe and Asia and those that wish to expand their overseas operations need to meet ISO standards. Some companies supply both ISO and ASTM test results to potential customers. Other companies are planning for the introduction of ISO methods.

Table 6-1 identifies a number of common tests for plastics along with corresponding ISO and ASTM methods.

Table 6-1. Summary of ISO and ASTM Testing Methods

Property	ISO Method	ASTM Test Method*	SI Units
Apparent density		D-1895	g/cm^3
Free flowing	60		
Non-pouring	61		
Arc resistance			
High voltage	1325	D-495	s
Low current			
Brittleness temperature	974	D-746	°C at 50%
Bulk factor	171	D-1895	Dimensionless
Chemical resistance	175	D-543	Changes recorded
Compression set	1856	D-395	Pa
Compressive strength	604	D-695	Pa
Conditioning procedure	291	D-618	Metric units
Creep	899	D-2990	Pa
Creep rupture		D-2990	Pa
Deflection temperature	75	D-648	°C at 18.5 MPa
Density	1183	D-1505	g/cm^3
Dielectric constant	1325	D-150	Dimensionless
Dissipation factor at 60 Hz, 1 KHz, 1 MHz			

Table 6-1. Summary of ISO and ASTM Testing Methods (Continued)

Property	ISO Method	ASTM Test Method*	SI Units
Dielectric strength	3915	D-149	V/mm
Short time			
Step by step			
Dynamic mechanical properties		D-2236	Dimensionless
Logarithmic decrement			
Elastic shear modulus			
Elasticity modulus			
Compressive	4137	D-695	Pa
Tangent, flexural		D-790	Pa
Tensile		D-638	Pa
Elongation	R527	D-638	%
Fatigue Strength	3385	D-671	Number of cycles
Flammability	181, 871, 1210	D-635	cm/min (burning), cm/s
Flexural strength	178	D-790	Pa
Flexural stiffness		D-747	Pa
Flow temperature		D-569	°C
Rossi-Peakes			
Gel time & peak exothermic temperature	2535	D-2471	
Hardness			
Durometer	868	D-2240	Read dial
Rockwell	2037/2	D-785	Read dial
Haze		D-1003	%
Impact resistance			
Dart		D-1709	Pa @ 50% failure
Charpy	179		
Izod	180	D-256	J/m
Indention Hardness		D-2583	Read dial
Barcol Impressor			
Linear coefficient of thermal expansion		D-696	mm/mm/°C
Load deformation		D-621	%
Luminous transmittance		D-1003	%
Melt-flow rate, thermoplastics	1133	D-1238	g/10 min.
Melting point	1218, 3146	D-2117	°C
Mold shrinkage	3146	D-955	mm/mm
Molding index		D-731	Pa
Notch sensitivity		D-256	J/m
Oxygen index		D-2863	%
Particle size		D-1921	Micrometre
Physical property changes		D-759	Changes recorded
Subnormal temperature			
Supernormal temperature	1137, 2578		
Refraction index	489	D-542	Dimensionless
Relative density	1183	D-792	Dimensionles
Shear Strength		D-732	Pa
Solvent swell		D-471	J
Surface abrasion resistance		D-1044	Changes recorded
Tear resistance		D-624	Pa
Tensile strength	R527	D-638	Pa
Thermal conductivity		C-177	W/K · m
Vicat softening point	306	D-1525	ohm-cm
Volume resistivity 1 min. at 500V		D-257	%
Water absorption			
24 hour immersion	62, 585, 960	D-570	%
Long-term immersion			
Water vapor		E-96	g/24 h
Permeability		E-42	
Weathering	45, 85, 877 4582, 4607	D-1435	Changes

*Use the latest version of any ISO and ASTM method referenced.

Table 6-2. SI Base Units

Quantity	Unit	Symbol
Length	metre	m
Mass	kilogram	kg
Time	second	s
Thermodynamic temperature	kelvin	K
Electric current	ampere	A
Luminous intensity	candela	cd
Amount of substance	mole	mol

Although the ASTM specifications use both metric measurements and English units, ISO methods use only SI, *System International d'Unites*, metric units. In the United States, both systems are used, and students must be able to work quickly and easily in both SI units and English units. This chapter will follow the publication guidelines of the Society of Plastics Engineers, namely that SI units are always used, occasionally followed by English units in parentheses.

SI Units

The SI metric system consists of seven *base units*, shown in Table 6-2. To simplify large and small numbers, the SI system uses a set of prefixes listed in Table 6-3. When the base units are combined or when additional measures are needed, derived units are used. Table 6-4 lists selected *derived units* which are frequently used in the plastics industry.

Table 6-3. Prefix and Numerical Expression

Symbol	Prefix	Phonic	Decimal Equivalent	Factor	Prefix Origin	Original Meaning
E	exa	*x'*a	1000000000000000000	10^{18}	Greek	colossal
P	peta	*pet'*a	1000000000000000	10^{15}	Greek	enormous
T	tera	*ter'*a	1000000000000	10^{12}	Greek	monstrous
G	giga	*ji'* ga	1000000000	10^{9}	Greek	gigantic
M	mega	*meg'*a	1000000	10^{6}	Greek	great
k	kilo	*kil'*o	1.000	10^{3}	Greek	thousand
h	hecto	*hek'*to	100	10^{2}	Greek	hundred
da	deka	*dek'*a	10	10^{1}	Greek	ten
d	deci	*des'*i	0.1	10^{-1}	Latin	tenth
c	centi	*cen'*ti	0.01	10^{-2}	Latin	hundredth
m	milli	*mill'*i	0.001	10^{-3}	Latin	thousandth
µ	micro	*mi'*kro	0.000001	10^{-6}	Greek	small
n	nano	*nan'*o	0.000000001	10^{-9}	Greek	very small
p	pico	*pe'*ko	0.000000000001	10^{-12}	Spanish	extremely small
f	femto	*fem'*to	0.000000000000001	10^{-15}	Danish	fifteen
a	atto	*at'*to	0.000000000000000001	10^{-18}	Danish	eighteen

Table 6-4. Selected Derived SI Units

Quantity	Unit	Symbol	Formula
absorbed dose	gray	Gy	J/kg
area	square metre	m^2	
volume	cubic metre	m^3	
frequency	hertz	Hz	s^{-1}
mass density (density)	kilogram per cubic metre	kg/m^3	
speed, velocity	metre per second	m/s	
acceleration	metre per second squared	m/s^2	
force	newton	N	$kg \cdot m/s^2$
pressure (mechanical stress)	pascal	Pa	N/m^2
kinematic viscosity	square metre per second	m^2/s	
dynamic viscosity	pascal second	$Pa \cdot s$	$N \cdot s/m^2$
work, energy, quantity of heat	joule	J	$N \cdot m$
power	watt	W	J/s
quantity of electricity	coulomb	C	$A \cdot s$
electric potential, potential difference, electromotive force	volt	V	W/A
electric field strength	volt per metre	V/m	
electric resistance	ohm	Ω	V/A

Mechanical Properties

The *mechanical properties* of a material describe how it responds to the application of a force or load. There are only three types of mechanical force that can effect materials. These are *compression, tension,* and *shear.* Figure 6-1 shows these as push (Fig. 6-1A), pull (Fig. 6-1B) , and opposing pull which threatens to shear off the bolt in (Fig. 6-1C). The mechanical tests consider these forces separately, and in combinations. Tensile, compression, and shear tests measure only one force; while flexural, impact, and hardness tests involve two or more simultaneous forces.

Brief discussions of selected tests for mechanical properties follow. The tests treated are: tensile strength, compressive strength, shear strength, impact strength, flexural strength, fatigue, hardness and abrasion resistance.

Tensile Strength (ISO 527, ASTM D-638)

Calculation of tensile force requires the SI base unit for mass and the derived unit of acceleration. By definition,

$$\text{force} = \text{mass} \times \text{acceleration.}$$

The unit of *mass* is the kilogram, and the unit of *acceleration* is meters per second squared. The standard value for the acceleration caused by gravity on earth is 9.806 65 meters per second squared. This value, 9.807 m/s^2, is called the *gravity constant.* The SI unit of force is the *newton,* which is the force of gravity acting on one kilogram.

$$1 \text{ newton} = 1 \text{ kilogram} \times 9.807 \text{ m/s}^2$$

Stress. Pressure is force applied over an area. The technical term for pressure is *stress.* The metric unit for stress is the *pascal* (Pa). One pascal equals the force of one newton exerted on the area of one square meter. In the English system, the unit is pounds per square inch (psi). Tensile strength is measured in pascals and is the ratio of the pulling force in newtons and the original cross-sectional area of the sample in square meters.

$$\text{Tensile strength (Pa)} = \frac{\text{pulling force (N)}}{\text{cross-section } (\text{m}^2)}$$

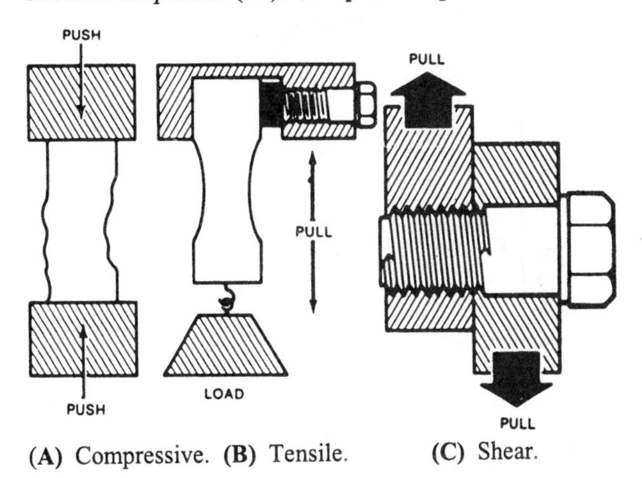

(A) Compressive. (B) Tensile. (C) Shear.

Fig. 6-1. Three types of stress.

Strain. Pulling stress usually causes material to deform by thinning in width and stretching in length. As shown in Figure 6-2, the change in length, in relation to the original length, is called *strain.*

Strain is measured in millimeters per millimeter (inches per inch). It can be expressed as a percent, and is then called *percent elongation.* To convert strain in meters per meter to a percent, merely multiply by 100 and report it as a percent. Strain is apparent when tensile testing plastics deform readily. Figure 6-3 shows typical deformation in an unreinforced plastic.

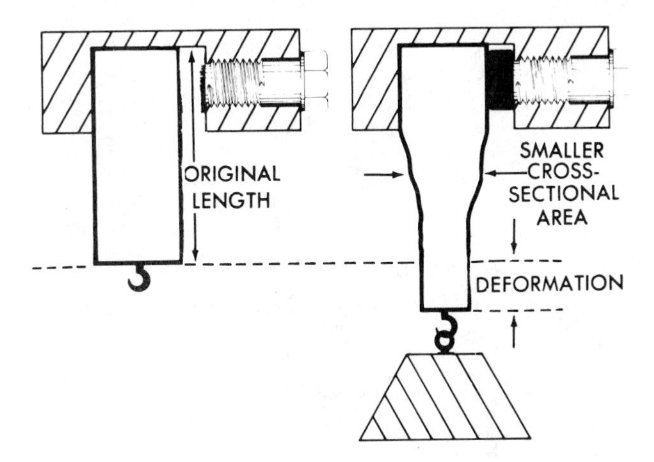

Fig. 6-2 . *Strain* is deformation due to pulling stress.

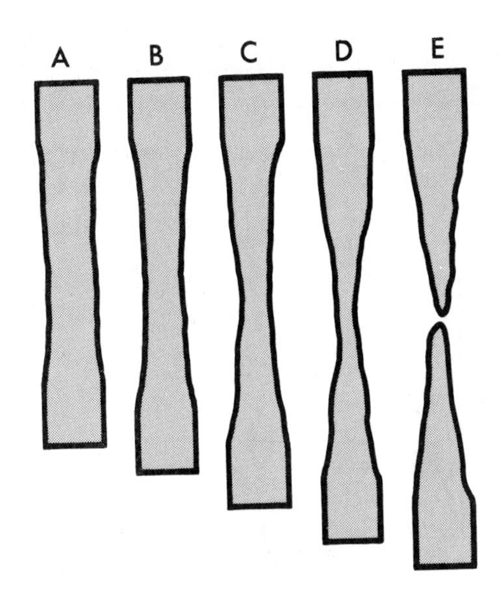

Fig. 6-3. Stages of deformation in unreinforced plastics.

Stress-Strain Diagrams. Modern tensile-testing apparatus create stress-strain diagrams. These plots accurately document the stress on the sample, and the resulting strain at all levels of load. Figure 6-4 shows a tensile-testing machine and peripheral equipment.

This tensile testing system includes a monitor, which displays stress-strain curves and numeric data; a printer for generating hard copies; and a plotter, which draws the stress-strain curves on graph paper. The computer does mathematical calculations and stores data for quality control reports.

Figure 6-5 shows a stress-strain curve generated by a plotter. The material tested was PC (polycarbonate).

Understanding stress-strain curves requires familiarity with a few technical terms. The *yield point A* is the point on the stress-strain curve, also called the load/extension curve, at which the extension increases without an increase in load (stress). Up to the yield point, the resistance of the PC to the applied force was linear. After point *A*, the relation between stress and strain was no longer linear. Calculation can provide the strength at yield and the elongation to yield.

At the *break point B*, the material failed completely and broke into two pieces. Calculations can readily provide the strength at break and the elongation to break. The *ultimate strength* measures the greatest resistance of the material to the stress. On a stress-strain curve, it corresponds to the highest point *C*.

Figure 6-6 shows a typical stress-strain curve for ABS. This curve shows that the ABS reached its ultimate strength at the yield point (*A* and *C* together.)

Figure 6-7 is a typical stress-strain curve for LDPE.

This curve exhibits no clear yield point. However, to determine strength or elongation at yield, a yield point must be located. The *offset yield*, used when the curve is not conclusive, is the point where a line parallel to the linear portion and offset by a specified amount, crosses the curve. Figure 6-8 shows the offset line, and the location of its intersection with the stress-strain curve (point *A*).

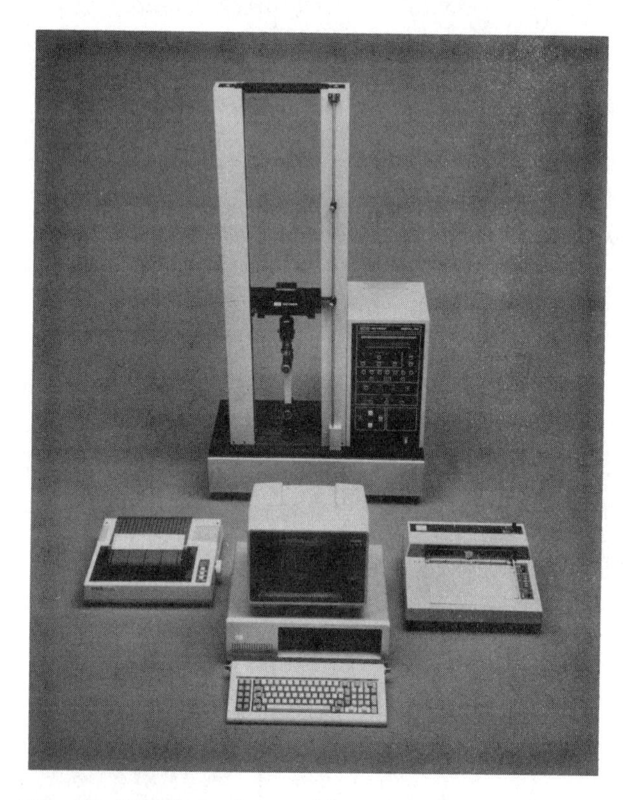

Fig. 6-4. Tensile-testing machine, printer, computer, and plotter. (Photo Courtesy of Instron Corporation)

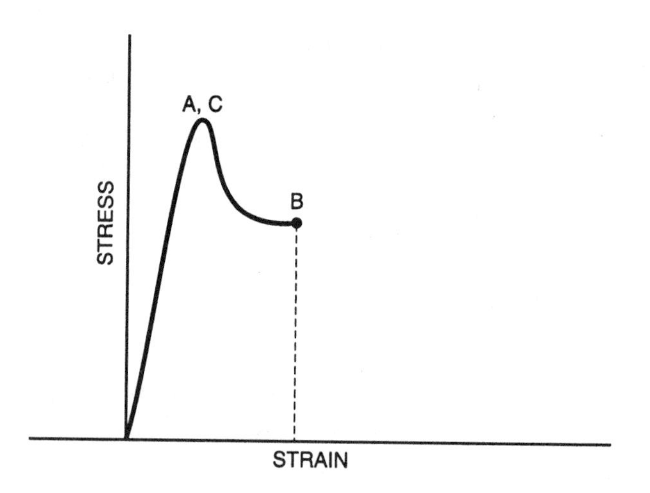

Fig. 6-6. Typical stress-strain curve for ABS.

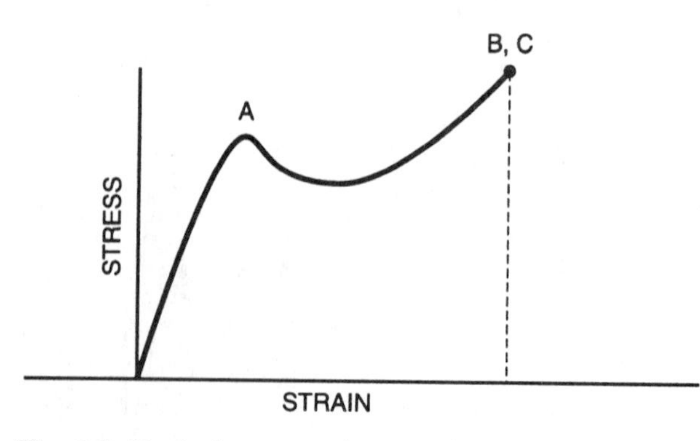

Fig. 6-5. Typical stress-strain curve for polycarbonate.

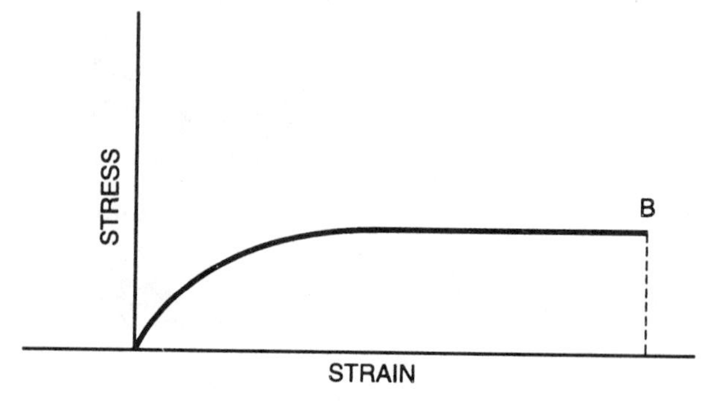

Fig. 6-7. Typical stress-strain curve for LDPE.

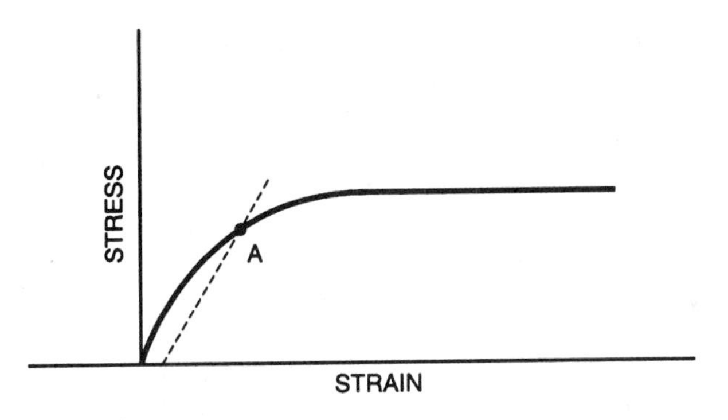

Fig. 6-8. Stress-strain curve with offset yield point shown at point A.

Toughness. A generalization about stress-strain curves is that brittle materials are often stronger and less extensible than soft materials. The weaker plastics often exhibit high elongation and low strength. A few materials are both strong and elastic. The area under the curve represents the energy required to break the sample. This area is an approximate measure of *toughness*. In Figure 6-9, the toughest sample has the largest area under the stress-strain curve.

Modulus of elasticity (tensile modulus). The *modulus of elasticity*, also called tensile modulus or Young's modulus, is the ratio between the stress applied and the strain, within the linear range of the stress-strain curve. Young's modulus has no meaning at stress beyond the yield point. It is calculated by dividing the stress (load) in pascals by the strain (mm/mm). Mathematically, Young's modulus is identical to the slope of the linear portion of the stress-strain curve. When the linear relation remains constant until yield, then dividing the yield strength (Pa) by the elongation to yield (mm/mm) results in the modulus of elasticity.

Young's modulus of elasticity = stress (Pa)/ stain(m/m)

The ratio of tensile force to elongation is useful in predicting how far a part will stretch under a given load. A large tensile modulus indicates that the plastic is rigid and resists elongation.

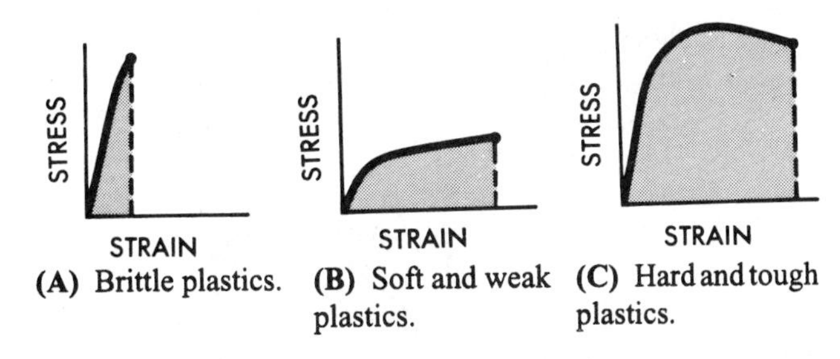

(A) Brittle plastics. **(B)** Soft and weak plastics. **(C)** Hard and tough plastics.

Fig. 6-9. Toughness is a measure of the amount of energy needed to break a material. It is often defined as the total area under the stress-strain curve.

Compressive Strength (ISO 604, ASTM D-695)

Compressive strength is a value that shows how much force is needed to rupture or crush a material.

Compressive strength values may be useful in distinguishing between grades of plastics and in comparing plastics to other materials. Compressive strength is especially significant in testing cellular or foamed plastics.

When calculating the compressive strength, the units needed are multiples of the pascal, such as kPa, MPa and GPa. To determine the compressive strength, divide the maximum load (force) in newtons by the area of the specimen in square meters.

Compressive strength (Pa) =
force (N)/cross-sectional area (m^2)

If 50 kg is required to rupture a 1.0 mm^2 plastics bar:

Force (N) = 50 kg × 9.8 m/s^2

where 9.8 m/s^2 is the gravity constant

Compressive strength (Pa) = (50 × 9.8) N/1 mm^2
490 N/ 1mm^2
= 490 N/0.000 001 m^2
= 490 MPa or
490 000 kPa (71 076 psi)

Shear Strength (ASTM 732)

Shear strength is the maximum load (stress) needed to produce a fracture by a shearing action. To calculate shear strength, divide the applied force by the cross-sectional area of the sample sheared.

$$\text{Shear strength (Pa)} = \frac{\text{force (N)}}{\text{cross-sectional area } (m^2)}$$

To shear a sample, several methods are common. Figure 6-10 shows three of them.

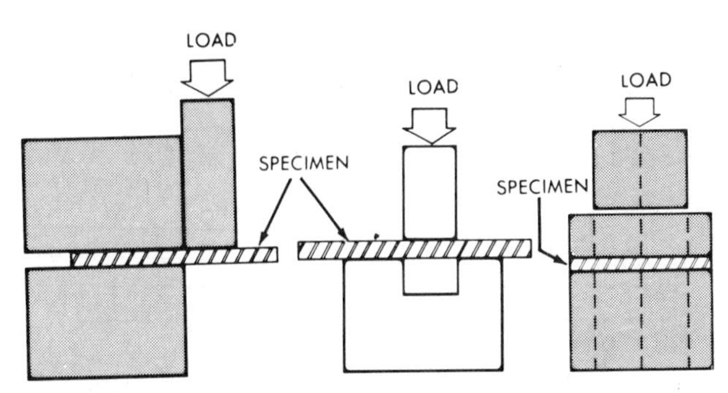

Fig. 6-10. Various methods used to test shear strength.

Impact Strength

Impact strength is not a measure of stress needed to break a sample. It does, however, indicate the energy absorbed by the sample prior to its fracture. There are two basic methods for testing impact strength: (a) falling mass tests, and (b) pendulum tests.

Falling mass test. (ASTM D-1709). Falling mass tests involve dropping a ball-shaped mass from a given height onto the plastic surface. Containers, dinnerware, and helmets are often tested in this manner. Figure 6-11 includes two variations.

When testing films, a blunt dart is used in place of a heavier mass, as shown in Figure 6-11B. Sometimes the sample is allowed to slide down a trough and strike a metal anvil (Fig. 6-12). This test may be repeated from various heights. If damaged, the sample will show cracks, chips, or other fractures.

Pendulum test (ISO 179, 180, ASTM D-256, D-618). Pendulum tests use the energy of a swinging hammer to strike the plastics sample. The result is a measure of energy or work absorbed by the specimen.

The basic formula is:

$$\text{Energy (J)} = \text{force (N)} \times \text{distance (m)}$$

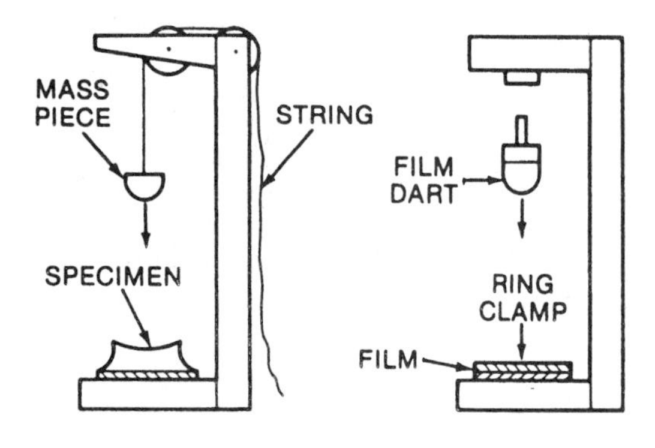

Fig. 6-11. Falling mass tests.

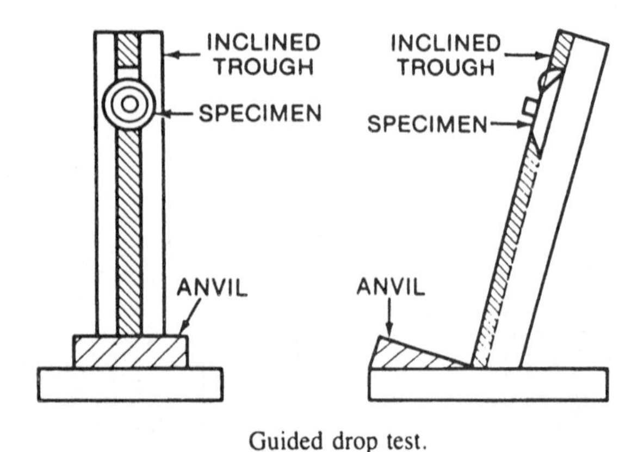

Guided drop test.

Fig. 6-12. Guided drop test.

The hammers of most plastics-testing machines have a kinetic energy of 2.7–22 J [2–16 ft-lb]. Figures 6-13D and 6-13E show two impact-testing machines.

In the *Charpy* (simple beam) method, the test piece is supported at both ends but not held down. The hammer strikes the sample in the center. (See Figures 6-13A and 6-13B). In the *Izod* (cantilever beam) method, the hammer strikes a specimen supported at one end.

Impact tests may specify notched or unnotched samples. In the Charpy test, the notch is on the side away from the striker. In the Izod test, the notch is on the same side as the striker, as shown in Figure 6-13C. In both tests, the depth and radius of the notch can dra-

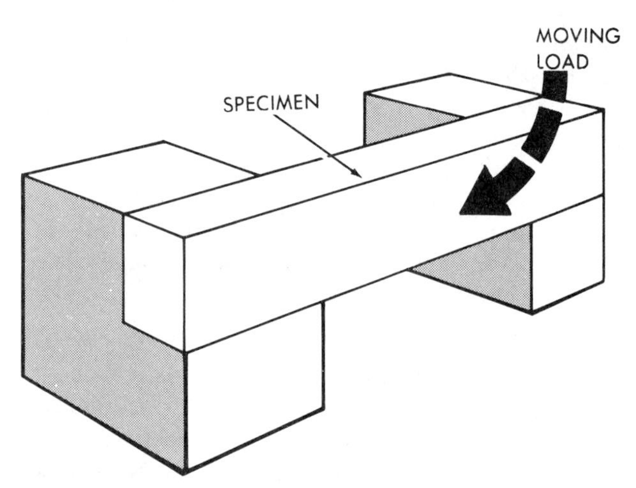

(A) Charpy pendulum method.

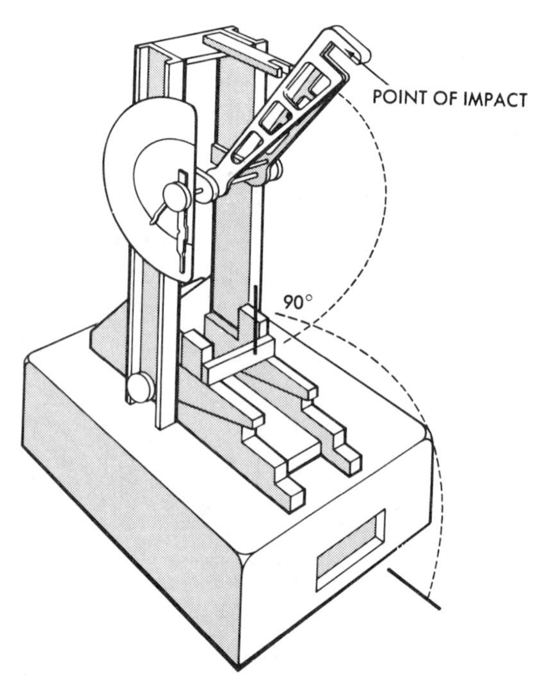

(B) Charpy simple beam impact machine. (Tinius Olsen Testing Machine Co., Inc.)

Fig. 6-13. Charpy and Izod method testing equipment.

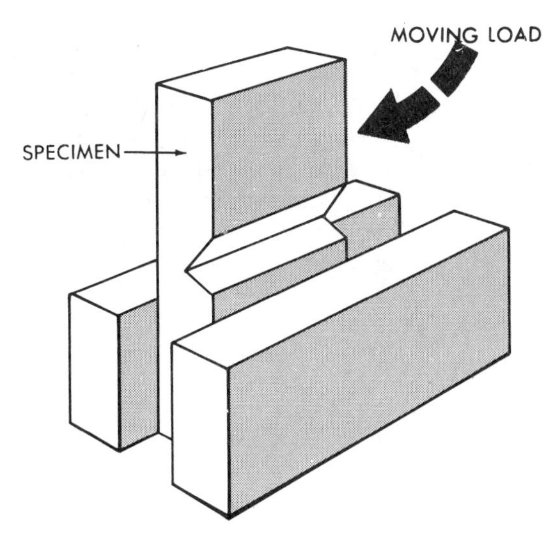

(C) Izod pendulum method.

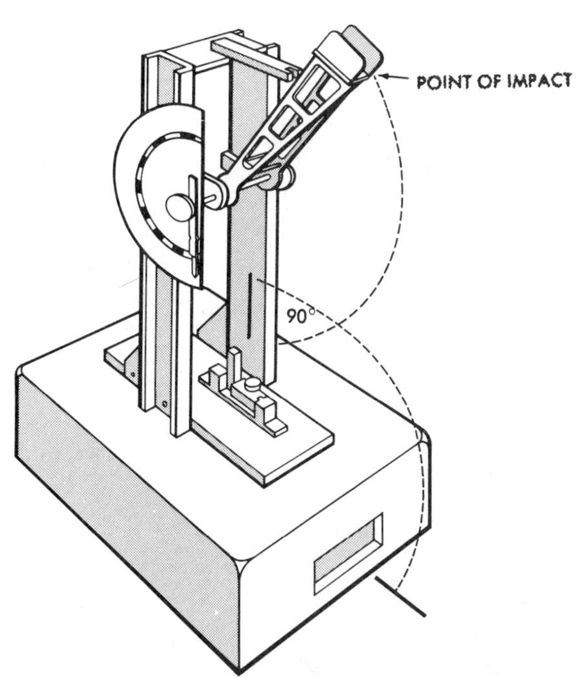

(D) Izod cantilever beam impact machine. (Tinius Olsen Testing Machine Co., Inc.)

matically alter the impact strength, especially if the polymer exhibits notch sensitivity.

PVC is a rather notch-sensitive material. If prepared with a blunt notch, with a radius of 2 mm, PVC has a higher impact strength than ABS. If the samples have sharp notches, with a radius of 0.25 mm, then the impact strength of PVC drops below that of ABS. Other materials that are notch brittle are acetals, HDPE, PP, PET and dry PA.

Moisture in plastics can also influence the impact strength. Polyamides (nylons) exhibit a large difference, having impacts of 5 kJ/m^2 [50 ft-lbs] when thoroughly dried, and strengths over 20 kJ/m^2 [200 ft-lbs] when they contain moisture.

(E) Impact tester for Charpy and Izod testing. (Tinius Olsen Testing Machine Co., Inc.)

Fig. 6-13. Charpy and Izod method testing equipment. *(Continued)*

Since impact measurements must take the thickness of the sample into account, impact-strength values are expressed in joules per square meter (J/m^2) or ft-lbs per inch of notch.

Flexural Strength (ISO 178, ASTM D-790 and D-747)

Flexural strength is a measure of how much stress (load) can be applied to a material before it breaks. Both tensile and compressive stresses are involved in bending the sample. The ASTM sample is supported on test blocks 4 in. [100 mm] apart. The ISO procedure varies the span according to the thickness of the sample. The load is applied in the center (Fig. 6-14).

Because most plastics do not break when deflected, the flexural strength at fracture cannot be calculated easily. In the ASTM method, most thermoplastics and elastomers are measured when 5 percent strain occurs in the samples. This is found by measuring the load in pascals that causes the sample to stretch 5 percent. In the ISO procedure, the force is measured when the deflection equals 1.5 times the thickness of the sample.

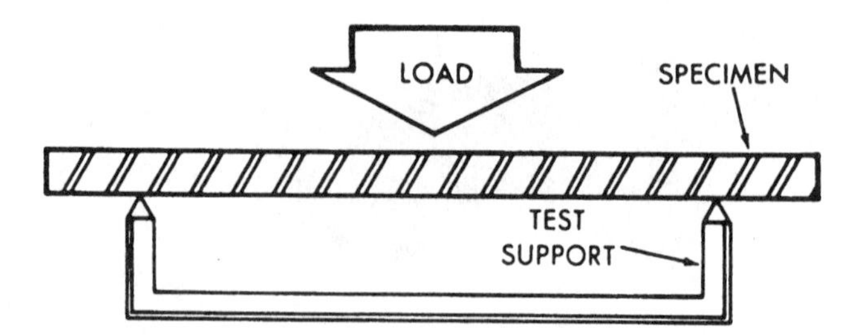

Fig. 6-14. A method used to test flexural strength (flexural modulus).

Fatigue and Flexing (ISO 3385, ASTM D-430 and D-813)

Fatigue is a term used to express the number of cycles a sample can withstand before it fractures. Fatigue fractures are dependent on temperature, stress, and frequency, amplitude, and mode of stressing.

If the load (stress) does not exceed the yield point, some plastics may be stressed for a great many cycles without failure. In producing integral hinges and one-piece box-and-lid containers, the fatigue characteristics of the plastics must be considered. Figure 6-15 shows two integral hinges and the apparatus for testing folding endurance.

Damping

Plastics can absorb or dissipate vibrations. This property is called *damping*. On an average, plastics have ten times more damping capacity than steel. Gears, bearing, appliance housings, and architectural applications of plastics make effective use of the vibration-reducing property.

Hardness

The term *hardness* does not describe a definite or single mechanical property of plastics. Scratch, mar and abrasion resistance are closely related to hardness. Surface wearing of vinyl floor tile and marring of PC optical lenses are affected by several factors. However, a widely accepted definition of hardness is resistance to compression, indentation, and scratch.

There are several types of instruments used to measure hardness. Since each instrument has its own scale for measurement, the values must also identify the scale used. Two tests of rather limited use in testing plastics are the *Mohs scale* and the *scleroscope*.

The Mohs hardness scale is used by geologists and mineralogists. It is based on the fact that harder materials scratch softer ones. The scleroscope (Fig. 6-16) is a nondestructive hardness test. The instrument measures the rebound height of a free-falling hammer called a *tup*.

Indentation test instruments (ASTM D-2240) are used for more sophisticated quantitative measurements. Rockwell, Wilson, Barcol, Brinell, and Shore are well-known testing tools. Figure 6-17 shows the basic differences in hardness tests and scales. Table 6-5 provides details about the various hardness scales. In these tests, either the depth or area of indentation is the measure of hardness.

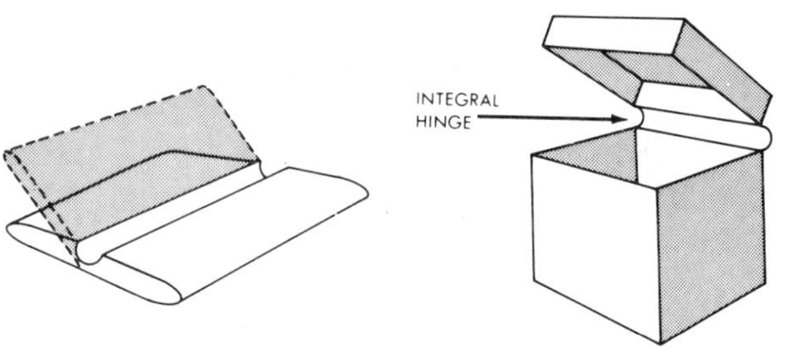

(A) Flexible hinge. (B) One-piece box and lid.

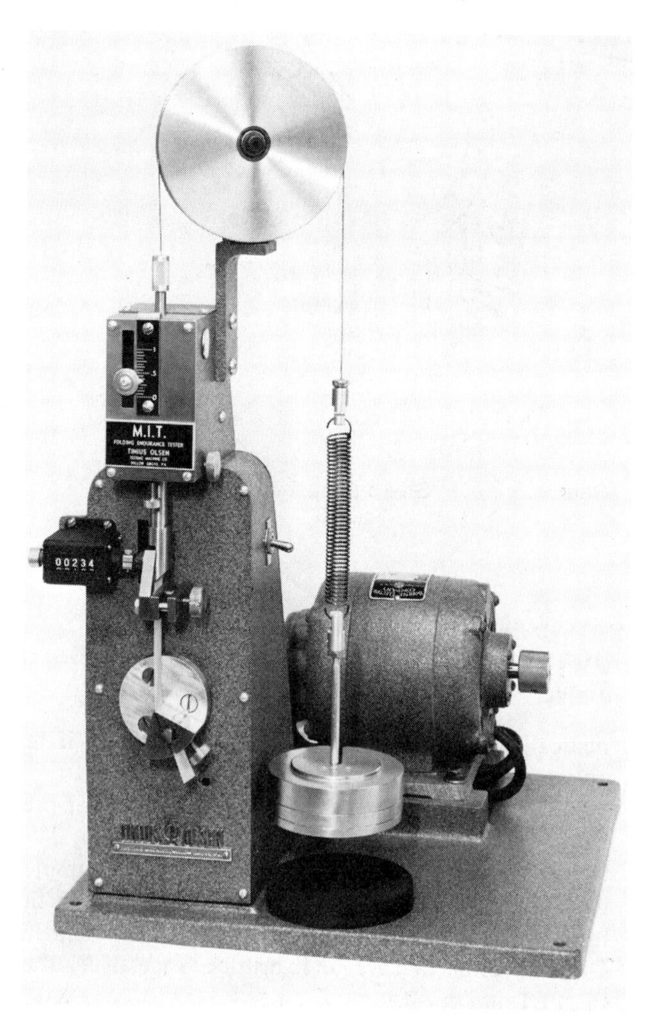

(C) A folding endurance tester, which records on a dial the number of flexings that take place before a plastics sample breaks. (Tinius Olsen Testing Machine Co., Inc.)

Fig. 6-15. Fatigue testing.

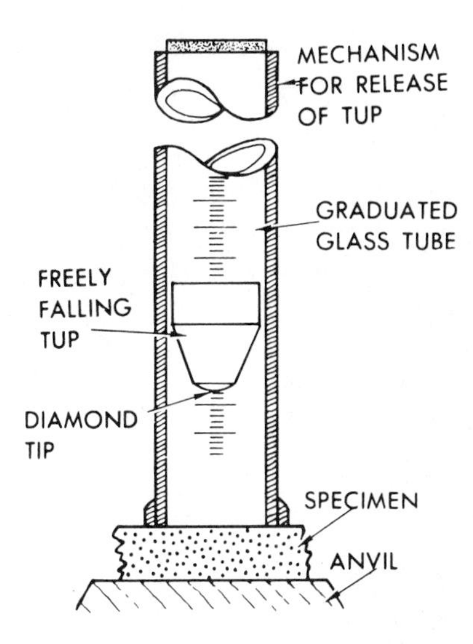

Fig. 6-16. Scleroscope for making hardness tests.

The Brinell test relates hardness to the area of indention. Typical Brinell numbers for selected plastics are acrylic, 20; polystyrene, 25; polyvinyl chloride, 20; and polyethylene, 2. Figure 6-18 shows a Brinell hardness tester.

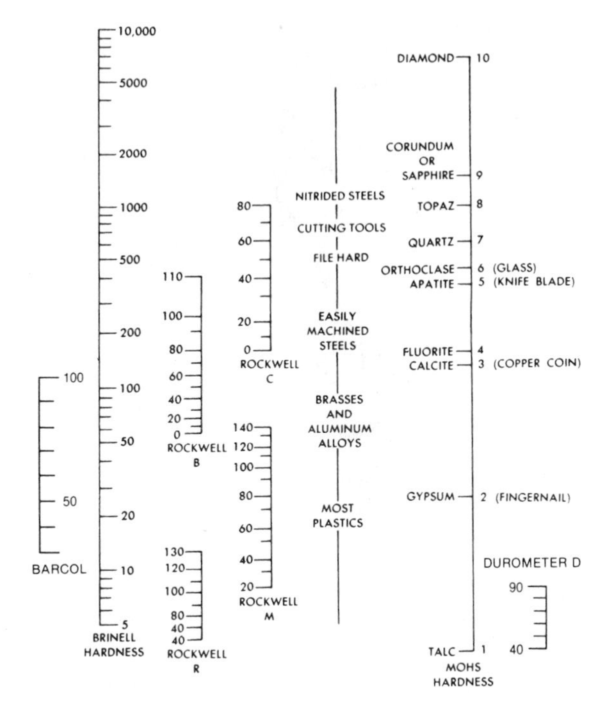

Fig. 6-17. Comparison of various hardness scales.

Table 6-5. Comparison of Selected Hardness Tests

Instrument	Indentor	Load	Comments
Brinnell	Ball, 10 mm diameter	500 kg 3 000 kg	Average out hardness differences in material. Load applied for 15–30 seconds. View through Brinnell microscope shows and measures diameter (value, of impression). Not for materials with high creep factors.
Barcol	Sharp-point, rod 26° 0.157 mm flat tip	Spring loaded. Push against specimen with hand 5–7 kg.	Portable. Readings taken after 1 or 10 s.
Rockwell C	Diamond cone	Minor 10 kg Minor 150 kg	Hardest materials, steel. Table model.
Rockwell B	Ball 1.58 mm (1/16 in)	Minor 10 kg Minor 100 kg	Soft metals and filled plastics.
Rockwell R	Ball 12.7 mm (½ in)	Minor 10 kg Minor 60 kg	Within 10 seconds after applying minor load, apply major load. Remove major load 15 s after application. Read hardness scale 15 s after removing major load or Apply minor load and zero within 10 s. Apply major load immediately after zero adjust. Read number of divisions pointer passed during 15 s of major load.
Rockwell L	Ball 6.35 mm (¼ in)	Minor 10 kg Minor 60 kg	
Rockwell M	Ball 6.35 mm (¼ in)	Minor 10 kg Minor 100	
Rockwell E	Ball 3.175 mm (1/8 in)	Minor 10 kg Minor 100 kg	
Shore A	Rod, 1.40 mm diameter, sharpened to 35° 0.79 mm.	Spring loaded. Push against specimen with hand pressure.	Portable. Readings taken in soft plastics after 1 or 10 s.
Shore D	Rod, 1.40 mm diameter. Sharpened to 30° point with 0.100 mm radius.	As above	As above

Fig. 6-18. This Brinell hardness tester is air-operated. (Tinius Olsen Testing Machine Co., Inc.)

The Rockwell test relates hardness to the difference in the depth of penetration of two different loads. The minor load (usually 10 kg) and the major load (from 60–150 kg) are applied to a ball-shaped indentor (Fig. 6-19). Typical Rockwell numbers for certain plastics are acrylic, M 100; polystyrene, M 75; polyvinyl chloride, M 115; and polyethylene, R 15. Figure 6-20 shows Rockwell tests in progress.

For soft or flexible plastics, the Shore durometer instrument may be used. There are two ranges of durometer hardness. Type A uses a blunt rod-shaped indentor to test soft plastics. Type D uses a pointed rod-shaped indentor to measure harder materials. The reading or value is taken after 1 or 10 seconds of applying pressure by hand. The scale range is 0 to 100.

The Barcol tester is similar to the Shore durometer, D type. It also uses a sharp-pointed indentor. Figure 6-21 shows a drawing of a Barcol tester.

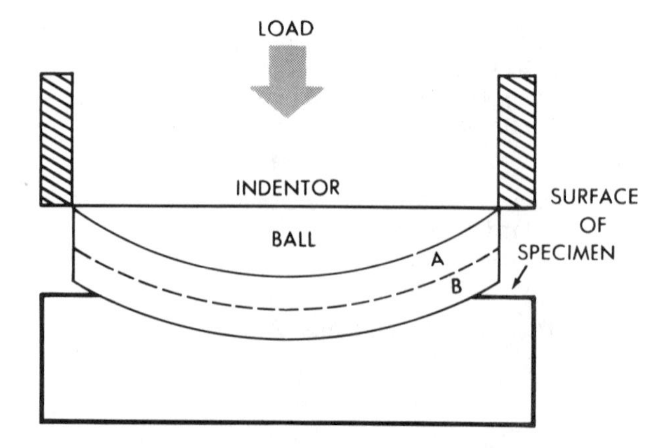

Fig. 6-19. The distance between Line *A* (minor load) and Line *B* (major load) is the basis for Rockwell hardness readings.

(A) Recording results of a Rockwell hardness test on an ABS bar sample. (Wilson Instrument Division of AACO)

(B) Molded bar of ABS, positioned under the indentor of a Rockwell hardness tester. (Wilson Instrument Division of AACO)

Fig. 6-20. Rockwell testing.

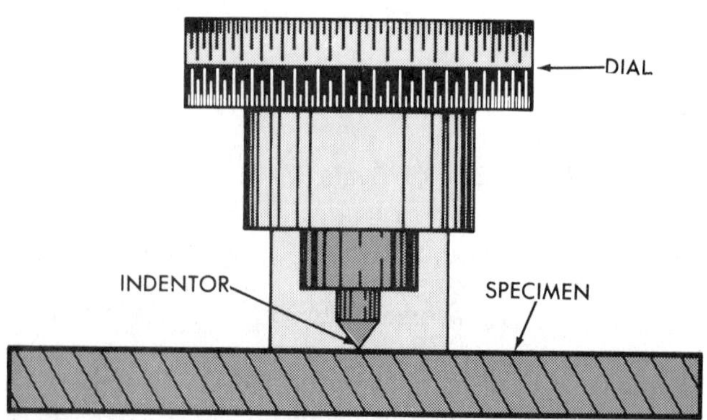

Fig. 6-21. A sharp-pointed indentor is used on the Barcol instrument (ASTM D-2583).

Abrasion Resistance (ASTM D-1044)

Abrasion is a process of wearing away the surface of a material by friction. The Williams, Lambourn and Tabor abraders measure the resistance of plastics materials to abrasion. In each test, an abrader rubs against the sample, removing some material. The amount of material loss (mass or volume) indicates how well the sample resists the abrasive treatment.

$$\text{abrasion resistance} = \frac{\text{original mass} - \text{final mass}}{\text{relative density}}$$

Physical Properties

In contrast to the mechanical properties, which require the basic forces of tension, compression, and shear, the physical properties of plastics may not involve these forces. The molecular structure of the material often effects the physical properties. A few properties will receive attention. They are relative density, mold shrinkage, tensile creep and viscosity.

Density and Relative Density (ISO 1183, ASTM D-792 and D-1505)

Density is mass per unit volume. The proper SI derived unit of density is kilograms per cubic meter, although it is commonly expressed as grams per cubic centimeter.

Example:

Density = mass (kg)/ volume (m^3)

For PVC:

Density = 1 300 kg/1 m^3 or 1.3 g/cm^3

Relative density is the ratio of the mass of a given volume of material to the mass of an equal volume of water at 23 °C. [73 °F]. Relative density is a dimensionless quantity, and will be the same in any measurement system.

Example:

relative density of PVC

$$= \frac{\text{density of PVC}}{\text{density of water}}$$

$$\frac{1300\ \text{kg/m}^3}{1000\ \text{kg/m}^3} = 1.3$$

The relative densities of a number of selected materials is given in Table 6-6. Notice that the polyolefins have densities less than 1.0, which means they float in water.

A simple method for determining relative density is to weigh the sample in air and in water. (ASTM D-792) A fine wire may be used to suspend the plastics sample in the water from a laboratory balance, as shown in Figure 6-22. You may calculate the relative density by the following formula:

$$D = \frac{a - b}{a - b + c - d}$$

D = density at 20 °C [68 °F]

a = mass of specimen and wire in air

b = mass of wire in air

c = mass of wire with end immersed in water

d = mass of wire and specimen immersed in water

Another method, given in ASTM D-1505, is a *density-gradient column*. This column is composed of

Table 6-6. Relative Densities of Selected Materials

Substance	Relative Density
Woods (based on water)	
Ash	0.73
Birch	0.65
Fir	0.57
Hemlock	0.39
Red oak	0.74
Walnut	0.63
Liquids	
Acid, muriatic	1.20
Acid, nitric	1.217
Benzine	0.71
Kerosene	0.80
Turpentine	0.87
Water 20 °C	1.00
Metals	
Aluminum	2.67
Brass	8.5
Copper	8.85
Iron, cast	7.20
Iron, wrought	7.7
Steel	7.85
Plastics	
ABS	1.02–1.25
Acetal	1.40–1.45
Acrylic	1.17–1.20
Allyl	1.30–1.40
Aminos	1.47–1.65
Casein	1.35
Celullosics	1.15–1.40
Chlorinated polyesters	1.4
Epoxies	1.11–1.8
Fluoroplastics	2.12–2.2
Ionomers	0.93–0.96
Phenolic	1.25–1.55
Phenylene oxide	1.06–1.10
Polyamides	1.09–1.14
Polycarbonate	1.2–1.52
Polyester	1.01–1.46
Polyolefins	0.91–0.97
Polysytrene	0.98–1.1
Polysulfone	1.24
Silicones	1.05–1.23
Urethanes	1.15–1.20
Vinyls	1.2–1.55

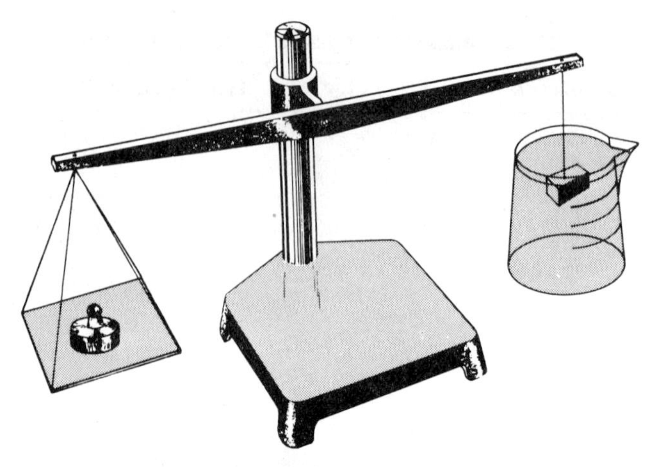

Fig. 6-22. An analytical balance is used as shown to determine the relative density of plastics samples.

liquid layers which decrease in density from bottom to top. The layer to which a sample sinks shows its density. A density-gradient column is rather complex and requires periodic maintenance to clean the column and to verify that the layers are of specified density.

A simpler approach is to create one or more mixtures of known density as shown in Figure 6-23. For densities greater than water, prepare a solution of distilled water and calcium nitrate and measure it with a technical-grade hydrometer. Add calcium nitrate until a desired density is obtained. For densities less than that of water, mix water and isopropyl alcohol to achieve a selected density.

When conducting density tests, remember that dirt, grease, and machining scratches may entrap air on the sample and cause inaccurate results. The presence of fillers, additives, reinforcements and voids or cells will also alter the relative density.

Mold Shrinkage (ISO 2577, ASTM D-955)

The mold (linear) shrinkage influences the size of molded parts. Typical mold cavities are larger than the desired finished parts. When the shrinkage of the parts is complete, they should then meet dimensional specifications.

Mold parts shrink when they crystallize, harden, or polymerize in a mold. The shrinkage also continues for some time after molding. To allow for complete post-mold shrinkage, do not take measurements until 48 hours have passed.

Mold shrinkage is the ratio of the decrease in length to the original length. The result is reported as mm/mm [in./in.]. The formula is:

$$\text{Mold shrink} = \frac{\text{length of cavity} - \text{length of molded bar}}{\text{length of cavity}}$$

Tensile Creep (ISO 899, ASTM D-2990)

When a mass suspended from a test sample causes the sample to change shape over a period of time, the strain is called *creep*. When creep occurs at room temperature, it is called *cold flow*.

Figure 6-24 depicts cold flow. The time duration required to go from *A*, the beginning of the test, to *E*, failure of the specimen, may be well over 1000 hours. Tensile creep test results report the strain in millimeters, as a percent, and as a modulus.

Creep and cold flow are very important properties to consider in the design of pressure vessels, pipes, and beams, where a constant load (pressure or stress) may cause deformation or dimensional changes. PVC pipes undergo specialized creep tests to measure their ability to withstand given pressures over time and to determine the burst or rupture strength. Figure 6-25 shows a section of pipe being tested for burst strength. This sample ruptured while under 5.85 MPa (848 psi) of pressure.

Viscosity

The property of a liquid that describes its internal resistance to flow is called *viscosity*. The more sluggish the

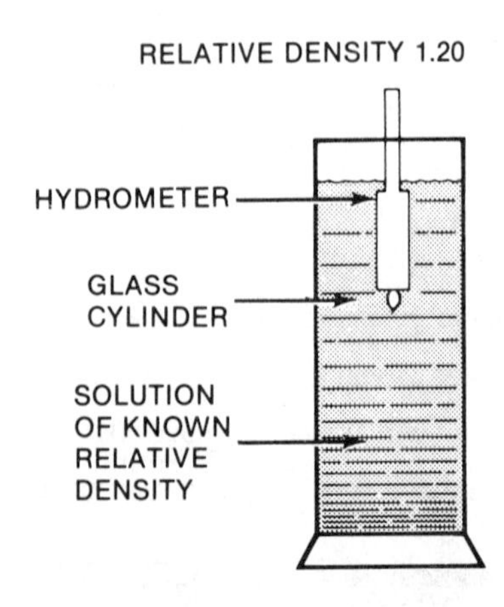

Fig. 6-23. An arrangement for measuring density.

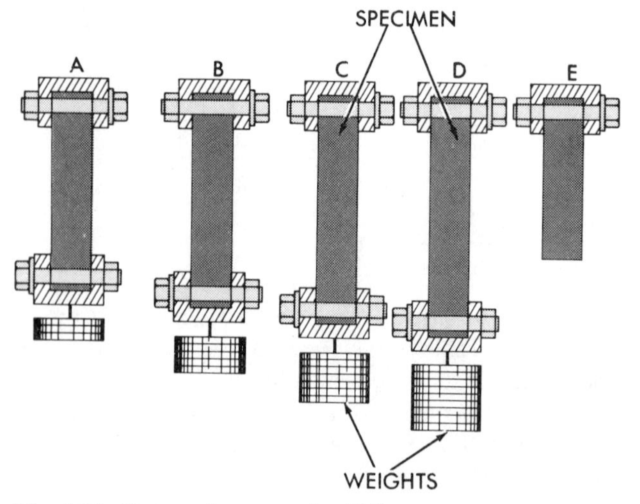

Fig. 6-24. Stages of creep and cold flow.

Fig. 6-25. Testing the bursting strength of pipe. (Schloemann-Fellows)

liquid, the greater its viscosity. Viscosity is measured in pascal-seconds (Pa × s) or units called poises. (See Table 6-7.)

Viscosity is an important factor in transporting resins, injecting plastics in a liquid state, and obtaining critical dimensions of extruded shapes. Fillers, solvents, plasticizers, thixotropic agents (materials that are gel-like until shaken), degree of polymerization, and density all may affect viscosity. The viscosity of a resin such as polyester ranges from 1 to 10 Pa's [1000 to 10000 centipoises]. One centipoise equals 0.01 poise. In the metric system, one centipoise equals 0.001 pascal-second. For a complete definition of poise, see any standard physics text or other reference work in which viscosity is described.

Thermal Properties

The important thermal properties of plastics are: thermal conductivity, specific heat, coefficient of thermal expansion, heat deflection, resistance to cold, burning rate, flammability, melt index, glass transition point, and softening point.

Table 6-7. Viscosity of Selected Materials

Material	Viscosity, Pa · s	Viscosity, centipoises
Water	0.001	1
Kerosene	0.01	10
Motor oil	0.01–1	10–100
Glycerine	1	1000
Corn syrup	10	10000
Molasses	100	100000
Resins	<0.1 to >10^3	<100 to >10^6
Plastics (hot, viscoelastic state)	<10^2 to >10^7	<10^5 to >10^{10}

As thermoplastics are heated, molecules and atoms within the material begin to oscillate more rapidly. This causes the molecular chains to lengthen. More heat may cause slippage between molecules held by the weaker van der Waals' forces. The material may become a viscous liquid. In thermosetting plastics, bonds are not easily freed. They must be broken or decomposed.

Thermal Conductivity (ASTM C-177)

Thermal conductivity is the rate of transmission of heat energy from one molecule to another. For the same molecular reasons that plastics are electrical insulators, they are also thermal insulators.

Thermal conductivity is expressed as a coefficient. Thermal conductivity is called the *k* factor. It should not be confused with the symbol K that indicates Kelvin temperature scale. Aluminum has a *k* factor of 122 W/K · m Some foamed or cellular plastics have *k* values of less than 0.01 W/K · m (Table 6-8). The *k* values for most plastics show that they do not conduct heat as well as an equal amount of metal.

The flow of heat energy should be measured in watts, not calories per hour or Btu per hour. The watt (W) is the same as the joule per second (J/s). It is best to remember that the joule is a unit of energy. The watt is a unit of power.

Specific Heat (Heat Capacity)

Specific heat is the amount of heat required to raise the temperature of a unit of mass by one kelvin, or one degree Celsius. See Figure 6-26. It should be expressed in joules per kilogram per kelvin (J/kg · K). The specific heat at room temperature for ABS is 104 J/kg · K; for polystyrene, 125 J/kg · K; and for polyethylene, 209 J/kg · K. This indicates that it will take more thermal energy to soften the crystalline plastics polyethylene than to soften ABS. The values for most plastics indicate that they require a greater amount of heat energy to raise their temperature than does water since the specific heat of water is 1. The amount of heat may be expressed in joules per gram per degree Celsius (J/g · °C).

Thermal Expansion

Plastics expand at a much greater rate than metals. This makes it difficult to join metals and plastics. Figure 6-27 shows the differences in coefficients of expansion of selected materials. The coefficient of expansion is used to determine thermal expansion in length, area, or volume per unit of temperature rise. It is expressed as a ratio per degree Celsius.

Table 6-8. Thermal Conductivity of Selected Materials

Material	Thermal Conductivity (*k*-factor), W/K · m	Thermal Resistivity (R-factor), K · m/W	Thermal Conductivity, Btu · in/hr · ft² °F
Acrylic	0.18	5.55	1.3
Aluminum (alloyed)	122	0.008	840
Copper (beryllium)	115	0.008	800
Iron	47	0.021	325
Polyamide	0.25	4.00	1.7
Polycarbonate	0.20	5.00	1.4
Steel	44	0.022	310
Window glass	0.86	1.17	6.0
Wood	0.17	5.88	1.2

*R-factor is the reciprocal of the *k* factor.

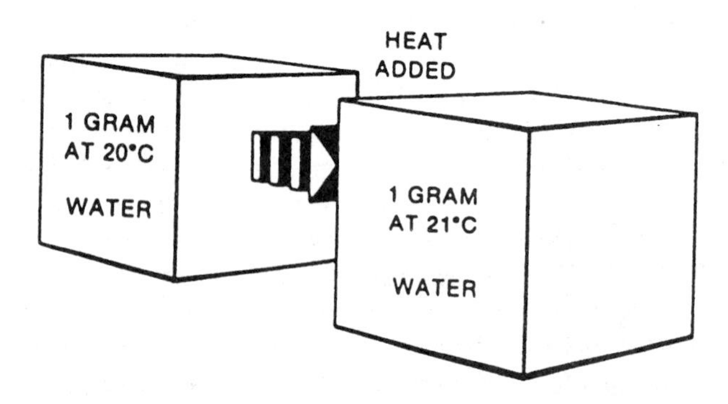

Fig. 6-26. How much heat was added?

Table 6-9. Thermal Expansion of Selected Materials

Substance	Coefficient of Linear Expansion $\times 10^{-6}$, mm/mm * °C
Nonplastics	
Aluminum	23.5
Brass	18.8
Brick	5.5
Concrete	14.0
Copper	16.7
Glass	9.3
Granite	8.2
Iron, cast	10.5
Marble	7.2
Steel	10.8
Wood, pine	5.5
Plastics	
Diallyl phthalate	50–80
Epoxy	40–100
Melamine-formaldehyde	20–57
Phenol-formaldehyde	30–45
Polyamide	90–108
Polyethylene	110–250
Polystyrene	60–80
Polyvinylidene chloride	190–200
Polytetraluoroethylene	50–100
Poly methyl methacrylate	54–110
Silicones	8–50

If a PVC rod 2 m in length is heated from –20 °C to 50 °C, it will change 7 mm in length.

Example:

change in length = coefficient of linear expansion × original length × change in temperature

$$= \frac{0.000\,050}{°C} \times 2\ m \times 70°$$

= 0.007 m or 7 mm

Since area is a product of two lengths, value of the coefficient must double. Similarly, we must triple the coefficient value to obtain thermal expansion for volume. Table 6-9 shows the thermal expansions of selected materials.

Deflection Temperature (ISO 75, ASTM D-648)

Deflection temperature (formerly called heat distortion) is the highest continuous operating temperature that the material will withstand. In general, plastics are not often used in high heat environments. However, some special phenolics have been subjected to temperatures as high as 2760 °C [4032 °F].

A device which provides heat, pressure, linear measurement, and a print-out of results is shown in Figure 6-28. In the ASTM test, a specimen (3.175 mm × 140 mm) is placed on supports 100 mm apart, then a force of 455–1820 kPa is pressed on the sample. The temperature is raised 2 °C per minute. The temperature at which the sample will deflect 0.25 mm is reported as deflection temperature.

In addition to the standard test, a number of nonstandard tests provide information about the heat deflection of various plastics. The materials may be tested in an oven. The temperature is raised until the material chars, blisters, distorts, or loses appreciable strength. Sometimes boiling water provides the heat and temperature level. Figure 6-29 shows a deflection experiment using an infrared radiant heater.

Test bars of glass-reinforced polycarbonate, polysulfone, and thermoplastic polyester were locked in a laboratory vice. Equal 175 g loads were applied. After only 1 minute under 155 °C [311 °F] heat from an infrared radiant heater, the polycarbonate bar began to deflect. One minute later, the polysulfone bar followed suit. The thermoplastic polyester bar was still not bent after 6 minutes at 185 °C [365 °F].

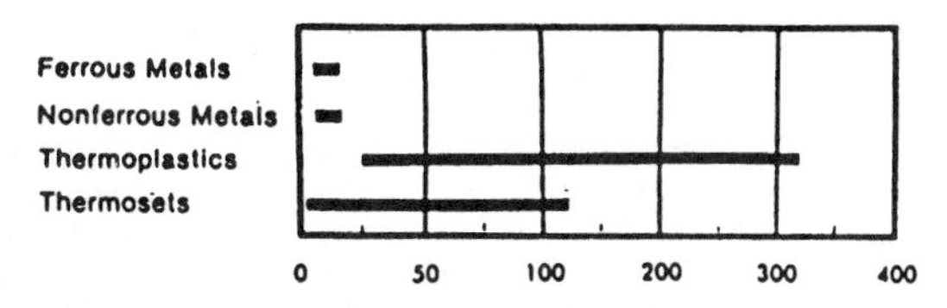

Fig. 6-27. Coefficient of expansion (per °C × 10^{-6})

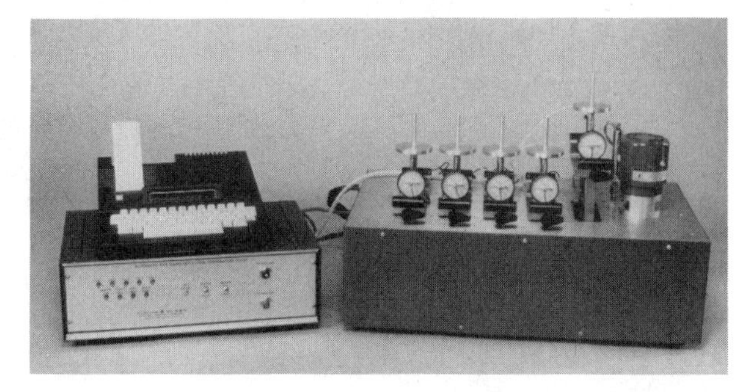

Fig. 6-28. Deflection Temperature/Vicat: The Tinius Olsen Automated Deflection Temperature/Vicat Tester fitted with a DS-5 offers testing of up to five specimens independently... or simulatneously.

Ablative Plastics

Ablative plastics have been used for spacecraft and missiles. On re-entry, the temperature of the outer surface of a heat shield may be greater than 13 000 °C [23 800 °F], while the inner surface is no more than 95 °C [203 °F]. Ablative plastics may be composed of a phenolic or epoxy resin and graphite, asbestos, or silica matrixes.

In ablative materials, heat is absorbed through a process known as pyrolysis. This takes place in the near-surface layer exposed to heat energy. Much of the plastics is consumed and sloughed off as large amounts of heat energy are absorbed.

(A) Before heat was applied. (Celanese Plastic Materials Co.)

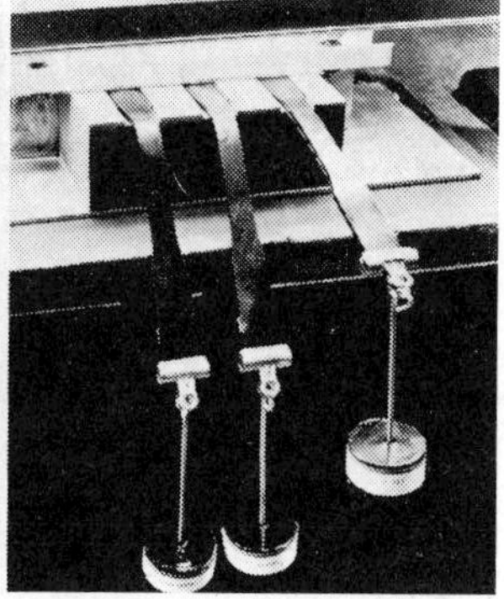

(B) After two minutes of heating. (Celanese Plastic Materials Co.)

Fig. 6-29. Heat deflection testing.

Resistance to Cold

As a rule, plastics have good resistance to cold. Food packages made of polyethylene routinely withstand temperatures of –60 °F [–51 °C]. Some plastics can withstand the extreme low temperature of –19 °F [–196 °C] with little loss of physical properties.

Flammability (ISO 181, 871, 1210, ASTM D-635, D-568, and E-84)

Flammability, also called *flame resistance*, is a term indicating a measure of the ability of a material to support combustion. Several tests measure this property. In one test, a plastics strip is ignited and the heat source (flame) removed. The time and amount of material consumed are measured, and the result is expressed in mm/min. Highly combustible plastics, such as cellulose nitrate, will have high values.

A rather loosely used term related to flammability is *self-extinguishing*. This indicates that the material will not continue to burn once a flame is removed. Figure 6-30A shows a flame-resistant material and Figure 6-30B a self-extinguishing plastics. Nearly all plastics may be made self-extinguishing with the proper additives.

Table 6-10 indicates that plastics will burn when exposed to direct flame. To cause self-ignition, the temperature must be higher than the temperature of ignition from a direct flame.

Melt Index (ISO 1133, ASTM D-1238)

Viscosity and flow properties effect both the processing of plastics and the design of molds. Melt viscosity provides the most accurate data, but melt-index values are common, often because the test for melt index requires little time.

Melt index is a measure of the amount of material in grams that is extruded through a small orifice in 10 minutes at a given pressure and temperature. A common load is 43.5 psi [300 kPa]. The ASTM procedure specifies temperatures of 190 °C [374 °F] for polyethylene and 230 °C [446 °F] for polypropylene. The ISO method specifies die diameter, temperature, die factor, reference time and nominal load. Figure 6-31 shows a melt-index measuring device.

A high melt-index value indicates a low-viscosity material. Usually, low-viscosity plastics have relatively low molecular mass. In contrast, high molecular mass materials are resistant to flow and have lower melt-index values.

Glass Transition Temperature

At room temperature, the molecules in amorphous plastics are in motion, but this motion is rather limited.

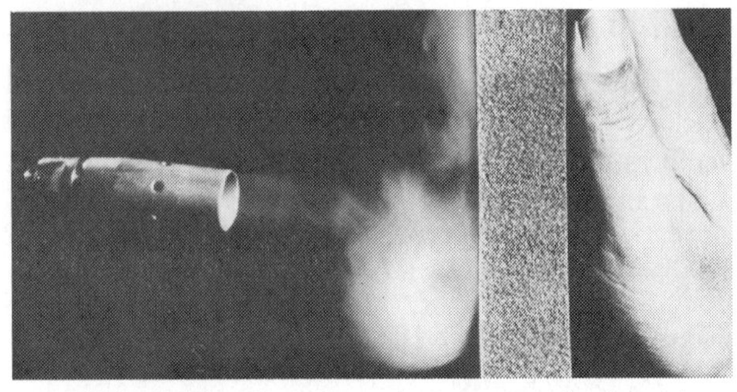

(A) This cellular polyurethane demonstrates the thermal insulating ability and flame resistance of special plastics formulations.

(B) When the flame is removed from a self-extinguishing plastics, burning ceases. (Henkel Corp.)

Fig. 6-30. Testing for flammability of plastics.

As an amorphous material heats up, the relative motion of the molecules increases. When the material reaches a certain temperature, it loses its ridigity, and becomes leathery. That temperature is identified as the *glass transition temperature,* T_g. Often glass transition temperatures are reported as a range of temperature, because the transition does not occur at one specific temperature. The glass transition points for several amorphous plastics are shown in Table 6-11.

Crystalline plastics contain both crystalline regions and amorphous regions. Consequently, they exhibit two changes upon heating. When the temperature is high enough, the amorphous regions alter from glass-like to flexible. As the temperature continues to rise, the energy disrupts the crystalline regions, causing the material to become a viscous liquid throughout. This transition occurs over a limited temperature range. It is identified as T_m, the melting temperature. Table 6-12 shows both the T_g and T_m for selected crystalline plastics.

Figure 6-32 shows the difference between amorphous and crystalline materials graphically. Note the two points of inflection on the curve for crystalline materials.

Softening Point (ISO 306, ASTM D-1525)

In the Vicat softening point test, a sample is heated at a rate of 50 °C [120 °F] per hour. The temperature at which a needle penetrates the sample—0.039 in. [1 mm]—is the Vicat softening point.

Table 6-10. Ignition Temperatures and Flammability of Various Materials

Material	Flash-Ignition Temperature, °C	Self-Ignition Temperature, °C	Burning Ratio, mm/min
Cotton	230–266	254	SB
Paper, newsprint	230	230	SB
Douglas fir	260		SB
Wool	200		SB
Polyethylene	341	349	7.62–30.48
Polypropylen, fiber		570	17.78–40.64
Polytetrafluoro-ethylene		530	NB
Polyvinyl chloride	391	454	SE
Polyvinylidene chloride	532	532	SE
Polystyrene	345–360	488–496	12.70–63.5
Polymethyl methacrylate	280–300	450–462	15.24–40.64
Acrylic, fiber		560	SB
Cellulose nitrate	141	141	Rapid
Cellulose acetate	305	475	12.70–50.80
Cellulose triacetate, fiber	540	SE	
Ethyl cellulose	291	296	27.94
Polyamide (Nylon)	421	424	SE
Nylon 6,6, fiber		532	SE
Phenolic, glass fiber laminate	520–540	571–580	SE-NB
Melamine, glass fiber laminate	475–500	623–645	SE
Polyester, glass fiber laminate	346–399	483–488	SE
Polyurethane, polyether, rigid foam	310	416	SE
Silicone, glass fiber laminate	490–527	550–564	SE

NB—Nonburning
SE—Self-extinguishing
SB—Slow burning

Table 6-11. Glass Transition of Selected Amorphous Plastics

Plastic	T_g °C
ABS	110
PC	150
PMMA	105
PS	95
PVC	85

Table 6-12. Glass Transition of Selected Crystalline Plastics

Plastics	T_g °C	T_m °C
PA	50	265
PE	-35	130
PET	65	265
PP	-10	165

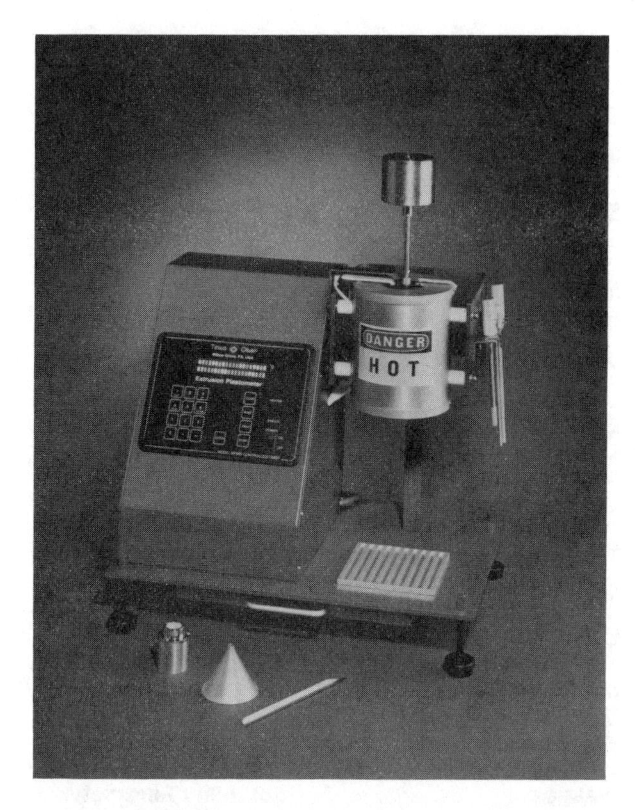

Fig. 6-31. Extrusion Plastometer: The Basic, Tinius Olsen Extrusion Plastometer (Melt Indexer) incorporates microprocessor-based MP 993 Controller/Timer (Procedure A - Manual Operations capability) for determining the Flow Rate (Melt Index) of thermoplastics.

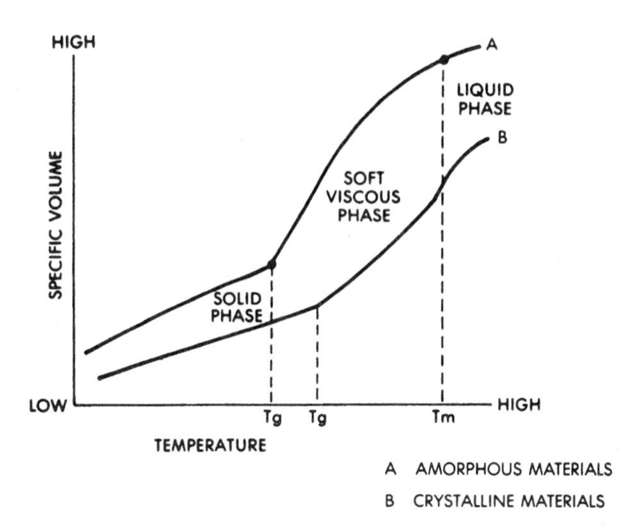

Fig. 6-32. Specific volume versus temperature for an amorphous and a crystalline plastics.

Environmental Properties

Plastics are found in nearly every environment. They are used as containers for chemicals, packages for food storage, and as medical implants inside the human body. Before a product is designed, plastics must be tested for endurance under the expected environmental extremes. The environmental properties of plastics include: chemical resistance, weathering, ultraviolet resistance, permeability, water absorption, biochemical resistance, and stress cracking.

Chemical Properties

The statement that "most plastics resist weak acids, alkalies, moisture, and household chemicals" must be used only as a broad rule. Any statement about the response of plastics to chemical environments must be only a generalization. It is best to test each plastics to determine how it can be specifically applied and what chemicals each is expected to resist.

The chemical resistance of plastics depends, to a large degree, on the elements combined into molecules and on the types and strengths of the chemical bonds. Some combinations are very stable; others are quite unstable. Polyolefins are exceptionally inert, non-reactive, and resistant to chemical attack. This is due to the C—C bonds in the backbone of the molecules, which are very stable. In contrast, polyvinyl alcohol contains hydroxyl groups (—OH) attached to the carbon backbone of the molecule. The bonds holding the hydroxyl groups onto the main chain break down in the presence of water.

Table 6-13 lists the chemical resistance of a number of plastics. This table provides information only about the natural materials. However, fillers, plasticizers, stabilizers, colorants, and catalysts can affect the chemical resistance of plastics.

The resistance of plastics to organic solvents can provide information for the identification of unknown materials. (See Appendix D on Material Identification) The reactivity of both plastics and organic solvents has been assigned a *solubility parameter*. In principle, a polymer will dissolve in a solvent with a similar or lower solubility parameter. This general principle may not apply in all cases, due to crystallization, hydrogen bonding, and other molecular interactions. Table 6-14 contains the solubility parameters of selected solvents and plastics.

Weathering (ISO 45, 85, 4582, 4607, ASTM D-1435 and G-23)

Many weathering tests are conducted in Florida, where samples receive considerable exposure to heat, moisture, and sunlight. Exposed samples are rated on color change, gloss change, cracks and crazing, and loss of physical properties. Since the weathering tests require long time periods, accelerated tests try to provide similar exposure in shorter time. Figure 6-33 shows an

Table 6-13. Chemical Resistance of Selected Plastics at Room Temperature

Plastics	Strong Acids	Strong Alkalies	Organic Solvents
Acetal	Attacked	Resistant	Resistant
Acrylic	Attacked	Slight	Attacked
Cellulose acetate	Affected	Affected	Attacked
Epoxy	Slight	Slight	Slight
Ionomer	Slight	Resistant	Resistant
Melamine	Slight	Slight	Resistant
Phenolic	Resistant	Attacked	Affected
Phenoxy	Resistant	Resistant	Attacked
Pollallomer	Resistant	Resistant	Resistant
Polyamide	Attacked	Slight	Resistant
Polycarbonate	Resistant	Attacked	Attacked
Polychlorotri-fluoroethylene	Resistant	Resistant	Resistant
Polyester	Slight	Affected	Affected
Polyethylene	Resistant	Resistant	Affected
Polyimide	Affected	Attacked	Resistant
Polyphenylene oxide	Resistant	Resistant	Slight
Polypropylene	Resistant	Resistant	Resistant
Polysultone	Resistant	Resistant	Affected
Polystyrene	Affected	Resistant	Affected
Polytetrafluoro-ethylene	Resistant	Resistant	Resistant
Polyurethane	Resistant	Affected	Slight
Polyvinyl chloride	Resistant	Resistant	Affected
Silicone	Slight	Affected	Slight

accelerated weathering tester. These machines cycle samples through moisture and temperature changes, and simulate sunlight with a variety of lamps that produce ultraviolet light.

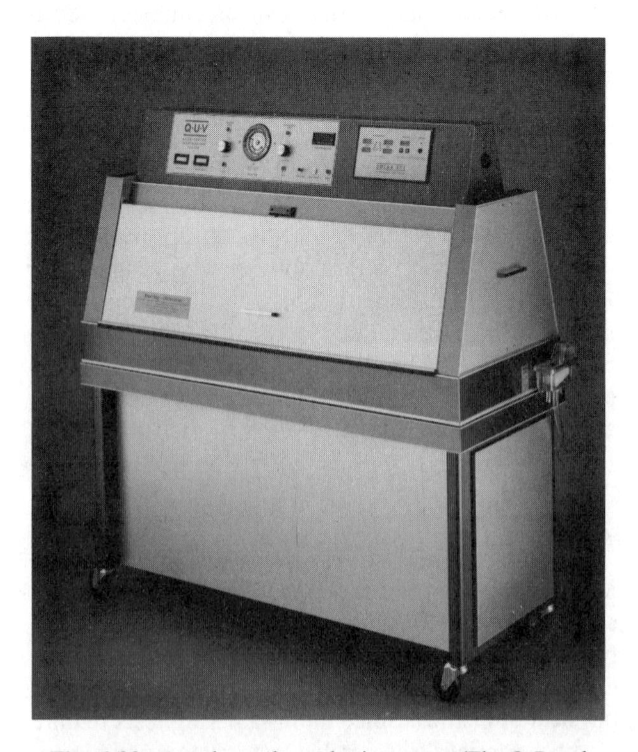

Fig. 6-33. Accelerated weathering tester. (The Q-Panel Company.)

Table 6-14. Solubility Parameters of Selected Solvents and Plastics

Solvent	Solubility Parameter
Water	23.4
Methyl alcohol	14.5
Ethyl alcohol	12.7
Isopropyl alcohol	11.5
Phenol	14.5
n-Butyl alcohol	11.4
Ethyl acetate	9.1
Chloroform	9.3
Trichloroethylene	9.3
Methylene chloride	9.7
Ethylene dichloride	9.8
Cyclohexanone	9.9
Acetone	10.0
Isopropyl acetate	8.4
Carbon tetrachloride	8.6
Toluene	9.0
Xylene	8.9
Methyl isopropyl ketone	8.4
Cyclohexane	8.2
Turpentine	8.1
Methyl amyl acetate	8.0
Methyl cyclohexane	7.8
Heptane	7.5
Plastics	**Solubility Parameter**
Polytetrafluoroethylene	6.2
Polyethylene	7.9–8.1
Polypropylene	7.9
Polystyrene	8.5–9.7
Polyvinyl acetate	9.4
Polymethyl methacrylate	9.0–9.5
Polyvinyl chloride	9.38–9.5
Bisphenol A polycarbonate	9.5
Polyvinylidene chloride	9.8
Polyethylene terephthalate	10.7
Cellulose nitrate	10.56–10.48
Cellulose acetate	11.35
Epoxide	11.0
Polyacetal	11.1
Polyamide 6, 6	13.6
Coumarone indene	8.0–10.6
Alkyd	7.0–11.2

Ultraviolet Resistance (ASTM G-23 and D-2565)

Linked with weatherability is the resistance of plastics to the effects of direct sunlight or artificial weathering devices. Ultraviolet radiation (combined with water and other environmental oxidizing conditions) may cause color fading, pitting, crumbling, surface cracking, crazing, and brittleness. The Altas 181 Fade-O-meter is used to check color stability. For artificial weathering, the Water-Cooled Zenon-Arc Light and Water Exposure Apparatus is commonly utilized.

Permeability (ISO 2556, ASTM D-1434 and E-96)

Permeability may be described as the volume or mass of gas or vapor penetrating an area of film in 24 hours. Permeability is an important concept in the food packaging industry. In some applications, a packaging film must allow the passage of oxygen, which keeps meats and vegetables looking fresh. Other applications need to selectively prevent gases, moisture, and other agents from contaminating the package contents. Frequently, packages contain several layers of different materials to achieve the desired control of permeability.

Water Absorption (ISO 62, 585, 960, ASTM D-570)

Some plastics are *hygroscopic*. That means that they absorb moisture, usually taking on water from humid air. Table 6-15 contains water absorption data for selected hygroscopic plastics. These materials require drying before entering any processes that involve heat or melting. If not properly dried, the moisture in these plastics will turn to steam, which may cause surface defects and voids in the material. To verify that drying equipment is functioning properly, many companies periodically test samples for moisture content.

A simple test involves accurately weighing a sample, heating it in an oven for a period of time, and reweighing to determine the weight loss. Some instruments provide rapid, accurate results based on this thermogravimetric principle. One such instrument is pictured in Figure 6-34.

The thermogravimetric method makes the assumption that all weight loss represents moisture. That assumption is not always accurate because some materials also loose lubricants, oils, and other volatiles when heated. To provide extremely accurate measures of moisture content, a moisture-specific apparatus is needed. Figure 6-35 shows a moisture meter that heats a sample, and draws the evolved gases into an analysis cell that traps only water vapor. The result is a test that accurately measures moisture in a specimen.Two simple, low-cost methods of checking moisture content are the Tomasetti's Volatile Indicator (TVI) and the Test Tube/Hot Block (TTHB) techniques. Follow the procedure in Figure 6-36 for the TVI technique.

1. Place two glass slides on a hot plate and heat from 1 to 2 minutes at 275 ±15 °C [526 ± 59 °F] (Fig. 34).
2. Place four pellets or granular plastics samples on one of the glass slides.
3. Place the second hot slide on top of the sample and press the sandwiched pellets to about 10 mm [0.393 in.] diameter.

Table 6-15. Water Absorption

Material	Water Absorbed, percent (24-Hour-Immersion)
Polychlorotrifluoroethylene	0.00
Polyethylene	0.01
Polystyrene	0.04
Epoxy	0.10
Polycarbonate	0.30
Polyamide	1.50
Cellulose acetate	3.80

4. Remove the slides from the hot-plate platen and allow then to cool.
5. The number and size of bubbles seen in the plastics samples indicate percentage of moisture absorbed. Some bubbles may be the result of trapped air, but numerous bubbles indicate moisture-laden material. There will be a direct correlation between the number of bubbles and moisture content.

The TTHB procedure is (Fig. 6-37):

1. Heat a hot block with test tube holes to 26 ± 10 °C (500 ± 50 °F).
2. Place 5.0 g of plastics in a 20 × 150 mm Pyrex test tube.
3. Place a stopper in the test tube, then carefully place the test tube in the hot block.
4. Allow the material to melt (about 7 minutes).
5. Remove the tube and sample from the hot block and allow it to cool for ten minutes.
6. Observe and record the correlation of moisture content with the surface area of the condensation in the test tube.

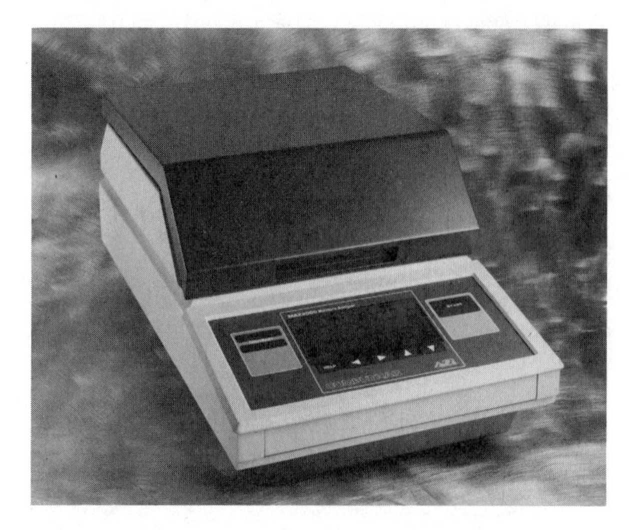

Fig. 6-34. A thermogravimetric moisture tester. (Arizona Instrument Corp.)

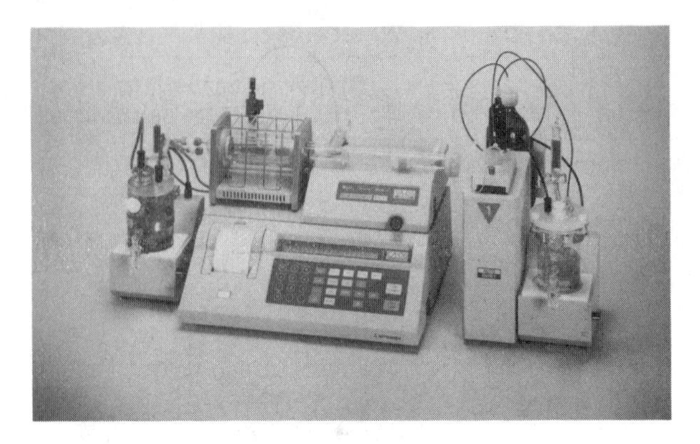

Fig. 6-35. A moisture specific moisture tester. (Mitsubishi Kasei Corp.)

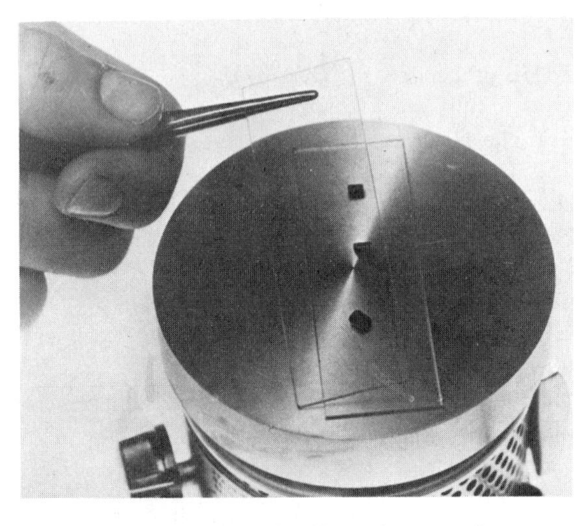

(A) Plug in hot plate and calibrate it to a surface temperature of 518° F(270°C ±10°). Be sure surface is clean; place two glass slides on surface for 1–2 min.

(B) When the glass surface temperature reaches 446–500 °F (230–250 °C) place four or five pellets on one glass slide using tweezers.

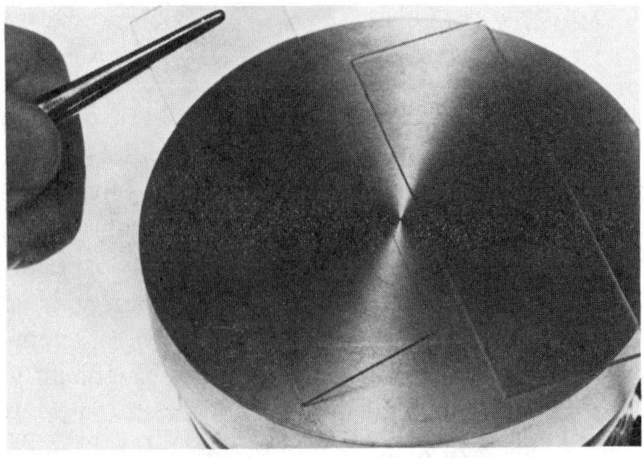

(C) Place the second hot slide over the pellets to form a sandwich.

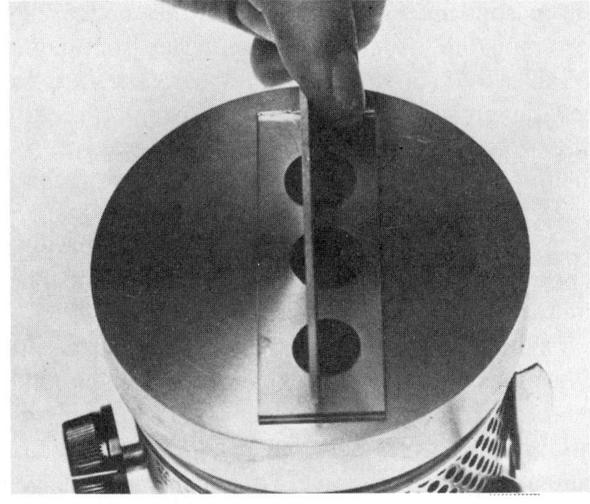

(D) Press on the top slide with a tongue depressor until the pellets flatten to about 10 mm diameter.

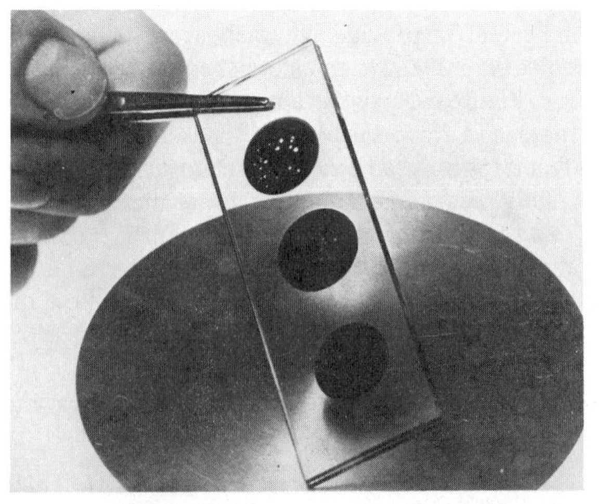

(E) Remove sandwich and allow to cool. Amount and size of bubbles indicate percentage of moisture.

Fig. 6-36. The six simple steps of the Resin Moisture Test.

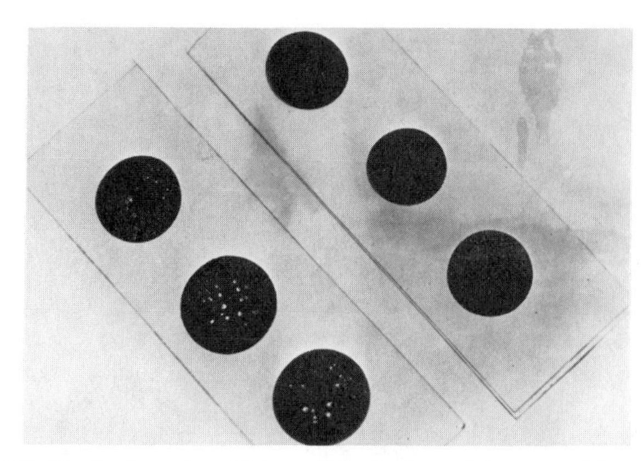

(F) Typical results. Slide at right indicated dry material; slide at left moisture-laden material. One or two bubbles may be only trapped air.

Fig. 6-36. The six simple steps of the Resin Moisture Test (continued).

Biochemical Resistance (ASTM G-21 and G-22)

Most plastics are resistant to bacteria and fungi, however, some plastics and additives are not. They may not be approved by the U.S. Food and Drug Administration (FDA) for use in packaging and as containers for foods or drugs. Various preservatives or antimicrobial agents may be added to plastics to make them resistant.

Stress Cracking (ISO 4600, 6252, ASTM D-1693)

Environmental stress cracking of plastics may be caused by solvents, radiation, or constant strain. There are several tests that expose the sample to a surface agent. One such test is shown in Figure 6-38.

In this test, a glass-reinforced polysulfone test bar at the rear cracked apart violently under a spray of acetone. This reaction broke an electrical connection, triggering the camera to take this picture.

The acetone did not affect the similarly stressed test bar of thermoplastic polyester in the foreground. Thermoplastic polyester withstands even higher stresses in the presence of carbon tetrachloride, methyl ethyl ketone, and other aromatic chemicals.

Optical Properties

Optical properties are closely linked with molecular structure. Because of this, the electrical, thermal, and optical properties of the plastics are interrelated. Plastics exhibit many optical properties. Among the most important are gloss, transparency, clarity, haze, color, and refractive index.

Specular Gloss (ASTM D2457)

Specular gloss is the relative luminous reflectance factor of a plastics sample. A glossmeter directs light onto the sample at incidence angles of 20°, 45°, and 60°. The light that reflects off the surface is collected and measured by a photosensitive device. A perfect mirror is used as a standard and yields values of 1 000 for the 20° and 60° incidence angles. Test results on plastics samples provide comparative data that can rate samples and estimate surface flatness. Comparisons should only be made between similar types of samples. For example, opaque films should not be compared to transparent films.

Luminous Transmittance (ASTM D-1003)

A cloudy or milky appearance in plastics is known as *haze*. A plastics termed *transparent* is one that absorbs very little light in the visible spectrum. *Clarity* is a measure of distortion seen when viewing an object through transparent plastics. All these terms relate to the test for luminous transmittance.

(A) Plastics samples being heated to drive off moisture.

Fig. 6-37. The Test Tube/Hot Block Method of Moisture Measurement. (General Electric Co.)

(Continued)

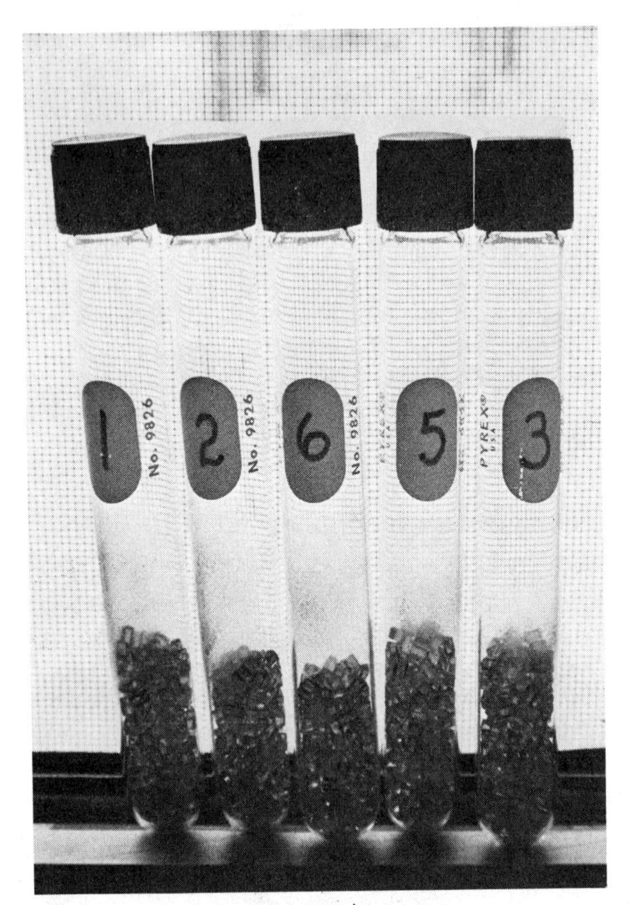

(B) Moisture condensed within test tubes.

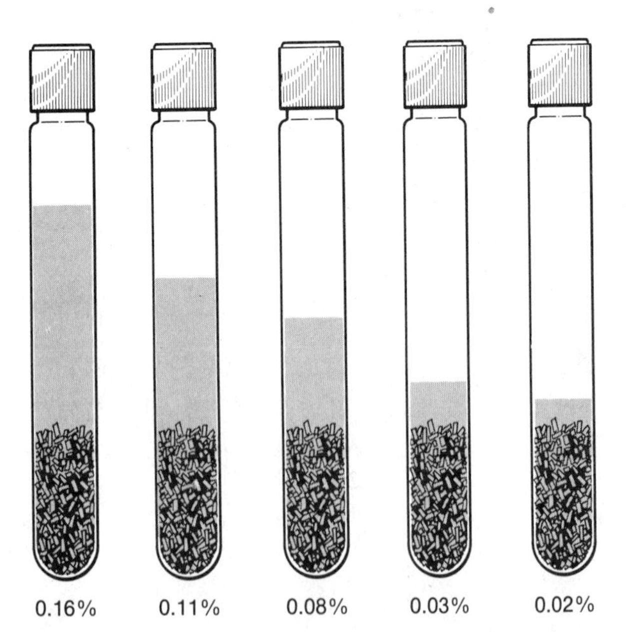

(C) Area of condensation on test tube surface shows the percentage of moisture in each plastics sample.

Fig. 6-37. The Test Tube/Hot Block Method of Moisture Measurement. (General Electric Co.) *(Continued).*

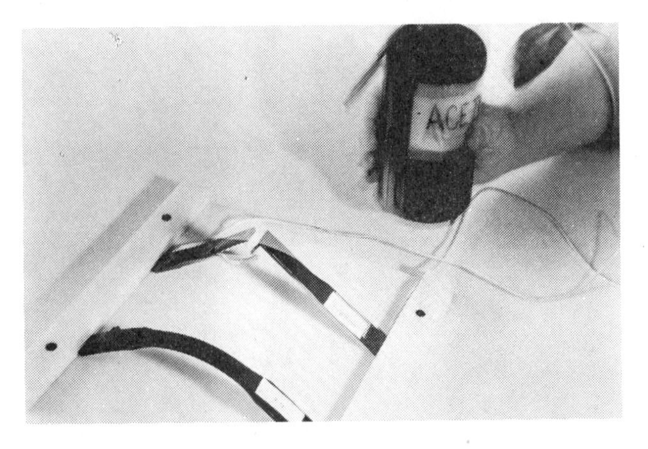

Fig. 6-38. An acetone spray causes a glass-reinforced polysulfone test bar to break. (Celanese Plastic Materials Co.)

Luminous transmittance is the ratio of transmitted light to incident light. In this test, a beam of light passes through air and into a receptor. This measures the incident light. After positioning a sample, the light then shines through it and into the receptor. The ratio of the reading through a sample to the reading through air provides a measure of total transmittance.

Unfilled amorphous plastics are the most transparent of the plastics. Fillers, colorants, and other additives, even in small amounts, interfere with the passage of light.

Color

The selective absorption of light results in *color*. A problem related to colored plastics parts is color match. When colored parts made by one manufacturer need to match the color of parts coming from another source, color measurement is crucial.

Currently, color measurement systems use three components, delta L^* (lightness), delta C^* (chroma) and delta H^*. When manufacturing companies agree on the color measurements need for parts, and when they have identical or similar color-measuring equipment, then color matches are frequently acceptable.

The color-measurement equipment utilizes computers to store and compare data, and photo cells to acquire color readings from parts or color samples. Figure 6-39 shows a portable color measurement device.

Index of Refraction (ISO 489, ASTM D-542)

When light enters a transparent material, part of that light is reflected and part is refracted (Fig. 6-40). The index of refraction n may be expressed in terms of the angle of incidence i and the angle of refraction r

$$n = \frac{\sin i}{\sin r}$$

where i and r are taken relative to the perpendicular to the surface at the point of contact. The index of refraction for most transparent plastics is about 1.5. This is not greatly different from most window glass. Table 6-16 gives indexes of refraction for selected plastics.

Table 6-16. Optical Properties of Plastics

Material	Refractive Index	Light Transmission, %
Methyl methacrylate	<1.49	94
Cellulose acetate	1.49	87
Polyvinyl chloride acetate	1.52	83
Polycarbonate	1.59	90
Polystyrene	1.60	90

Electrical Properties

The five basic properties that describe the electrical behavior of plastics are: resistance, insulation resistance, dielectric strength, dielectric constant, and dissipation (power) factor. The predominantly covalent bonds of polymers limit their electrical conductivity and cause most plastics to be electrical insulators. With the addition of fillers such as graphite or metals, plastics can be made conductive or semiconductive.

Arc Resistance (ISO 1325, ASTM D-495)

Arc resistance is a measure of the time needed for a given electrical current to render the surface of a plastics conductive because of carbonization. The measurement is reported in seconds. The higher the value, the more resistive the plastics is to arcing. The breakdown in arc resistance may be a result of corrosive chemicals. Ozone, nitric oxides, or a build-up of moisture or dust may also lower values.

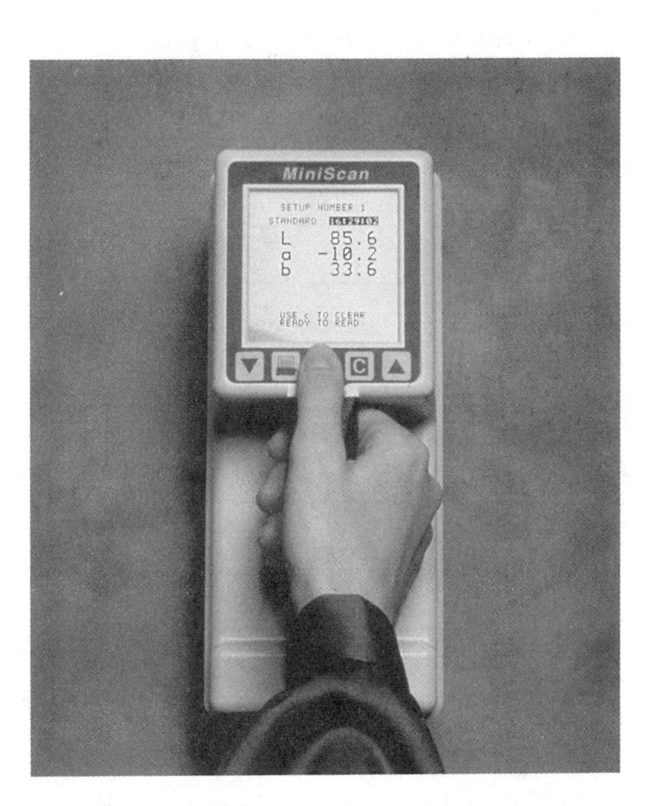

Fig. 6-39. A portable color meter. (Hunter Associates Laboratory, Inc.)

Resistivity (ISO 3915, ASTM D-257)

Insulation resistance is the resistance between two conductors of a circuit or between a conductor and ground when they are separated by an insulator. The insulation resistance is equal to the product of the resistivity of the plastics and the quotient of its length divided by its area:

$$\text{insulation resistance} = \frac{\text{resistivity} \times (\text{length})}{\text{area}}$$

Resistivity is expressed in ohm-centimeters. Table 6-17 gives resistivities for certain plastics.

Dielectric Strength (ISO 1325, 3915, ASTM D-149)

Dielectric strength is a measurement of electrical voltage required to break down or arc through a plastics material. The units are reported as volts per millimeter of thickness (V/mm). This electrical property gives an indication of the ability of a plastics to act as an electrical insulator. See Figure 6-41 and Table 6-17.

Dielectric Constant (ISO 1325, ASTM D-150)

The *dielectric constant* of a plastics is a measure of the ability of the plastics to store electrical energy, as shown in Figure 6-42. Plastics are used as dielectrics in the production of capacitors, which are used in radios

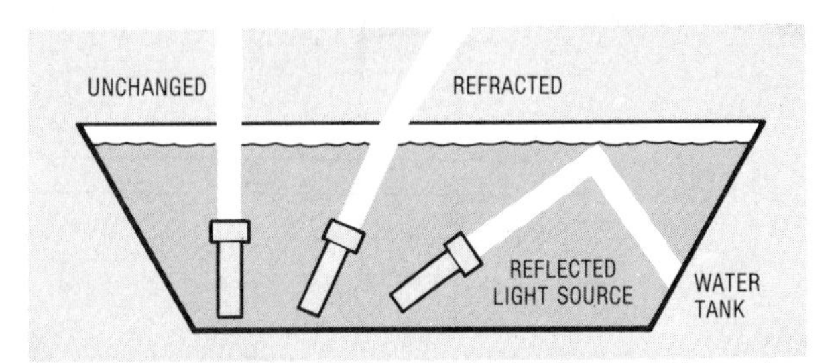

Fig. 6-40. Reflection and refraction of light.

Table 6-17. Electrical Properties of Selected Plastics

Plastics	Resistivity, ohm-cm	Dielectric Strength, V/mm	Dielectric Constant		Dissipation (Power) Factor	
			At 60 Hz	At 10^6 Hz	At 60 Hz	At 10^6 Hz
Acrylic	10^{16}	15 500–19 500	3.0–4.0	2.2–3.2	0.04–0.06	0.02–0.03
Cellulosic	10^{15}	8 000–23 500	3.0–7.5	2.8–7.0	0.005–0.12	0.01–0.10
Fluoroplastics	10^{18}	10 000–23 500	2.1–8.4	2.1–6.43	0.000 2–0.04	0.000 3–0.17
Polyamides	10^{15}	12 000–33 000	3.7–5.5	3.2–4.7	0.020–0.014	0.02–0.04
Polycarbonate	10^{16}	13 500–19 500	2.97–3.17	2.96	0.000 6–0.000 9	0.009–0.010
Polyethylene	10^{16}	17 500–39 000	2.25–4.88	2.25–2.35	<0.000 5	<0.000 5
Polystyrene	10^{16}	12 000–23 500	2.45–2.75	2.4–3.8	0 000 1–0.003	0.000 1–0.003
Silicones	10^{15}	8 000–21 500	2.75–3.05	2.6–2.7	0.007–0 001	0.001–0 002

and other electronic equipment. The dielectric constant is based upon air, which has a value of 1.0. Plastics with a dielectric constant of 5 will have five times the electricity-storing capacity of air or of a vacuum.

Nearly all electrical properties of plastics will vary with time, temperature, or frequency. For example, the values may vary as frequency is increased. (See Table 6-17, for dielectric constant and dissipation factor.)

Dissipation Factor (ASTM D-150)

Dissipation (power) factor or *loss tangent* also varies with frequency. It is a measure of the power (watts) lost in the plastics insulator. A test like those used in determining the value of the dielectric constant is used to measure this power loss. As a rule, measurements are made at one million hertz. They indicate the percent of alternating current lost as heat within the dielectric material. Plastics with low dissipation factors waste little energy and do not become overheated. For some plastics, this is a disadvantage since they cannot be preheated or heat sealed by high-frequency methods of heating. (See Table 6-17 for various dissipation factors.)

The relationship between heat, current, and resistance is shown in the power equation:

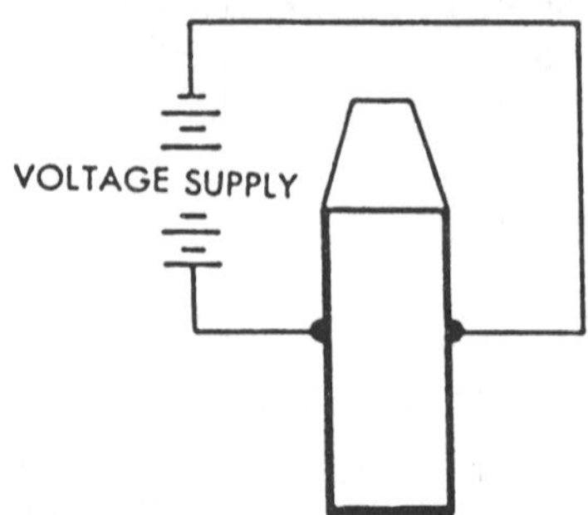

Fig. 6-41. Testing dielectric strength, an important characteristic of plastics for insulating applications.

$$P = I^2R$$

The power P used to perform wasted work is lost or dissipated power. In this formula, the amount of power can be decreased by lowering either the current I or the resistance R. In electrical appliances designed to produce heat, a low dissipation factor is not considered desirable.

Vocabulary

The following vocabulary words are found in this chapter. Look up the definition of unfamiliar words in the glossary, Appendix A.

Brittleness temperature
Centipoise
Cold flow
Compressive strength

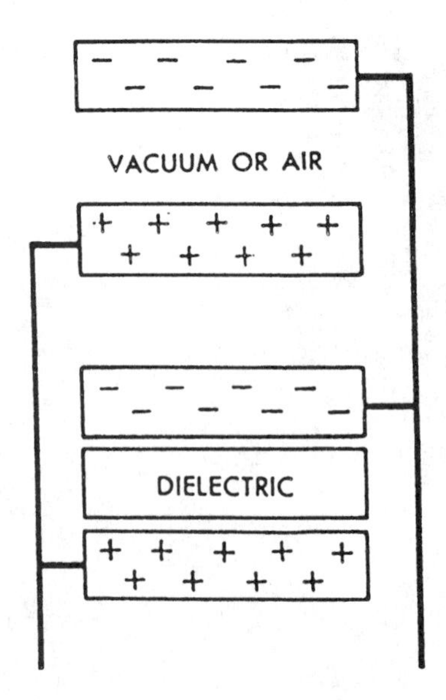

Fig. 6-42. The dielectric constant is the amount of electricity stored across an insulating material, divided by the amount of electricity stored across air or a vacuum.

Creep
Damping
Density
Density gradient column
Dimensional stability
Elongation percent
Dielectric strength
Fatigue
Flexural modulus
Flexural strength
Glass transition temperature
Gravity constant
Hardness
Haze
Hygroscopic
Impact strength
Index of refraction
Melt index
Offset yield
Plastic strain
Poise
Proportional limit
Relative density
Scleroscope
Solubility parameter
Solvent resistance
Specular gloss
Stiffness
Strain
Thixotropy
Toughness
Vicat softening point
Viscosity

Questions

6-1. Name the seven base units of the SI metric system.

6-2. A gigahertz is equal to ___?___ Hz.

6-3. Identify the SI metric unit for force and its formula.

6-4. Tensile strength, modulus of elasticity, and air pressure are measured in ___?___ .

6-5. In the SI metric system, temperatures are measured in ___?___ .

6-6. Two international, technical societies that develop standard and specifications for testing of plastics are ___?___ and ___?___ .

6-7. T or F. In testing mechanical properties, it is generally important to apply force at a specified rate.

6-8. Young's modulus is the ratio of ___?___ to ___?___ .

6-9. To choose a tougher plastic, choose one having a ___?___ area under the stress-strain curve.

6-10. The pendulum test measures ___?___ .

6-11. A plastics hinge relies on which property?

6-12. Resistance to transmitting vibration is called ___?___ .

6-13. Viscosity is defined as a measure of the ___?___ of a fluid.

6-14. Elongation over time due to a constant force is called ___?___ .

6-15. Plastics for space-craft heat shields are chosen for their ___?___ properties.

6-16. As the melt-index value of a plastic goes up, the viscosity goes ___?___ .

6-17. Below the glass transition temperature, a plastic becomes ___?___ .

6-18. Name a plastic that is hygroscopic.

6-19. Fillers used to make plastics conduct electricity are ___?___ and ___?___ .

6-20. In the arc resistance test, the surface of a sample becomes conductive because of ___?___ .

6-21. If the resistivity of a material is high, insulation resistance will be ___?___ .

6-22. Dielectric strength indicates suitability of a plastic for service as ___?___ .

6-23. Plastics used in electrical capacitors must have a high ___?___ .

6-24. For heat-sealing a plastic film by high-frequency methods, the ___?___ must not be low.

6-25. A viscosity of 1 pascal-second is equal to ___?___ poise.

6-26. Viscosity is defined as a measure of the ___?___ of a fluid.

6-27. Handles of pots and pans are often made of plastics because of the property of low ___?___ .

6-28. The plastics with the lowest self-ignition temperature is ___?___ .

6-29. Stress-cracking tests combine physical stress and ___?___ stress.

Activities

Tensile Testing

Materials and equipment. Constant speed tensile tester, stress-strain plotter, calipers, tensile test bars. Acquire standard ISO or ASTM test bars if available. Samples can also be cut from sheet materials.

6-1. Acquire or prepare 10 sample pieces and measuring:

overall length
gage length
gage width and thickness

(record dimensions in meters and inches)

6-2. Pull samples to failure at constant speed. Calculate strength at yield and break, elongation to yield and break, and modulus of elasticity in SI and English systems.

6-3. Calculate mean and standard deviations on stress (load) and elongation (strain) at yield and break.

6-4. Prepare an additional 10 bars and pull them to failure at a significantly different strain rate than the first group. For example, use 25 mm/minute

[1 in./minute] on one group and 500 mm/minute [20 in./minute] on the second group.

6-5. Calculate mean and standard deviations as in 3.

6-6. Sketch bell curves comparing strengths at yield and elongations at yield.

6-7. How great an effect did changing the strain rate have?

6-8. Write up a report summarizing the findings.

Additional activity for tensile testing

If samples are available with differing gating arrangements, create groups based on gate location. Test to see what effect the gate location has on strength and elongation. Molds that provide bars gated from one end, and others gated at both ends are extremely useful. The double gate yields a weld line at the center of the part. This allows comparsions between parts with weld lines and parts with no weld line.

Hardness testing

Equipment. Rockwell hardness tester, strip heater, temperature measuring device.

Procedure

6-1. Cut a piece of sheet material (acrylic or polycarbonate) into a square 75mm × 75 mm. The material should be at least 3 mm thick.

6-2. Test for hardness at 10 locations on the sample.

6-3. Place sample on strip heater and heat until soft enough to bend. Measure the highest temperature achieved by the sample. Do not bend, but cool sample preserving a flat surface.

6-4. After letting cool, test 10 locations within the "heat effected zone."

6-5. Calculate mean and standard deviations for the heated group and the unheated group. Sketch bell curves.

6-6. How much effect did the heating have on the hardness?

6-7. Summarize the findings in a short report.

Additional activities

Systematically alter the temperature achieved in sample pieces. Find the temperature range that causes the largest or the smallest change in hardness.

Impact testing

Equipment. Izod or Charpy tester, samples of appropriate dimensions.

Procedure

6-1. Impact 10 pieces and record results.

6-2. Expose 10 pieces of the same material to cold. Remove the samples one at a time from the cold and impact as rapidly as possible.

6-3. Calculate means and standard deviations. Sketch bell curves.

6-4. How great an effect did the cold have on the impact strength?

Additional activities

Expose samples to extreme cold. Let them return to room temperature before impact testing. Did the exposure to cold produce a lasting effect?

Linear Thermal Expansion Tests

6-1. If thermal expansion apparatus is available, follow the manufacturer's instructions for use.

6-2. To obtain a relative measurement of thermal expansion, carefully measure the length of a sample. Acquire 1 L of water that has a temperature of 20 °C.

6-3. Place the sample in the water and heat the water to 40 °C. Remove the sample and quickly measure length.

CAUTION: Do not exceed 40 °C.

6-4. Calculate the theoretical thermal expansion with the following formula:

Theoretical Thermal expansion (mm) =

difference in temperature (°C) × coefficient of thermal expansion (1/°C) × original length (mm)

The coefficients for selected plastics are shown in Table 6-10.

6-5. Calculate the observed thermal expansion with the following formula:

Observed Thermal Expansion (mm) =

Length hot – Length cool

6-6. Compare the observed and the theoretical values.

Chapter 7

Ingredients of Plastics

Introduction

Most plastics products consist of a polymeric material that has been altered to change or improve selected properties. This chapter focuses on those special ingredients used to alter and enhance plastics. For information of the processes used to mix these special materials with selected plastics, see Chapter 11 on extrusion. There are three large categories of these ingredients included in the chapter outline below:

I. Additives
   A. Antioxidants
   B. Antistatic agents
   C. Colorants
   D. Coupling agents
   E. Curing agents
   F. Flame retardants
   G. Foaming/blowing agents
   H. Heat stabilizers
   I. Impact modifiers
   J. Lubricants
   K. Plasticizers
   L. Preservatives
   M. Processing aids
   N. UV stabilizers
II. Reinforcements
   A. Lamina
III. Fillers

Some of the reasons for including additives, reinforcements, and fillers are:

- to improve processability
- to reduce material costs
- to reduce shrinkage
- to permit higher curing temperatures by reducing or diluting reactive materials
- to improve surface finish
- to change the thermal properties such as expansion coefficient, flammability, and conductivity
- to improve electrical properties including conductivity or resistance
- to prevent degradation during fabrication and service
- to provide desirable color or tint
- to improve mechanical properties such as modulus, strength, hardness, abrasion resistance, and toughness
- to lower the coefficient of friction.

A host of chemicals found usage in plastics materials because they produce desired changes in properties. However, some of the most successful chemicals were also dangerous and even toxic.

The environmental movement has had a significant effect on the use of chemicals in the plastics industry. Public concern about water and air pollution caused many significant changes in plastics materials and manufacturing processes. Agencies overseeing packaging for food, drugs, and cosmetics have sought to eliminate the use of toxic and dangerous chemicals. A potent approach used by several agencies is the banning of selected chemicals. This chapter includes descriptions of the efforts of the plastics industry to comply with environmental regulations.

Additives

The term *additives* covers a wide range of chemicals that are *added* to plastics. The major categories of additives are: antioxidants, antistatic agents, colorants, coupling agents, curing agents, flame retardants, foaming/blowing agents, heat stabilizers, impact modifiers, lubricants, plasticizers, preservatives, processing aids, and UV stabilizers.

Antioxidants

Oxidation of plastics involves oxygen in a series of chemical reactions that result in the breaking of bonds in polymers. Long-chain molecules are cut into shorter chains. If the oxidation continues, the chain cutting, often called *chain scission*, progresses to the point where the material becomes very weak and will disintegrate into a powder (Fig. 7-1). At elevated temperatures, oxidation generally occurs much more rapidly than at room temperature. Consequently, tests for oxidation usually expose samples to heat.

To combat oxidation, chemical substances that slow down or stop oxidation are added to the plastics. These substances are named *antioxidants*. Since the chemical reactions that occur in oxidation are rather complex, *antioxidant packages* combine two or more chemicals to increase the resistance to oxidation. Most antioxidant packages contain a *primary antioxidant*, and a *secondary antioxidant*. The primary antioxidant works to stop or terminate oxidative reactions. The secondary antioxidants work to neutralize reactive materials that cause new cycles of oxidation. When properly selected, the primary and secondary antioxidants may work together with a synergistic effect that enhances the results.

The major types of antioxidants are:

1. Phenolic
2. Amine
3. Phosphite
4. Thioesters

The phenolic and amine materials are often used as primary antioxidants, while the phosphite and thioesters serve as secondary antioxidants.

Some plastics are more susceptible to breakdown through oxidation than others. Polypropylene and polyethylene oxidize readily. Because of this tendency, chemical companies that manufacture polypropylene usually add a small amount of primary antioxidant to their polypropylene to prevent it from oxidizing during the extrusion processes needed to pelletize it.

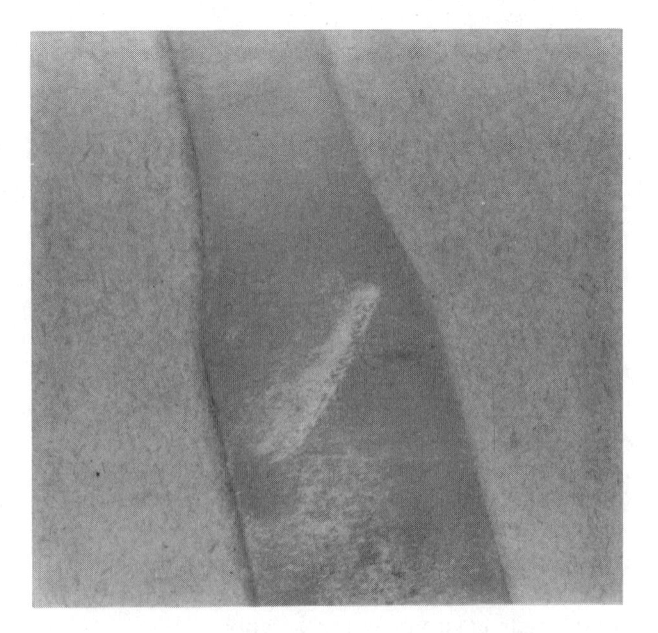

Fig. 7-1. Oxidative degradation of unstabilized polypropylene. This damage occurred during 50 hours at 180 °C. A fingernail made the diagonal scratch.

Antistatic Agents

Antistatic agents may be compounded into the plastics or applied to the product surface. These agents attract moisture from the air making the surface more conductive, which in turn dissipates static charges.

The most common antistatic agents are amines, quaternary ammonium compounds, organic phosphates, and polyethylene glycol esters. Concentrations of antistatic agents may exceed 2 percent, but application and FDA approval are the prime considerations in their use.

Colorants

Plastics can possess a wide range of colors. Plastics designers have exploited this property of plastics. In fact, some uses for plastics rely completely on their availability in a multitude of colors.

When making colored products, processors use precolor, dry color, liquid color, or color concentrates. *Precolor* is material already compounded to a desired color. *Dry color* is powdered colorant. It is frequently difficult to handle and leads to dust problems. *Liquid color* is a color in a liquid base. It requires special pumps. A *color concentrate* is a high loading of a colorant carried in a base-resin. It comes in pelletized and diced forms.

There are four basic types of colorants used in these various forms:

1. Dyes
2. Organic pigments
3. Inorganic pigments
4. Special-effect pigments

Dyes. *Dyes* are organic colorants. In contrast to pigments, dyes are soluble in plastics and color the material by forming chemical linkages with molecules. They are often brighter and stronger than inorganic colorants. Dyes are the best choice for a totally transparent product. Although some dyes have poor thermal and light stability, thousands of dyes are currently used in plastics.

Because dyes are soluble in plastics, they may move or migrate. A red dye may migrate into a white parts, causing it to turn pink.

Organic pigments. *Pigments* are not soluble in common solvents or in the resin; therefore, they must be mixed and evenly dispersed within the resin. Organic pigments provide the most brilliant opaque colors available. However, the translucent and transparent colors achieved with organic pigments are not as brilliant as those produced with dyes. Organic pigments can be hard to disperse. They tend to form *agglomerates*, which are clumps of pigment particles that cause spots and specks in products.

Inorganic pigments. Most *inorganic pigments* are based on metals. Oxides and sulfides of titanium, zinc, iron, cadmium, and chromium are common. Some of the colorants rely on the heavy metals (Fig. 7-2).

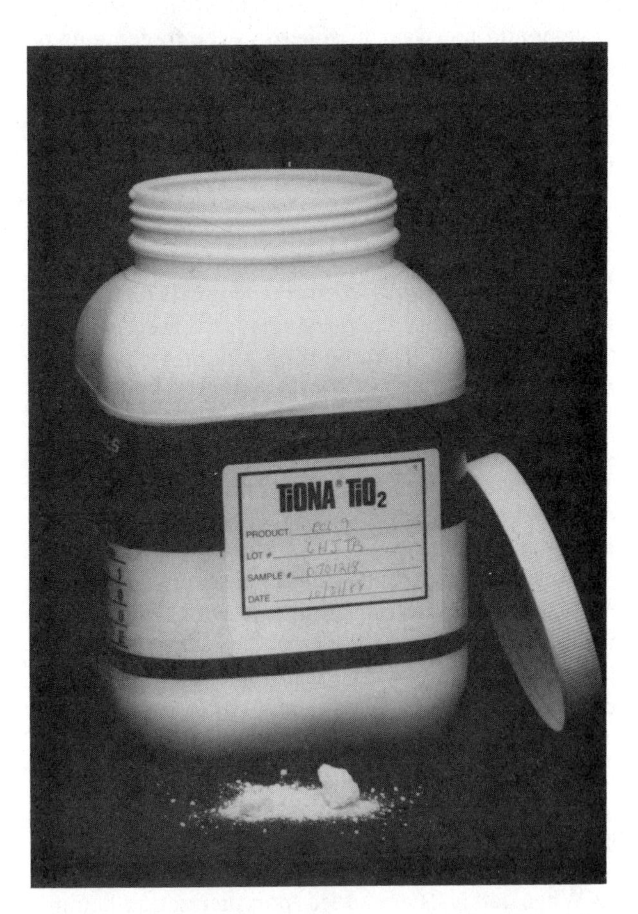

Fig. 7-2. This inorganic pigment is a powdered form of titanium dioxide. Titanium white is both bright and stable.

Environmental agencies have analyzed the health effects of heavy metals and recommend a ban on a number of them. By 1993, 11 states had banned or restricted heavy metals in packaging applications. The metals of chief concern are lead, mercury, cadmium and hexavalent chromium. The use of these materials needs to be less than 100 parts per million (ppm) within a few years after the passage of legislation. The EPA has also proposed legislation concerning the amount of cadmium and lead permitted in incinerator ash.

Selected metals in order of weight are:

Metal	Weight in grams per mole
Lead	207
Mercury	201
Gold	197
Tungsten	184
Barium	137
Cesium	133
Iodine	127
Tin	119
Cadmium	112
Silver	108
Bromine	80
Chromium	52

The use of some of these metals is restricted. A major concern is that when heavy metals leach out of landfills and enter ground water, they pose a health hazard. Also, when heavy metals are incinerated, the metal residue left in the ash is significant. The incinerator ash cannot go to a traditional landfill; consequently, its handling, storage, or disposal is a major problem.

Pigments containing lead, mercury, cadmium and hexavalent chromium are banned or under scrutiny. Many companies are developing and marketing heavy-metal-free (HMF) colorants. Some companies expect similar restrictions to affect the use of barium.

Other inorganic pigments pose no environmental and health danger. They include quite simple chemicals, such as carbon (black), iron oxide (red), and cobalt oxide (blue). Although lead sulfate (white) and cadmium sulfide (yellow) were popular for years, these pigments have lost sales in recent years.

These metallic oxides are easily dispersed in the resin. They do not produce colors as brilliant as those from organic pigments and dyes, but because of their inorganic structure, they resist light and heat more effectively. Most inorganic pigments are used in high concentrations to produce opaque-colored plastics. Low concentrations of iron-oxide pigment will produce a translucent color.

Special-effects pigments. *Special-effects pigments* may be either organic or inorganic compounds. Colored glass is used in a fine powdered form, and is a heat- and light-stable pigment for plastics. Colored glass powder is effective in exterior uses because of its color stability and chemical resistance.

Flakes of aluminum, brass, copper, and even gold may be used to produce a striking metallic sheen. Iridescent plastics are used by the automotive industry in producing metallic finishes. When metallic powders are mixed with a colored plastics, a finish varying in highlights and reflective hues is produced. Pearl essence, either natural or synthetic, may be used to produce a brilliance where pearl luster is desired.

When energy is absorbed by a material, a portion of that energy may be released in the form of light. This light is radiated when the molecules and atoms have their electrons excited to such a state that they begin to lose energy in the form of *photons*, or light particles. If heat causes the electrons to release photons of light energy, the radiation is called *incandescence*.

When chemical, electrical, or light energy excites electrons, the radiation of light is called *luminescence*. Luminescent materials are often added to plastics for special effects. Luminescence is categorized into fluorescence and phosphorescence (Fig. 7-3). *Fluorescent* materials emit light only when their electrons are being excited. These materials cease to emit light when the energy source exciting their electrons is removed. Fluorescent materials are made from sulfides of zinc, calcium, and magnesium. To be environmentally safe, some companies are offering fluorescent colors that are formaldehyde free. Fluorescent paint on instrument dials allows a pilot to read instruments with little visible light being emitted. Other uses of fluorescent materials include hunting jackets, protective helmets, gloves, life preservers, rain slickers, bicycle stripes, and road warning signs.

Phosphorescent pigments possess an afterglow; that is, they continue to emit light for a limited time after the exciting force has been removed. The most common example of phosphorescence is the television picture tube which emits light when electrical energy excites the phosphorescent materials coating the inside of the face of the tube. Phosphorescent pigments used in plastics and paints are made from calcium sulfide or strontium sulfide.

Mesothorium and radium compounds are radioactive materials sometimes used for special luminescence. Note that there may be harmful effects from prolonged exposure to radioactive materials.

(A) Illuminated signs.

(B) Nonilluminated signs.

Fig. 7-3. Phosphorescent pigments glow in the dark after exposure to light.

Coupling Agents

Coupling agents are sometimes called promotors (not to be confused with promoters). They are especially important in processing composites. Coupling agents are used as surface treatments to improve the interfacial bond between the matrix, reinforcements, fillers, or laminates. Without this treatment, many resins and polymers do not want to adhere to reinforcements or other substrates. Good adhesion is essential if the polymer matrix is to transfer stress from one fiber, particle, or laminar substrate to the next. Silane and titanate coupling agents are commonly used.

Curing Agents

Curing agents are a group of chemicals that cause cross-linking. These chemicals cause the ends of the monomers to join, forming long polymer chains and crosslinkages.

Because resins may be partially polymerized systems (e.g. B-stage resins), other forms of energy may cause premature polymerization. *Inhibitors* (stabilizers) may be used to prolong storage and block polymerization.

Catalysts, sometimes called hardeners (more correctly initiators), are the chemicals which help cause the monomers to join and/or cross-link. Organic peroxides are used to polymerize and cross-link thermoplastics (PVC, PS, LDPE, EVA, and HDPE) as well as the familiar thermosetting polyesters.

The most widely used initiators are unstable peroxides or *azo compounds*. Benzoyl peroxide and methyl ethyl ketone peroxide are widely used organic initiators.

When catalysts are added, polymerization begins. Catalysts are little influenced by the inhibitors in the resin. As organic peroxides are added to polyester resin, the polymerization reaction begins and yields exothermic heat. This formation of heat further speeds up the cross-linking and polymerization. *Promoters* (or *accelerators*) are additives that react in a manner opposite that of inhibitors, and they are often added to resins to aid in polymerization. Promoters react only when the catalyst is added. This reaction, which causes the polymerization, produces heat energy. A common promoter used with the catalyst methyl ethyl ketone peroxide is cobalt naphthanate. All promoters and peroxides should be handled with caution.

CAUTION: Peroxides may cause skin irritation and acid burns. If promoters and catalysts are added at the same time, a violent reaction may occur. Always thoroughly mix in the promoter, then add the desired amount of catalyst to the resin. Be sure that there is adequate ventilation and use personal protective wear.

Resins that have not been prepromoted generally have a longer shelf life. Remember that other forms of energy also cause polymerization. Heat, light or electrical energy may initiate this reaction. Always store curing agents at their recommended storage temperature and in their original containers.

Flame Retardants

Most commercial flame-retarding chemicals are based on combinations of bromine, chlorine, antimony, boron, and phosphorus. Many of these retardants emit a fire-extinguishing gas (halogen) when heated. Others react by swelling or foaming, forming an insulation barrier against heat and flame (Fig. 7-4).

Some of the most common chemicals used for retarding combustion are alumina trihydrate (ATH), halogenated materials, and phosphorus compounds.

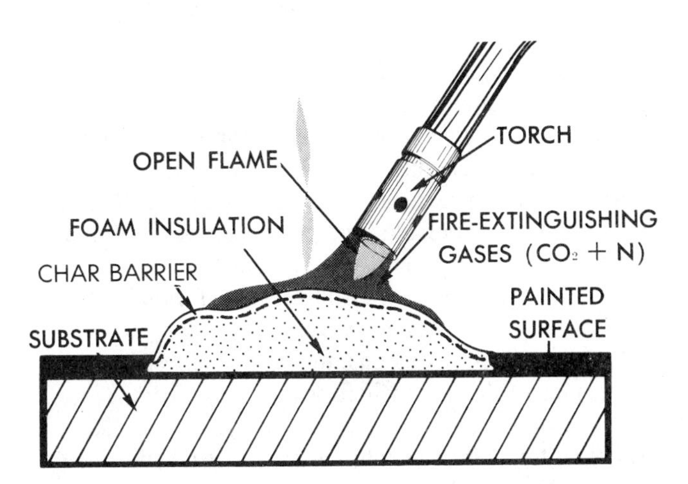

Fig. 7-4. This protective finish swells to form an insulating char barrier when heated. It also emits a fire-extinguishing gas to retard burning.

ATH cools the flame area by producing water. Halogenated materials release inert gases that reduce combustion. Several phosphorus materials form char barriers, which insulate combustibles.

Recent concern about brominated flame-retardant systems, particularly compounds containing polybrominated diphenyloxide (PBDPO), has led companies to offer halogen-free flame retardants. Development work on the halogen-free systems continues, since they are not as efficient as halogenated products.

Foaming/Blowing Agents

The terms *foaming, blowing, frothed, cellular,* and *bubble* are sometimes used to cover a wide variety of compounds and processing techniques to make polymers with a cellular structure. (See Chapter 16, Expansion Processes) There are two major types of foaming agents, the physical types and the chemical types. *Physical foaming agents* decompose at specific temperatures, releasing gases. The gases cause cells or voids in the plastics. The *chemical foaming agents* release gases due to a chemical reaction.

One of the biggest uses of foaming agents occurs in the manufacture of foamed polyurethane pads and seats for cars, trucks, sofas, and other furniture. Chlorinated fluorocarbons (CFC) are a very efficient physical foaming agent for polyurethane and received wide use for many years. However, research indicated that CFCs were causing damage to the ozone layer in the upper atmosphere. Starting on May 15, 1993, products made with CFCs must have a clearly visible warning that it contains a material that "harms human health and the environment by destroying ozone in the upper atmosphere." To provide additional incentive to reduce or eliminate the use of CFCs, a special tax on them began in 1994.

In response to these concerns, many foam makers have switched to hydrochlorofluorocarbon (HCFC), which is much less ozone-depleting than traditional CFCs. Compared to CFCs, HCFCs have only 2–10 percent the ozone depletion potential. In contrast to CFCs, a difficulty with the new foaming agents is that they yield foam that is more dense, and therefore less efficient as an insulating material. Researchers are working on two fronts; first to improve the effectiveness of the HCFCs, and second, to develop foaming agents that contain no chlorine and have no ozone-depleting potential.

Chemical blowing agents such as azodicarbonamide are widely used to produce cellular HDPE, PP, ABS, PS, PVC, and EVA. This chemical has several advantages, including efficient gas yield, some FDA approvals for food contact applications, and its ease of modification for various plastics (Fig. 7-5).

Fig. 7-5. A pelletized azo-type blowing agent.

Heat Stabilizers

Heat stabilizers are additives that retard the decomposition of a polymer caused by heat, light energy, oxidation, or mechanical shear. PVC has poor thermal stability, and has been the focus of most heat stabilizers. In the past, heat stabilizers were compounds based on lead and cadmium. Lead has been the predominant additive for wire and cable coatings. Because of environmental concerns surrounding heavy metals, non-cadmium stabilizers have taken over many applications previously held by cadmium stabilizers.

This change away from cadmium heat stabilizers will have a large effect. In 1993, PVC stabilizers accounted for about 15 percent of the cadmium found in municipal solid wastes in the U.S. When combined with all other plastics, 28 percent of the cadmium came from plastics. Reducing or eliminating this source of cadmium will help clean-up the environment.

To replace lead and cadmium, suppliers have developed compounds using barium-zinc, calcium-zinc, magnesium-zinc, and magnesium-aluminum-zinc, and phosphite formulations.

Impact Modifiers

One or more monomers (usually elastomers) in varying amounts may be added to rigid plastics to improve (modify) impact properties, melt index, processability, surface finish, and weather resistance. PVC is toughened by modification with ABS, CPE, EVA, or other elastomers. (See Alloys, Blends, Ethylene-ethyl acrylate, and Styrene-butadiene.)

Lubricants

Lubricants are needed for making plastics. During the making of polymers, lubricants are added for three basic reasons. First, they help get rid of some of the friction between the resin and the manufacturing equipment. Second, lubricants aid in emulsifying other ingredients and provide internal lubrication for the resin. Third, some lubricants prevent the plastics from sticking to the mold surface during processing. After the products are taken from the mold, lubricants may exude from the plastics and prevent the products from adhering to each other, and may provide a nonsticking or slippery quality to the plastics surface.

Many different lubricants are used as ingredients in plastics. Some examples are: waxes, such as montan, carnauba, paraffin, and stearic acid. Metallic soaps such as the stearates of lead, cadmium, barium, calcium and zinc are also used as lubricants (Table 7-1). Most of the lubricant is lost during the process of manufacturing the resin. Excess lubricant may slow polymerization or cause a *lubrication bloom,* seen as an irregular, cloudy patch on the plastics surface.

Some plastics exhibit nonstick and self-lubrication properties. Examples are fluorocarbons, polyamides, polyethylene, and the silicone plastics. They are sometimes used as lubricants in other polymers. Remember, all additives must be carefully selected for toxic effects and desired service use.

Plasticizers

Plasticity refers to the ability of a material to flow or become fluid under force. A *plasticizer* is a chemical agent added to plastics to increase flexibility; reduce melt temperature; and lower viscosity. All of these properties aid in processing and molding. Plasticizers act much like solvents by lowering the viscosity. However, they also act like lubricants by allowing slip to occur between molecules.

Remember, van der Waals' bonds are only physical attractions and not chemical bonds, and plasticizers help neutralize most of these forces. Plasticizers, much like solvents, produce a more flexible polymer. However they are not designed to evaporate from the polymer during normal service life.

Plasticizer leaching or loss is an important consideration. It is undesirable when in contact with food, pharmaceutical or other products for consumption. Leaching and degassing may cause PVC hoses, upholstery, and other products to become stiff or brittle and to crack. For best results, the plasticizer and polymer must have similar solubility parameters.

Over 500 different plasticizers are formulated to modify polymers. Plasticizers are vital ingredients in plastics coatings, extrusions, moldings, adhesives, and films. One of the most widely-used plasticizers is dioctyl phthalate. Some plasticizers may be hazardous. The EPA found di-2-ethylhexyl phthalate plasticizer as carcinogenic in animals in lab tests. It is currently

Table 7-1. Lubricants Chart

	Lubricant								
Plastics	**Alcohol Esters**	**Amide Waxes**	**Complex Esters**	**Comb. Blends**	**Fatty Acids**	**Glycerol Esters**	**Metallic Stearates**	**Paraffin Waxes**	**Poly-ethylene Waxes**
ABS		X			X	X			
Acetals	X		X						
Acrylics	X			X					
Alkyd							X		
Cellulosics	X	X			X	X			
Epoxy				X		X			
Ionomers		X							
Melamines			X		X				
Phenolics				X	X	X			
Polyamides	X			X					
Polyester			X	X	X				
Polyethylene		X							
Polypropylene		X			X	X			
Polystyrene		X	X		X		X		
Polyurethanes			X						
Polyvinyl chloride	X	X	X	X	X	X	X	X	X
Sulfones			X						

Table 7-2. Compatibility of Selected Plasticizers and Resins

	Resin											
Plasticizer	Polyvinyl Acetate	Polyvinyl Chloride	Polyvinyl Butyral	Polystyrene	Cellulose Nitrate	Cellulose Acetate	Cellulose Acetate Butyrate	Ethyl Cellulose	Acrylic	Epoxy	Urethane	Polyamide
Butyl benzyl phthalate	C	C	C	C	C	P	C	C	C	C	C	C
Butyl cyclohexyl phthalate	C	C	C	C	C	P	C	C	C	C	C	C
Didecyl phthalate	I	C	C	C	C	C	C	C	C	P	C	P
Butyl octyl phthalate	I	C	P	C	C	I	C	C	C	P	C	C
Dioctyl phthalate	I	C	P	C	C	I	C	C	C	I	C	C
Cresyl diphenyl phosphate	C	C	C	P	C	C	C	C	C	C	C	C
N-Ethyl-*o,p*-toluenesulfonamide	C	I	C	P	C	C	C	C	C	C	P	C
o, p-Toluenesulfonamide	C	I	C	P	C	C	C	C	C	C	P	C
Chlorinated paraffins	C	P	P	C	P	I	P	C	P	P	C	C
Didecyl adipate	I	C	I	C	C	I	C	C	I	I	P	C
Dioctyl adipate	I	C	C	C	C	I	C	C	I	I	P	C
Dioctyl sebacate	I	C	P	C	C	I	P	C	I	I	P	C

Notes:
C—Compatible
P—Partially compatible
I—Incompatible

labeled as a potential carcinogen. Some plasticizers are listed in Table 7-2.

Preservatives

Elastomers and heavily plasticized PVC are most susceptible to attack by microorganisms, insects, or rodents. As moisture stays or condenses on shower curtains, automobile tops, pool liners, waterbed liners, cable coatings, tubing, etc., microbiological deterioration may occur. Antimicrobials, mildewicides, fungicides, and rodenticides may be used to provide the necessary protection in many polymers. The EPA and FDA carefully regulate the use and handling of all antimicrobials.

Processing Aids

There are a variety of additives used to improve processing behavior, increase production rates, or improve surface finish. *Antiblocking* agents such as waxes exude to the surface and prevent two polymer surfaces from adhering. *Emulsifiers* are used to lower the surface tension between compounds. They act as detergents and wetting agents. Wetting agents used to lower viscosity are called *viscosity depressants*. They are used in plastisol compounds to assist in processing heavily filled materials or those that become too thick with age.

Solvents are added to resins for several reasons. Many natural resins are very viscous or hard; therefore, they must be diluted or dissolved before processing. Resinous varnish and paints must be thinned with solvents for proper application.

Solvents may be considered a processing aid. In solvent molding, the solvent holds the resin in solution while it is being applied to the mold. The solvent rapidly evaporates, leaving a layer of plastics film on the mold surface. Solvents dissolve many thermoplastics; therefore, they are used for both identification and cementing purposes. Solvents are also useful for cleaning resins from tools and instruments. Benzene, toluene, and other aromatic solvents will dissolve the natural oils of the skin. All chlorinated solvents are potentially toxic. Avoid breathing the fumes or allowing skin contact when using plastics additives. (See Chapter 4, Health and Safety.)

UV Stabilizers

Polyolefins, polystyrene, polyvinyl chloride, ABS polyesters, and polyurethanes are susceptible to ultraviolet solar breakdown. Solar radiation of polymers may result in crazing, chalking, color changes; or loss of physical, electrical, and chemical properties. This weathering damage is caused when the polymer absorbs light energy. Ultraviolet light is the most destructive portion of the solar radiation striking plastics products. There may be enough energy involved to break chemical bonds between atoms.

To reduce the damage done by exposure to UV light, compounding processes add UV stabilizers to plastics. Carbon black is of some use as a UV stabilizer, but of limited use due to its color. In the past, the most widely used ultraviolet absorbers were 2-hydroxybenzophenones, 2-hydroxyphenylbenzotriazoles and 2-cyanodiphenyl acrylates. Currently, most developments involve hindered-amine light stabilizers (HALS).

HALS often contain reactive groups, which chemically bond onto the backbone of the polymer molecules. This reduces migration and volatility. The combination of a HALS and phosphite or phenolic antioxidant increase the UV resistance.

Reinforcements

Reinforcements are ingredients added to resins and polymers. These ingredients do not dissolve in the polymer matrix. Consequently the material becomes a composite. Among many reasons for adding reinforcements, one important reason is to produce dramatic improvements in the physical properties of the composite.

Reinforcements are often confused with fillers. Fillers, however, are only small particles and contribute only slightly to strength. Reinforcements, on the other hand, are ingredients that increase strength, impact resistance, and stiffness. One major reason for confusing the two is that some materials—glass, for example—may act as a filler, reinforcement, or both.

There are six general variables that influence the properties of the reinforced composite materials and structures.

1. *Interface bond between matrix and reinforcements.* The matrix functions to transfer most of the stress to the (much stronger) reinforcements. In order to accomplish this task, there must be an excellent adhesion between the matrix and the reinforcement.

2. *Properties of the reinforcement.* It is assumed that the reinforcement is much stronger than the matrix. The actual properties of each reinforcement may vary by composition, shape, size, and number of defects. The production, handling, processing, surface enhancement, or hybridization can also determine properties for each type of reinforcement.

3. *Size and shape of the reinforcement.* Some shapes and sizes may help provide superior handling, loading, processing, packing orientation, or adhesion in the matrix. Some fibers are so small that they are handled in bundles, while others are woven into cloth. Particulates are more likely to be randomly distributed than long fibers.

4. *Loading of the reinforcement.* Generally, mechanical strength of the composite depends on the amount of reinforcing agent it contains. A part containing 60 percent reinforcement and 10 percent resin matrix is almost six times stronger than a part containing the opposite amounts of these two materials. Some glass filament-wound composites may have up to 80 percent (by weight) loading by unidirectional orientation of the filament. Most reinforced thermoplastic composites contain less than 40 percent (by weight) reinforcements.

5. *Processing technique.* Some processing techniques allow the reinforcements to be more carefully aligned or oriented. During processing, reinforcements may be broken or damaged, resulting in lower mechanical properties. Depending on processing technique, particulate reinforcements and short fibers are more likely to have random rather than oriented placement in the matrix.

6. *Alignment or distribution of the reinforcement.* Alignment or distribution of the reinforcement allows versatility in composites. The processor can align or orient the reinforcements to provide directional properties. In Figure 7-6, the parallel (anisotropic) alignment of continuous strands provided the highest strength; bidirectional (cloth) alignment provides a middle strength range, and random (mat) gives the lowest.

Reinforcements may be divided into two major groups of materials: 1) lamina, and 2) fibrous. The basic structural element of laminar composites is the *lamina.*

Lamina

Lamina may be unidirectional fibers, woven cloth, or sheets of material. Because the individual layers act as a reinforcement, they may be included as an ingredient or additive. These laminar layers are more than just a processing technique. (See Laminating) The lamina selection, alignment, and composition constitute the performance properties of the laminar composite. (See Sandwich Composites)

It should become clear that alignment of the reinforcements is the key to designing the composite with anisotropic or isotropic properties. As a rule, if all the reinforcements are placed parallel to each other (0° lay-up), the composite will be directional. (See Pultrusion) The directional tensile strength properties achieved with different fiber reinforcement alignment is illustrated in Figure 7-7. The random-chopped strand mat provides equal strength properties in all directions. The unidirectional fiber alignment has the highest strength parallel to (or in the direction of) the fiber. As this angle varies from 0° to 90°, the strength varies proportionally. Remember that the matrix must securely adhere to the reinforcement and prevent the reinforcement from buckling to transfer the applied stress.

In the fibrous group of reinforcements, there are six sub-classes. They are:

1. glass
2. carbonaceous
3. polymer
4. inorganic
5. metal
6. hybrids

Glass fibers. One of the most important reinforcing materials is fibrous glass (Tables 7-3 and 7-4). Because the strength of glass reinforces plastics, many parts previously made from metals have been replaced with plastics. Figure 7-8 shows a reinforced-plastics gear housing. It is lighter and stronger than the metal housing it replaces.

Glass fibers are produced by several different methods. One common method of production involves pulling a strand of molten glass after it has been formed by a small orifice. The diameter of the strand is controlled by the pulling action.

The major constituent of glass is silica but other ingredients allow the production of many types of

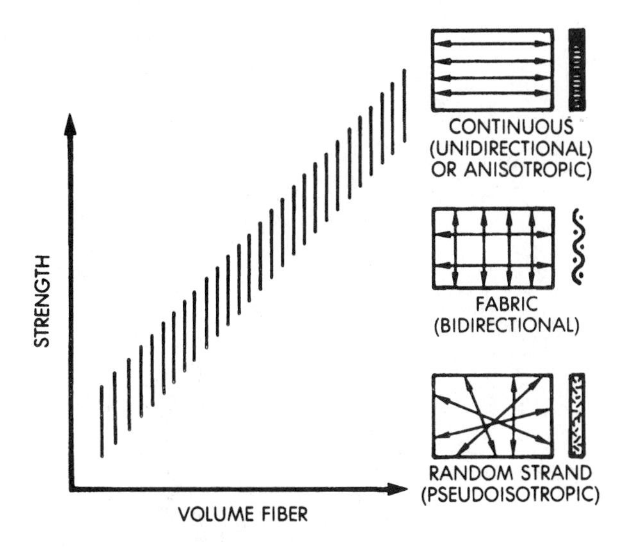

Fig. 7-6. Strength relation to reinforcement alignment and volume of fiber.

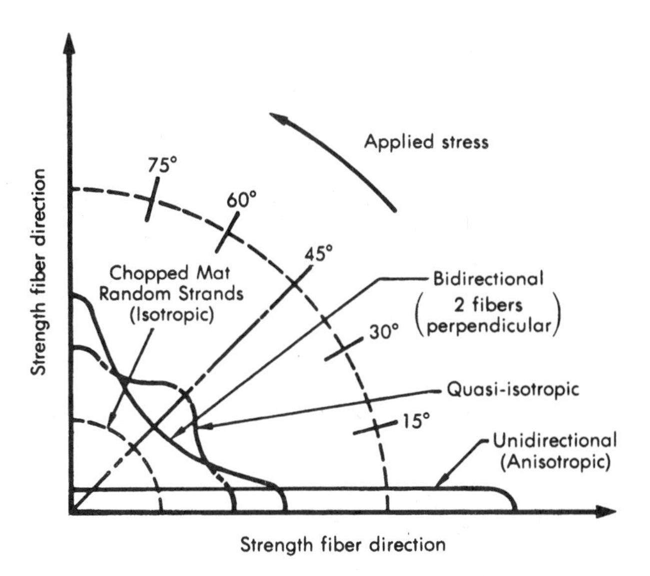

Fig. 7-7. Effect of alignment or distribution of the reinforcement.

fibrous glass. The most common type is E glass fiber, which has good electrical (E) properties and high strength. For chemical resistance, C glass is used. Both E and C glass have tensile strength exceeding 3.4 GPa [493 183 psi]. For low dielectric constant and density, D glass is used. For radiation protection, I glass contains lead oxide. For high-strength uses, S glass is selected. It is about 20 percent stronger and stiffer than E-glass. S glass has a tensile strength of more than 4.8 GPa [696258 psi].

Plastics processors purchase these differing types of glass in several forms. *Rovings* are long strands of fibrous glass that may be easily cut and applied to resins. A roving is made up of many strands of glass resembling a loosely twisted or stranded rope (Fig. 7-9). *Chopped fibers* (Fig. 7-10) are among the least costly forms of glass reinforcement. Chopped strands range in length from 3 to 50 mm [0.125 to 2 in.] Figure 7-11 shows the production of chopped strands from rovings. *Milled fibers* are less than 1.5 mm [0.062 in.] in length, and are produced by hammer milling glass strands (Fig. 7-12). Milled fibers are added to a resin as

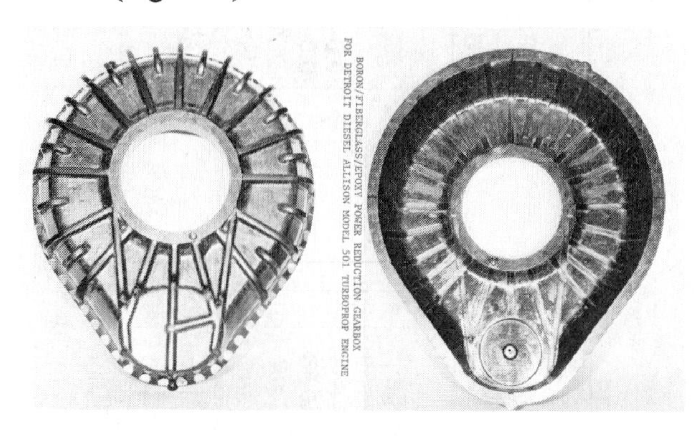

Fig. 7-8. This boron-epoxy gear housing has replaced a metal housing. (Allison Division, Detroit Diesel)

Table 7-3. Properties of Thermoplastics: Unreinforced and Reinforced Materials

	Polyamide		Polystyrene*		Polycarbonate		Styrene-Acrylonitrile†		Polypropylene		Acetal		Linear Polyethylene	
Property	**U**	**R**	**U**	**R**	**U**	**R**	**U**	**R**	**U**	**R**	**U**	**R**	**U**	**R**
Tensile strength, MPa	82	206	59	97	62	138	76	124	35	46	69	86	23	76
Impact strength, notched, J/mm														
At 22.8 °C	0.048	0.202	0.016	0.133	0.106§	0.213§	0.024	0.160	0.069-0.112	0.128	3.20	0.160	—	0.240
At –40 °C	0.032	0.224	0.010	0.170	0.080§	0.213§	—	0.213	—	0.133	—	0 160	—	0.266
Tensile strength, MPa	2.75	—	2.75	8.34	2.2	11.71	3.58	10.34	1.37	3.10	2.75	5.58	0.82	6.20
Shear strength, MPa	66	97	—	62	63	83	—	86	33	34	65	62	—	38
Flexural strength, MPa	79	255	76	138	83	179	117	179	41 to 55	48	96	110	—	83
Compressive strength, MPa	34††	165	96	117	76	130	117	151	59	41	36	90	19 to 24	41
Deformation (27.58 MPa), %	2.5	0.4	1.6	0.6	0.3	0.1	—	0.3	—	6.0	—	1.0	—	0.4‡
Elongation, %	60.0	2.2	2.0	1.1	60-100	1.7	3.2	1.4	>200	3.6	9-15	1.5	60.0	3.5
Water absorption in 24 hr, %	1.5	0.6	0.03	0.07	0.3	0.09	0.2	0.15	0.01	0.05	0.20	1.1	0.01	0.04
Hardness, Rockwell	M79	E75 to 80	M70	E53	M70	E57	M83	E65	R101	M50	M94	M90	R64	R60
Relative density	1.14	1.52	1.05	1.28	1.2	1.52	1.07	1.36	0.90	1.05	1.43	1.7	0.96	1.30
Heat distortion temp. (at 1.82 MPa), °C	65.6	261	87.8	104.4	137.8	148.9	933	107	683	137.8	100	168.6	52.2	126.7
Coef. of thermal expansion, per °C × 10^{-6}	90	15	60	35	60	15	60	30	70	40	65	30	85	25
Dielectric strength (short time), V/mm	15157	18898	19685	15591	15748	18976	17717	20276	29528	—	19685	—	—	2362 2
Volume resistivity, ohm-cm × 10^{15}	450	2.6	10.0	36.0	20.0	1.4	10^{16}	43.5	17.0	15.0	0.6	38.0	10^{15}	29.0
Dielectric constant at 60 Hz	4.1	4.5	2.6	3.1	3.1	3.8	3.0	3.6	2.3	—	—	—	2.3	2.9
Power factor at 60 Hz	0.0140	0.009	0.0030	0.0048	0.0009	0.0030	0.0085	0.005	—	—	—	—	—	0.001
Approximate cost, ¢/cm³	0.256	0.70	0.04	0.21	0.31	0.56	0.08	0.30	0.05	0.18	0.28	0.67	0.06	0.26

Notes: Columns marked "U" unreinforced, "R" reinforced. *Medium-flow, general-purpose grade.
†Heat-resistant grade. §Impact values for polycarbonates are a function of thickness, ‡6.8 mPa load.
††At 1% deformation. Source: *Machine Design*, Plastic Reference Issue.

Table 7-4. Properties of Thermosetting Plastics: Glass-Fiber Reinforced Resins

	Base Resin				
Property	**Polyester**	**Phenolic**	**Epoxy**	**Melamine**	**Polyurethane**
Molding quality	Excellent	Good	Excellent	Good	Good
Compression molding					
Temperature, °C	76.7–160	137.8–176.7	148.9–165.6	137.8–171.1	148.9–204.4
Pressure, MPa	1.72–13.78	13.78–27.58	2.06–34.47	13.78–55.15	0.689 34.47
Mold shrinkage, mm/mm	0.0–0.05	0.002–0.025	0.025–0.05	0.025–0.100	0.228–0.762
Relative density	1.35–2.3	1.75–1.95	1.8–2.0	1.8–2.0	1.11–1.25
Tensile strength, MPa	173–206	35–69	97–206	35–69	31–55
Elongation, %	0.5–5.0	0.02	4	—	10–650
Modulus of elasticity, Pa	0.55–1.38	2.28	2.09	1.65	—
Compressive strength, MPa	103–206	117–179	206–262	138–241	138
Flexural strength, MPa	69–276	69–414	138–179	103–159	48–62
Impact, Izod, J/mm	0.1–0.5	0.5–2.5	0.4–0.75	0.2–0.3	No break
Hardness, Rockwell	M70–M120	M95–M100	M100–M108	—	M28–R60
Thermal expansion, per °C	$5–13(\times 10^{-4})$	4×10^{-4}	$2.8–7.6(\times 10^{-4})$	3.8×10^{-4}	$25–51(\times 10^{-4})$
Volume resistivity (at 50% RH, 23 °C), ohm-cm	1×10^{14}	7×10^{12}	3.8×10^{15}	2×10^{11}	$2 \times 10^{11}–10^{14}$
Dielectric strength, V/mm	13780–19685	5512–14567	14173	6693–11811	12992–35433
Dielectric constant					
At 60 Hz	3.8–6.0	7.1	5.5	9.7–11.1	5.4–7.6
At 1 kHz	4.0–6.0	6.9	—	—	5.6–7.6
Dissipation factor					
At 60 Hz	0.01–0.04	0.05	0.087	0.14–0.23	0.015–0.048
At 1 kHz	0.01–0.05	0.02	—	—	0.043–0.060
Water absorption, %	0.01–1.0	0.1–1.2	0.05–0.095	0.9–21	0.7–0.9
Sunlight (change)	Slight	Darkens	Slight	Slight	None to slight
Chemical resistance	Fair*	Fair*	Excellent	Very good†	Fair
Machining qualities	Good	—	Good	Good	Good

*Attacked by strong acids or alkalies. †Attacked by strong acids. Source: *Machine Design*, Plastics Reference Issue.

a premix to increase viscosity and product strength *Yarns* resemble rovings but are twisted like a rope. (Fig. 7-13) Reinforcing yarns are used in the fabrication of large liquid tank containers.

Fiberglass yarn product nomenclature is based on a letter-number system. For example, a yarn designated as ECG 150 2/2 2.8 would be:

E = Electrical glass

C = Continuous filament

G = Filament diameter of 9 μm (See Table 7-5.)

150 = 1/100 of the total approximate bare yardage in a pound or 1500 yards.

2/2 = single strands were twisted and two of the twisted strands were plied together (S or Z may be used to designate the type of twist) (Fig. 7-14)

2.8 = number of turns per inch in the twist of final yarn with S twist

There were two basic strands in the yarn and two twisted strands. Thus,

15000/2x2 = 3750 yards per pound of yarn

Fig. 7-9. Fibrous glass roving. (Reichhold)

Fig. 7-10. Chopped strands of fibrous glass. (PPG)

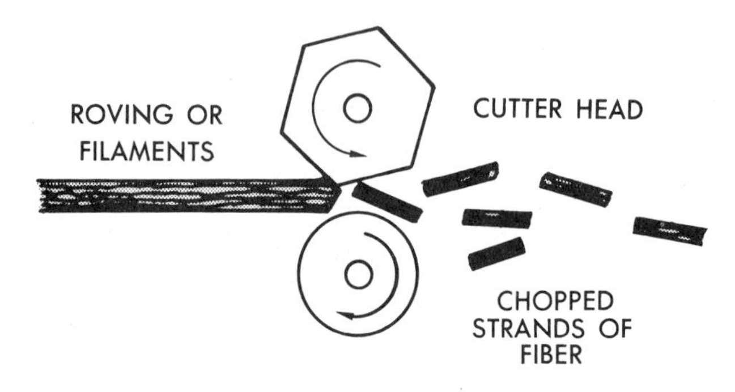

Fig. 7-11. Production of chopped glass strands.

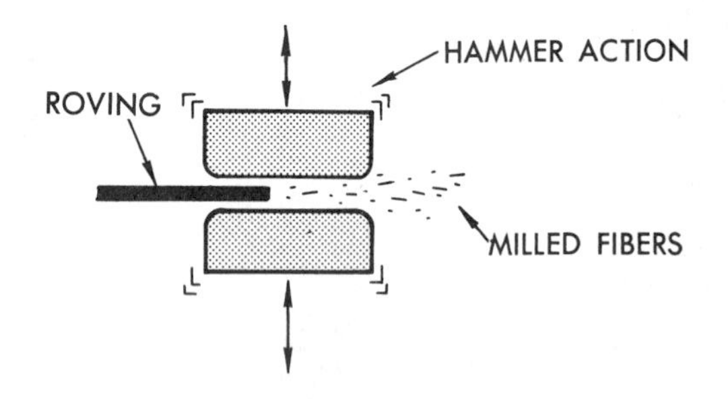

Fig. 7-12. Production of milled glass fibers.

Table 7-5. Glass Fiber Diameter Designations

Filament Designation	Filament Diameter in μm	(inches)
C	4.50	0.000 175
D	5.00	0.000 225
DE	6.00	0.000 250
E	7.00	0.000 275
G	9.10	0.000 375
H	11.12	0.000 425
K	13.14	0.000 525

(A) Monofilament yarn.

(B) Multifilament yarn.

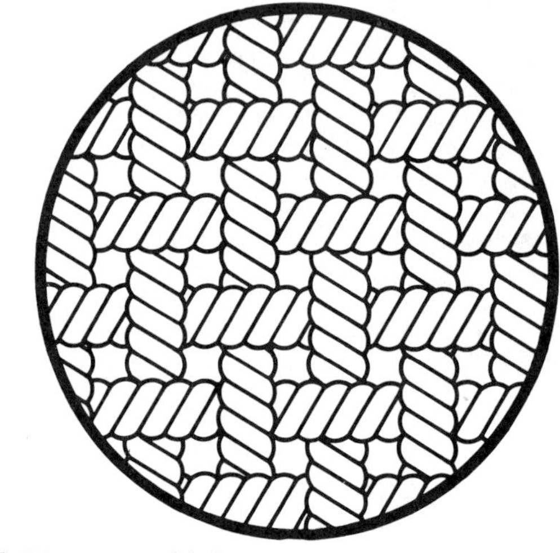

(A) Woven yarn fabric.

Fig. 7-13. Yarns.

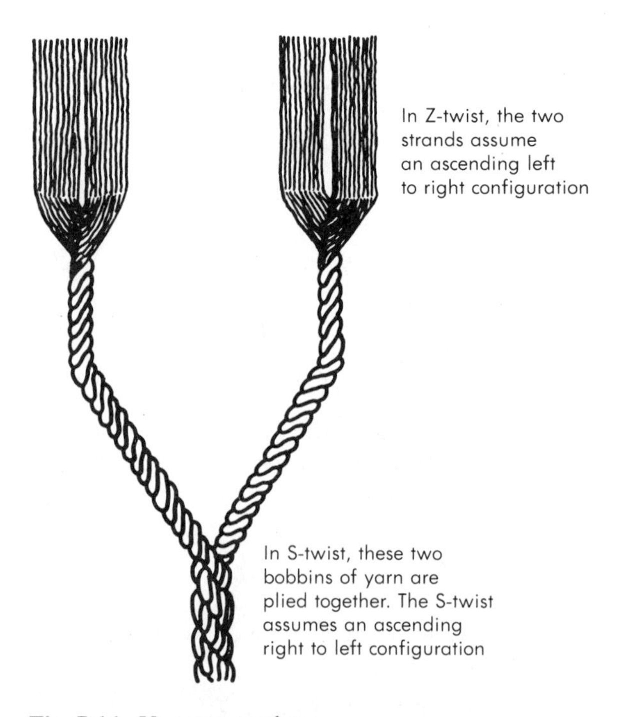

Fig. 7-14. Yarn nomenclature

In addition to fiber and yarn forms, glass reinforcements are also available in cloth and mat forms. *Mats* consist of nondirectional pieces of chopped strands. They are held together by a resinous binder or by mechanical stitching called needling (Fig. 7-15).

Woven cloth can provide the greatest physical strength of all the fibrous forms but is about 50 percent more costly than other forms. Standard rovings may be woven into fabric form (known as woven roving) and used for thick reinforcements.

There are several types of woven glass fabrics. Glass fiber yarns are woven into several basic patterns as shown in Figure 7-16. Figure 7-17 shows three different forms of fibrous glass reinforcements.

Carbonaceous fibers. Carbon fibers are usually made by oxidizing, carbonizing and graphitizing an organic fiber. Rayon and polyacrylonitrile (PAN) are currently used. Pitch fibers can also be produced directly from oil and coal. Ordinary pitches are isotropic in character and must be oriented to be as useful as reinforcing agents. The terms carbon and graphite are used interchangeably, but there is a distinction. Carbon (PAN) fibers are about 95 percent carbon, while graphite fibers are

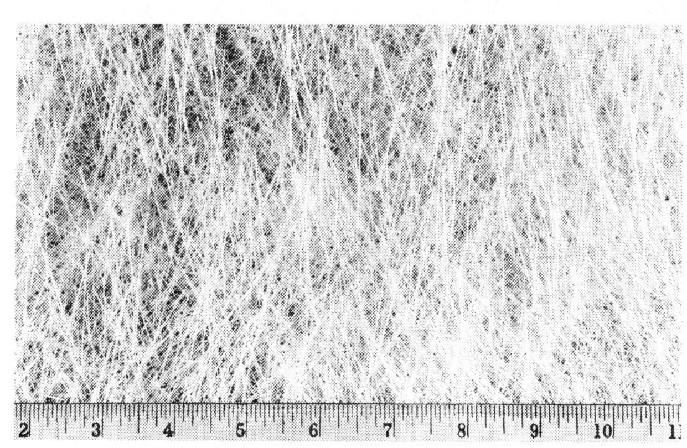

(A) Resin-bonded.

(B) Stitched (needled)

Fig. 7-15. Fibrous glass mats. (Owens-Corning Fiberglas Corp.)

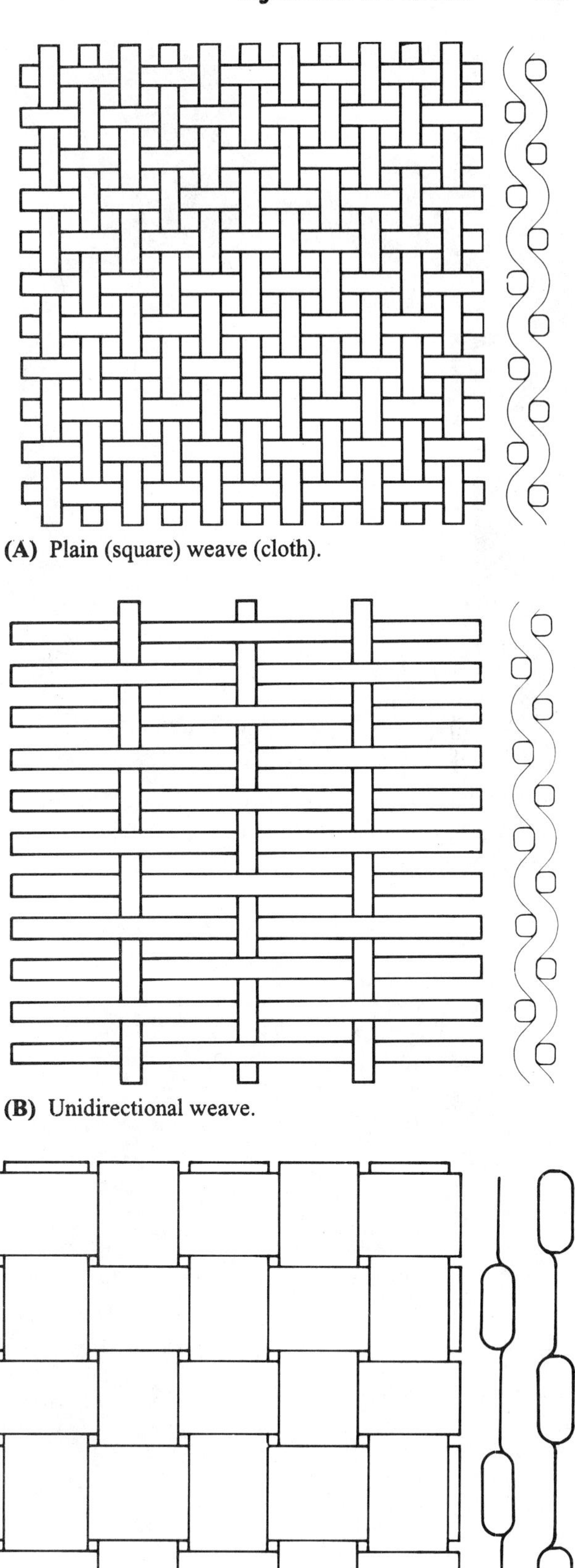

(A) Plain (square) weave (cloth).

(B) Unidirectional weave.

(C) Square-weave woven rovings.

Fig. 7-16. Weave patterns and roving. (Owens-Corning Fiberglas Corp.)

(D) Multifilament wound or twisted roving used in making heavy woven fibrous glass products. (Owens-Corning Fiberglass Corp.)

Fig. 7-16. Weave patterns and roving. (Owens-Corning Fiberglas Corp.) *(Continued)*

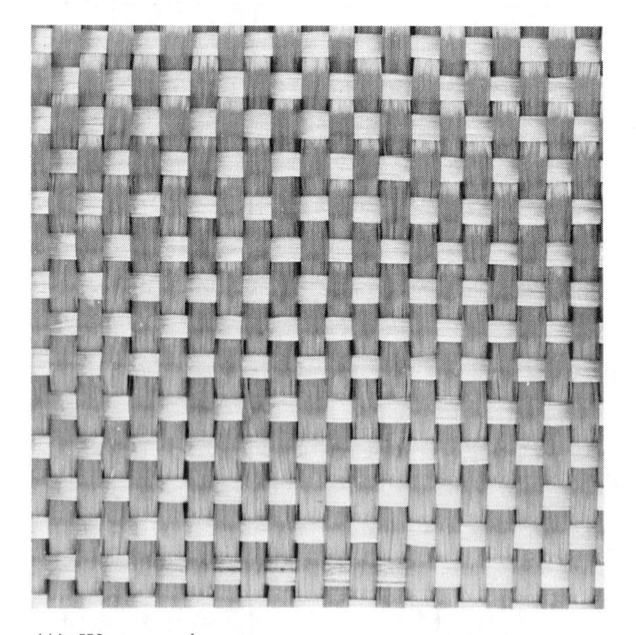

(A) Woven roving.

graphitized at much higher temperatures and result in a carbon elemental analysis of 99 percent. Once the organic materials have been driven off (pyrolized and stretched into filaments), the result is a high-strength, high-modulus, low-density fiber (Fig. 7-18).

(B) Fine strand mat.

(C) Combination woven roving and mat.

Fig. 7-17. Some of the many forms of fibrous glass reinforcements. (PPG)

Polymer fibers. For years cotton and silk filaments were used as reinforcements in belting, tires, gears, and other products. Synthetic polymers of polyester, polyamide (PA), polyacrilonitrile (PAN), polyvinyl acetate (PVA), cellulose acetate (CA), and others are currently used. Kevlar aramid is an aromatic polyamide polymer fiber with nearly twice the stiffness and about half the density of glass. Kevlar is a trademark of DuPont. Aramid is the generic name for a series of Kevlar fibers. Unlike carbon fibers, Kevlar fibers do not conduct electricity, nor are they electrically opaque to radio

Fig. 7-18. This reinforcing cloth contains both glass and carbon fibers. The glass is lighter in color, the carbon darker.

waves. Kevlar 29 fibers are used for ballistic protection, ropes, army helmets, coated fabrics, and a variety of composite applications. Kevlar 49 is used in boat hulls, flywheels, v-belts, hoses, composite armor, and aircraft structures. It is of equal strength but has a much higher modulus than Kevlar 29.

A common high-strength polymer matrix is epoxy. Polyesters, phenolic polyimide and other resin and polymer systems are used.

Polyester and polyamide-based thermoplastic fibers find applications in bulk molding compounds (BMC), sheet molding compounds (SMC), thick molding compounds (TMC), layup, pultrusion, filament winding, resin transfer molding (RTM), reinforced reaction injection molding (RRIM), thermal expansion resin transfer molding (TERTM), and injection molding operations.

Inorganic fibers. Inorganic fibers are a class of short crystalline fibers. They are sometimes called crystal whisker fibers. Crystal whisker fibers are made of aluminum oxide, beryllium oxide, magnesium oxide, potassium titanate, silicon carbide, titanium boride, and other materials (Fig. 7-19). Potassium titanate whiskers are used in large quantities to strengthen composites in thermoplastic matrices. Inorganic continuous boron fibers are stronger than carbon and may be used in a polymer and aluminum matrix. Boron in an epoxy matrix is used to make many composite parts for military and civilian aircraft.

These fibers are very costly to make with present technologies; however, they display tensile strengths greater than 40 GPa [5 802 146 psi]. Research into use of these reinforcements in dental plastics fillings, turbine compressor blades, and special deep-water equipment has shown encouraging results. Figure 7-20 shows a helicopter tail rotor shaft made of boron-reinforced epoxy.

Fig. 7-19. Submicron ceramic whiskers grown in a fibrous ball. There is a higher concentration of fibers near the center of the ball. The fibers range from as small as two billionths of a meter up to 50 billionths. The minute diameter and length of these fibers are advantageous in injection molding, permitting greater processing speed with minimal fiber damage. (J. M. Huber Corp).

Fig. 7-20. Rotor driveshaft for helicopter, with end fittings and bearing supplements. (Bell Helicopter Co.)

Carbon and graphite fibers may exceed glass in strength. They are finding many uses as self-lubricating materials, heat-resistant reentry bodies, blades for turbines and helicopters, and valve-packing compounds. Figure 7-21 shows carbon and glass fibers used together to reinforce an injection-molded nylon racquetball racquet.

Ceramic fibers have high tensile strengths and low thermal expansion. Some fibers may reach a tensile strength of 14 GPa (2 030 750 psi). Present applications for ceramic fibers included dental fillings, special electronics, and spacecraft research. (See tensile strength of whisker fibers in Figure 7-22.)

Metal fibers. Steel, aluminum, and other metals are drawn into continuous filaments. They do not compare with the strength, density, and other properties exhibited by other fibers. Metal fibers are used for added strength, heat transfer, and electrical conductivity.

Hybrid fibers. Hybrid fibers refer to a special available form of fibers. Two or more fibers may be combined (hybridization) to tailor the reinforcement to the needs of the designer. Hybrid fibers provide diverse properties and many possible material combinations. They can maximize performance, minimize cost, or improve any deficiency of the other fiber component (a synergistic effect). Glass and carbon fibers are used together to increase impact strength and toughness; prevent galvanic action; and reduce cost of a 100 percent carbon composite. When these fibers are placed in a matrix, the composite—not the fibers—is called a hybrid. A composite of metal foils or metal composite plies stacked in a specified orientation and sequence is called a super hybrid. (See Laminar)

Properties of the most commonly used fiber reinforcing agents are shown in Table 7-6.

Fig. 7-21. A combination of glass fiber and carbon fiber is used to reinforce this injection-molded nylon racquet-ball racquet.

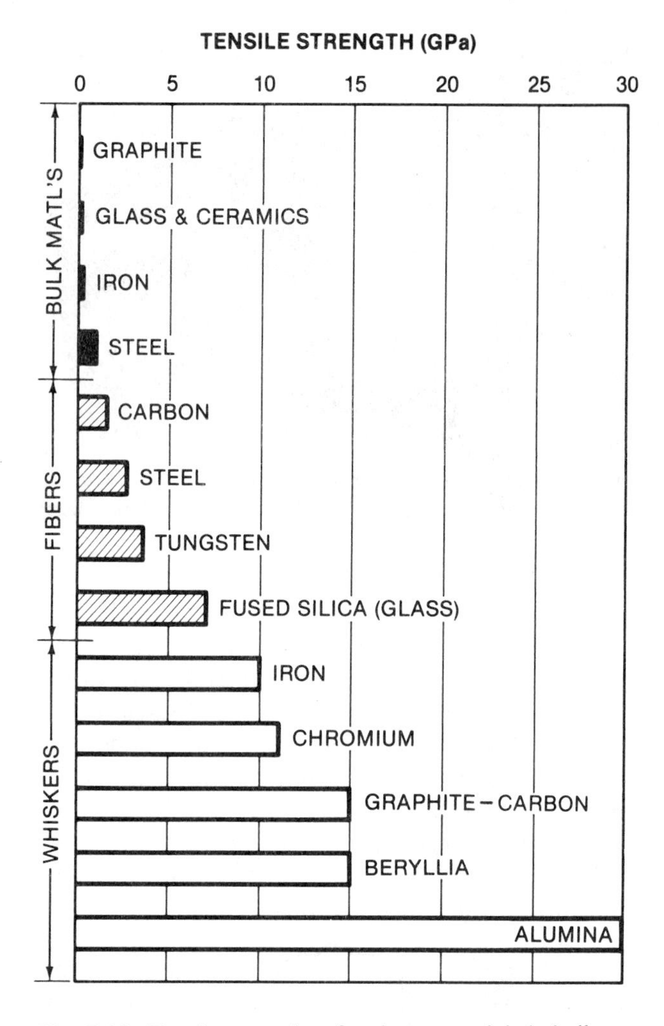

Fig. 7-22. Tensile strengths of various materials in bulk, fiber, and whisker forms.

Fillers

The term *filler* is often rather confusing. Filler was originally selected to describe any additive used to *fill* space in the polymer and lower cost. Because some fillers are more expensive than the polymer matrix, the word extender can be misleading. The terms *dilutant* and *enhancer* are sometimes used to describe the addition of fillers. Ambiguity of terms and overlap of function add to the problem. In this book, the term filler will mean any minute particle from various sources, functions, composition, and morphology. Fillers can be saucer-, sphere-, needle-shaped, or irregular in shape. (Fig. 7-23)

According to the ASTM, a filler is a relatively inert material added to a plastic to modify its strength, permanence, working properties or other qualities, or to lower costs.

Fillers may be either organic or inorganic ingredients of plastics or resins. They may increase bulk or viscosity, replace more costly ingredients, reduce mold

Table 7-6. Properties of the most Commonly Used Fiber Reinforcing Agents (Metallic and Nonmetallic)

Fiber	Relative Density	Tensile Strength: Ultimate (MPa)	Tensile Modulus of Elasticity: Modulus (GPa)
Aluminum	2.70	620	73.0
Aluminum oxide	3.97	689	323.0
Aluminum silica	3.90	4130	100.0
Aramid (Kevlar 49)	1.4	276	131.0
Asbestos	2.50	1380	172.0
Beryllium	1.84	1310	303.0
Beryllium carbide	2.44	1030	310.0
Beryllium oxide	3.03	517	352.0
Boron-Tungsten boride	2.30	3450	441.0
Carbon	1.76	2760	200.0
Glass, E-glass	2.54	3450	72.0
S-glass	2.49	4820	85.0
Graphite	1.50	2760	345.0
Molybdenum	10.20	1380	358.0
Polyamide	1.14	827	2.8
Polyester	1.40	689	4.1
Quartz (fused silica)	2.20	900	70.0
Steel	7.87	4130	200.0
Tantalum	16.60	620	193.0
Titanium	4.72	1930	114.0
Tungsten	19.30	4270	400.0

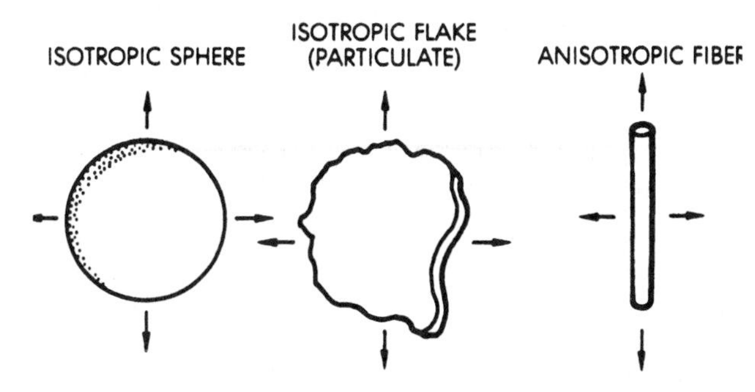

Fig. 7-23. Spheres are isotropic but have no aspect ratio. Isotropic particulates have uniform mechanical properties in the plane of the flake. Fibers have lower aspect ratios but are anisotropic.

shrinkage, and improve the physical properties of the composite item. The size and shape of the filler greatly influence the composite. The *aspect ratio* of a filler is the ratio of length to width. Flakes or fibers have aspect ratios which make them resist movement or realignment; thus, they improve strength. Spheres have no aspect ratio and produce composites with isotropic properties. Metallic flakes are used in particulate composites to form an electrical barrier or layer in the polymer matrix. The main types of fillers and their functions are shown in Table 7-7.

Fillers may improve processability, product appearance, and other factors. One filler is wood flour, which is obtained by grinding waste wood stock into a fine granular state. This powdered filler is often added to phenolic resins to reduce brittleness and resin cost, and improve product finish.

In most molding operations, the volume of fillers does not exceed 40 percent; but as little as 10 percent resin may be used for molding large desk tops, trays, and particle boards. In the foundry industry, as little as 3 percent resin is used to adhere sand together in the shell-molding process.

One product, called cultured marble, uses inorganic marble dust and polyester resin to produce items that have the appearance of genuine marble. This product has the advantage of being stain resistant, and it is easily produced in many colors, shapes, or sizes.

Many organic fillers cannot withstand high temperatures, that is, they have low heat resistance. To improve heat resistance, a silica filler is used such as sand, quartz, tripoli, and diatomaceous earth.

Diatomaceous earth consists of the fossilized remains of microscopic organisms (diatoms). This filler provides improved compressive strength in rigid polyurethane foam.

A filler as fine as cigarette smoke (0.007 to 0.050 μm) is fumed silica. This submicroscopic silica is added to resins to achieve thixotropy. *Thixotropy* is a state of a material that is gel-like at rest but fluid when agitated. Other thixotropic fillers may be made from very fine powders of polyvinyl chloride, china clay, alumina, calcium carbonate and other silicates. Cab-O-Sil and Sylodex are trade names for two commercial thixotropic fillers. Thixotropic fillers may be added to either thermosetting or thermoplastic resins. They are

Table 7-7. Principle Types of Fillers

Filler	Function														
	Bulk	Processibility	Thermal Resistance	Electrical Resistance	Stiffness	Chemical Resistance	Hardness	Reinforcement	Electrical Conductivity	Thermal Conductivity	Lubricity	Moisture Resistance	Impact Strength	Tensile Strength	Dimensional Stability
Organic															
Wood flour	x	x												x	x
Shell flour	x	x										x		x	x
Alpha cellulose (wood pulp)	x			x	x									x	
Sisal fibers	x			x	x	x	x	x				x	x	x	x
Macerated paper	x			x									x		
Macerated fabric	x				x								x		
Lignin	x	x													
Keratin (feathers, hair)	x				x								x		
Chopped rayon		x	x	x		x	x	x				x	x	x	x
Chopped Nylon		x	x	x	x	x	x	x			x		x	x	x
Chopped Orlon		x	x	x	x	x	x	x				x	x	x	x
Powdered coal	x		x		x	x						x			
Inorganic															
Mica	x		x	x	x	x	x				x	x			x
Quartz			x	x	x		x					x	x		
Glass flakes		x	x	x	x	x	x	x			x	x	x		
Chopped glass fibers			x	x	x	x	x	x			x	x	x	x	
Milled glass fibers	x	x	x	x	x	x	x	x			x	x	x	x	x
Diatomaceous earth	x	x	x	x	x		x						x		x
Clay	x	x	x	x	x		x						x		x
Calcium silicate		x	x		x		x					x	x		x
Calcium carbonate		x	x		x		x								
Alumina trihydrate		x		x	x		x					x			
Aluminum powder					x		x	x	x	x			x		
Bronze powder					x		x	x	x	x			x		
Talc	x	x	x	x	x	x	x			x		x			x

This table does not indicate the degree of improvement of function. The prime function will also vary between thermosetting and thermoplastic resins. This table is to be used only as a guide to the selection of fillers.

used to thicken the resin, improve strength, suspend other additives, improve the flow properties of powders, and decrease costs. (See Figure 7-24.)

Thixotropic fillers increase viscosity (internal resistance to flow), and thus are desirable in paints, adhesives, and other compounds applied to vertical surfaces. These fillers may be added to polyester resin in fabrication of inclined or vertical surfaces. Thixotropic fillers can also act as emulsifiers to prevent separation of two or more liquids. Both oil- and water-based additives may be added and held in an emulsified state.

Fillers such as steel, brass, graphite, and aluminum are added to resins to produce electrically conductive moldings, or to give added strength. Plastics with these fillers may be electroplated. Plastics containing powdered lead are used as neutron and gamma-ray shields.

Wax, graphite, brass, or glass are sometimes added to provide self-lubricating qualities to plastics gears, bearings, and slides.

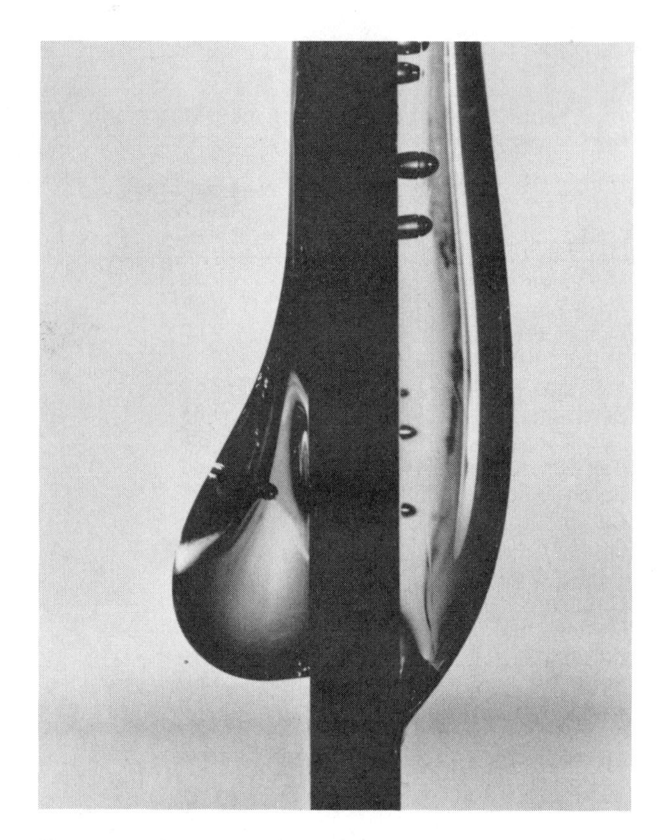

Fig. 7-24. The resin on the right contains a thixotropic agent. The one on the left will sag and run without this ingredient. (Cabot Corp.)

Glass, especially, is used in plastics for several reasons. It is easily added, relatively inexpensive, improves physical properties, and may be colored. Colored glass has optical advantages (especially color stability) over other chemical colorants. Tiny, hollow glass spheres called *microballons* (also called microspheres) are used as a filler in producing low-density composites.

Vocabulary

The following vocabulary words are found in this chapter. Look up the definition of any of these words you do not understand as they apply to plastics, in the glossary, Appendix A.

Antiblocking
Antioxidant
Antistatic
Aspect ratio
Carbonaceous
Catalysts
Char barrier
Colorants
Coupling agents
Curing agents

Exothermic
Filler
Flame retardants
Fluorescence
Foaming/blowing agents
Glass, type C
Glass, type E
Impact modifiers
Inhibitor
Initiator
Lamina
Lubrication bloom
Luminescence
Microballoons (microspheres)
Phosphorescence
Plasticizer
Promoters
Reinforcements
Roving
Stabilizer
Strand
Thixotropy
Whiskers
Yarn

Questions

7-1. To improve or extend the properties of plastics, ___?___ are used.

7-2. Static charges may be dissipated by the addition of ___?___ agents.

7-3. Name four types of colorants.

7-4. The major difference between dyes and pigments is that dyes ___?___ in the plastics material.

7-5. A major disadvantage of organic dyes is their poor ___?___ and ___?___ stability.

7-6. Polyvinyl chloride commonly has plasticizers added to provide ___?___.

7-7. What name is given to chemical-resistant glass fibers.

7-8. Single crystals used as reinforcements are called ___?___.

7-9. Thixotropic fillers increase ___?___.

7-10. One of the most widely used plasticizers is ___?___.

7-11. Name the type of light that is destructive to plastics.

7-12. Polymerization is initiated by the use of ___?___.

7-13. Chemical ___?___ are sometimes used to produce cellular plastics.

7-14. Identify the special stabilizers that retard or inhibit degradation through oxidation.

7-15. Radical initiators such as ___?___ are curing agents used to cure unsaturated polyester.

7-16. Name three functions fillers provide in plastics.

7-17. A disadvantage of ___?___ fillers is that they can not withstand high processing temperatures.

7-18. Long fibrous ingredients that increase strength, impact resistance and stiffness are called ___?___.

7-19. Name four reasons why glass fiber is often selected as a reinforcement.

7-20. Long rope-like strands of fibrous glass are called ___?___.

7-21. To lower viscosity and aid in processing, ___?___ and ___?___ are added to resins.

7-22. Zinc stearate is a ___?___ additive that acts as a lubricant in molding.

7-23. Small particles that contribute only slightly to strength are called ___?___.

7-24. To help prevent discoloration and decomposition of resins and plastics ___?___ are added.

7-25. Crazing, chalking, color changes and loss of properties may be caused by ___?___.

7-26. Name the four types of colorants used in plastics.

7-27. When chemical, electrical or light energy excites electrons, the radiation of light from the plastics is called ___?___.

7-28. Lubrication bloom is caused by excess ___?___.

7-29. Because the color is distributed throughout the product, ___?___ plastics are superior to painted plastics.

7-30. Which additive will produce the stronger composite, carbon black or graphite fibers?

7-31. Which is stronger, a composite made by curing SMC or a filament wound structure?

7-32. Pre-promoted polyester resins should not be stored for a period of time, in a ___?___ container or in a ___?___ place.

Activities

Antioxidants

Antioxidants are chemicals that reduce oxidative degradation of plastics. Without antioxidants, many common plastics could not remain useful for a long period of time.

Equipment. Polypropylene reactor flake, natural PP pellets, injection molding grade, any injection molding equipment, an oven (preferably a air-recirculating type).

7-1. Make several parts by injecting any shape or object. Use PP reactor flake, which should contain no additives, fillers, or reinforcements. Also use pelletized PP, which contains some antioxidant.

7-2. Hang parts in oven at 180 °C. Don't use wire or metal clips to hang parts, because the metal may promote degradation. Monofilament fishing line may work if parts are not very heavy.

7-3. Leave parts in oven until some degradation is apparent on the parts made from reactor flake. Remove those parts at regular time intervals to document the progression of the degradation. Leave the last one or two parts until they are fully degraded, and almost crumble when handled. Continue to expose parts made with pelletized material to heat. They should withstand the heat longer before the onset of degradation.

7-4. Write a report summarizing findings.

7-5. To speed up this test, mold PP pellets and reactor flake into thin sheets. Try to produce sheets less than 1 mm thick. Expose samples to heat as in step 7-2. Noticeable degradation should occur in a few hours. Figure 7-25 shows a sample that exhibited degradation after 4 hours at 180 °C.

Fig. 7-25. This piece of unstabilized polypropylene film exhibited brittle fracture after 2 hours at 180° C.

Coupling Agents

Coupling agents promote adhesion between a plastics material and fillers or reinforcements. Fiberous reinforcements can not provide their strength if they easily slip out of the plastics matrix.

Equipment. Natural PP, extrusion or injection molding grade, two types of chopped glass fibers—one type compatible with styrene, and another type compatible with PP, an extruder, an injection molder, and a tensile tester.

Procedure

7-1. Compound a batch of PP with a selected loading of PS compatible glass fibers. Determine loading percent to make it feasible for the extruder available. If the extruder does not handle glass well, keep loadings less than 10 percent. Compound a batch of material with the same loading of glass compatible with PP. If the glass fibers are not well dispersed after the first pass through the extruder, run both types through the extruder a second, and perhaps a third time.

7-2. Injection mold parts. If available, use a mold that creates tensile test "dog-bones." Pull parts on tensile tester. Determine strengths and elongations at yield and break. Calculate means and deviations, sketch curves comparing PS compatible glass to PP compatible glass.

7-3. Write a report summarizing findings.

Chapter 8

Characterization and Selection of Commercial Plastics

Introduction

Every new plastics product represents a host of decisions made by designers, engineers, processors and marketing specialists. They determine the shape, color, function, strength, style, appearance, and reliability of the product. One aspect of new product development involves identifying the material to use. The purpose of this chapter is to provide an introduction to several facets of material characterization and selection. The content outline for this chapter is:

I. Basic materials
   A. Polymerization types
   B. Melt index
II. Selection of material grade
III. Computerized databases for material selection

Basic Materials

Picking between ABS, PC, PS, or PP is often not simple. In addition to cost factors, the decision involves strength and flexural characteristics, use-environment, and surface appearance. In a few cases, the choices are well established.

A thermoformed packaging container for a round, decorated cake does not need to be air tight or have oxygen barrier properties. It does not need resistance to ultraviolet radiation. It does need to protect the decorations in the frosting and allow some stacking. It will usually consist of polystyrene which is both transparent and inexpensive. Although polystyrene is rather brittle, and will probably crack during use, it will often withstand the two or three openings required in its service life.

Plastics lenses for glasses have much more rigorous requirements. They must be highly transparent, scratch and impact resistant. They must be moldable, and yet easily worked with sanding, grinding, and polishing equipment. A typical choice of material is polycarbonate. It will exceed most impact requirements, does not shatter, and can be rather scratch resistant.

However, many choices of the basic plastics material are not obvious. Improvements in the commodity resins have made selections even more difficult. For example, reinforced polypropylene has found use in some applications previously reserved for engineering materials such as nylon.

It is far beyond the scope of this chapter to enumerate common uses for a host of plastics. Appendix E and F list many examples of applications for common plastics. However, there are some considerations that reoccur in material selection processes. One consideration is understanding the characteristics of the basic plastics material.

Determining the basic plastics is not equal to material selection because, although some commercial products use natural, unmodified, and unreinforced plastics, such products are never 100 percent polymer. They always contain some additives. A key to understanding various plastics is the melt flow index (MFI), often called the melt index. (See Chapter 6.) In order to clearly understand the importance of the MFI, an introduction to techniques of polymerization is helpful.

Polymerization Techniques

In polymerization, tiny hydrocarbon molecules (mers) combine to form huge molecules, often called macromolecules. For this to occur efficiently in a chemical reactor, the monomer needs to be in a form that provides high surface area and low volume. Four differing processes ensure a high ratio between surface area and volume, namely bulk, solution, suspension, and emulsion polymerization. In bulk polymerization, narrow, tubular reactors guarantee the ratio. Solution polymerization requires small drops of monomer added to a large bath of solvent. For example, in polymerizing polystyrene from styrene monomer, the monomer, approximately 20 percent by weight, is dissolved in about 80 percent benzene. To avoid the use of solvents, emulsion and suspension polymerization processes use water to surround the tiny droplet of monomer. Close examination of a polymer that is in the form of reactor flake or sphere can provide a feel for the small physical scale in which the polymerization reactions take place.

Regardless of the process, polymerization proceeds according to two major types of chemical reactions, chain growth (sometimes called addition polymerization) or stepwise growth (also called condensation polymerization).

In chaining, polymerization begins at one location by the action of a chemical initiator. Almost instantaneously, the complete chain forms without yielding chemical by-products. The image of a train forming when hundreds of box cars join together is appropriate. When complete, the train leaves behind no excess parts. The chain stops growing in length due to the effect of chemicals which cause termination of the chain. Termination is also effected by probability, monomer purity and type.

In stepwise polymerization, monomers combine to form blocks two units long. These in-turn combine to form blocks four units long. This process continues until the process is terminated. Stepwise polymerization usually requires some chemical alteration of the monomer. This results in by-products. If the by-products are not removed continuously, they will slow or inhibit the polymerization process. The most common by-products are water, acetic acid and hydrogen chloride. For example, in reaction-injection molding, some materials polymerize by the stepwise reaction, producing water within the mold. Molds for this manufacturing process must be plated with nickel to avoid continuous rusting.

In both chain polymerization and stepwise polymerization, the process is not perfect. Some molecules grow beyond the desired length, some are too short. This results in a distribution of molecular lengths. Chapter 5 provided instruction in calculating the mean and standard deviation for a distribution of normally distributed values. Distributions of polymer molecules

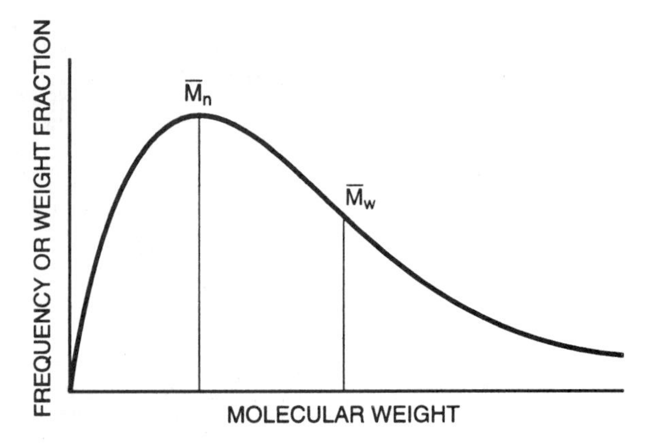

Fig. 8-1. This distribution of molecular weights is typical for the polymers in many commercial plastics.

are more complex, and cannot be fully described with only the mean and standard deviation. The reason for this is that most distributions of molecular lengths are not *normal*. In contrast, they tend to have the shape shown in Figure 8-1.

The shape of this curve (Figure 8-1) is very important. As with a normal distribution, this curve has a peak value, identified as M_n. In contrast to normal distributions, this curve is not symmetrical. It has a long *tail* extending toward the right. That means that there are some very long molecules in the polymer. If these long molecules were rather unimportant, it might be reasonable to pretend that the distribution was normal. However, the long molecules are very significant because they alter the physical properties of the material.

To achieve a more adequate characterization of the distribution of molecular weights in a polymer, it is necessary to use both the number average molecular weight and the weight average molecular weight. The *number average molecular weight* is based on the frequency of various molecular lengths in a sample. It implies that the short molecules are *just as important* as the long ones. In contrast, the *weight average molecular weight* counts not the frequency of various molecular lengths, but the contribution of the various molecules to the total weight of the sample. This approach gives the longer molecules greater importance than only counting their frequency.

The ratio of the M_w to M_n is the polydispersity index (PI). This number is somewhat similar to a standard deviation for a normal distribution. Both the standard deviation and the PI indicate the spread of the values in the distribution. A PI of 1.0 is a theoretically perfect polymer in which all the molecules are of exactly the same length. As the PI value increases, the difference between the shortest and longest molecules in a sample increases. Commercial polymers range from a PI of about 2 to 40.

Determining number average molecular weight and weight average molecular weight is not easy. It

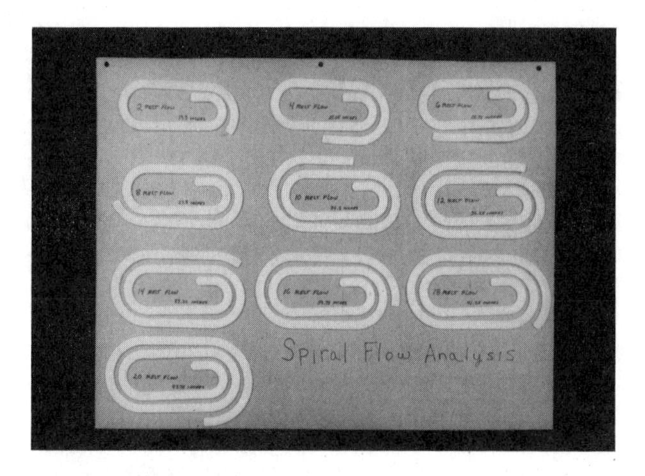

Fig. 8-2. Results of sprial flow analysis of polypropylene homopolymer.

requires special instruments and trained personnel. Consequently, many manufacturing companies turn to melt index values for a "quick and dirty" estimation of the average molecular lengths in a sample.

Melt Index

Figure 8-2 shows the influence of various melt flow rates on spiral flow. Spiral-flow molds for injection molding have very long cavities, so that a *shot* of plastics will *freeze-up* long before it reaches the end of the cavity. During spiral flow analysis, the set-up on an injection-molding machine provides uniform temperature for the mold and melted plastics. Set-up also ensures uniform injection pressure and injection speed. Each shot has excess material available, which means that the flow does not stop because the material was used up, but because—given the conditions of molding—it would not flow any further.

The material used for the spiral flow shown in Figure 8-2 was natural polypropylene homopolymer. The differences resulted from selection of differing melt index materials. The shortest flow was a melt index of 2, the longest—a melt index of 20. The melt index values indicate the number of grams of material extruded through a standard orifice in a 10 minute period. The flow lengths for the various melt index values are shown in Table 8-1.

As is apparent, when the melt flow index (MFI) increased, the flow length also increased. The reason the flow length increases is that the average length of the polymer chains decrease as the MFI increased. That means that the higher melt material was "runnier" than the lower melt material. The changes in the MFI are inversely correlated with the average chain lengths (the average molecular weight).

Table 8-2 presents these relationships in data of selected types of polyproplyene homopolymer.

Table 8-1. Relation of MFI to Sprial Flow

MFI	Flow length (inches)
2	19.5
4	23.25
6	25.75
8	27.5
10	34.5
12	36.25
14	37.25
16	39.75
18	41.25
20	43.75

Table 8-2. Relation of MFI to Molecular Weights

Melt Flow Range	M_n	M_w	PI(M_w/M_n)
0.3–0.6	90000	850000	9.5
1–3	65000	580000	9
2.6	61500	375000	6.1
3–5	60000	450000	8
4.6	57000	333000	5.9
5–8	35000	350000	10
7.5	51000	296000	5.8
8.5	50000	305000	6.1
8–16	30000	300000	10

The data in Table 8-2 show that as the MFI increases, both the M_n and M_w decrease. Although estimates of molecular weight based on MFI values are not highly accurate, the general trends are very consistent. Understanding these relations can assist in the process of material selection.

Selection of Material Grade

The selection of a grade of material is also rather complicated. Because several companies produce comparable materials, the choices are numerous. For example, one major petrochemical company offers five grades of high heat polycarbonate; three general-purpose (GP) grades; three flame-retardant grades; two glass-reinforced graded; one extrusion grade; one impact-modified grade; and four specialty grades, which include medical applications, one lighting grade, three optical types, and three types of automotive lens materials. Another major company offers 12 grades of general purpose polycarbonate, three high-flow grades, eight healthcare-product grades, six flame-retardant grades, five glass-reinforced grades, seven wear-resistant grades, seven optical-quality grades, four grades for blow molding, and five grades of high-heat material.

Most petrochemical companies provide MFI values on material data sheets. These numbers imply the usually unstated values for the M_n, M_w and PI. Beside differences in polydispersity and molecular weight, the other differences stem from various fillers and additives. The flame-retardant grades include chemicals that inhibit combustion. High-flow grades often include internal lubricants to promote flow. The search for the optimum material for a specific application can be overwhelming.

Computerized Databases for Material Selection

Because there are hundreds of companies producing plastics and thousands of grades, selecting the *best* material for a product is very difficult. To assist in this task, computerized databases assemble material data. These databases collect and organize material data. They typically include such information as trade name, chemical name, manufacturer name, and most physical properties. The properties include strength, impact, flexural and tensile modules, hardness, filler content, brittleness temp, heat deflection temperature, refractive index, water absorption, melt flow, linear mold shrinkage, and others. The most thorough databases include rheological data, creep curves, aging data, chemical resistance, and weatherability.

There are two basic types of material databases, those produced by software companies, and those created by plastics-producing companies. The software companies gather huge numbers of materials into databases, and then sell compact disks or on-line hookup at a charge of about $1000.00 per year. One such database includes about 17000 thermoplastic materials. The resin manufacturers, such as GE, DuPont, Mobay, and BASF generally focus only on their own materials, and consequently have databases on about 200 to 600 materials.

The advantages of a company-specific database is that it generally includes more information on the physical properties than gathered by the software companies. The smaller databases may also contain more accurate information.

The databases include sort features, which allow a user to search the database for materials that match a set of specified characteristics. The sort will then return to the user a list of materials that fit the conditions.

Inherent in this process is the assumption that direct comparisons between materials are basically accurate. However, the validity of such comparisons rests on the comparability of the data included. In many cases, tests performed to ASTM standards may not be comparable due to varying size and shape of test specimens, varying mold designs, and varying molding conditions. Some of the company-specific databases have moved to ISO tests in the hope of more accurate comparisons. However, even within one plastics-producing company, the same problem arises. Large companies have multiple testing laboratories in various geographic locations. Often, the machines used to mold test samples are from differing manufacturers, the molds are not identical in size or cooling, and the molding conditions are not identical from one location to the next. All of these factors introduce variability into the testing results.

Summary

In material selection, awareness of the characteristics of the basic polymer is critical. The average lengths of molecular chains and the distribution of lengths impact physical properties and flow characteristics. In addition to the characteristics of the polymeric molecules; additives, fillers, reinforcing agents, and colors tailor plastics for specific applications. Correct decisions about materials enhance product life and reliability. Incorrect decisions often lead to product failures.

Vocabulary

Additive polymerization
Bulk polymerization
Chain growth polymerization
Condensation polymerization
Emulsion polymerization
Melt flow index
Number average molecular weight (M_n)
Polydispersity index (PI)
Solution polymerization
Stepwise polymerization
Suspension polymerization
Weight average molecular weight (M_w)

Questions

8-1. Which polymerization processes use water surrounding droplets of monomer?

8-2. How is a high surface area to volume ratio created in bulk polymerization?

8-3. What is a common solvent used in the solvent polymerization of styrene?

8-4. How is chain growth different from stepwise growth in polymerization reactions?

8-5. Which type of polymerization reaction produces by-products?

8-6. What does a polydispersity index of 1.0 mean?

8-7. What is the approximate range of PI values common in commercial plastics?

Activities

Introduction

A practical approach to learning about molecular weight distributions involves investigation of melt mixtures. Melt mixtures are blends of plastics in which the mixing is done after the initial polymerization. Many compounding companies need to provide customers with specific melt flow plastics. To achieve the desired flow rate, compounders mix various percentages of readily available melt flow materials. The difficulty is that predicting a resultant melt flow from the flow rates of the ingredients requires a practical understanding of the differences between number average molecular weights and weight average molecular weights.

Equipment. Compounding extruder, pelletizer, extrusion plastometer or other type of melt flow measurement device, selected plastics, and personal safety equipment.

Procedure

8-1. Acquire two or three natural homopolymers with differing melt flow rates. If possible, get information on the number average molecular weight and weight average molecular weight for each type.

8-2. Use melt index equipment to verify the melt flow of the selected materials. For accuracy, run several tests on each type.

8-3. Determine the relative percentages of the ingredients and weigh them out. For example, select a 2 melt type and a 20 melt type. For a 500 gram batch, use 50 percent (250 g) of the 2 melt material and 50 percent (250 g) of the 20 melt. If the differences in flow rates are rather extreme, the difficulty with prediction will be most obvious.

8-4. Melt mix the two melt flow rates with an extruder, and pelletize the extrudate.

8-5. Predict the melt flow of the "new" plastics. If the predicted value is 11, based on adding 20 and 2, then dividing 22 by 2. Review Figure 8-2. The typical distribution of molecular weights is not symmetrical, but is skewed to the right. Since the weight average molecular weight is also to the right of the peak, its influence will also be skewed.

8-6. Measure the melt flow of the compounded material.

8-7. Use the following technique to create an estimate of the melt flow of the new material.

- **a.** Determine the log of one melt index value. The log of 2 is 0.3.
- **b.** Multiply the percentage of the 2 melt material by the log value ($0.5 \times 0.3 = .15$)
- **c.** Determine the log of the other melt index value and multiply by the percentage. ($0.5 \times 1.3 = .65$)
- **d.** Add the two values and take the inverse log of the sum. Inverse log of 0.8 is 6.3. This 6.3 value should approximate the measured melt index.

8-8. Melt mix unequal percentages, such as 15 percent 2 melt and 85 percent 20 melt. Predict the resulting melt flow and then determine it experimentally.

8-9. Use more than two ingredients. Does the mathematical procedure accurately predict the mixture of 3 or more melt index materials?

8-10. Write a report summarizing the results.

Chapter 9

Machining and Finishing

Introduction

In this chapter, you will learn how plastics and composites are machined and finished. Molded or formed plastics parts often require further processing, including such common operations as flash removal, slot cutting, polishing, and annealing. Many of the operations are similar to those used for the machining and finishing of metal or wood products.

The vast numbers of machines and processes used in the shaping and finishing of plastics do not allow discussion in detail, although certain basics apply to all machining and finishing processes. Additives, fillers, and the plastics of each family require different shaping and finishing techniques. Few plastics pieces are made solely by machining, though many molded parts must be finished or fabricated into useful items.

All machining and finishing operations present potential physical hazards. Fine dusts or particles are produced when sawing, laser cutting, or water-jet cutting. Eye- and face-mask protection should be worn by the operator to prevent injury or inhalation of particles. (See Chapter 4 on Health and Safety)

Processing techniques for plastics are based on those used for wood and metal. Nearly all plastics may be machined (Fig. 9-1). As a rule, thermosets are more abrasive to cutting tools than thermoplastics.

Machining techniques for composite materials such as high-pressure laminates, filament-wound parts, and reinforced plastics, attempt to prevent fraying and delamination of the composite. The reinforcing agents used with various matrix compounds are abrasive. Most cutting tools must be made from tungsten carbide or coated (with titanium diboride, for example). High-speed steel (M2) or diamond-tipped cutters are also used. Boron/epoxy composites are generally cut using diamond-tipped tools. The lower thermal conductivity and the lower modulus of elasticity (softness, flexibility) of most thermoplastics mean that tools should be kept properly sharp to allow them to cut cleanly, without burning, clogging, or causing frictional heat.

Elastic recovery causes drilled or tapped holes to become smaller than the diameter of the drills used, and turned diameters often become larger. The low melting points of some thermoplastic materials tend to make them gum, melt, or craze when machined. Plastics expand more than most materials when heated.

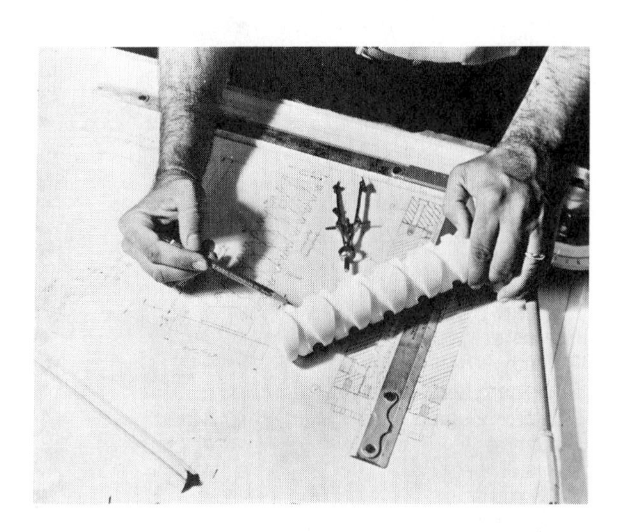

Fig. 9-1. This part was machined from bar stock. (The Polymer Corp.)

The coefficient of thermal expansion for plastics is roughly ten times greater than that of metals. Cooling agents (liquids or air) may be needed to keep the cutting tool clean and free of chips. The benefits of cooling include increased cutting speed, smoother cuts, longer tool life, and elimination of dust. Because the polymer matrix has a high coefficient of expansion, even small variations of temperature may cause dimensional control problems.

Topics covered in this chapter include:

I. Sawing
II. Filing
III. Drilling
IV. Stamping, blanking, and die cutting
V. Tapping and threading
VI. Turning, milling, planing, shaping, and routing
VII. Laser cutting
VIII. Induced fracture cutting
IX. Thermal Cutting
X. Hydrodynamic cutting
XI. Smoothing and polishing
XII. Tumbling
XIII. Annealing and posturing

Sawing

Nearly all types of saws have been adapted to cutting plastics. Backsaws, coping saws, hacksaws, saber saws, hand saws, and jeweler's saws may be used for hobbycraft or short-run cutting. The shape of the tooth is important for proper cutting of plastics.

Circular blades should have plenty of *set* or be *hollow ground.* Blades should have a deep, well-rounded *gullet* (Fig. 9-2A). The *rake* (or hook) angle should be zero (or slightly negative). The *back clearance* should be about 30°. The preferred number of teeth per centimetre varies with the thickness of the material to be cut. Four or more teeth per centimetre should be used for cutting thin materials. Fewer than four teeth per centimetre are needed for plastics over 25 mm [1 in.] thick.

A *skip-tooth* bandsaw blade is preferred (Fig. 9-3A). The wide gullet in this blade provides ample space for plastics chips to be carried out of the *kerf* (cut made by the saw). For best results, the teeth should have zero rake and some set.

Bandsaw blades may be reversed to accomplish a zero or negative rake. Abrasive carbide or diamond grit blades may be used to cut graphite and boron/epoxy composites. In all cutting operations, it is best to back up the work with a solid material to reduce chipping, fraying, and delamination of composites. Table 9–1 suggests the number of teeth per centimetre for various speeds and thicknesses of material.

***Note:** Fewer teeth per centimetre are needed for cutting pastics over 0.23 in. [6 mm] thick. Thin or flexible plastics may be cut with shears or blanking dies. Foams require cutting speeds above 8000 fpm [40 m/s]*

Table 9-1. Power Saws for Cutting Plastics

	Circular Saws			Band Saws		
	Teeth per cm		Speed,	Teeth per cm		Speed, m/s
Plastics	(<6 mm)	(>6 mm)	m/s	(<6 mm)	(>6 mm)	(>6 mm)
Acetal	4	3	40	8	5	7.5–9
Acrylic	3	2	15	6	3	10–20
ABS	4	3	20	4	3	5–15
Cellulose acetate	4	3	15	4	2	7.5–15
Diallyl phthalate	6	4	12.5	10	5	10–12.5
Epoxy	6	4	15	10	5	7.5–10
Ionomer	6	4	30	4	3	7.5–10
Melamine-formaldehyde	6	4	25	10	5	12.5–22.5
Phenol-formaldehyde	6	4	15	10	5	7.5–15
Polyallomer	4	3	45	3	2	5–7.5
Polyamide	6	4	25	3	2	5–7.5
Polycarbonate	4	3	40	3	2	7.5–10
Polyester	6	4	25	10	5	15–20
Polyethylene	6	4	45	3	2	7.5–10
Polyphenylene oxide	6	4	25	3	2	10–15
Polypropylene	6	4	45	3	2	7.5–10
Polystyrene	4	3	10	10	5	10–12.5
Polysulfone	4	3	15	5	3	10–15
Polyurethane	4	3	20	3	2	7.5–10
Polyvinyl chloride	4	3	15	5	3	10–15
Tetrafluoroethylene	4	3	40	4	3	7.5–10

See Appendix G. Useful Tables.

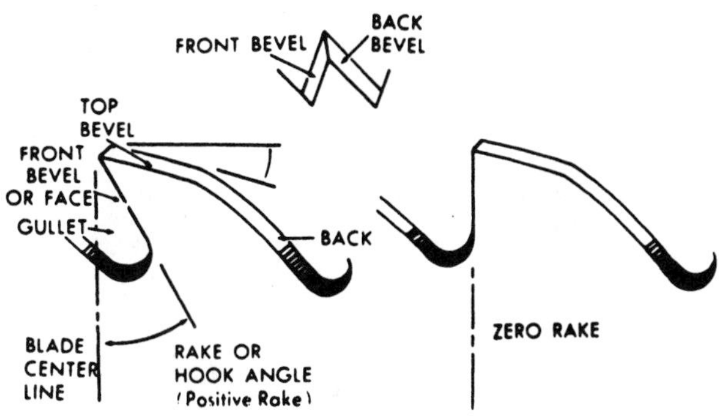

(A) Parts of the circular saw tooth.

(B) Zero rake of saw tooth.

(C) Zero rake where line of tooth face crosses center of blade.

(D) Negative rake.

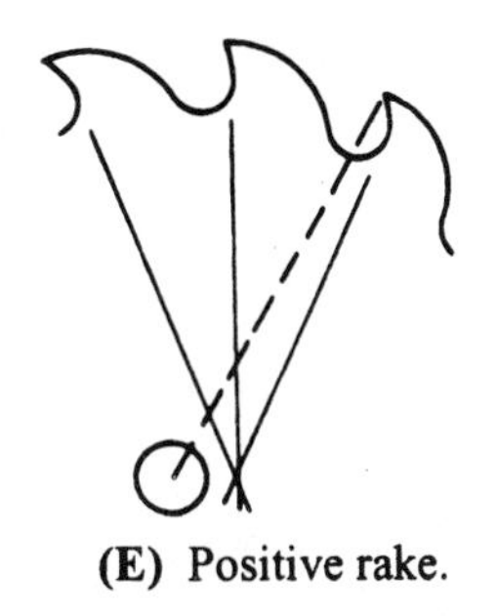

(E) Positive rake.

Fig. 9-2. Tooth characteristics of a circular saw blade.

For cutting reinforced or filled plastics, and for many thermosetting plastics, carbide-tipped blades are recommended (Fig. 9-4). They provide accurate cuts and long blade life. Abrasive-or diamond-tipped blades may be used also. A liquid coolant is advised to prevent clogging or overheating, although jets of CO_2 are often used as a coolant while thermoplastics are being machined. All cutting tools must have protective shields and safety devices.

The feeds and speeds for cutting composites vary greatly with thickness and material, but are similar to those for nonferrous materials. (See Table 9–2.)

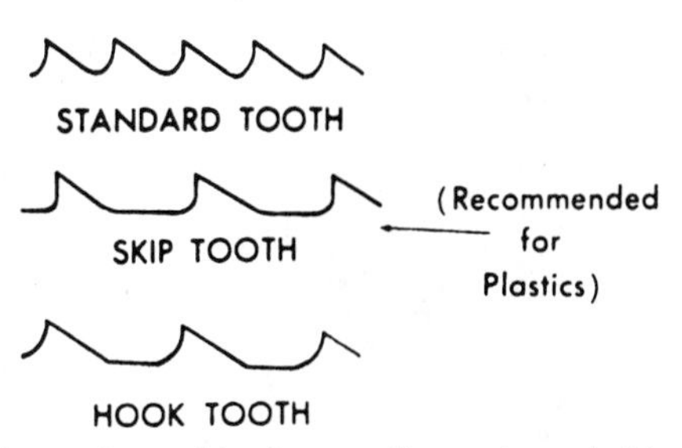

(A) Skip tooth provides large gullet and good chip clearance. The hook tooth is sometimes preferred for glass-filled thermosets.

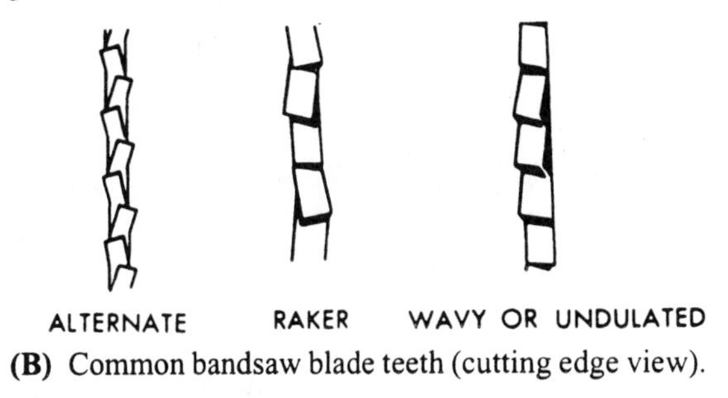

(B) Common bandsaw blade teeth (cutting edge view).

Fig. 9-3. Bandsaw blade teeth.

Filing

Thermosetting plastics are quite hard and brittle, therefore filing removes material in the form of a light powder. Aluminum type A, sheartooth, or other files that have coarse, single-cut teeth with an angle of 45° are preferred (Fig. 9-5). The deep-angled teeth help the file to clear itself of plastics chips. Many thermoplastics tend to clog files. Curved-tooth files like those used in auto body shops are good because they clear themselves of plastics chips. Special files designed for plastics should be kept clean and not used for filing metals.

Drilling

Thermoplastic and thermosetting materials may be drilled with any standard twist drill; however, special drills designed for plastics produce better results. Carbide-tipped drills will give long life. Holes drilled in most thermoplastics and some thermosets are usually 0.002 to 0.004 in. [0.05 to 0.10 mm] undersize. Thus, a 0.23 in. [6 mm] drill will not produce a hole wide enough for a 6 mm rod. Thermoplastics may need a coolant to reduce frictional heat and gumming during drilling.

For most plastics, drills should be ground with a 70° to 120° point angle and a 10° to 25° lip-clearance

(A) Diamond cutoff used to cut boron-epoxy reinforced tube. (Advanced Structures Division, TRE Corp.)

(B) This machine skives (pares) sheets of plastics from slab stock. (McNeil Akron Corp.)

(C) Space-age composites like Kevlar are quickly and cleanly cut by water jet. (Flow Systems Inc.)

Fig. 9-4. Various methods of cutting plastics.

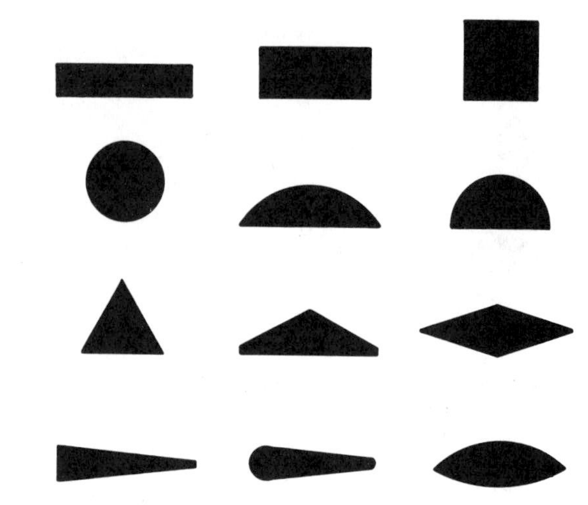

(A) Various file profiles.

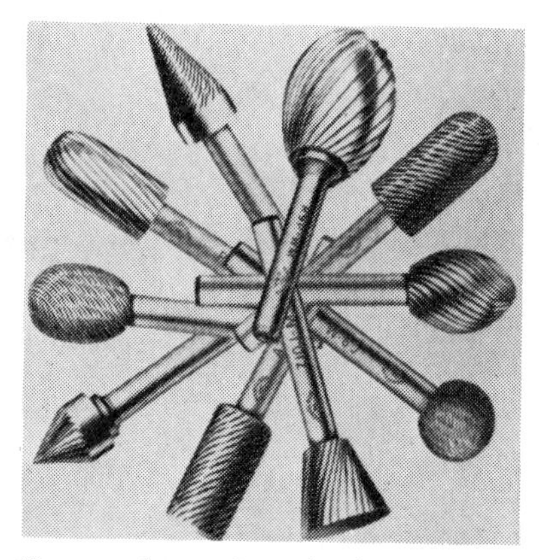

(B) Rotary files may be used on plastics.

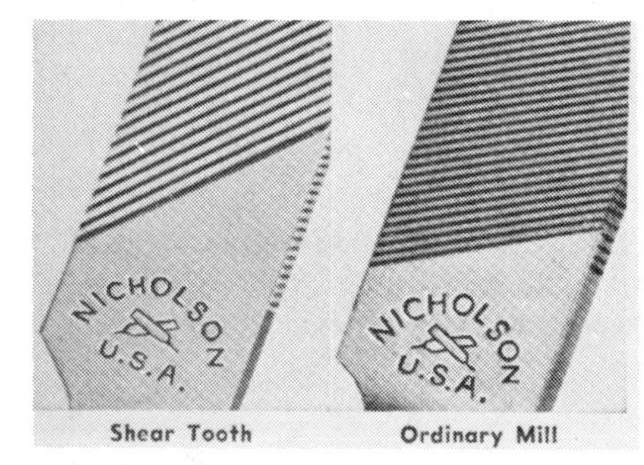

(C) Comparison of shear-tooth and ordinary mill files.

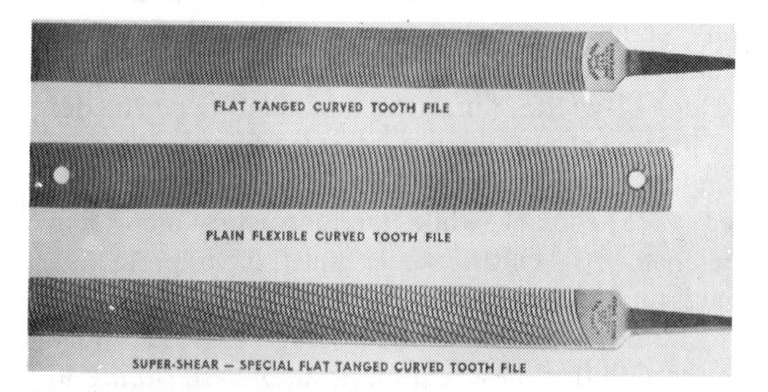

(D) Curved-tooth files used on plastics.

Fig. 9-5. Files used on plastics.

Table 9-2. Machining Composites

Operation	Material	Cutting Tool	Speeds	Feeds (<0.250 thick)
Drilling	Gl-Pe	0.250-Diamond	20000 RPM	0.002/rev
	B-Ep	0.250-Diamond core (60–120 grit)	100 SFPM	0.002/rev
	B-Ep	0.250 2–4 flute (HSS)	25 SFPM	0.002/rev
	Kv-Ep	Spade-carbide	>25000 RPM	0.002/rev
	Kv-Ep	Brad point-carbide	>6000 RPM	0.002/rev
	Gl-Ep	Tungsten Carbide	<2000 RPM	<0.5 ipm
	Gr-Ep	Tungsten Carbide	>5000 RPM	<0.5 ipm
Band Saw	Kv-Ep	14 teeth, honed saber	3000–6000 SFPM	<30 ipm
	B-Ep	Carbide or 60 gut Diamond	2000–5000 SFPM	<30 ipm
	Hybrids		3000–6000 SFPM	<30 ipm
	G-Pe	14 teeth, honed saber	3000–6000 SFPM	<30 ipm
Milling	most	Four-flute carbide	300–800 SFPM	<10 ipm
Circular Saw	Gr-Ep, B-Ep	60 grit diamond	6000 SFPM	<30 ipm
	Gl-Pe	60 tooth carbide or	5000 SFPM	<30 ipm
		60 grit diamond	5000 SFPM	<30 ipm
Lathe	Kv-Ep	carbide	250–300 SFPM	0.002/rev
	Gl-Pe	carbide	300–600 SFPM	0.002/rev
Shears	Kv-Ep	HSS or Carbide	—	<30 ipm
Countersink or counterbore	most	Diamond grit or carbide	20000 RPM 6000 RPM	<0.5 ipm
Laser (10kW)	most <0.250 thickness	CO_2 cooling	—	<30 ipm, depending upon material
Water jet	most <0.250 thickness	60000 psi–0.10 in orifice	—	<30 ipm, depending upon material
Abrasive (sanding-grinding)	most	Silicone carbide or alumina grit (wet)	4000 SFPM	—
Router	most	Carbide or Diamond grit	20000 RPM	—

angle (Fig. 9-6). The rake angle on the cutting edge should be zero or several degrees negative. Table 9–3 gives rake and point angles for many plastics materials. Highly polished, large, slowly twisting flutes (large helix or helix rake angle) are desirable for good chip removal.

Four factors that affect cutting speeds are as follows:

1. type of plastics
2. tool geometry
3. lubricant or coolant
4. feed and depth of cut

The cutting speed of plastics is given in surface feet per minute (fpm) or metres per second (m/s). Metres per second refers to the distance the cutting edge of the drill travels in one second when measured on the circumference of the cutting tool. The following formula

Table 9-3. Drill Geometry

Material	Rake Angle	Point Angle	Clearance	Rake
Thermoplastic				
Polyethylene	10°–20°	70°–90°	9°–15°	0°
Rigid polyvinyl chloride	25°	120°	9°–15°	0°
Acrylic (polymethyl methacrylate)	25°	120°	12°–20°	0°
Polystyrene	40°–50°	60°–90°	12°–15°	0° to neg. 5°
Polyamicle resin	17°	70°–90°	9°–15°	0°
Polycarbonate	25°	80°–90°	10°–15°	0°
Acetai resin	10°–20°	60°–90°	10°–15°	0°
Fluorocarbon TFE	10°–20°	70°–90°	91–15°	0°
Thermosetting				
Paper or cotton base	25°	90°–120°	10°–15°	0°
Fibrous glass or other fillers	25°	90°–120°	10°–15°	0°

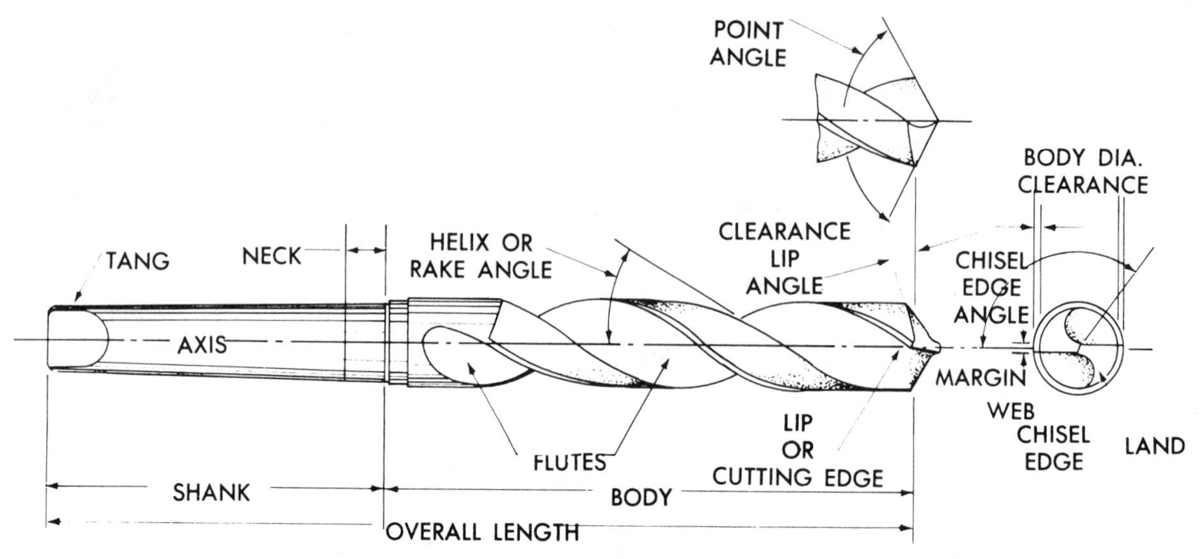

(A) Selected nomenclature for tapered-shank twist drill. Drills of 1/2 inch (12.5 mm) or less diameter usually have straight shanks. (Morse Twist Drill Machine Co.)

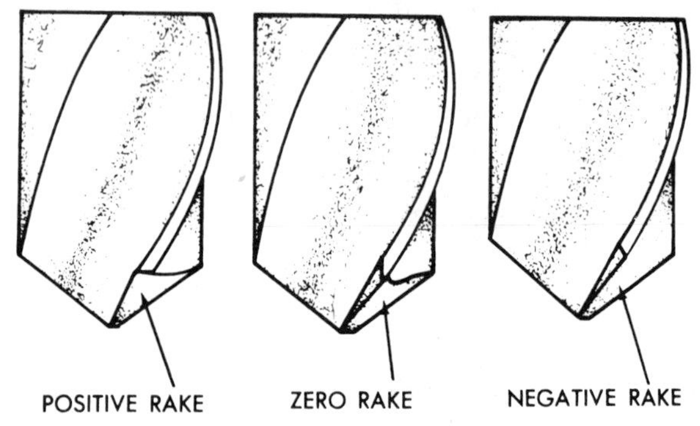

(B) A zero rake is usually preferred for plastics, but a negative rake is sometimes used with polystyrene.

Fig. 9-6. Drill nomenclature

is used to determine surface metres per second. This information may be obtained from handbooks:

$$m/s = \pi D \times rpm$$

$$r/s = (m/s) \div (\pi D)$$

where rpm = revolutions per minute

r/s = revolutions per second

m/s = surface metres per second

D = diameter of cutting tool in metres

$\pi = 3.14$

As a rule of thumb; plastics have a cutting speed of 200 fpm [1 m/s]. A guide for drilling thermoplastics and thermosets is shown in Table 9–4.

The rate at which the drill or cutting tool moves into the plastics is crucial. The distance the tool is fed into the work each revolution is called the *feed.* Feed is measured in inches or millimetres. Drill feed ranges from 0.001 to 0.0031 in. [0.25 to 0.8 mm] for most plastics, depending on thickness of the material. (Table 9-5)

Many of these same principles of speed and feed apply to reaming, countersinking, spotfacing, and counterboring (Fig. 9-7). In many thermoplastics, holes may be made with hollow punches. Warming the stock may help the punching operation.

Sintered-diamond core drills, countersinks, reamers, or counterbores may be used with ultrasonic energy to bore into some composites. Boron/epoxy, graphite/boron/epoxy, and other hybrid composite materials may require ultrasonic techniques.

Stamping, Blanking, and Die Cutting

Many thermoplastics and thin pieces of thermosets may be cut using rule, blanking, piercing, or matched

Table 9-4. Guide to Speeds for Drilling Plastics

Drill Size	Speed for Thermoplastics, r/s	Speed for Thermosets, r/s
No. 33 and smaller	85	85
No. 17 through 32	50	40
No. 1 through 16	40	28
1.5 mm	85	85
3 mm	50	50
5 mm	40	40
6 mm	28	28
8 mm	28	20
9.5 mm	20	16
11 mm	16	10
12.5 mm	16	10
A–C	40	28
D–O	20	20
P–Z	20	16

Table 9-5. Drilling Feeds of Plastics

Material	Speed, m/s	Feed, mm/revolution					
		1.5	3	6	12.5	19	25
Thermoplastics							
Polyethylene Polypropylene TFE fluorocarbon	0.75–1.0	0.05	0.08	0.13	0.25	0.38	0.5
Butyrate High impact styrene Acrylonitrile-butadiene-styrene	0.75–1.0	0.05	0.1	0.13	0.15	0.15	02
Modified acrylic Nylon Acetals Polycarbonate	0.75–1.0	0.05	0.08	0.13	0.2	0.25	0.3
Acrylics	0.75–1.0	0.02	0.05	0.1	0.2	0.25	0.3
Polystyrenes	0.75–1.0	0.02	0.05	0.08	0.1	0.13	0.15
Thermosets							
Paper or cotton base	1.0–2.0	0.05	0.08	0.13	0.15	0.25	0.3
Homopolymers	0.75–1.5	0.05	0.08	0.1	0.15	0.25	0.3
Fiber glass, graphitized, and asbestos base	1.0–1.25	0.05	0.08	0.13	0.2	0.25	0.3

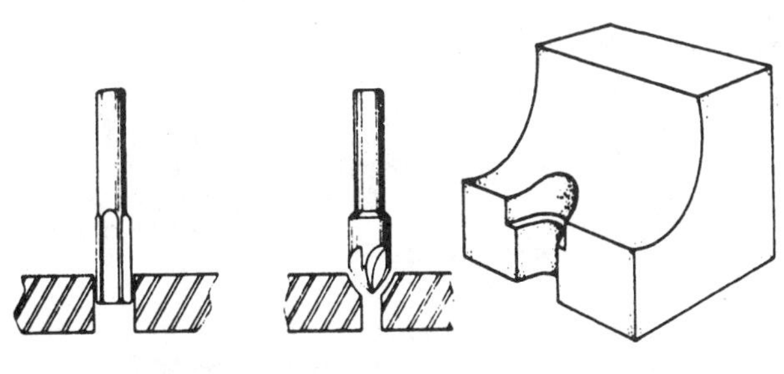

(A) Reaming. (B) Countersinking. (C) Spotfacing.

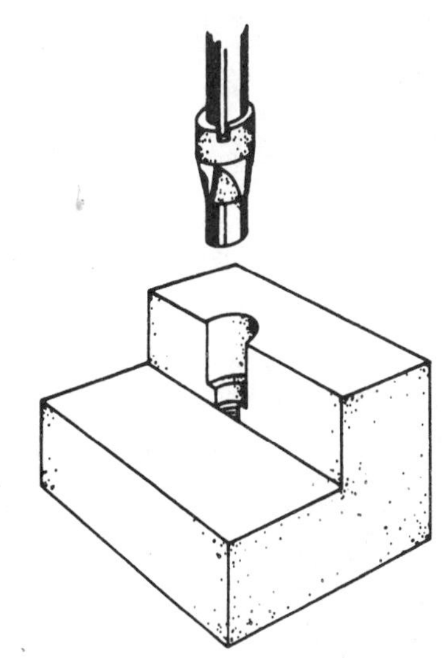

(D) Counterboring.

Fig. 9-7. Drilling operations in plastics.

molding dies (Fig. 9-8). It is done on flat parts less than 0.23 in. [6 mm] thick. Holes may be either drilled or die cut, and heating the plastics stock may aid in these operations.

Punching and shearing of laminar composite materials usually results in some delamination, edge raging, or fiber tearing. It is recommended that the part be abrasively ground to the final dimension.

Tapping and Threading

Standard machine-shop tools and methods may be used for tapping and threading. To prevent overheating, taps should be finish ground and have polished flutes. Lubricants may also be used to help clear chips from the hole. If transparency is needed, a wax stick may be inserted in the drilled hole before tapping. The wax lubricates, helps expel chips, and makes a more transparent thread.

Because of the elastic recovery of most plastics, oversized taps should be used. Oversized taps are designated as follows:

H1: Basic size to basic + 0.012 mm
H2: Basic + 0.012 mm to basic + 0.025 mm
H3: Basic + 0.025 nun to basic + 0.038 mm
H4: Basic + 0.038 mm to basic + 0.050 mm

The cutting speed for machine tapping should be less than 9.842 in./s [0.25 m/s]. The tap should be backed out often to clear chips. Usually, not more than 75 percent of the full thread is cut into the plastics. Sharp V-threads are not advised. Acme threads (Fig.

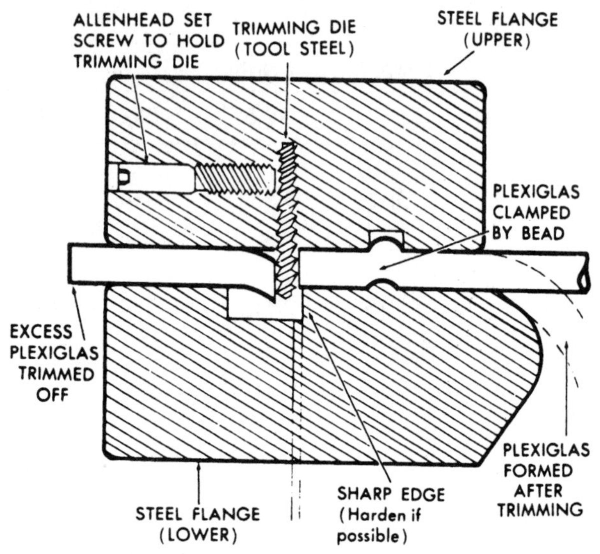

(A) Blank die and clamp for cutting plexiglas.

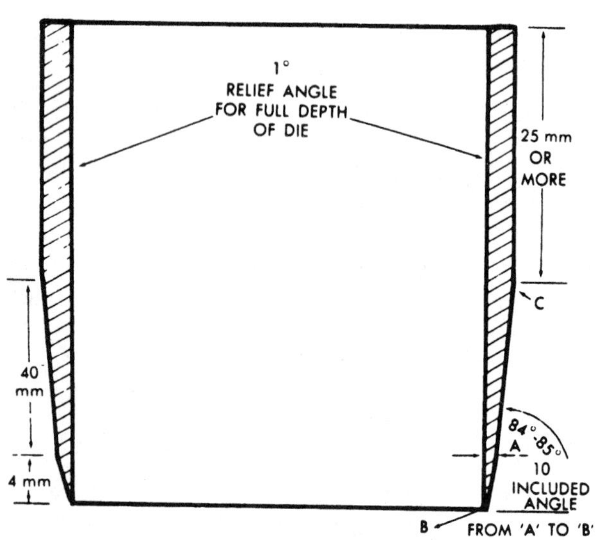

(B) Modified shoemaker die.

Fig. 9-8. Dies used for cutting plastics. (Rohm & Haas Co.)

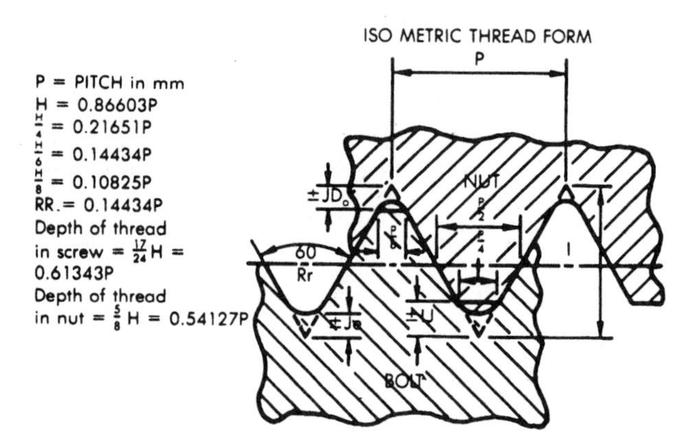

Fig. 9-9. Simplified ISO metric thread forms.

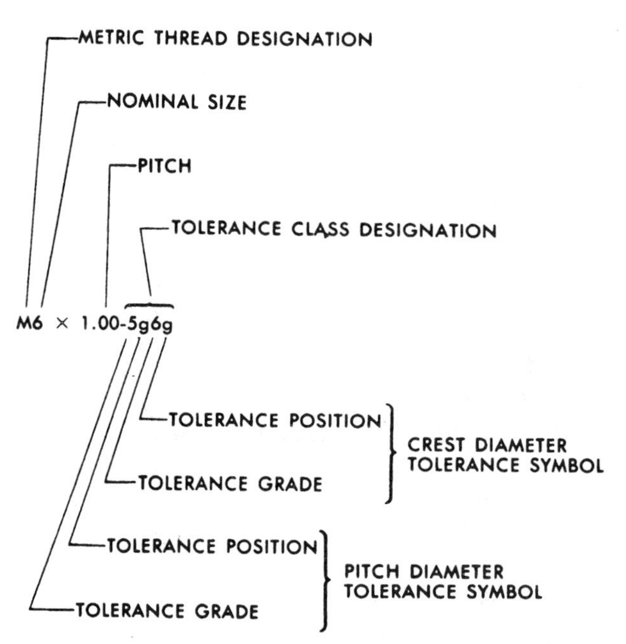

(A) Metric screw thread designation.

CLASSES OF FIT	INTERNAL THREADS (NUTS)	EXTERNAL THREADS (BOLTS)
CLOSE (close accuracy required)	5H	4h
MEDIUM (general purposes)	6H	6g
FREE (easy assembly)	7H	8g

(B) Classes of fit.

Fig. 9-10. Metric screw thread designation and classes of fit.

9-9) and ISO metric threads are preferred. Figure 9-10 displays the metric screw thread designation. Selected ISO metric threads are shown in Table 9-6. (ISO stands for the International Organization for Standardization.) To obtain tap drill size, subtract the pitch from the diameter. National coarse and national fine thread and tap drill sizes are shown on Table 9-7.

Plastics may be tapped and threaded on lathes and screw machines. (See Mechanical Fastening in Chapter 18.)

Turning, Milling, Planing, Shaping, and Routing

High-speed steel or carbide cutting tools used for machining brass and aluminum are advised for machining plastics. (Fig. 9-11A). The feeds and speeds are

Table 9-6. Selected ISO Metric Threads—Coarse Series

Diameter, mm	Pitch, mm	Tap Drill, mm	Depth of Thread, mm	Area of Root, mm^2
M 2	0.40	1.60	0.25	1.79
M 2.5	0.45	2.05	0.28	2.98
M 3	0.50	2.50	0.31	4.47
M 4	0.70	3.30	0.43	7.75
M 5	0.80	4.20	0.49	12.7
M 6	1.00	5.00	0.61	17.9
M 8	1.25	6.75	0.77	32.8
M 10	1.50	8.50	0.92	52.3
M 12	1.75	10.25	1.07	76.2
M 16	2.00	14.00	1.23	144

Table 9-7. National Coarse, and National Fine Threads and Tap Drills

Size	Threads Per Inch	Major Dia.	Minor Dia.	Pitch Dia.	Tap Drill 75% Thread	Decimal Equivalent	Clearance Drill	Decimal Equivalent
2	56	.0860	.0628	.0744	50	.0700	42	.0935
	64	.0860	.0657	.0759	50	.0700	42	.0935
3	48	.099	.0719	.0855	47	.0785	36	.1065
	56	.099	.0758	.0874	45	.0820	36	.1065
4	40	.112	.0795	.0958	43	.0890	31	.1200
	48	.112	.0849	.0985	42	.0935	31	.1200
6	32	.138	.0974	.1177	36	.1065	26	.1470
	40	.138	.1055	.1218	33	.1130	26	.1470
8	32	.164	.1234	.1437	29	.1360	17	.1730
	36	.164	.1279	.1460	29	.1360	17	.1730
10	24	.190	.1359	.1629	25	.1495	8	.1990
	32	.190	.1494	.1697	21	.1590	8	.1990
12	24	.216	.1619	.1889	16	.1770	1	.2280
	28	.216	.1696	.1928	14	.1820	2	.2210
1/4	20	.250	.1850	.2175	7	.2010	G	.2610
	28	.250	.2036	.2268	3	.2130	G	.2610
5/16	18	.3125	.2403	.2764	F	.2570	21/64	.3281
	24	.3125	.2584	.2854	1	.2720	21/64	.3281
3/8	16	.3750	.2938	.3344	5/16	.3125	25/64	.3906
	24	.3750	.3209	.3479	Q	.3320	25/64	.3906
7/16	14	.4375	.3447	.3911	U	.3680	15/32	.4687
	20	.4375	.3725	.4050	25/64	.3906	29/64	.4531
1/2	13	.5000	.4001	.4500	27/64	.4219	17/32	.5312
	20	.5000	.4350	.4675	29/64	.4531	33/64	.5156
9/16	12	.5625	.4542	.5084	31/64	.4844	19/32	.5937
	18	.5625	.4903	.5264	33/64	.5156	37/64	.5781
5/8	11	.6250	.5069	.5660	17/32	.5312	21/32	.6562
	18	.6250	.5528	.5889	37/64	.5781	41/64	.6406
3/4	10	.7500	.6201	.6850	21/32	.6562	25/32	.7812
	16	.7500	.6688	.7094	11/16	.6875	49/64	.7656
7/8	9	.8750	.7307	.8028	49/64	.7656	29/32	.9062
	14	.8750	.7822	.8286	13/16	.8125	57/64	.8906
1	8	1.0000	.8376	.9188	7/8	.8750	1- 1/32	1.0312
	14	1.0000	.9072	.9536	15/16	.9375	1- 1/64	1.0156
1-1/8	7	1.1250	.9394	1.0322	63/64	.9844	1- 5/32	1.1562
	12	1.1250	1.0167	1.0709	1- 3/64	1.0469	1- 5/32	1.1562
1-1/4	7	1.2500	1.0644	1.1572	1- 7/64	1.1094	1- 9/32	1.2812
	12	1.2500	1.1417	1.1959	1- 11/64	1.1719	1- 9/32	1.2812
1-1/2	6	1.5000	1.2835	1.3917	1- 11/32	1.3437	1-17/32	1.5312
	12	1.5000	1.3917	1.4459	1- 27/64	1.4219	1-17/32	1.5312

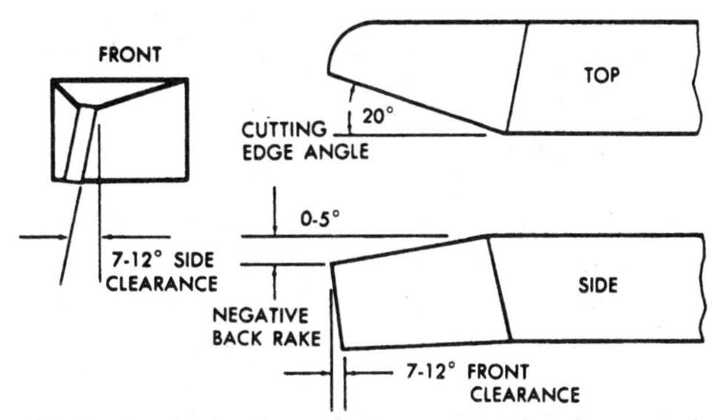

(A) Cutting tool rake and clearance angles for general-purpose turning of plastics. Note the 0-5° negative back rake angle. A sharp-pointed tool with a +20° rake is used in turning polyamides.

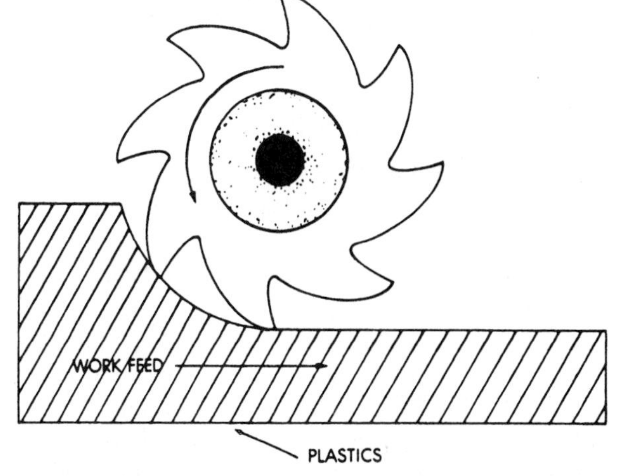

(B) Climb milling, or down milling, a technique in which the work moves in the same direction as the rotating cutter.

Fig. 9-11. Machining plastics.

Table 9-8. Turning and Milling Plastics

Material	Turning Single Point (H-S Steel)			Milling Tool Per Tooth (H-S Steel)		
	Depth of cut, mm	Speed, m/s	Feed, mm/r	Depth of cut, mm	Speed, m/s	Feed, mm per tooth
Thermoplastics						
Polyethylene	3.8	0.8–1.8	0.25	3.8	2.5–3.8	0.4
Polypropylene	0.6	1.5–2	0.05	3.8	2.5–3.8	0.4
TFE-fluorocarbon				1.5	3 8–5	0.1
Butyrates				3.8	2.5–3.8	0.4
ABS	3.8	1.2–1.8	0.38	3.8	2.5–3.8	0.4
Polyamides	3.8	1.5–2	0.25	3.8	2.5–3.8	0.4
Polycarbonate	0.6	2–2.5	0.05	1.5	3.8–5	0.1
Acrylics	3.8	1.2–1.5	0.05	1.5	3.8–5	0.1
Polystyrenes, low	3.8	0.4–0.5	0.19	3.8	2.5–3.8	0.4
and medium impact	0.6	0.8–1	0.02	3.8	2.5–3 8	0.4
Thermosets						
Paper	3.8	2.5–5	0.3	1.5	2.0–2.5	0.12
and cotton base	0.6	5–10	0.13	1.5	2.0–2.5	0.12
Fiber glass	3.8	1–2.5	0.3	1.5	2.0–2.5	0.12
and graphite base	0.6	2.5–5	0.13	1.5	2.0–2.5	0.12
Asbestos base	3.8	3.2–3.8	0.3	1.5	2.0–2.5	0.12

similar. For many plastics, a surfac speed of 492 ft/min. [2.5 m/s] with feeds (depth of cut) of 0.02 to 0.005 in./r [0.5 to 0.12 mm] per revolution will produce good results. On cylindrical stock, a 0.049 in. [1.25 mm] cut will reduce the diameter by 0.098 in. [2.5 mm].

Climb-cutting (or down-cutting) a milling operation using lubrication, gives a good machined finish on plastics (Fig. 9-11B). In climb milling, the work moves in the same direction as the rotating cutter. The feed rate on multiple-edged milling cutters is expressed in millimetres of cut per cutting edge per second. The feed of a milling machine is expressed in millimetres of table movement per second rather than millimetres per spindle rotation. The formula below is used to determine the amount of feed in inches per minute or millimetres per second:

$$\text{mm/s} = t \times \text{fpt} \times \text{r/s}$$

where

t = number of teeth

mm/s = feed in millimetres per second

fpt = feed per tooth (chip load)

r/s = revolutions per second (spindle or work)

Table 9-8 gives turning and milling data for various plastics materials. Table 9-9 gives side and end relief angles and back rake angles for cutting tools used with different plastics.

For all milling, planing, shaping, and routing work, carbide-tipped cutters are advised. Conventional high-speed steel shapers, planers, and routers used for wood-working may be employed with plastics if tools are sharpened in a proper manner. Routers and shapers are useful for cutting beads, rabbets, and flutes and for trimming edges. Carbide or diamond-tipped tools are vital for long runs, uniformity of finish, and accuracy.

Table 9-9. Design of Turning Cutting Tool

Work Material	Side Relief Angle, Deg.	End Relief Angle, Deg.	Back Rake Angle, Deg.
Polycarbonate	3	3	0–5
Acetal	4–6	4–6	0–5
Polyamide	5–20	15–25	neg. 5–0
TFE	5–20	0.5–10	0–10
Polyethylene	5–20	0.5–10	0–10
Polypropylene	5–20	0.5–10	0–10
Acrylic	5–10	5–10	10–20
Styrene	0–5	0–5	0
Thermosets:			
Paper or Cloth	13	30–60	neg. 5–0
Glass	13	33	0

Laser Cutting

A CO_2 laser (light amplification by stimulated emission of radiation) can deliver powerful radiation at a wavelength of 10.6 μm (microns). A laser may be used to make intricate holes and complex patterns in plastics (Fig. 9-12). The laser power can be controlled to merely etch the plastics surface or actually vaporize and melt it. Holes and cuts made by a laser have a slight taper, but the cuts are clean with a finished appearance. Cuts made by a laser are more precise, and tolerances are held more closely than those made with conventional machining operations. There is no physical contact between the plastics and the laser equip-

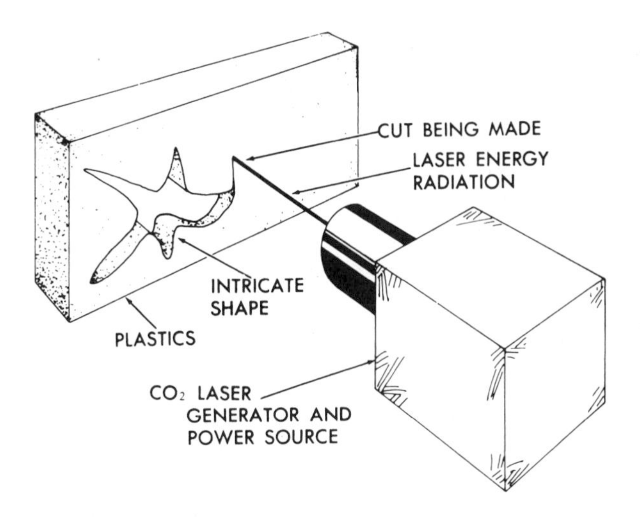

(A) Basic concept.

(B) A five axis robotic laser is used to trim the three-dimensional molded Kevlar part. (Russel Plastics Technology)

Fig. 9-12. Light energy from a laser can be used to cut intricate shapes in plastics, or to trim to final shape.

ment, therefore no chips are produced. Laser cutting does produce a residue of fine dust; however, this is easily removed by vacuum systems. Most polymers and composites may be laser machined. Some laminar composites tend to heat up, bubble and char.

Induced Fracture Cutting

Acrylics and several other plastics including some composites, may be cut to shape by *induced fracture* methods. The methods are similar to cutting glass. A sharp tool or cutting blade is used to score, or scratch, the plastics surface. On thick pieces, both sides are scored. Pressure is applied along the scratch line, and the plastics fractures. The fracture will follow the score line (Fig. 9-13).

(A) Score plastics with tool.

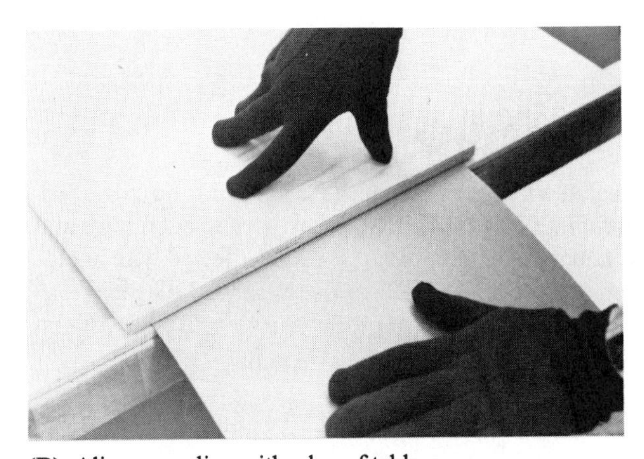

(B) Align score line with edge of table.

(C) Press down on plastics piece to induce fracture.

Fig. 9-13. The induced-fracture cutting method.

(continued)

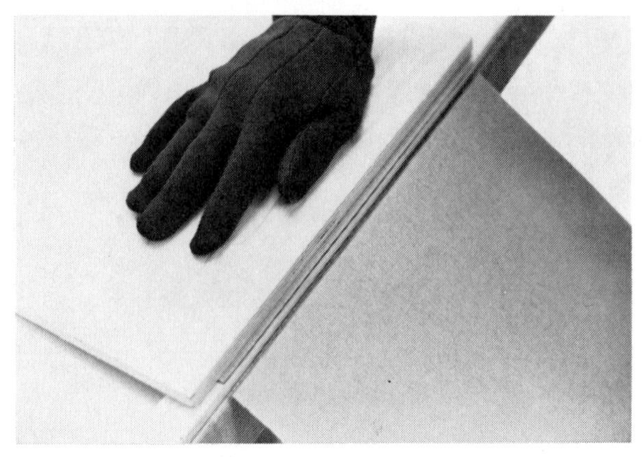

(D) Completed cutting.

Fig. 9-13. The induced-fracture cutting method. *(Continued)*

Thermal Cutting

Heated wires or dies are used to cut solid and expanded, or foamed, plastics. Hot dies are used to cut fabrics and silhouette-shaped products, while a heated wire or ribbon is commonly used to cut expanded plastics (Fig. 9-14). Thermal cutting produces a smooth edge with no chips or dust.

Hydrodynamic Cutting

High-velocity fluids may be used to cut many plastics and composites (Fig. 9-4C). Pressures of 320 MPa [46417 psi] are used. Foamed or cellular plastics are reinforced and filled plastics have been successfully machined by this method.

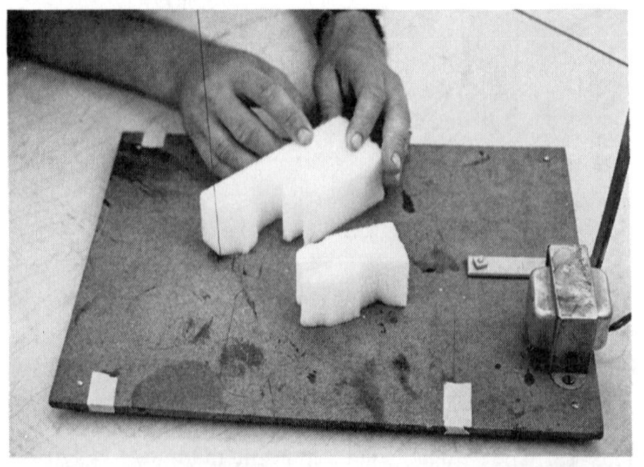

Fig. 9-14. Expanded polystyrene is easily cut using a hot Nichrome wire. The wire melts a path through the cellular material.

Smoothing and Polishing

Smoothing and polishing techniques for plastics are like those used on woods, metals, and glass.

Because of the elastic and thermal properties of thermoplastics, many are difficult to grind. Abrasive grinding is more easily accomplished on thermosetting materials, reinforced plastics, and most composites. Grinding is not advised unless open-grit wheels are used with a coolant. Hand and machine sanding is an important operation. *Open-grit sandpaper* is used on machines to prevent clogging (loading). A number 80-grit siliconcarbide abrasive is advised for rough sanding. In any machine sanding, light pressure is used to prevent overheating the plastics.

Disc sanders (working at 30 r/s) and belt sanders (working at a surface speed of 18 m/s [59 ft/sl) are used for dry sanding. If water coolants are used, the abrasive lasts longer and cutting action is increased. Progressively finer abrasives are used; that is, the first rough sanding using 80-grit paper should be followed by 280-grit silicon-carbide wet or dry sandpaper. The final sanding may be with 400- or 600-grit sandpaper. After the sanding is finished and the abrasives removed, further finishing operations are used.

Ashing, buffing, and polishing are done on abrasive-charged wheels. These wheels may be made of cloth, leather, or bristles. A different wheel is used for each abrasive grit. Finishing wheel speeds should not exceed 10 surface m/s [32.8 ft/s]. With the use of coolants, surface speed may be increased.

Never finish plastics on wheels used for metals. Small metal particles may be left in the wheel and these will damage the plastics surface. Machines should be grounded. Static electricity is generated by the movement of the wheels over the plastics. Remove coarse tool marks before the finishing wheels are used.

Ashing is a finishing step in which a wet abrasive is applied to a loose muslin wheel. Number 00 pumice is commonly used (Fig. 9-15). A hood or shield is used over the wheel because the operation is wet. Surface speeds of over 20 m/s [65.6 ft/s] may be used. Overheating is avoided in this process and the loose muslin wheel is fast-cutting on irreguar surfaces.

Buffing is an operation in which grease- or wax-filled abrasive bars or sticks are applied to a loose or sewn muslin wheel.

Loose buffs are used for more irregular shapes or for entering crevices. Hard buffing wheels should be avoided.

The buffing wheels are charged by holding the bars or sticks against them as they revolve, producing frictional heat that leaves the wax-filled abrasive on the wheel. The most common buffing abrasives are tripoli, rouge, or other fine silica.

Polishing, sometimes called luster buffing or burnishing, employs wax compounds containing the finest

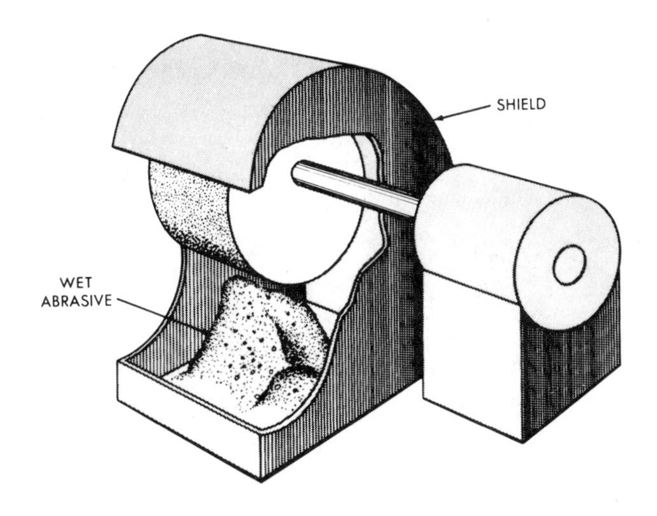

Fig. 9-15. Ashing with wet pumice is a faster cutting method than the use of grease-or wax-based compounds. Cooling action is also better.

abrasives such as levigated alumina or whiting. Polishing wheels are generally made of loose flannel or chamois. A final polishing is sometimes done with clean, abrasive-free waxes on a flannel or chamois wheel. The wax fills many imperfections and protects the polished surface.

Never let the finishing wheel rotate to the edge of a part because it may be jerked from your hands. The wheel may pass over an edge but never to it. Always keep the work piece below the center of the buffing wheel. The preferred procedure is to do about half of the surface and then turn it around and finish the final portion. The part should be moved or pulled toward the operator in rapid, even strokes (Fig. 9-16). Do not spend very much time at the finishing wheels. Move stock around. If you hold a piece in one place, the heat generated by friction between the wheel and the workpiece will melt many thermoplastics.

Solvent-dip polishing of cellulosic and acrylic plastics may be used to dissolve minor surface defects (Fig. 9-17A). The parts are either dipped into or sprayed with solvents for about one minute. Solvents are sometimes used to polish edges or drilled holes. All solvent-polished parts should be annealed to prevent crazing.

Surface coatings may be used on most plastics to produce a high surface gloss, which may cost less than other finishing operations.

Flame polishing with an oxygen-hydrogen flame may be used to polish some plastics (Fig. 9-17B).

Tumbling

The tumbling-barrel process is one of the least costly ways to finish plastics molded parts rapidly. It produces a smooth finish on plastics parts by rotating them in a drum with abrasives and lubricants, causing the parts and abrasives to rub against each other with a smoothing effect (Fig. 9-18A). The amount of material removed depends on the speed of the tumbling barrel, the abrasive grit size, and the length of the tumbling cycle.

In another tumbling process, abrasive grit is sprayed over the parts as they tumble on an endless rubber belt. Figure 9-18B shows parts being tumbled while being doused with abrasive grit.

Dry ice is sometimes used in tumbling to remove molding flash. The dry ice chills the thin flash and makes it brittle; then tumbling breaks it free in a very short time.

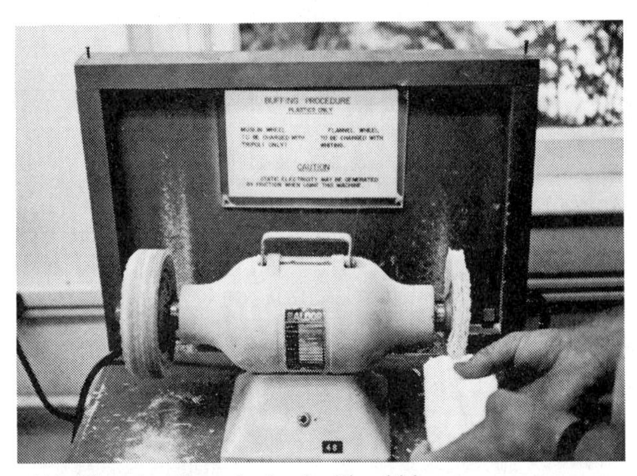

(A) Charging flannel wheel with whiting.

(B) Abrasive wheel passing over an edge. About half of each side is done by pulling work toward the operator.

Fig. 9-16. Buffing plastics materials. Note: Guards have been removed from equipment for this illustration. Protective devices and guards should *always* be used when working with such equipment.

Annealing and Postcuring

During the molding, finishing, and fabrication processes, the plastics or composite part may develop internal stresses. Chemicals may sensitize the plastics and cause crazing.

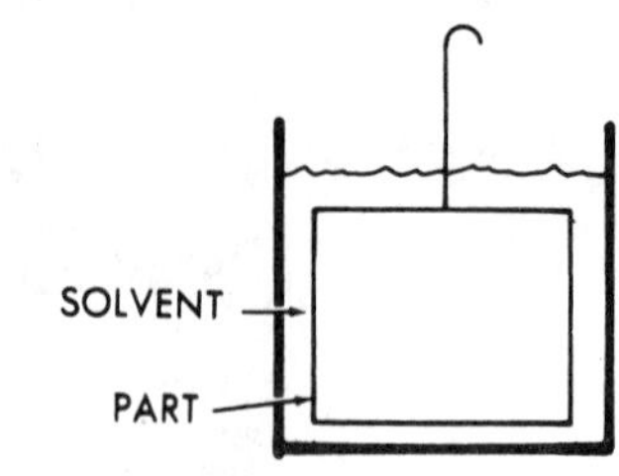

(A) Solvent-dip polishing.

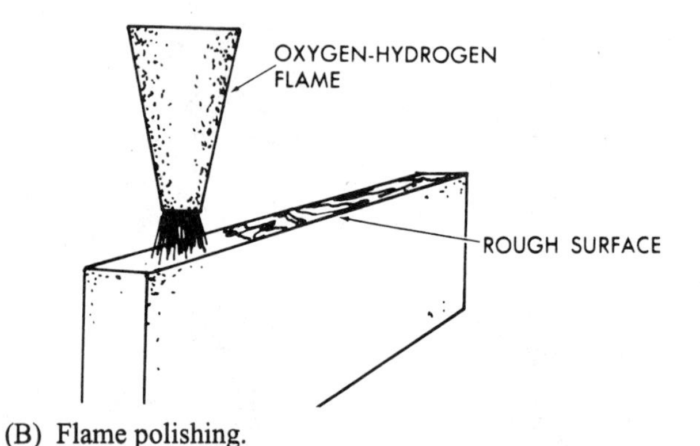

(B) Flame polishing.

Fig. 9-17. Two methods of polishing.

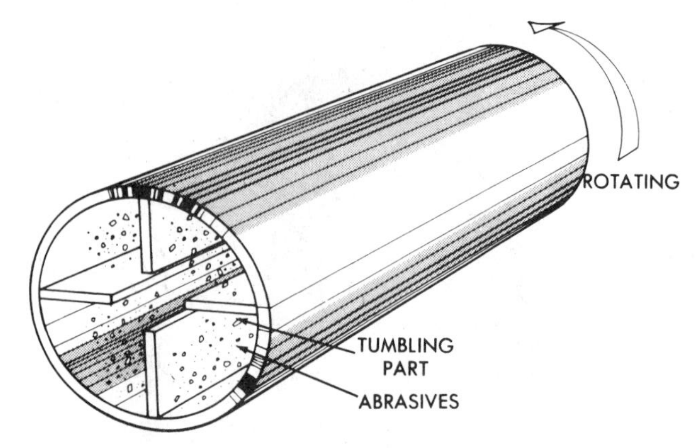

(A) Parts being tumbled in a revolving drum.

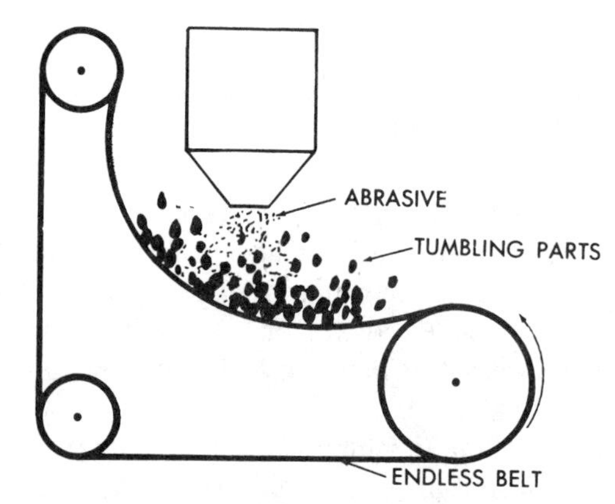

(B) Tumbling of parts on an endless revolving belt.

Fig. 9-18. Two tumbling methods.

Many parts develop these internal stresses as a result of cooling immediately after molding or post-mold curing, as the chemical reactions continue during complete polymerization. Composites are sometimes left in the mold or placed in a curing jig until the curing process has been completed and all chemical activity and temperatures are brought to ambient levels. For some plastics and composites, the internal stresses may be reduced or eliminated by annealing. Annealing consists of prolonged heating of the plastics part at temperature lower than molding temperatures. The parts are then slowly cooled. All machined parts should be annealed before cementing.

Tables 9-10 and 9-11 give heating and cooling times for the annealing of Plexiglas. Figure 9-19 shows a large oven that may be used in the process.

Table 9-10. Heating Times for Annealing of Plexiglas

	Time in a Forced-Circulation Oven at the Indicated Temperature, h									
	Plexiglas G, 11, and 55					Plexiglas I-A				
Thickness (mm)	110 °C*	100 °C*	90 °C*	80 °C	70 °C**	90 °C*	80 °C*	70 °C*	60 °C	50 °C
1.5 to 3.8	2	3	5	10	24	2	3	5	10	24
4,8 to 9.5	2½	3½	5½	10½	24	2½	3½	5½	10½	24
12.7 to 19	3	4	6	11	24	3	4	6	11	24
22.2 to 28.5	3½	4½	6½	11½	24	3½	4½	6½	11½	24
31.8 to 38	4	5	7	12	24	4	5	7	12	24

Note: Times include period required to bring part up to annealing temperature, but not cooling time. See Table 9–9.
*Formed parts may show objectional deformation when annealed at these temperatures.
**For Plexiglas G and Plexiglas 11 only. Minimum annealing temperature for Plexiglass 55 is 80 °C.
Source: *Rohm & Haas Co.*

Table 9-11. Cooling Times for Annealing of Plexiglas

Thickness (mm)	Rate (°C)/h	Time to Cool from Annealing Temperature to Maximum Removal Temperature							
		Plexiglas G, 11, and 55				Plexiglas I-A			
		230 (110 °C)	212 (100 °C)	184 (90 °C)	170 (80 °C)	194 (90 °C)	176 (80 °C)	158 (70 °C)	140 (60 °C)
1.5 to 3.8	122 (50)	¾	½	½	¼	¾	½	½	¼
4.8 to 9.5	50 10	1½	1¼	¾	½	1½	1¼	¾	½
12.7 to 19	22 – 5	3¼	2¼	1½	¾	3	2¼	1½	¾
22.2 to 28.5	18 – 8	4¼	3	2	1	4	3	2¼	1
31.8 to 38	14–10	5¾	4½	3	1½	5¾	4½	3	1½

Note: Removal temperature is 70 °C for Plexiglas G and II, 80 °C for Plexiglas 55, and 50 °C for Plexiglas 1-A.
Source: *Rohm & Haas Co.*

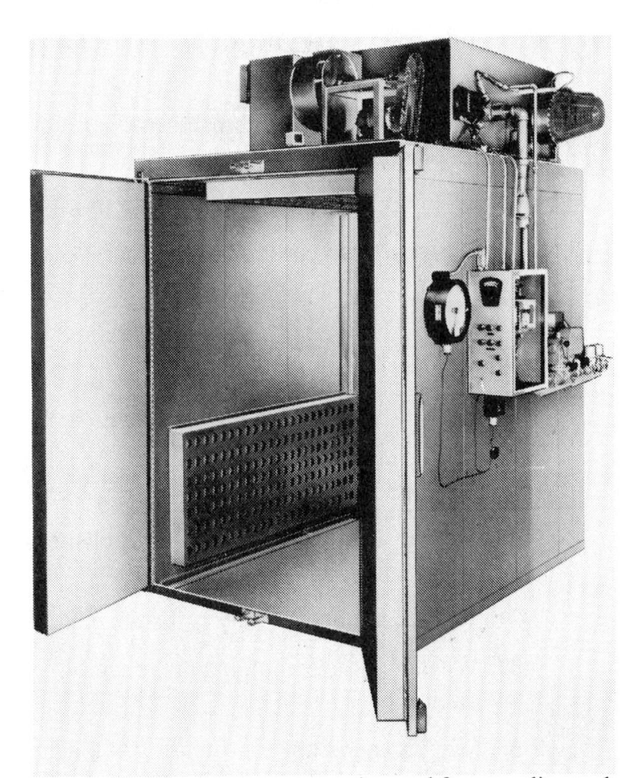

Fig. 9-19. This large oven may be used for annealing and posturing plastics. (Precision Quincey Corp.)

Vocabulary

The following vocabulary words are found in this chapter. Look up the definition of any of these words you do not understand as they apply to plastics in the glossary, Appendix A.

Annealing
Ashing
Blanking
Charging
Die cutting
Feed
Hollow ground
Kerf
Laser cutting
m/s
Negative rake
Open-grit sandpaper
rpm
r/s
Tripoli
Tumbling
Whiting

Questions

9-1. Name the process of slowly cooling plastics to remove internal stresses.

9-2. Identify the number of teeth per centimetre for cutting 6 mm thick polycarbonate on the band-saw.

9-3. Name the rake angle of a circular saw when the hook angle of the tooth and the center of the blade are in line.

9-4. How many r/s are required for drilling with an 8 mm drill in thermoplastics?

9-5. The distance the cutting tool is fed into the work each revolution is called __?__ .

9-6. What is the abbreviation or symbol for the International Organization for Standardization?

9-7. Name the slit or notch made by a saw or cutting tool.

9-8. Name the burnishing agent sometimes used to polish the edges of acrylic plastics.

9-9. Name the finishing operation in which wet abrasives are used.

9-10. Name the cutting operation that uses high velocity fluids to cut plastics.

9-11. What type of cutting edges or teeth on tools are essential for long runs, uniform finishing and accuracy?

9-12. What does the 1.00 show in M6X1.00-5g6g thread designation?

9-13. Name a popular silica abrasive used in some finishing operations.

9-14. Frictional __?__ is a major problem in machining most plastics.

9-15. Name the operation that is preferred over sawing, for many plastics, because it produces a smoother edge.

9-16. How many teeth per centimetre should a circular saw have for cutting thin plastics materials? Thick materials?

9-17. What type of bandsaw blade and what cutting speed should you use to cut 0.118 in. [3 mm] thick acrylic plastics?

9-18. What is a skip-tooth blade?

9-19. Name some-saws that may be used to cut plastics. Indicate the kinds of jobs for which each may be used.

9-20. What is drill feed? What is its range in millimetres for most plastics?

9-21. What factors affect the cutting speed of drills in plastics materials?

9-22. What precaution must be observed in drilling a hole of a given size in a plastic rod?

9-23. Why are oversized taps necessary for plastics materials?

9-24. Name the preferred thread forms for taps used on plastics.

9-25. What is climb milling? Why is it used?

9-26. What is laser cutting of plastics? Where is it used?

9-27. What grit number of sandpaper should be used for final finishing of plastics?

9-28. What is ashing? Buffing? Charging a wheel?

9-29. Which abrasives are used in buffing? In burnishing?

9-30. Briefly describe solvent-dip polishing and flame polishing.

9-31. What is tumbling? Why is it so named?

9-32. What does the smoothing in the tumbling operation?

9-33. What is annealing or postcuring of machined or molded parts? Why is this process done?

9-34. What machining operation would you select for shaping or finishing of the following products?

a. 0.23 in. [6 mm] thick polycarbonate window glazings.

b. Christmas tree decoration shapes made of thin sheet or films.

c. Removing flash from radio cabinet housing.

d. Making the edges of a plastics part smooth and glossy.

Activities

Drilling

Introduction. Drill geometry dramatically effects the drilling process. Figure 9-20 shows the start of a hole in a piece of acrylic. Notice the surface roughness. A standard twist drill caused this result. Compare Figure 9-20 to Figure 9-21, which shows another drill dimple in acrylic. The surface in Figure 9-21 is much smoother because the drill used had a geometry appropriate for acrylics.

Equipment. Drill press, selected drills, sheet acrylic plastics, and safety glasses.

Procedure

9-1. Acquire thick acrylic sheet, at least 6 mm (0.25 inch thick). Thicker sheets permit better examination of the walls of the hole.

9-2. Using appropriate safety equipment and safe practices, drill dimples, partial holes, and through holes using a drill at least 12 mm (0.5 inch) in diameter. (Smaller holes exhibit the same characteristics, but they are harder to thoroughly inspect.) Save all chips and cuttings.

9-3. Examine the start of a hole as shown in Figure 9-20 and 9-21. Examine the inside walls of the holes. Examine the exit of the drill on the back of the piece.

9-4. Purchase or grind a drill to the geometry recommended for acrylic in Table 9-3.

9-5. Drill a second set of dimples, partial holes, and through holes. Examine them carefully.

9-6. Compare the chips made by the differing drill geometries.

9-7. Did the standard drill cause chips when exiting the piece? Did the special drill also cause chips.

9-8. Drill holes with standard and special drills at various RPMs and various feed rates.

9-9. At high RPMs and/or high fed rates, does the heat generate melt the acrylic? Does the special drill generate more or less frictional heat than the standard bit?

9-10. Record observations and offer explanations for differences caused by the drill geometries.

Milling

Introduction. Machining a groove with a ball-end mill or a slot with a flat-end mill provides a direct comparison between climb milling and conventional milling. The effects of speed and feed on the quality of a machined surface are also observable.

Equipment. Milling machine, selected end mills, plastics sheet or moldings thick enough for ease of clamping, and safety glasses.

Procedure

1-1. Use a ball-end mill to cut a groove in a selected material using the parameters suggested in Table 9-8. Tools ranging from 12 to 25 mm (0.5 to 1 inch) in diameter provide ease of inspection.

9-2. Is the side of the cut that experienced climb milling smoother or rougher than the side that experienced conventional milling?

9-3. Does the surface quality change significantly with changes in speeds and feeds?

9-4. Record and explain your observations.

Fig. 9-20. Beginning of a drilled hole, using standard twist drill.

Fig. 9-21. Beginning of a drilled hole, using a drill with special geometry.

Flame Polishing

Introduction. Flame polishing is a common technique to finish the edges of thermoformed or fabricated acrylic products. Department store displays often exhibit flame-polished edges. Flame polishing is popular because it is several times faster than buffing.

Equipment. Oxy-hydrogen torch, selected plastics, gloves, safety glasses. Oxy-acetylene welding or cutting equipment can provide much of the necessary components. A hydrogen regulator is necessary, because an acetylene regulator can not be used on a hydrogen tank.

Procedure

9-1. Cut samples of acrylic or other plastics.

9-2. Use an oxy-acetylene flame to polish the edges. Try excess oxygen flames and excess acetylene flames. How do these flames effect the plastics? Can the oxy-acetylene be adjusted so that it does not discolor the samples?

9-3. Prepare the oxy-hydrogen flame.

CAUTION: An oxy-hydrogen flame is almost invisible. Exercise great care to avoid burns when manipulating the oxy-hydrogen flame.

9-4. Experiment with the rate of travel of the flame along the edge of a sample.

9-5. Does the flame remove the scratches caused by sawing?

9-6. Intentionally scratch a sample and then polish it. How deep do the scratches need to be to prevent them from being polished out?

9-7. Experiment with the relative amounts of oxygen and hydrogen. Is the neutral flame the best for polishing?

9-8. Record and explain your observations.

Chapter 10

Molding Processes

Introduction

Molding processes convert plastics resins, powders, pellets and other forms into useful products. A characteristic common to the molding processes is that they all involve some level of force. In processing powders and pellets, the required force can be enormous. Although driving liquid resins into molds takes much less force than needed to move melted pellets, some level of pressure is essential.

A host of molding processes are available to processors. This chapter will not attempt to discuss all these techniques. It will focus on three main areas of molding, namely injection molding, compression and transfer molding, and techniques for liquid resins. The outline for this chapter is:

I. Injection molding
   A. Injection unit
   B. Clamping unit
   C. Injection-molding safety
   D. Specification of molding machines
   E. Elements of molding cycles
   F. Advantages of injection molding
   G. Disadvantages of injection molding
   H. Injection molding thermosets
   I. Coinjection molding
III. Molding liquid materials
   A. Reaction injection molding
   B. Reinforced reaction injection molding
   C. Liquid resin molding
IV. Molding granular thermoset materials
   A. Compression molding
   B. Transfer molding

Injection Molding

Injection molding is a principal process for converting plastics into products. The list of injection-molded products that influence daily life is almost endless: TV, VCR, and computer housings, CDs, CD players, glasses, toothbrushes, automobile parts, athletic shoes, ball-point pen barrels, and office furniture.

Injection molding is appropriate for all thermoplastics except polytetraflouroethylene (PTFE) fluoroplastics, polyimides, some aromatic polyesters, and some specialty grades. Injection molding machines, hereafter IMM, for thermosets provide processing for phenolics, melamine, epoxy, silicone, polyester and numerous elastomers. In all cases, pelletized or granular materials absorb sufficient heat to make them "flowable." The machine injects the hot plastics into a closed mold which creates a desired shape. After cooling or chemical transformation, an ejector system removes the parts from the mold.

Figure 10-1 shows two modern injection-molding machines; one small model, and one of medium size. Both machines have the same basic design, namely the reciprocating screw style. Although there are a number of other types of molding machines, the reciprocating screw type is dominant. In the reciprocating screw machines, granular material is quickly made molten by the heat from the barrel and by frictional heat created when the screw turns. The screw has the tasks of heating, convening pellets, and acting like a plunger. When the material is uniformly fluid, the screw moves forward, forcing hot melt through the runner system into the mold cavities.

A closer look at IMMs reveals that they contain two major components, the injection unit and the clamp unit.

(A) Boy 50m. 55 US tons with "Procan Control®" fully closed loop for all machine functions. (Boy Machine Inc.)

Injection Unit

The injection unit has the task of melting and injecting the materials. Within this unit, the major parts are: the barrel, the end cap on the barrel, the nozzle, the screw and non-return valve, heater bands, a motor to rotate the screw, and a hydraulic cylinder to move the screw forward and backward. Control systems keep the temperatures at selected levels, as well as initiate and time screw rotation and injection strokes. Figure 10-2 is a simple schematic of a typical plasticating unit.

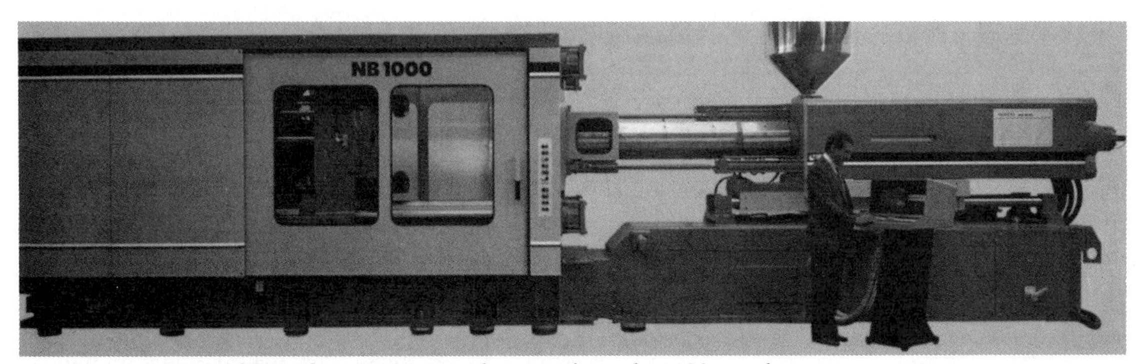

(B) This medium sized IMM has a 200 ounce shot capacity and a 1120 ton clamp.

Fig. 10-1. Two sizes of modern injection-molding machines.

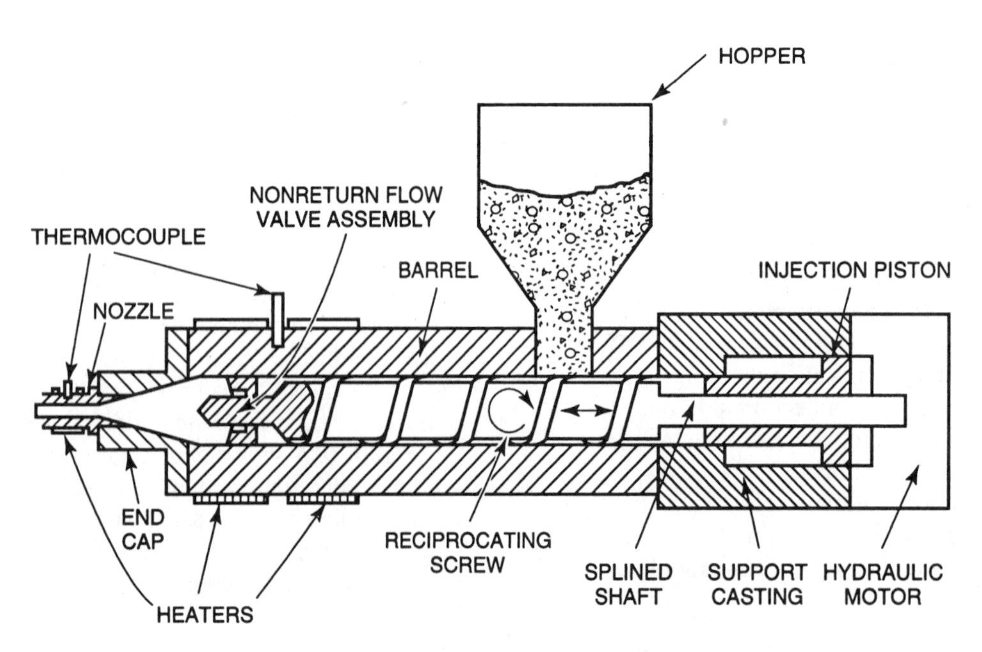

Fig. 10-2. Simplified schematic of an injection unit.

The action of the screw determines the speed and efficiency of plasticizing pellets. Figure 10-3 shows a small injection screw. Note that the depth of the screw flights near the rear of the screw is greater than near the front end.

A typical screw consists of three major sections: the feed zone, the transition zone, and the metering zone. The feed zone contains deep flights and accounts for approximately one-half of the total length. The transition zone is about one-fourth of the total length. During transition, the flight depth reduces and the resulting compression and friction cause most of the melting of pellets. In the metering zone, which has shallow flight depth, any unmelted pellets should melt, thus providing fully melted material to pass through the non-return valve and be available for injection into a mold.

Non-return valves have two common styles, as shown in Figure 10-4. The style shown in Figure 10-3 corresponds to Figure 10-4A. The purpose of the non-return valve is to prevent backflow of material during injection. If the non-return valve does not function properly, the pressure on the melted plastics may be insufficient to cause the desired flow into the recesses of the mold.

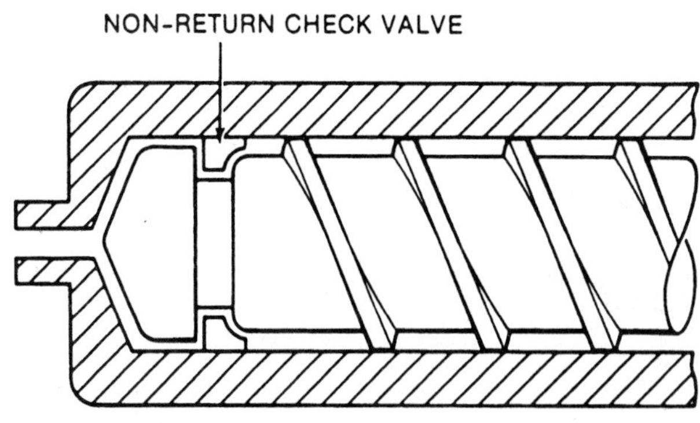

(A) Ring valve.

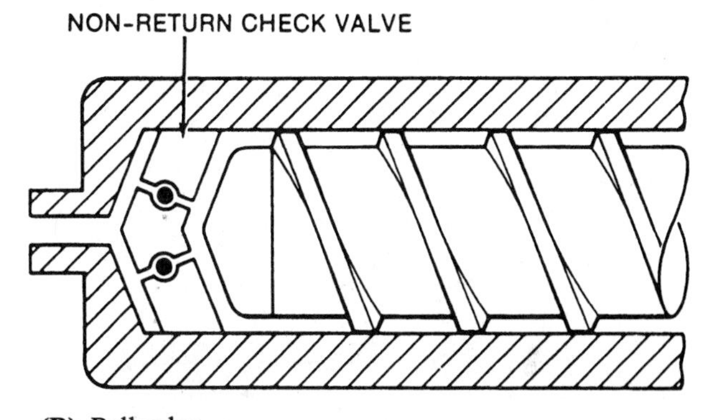

(B) Ball valve.

Fig. 10-4. Two non-return valve assemblies used to prevent the backflow of melted plastics during injection.

Clamping Unit

The clamping unit has the task of opening and closing the mold, and ejecting the parts. The two most common methods to generate clamping forces are direct hydraulic clamps, and toggle clamps actuated by hydraulic cylinders. Figure 10-5 shows the toggle style in closed and open position. Figure 10-6 portrays the hydraulic clamp.

Both styles are available in a wide range of sizes. Toggle clamps generate force mechanically, thus requiring smaller clamping cylinders. Hydraulic clamps eliminate mechanical linkages, but require much larger clamp cylinders. Very large machines may utilize a combination of hydraulic and mechanical clamping mechanisms, as seen in Figure 10-7.

In addition to the plasticating clamp units, a typical IMM also includes a hydraulic pump to move and pressurize hydraulic oil, and an oil reservoir. Guards cover

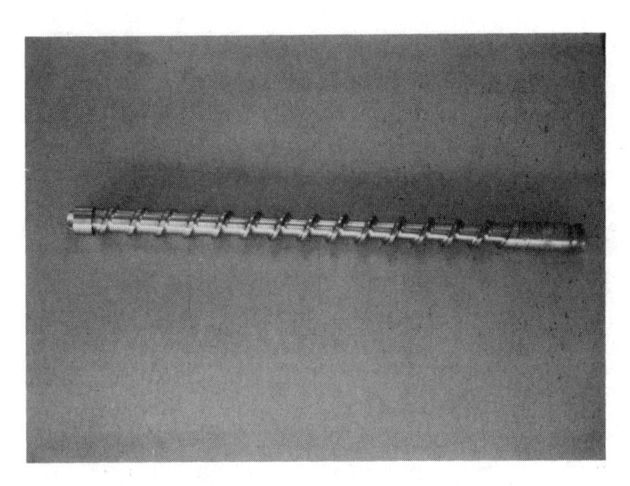

Fig. 10-3. This 30 mm IMM screw shows a ring-type non-return valve. Notice the differing flight depths.

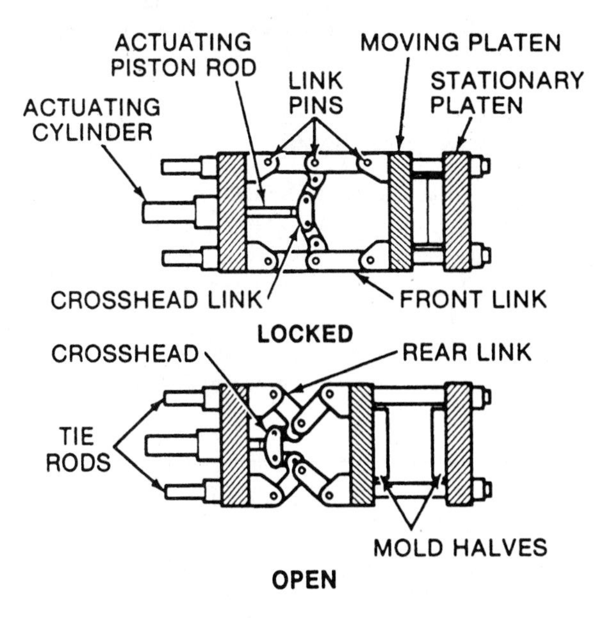

Fig. 10-5. The toggle clamp design in open and closed positions.

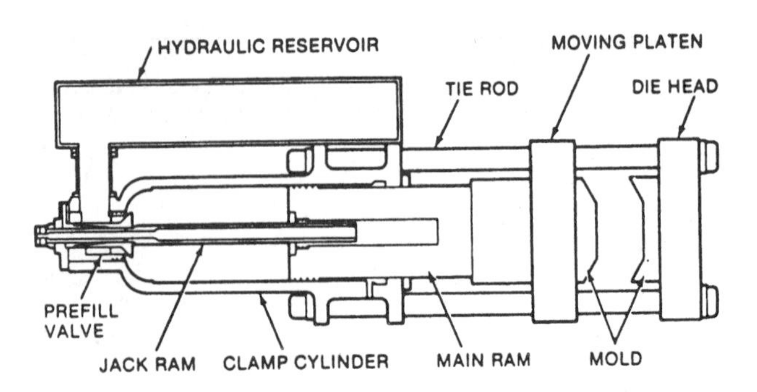

Fig. 10-6. The straight hydraulic style of mold clamp. *(Modern Plastics Encyclopedia)*

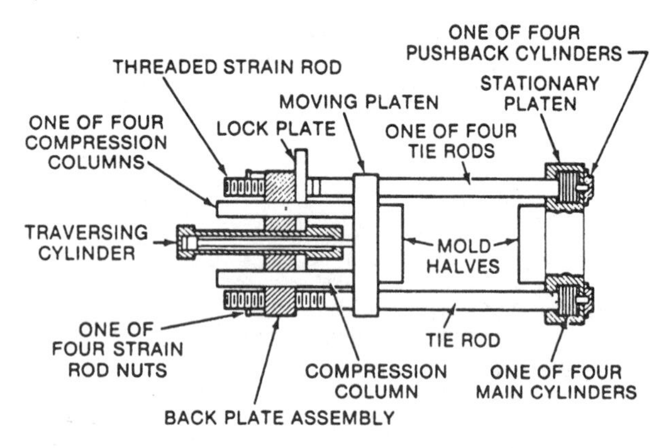

Fig. 10-7. The hydromechanical clamp is used most frequently on very large machines.

the hot barrel and prevent contact with heater bands and electrical terminals. The front door allows operators to take parts out of a mold, place inserts in molds, and keep molds clean. The most dangerous portions of a molding cycle, namely closing the mold and injection of hot plastics, occur only when guards are in place and safety doors are closed. The safety systems prevent operations if the guards and doors are not all closed.

Injection-Molding Safety

Machinery manufacturers include devices to protect both the machine operators and the machine itself. Operator and technician protection relies upon guards, doors, mold-closing safety systems, purge guards, and rear door systems. Mold-closing safety systems have, as required by law, three separate systems: a mechanical drop bar, electrical interlocks, and hydraulic interlocks.

Mechanical drop bars. The purpose of a drop bar is to prevent the closing of a mold when an operator may have hands and arms between the mold halves. If electrical and hydraulic safety systems failed and a mold began to close on an operators arm, the drop bar must stop the closing of the mold. All persons working on or around injection molders should recognize drop bars and their proper adjustments.

Drop bars respond to the motion of the front gate of an IMM. If the door is open, the bar must drop. Drop bars come in many sizes and shapes, but two styles predominate: a straight rod and a lobed rod. In the first style, a rod passes through a hole in the stationary platen of the machine. The stationary plates hold the half of the mold that does not move. When the door is open, a flap or bar drops in front of the hole. If the mold begins to close, the rod bumps into the bar and prevents mold closing.

The safety rod requires adjustment upon every mold change. The distance between the end of the rod and the bar should be adequate to allow the bar to drop easily. It should also be short enough that the moving platen has little distance to move before hitting the stop. That will guarantee that the momentum of the moving platen is limited.

Figure 10-8 shows a drop bar in the down position. Figure 10-9 shows the same drop bar in up position, allowing the rod to move forward through the hole in the block attached to the platen. Figure 10-10 shows

Fig. 10-8. A drop bar in down position. Notice the dents on the bar caused when the bar prevented mold closing.

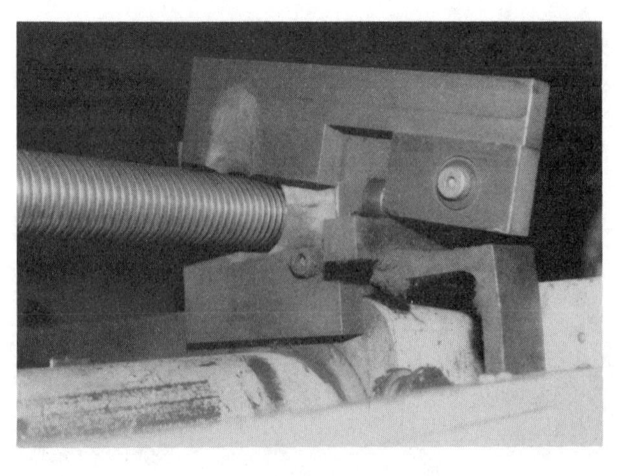

Fig. 10-9. A drop bar in up position. The large threaded rod can move forward and permit closing of the mold.

another style of drop bar. The bar is down, which means the door on the machine was open.

The adjustment of this style device is either by a lock nut on the threaded rod, or by machined grooves in the rod. The threaded type is visible in Figures 10-8 and 10-10. The grooved type appears in Figure 10-11.

A problem with the straight rod type is that if the bar does not drop, no protection is available. The bar won't drop if the rod is not properly adjusted. It also won't drop if the mold does not open fully. Please carefully examine Figure 10-12. It shows the front door open, but the bar still up. This dangerous situation should never be permitted.

Fig. 10-10. This safety rod has a threaded portion for length adjustments.

Fig. 10-11. The grooves in this style safety rod control length adjustments.

Fig. 10-12. Recognize the horrible danger shown here. The front door of the IMM is open, but the bar has not dropped. It is misadjusted or the mold is not fully open.

To avoid the possible failures of the straight rod type, machine manufacturers developed lobed style safety rods. These lobes, machined into the rod, allow the mold to open, but prevent its close unless the bar that engages the lobes is up. Figure 10-13 shows a lobed style. Figure 10-14 shows another style of lobed safety rod. The advantage of the lobed style is that even if the mold is not fully open, the stop will engage after a short distance if the mold begins to close.

Electrical interlock. The electrical interlock should disable the electrical circuit that controls mold closing when the door is open. Many machines feature a small rod attached to the door. When the door closes, the rod trips a limit switch and allows the machine to initiate an injection cycle. The electrical interlocks can fail if the limit switch does not function or if critical parts become loose and do not meet properly.

Figure 10-15 shows a push rod and the hole that allows the rod to actuate a limit switch.

One way to check on the electrical interlock is to close the door without tripping the electrical interlock. Removing or rotating the trip pin for the electrical interlock can achieve this goal. Figure 10-16 shows such a set-up. With the door closed, the platen should not move. If it does close, seek maintenance on the electrical interlock immediately.

This procedure of checking the electrical interlock involves tampering with the existing safety equipment.

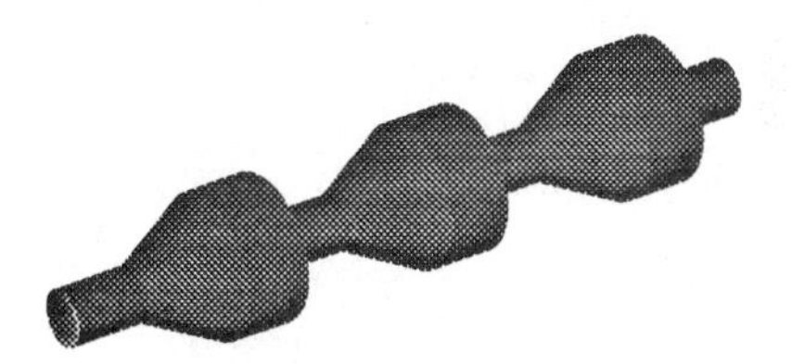

Fig. 10-13. The lobed style safety rod avoids the danger shown in Figure 10-12.

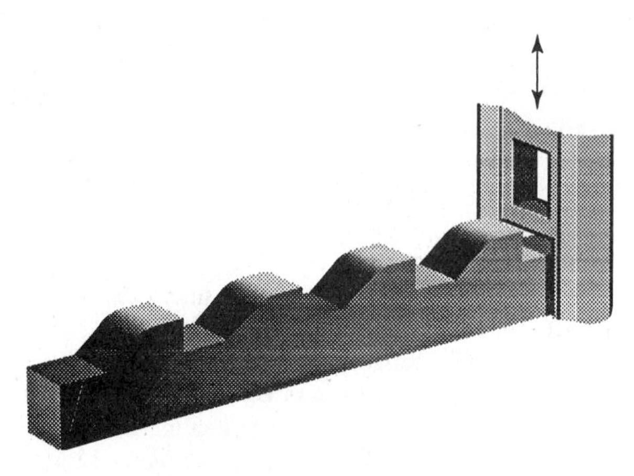

Fig. 10-14. This style of lobed safety rod appears on some medium and large size IMMs.

Fig. 10-15. The push rod trips the limit switch, signaling that the door is closed.

Fig. 10-16. The push rod shown in 10-15 has been rotated so that the door closes without tripping the limit switch. Return it to correct position immediately after testing the electrical safety system.

Fig. 10-17. This arm prevents or allows the flow of hydraulic oil to the mold closing cylinder.

Fig. 10-18. This hydraulic safety arm is directly connected to a switch controlling the flow of hydraulic oil.

Follow this procedure only with great caution and immediately replace the trip pin when complete.

Hydraulic interlock. Hydraulic interlock devices should prevent mold closing when the door is open. They often consist of a hydraulic switch and an actuating arm. Figure 10-17 shows an arm that raises when the door is closed and lowers when the door is open. Another style appears in Figure 10-18. Only when the door is closed can hydraulic oil enter the mold closing cylinder.

In comparison to the electrical interlock, which generally functions correctly or not at all, the hydraulic safety system may function partially, due to misadjustment. To check the hydraulic safety, open the door, trip the limit electrical system safety switch manually, and attempt to close the mold manually. The platen should not move.

It may be advantageous to manually lift the drop bar before checking the hydraulic safety. If the hydraulic safety system fails, the mold will begin to close and hit the mechanical stop. It is undesirable to stop the platen with the drop bar, because the platen may wedge on the tie rods. Some machines have two drop bars. On these machines, the platen will usually not wedge on the tie rods. However, for machines with one safety rod and drop bar, the platen may wedge on the tie rods. This may cause great difficulty in freeing the platen. Try to avoid this potential problem.

If the hydraulic safety is misadjusted, the clamping cylinder may receive enough oil to slowly move the mold. If this occurs, adjust the linkages to prevent any closing motion of the mold.

Purge guards. During purging, hot plastics may spray on nearby personnel. To prevent such accidents, injection-molding machines have purge guards. They are box-like metal guards that enclose the nozzle of the machine. Figure 10-19 shows one style of purge guard.

This guard attaches to the stationary platen with hinges, which allow it to swing up. The switch on the front of the guard is a mercury switch. When the guard is up, the switch prevents injection of hot plastics. Another type of purge guard has a hinged panel, which swings open like a door. A switch senses when the panel is open, and prevents injection.

Rear door safety systems. Many molding machine operators open and close the front door every shot. They have three safety systems to provide them protection. The rear door, however, does not activate similar redundant safety systems. Most molding machines have a switch attached to the rear door that completely shuts off the machine if the door is open. Figure 10-20 shows a type of rear-door safety device. In order to open the rear-door, the cap must come off. When the cap is off, as shown in Figure 10-21, the main motors will not run.

Fig. 10-19. If this purge guard is up and the nozzle is exposed, the machine will not initiate an injection cycle.

Fig. 10-20. This safety switch is attached to the rear door of the molding machine.

Safe molding practices. Properly guarded machines do provide protection for operators and technicians. However, guards cannot replace safe molding practices. Plastics can degrade in a molding machine and generate immense forces in the barrel and nozzle. Although uncommon, extreme pressures in the barrel have blown off end caps. For that to happen, the internal pressure must cause the many bolts that hold the end cap onto the barrel to shear. On a few occasions, barrels have exploded due to internal pressures.

To prevent any build up of pressure inside the barrel, always back off the carriage when a machine will be out of cycle for more than a few minutes. When the nozzle is not in contact with a mold, excess plastics can escape by drooling out the nozzle. This prevents the build-up of pressure.

The throat area of a molding machine should be cool enough to not melt pellets in the throat or in the bottom of the hopper. If the cooling in the throat area is inadequate, or the heater bands in the rear heating zone overheat—pellets may melt in the feed zone of

Fig. 10-21. Removing the cap stops the machine pumps before the rear door can be opened.

the screw, in the throat, and in the bottom of the hopper. This can be a very dangerous situation.

Pellets may melt together into a *bridge* in the bottom of the hopper. This bridge will prevent the flow of fresh pellets into the feed zone. Molding technicians may break through the bridge with a bar or rod. This can be extremely dangerous. If hot plastics under pressure are below the bridge, breaking down the bridge may cause them to blow upward. Molding technicians have died from severe burns sustained when hot plastics blew upward from the throat of a machine.

Safety for the machine. Machine safety involves systems to protect the machine from damage. Most IMMs have shear pins or shear keys connecting the screw motor to the screw. If the barrel is not hot enough or if a hard object blocks the screw, the shear pin should break before severe damage occurs to the screw.

Low mold pressure protection devices protect the mold from damage if some foreign object gets pinched between the mold halves. The key to this protection is preventing the application of full clamping pressure until the mold is almost fully closed. Only then does the machine exert full pressure against the mold.

A limit switch signals the machine controls when the mold is almost closed. By carefully adjusting the limit switch, molding technicians can control the initiation of full clamping pressure. Some molding technicians use pieces of cardboard as spacers to check the setting of the low mold switches. One or two pieces of cardboard between the mold halves should prevent the application of full clamping pressure. Without the cardboard pieces, the mold should clamp tightly.

When the switches controlling the mold pressure are correctly set, the presence of an object between the mold halves will inhibit clamping and interrupt the molding cycle. This is extremely important when operators or robots load inserts into a mold. It is possible for metallic inserts to jiggle or fall out of correct position. Unless the mold pressure protection is adjusted, severe mold damage could ensue.

Specification of Molding Machines

Molding machines have dozens of characteristics, but two of them provide a quick method to describe a machine. These two capabilities are shot size and clamp tonnage.

Shot size. Shot size is the maximum amount of material the machine will inject per cycle. Because of the great variations in the density of commercial plastics, a standard for comparison is needed. The accepted standard for shot size measurements is polystyrene. A small laboratory machine may have a maximum shot size of 20 grams [0.70 oz]. Large capacity machines may have a shot size of more than 9000 grams (g) or 9 kilograms (kg) [19.8 lbs].

Clamp tonnage. Clamp tonnage is the maximum force a machine can apply to a mold. One method of categorizing molding machines is to distinguish small, medium, and jumbo sizes. Generally, small machines have clamp tonnages of 99 tons or less; medium-sized machines run from 100 to 2000 tons; and jumbo machines over 2000 tons. Jumbo machines are available up to 10000 tons as a standard machine. Larger machines require special orders.

Sales figures help describe the utilization of the various sizes. During one year in the early 1990s, U.S. sales of small machines were slightly less than 400. Sales of medium sized machines were about 1200 machines. The jumbo machines sold about 50. It is clear that the medium range was dominant. In that range, the most common press size was about 300 tons.

As a rule of thumb, 3.5 kN [786 pounds-force] of force is needed for each square centimeter of mold cavity area. A machine with a clamp force of 3MN [337 tons-force] should be capable of molding a polystyrene plastics part 250 × 325 mm [9.8 × 12.8 in.]. This part would have a surface area of 812.5 cm^2 [126 $in.^2$]. Using the rule of thumb:

$$3000\ kN / 3.5\ kN/cm^2 = 857\ cm^2$$

Elements of Molding Cycles

Injection molding consists of five basic steps:

1. The mold closes.
2. As the screw begins to move forward, the non-return valve on the front end of the screw prevents the plasticated material from moving backward along the screw flights. Consequently, the screw functions as a ram and forces the hot materials into the mold cavity.
3. The screw maintains pressure through the nozzle until the plastics is cooled or set. In thermoplastic molding, timers maintain pressure on the plastics until the gates freeze. Gate freeze effectively separates the molded parts from the injection pressure. Maintaining further pressure on the plastics wastes time.
4. Timers stop injection pressure and the screw turns to draw fresh material from the feed hopper. The screw backs up until a limit switch signals the completion of the shot size. A decompression stroke pulls the screw backward a short distance. The purpose of decompression is to prevent the drooling of hot plastics into the sprue.
5. The mold opens and ejector pins remove the molded part.

It is common to group these basic steps as a *time cycle*. All injection systems have the following four elements in a time cycle:

1. *Fill time* is the time it takes to displace the air in the mold cavity with plastics material.
2. *Pack time* is the time required to maintain enough pressure to fill out the part and to achieve gate freeze.
3. *Cooling* or *dwell time* is the time required to cool or set enough for safe removal from the mold cavity.
4. *Dead time* is the time required to open the mold, remove the molded part and close the mold.

Advantages of Injection Molding

Injection molding is popular because metal inserts may be used; output rates are high; surface finish can be controlled to produce any desired texture; and dimensional accuracy is good. For thermoplastics, gates, runners, and rejected parts may be ground and reused. The following list enumerates eight advantages.

1. High output rates.
2. Fillers and inserts may be used.
3. Small, complex parts with close dimensional tolerances can be molded.
4. More than one material may be injection molded. (coinjection molding).
5. Parts require little or no finishing.
6. Thermoplastics scrap may be ground and reused.
7. Self-skinning structural foams may be molded (reaction injection molding).
8. Process may be highly automated.

Disadvantages of Injection Molding

Injection molding is not practical for short production runs. Molding machines are costly; consequently, hourly costs to operate the machines are considerable. Even small injection molds often cost many thousands of dollars. To make molding cost-effective, the numbers of parts must be high. Because molding is such a popular process, many companies compete for contracts. Some companies are unable to maintain profits and fall into bankruptcy.

The process of injection molding is complicated. Occasionally, due to poor part or mold design, molders find it very difficult to make acceptable parts. When processes are not under contol, scrap rates increase and part rejections by customers may cause major financial loses. Table 10-1 lists some of the problems associated with injection molding.

Table 10-1. Problems In Injection Molding

Difficulty	Cause	Possible Remedy
Black specks, spots, or streaks	Flaking off of burned plastics on cylinder walls	Purge heating cylinder
	Air trapped in mold causing burning	Vent mold properly
	Frictional burning of cold granules against cylinder walls	Use lubricated plastics
Bubbles	Moisture on granules	Dry granules before molding
Flashing	Material too hot	Reduce temperature
	Pressure too high	Lower pressure
	Poor parting line	Reface the parting line
	Insufficient clamp pressure	Increase clamp pressure
Poor finish	Mold too cold	Raise mold temperature
	Injection pressure too low	Raise injection pressure
	Water on mold face	Clean mold
	Excess mold lubricant	Clean mold
	Poor surface on mold	Polish mold
Short moldings	Cold material	Increase temperature
	Cold mold	Increase mold temperature
	Insufficient pressure	Increase pressure
	Small gates	Enlarge gates
	Entrapped air	Increase vent size
	Improper balance of plastics flow in multiple cavity molds	Correct runner system
Sink marks	Insufficient plastics in mold	Increase injection speed, check gate size
	Plastics too hot	Reduce cylinder temperature
	Injection pressure too low	Increase pressure
Warping	Part ejected too hot	Reduce plastics temperature
	Plastics too cold	Increase cylinder temperature
	Too much feed	Reduce feed
	Unbalanced gates	Change location or reduce gates
Surface marks	Cold material	Increase plastics temperature
	Cold mold	Increase mold temperature
	Slow injection	Increase injection speed
	Unbalanced flow in gates and runners	Rebalance gates or runners

Injection Molding Thermosets.

Both machine and mold designs differ when molding thermosetting materials. A non-return check valve is not required since the material is very viscous and little material is left in the barrel after injection. Screw flights are shallow with a one-to-one compression ratio. Bulk-molding compounds or other heavily filled or reinforced materials generally use a plunger machine. The length-to-diameter (L/D) ratio in these machines generally ranges from 12:1 to 16:1; compared to higher ratios for injection molding of thermoplastics (Fig. 10-22).

Coinjection Molding.

Coinjection molding is a process in which two or more materials are injected into the mold cavity (Fig. 10-23).

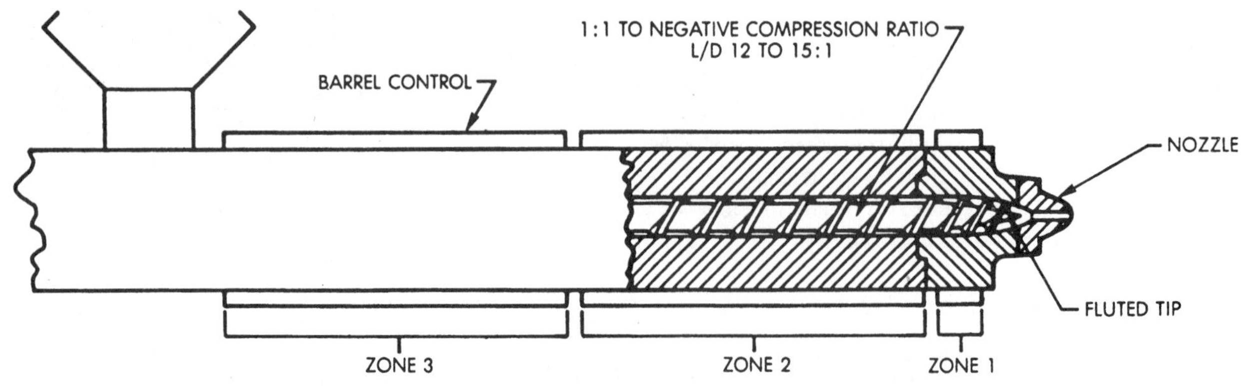

Fig. 10-22. Schematic of basic thermoset screw and barrel assembly.

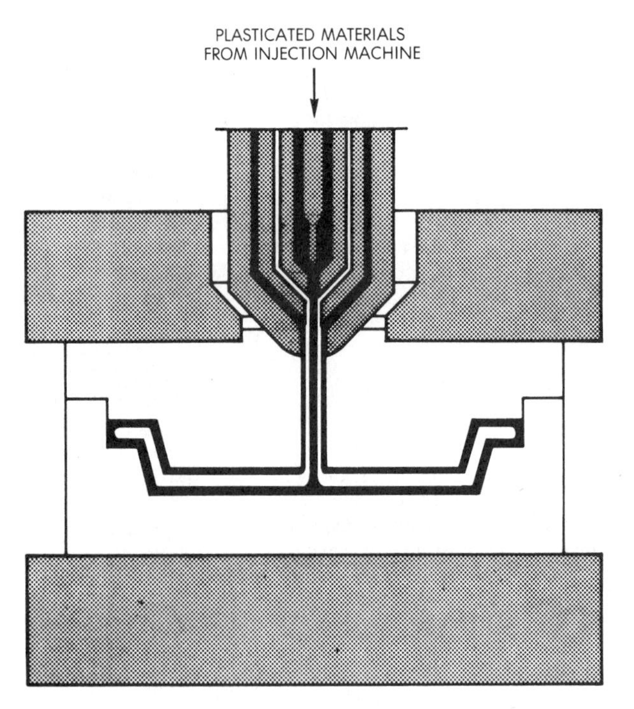

Fig. 10-23. This coinjection molding machine feeds three separate melt channels to the mold.

This usually produces a skin of material on the mold surfaces and a cellular center core. The core material includes blowing agents to produce the desired cellular densities. The process is sometimes incorrectly called *sandwich molding,* because of the composite layered effect.

Coinjection molding may use differing families of plastics for the skin or core layer. Fiber reinforcements provide greater strength, but the flow patterns may cause unwanted orientation of fibers. The selection of materials and additives is limited only for the exposed skin surfaces. They generally need to be platable or pigmented.

Items that use the coinjection molding process include: automotive parts, office machine housings, furniture components, and appliance housings.

Molding Liquid Materials

Several processes convert liquid resins into finished plastics parts. An important reason to use liquid materials is that they flow into mold cavities with much less force than melted thermoplastic pellets. Liquid materials flow around fibrous reinforcements and do not damage delicate inserts. The most significant processes relying on liquid materials are reaction injection molding (RIM), reinforced reaction injection molding (RRIM), and liquid resin molding or resin transfer molding (RTM).

Reaction Injection Molding

Reaction injection molding (RIM) is also known as liquid reaction molding or high-pressure impingement mixing. It is a process in which several reactive chemical systems are mixed and forced into the molding cavity where the polymerization reaction occurs. Although most of the current RIM components are polyols and isocyanates, other modified polyurethanes; polyester, epoxies, and polyamide monomers are used.

The process involves impingement-atomized mixing of two or more liquids in a mixing chamber. This mixture is immediately injected into a closed mold, and a rigid, structural-foamed or cellular product results (Fig. 10-24).

The automotive and furniture industries are the major users of RIM parts. Bumpers, belts, shock-absorbing parts, fender components, and cabinet elements are familiar examples (Fig. 10-25.) Reaction injection molding is limited only by mold and equipment size. Present machines are capable of molding 300 kf (135 lbs) of mixture in one shot. Clamp capacity requirements are much lower than those of conventional injection molding. Clamps are often designed to be opened and closed like a book. This allows easy parts removal and operator access to the mold (Fig. 10-26). Seven advantages and four disadvantages of reaction injection molding are listed in Figures 10-25 and 10-26.

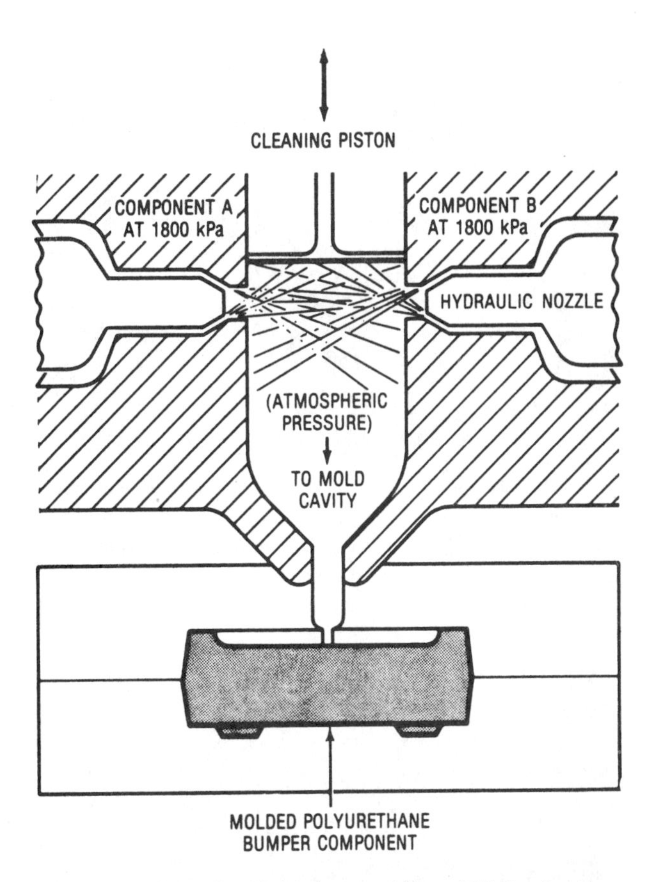

Fig. 10-24. Reaction injection molding (RIM), showing impingement mixing. Components are atomized to a fine spray by a pressure drop from 2 500 psi (1 800 kPa) to atmospheric pressure.

Fig. 10-25. RIM equipment producing automobile components. (General Motors)

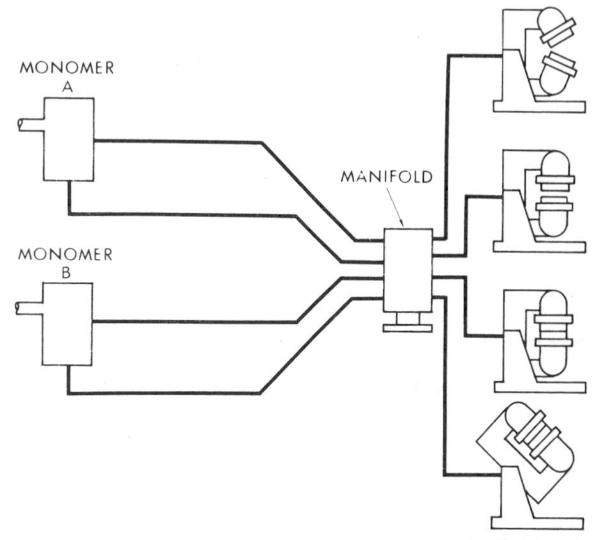

(A) The C-frame stations (clamps) open in book-fashion for fast part removal and rock back for venting.

(B) RIM molding machine with mold open.

Fig. 10-26. Molding machine system for RIM. (Cincinnati Milacron)

Reinforced Reaction Injection Molding (RRIM)

When short fibers or flakes (particulates) are used to produce a more isotropic product, the process is called reinforced reaction injection molding (RRIM). Fiber loading increases monomer viscosities and abrasive wear on all flow surfaces.

Polyurethane/urea hybrid, epoxy, polyamide, polyurea, polyurethane/polyester hybrid, polydicyclopentadiene, and other resin systems have been used for RIM and RRIM. RRIM applications include: automobile fenders, panels, bumpers, shields, radomes, appliance housings, and furniture components.

Advantages of Reaction Injection Molding (RIM)

1. Cellular core and integral skin for durable products
2. Fast cycle times for large products
3. Good finishes that are paintable
4. Less cost than castings
5. Polymers may be reinforced
6. Reduced tooling and energy cost (compared with injection molding)
7. Lower equipment cost, due to low pressures

Disadvantages of Reaction Injection Molding

1. New technology requiring investment in equipment
2. System requires four or more chemical-component tanks
3. System requires handling of isocyanates
4. Releasing agents required

Liquid Resin Molding

Liquid resin molding (LRM) is a term used to describe products produced by a variety of low-pressure methods in which mixing is often mechanical rather than by impingement. The term once described a very specialized process for potting and encapsulating components. LRM is used to describe a group of processing methods including resin transfer molding (RTM), vacuum injection molding (VIM), and thermal expansion resin transfer molding (TERTM) in which resins are forced under low pressure into the molding cavity and rapidly cured. Epoxies, silicones, polyesters, and polyurethanes lend themselves to use in liquid resin molding processes.

Resin transfer molding. *Resin transfer molding,* also called *resin injection molding,* is a process whereby catalyzed resin is forced into a mold in which fragile parts or reinforcements have been placed. Low pressure does not distort or move the desired fiber orientation of preforms or other materials (Fig. 10-27). Boat hulls, hatches, computer housings, fan shrouds, or other large composite structures may be produced by this technique. The basic concept of RTM is illustrated in Figure 10-28.

Advantages of RTM

1. Eliminates the plasticizing stage necessary with dry compounds
2. Permits encapsulation of delicate, or fragile parts
3. No hand mixing
4. Eliminates preheating and preforming
5. Lower pressures are needed
6. Minimum material waste
7. Rapid cure of resins at low temperatures
8. Improved reliability and dimensional stability
9. Reduction of material handling

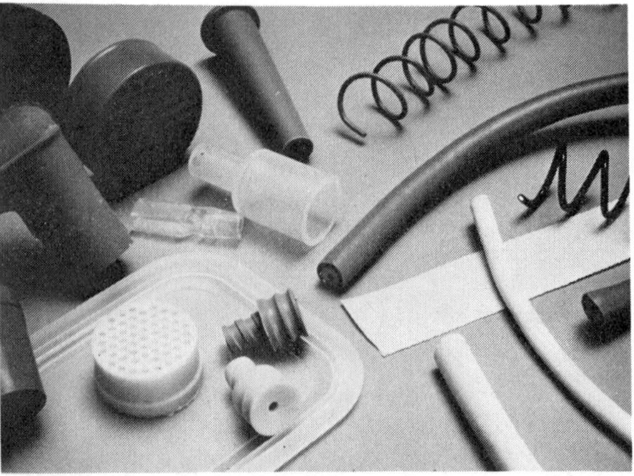

(A) RTM is recommended for such parts as electrical connector inserts, diaphragms, valves, O-rings, stoppers, and plungers as shown above. This molding technique does not distort windings or shift delicate devices in the molding cycle. (Plastics Design Forum)

(B) RTM bathtub with glossy gelcoat finish. (Molded Fiber Glass Co.)

Fig. 10-27. RTM products.

Vacuum injection molding. In a process similar to RTM, preforms are placed on a male mold and the female mold is closed. A vacuum is drawn pulling the reactive resin system into the mold cavity. This technique (VIM) is illustrated in Figure 10-29.

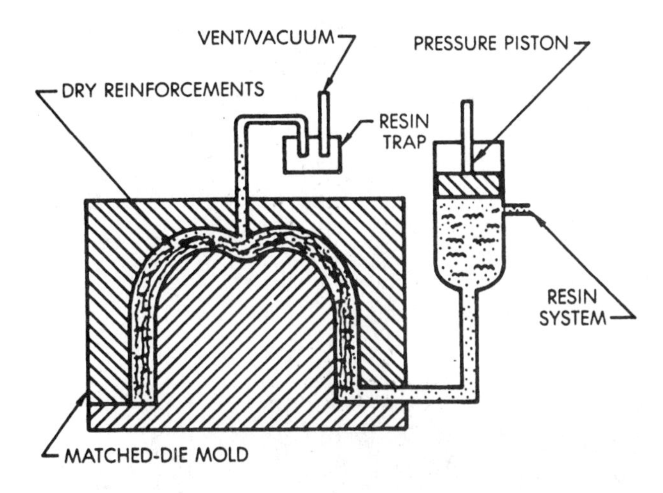

Fig. 10-28. Concept of RTM. Preform reinforcements are loaded into a matched-die mold. After mold closing, a liquid resin system is forced into and around the preform.

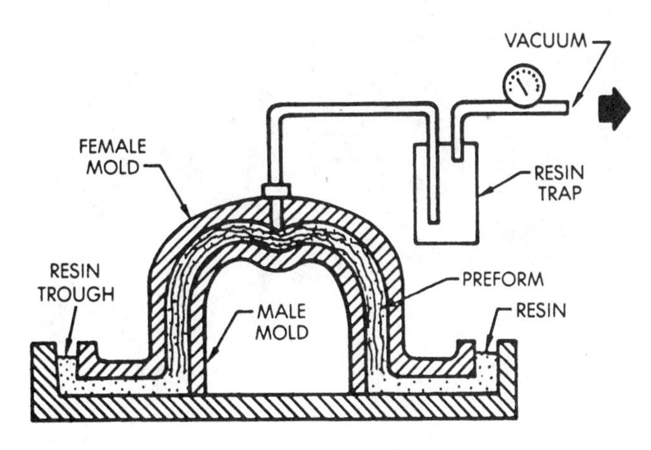

Fig. 10-29. Concept of VIM.

Thermal-expansion resin transfer molding. TERTM is a variation of the RTM process. A cellular mandrel of PVC or PU is wound or wrapped with reinforcements and placed into a matched die. Epoxy or other resin systems are injected to impregnate the reinforcements. The heated die causes the cellular material to further expand, forcing the impregnated reinforcements against the mold walls. The tooling is vented to allow excess matrix or entrapped air to escape.

Molding Granular Thermoset Materials

There are two processes in widespread use for the molding of granular or pelletized thermoset materials. They are compression molding and transfer molding.

Compression Molding

One of the oldest known molding processes is compression molding. The plastics material is placed in a mold cavity and formed by heat and pressure. As a rule, thermosetting compounds are used for compression molding, but thermoplastics may be used. The process is somewhat like making waffles. Heat and pressure force the materials into all areas of the mold. Then, after the heat hardens the substance, the part is removed from the mold cavity (Fig. 10-30).

To reduce pressure requirements and production (cure) time, the plastics material is usually preheated with infrared, induction, or other heating methods before it is placed in the mold cavity. A screw extruder is sometimes used to reduce cycle time and increase output. The screw extruder is often used to make preformed slugs that are loaded into the molding cavity. The screw-compression process greatly reduces cycle time, eliminating the greatest drawback of compression molding. Compression-molded parts with heavy wall thickness may be produced with up to 400 percent

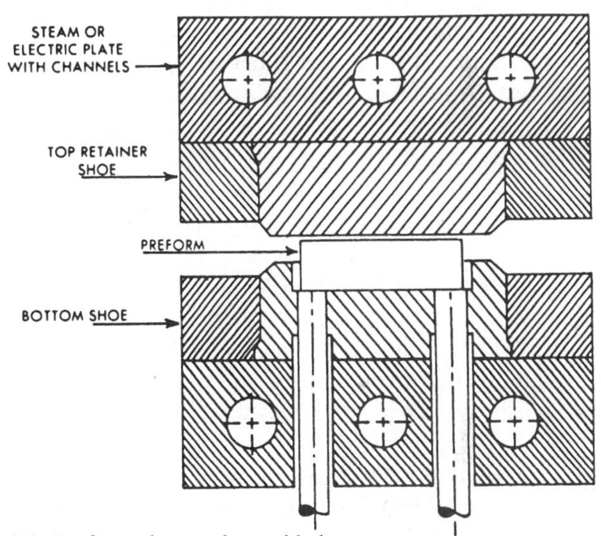

(A) Preform about to be molded.

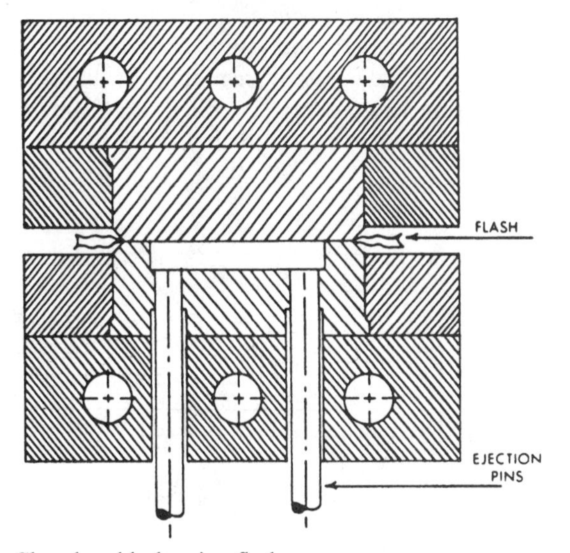

(B) Closed mold, showing flash.

Fig. 10-30. Principle of compression molding

greater product output per mold cavity when this method is used.

Thermosetting bulk molding compounds (BMC) are used in forming polyester compounds. BMC is a mixture of fillers, resins, hardening agents, and other additives. Hot extruded preforms of this material may be loaded straight into the cold cavity or feed chute (Fig. 10-31).

Phenolic plastics, urea-formaldehyde, and melamine compounds are also popular molding materials. Like BMCs, they are usually preformed for automation and speed. Heavily filled and reinforced sheet-molding compounds are used. They may be placed in alternating layers for more isotropic properties, or placed in one direction for more anisotropic properties.

Most compression-molding equipment is sold by the press or platen rating. A force of 20 MPa or 2 900 psi [2 kN/cm^2] is usually required for moldings up to 1 in. [25 mm] thick. An added 725 psi [0.5 kN/cm^2] should be provided for each 1 in. [25 mm] increase. Hydraulic action provides this force (Fig. 10-32).

Steam, electricity, hot oil, and open flame are some of the means for heating the molds, platens, and related equipment. Hot oil is common because it may be heated to high temperatures with little pressure. Electricity is clean, but limited by wattage.

During preforming and the actual molding process, heat and various catalysts begin cross-linking the molecules. During the cross-linking reaction, gases, water, or other by-products may be freed. If they are trapped in the mold cavity, they may affect the plastics part and the part may be damaged, of poor quality, or marked by surface blisters. Molds are usually vented to allow the escape of these by-products.

In molding, the thermosetting plastics compound becomes cross-linked and infusible, thus these products may be removed from the molding cavity while hot. Thermoplastic materials, because they do not cross-link to any extent, must be cooled before removal. Many elastomers are molded by this process.

Long runs of moderately complex parts are often produced by compression molding. Mold maintenance and starting costs are low; material waste is rather low; and large bulky parts are practical. Very complex parts are hard to mold, however. Inserts, undercuts, side draws, and small holes are not practical with this method when it is necessary to maintain close tolerances.

The sequence of compression molding may include the following six steps:

1. Clean mold and apply mold release (if required).
2. Load preform into cavity.

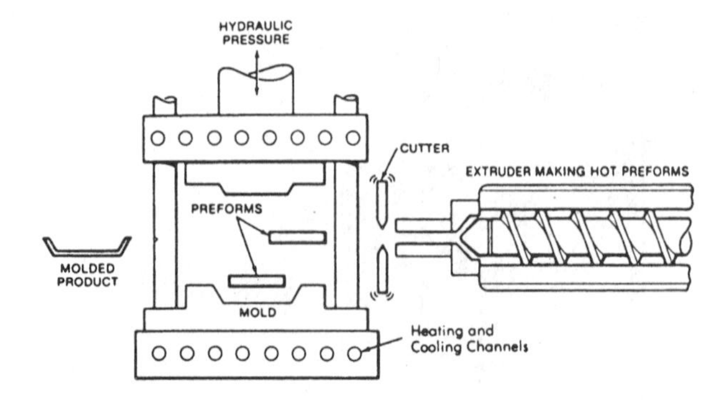

Fig. 10-31. Compression molding, showing hot preforms being fed to mold cavity.

(A) Large compression-molding press being tested. (Hull Corp.)

(B) Stokes press used to manufacture electrical plate covers.

Fig. 10-32. Compression-molding presses.

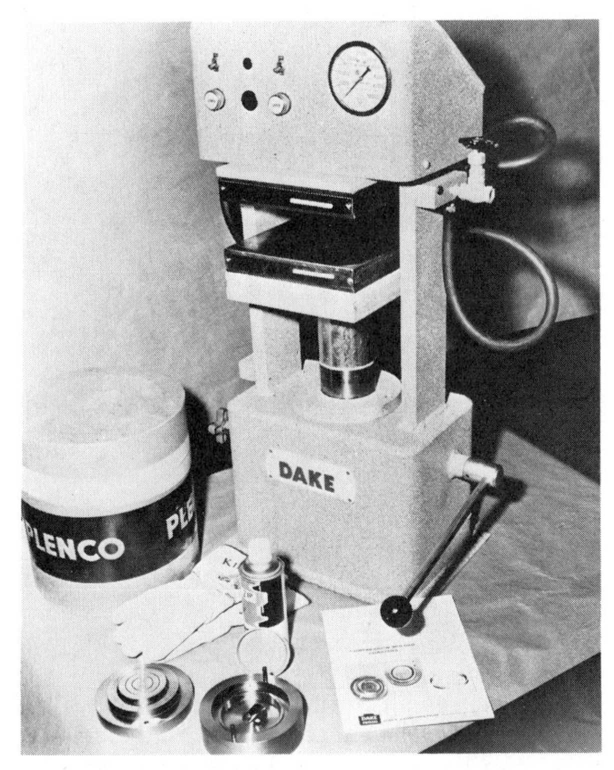

(C) Small laboratory press and mold. (Dake Corp.)

Fig. 10-32. Compression-molding presses.

3. Close mold.
4. Open mold briefly to release trapped gases (breathe mold).
5. Apply heat and pressure until cure is complete (dwell).
6. Open mold and place hot part in cooling fixture.

Six advantages and eight disadvantages of compression molding are given here.

Advantages of Compression Molding

1. Little waste (no gates, sprues, or runners in many molds)
2. Low tooling costs
3. Process may be automated or hand-operated
4. Parts are true and round
5. Material flow is short; less chance of disturbing inserts, causing product stress, and/or eroding molds
6. Multiple cavity placement in tooling is not dependent on a balanced feeder system

Disadvantages of Compression Molding

1. Hard to mold complex parts
2. Inserts and fine ejector pins are easily damaged
3. Complex shapes are sometimes hard to achieve
4. Long molding cycles may be needed
5. Unacceptable parts cannot be reprocessed
6. Trimming flask can be difficult
7. Some part dimensions are controlled by material charge rather than tooling
8. Require external loading and unloading equipment for automation.

Compression-molded products include dinnerware, buttons, buckles, knobs, handles, appliance housings, drawers, parts bins, radio cases, large containers, and many electrical parts (Fig. 10-33).

Two variations of the compression-molding processes deserve special attention. They are *cold molding* and *sintering*.

Cold molding. In cold molding, plastics compounds (mostly phenolics) are formed in unheated molds. After forming, a part is hardened into an infusible mass in an oven (Fig. 10-34). Electrical insulator parts, utensil handles, battery boxes, and valve wheels are examples of products made in this manner.

Sintering. *Sintering* is the process of compressing a powdered plastics in a mold at temperatures just below its melting point for about one-half hour (Fig. 10-35). The powdered particles are fused (sintered) together,

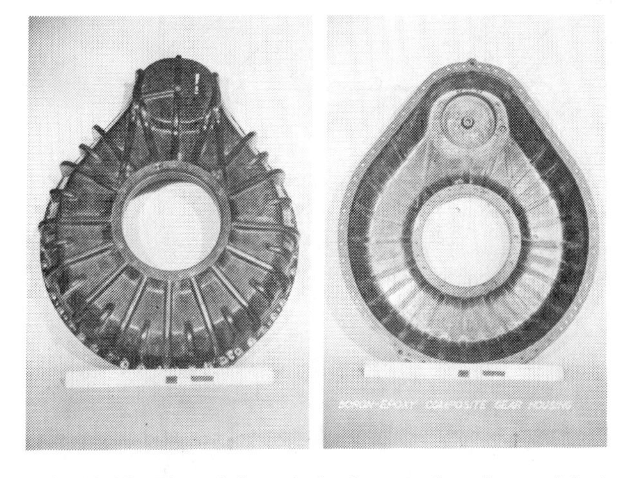

Fig. 10-33. The reinforced plastics gear housing at right is lighter and stronger than the metal housing it replaces, shown at left. Many similar parts are compression-molded. (Allison Division, Detroit Diesel)

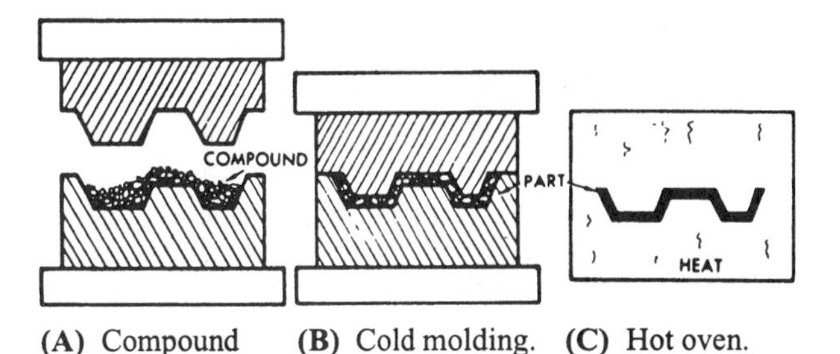

(A) Compound charged into mold. (B) Cold molding. (C) Hot oven.

Fig. 10-34. Principle of cold molding.

but the mass as a whole does not melt. Bonding is done by the exchange of atoms between the individual particles. After the fusion process, the material may be postformed under heat and pressure to the needed dimensions.

The three major variables governing the sintering process are temperature, time, and the composition of the plastics.

The process is adapted from the sintering operations of powder metallurgy. Sintering can be used to

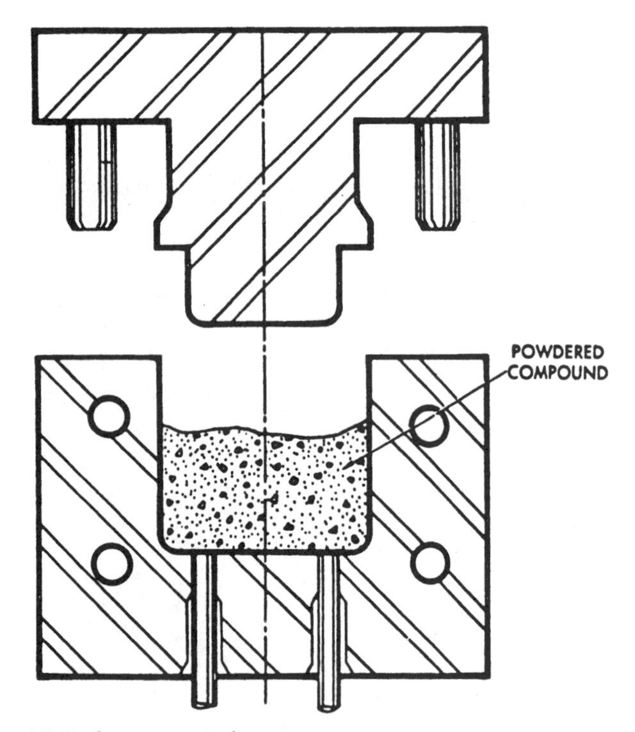

(A) Before compression.

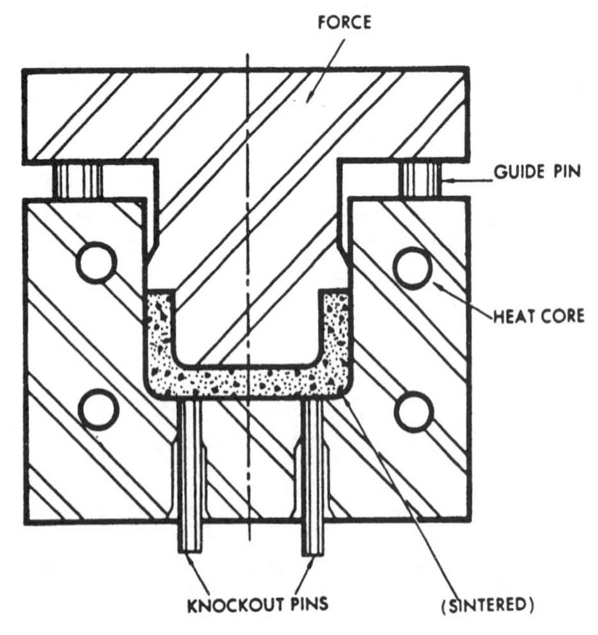

(B) Compression and heating.

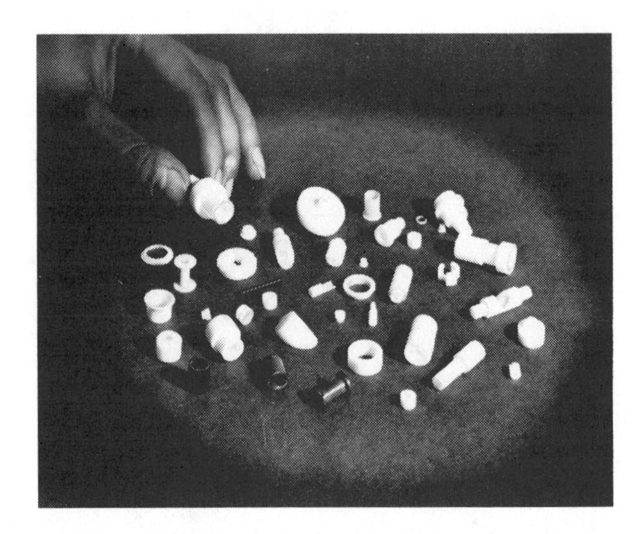

(C) PTFE parts have been sintered to basic shape before machining. (Chemplast, Inc.)

Fig. 10-35. Sintering plastics parts.

process polytetrafluoroethylene, polyamides, and other specially filled plastics, and is the primary method by which polytetrafluoroethylene is processed. Dense parts with first rate electrical and mechanical properties may be produced. The cost of tools and production is high, and parts with thin walls or variations in cross-sectional thickness are hard to form.

Transfer Molding

Transfer molding has been known and practiced since World War II. The process is sometimes called plunger molding, duplex molding, reserve-plunger transfer molding, step molding, injection transfer molding, or impact molding. It is actually a variation of compression molding but differs in that the material is loaded in a chamber outside the mold cavity. An advantage of transfer molding is that the molten mass is fluid when entering the mold cavity. Fragile, complex shapes with inserts or pins may be formed with accuracy. Transfer-molding techniques are much like those of injection molding, except that thermosetting compounds are normally used.

The American Society of Tool and Manufacturing Engineers, recognizes two basic types of transfer molds:

1. Pot or sprue molds
2. Plunger molds

Plunger molds (Fig. 10-36) differ from the sprue molds (Fig. 10-37) in that the plunger or force is pushed to the parting line of the mold cavity when inserting the plastics material. In sprue molds the plastics is fed by gravity down a hole (sprue). With plunger

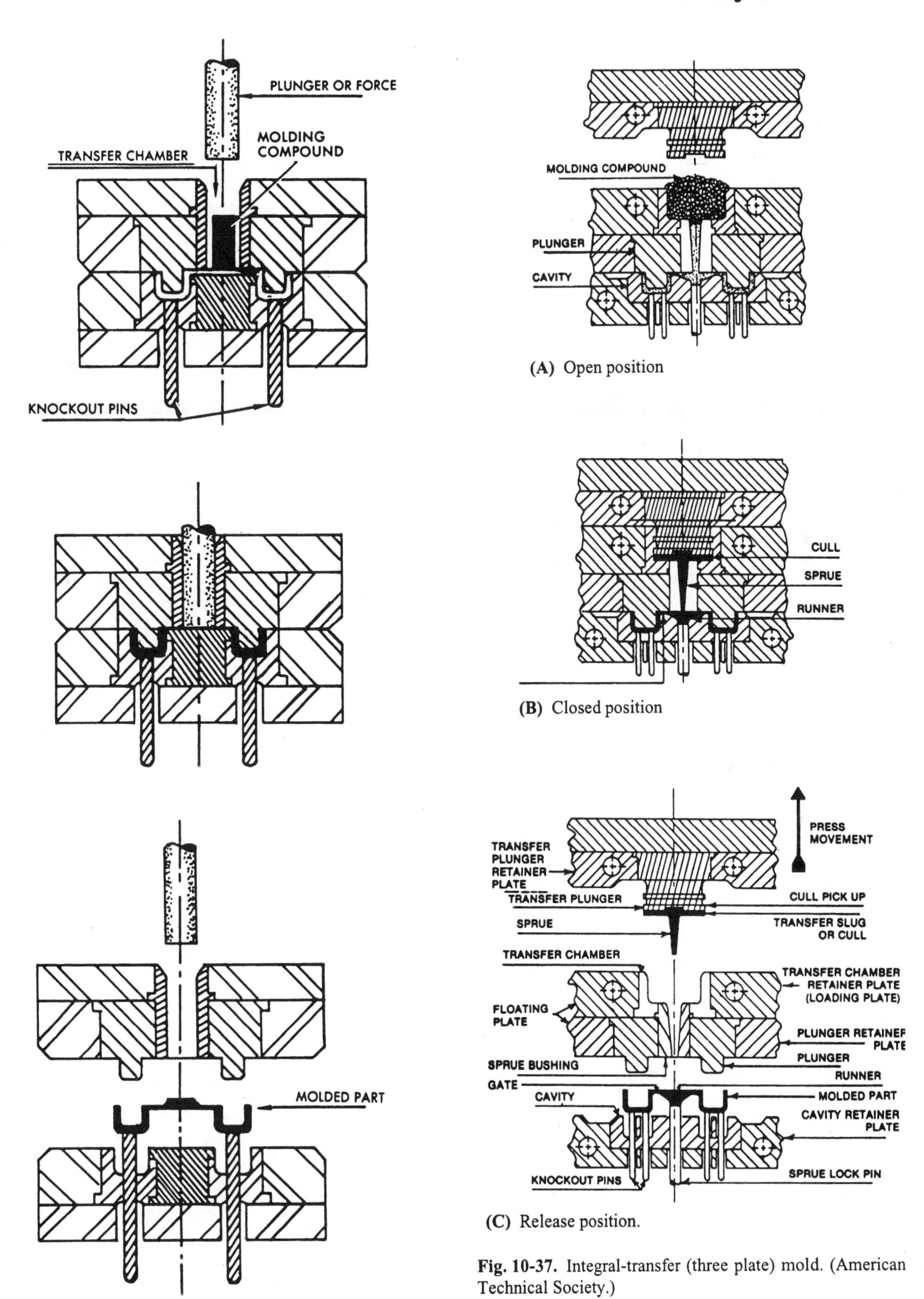

Fig. 10-36. Plunger (two-plate) transfer mold.

(A) Open position

(B) Closed position

(C) Release position.

Fig. 10-37. Integral-transfer (three plate) mold. (American Technical Society.)

molds, only runners and gates are left as waste on the molded part.

A third type of mold may also be included in transfer molding. In this type, molding compound is preplasticized by extruder action, then a plunger forces the melt into the mold (Fig. 10-38).

The cost of detailed mold designs and high waste from culls, sprues, and flash are two major limitations of transfer molding.

Although most parts are limited in size, there are numerous applications including distributor caps, camera parts, switch parts, buttons, coil forms, terminal block insulators; and complex shapes such as cups and caps for cosmetic containers.

Problems in compression and transfer molding are found in Table 10-2. Six advantages and four disadvantages of transfer molding are shown below.

Advantages of Transfer Molding

1. Less mold erosion or wear
2. Complex parts (small diameter holes or thin wall sections) and inserts may be molded and used
3. Less flash than in compression molding
4. Densities are more even than in compression molding
5. Multiple pieces may be molded
6. Shorter mold and loading times than most compression molding processes

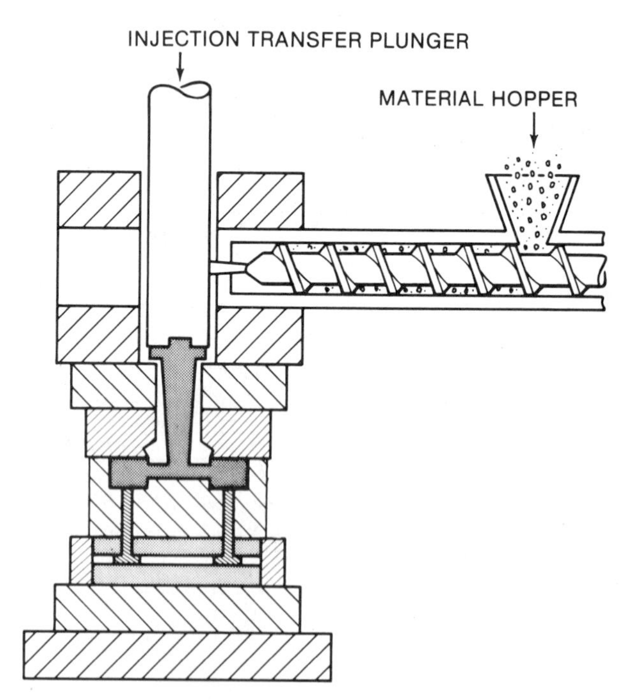

Fig. 10-38. Injection-transfer molding, showing the hot extruded compound being forced into the mold cavity by the transfer plunger.

Table 10-2. Problems In Compression and Transfer Molding

Defect	Possible Remedy
Cracks around inserts	Increase wall thickness around inserts Use smaller inserts Use more-flexible material
Blistering	Decrease cycle and/or mold temperature Vent mold—breathe mold Increase cure—increase pressure
Short and porous moldings	Increase pressure Preheat material Increase charge weight of material Increase temperature and/or cure time Vent mold—breathe mold
Burned marks	Reduce preheating and molding temperature
Mold sticking	Raise mold temperature Preheat to eliminate moisture Clean mold—polish mold Increase cure Check knockout pin adjustments
Orange peel surface	Use a stiffer grade of molding material Preheat material Close mold slowly before applying high pressure Use finer ground materials Use lower mold temperatures
Flow marks	Use stiffer material Close mold slowly before applying high pressure Breathe mold Increase mold temperature
Warping	Cool on jig or modify design Heat mold more uniformly Use stiffer material Increase cure Lower temperature Anneal in oven
Thick flash	Reduce mold charge Reduce mold temperature Increase high pressure Close slowly—eliminate breathe Increase temperature Use softer grade material Increase clamping pressure

Disadvantages of Transfer Molding

1. More waste from runners and sprues
2. More costly equipment and molds
3. Molds must be vented
4. Runners and gates must be removed

Vocabulary

The following vocabulary words are found in this chapter. Look up the definition of any of these words you do not understand as they apply to plastics in the glossary, Appendix A.

- Automatic cycle
- Barrel
- Breathing
- Cold molding
- Compression mold
- Compression molding
- Dwell time
- Flash
- Impingement
- Injection molding
- LRM
- Plasticate
- Platens
- Purging
- RIM
- RRIM
- RTM
- Sintering
- TERTM
- Transfer molding
- VIM

Questions

10-1. The excess material left on a product after compression molding is called ___?___.

10-2. The amount of material used to fill a mold during an injection-molding process is called ___?___.

10-3. The chief advantage of injection molding over other molding processes is ___?___.

10-4. Name three ways to reduce production time in compression molding.

10-5. Name three ways to reduce the cycle time in compression molding.

10-6. Opening the compression mold to allow gasses to escape during the molding cycle is called ___?___.

10-7. The mark on a molding where the halves of the mold meet in closing is called ___?___.

10-8. The process of molding or forming items from pressed powders at a temperature just below the plastics' melting point is called ___?___.

10-9. Name three major disadvantages of transfer molding.

10-10. The time it takes to close a mold, form a part, open the mold and remove the cooled part is called the ___?___ time, in injection molding.

10-11. In ___?___ molding, the material is loaded in a chamber outside the mold cavity before it is made fluid and forced into the mold cavity.

10-12. The two factors that rate the capacity of an injection-molding machine are ___?___ and ___?___.

10-13. Thermosetting materials with fragile, intricate shapes, and with inserts or pins, may be ___?___ or ___?___ molded.

10-14. Name the two molding processes that use thermosetting and selected thermoplastic materials in preform or bulk molding compound forms.

10-15. Name three processes that are similar to transfer-molding processes.

10-16. Name four commonly used polymers in liquid resin molding.

10-17. A common expression for the time during which a molding machine is not operating is ___?___.

10-18. Reaction injection molding is also known as ___?___ or high-pressure impingement mixing.

Activities

Compression Molding

Introduction. The compression-molding process is one of the oldest and simplest plastics processing techniques. It normally utilizes thermosetting resins.

Equipment. Compression molder, mold, resin, 35 g phenolic, silicone mold release, safety glasses, gloves to resist high temperatures.

Procedure

10-1. Adjust temperature to 190 °C [356 °F] and connect cooling hoses to both platens.

10-2. Inspect and clean mold with wooden scraper.

CAUTION: Steel instruments may damage mold surfaces.

10-3. Place all parts of the mold on lower platen and close press platens. Heat for 10 minutes.

10-4. Carefully measure all phenolic molding materials in a container. (Fig. 10-39) Too much material will produce a thick product, too little will produce a thin product.

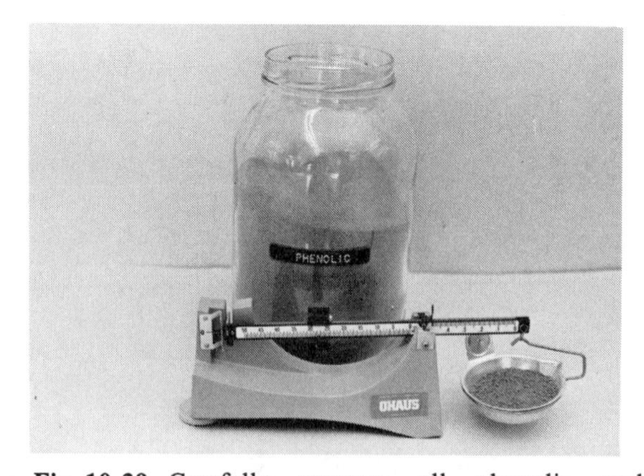

Fig. 10-39. Carefully measure all phenolic molding materials.

10-5. Remove the mold from the press using insulated gloves and set it on a heat resistant surface, as seen in Figure 10-40.

CAUTION: Mold and platens are hot. Some practice is needed to handle hot molds without wasting time and cooling the molds.

10-6. Assemble the mold and pour measured phenolic charge into mold. (Fig. 10-41)

10-7. Close the mold and return to the press. Be certain to center the mold on the platen. (Fig. 10-42.)

10-8. Apply 13 500 kPa [1 950 psi] on molding surface. After about 10 seconds, release pressure to allow gas to escape, then apply pressure again. (Fig. 10-43) A MSDS on phenolic molding compounds will indicate that processing releases small amounts of ammonia, formaldehyde, and phenol. Formaldehyde has the lowest TLV of those three gases. It is rated at 0.3 ppm ceiling by the ACGIH. Use adequate ventilation to remove these gases from the work area.

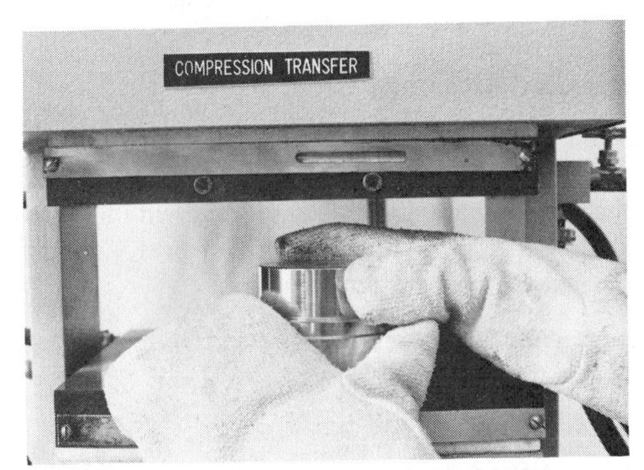

Fig. 10-40. A hot mold is removed from compression molding platens.

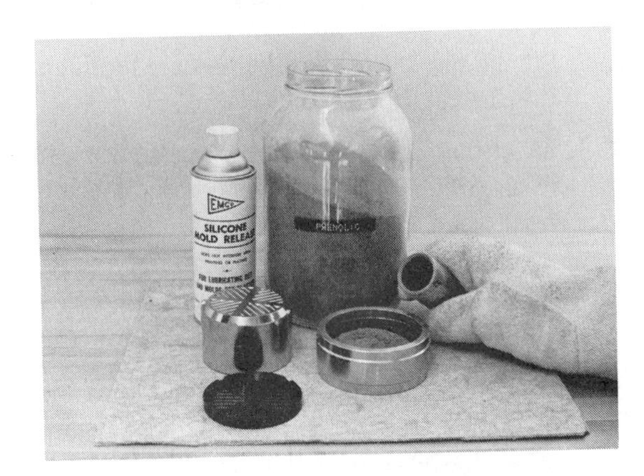

Fig. 10-41. Place measured amount of phenolic material in the hot mold.

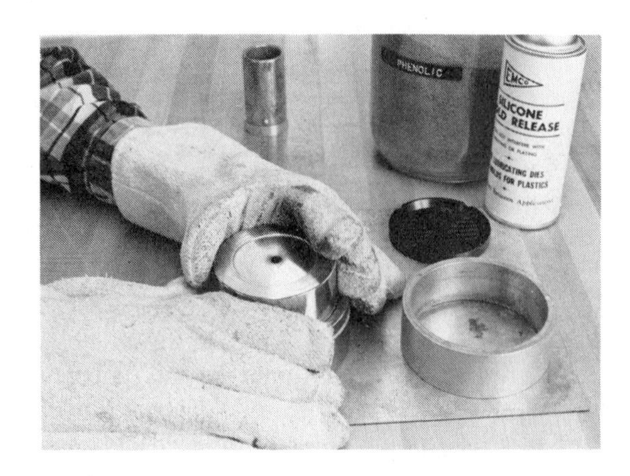

Fig. 10-42. Assembly hot mold, and return it to hot platens.

Fig. 10-43. Pressure is maintained to complete the compression molding.

10-9. Hold pressure for 5 minutes. Be sure pressure is maintained.

10-10. To cool platens, turn off power and slowly open water valve.

CAUTION: Steam can cause severe burns.

10-11. Release pressure and remove mold.

CAUTION: Mold will still be hot.

10-12. Carefully remove part from mold and allow to cool. Trim flash from part. If part exhibits a dull surface, repeat the process at higher temperature and/or pressure.

Injection Molding

Introduction. The finished size of injection molded parts is controllable, within limits, by the processing conditions. To investigate the relation between part size and process parameters, accurate measurement is essential. Figure 10-44 shows a devise for rapid and accurate length measurement.

Equipment. Injection-molding machine, mold that produces a rectangular part such as an impact or tensile bar, accurate measuring device, crystalline thermoplastics (HDPE, LDPE, or PP).

Procedure

10-1. Establish a *control* condition which is capable of producing acceptable parts with little variation in length. This will require holding pressure on the mold until the gate(s) freeze.

To determine this time, step the injection times from a short time to a long time. A short time will be slightly longer than the time needed to fill the part. A long time is over 30 seconds. Carefully cut the parts off the runner system and weigh the parts individually. Longer injection time will cause weight gain. At some point in time, the parts will no longer gain weight. That indicates gate freeze.

Some parts with large gates, or sprue gated parts will continue to gain weight even after 1 minute of injection hold pressure. Since that will delay the activity, select a mold that has gates small enough to freeze in about 10 seconds.

10-2. Select one process parameter, and vary it from one extreme to another. For example, select injection pressure, and vary it in increments of 100 psi in the hydraulic pressure. Reduce pressure until the parts are not completely filled or *short*, then increase incrementally until the mold flashes.

10-3. Number each shot in order and keep track of the parameter under investigation.

10-4. Allow 40 or more hours for part stabilization.

10-5. Measure the parts and create a graph of length on the Y axis against the increments of pressure (or other parameter) change on the X axis.

10-6. If time allows, investigate changes in injection speed, cooling time, mold temperature, back pressure, screw RPM, and other process parameters.

10-7. Arbitrarily select a length near the high or low extreme, and find a process which yields those parts with low variations in length.

Fig. 10-44. Three pins locate the tensile bar exactly and the dial indicator shows its length relative to a standard.

Characteristics of Polymer Flow

Introduction. Most new students of plastics do not intuitively grasp polymer flow. The rheological aspects of flow are difficult to observe without instrumentation or a spiral-flow mold. However, the nature of fountain flow is easy to observe.

Equipment. Injection-molding machine, plastics pellets, a mold, safety glasses.

Procedure

10-1. Position a small piece of facial or toilet tissue on a flat cavity surface. A tiny amount of water or grease will gently adhere the tissue to the mold.

10-2. Predict the location of the tissue after the melt fills the cavity.

10-3. Inject the part and observe the location of the tissue. Figure 10-45 shows a tissue piece on a tensile bar.

10-4. Put a piece of tissue at various locations in the cavity or runner system. Do the pieces move?

10-5. Try clamping a tissue over the entire part surface. Figure 10-46 shows that the force of the flow did not rip or significantly move the tissue in any location.

10-6. Describe the process of mold filling, taking into account the observations that the melt does not slide along the surface of the mold.

Fig. 10-45. This small piece of tissue did not move during the flow of plastics into the mold.

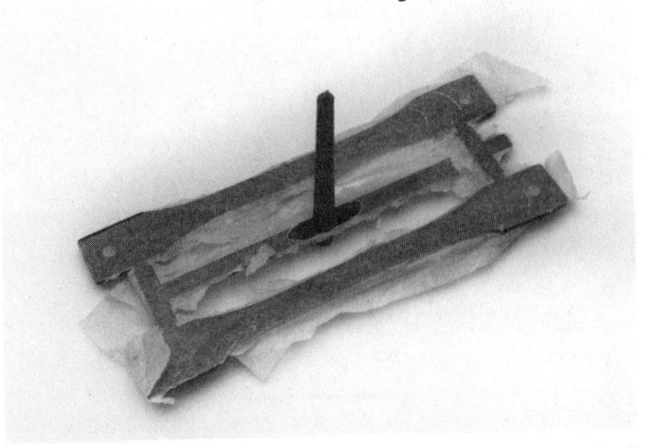

Fig. 10-46. The slight wrinkles in the tissue record the shape of the flow front as it advanced through the mold.

Chapter 11

Extrusion Processes

Introduction

The word extrusion is derived from the Latin word *extrudere* meaning (ex) out and (trudere) to push. In extrusion, dry powder, granular, or heavily reinforced plastics is heated and forced through an orifice in a die. Although ram extruders are still in use, particularly for products made of ultra-high molecular weight polyethylene (UHMWPE), screw extruders are the rule. The screw plasticates (melts and mixes) the material and forces it through the die.

This chapter deals with extrusion products and processes. The major topics are:

I. Extrusion equipment
II. Compounding
III. Major types of extrusion products
  A. Profile extrusion
  B. Pipe extrusion
  C. Sheet extrusion
  D. Film extrusion
  E. Blown-film extrusion
  F. Filament extrusion
  G. Extrusion coating and wire covering
IV. Blow Molding
  A. Injection blow molding
  B. Extrusion blow molding
  C. Blow-molding variations

Extrusion Equipment

Figures 11-1 and 11-2 show typical single-screw type extruders. Although computerized instrumentation has improved process control, the basic design of single-screw extruders has not changed for several decades. Screws are measured by the diameter, ranging from very small machines with 19 mm [0.75 in.] diameter screws up to very large machines with screws of 300 mm [12 in.] diameter. The most popular machines are the 64 to 76 mm [2.5 to 3 in.] size.

Beside the diameter of the screw, extruders are sold by amount of material they will plasticate per minute or hour. Extruder capacity with low-density polyethylene may vary from less than 2 kg [4.5 lb] to more than 5000 kg [11 000 lb] per hour.

Screws are characterized by their L/D ratios. A 20:1 screw could be 50 mm [1.96 in.] in diameter and 1000 mm [39.37 in.] long. Short screws, such as one with a L/D ratio of 16:1 would be appropriate for a profile extruder. Long screws, up to 40:1, provide more mixing of the materials than shorter screws. Some screw designs are shown in Figure 11-3.

The channel depth of the screw is greatest in the feed section. That permits the screw to draw in pellets and other forms of material. The channel depth decreases beginning at the transition section. This continuous reduction forces out air and compacts the material (Fig. 11-3A). In the metering section, the narrow flight depth completes the melting of the plastics. At the end of the barrel, a breaker plate acts as a mechanical seal between the barrel and die. The breaker plane also holds the screen pack in place. Figure 11-4 shows screens of varying mesh size. Several screens together are called a screen packs and filter out pieces of foreign material. As screens become plugged, back pressure increases.

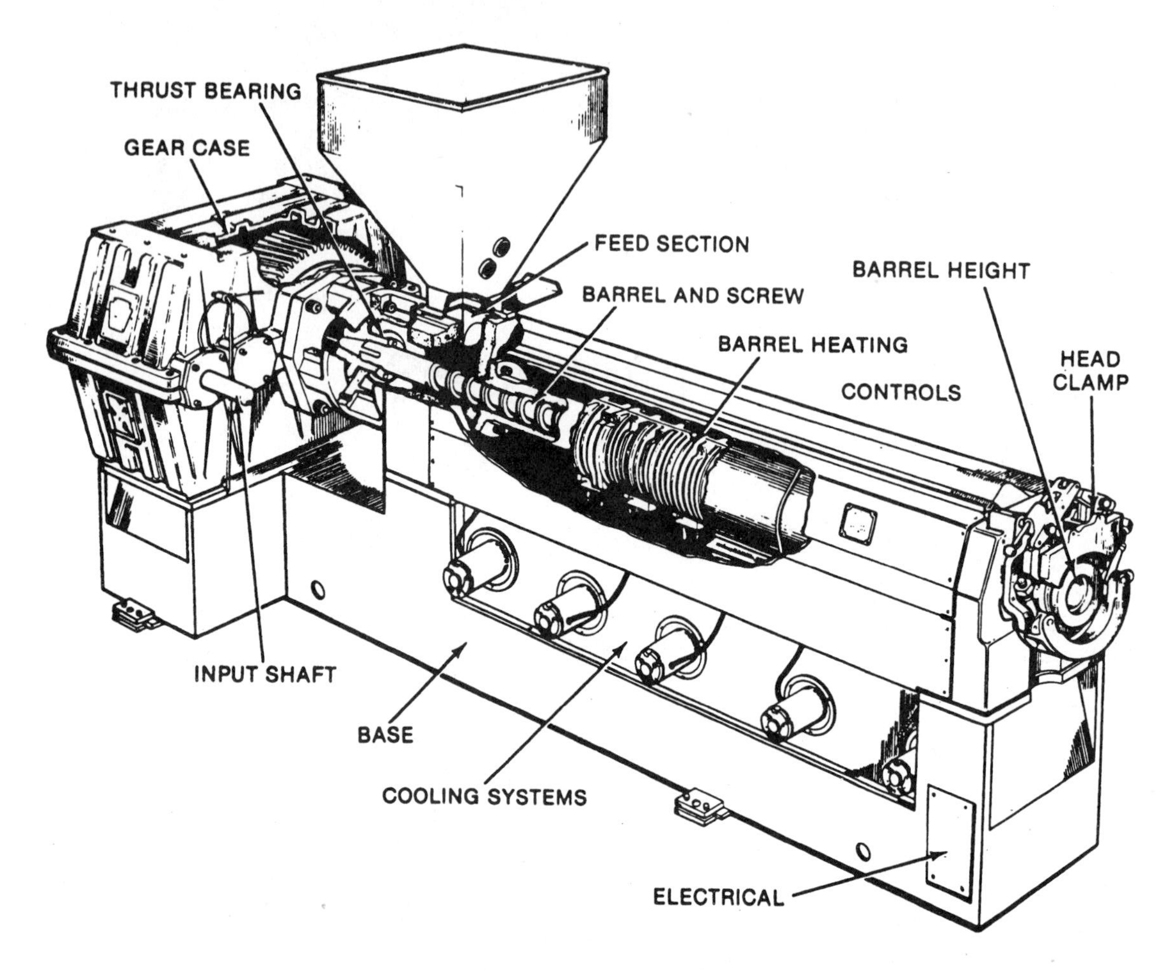

Fig. 11-1. Extruder, with parts labeled. (Davis-Standard Division, Crompton & Knowles Corp.)

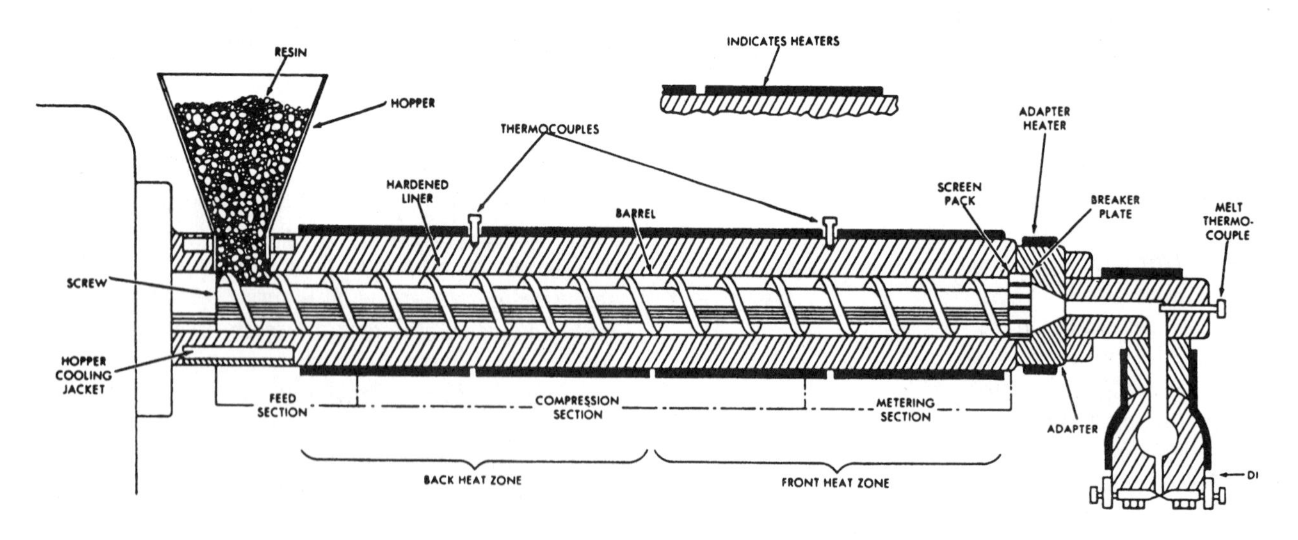

Fig. 11-2. Cross-section of a typical screw extruder, with the die turned down. (USI)

Most extruders are equipped with a screen changer. The most common is a plate type that moves from one side to the other. Sliding a clean screen pack into place exposes the contaminated screens, as seen in Figure 11-5. The dirty screens are removed, clean ones installed, and the changer is then ready. Some machines have a continuous ribbon of screen (sometimes rotary) that can be automatically controlled to maintain a steady head pressure despite varying levels of contamination in the polymer or other flow rate conditions.

After passing through the breaker plate and screens, the melt enters the die. The die actually forms

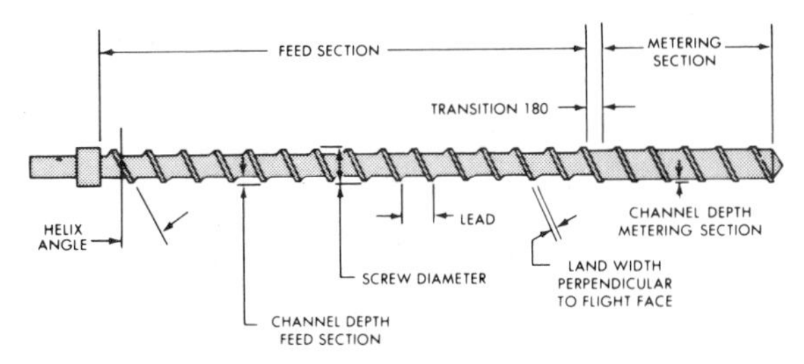

(A) Metering screw. (Processing of Thermoplastic Materials.)

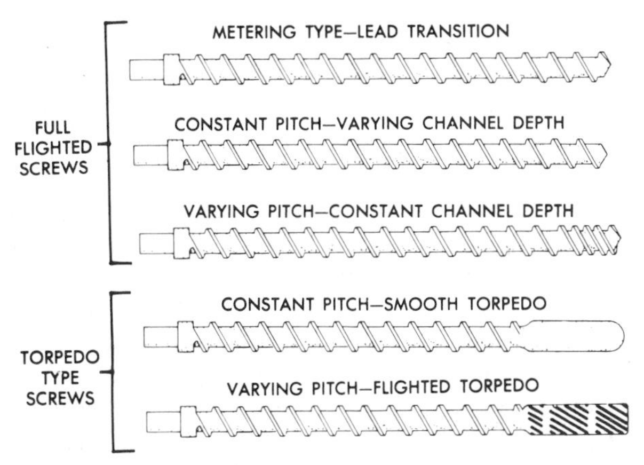

(B) Common extruder screws. (Processing of Thermoplastic Materials.)

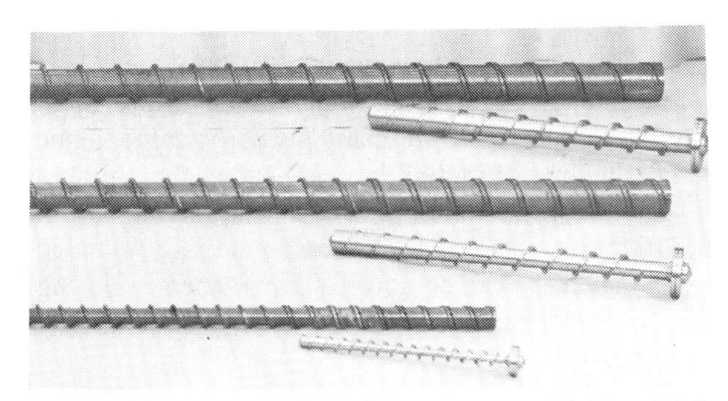

(C) Side view of extruder screws. (Cameron-Waldron Division, Midland-Ross Corp.)

(D) End view of extruder screws. (Cameron-Waldron Division, Midland-Ross Corp.)

Fig. 11-3. Common extruder screws. (Cameron-Waldron Division, Midland-Ross Corp.)

Fig. 11-4. Screens in varying mesh size. Mesh numbers refer to the number of openings per inch. 14 mesh is coarse, 200 mesh is very fine.

Fig. 11-5. The screen pack creates back-pressure on material in the extruder barrel. (Cameron-Waldron Division, Midland-Ross Corp.)

the molten plastics as it comes out of the extruder. The simplest die is a single strand die, which extrudes a strand slightly larger than the diameter of the die. Multiple-strand dies create many strands simultaneously. Sheet, tube, pipe, and profile dies force the material into a desired shape. Dies may be made of mild steel but should be made of chromium-molybdenum steel for long runs. Stainless alloys are used with corrosive materials.

Electric heaters are used around the barrel to help melt the plastics. Once the extruder is mixing, blending, and forcing the material through the die, frictional heat produced by the action of the screw may be enough to partly plasticate the material. External heaters are used to maintain a fixed temperature once the process is started.

Compounding

Compounding is the process of blending basic plastics with plasticizers, fillers, colorants, and other ingredients. Compounding companies used to rely on single-screw extruders, multistrand dies, long water-filled cooling tanks and pelletizers to chop the strands into pellets. Most compounders are now using twin-screw equipment. (See Figure 11-6.)

Twin-screw compounders are of two basic types, co-rotating and counter-rotating. In the co-rotating design, the material is forced through the gap between the two screws, which is also called the nip. The material that goes through the nip experiences intensive mixing, but some material may move forward down the barrel without passing through the nip.

In the counter-rotating design, the material is transferred from one screw to the other at the point of intermeshing. This created a figure-8 path for the material, and because this path is rather long, it generally yields more thorough mixing than the co-rotating design. Figure 11-7 shows the transfer of material from one screw to the other.

Counter-rotating twin-screw compounding extruders are available in L/D ratios from 12:1 to 48:1, with screw diameters ranging from 25 mm to 300 mm [1 to 12 in.] and outputs up to 27000 kg [about 60000 lbs]

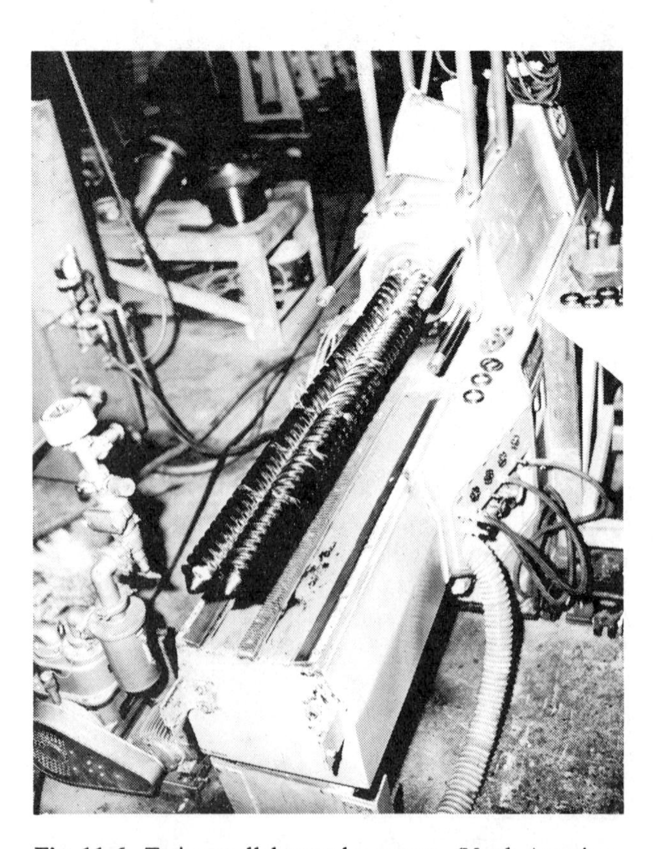

Fig. 11-6. Twin parallel extruder screws. (North American Bitruder Co.)

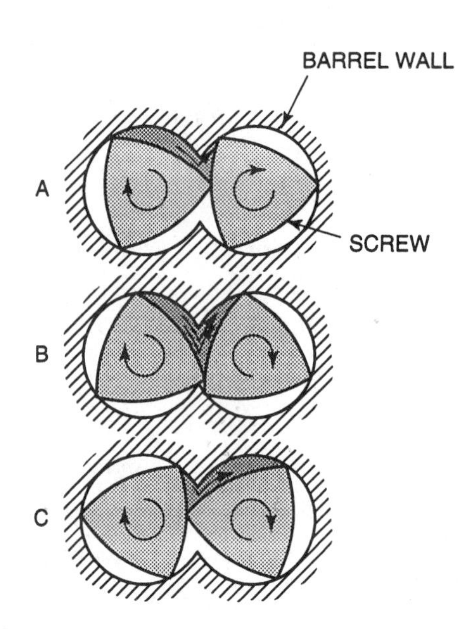

Fig. 11-7. In co-rotating twin screw machines, one screw wipes the material off the other one, allowing little material to pass between the two screws.

per hour. Some companies offer modular screw design. That means that the screw consists of a central splined rod and various mixing and conveying elements that slip over the rod. Kneeding and conveying elements can be right- or left-handed. Use of left-handed elements creates back pressure on the melt behind the element and reduces the pressure ahead. With careful arrangement of these elements, effective mixing and conveying of even difficult-to-handle materials is possible.

Typical compounding companies must store, measure, mix, and transport many types of ingredients. Batch blending of ingredients can lead to the problem of segregation, in which the heavier ingredients separate from the lighter ones. A homogeneously blended batch may become no longer homogeneous during transportation, particularly when passing through long pipe lines.

To avoid the problems of segregation due to varying bulk densities, many companies bring the raw materials directly to the extruder and use several weight-loss hoppers to feed precisely measured amounts of ingredients to the extruder feed throat. Well-equipped compounding extruders can handle up to 10 separate feed streams. Some liquids and fibers must be added to already melted plastics. Such *downstream* feeding requires special pumps for liquids and feeders for fibers and powders.

To save space, some compounders purchase under-water pelletizing equipment. Figure 11-8 shows an underwater pelletizing head with the covers open. Note the cutter blade just above the multistrand die head. Figure 11-9 shows a schematic of an underwater pelletizing system.

Fig. 11-8. Multistrand die and cutter inside an underwater pelletizing head. (North American Bitruder Co.)

Major Types of Extrusion Products

Because extrusion includes so many types, it is useful to break extrusion down into major types of products. These categories are profile, pipe, sheet, film, blown film, filament, and coating and wire covering.

Profile Extrusion

The term *profile extrusions* applies to most extruded products other than pipe, film, sheet, and filaments. Figure 11-10 shows just a few of these items. Such profiles are generally extruded horizontally. Achieving a desired shape requires equipment to support and shape the extrudate during cooling. The cooling comes from air jets, water troughs, water sprays, and cooling sleeves. The size control requires sizing dies, hold-down fingers, or sizing plates. Figure 11-11 shows sizing fingers and water jets.

Controlling the size or shape of such profiles can be difficult. When a material exits an extruder die, it changes shape due to a phenomenon called *die swell.* In the extruder and die, the melted plastics is under compression. The release of this compression causes the extrudate to expand in cross section. When the extrudates do not have uniform cross sections, the shrinkage upon cooling will also not be uniform.

To produce exact cross-sectional dimensions after the extrudate has cooled, allowances must be made in the orifice design. In complex cross sections where thin sections or sharp edges are formed, cooling occurs more quickly at such portions. These areas shrink first, therefore they are smaller than the rest of the section. This means that the die and the shape of the plastics (extrudate) may be different. To correct this problem, the orifice in the die is made larger at these points (Fig. 11-12).

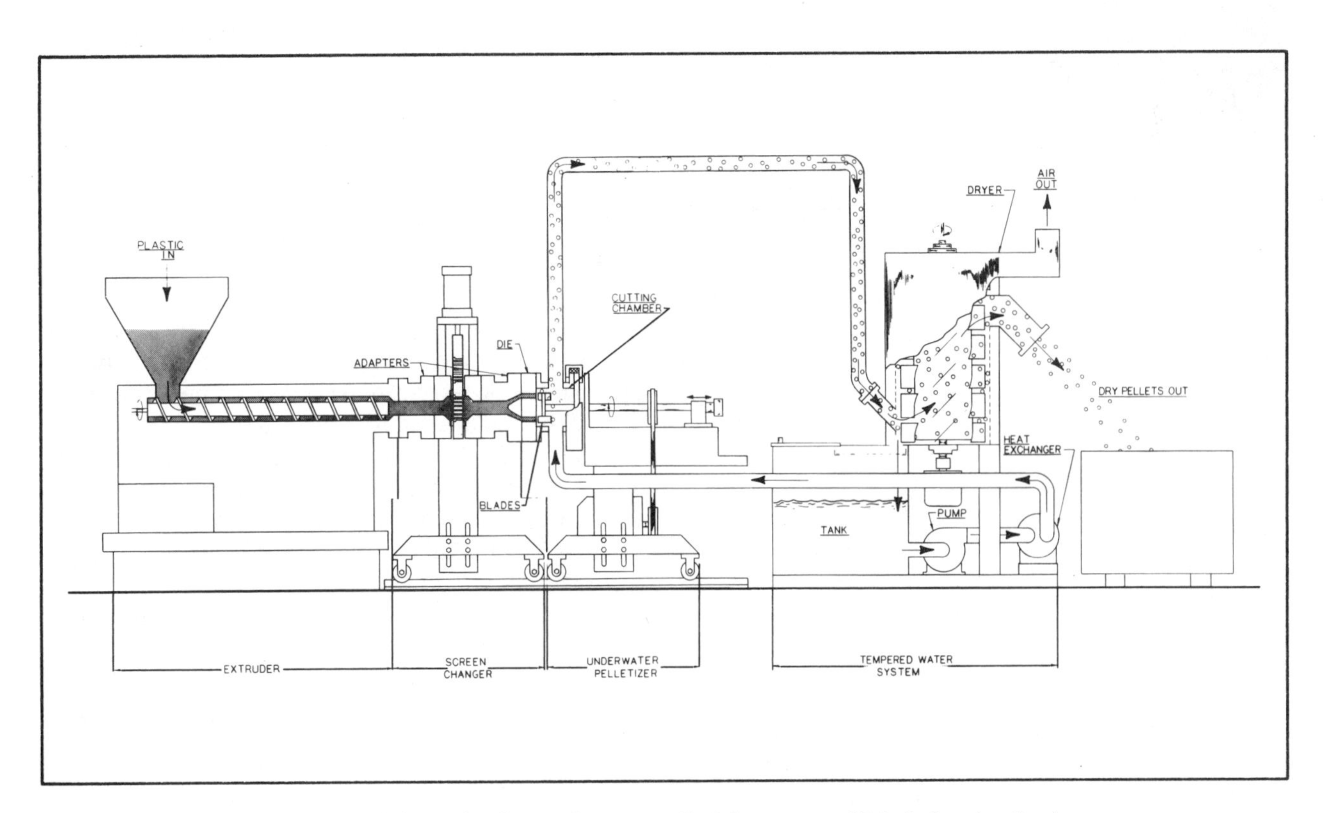

Fig. 11-9. Schematic of an underwater pelletizing system. (Gala Industries, Inc.)

Fig. 11-10. Some of the many profile extrusions of plastics. (Fellows Corp.)

Fig. 11-11. Note the fingers that help shape the material, and the water jets that cool the extruded shape. (Alma Plastics Co.)

To reduce problems caused by complex geometry of profile dies, some manufacturers choose postforming of simpler shapes. Postforming into different shapes requires the use of sizing plates, shoes, or rollers. A flat-tape shape may be postformed into a corrugated form. Round rods may be postformed into oval or other new shapes while the extrudate is still hot (Fig. 11-13).

Pipe Extrusion

Pipes (tubular forms) are shaped by the exterior dimensions of the orifice and by the mandrel (sometimes called a *pin*), which shapes inside dimensions (Fig. 11-14). The mandrel is held in place by thin pieces of metal called *spiders*.

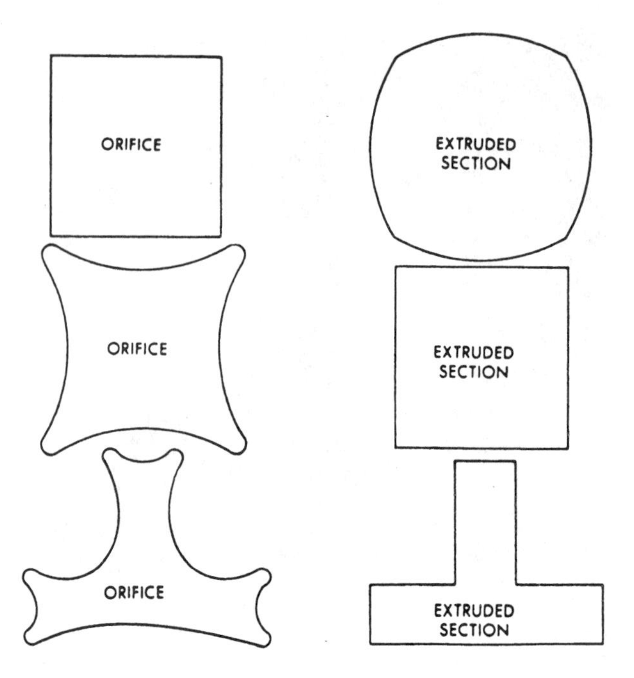

Fig. 11-12. Relationships between die orifices and extruded sections.

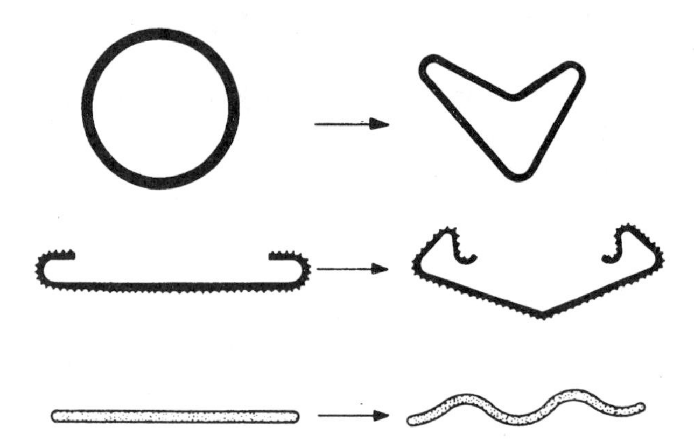

Fig. 11-13. The extrudates at left are postformed into the shapes at right by being passed through rollers.

The diameter of the pipe or tube is also controlled by the tension of the take-up mechanism. If the tube is pulled faster than speed of the extrudate melt, the product will be smaller and thinner than the die.

To prevent the tube from collapsing before cooling, it is pinched shut on the end and air is forced in through the die. This air pressure expands the pipe slightly. The hot tube may be pulled through a sizing ring or vacuum ring to hold the outside diameter to a close tolerance. The thickness of the pipe wall is controlled by the mandrel and die size.

Sheet Extrusion

The American Society for Testing and Materials (ASTM) has defined *film* as plastics sheeting 0.25 mm [0.01 in.] or less in thickness. Sheet materials thicker

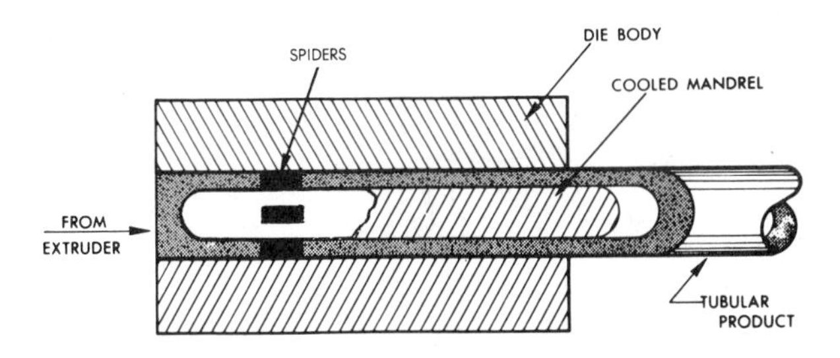

Fig. 11-14. In this pipe-forming operation, hot material is extruded around a cold mandrel or pin. (DuPont Co.)

than 0.25 mm are considered sheet. Sheet extrusion produces stock for use in most thermoforming operations.

Most sheet forms involve extruding molten thermoplastic materials through dies with a long horizontal slot, as seen in Figure 11-15. These dies come in two major styles, the *T-shape,* and the *coat hanger-shape*. In both styles, the molten material is fed to the center of the die. It is then formed by the die lands and adjustable jaw. The width may be controlled by the external deckle bars or by the actual die width. In Figure 11-16 an adjustable choke or restrictor bar is used in the extrusion of sheeting.

The extruded sheet goes over or through a set of rollers to provide desired surface finish or texture and

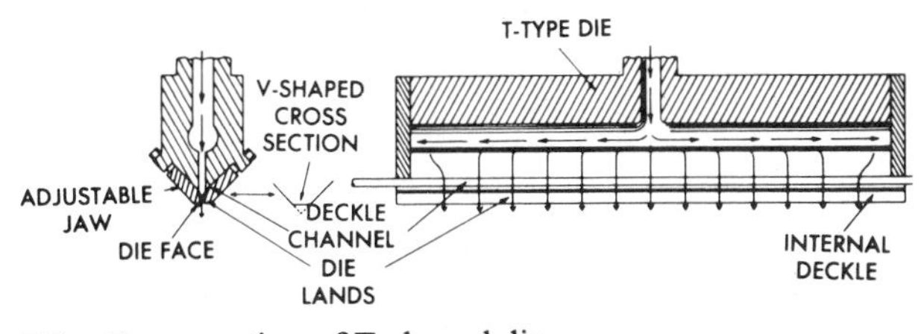

(A) Cross section of T-shaped die.

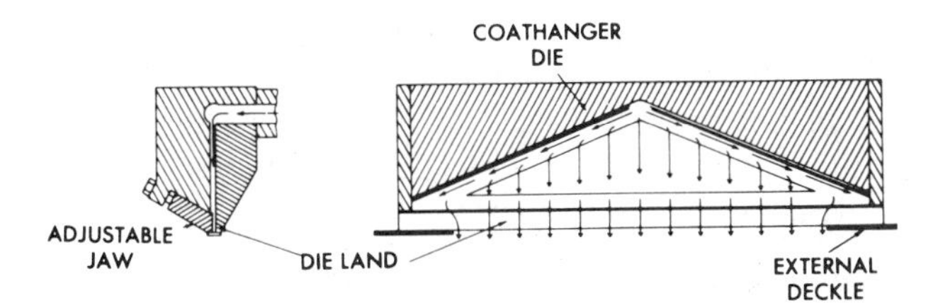

(B) Cross section of coathanger-shaped die.

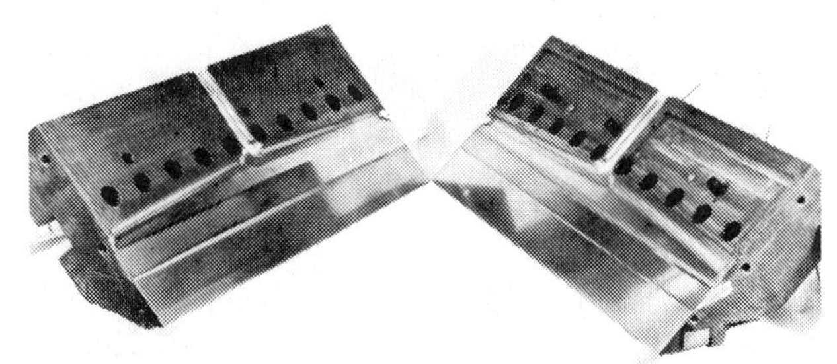

(C) T-shaped sheet dies. (Cameron-Waldron Division, Midland-Ross Corp.)

Fig. 11-15. Two types of extrusion dies.

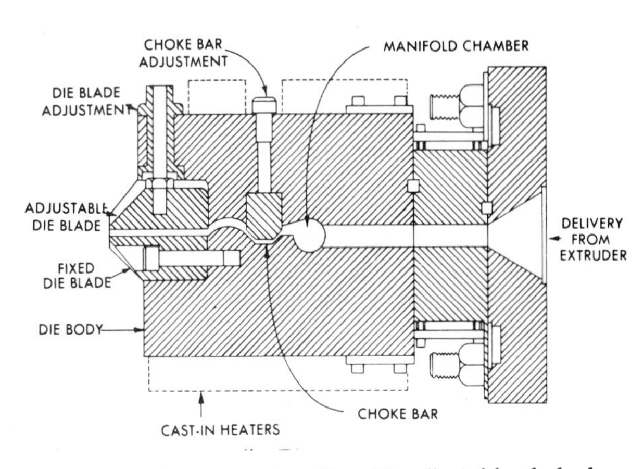

Fig. 11-16. Sheet extrusion die with adjustable choke bars. (Phillips Petroleum Co.)

to accurately size the thickness. Figure 11-17 shows a drawing of sheeting rollers and a photograph of the sheet leaving the die and entering the rollers.

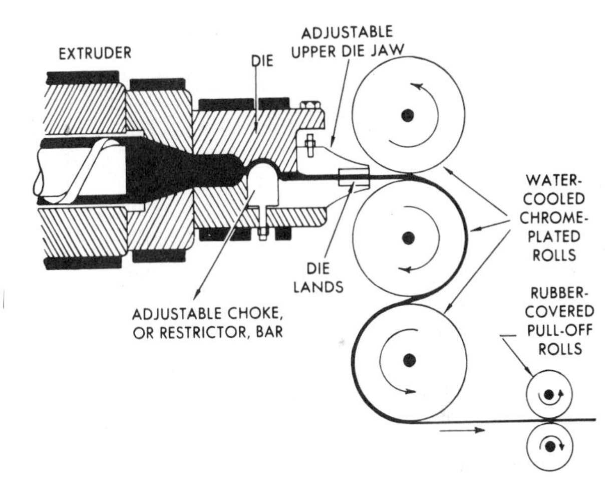

(A) Schematic of sheeting die and takeoff unit. (USI)

(B) Sheet extrusion from the die into the nip of chill rollers. (North American Bitruder Co.)

Fig. 11-17. Takeoff units with chill rollers.

Table 11-1 gives some common sheet-extrusion problems and their remedies.

Film Extrusion

Film extrusion and calendaring produce finished products that are rather similar. Although calendaring is not technically an extrusion process, it logically deserves treatment along with film extrusion.

Film extrusion is similar to sheet extrusion. Beside the difference in thickness, film-extrusion dies are lighter than sheeting dies and possess shorter die lands than sheet dies.

In Figure 11-18A the film is extruded into a quench tank, while in Figure 11-18B the film is being drawn over chill rollers in a process sometimes called *film casting*. Both chill-roller and water-tank film extrusions are used commercially. Temperature, vibration, and water currents must be controlled with care

Table 11-1. Troubleshooting Sheet Extrusion Equipment

Defect	Possible Remedy
Continuous lines in direction of extrusion	Repair or clean out die Die contamination or polish of rollers scored Reduce die temperatures Use properly dried materials
Continuous lines across sheet	Jerky operation—adjust tension on sheet Reduce polishing roll temperatures or increase roll temperatures Check back-pressure gauge for surging
Discoloration	Use proper die and screw design Minimize material contamination Temperature too high—too much regrind Repair and clean out die
Dimensional variation across sheet	Adjust bead at polishing rolls Balance die heats Reduce polishing roll temperatures Check temperature controllers Repair or clean out die
Voids in sheet	Balance extruder line conditions Use proper screw design Minimize material contamination Reduce stock temperature
Dull strip	Die set too narrow at this point Minimize material contamination Increase die temperature Repair or clean out die
Pits, craters	Balance extruder line conditions Minimize material contamination Use properly dried materials Control stock feed Reduce stock temperature

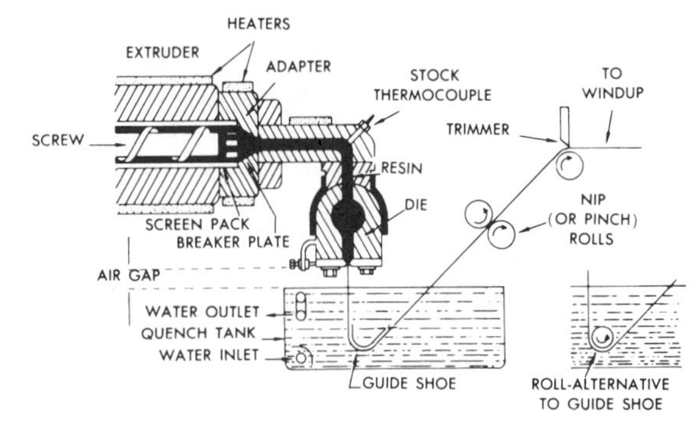

(A) Cross section of the front part of a flat-film extruder, the quench tank, and takeoff equipment. (USI)

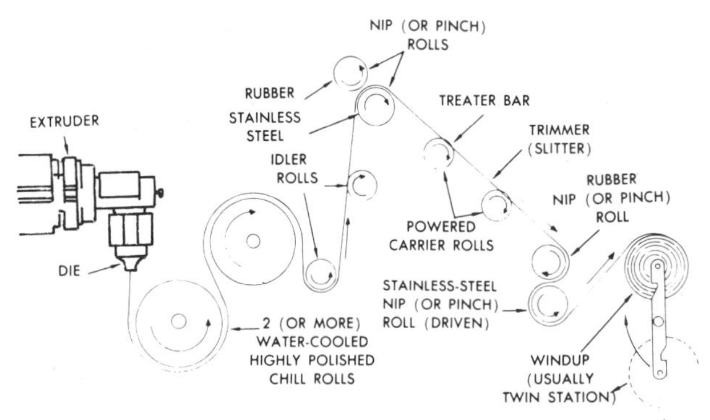

(B) Schematic of chill-roll film extrusion equipment.

Fig. 11-18. Types of film takeoff equipment.

when using the water-tank method to allow clear, defect-free film to be produced. Heavy-gauge films are currently being made by this high-speed method.

Film extrusion is more costly than blown-film, and consequently, is used only when the film quality must be higher than blown film. Some plastics are heat sensitive and tend to decompose or degrade due to high temperature. PVC is such a material. To make PVC film, slot extrusion or blown film techniques are generally not used. PVC film generally requires calendering.

Calendering. In calendering, thermoplastic materials are squeezed to final thickness by heated rollers (Fig. 11-19). Films and sheet forms with a glossy or embossed finish may be produced by this method (Fig. 11-20). Much calendered film is used in the textile industry. Embossed or textured film is used to produce leather-like apparel, handbags, shoes, and luggage.

The calendering process consists of blending a hot mix of resin, stabilizers, plasticizers, and pigments in a continuous kneader, or Banbury mixer. This hot mix is led through a two-roll mill to produce a heavy sheet stock. As the sheet passes through a series of heated, revolving rollers, it becomes progressively thinner until the desired thickness is reached. A pair of precision, high-pressure finish rollers are used for gauging

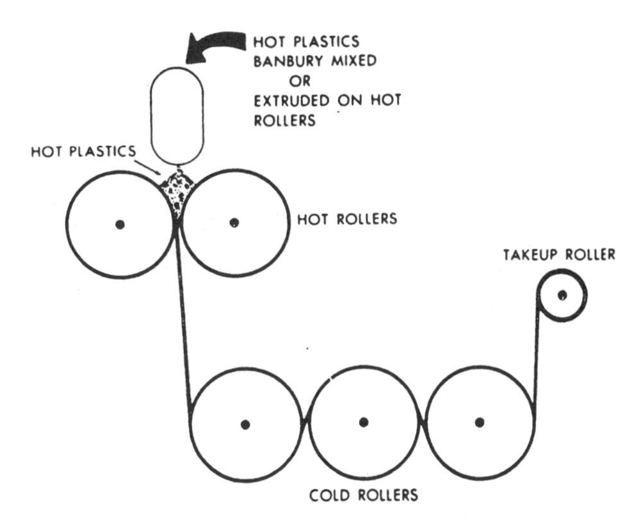

Fig. 11-19. The calendering process.

Fig. 11-20. Calendering a thermoplastic material. (Monsanto Co.)

and embossing. At the end, the hot sheet is cooled on a chill roller and is taken off in sheet or film form.

Calender rollers are very costly and are easily damaged by metal contaminants. For this reason, metal detectors are often used to scan the sheet before it enters the calender. The calendering equipment, together with accessory controls, is also very costly. Replacement costs for a calender line may exceed $1 million, which has discouraged the installation of much calendering equipment. Calendering has some advantages over extrusion and other methods when producing colored or embossed films and sheets. When colors are changed, a calender requires a minimum of cleaning, while an extruder must be purged and cleaned thoroughly.

Though calenders are expensive, calendering remains the preferred method of producing PVC sheets at high rates. Roughly 95 percent of all calendered products are PVC, and only about 15 percent are being used for rigid production.

Calendering equipment may involve multi-story complexes, because calender rollers are usually placed in an inverted L or a Z arrangement (Fig. 11-21). The rolls and supportive equipment are controlled by many sensing devices and by computers. Calenders are usually rated by the amount (mass) of material that they can produce per unit of time. This rate depends on the

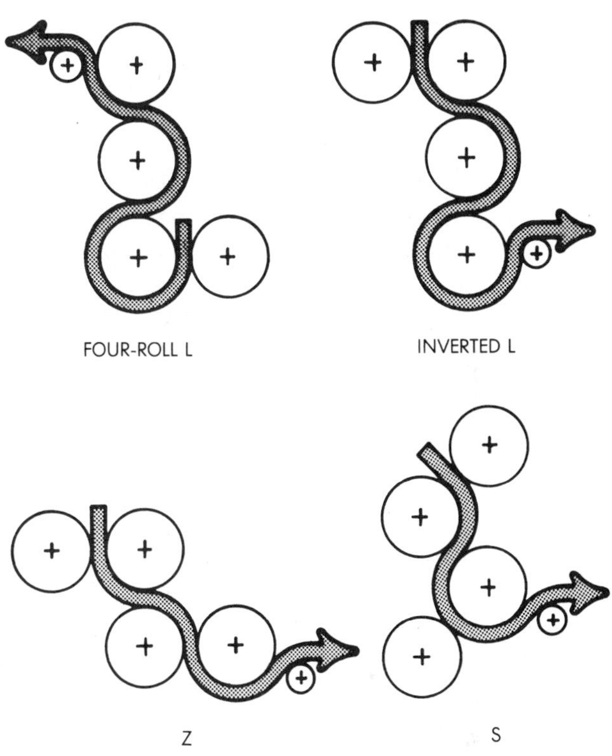

(A) Calender configurations.

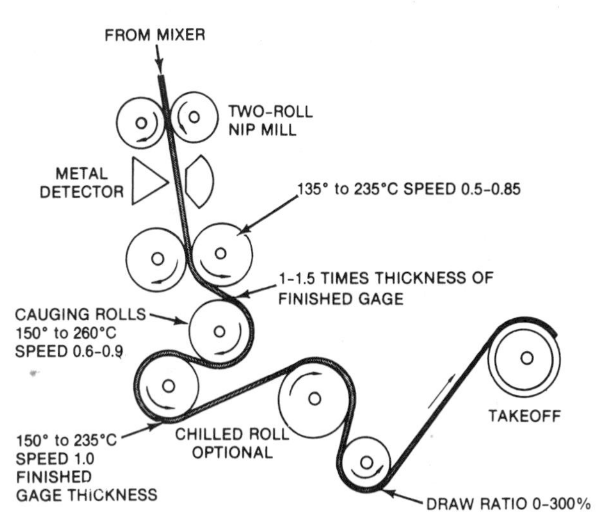

(B) A series of rollers arranged in a Z-form is used to calender thermoplastic sheet material.

Fig. 11-21. Common calender roll configurations.

material, plasticating rate, required surface finish, and take-up capacity. Large machines have an output rate of nearly 3000 kg/h [6662 lb/hr]. Most rollers are less than 2 m [6 ft. 7 in.] wide. With soft materials, widths over 3 m [10 ft] are possible. Roll forces may approach 350 kN [39.34 tons-force] for thin, rigid materials.

Several methods are used to compensate for roll deflection (bending or bowing): 1) force is applied to the outboard or inboard main bearings, 2) rolls are produced with slight crowns, or 3) one roll is skewed with respect to the other (roll crossing). (See Figure 11-22.) Table 11-2 gives some common calendering problems and remedies.

Blown-Film Extrusion

In blown-film extrusion, the film is produced by forcing molten material through a die and around a mandrel. It emerges from the orifice in tube form (Fig. 11-23). This process is like that used to make pipes or tubes. This tube, or bubble, is expanded by blowing air through the center of the mandrel until the desired film thickness is reached. This process is something like blowing up a balloon. The tube is usually cooled by air from a cooling ring around the die. (See Figure 11-24.) The *frost line* is the zone where the temperature of the tube has fallen below the softening point of the plastics. In polyethylene or polypropylene film extrusion, the frost zone is evident; it actually appears frosty. The frost zone shows the change taking place as the plastics cools from the melt (an amorphous state) to a crystalline state. With some plastics, there will be no visible frost line.

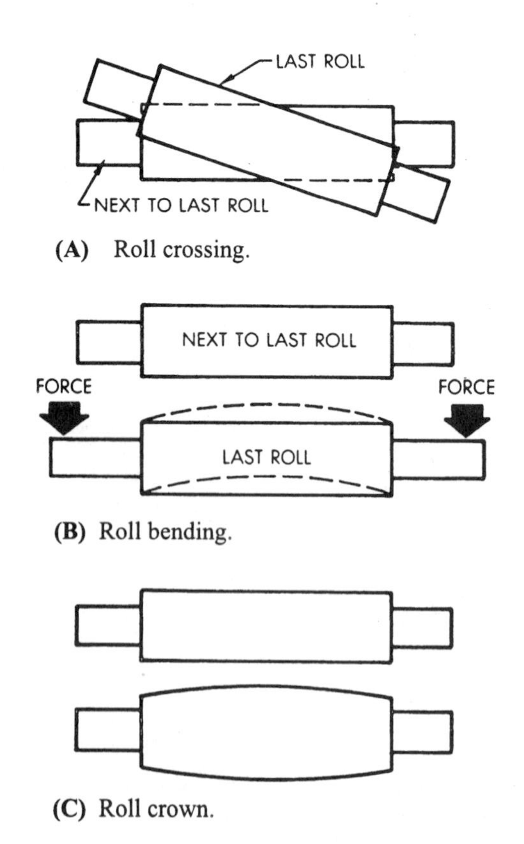

(A) Roll crossing.

(B) Roll bending.

(C) Roll crown.

Fig. 11-22. Methods used for correcting sheet profile.

Table 11-2. Troubleshooting Calendering

Defect	Possible Remedy
Blistering of film or sheet	Reduce melt temperature Reduce program speed of rolls Check for resin contamination Reduce temperature of chill rolls
Thick section in center, thin edges	Use crowned rolls Increase nip opening Check bearing load of rolls
Cold marks or crow's feet	Increase stock temperature Decrease program feed
Pin holes	Check for contamination in resin Blend plasticizers more thoroughly into resin
Dull blemishes	Check for lubricant or resin contamination Check roll surfaces Increase melt temperature Increase roll temperature
Rough finish	Raise roll temperature Excessive nip between rolls
Roll bank on sheet	Incorrect stock temperature—increase/decrease Provide constant takeoff speeds Reduce roll temperatures and nip clearance

The size and thickness of the finished film is controlled by several factors including extrusion speed, takeoff speed, die (orifice) opening, material temperature, and the air pressure inside the bubble or tube. Blow-up ratio is the ratio of the die diameter to the bubble diameter. Blow-extruded film is sold as seamless tubing, as fiat film, or as film folded in a number of ways. Film producers may slit the tubing on one edge during windup. If the tube is blown to a diameter of 2 m [6.5 ft] the flat film will have a width (slit and opened) of over 6 m [19 ft]. Slot dies of this size are not practical. Tubular films are desirable as low-cost packaging for some foods and garments. Only one heat seal is needed in the production of bags from blown tubing.

Blown films are semioriented; that is, they have less orientation of molecules in a single direction than film from slot dies. Blown films are stretched as the tube is expanded by air pressure. Such stretching results in a more balanced molecular orientation in two directions. Products are biaxially oriented; one in the direction of length and one across the diameter of the bubble. Improved physical properties are an asset of blown film. However, clarity, surface defects, and film thicknesses are harder to regulate than with slot extru-

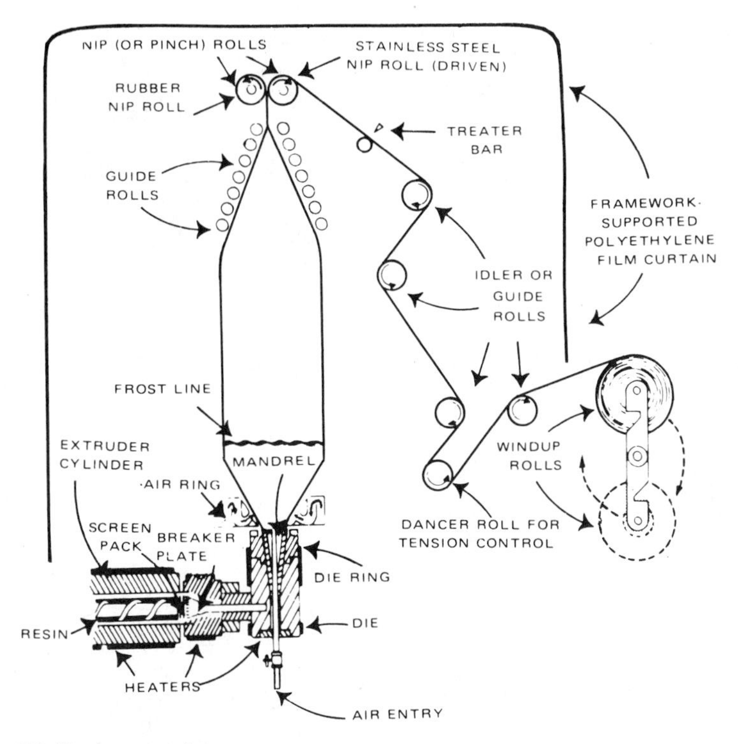

(A) Basic apparatus.

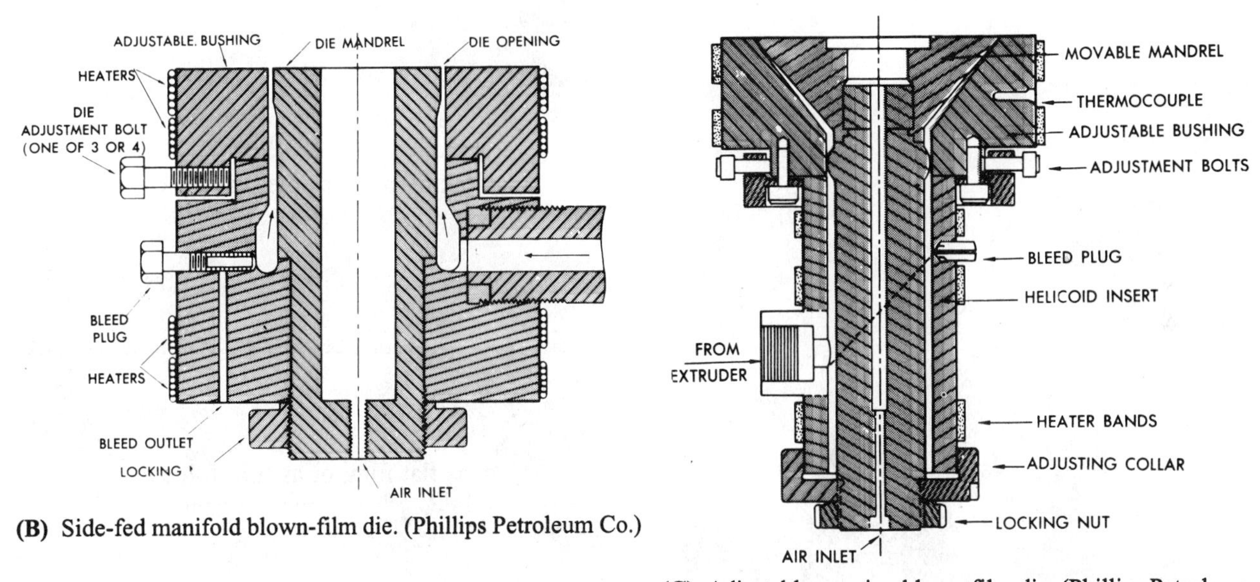

(B) Side-fed manifold blown-film die. (Phillips Petroleum Co.)

(C) Adjustable-opening blown-film die. (Phillips Petroleum Co.)

Fig. 11-23. Schematic drawings of blown-film extrusion procedures. (U.S. Industrial Chemicals)

sion. Table 11-3 gives troubleshooting information for blow-extruded film.

Filament Extrusion

Important terms for filament extrusion. A *filament* is a single, long, slender strand of plastics. The fisherman is probably most familiar with monofilament fishing line. This single filament of plastics may be made in any desired length. *Yarns* may be composed of either mono-filament or multifilament strands of plastics.

The term *fiber* is used to describe all types of filaments; natural or plastics, monofilament or multifilament. Fibers are first spun or twisted into yarns. They are then woven into finished fabrics, screening, or other products ready for consumer use.

The fineness of a fiber is expressed by a unit called *denier.* One denier equals the mass in grams of 9000 m

(A) Extruder, with blown film being taken off. (Chemplex Co.)

(C) Close-up view, looking down at extruder and gauge bars. (Chemplex Co.)

(B) Close-up view of extruded blown film. (Chemplex Co.)

(D) Note size of film after blowing. (BASF)

Fig. 11-24. Blown-film extrusion.

Table 11-3. Troubleshooting Blown Film

Defect	Possible Remedy
Black specks in film	Clean die and extruder Change screen pack Check resin for contamination
Die lines in film	Lower die pressure Increase melt temperature Polish all rough edges in film path Check nip rolls—make smooth
Bubble bounces	Increase screw rpm and nip roll speed Enclose tower or stop drafts Adjust cooling ring to obtain constant air velocity around ring
Poor optical and physical properties	Raise melt temperature Increase blow-up ratio increase frost line height Clean die lips, extruder, and rollers
Failures at fold	Decrease nip roll pressure
Failure at weld lines	If possible, bleed die at weld line Heat die spiders—insulate air lines there Increase melt temperature Check for contamination
Film won't run continuously	Clean die and extruder Lower melt temperature Increase film thickness

[29527 ft] of fiber. For example, 9000 m of a 10-denier yarn weighs 10 g [0.35 oz.]. The name of the unit may have evolved from the name for a sixteenth-century French coin that was used as a standard for measuring the fineness of silk fibers.

Plastics fibers of the same fineness vary in denier because of differences in density. Although two filaments may have the same denier, one may have a larger diameter because of a lower relative density.

To calculate the denier of the filaments in yarn, divide the yarn denier by the number of filaments:

$$\frac{\text{80-denier yarn}}{\text{40 filaments}} = \text{2 denier for each filament.}$$

Remember: 9000 m of 1-denier filament weighs 1 g.

9000 m of 2-denier filament weighs 2 g.

The International Organization for Standards (ISO) has developed a universal system for designating linear density of textiles called the *tex*. The fabric industry has adopted the tex as a measure of linear density. In the tex system the yarn count is equal to the mass of yarn in g/km.

Types of filaments. Not all of the synthetic filaments are used by the textile industry. Some monofilaments are used for bristles in brooms, toothbrushes, and paint brushes. The relative shapes of these filaments vary, as shown in Figure 11-25.

If the fiber is to be pliable and soft, a fine filament is needed. If the fiber is to withstand crushing and be stiff, a thicker filament is used. Clothing fibers range from 2 to 10 denier per filament. Fibers for carpeting range from 15 to 30 denier in fineness per filament.

The cross-sectional shape of the fiber helps determine the texture of the finished product. Triangular and trilobal shapes impart to the synthetic fiber many of the properties of the natural fiber, silk. Many of the ribbon- and bean-shaped filaments resemble cotton fibers.

Manufacturing of filaments. Monofilaments are produced much as are profile shapes, except that a multi-orifice die is used. These dies contain many small openings from which the molten material emerges. Such dies are used to produce granular pellets, monofilaments, and multifilament strands.

Filament shapes are made by forcing plastics through small orifices in a process referred to as *spinning*. The plastics is shaped by the opening in the die or *spinneret*. This process may have been named spinning from the method of spinning natural fibers. The small opening under the jaw of the silkworm is also called the spinneret.

Because these orifices are often much finer than the diameter of human hair, spinnerets are often made of such metals as platinum, which will resist acids and orifice wear. To be forced or extruded out of these small openings, the plastics must be made fluid.

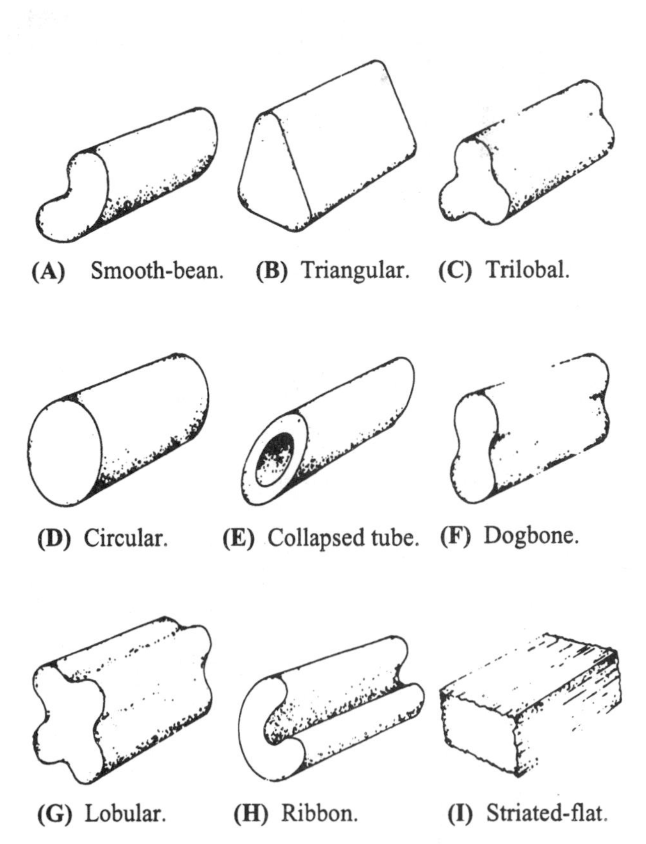
(A) Smooth-bean. (B) Triangular. (C) Trilobal.
(D) Circular. (E) Collapsed tube. (F) Dogbone.
(G) Lobular. (H) Ribbon. (I) Striated-flat.

Fig. 11-25. Cross-sectional shapes of selected fibers.

Acrylic fiber is produced in the spinning process shown in Figure 11-26. A thick chemical solution is extruded into a coagulation bath through the tiny holes of the spinneret. In the bath, the solution coagulates (becomes a solid) and becomes Acrilan acrylic fiber. The fiber is washed, dried, crimped, and cut into staple lengths. It is then baled for shipment to textile mills where it is converted into carpeting, wearing apparel, and many other products.

There are three basic methods of spinning fibers:

1. In *melt spinning*, plastics such as polyethylene, polypropylene, polyvinyl, polyamide, or thermoplastic polyesters are melted and forced out the spinneret. As the filaments hit the air, they solidify and are passed through other conditioners (Fig. 11-27A).
2. In *solvent spinning*, plastics such as acrylics, cellulose acetate, and polyvinyl chloride are dissolved by certain solvents. The solution is forced through the spinneret (Fig. 11-27B), and the filament then passes through a stream of hot air. The air aids in evaporating the solvents from the slender fibers. For economy, these solvents must be recovered for use again.
3. The first step of *wet spinning* (Fig. 11-27C) is like solvent spinning. The plastics is dissolved in chemical solvents. This fluid solution is forced out through the spinneret into a coagulating bath that makes the plastics gel into a solid filament form. Some members of the cellulosic, acrylic, and polyvinyl plastics families may be processed by wet spinning.

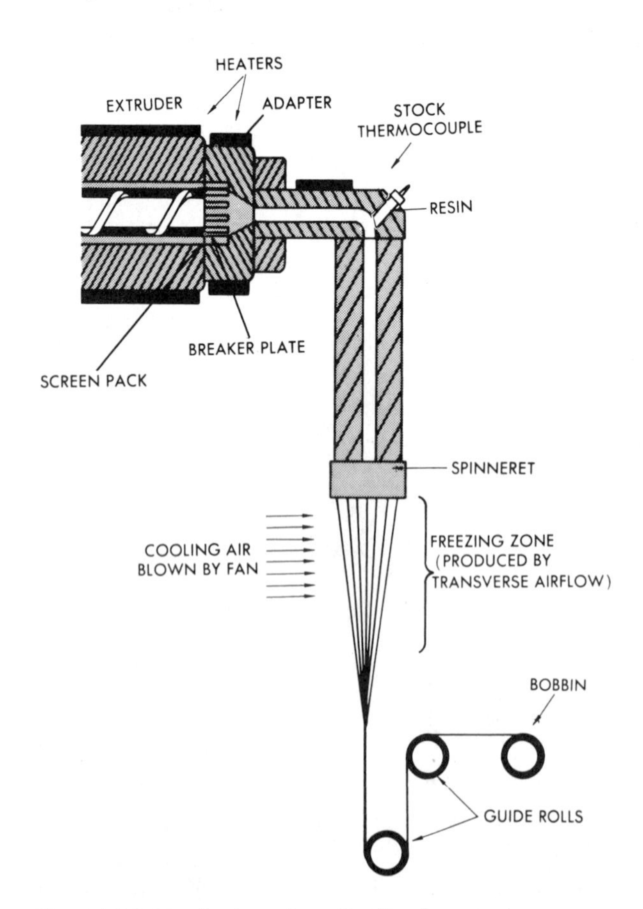

Fig. 11-26. Production of acrylic fiber by spinning.

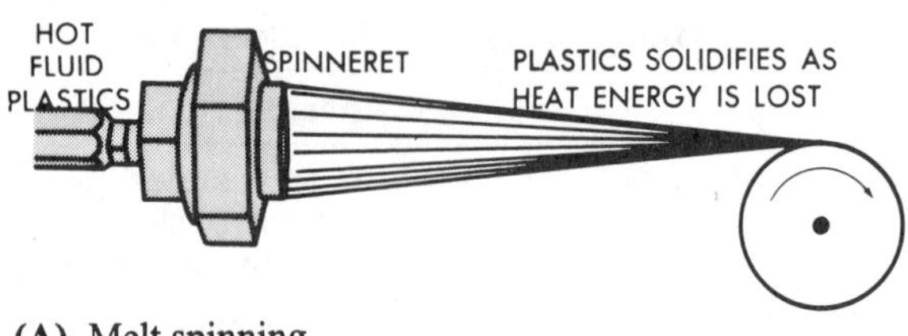

(A) Melt spinning.

(A) Melt spinning.

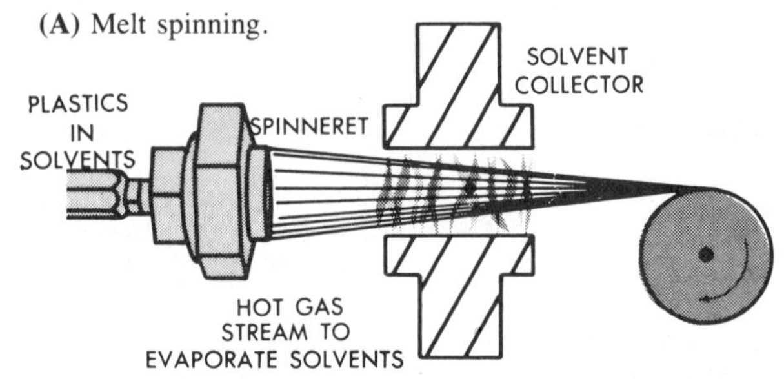

(B) Solvent spinning.

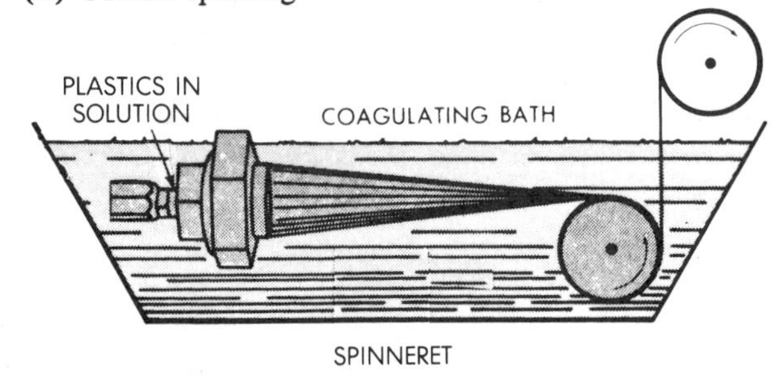

(C) Wet spinning.

Fig. 11-27. Three basic methods of spinning plastics fibers.

All three processes begin by forcing fluid plastics through a spinneret. They end by solidifying the filament through cooling, evaporation, or coagulation (Table 11-4).

The strength of the single filaments may be determined by several factors. Most of the selected filament fibers are linear and crystalline in composition. When groups of molecules lie together in long parallel molecular chains, there are additional strong bonding sites available. When the liquid plastics are forced through the spinneret, many of the molecular chains are forced closer together and parallel to the filament axis. This packing, arranging, and drawing provides increased strength throughout the filament. By mechanically working the filament, further molecular orientation and packing can be accomplished. This mechanical process is called *drawing*. The drawing of non-crystalline plastics also helps to orient molecular chains and thus improve strength.

Drawing or stretching of the plastics is done by running the filaments through a series of variable-speed rollers (Fig. 11-28). The drawing of crystalline plastics

Table 11-4. Selected Fibers and Production Processes

Fibers	Extrusion Spinning Process
Acrylics and modacrylics	
Acrilan	Wet
Creslan	Wet
Dynel (vinyl-acrylic)	Solvent
Orlon	Solvent
Verel	Solvent
Cellulose esters	
Acetate (Acele, Estron)	Solvent
Triacetate (Arnel)	Solvent
Cellulose, regenerated	
Rayon (viscose, cuprammonium)	Wet
Olefins	
Polyethylene	Melt
Polypropylene (Avisun, Herculon)	Melt
Polyamides	
Nylon 6,6, Nylon 6, Qiana	Melt
Polyesters	
Dacron, Trevira, Kodel, Fortrel	Melt
Polyurethanes	
Glospan	Wet
Lycra	Solvent
Numa	Wet
Vinyls and vinylidines	
Saran	Melt
Vinyon N	Solvent

is continuous. As the fiber goes through the drawing process, each roller is rotated at a faster speed, and roller speed determines the amount of stretching or drawing.

Monofilaments range from 0.12 mm to 1.5 mm [0.0047 to 0.0059 in.] in diameter. They may be handled individually or by takeoff machines.

High-bulk fibers and yarns may be processed from films. In this process, a composite film that has been coextruded or laminated is used. It is mechanically drawn and cut (fibrillated) into fine strands. The fibrillation is done during the stretching or drawing process as the film passes between serrating rollers rotating at different speeds. The teeth of these rollers cut the film into fibrous form (Fig. 11-29). The composite film develops internal stresses as it is extruded, drawn and fibrillated. This unequal stress orientation in the film layers makes the fibers curl and exhibit properties much like those of natural fibers.

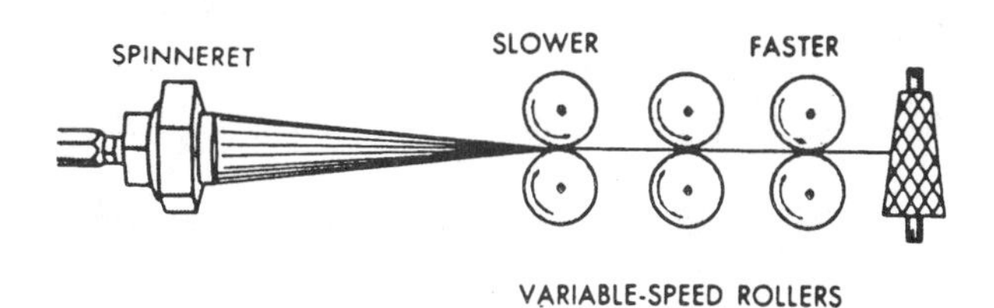

Fig. 11-28. Drawing plastics filaments.

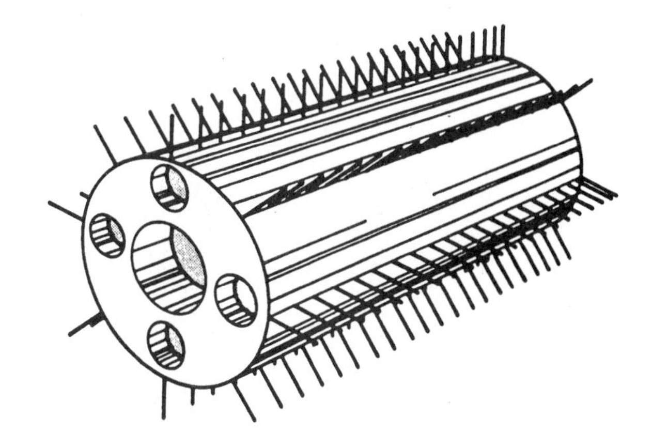

Fig. 11-29. This device has rows of pins that fibrillate yarn of very fine denier.

Extrusion Coating and Wire Covering

Paper, fabric, cardboard, plastics, and metal foils are common substrates for extrusion coating (Fig. 11-30). In extrusion coating, a thin film of molten plastics is applied to the substrate without the use of adhesives, while substrate and film are pressed between rollers. For special applications, adhesives may be needed to ensure proper bonding. Some substrates are preheated and primed with adhesion promoters using slot-extruder dies.

In wire and cable covering, the substrate for extrusion coating is a wire. The setup for extrusion coating of wire and cable is shown in Figure 11-31. During this process a molten plastics is forced around the wire or cable as it passes through the die. The die actually controls and forms the coating on the wire. Wires and cables are usually heated before coating to remove moisture and ensure adhesion. As the coated wire emerges from the crosshead die, it is cooled in a water bath. Two or more wires may be coated at one time. Television and appliance cords are common examples. Wooden strips, cotton rope, and plastics filaments may be coated by this process.

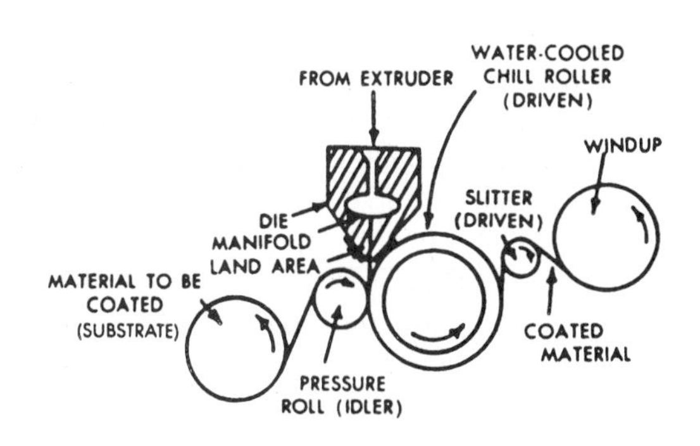

Fig. 11-30. Extrusion coating of substrates.

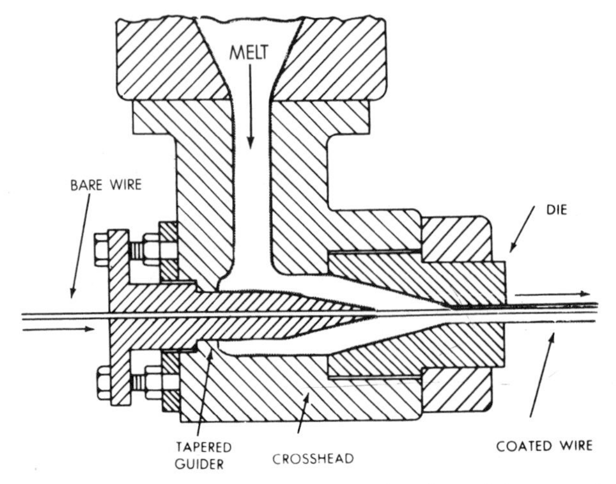

(A) A crosshead holds the wire-coating die and the tapered guide as the soft plastics flows around the moving wire.

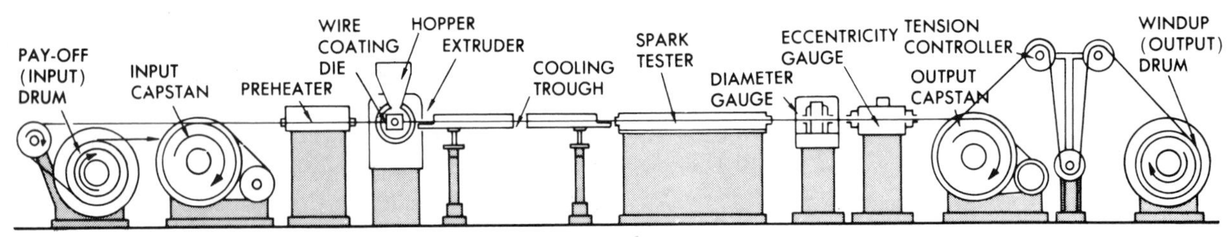

(B) A general component layout in a wire-coating extrusion plant. (U.S. Industrial Chemicals Co.)

Fig. 11-31. Extrusion coating of wire and cable.

Blow Molding

This process is sometimes listed as a molding technique because force is used to press the hot, soft tubular material against the mold walls. In this chapter, it is considered as a type of extrusion process, because extruders are required to create the tube shape that is then inflated into a hollow container.

Blow molding is a technique adopted and modified from the glass industry for making one-piece containers and other articles. The process has been used for centuries for making glass bottles. Blow molding of thermoplastics did not develop until the late 1950s. In 1880, blow molding was accomplished by heating and clamping two sheets of celluloid in a mold. Air was then forced in to form a blow-molded baby rattle. This may have been the first blow-molded thermoplastic article produced in the United States.

The basic principle of blow molding is simple (Fig. 11-32). A hollow tube (parison) of molten thermoplastic is placed in a female mold and the mold closed. It is then forced (blown) by air pressure against the walls of the mold. After a cooling cycle, the mold opens and ejects the finished product. The process is used to produce many containers, toys, packaging units, automobile parts, and appliance housings.

There are two basic blow molding methods:

1. Injection blowing
2. Extrusion blowing

The major difference is in the way that the hot, hollow tube, or *parison,* is produced.

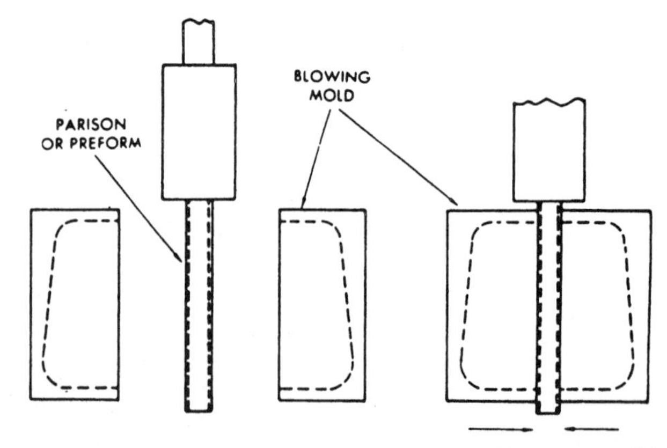

(A) Molded hollow tube (parison) is placed between mold halves, which then close.

Fig. 11-32. Blow molding sequence. (USI)

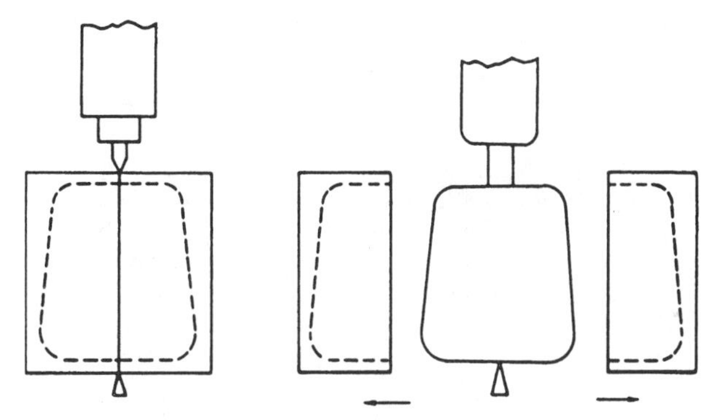

(B) The still-molten parison is pinched off and inflated by an air blast. The blast forces the plastics against the cold walls of the mold. Once the product has cooled, the mold opens and the object is ejected.

Fig. 11-32. Blow molding sequence. (USI) *(Continued)*

Injection Blow Molding

Injection blow molding can more accurately produce the desired material thickness in specified areas of the part. The major advantage is that any shape with varying wall thickness can be made exactly the same each time. There is no bottom weld or scrap to reprocess. The major disadvantage is that two different molds are required. One is used to mold the preform (Fig. 11-33A), and the other for the actual blowing operation (Fig. 11-33B). During the blow-molding operation, the hot injection-molded preform is placed in the blowing mold. Air is then forced into the preform, making it expand against the walls of the mold. The injection-blow process has been called *transfer blow* because the injected preform must be transferred to the blowing mold (Fig. 11-34).

Extrusion Blow Molding

In extrusion blowing, a hot tubular parison is extruded continuously (except when using accumulator or ram systems). The mold halves close, sealing off the open end of the parison (Fig. 11-35). Air is then injected, and the hot parison expands against the mold walls. After cooling, the product is ejected. Extrusion blow

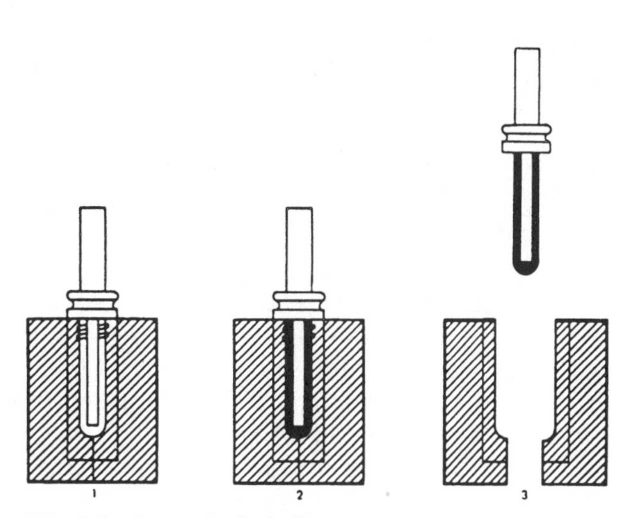

(A) Injection cycle (1, 2, 3).

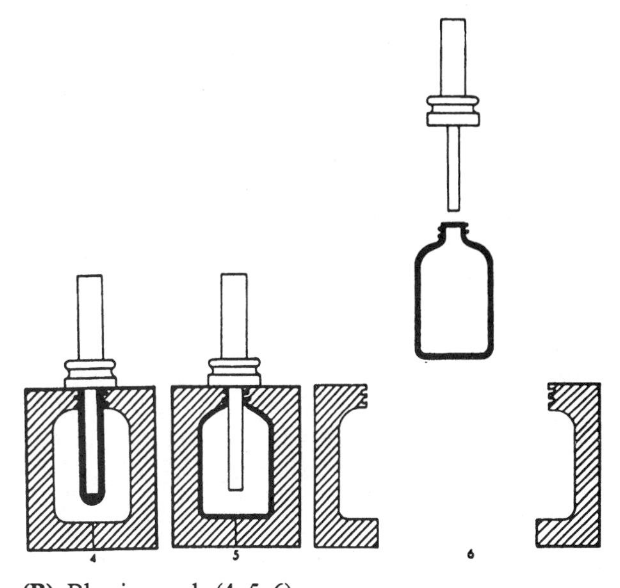

(B) Blowing cycle (4, 5, 6).

Fig. 11-33. The injection blow molding process.

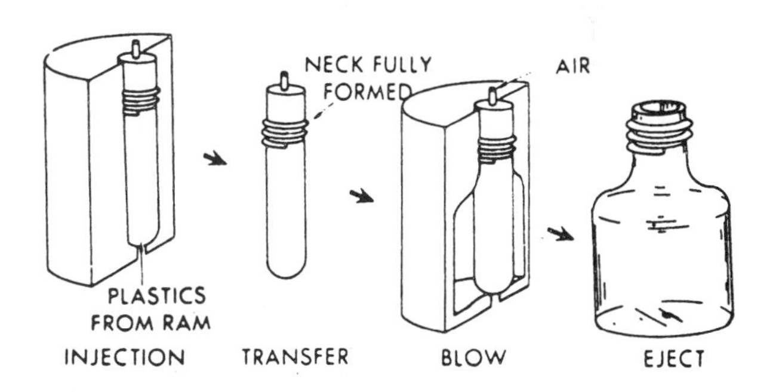

Fig. 11-34. Injection blow process. (Monsanto Co.)

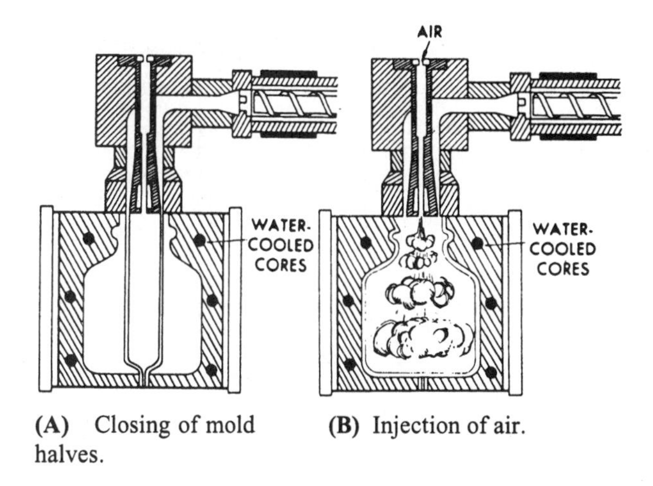

(A) Closing of mold halves.

(B) Injection of air.

Fig. 11-35. Extrusion blow molding.

molding can produce articles large enough to hold 10000 L [2646 gal.] of water; however preforms of this size are too costly. Blow extrusion offers strain-free articles at high production rate but scrap reprocessing is required.

Controlling wall thickness is the largest disadvantage. By controlling (sometimes called *programming)* the wall thickness of the extruded parison, thinning is reduced. For a part requiring an extremely large body but needing strength at the corners, a parison could be produced with the corner areas much thicker than the walls (Fig. 11-36).

In Figure 11-37, the arrangement of the extruder and die parts is shown. By this method, one or more continuous parisons may be extruded. In Figure 11-38, hot plastics is fed into an accumulator and then forced through the die. A controlled length of parison is produced when the ram or plunger operates. The extruder fills the accumulator and the cycle begins again.

The wall thickness of the tube or parison may be controlled (programmed) to suit the container configuration. This is done by using a die with a variable orifice, as shown in Figure 11-39.

Many different ways of forming the blow-molded product have been developed (Fig. 11-40), and each

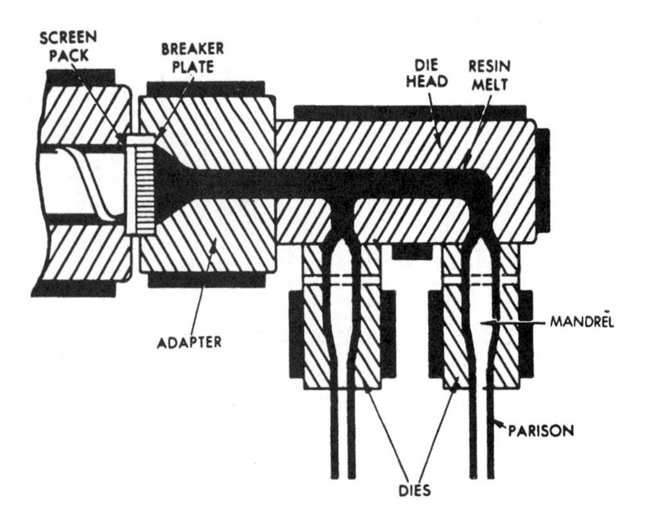

Fig. 11-37. Parts found in most extrusion blow molders.

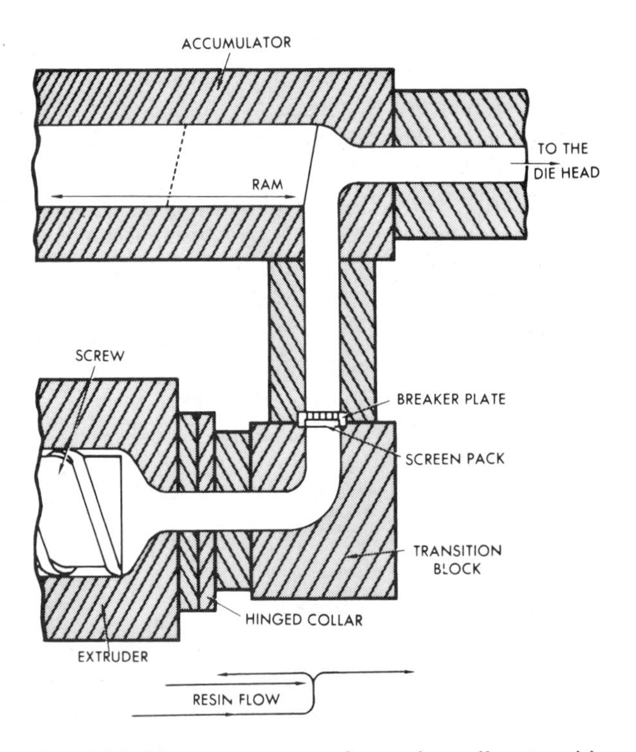

Fig. 11-38. The arrangement of extruder collar, transition block, screen pack, and breaker plate found on a blow-molding press with accumulator.

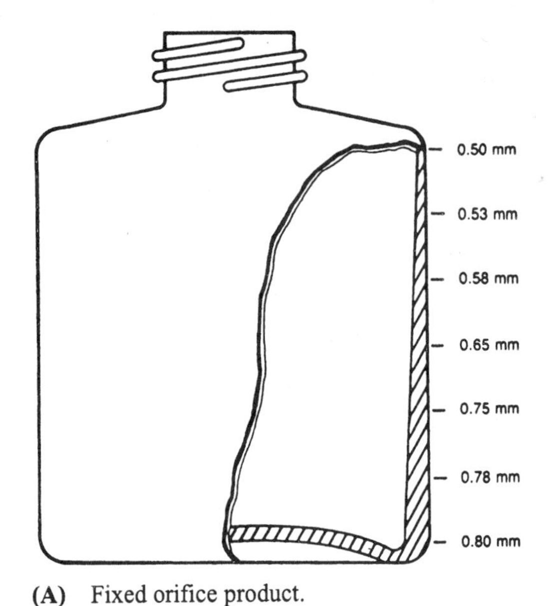

(A) Fixed orifice product.

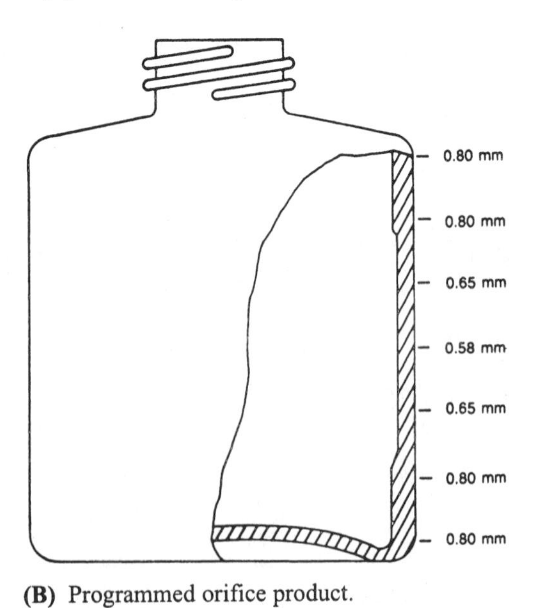

(B) Programmed orifice product.

Fig. 11-36. Parison programming with variable die orifice. Note wall thicknesses.

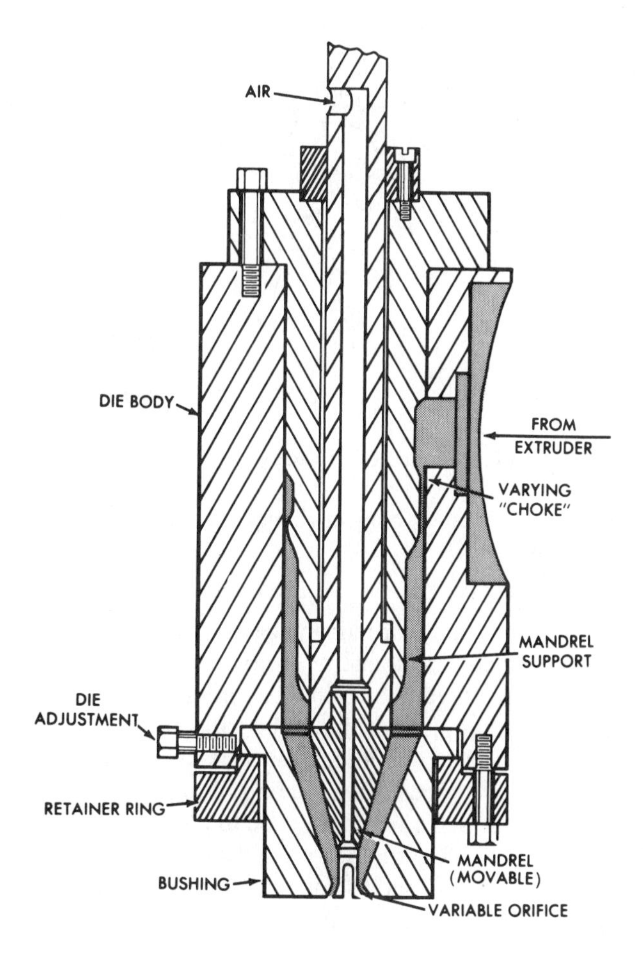

Fig. 11-39. Programming die used for blow molding. (Phillips Petroleum Co.)

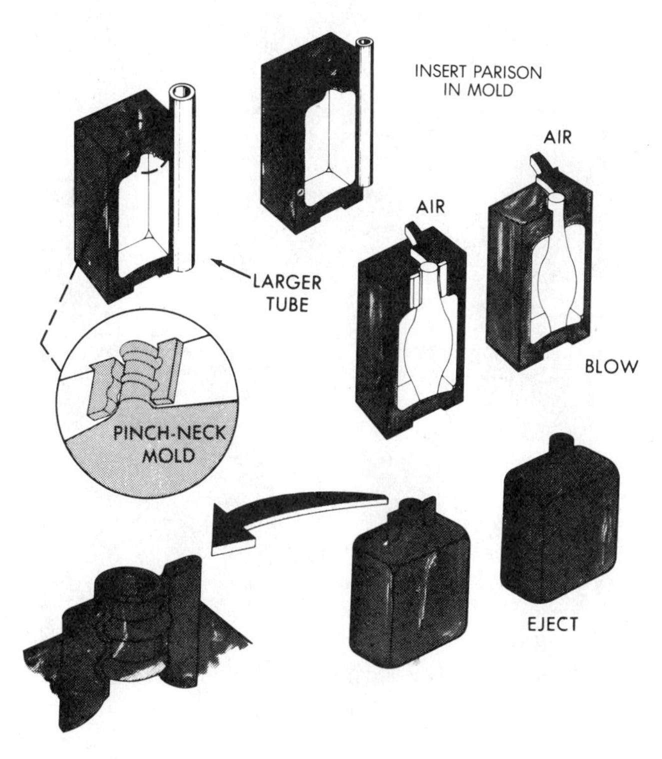

(A) Pinch-neck and regular processes.

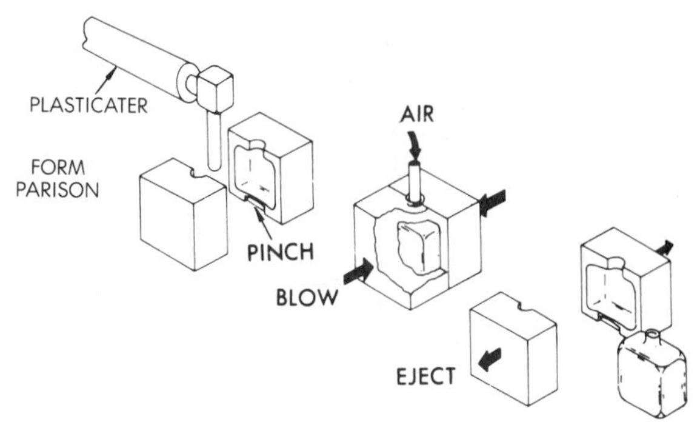

(B) Basic pinch-parison process.

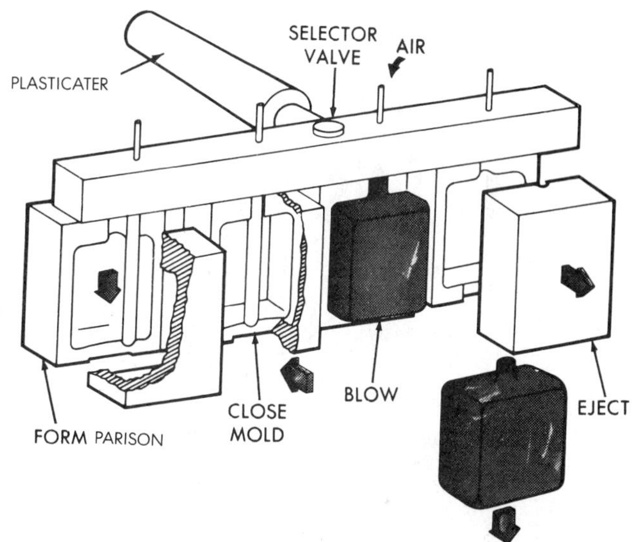

(C) In-place process.

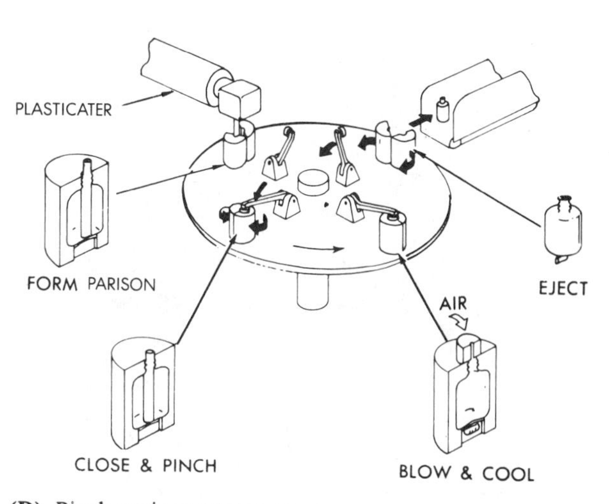

(D) Pinch-parison rotary process.

Fig. 11-40. Various blow-molding processes. (Monsanto Co.)
(continued)

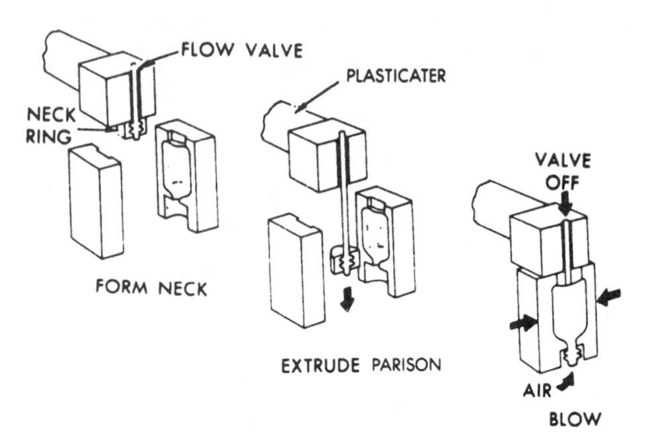

(E) Neck-ring process.

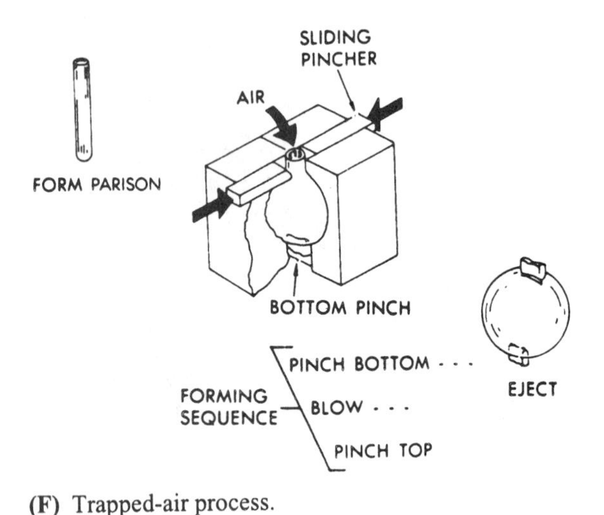

(F) Trapped-air process.

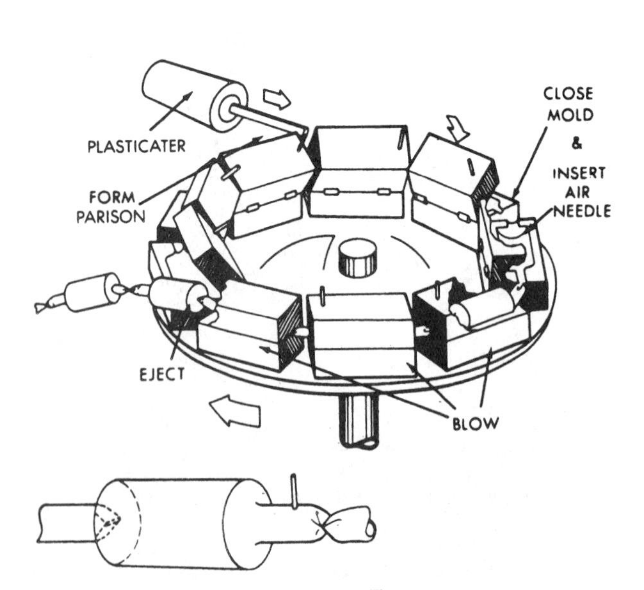

(G) Continuous-parison process (I).

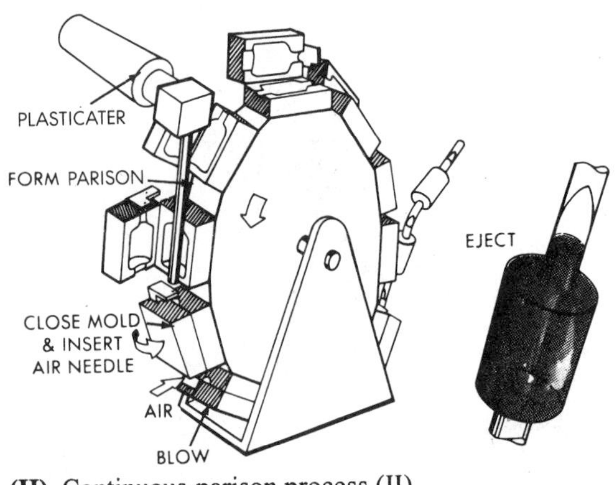

(H) Continuous-parison process (II).

Fig. 11-40. Various blow molding processes. (Monsanto Co.) *(Continued)*

process may have an advantage in molding a given product. One manufacturer forms the container and fills it in a single operation. The product, rather than compressed air, is forced into the parison.

Extrusion blow molding may be used to produce all forms of composites, including fibrous, particulate, and laminar. Short fibers are used to produce a variety of reinforced blow-molded products.

Figure 11-41 shows some blow-molded products and a mold. Note the pinched parison on the product in

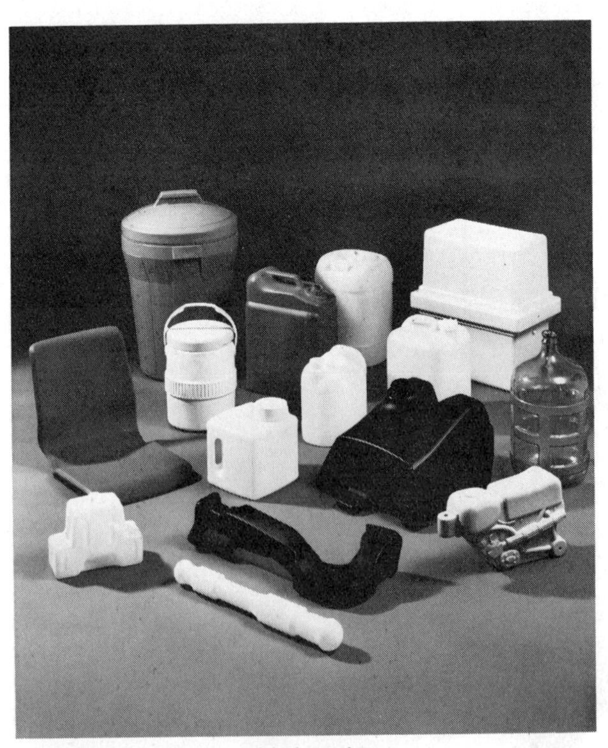

(A) Assorted blow-molded products.

Fig. 11-41. Some blow-molded products.

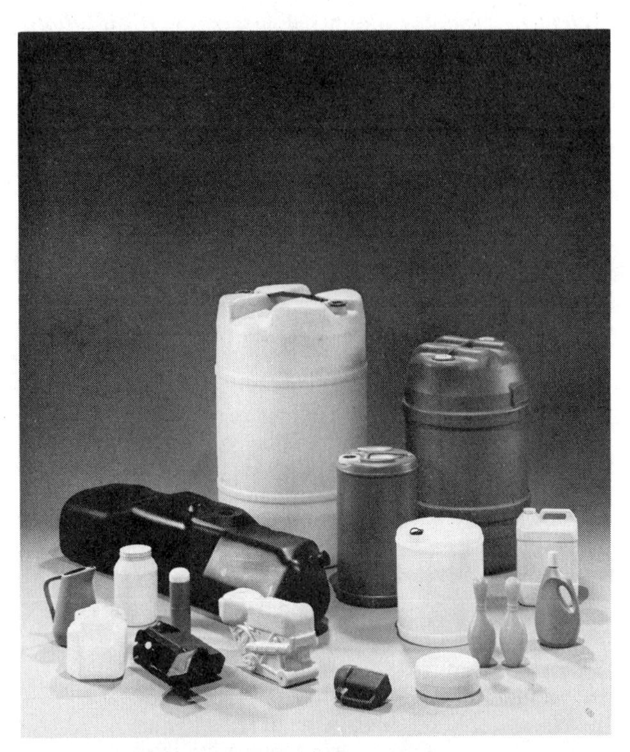

(B) More blow-molded products.

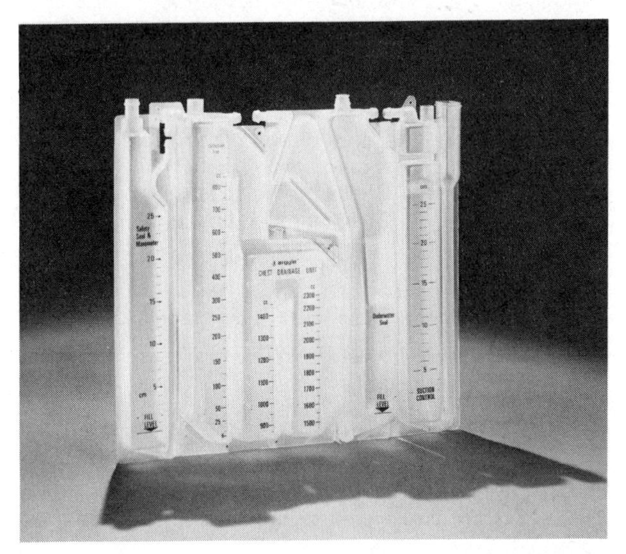

(C) Double-walled, blow-molded chest evacuator. Double-walled products are rigid, yet have some flexibility. (Geauga)

Fig. 11-41. Some blow-molded products. *(Continued)*

Figure 11-42A. Table 11-5 gives some common problems in blow molding and their remedies, while six advantages and five disadvantages of extrusion blow molding are shown below.

Advantages of Extrusion Blow Molding

1. Most thermoplastics and many thermosets may be used
2. Die costs are lower than those for injection molding

(A) Note the pinched parison and untrimmed, or unseparated, lids. (Hoover Universal)

(B) Note the pinchoff on the milk containers from this six-head blow molder.

Fig. 11-42. These containers require trimming.

(continued)

(C) A spin-off trimmer for wide-mouth jars.

Fig. 11-42. These containers require trimming. *(Continued)*

Table 11-5. Troubleshooting Blow Molding

Defect	Possible Remedy
Excess parison stretch	Reduce stock temperature Increase extrusion rate Reduce die tip heat
Die lines	Die surface poorly finished or dirty Blowing air orifice too small—more air Extrusion rate too slow—parison cooling
Uneven parison thickness	Center mandrel and die Check heater bands for uneven heating Increase extrusion rate Reduce melt temperature Program parison
Parison curls up	Excessive temperature difference between mandrel and die body Increase heating period Uneven wall thickness or die temperature
Bubbles (fisheyes) in parison	Check resin for moisture Reduce extruder temperatures for better melt control Tighten die tip bolts Reduce feed-section temperature Check resin for contamination
Streaks in parison	Check die for damage Check melt for contamination Increase back pressure on extruder Clean and repair die
Poor surface	Extrusion temperature too low Die temperature too low Poor tool finish or dirty tools Blowing air pressure too low Mold temperature too low Blowing speed too slow

Table 11-5. Troubleshooting Blow Molding *(Continued)*

Defect	Possible Remedy
Parison blowout	Reduce melt temperature Reduce air pressure or orifice size Align parison and check for contamination Check for hot spots in mold and parison
Poor weld at pinch-off	Parison temperature too high Mold temperature too high Mold closing speed too fast Pinch-off land too short or improperly designed
Container breaks on weld lines	Increase melt temperature Decrease melt temperature Check pinch-off areas Check mold temperature and decrease cycle time
Container sticks in mold	Check mold design—eliminate undercuts Reduce mold temperature and melt temperature Increase cycle time
Part weight too heavy	Parison temperature too low Melt index of resin too low Annular opening too large
Warpage of container	Check mold cooling Check for proper resin distribution Lower melt temperature Reduce cycle time for cooling
Flashing around container	Lower melt temperature Check blowing pressure and air start time Check for molds closing on parison Check air start time and pressure

3. Extruder compounds and blends materials well
4. Extruder plasticates material efficiently
5. Extruder is basic to many molding processes
6. Extrudates may be any practical length

Disadvantages of Extrusion Blow Molding

1. Costly secondary operations are sometimes needed
2. Machine cost is high
3. Purging and trimming produce waste
4. Limited programmed shapes and die configurations are available
5. Screw design must match material melt and flow characteristics for efficient operation

Blow-Molding Variations

Four blow molding variations should be mentioned:

1. Cold parison
2. Sheet
3. Stretched or biaxial
4. Multilayered (co-extrusion or co-injection)

In the cold-parison process, the parison is extruded by normal means (either injection or extrusion), then cooled and stored. The parison is later heated and blown to shape. The major advantage is that the parison can be shipped to other locations or stored in case of a breakdown or materials shortage.

Multilayer bottles may be produced by co-injection blow molding or co-extrusion methods. The three-layered product generally contains a barrier layer sandwiched between two body layers.

In the *sheet blow molding* process, hot extruded sheets are blow formed as they are pinched between mold halves. The edges are fused together by the pinching action of the mold. Two different colored sheets may be extruded and formed into a product with two separate colors (Fig. 11-43). The pinch-welded seams are the biggest disadvantage; in addition, two extruders are usually needed, and there is much scrap to be reprocessed.

In the *stretch or biaxial blow molding* process, molded preforms and extruded tubes may be stretched before blowing. This produces a blow-molded product with better clarity, reduced creep, higher impact strength, improved gas and water vapor barrier properties, and lower mass. Homopolymers may be used instead of more costly copolymers.

In injection-molded preforms, a rod stretches the hot preform during the blow cycle (Fig. 11-44). In parison or tube methods, the hot tube is stretched prior to the blow cycle (Fig. 11-45).

Co-extrusion blow molding actually produces a laminar bottle product. (See Laminating.) Several extruders are used to extrude the material into the manifold. The multilayered container is then blow molded from the emerging parison. For long shelf life, some food containers may have seven layers. A common food container consists of PP/adhesive/ EVOH/adhesive/PP (Fig. 11-46).

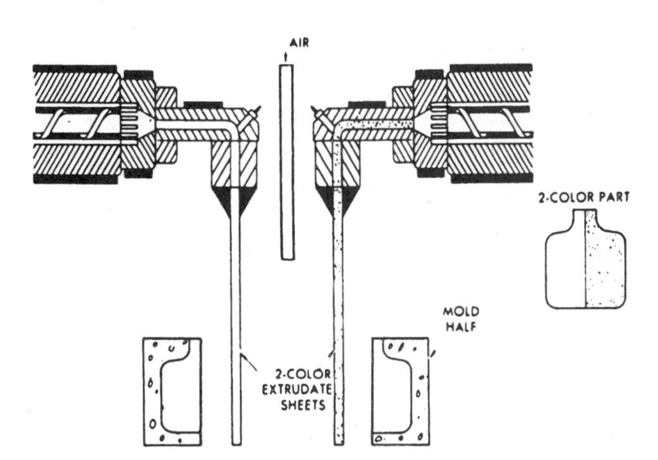

Fig. 11-43. Extrudates of two different colors are pinched between mold halves to form a sheet-blown two-color part.

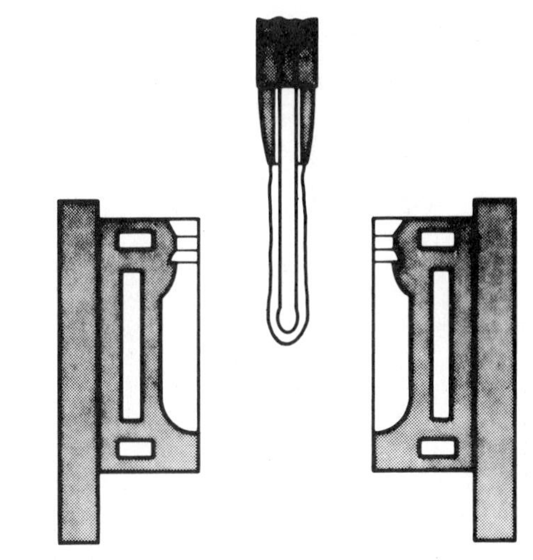

(A) Injection-molded preformed parison

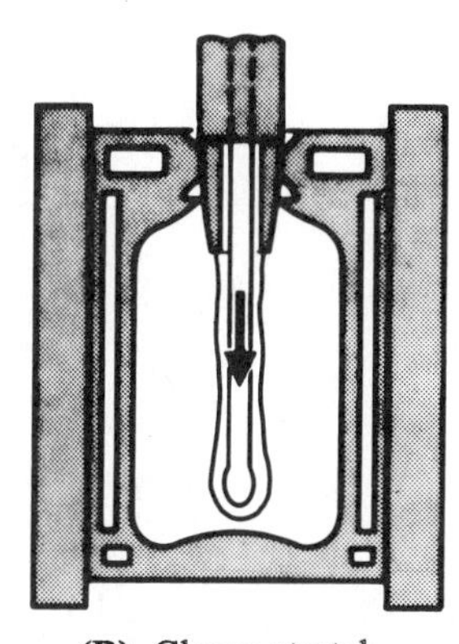

(B) Clamp-stretch

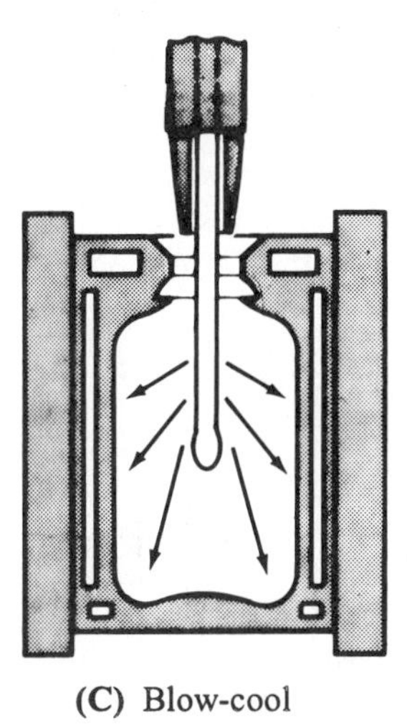

(C) Blow-cool

Fig. 11-44. The preform is stretched by action of the rod and the air pressure. This biaxial stretching improves properties.

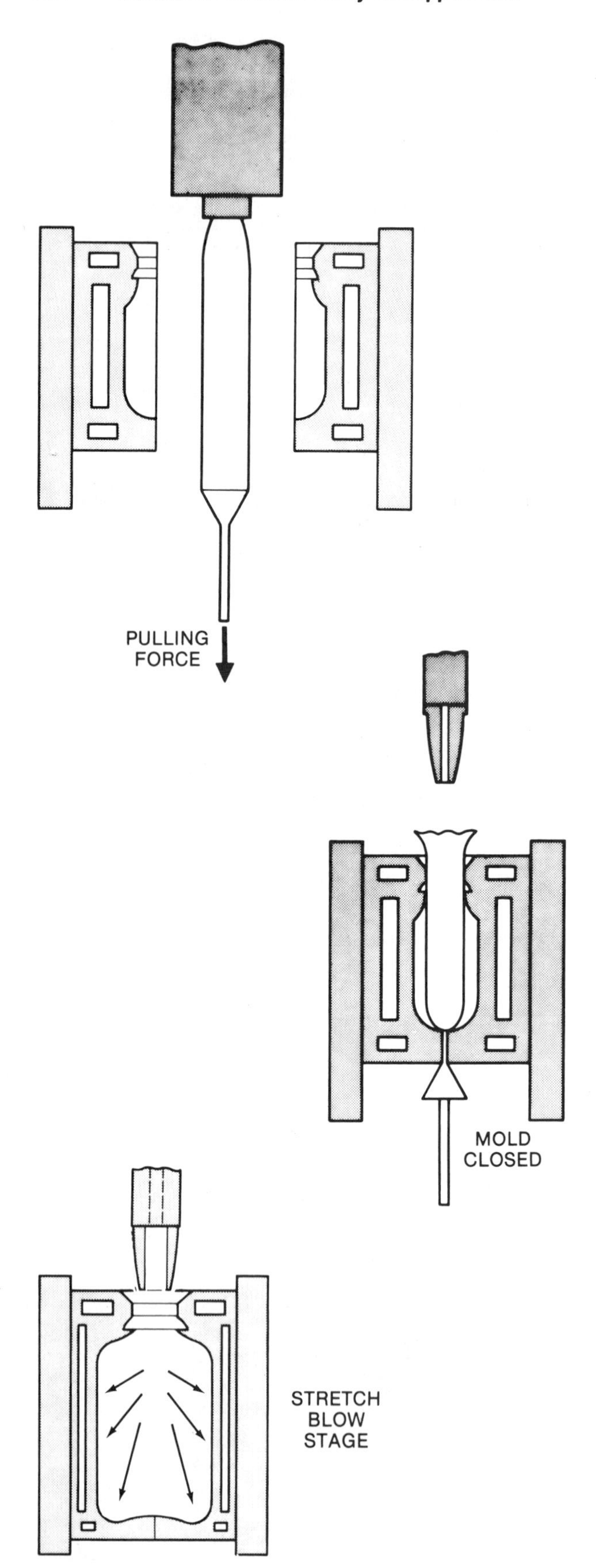

Fig. 11-45. Biaxially oriented product is produced by pulling the extruded tube, then blowing.

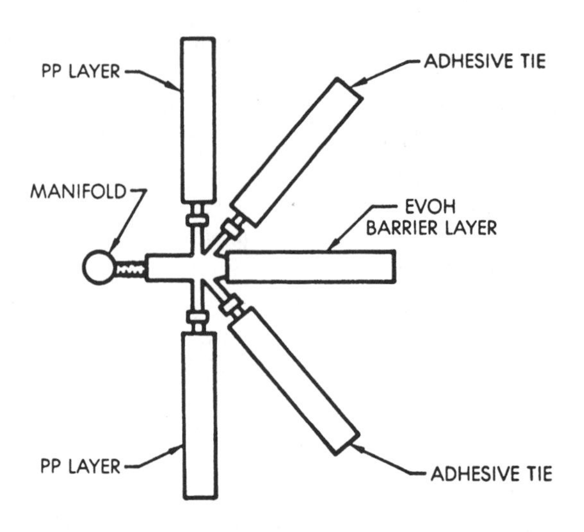

Fig. 11-46. Five extruders are used to produce a multilayered parison (for blow molded parts) or bubble (for films).

Vocabulary

The following vocabulary words are found in this chapter. Look up the definition of any of these words you do not understand as they apply to plastics in the glossary, Appendix A.

Back pressure
Banbury
Barrel
Barrel vent
Breaker plate
Blow molding
Blowup ratio
Calender or calendering
Coating
Denier
Drawing
Extrudate
Extrusion
Frost line
Hopper dryer
Parison
Pinch-off
Spinneret
Tex

Questions

11-1. The hollow plastics tube used in blow molding is called a __?__ .

11-2. The process that forces hot plastics through machine dies to form continuous shapes is called __?__ .

11-3. Which molding process would be used to make long sections of plastics pipe?

11-4. Name three extrusion-spinning processes.

11-5. The process of stretching a thermoplastic sheet, rod, or filament to reduce its cross-sectional area and change its physical traits is called __?__ .

11-6. Because of molecular orientation, __?__ film extrusion film is generally stronger than __?__ film.

11-7. The extruder screw moves the plastics materials through the __?__ of the machine.

11-8. The __?__ in the extruder helps support the screen pack and creates additional mixing of the plastics before it leaves the die.

11-9. The general term for the product or material delivered by an extruder is __?__ .

11-10. Name the two methods used to make films by the extrusion process.

11-11. The __?__ line is the name of the zone that appears frosty in the extrusion blow-film process.

11-12. Blow molding is a refinement of the ancient __?__ processes.

11-13. The two basic blow-molding methods are __?__ blowing and __?__ blowing.

11-14. In blow molding, a die with a variable orifice may be used to control __?__ and thickness.

11-15. In the __?__ process, hot material is pressed between two or more rotating rollers.

11-16. Patterns or textures may be placed on calendered films or sheets by passing the soft, hot plastics between __?__ rolls.

11-17. Spinnerets are used to produce __?__ and __?__ plastics forms.

11-18. Films are less than __?__ mm in thickness.

11-19. Sheets are greater than __?__ mm in thickness.

Activities

Extrusion Compounding

Introduction. Extrusion compounding is the process of mixing various ingredients into a base polymer to achieve a plastics material with desired properties.

Equipment. An extruder with pelletizer, polypropylene, chopped glass fibers, accurate scale, small furnace, small tongs, Pyrex dish or ceramic crucible, safety glasses, insulated gloves, special magnifying glass or microscope.

Procedure

11-1. Check the extruder to make sure that the rupture disk is good. Most extruders have a rupture disk in a hollow threaded fastener. The disk is located near the front end of the barrel. Rupture disks blow out if the pressure near the die is too great. This protects the bolts holding the die in place. Extreme pressures can break off the bolts holding the die, or cause their threads to shear. This can be extremely dangerous. To provide a release for extreme pressure, the rupture disk fails and the plastics flow out through the hollow fastener. Figure 11-47 shows a rupture disk. The disks fail at varying pressures. The one shown is a 5000 psi disk.

11-2. If the extruder is dirty, take off the die and push out the extruder screw. With many laboratory-sized machines, it is possible to push out the screw with a rod, if the barrel is at processing temperature. Use copper mesh, as seen in Figure 11-48 to clean the screw. This kind of mesh is supplied in a roll. To use it, cut a portion about 3 feet long, wrap it around the screw once or twice, and work it back and forth to remove the hot plastics. Clean a small portion, then push out another section and clean it. Pushing out a long section may allow the material in the screw time to stiffen before removal.

11-3. Measure the length of the glass fibers before any processing.

11-4. Calculate the amount of glass required for batches of 2, 4, 6, 8, 10, and 12 percent glass. A batch of 1.0 kg is usually sufficient to pro-

Fig. 11-47. This rupture disk will fail at 5 000 psi.

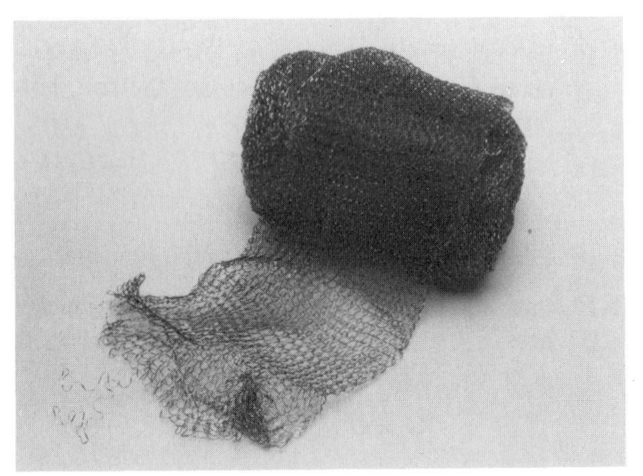

Fig. 11-48. This copper mesh is used to clean extruder screws of hot plastics.

duce many test samples. Measure out the glass and propylene and hand mix to ensure uniform distribution of the glass. Does the mixture tend to separate into glass rich and glass poor layers? How can this be avoided?

11-5. Compound the glass into the polypropylene. Select a temperature of about 204 °C [400 °F] at the die and the front zone. If the L/D ratio of the extruder is low, two passes through the extruder may be needed for adequate dispersion of the glass. The lower percentages should compound rather easily. As the glass content increases, the tendency to surge caused by non-uniform feeding into the screw, may increase. If the extruder has a screw no more than 25 mm [1.0 in.] diameter, the glass may hinder the feeding. To reduce this problem, agitate the material in the feed throat with a soft tool.

11-6. After compounding, carefully weigh a crucible or Pyrex container and then measure out 5 to 10 g of pellets. Use insulated gloves and small tongs to place the crucible in a furnace. Burn off the propylene in a furnace at a temperature of 482 to 538 °C [900 to 1 000 °F]. A small heat-treating furnace is appropriate for burn-outs. Make sure there is adequate ventilation to remove the smoke and fumes produced by the burning polypropylene.

11-7. When the container has cooled, carefully measure the container and glass. Calculate the percentage of glass content. Does the calculated percent match the percent of glass in the original batch? If the percents do not match, why is there an error?

11-8. Figure 11-49 shows a specialty magnifying glass with various scales printed on the stages. Such magnifiers are available in many industrial supply catalogs. Select an appropriate scale and measure the length of the glass fibers. How much shorter are they after compounding than before?

11-9. To compare the relative amounts of glass in the various batches, select one pellet from each batch. Arrange them on a microscope slide in order of glass content. See Figure 11-50 for such an arrangement. After burning off the polypropylene, the results give visual comparison of varying glass contents (Figure 11-51). Examine the glass under a microscope.

11-10. Mold test bars and tensile test. Figure 11-52 shows a close up of the broken edge of a tensile bar. Notice the glass fibers sticking out of the bar.

11-11. Compare test results to determine how great an effect on strength and elongation the levels of glass have.

11-12. Write a report summarizing the results.

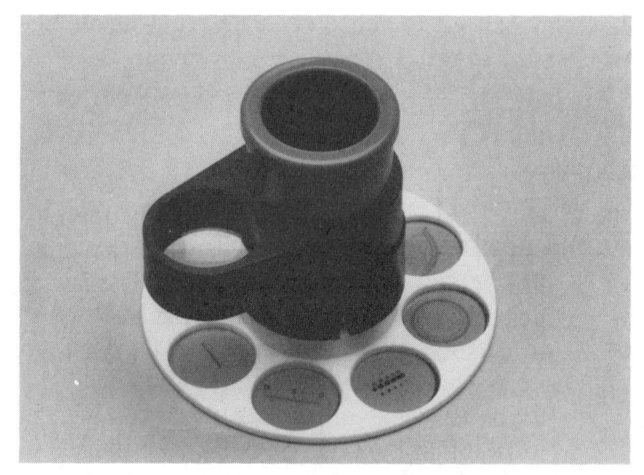

Fig. 11-49. This magnifying glass with graduated stages provides quick measurement of fiber lengths.

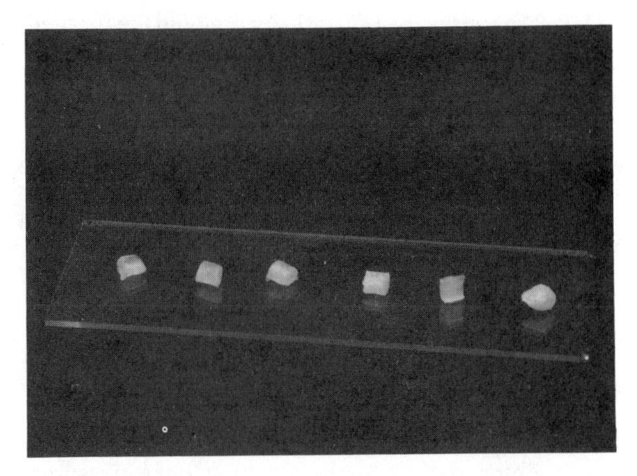

Fig. 11-50. These pellets are arranged in order of increasing glass content.

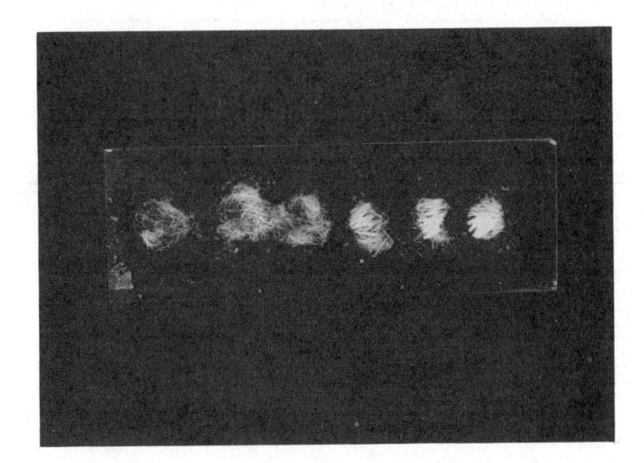

Fig. 11-51. After burnout, the remaining glass fibers show changes in percent of glass content.

Profile Extrusion

Introduction. Many laboratory extruders do not include required sizing fingers, cooling apparatus, and take-off equipment to support profile extrusion. However, output calculations on strand dies offer the opportunity to determine the efficiency of the extruder under various conditions.

Fig. 11-52. This close up of a broken tensile bar shows glass fibers. Some have broken during testing, others pulled out without breaking.

Equipment. Extruder, accurate scale, selected plastics.

Procedure

11-1. Clean the extruder screw and remove it.

CAUTION: Use insulated gloves when handling the hot screw. Never use steel tools on the screw. If direct gripping of the screw is required, use brass tools.

11-2. Figure 11-53 shows a single-stage extruder screw. Collect measurements of flighted length, screw diameter; length of feed, transition, and metering zones; and flight depth in feed and metering zones.

11-3. Set up the extruder and extrude enough material needed for the machine to arrive at a rather stable condition.

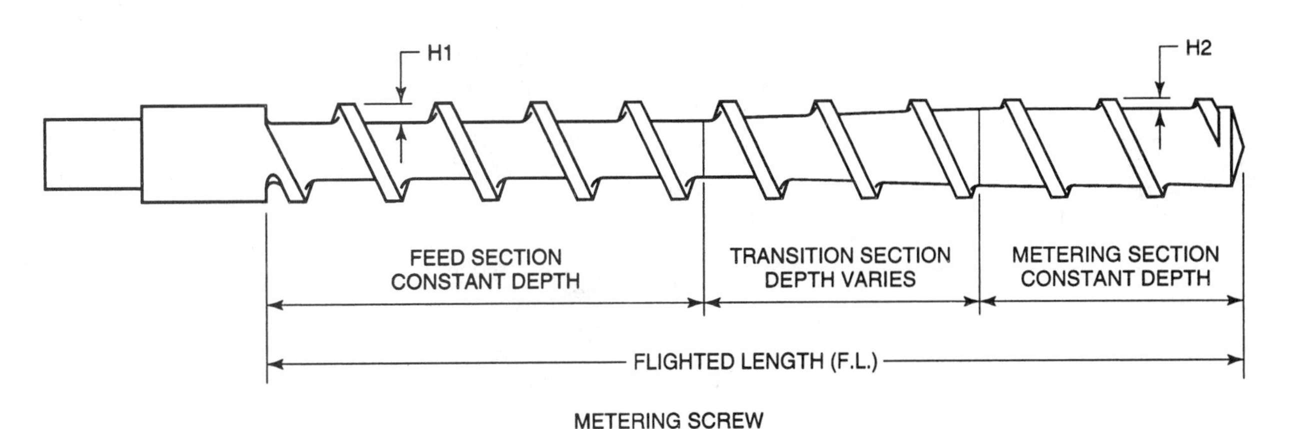

Fig. 11-53. Schematic of a single-state extruder screw.

11-4. Starting near the top of the RPM capabilities of the equipment, catch the extrudate on a piece of cardboard or sheetmetal for a period of 30 seconds.

11-5. Reduce the RPM by 10 and repeat step 2. Continue this process until reaching a low RPM, perhaps 10 or 20.

11-6. When the extrudate samples have cooled, accurately weigh them. Convert the results into output in pounds per hour.

11-7. Determine the compression ratio of the screw by dividing the depth of the feed zone by the depth of the metering zone.

11-8. Determine the L/D ratio by dividing the flighted length of the screw by its outside diameter.

11-9. Determine the theoretical output of the screw with this formula:

$R = 2.2\ D^2hgN$
R = output in pounds per hour
D = screw diameter in inches
h = depth of metering zone in inches
g = melt density
N = screw RPM

11-10. To determine the melt density, refer to Table 11-6.

11-11. Compare the theoretical output to the measured output.

Table 11-6. Material Data for Extrusion

Material	Melt Density Specific Gravity Grams/cm^3 at Processing Temperature	Extrusion Temperature °F
ABS - Extrusion	0.88	435
ABS - Injection	0.97	
Acetal - Injection	1.17	
Acrylic - Extrusion	1.11	375
Acrylic - Injection	1.04	
CAB	1.07	380
Cellulose Acetate - Extrusion	1.15	380
Cellulose Acetate - Injection	1.13	
Cellulose Proprionate - Extrusion	1.10	380
Cellulose Proprionate - Injection	1.10	
CTFE	1.49	
FEP	1.49	600
Ionomer - Extrusion	.73	500
Ionomer - Injection	0.73	
Nylon 6	0.97	520
Nylon 6/6	0.97	510
Nylon 6/10	0.97	
Nylon 6/12	0.97	475
Nylon 11	0.97	460
Nylon 12	0.97	450
Phenylene Oxide Based	0.90	480
Polyallomer	0.86	405
Polyarylene Ether	1.04	460
Polycarbonate	1.02	550
Polyester PBT	1.11	
Polyester PET	1.10	480
HD Polyethylene - Extrusion	0.72	410
HD Polyethylene - Injection	0.72	
HD Polyethylene - Blow Molding	0.73	410
LD Polyethylene - Film	0.77	350

Table 11-6. Material Data for Extrusion

Material	Melt Density Specific Gravity Grams/cm3 at Processing Temperature	Extrusion Temperature F
LD Polyethylene - Injection	0.76	
LD Polyethylene - Wire	0.76	400
LD Polyethylene - Ext. Coating	0.68	600
LLD Polyethylene - Extrusion	0.75	500
LLD Polyethylene - Intrusion	0.70	
Polypropylene - Extrusion	0.75	450
Polypropylene - Intrusion	0.75	
Polystyrene - Impact Sheet	0.96	450
Polystyrene - G. P. Crystal	0.97	410
Polystyrene - Injection Impact	0.96	
Polysulfone	1.16	650
Polyurethane (non-elastomer)	1.13	400
PVC - Rigid Profiles	1.30	365
PVC - Pipe	1.32	380
PVC - Rigid Injection	1.20	
PVC - Flexible Wire	1.27	365
PVC - Flexible Extruded Shapes	1.14	350
PVC - Flexible Injection	1.20	
PTFE	1.50	
SAN	1.00	420
TFE	1.50	
Urethane Elastomer (TPE)	0.82	390

Chapter 12

Laminating Processes and Materials

Introduction

This chapter concerns laminates and the processes required for their formation. Before discussing types of laminates and various production techniques, basic definitions are essential.

The verb *laminate* describes the process of bonding two or more layers of material by cohesion or adhesion. In the plastics industry, laminates contain layers held together by a plastics material. The layers bonded together often provide strength and reinforcement. This makes a clear distinction between laminates and reinforced plastics more difficult. A key characteristic of laminates is that they contain layers. In contrast, reinforced plastics generally get their strength from fibers contained within the plastics. Laminates are generally flat sheets, while reinforced plastics are often molded into complex shapes.

Laminates have numerous uses. They appear as structural materials in automobiles, furniture, bridges, and homes. They are very important in aircraft structures, helicopter blades, and aerospace applications (Figs. 12-1 and 12-2).

Laminates include metallic foils bonded to paper or fabric. In the textile field, layers of cloth, plastics foams and films are bonded into special-purpose fabrics.

In order to bring some order into the diverse applications of laminates, the sections in this chapter are organized according to the materials that form the layers in the laminates. These are:

I. Layers of differing plastics
II. Layers of paper
III. Layers of glass cloth or mat
IV. Layers of metal and metal honeycomb
V. Layers of metal and foamed plastics

Fig. 12-1. A radome of composite laminates protects the radar anenna of this capsule. (FMC Corporation)

Please note that this list does not include all laminates. Some laminates, bonded with plastics, will receive no attention, for example, plywood. Table 12-1 provides basic data on the types of plastics used in many laminates.

Layers of Differing Plastics

Although it is possible to start with finished film or sheet when making laminates of various plastics, it is

Fig. 12-2. This prototype turboprop aircraft has an all-composites primary structure. It has the strength of titanium, but is somewhat lighter than aluminum. (Plastics Design Forum)

usually more economical to extrude the various layers simultaneously. Engraving stock, which contains two or more different colors, is an example of continuous extrusion laminating. Engraving stock demonstrates the capability of extrusion laminating to make rather thick products.

Extrusion-laminated films are common in the packaging industry. Co-extruded composite films of polyethylene and vinyl acetate produce a tough, durable two-layer film that may be heat sealed. The plastics layers are brought together in the molten state and extruded through a single die opening to make the multilayered laminate film (Figs. 12-3 and 12-4).

A three-ply laminate film is used to wrap bread products. This film laminate is composed of an inner core (ply) or polypropylene with two outer layers (plies) of polyethylene (Fig. 12-5).

A vinyl plastics film laminate was developed specifically for packaging meat products by the Dow

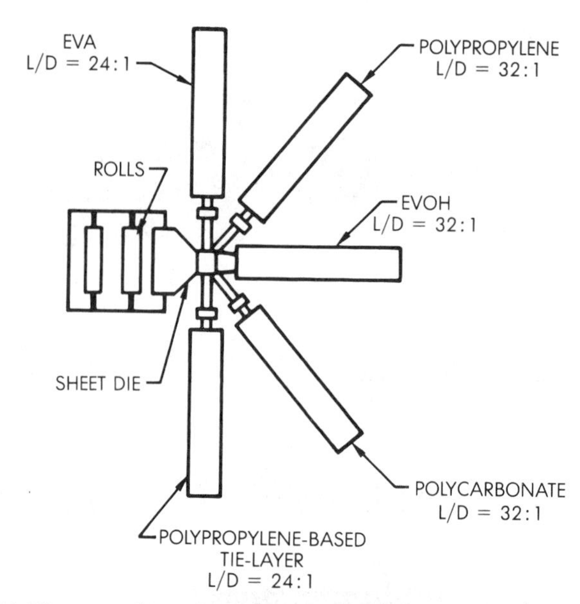

A) Five extruders are used to produce a true composite laminate.

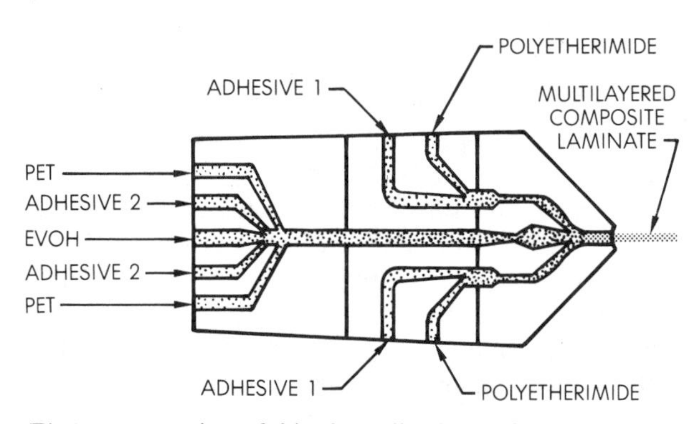

(B) A cross section of this sheet die shows that at least two extruders were used for the adhesives, (1 and 2), one for the outside polyetherimide layers, one for the two PET layers and one for the EVOH center barrier layer.

Fig. 12-3. Multilayer extrusion equipment.

Table 12-1. Selected Resins/Plastics and Materials Used in Lamination

Resin/Plastics	Paper	Cotton Fabric	Fibrous Glass Fabric/Mat	Metallic Foils	Composites, Honeycomb, etc.
Acrylic	LP	—	LP		—
Polyamide	LP	—	LP	LP	—
Polyethylene	LP	—	LP	LP	—
Polypropylene	LP	—	LP	LP	—
Polystyrene	—	—	LP		
Polyvinyl chloride	LP	—	—	LP	LP
Polyester	LP-HP	LP	LP	—	LP
Phenolic	LP-HP	LP-HP	HP-LP	HP	LP-HP
Epoxy	LP	LP-HP	LP-HP	LP	LP
Melamine	HP	HP		HP	
Silicone	—	—	HP	—	LP

LP—Low-pressure (laminate).
HP—High-pressure (laminate).
——Only limited amounts manufactured in this category.

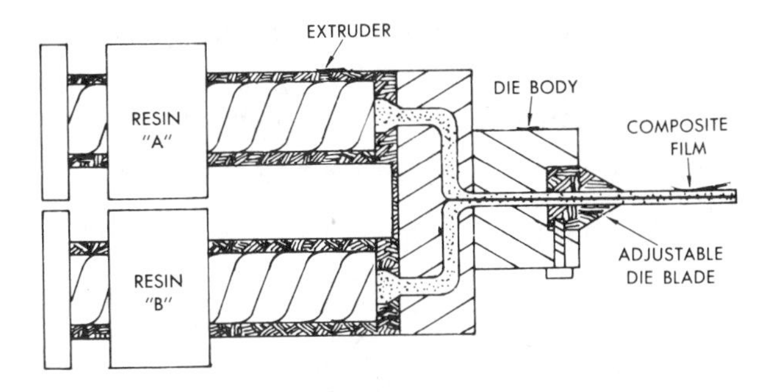

Fig. 12-4. Production of coextruded film.

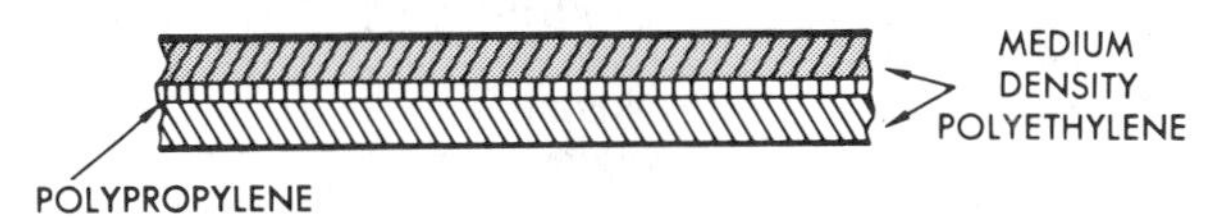

(A) Three-ply film for bread bag use.

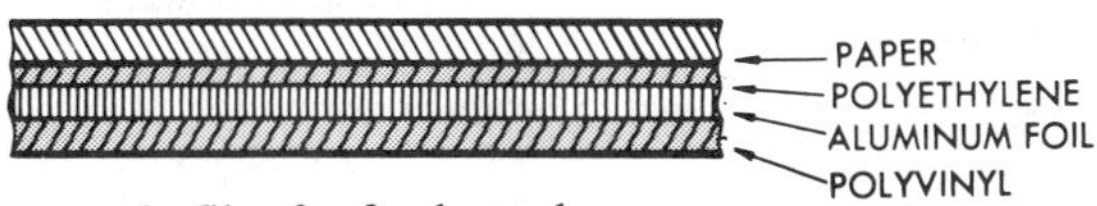

(B) Four-ply film for food pouch use.

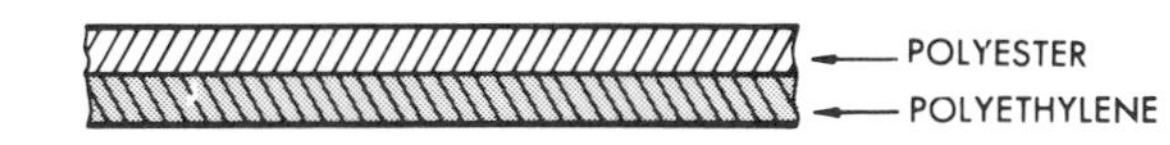

(C) Two-ply film for boil-in-bag use.

Fig. 12-5. Plastics laminates.

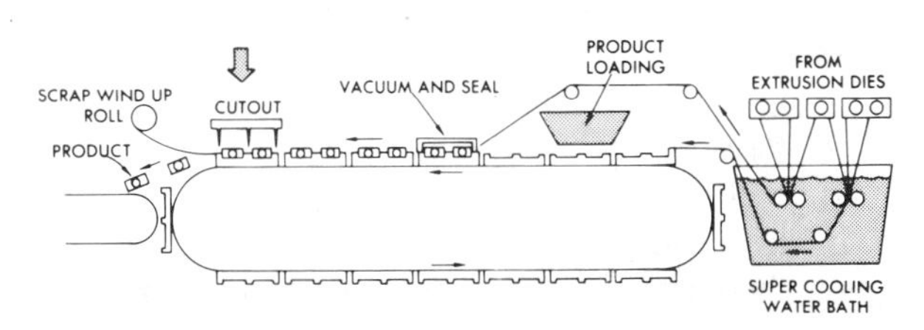

(A) Extrusion and packing.

(B) Three-ply laminate film is used to wrap bread loaves.

(C) Saranpac process is used for meat products.

(D) Nuts packaged in acetate-polyethylene film.

(E) Foil-paper-polyethylene packages for dry meats.

Fig. 12-6. Extrusion and packing technique, with some product examples.

Chemical Company and the Oscar Mayer Company. Three different plies are used in this laminate; Saran 18 (polyvinylidene chloride) for the outer layer, polyvinyl chloride 88 for the core, and Saran 22 for the inner sealing layer (Fig. 12-6).

In this *Saranpac* process, all three film layers are extruded and pressed together in a cooling tank. The laminated film is then formed to contain the meat product and vacuum sealed.

In another application, two different polymers are blow-film co-extruded with molecular orientation and then squeezed together into a laminate sheet. (See Figure 12-7.)

There are numerous applications of extruded film laminates. A few examples are listed in Table 12-2.

Layers of Paper

Paper appears in laminates in two distinct forms, as a not impregnated layer, or as a thoroughly impregnated material.

When the paper is not impregnated, one or more layers of plastic—generally an olefinic film—are adhered to the paper with heat and pressure. The purpose is to create a glossy, water-resistant finish. The cover of this book shows this application. Compare the inside of the cover to the outside. The gloss comes from a plastics film.

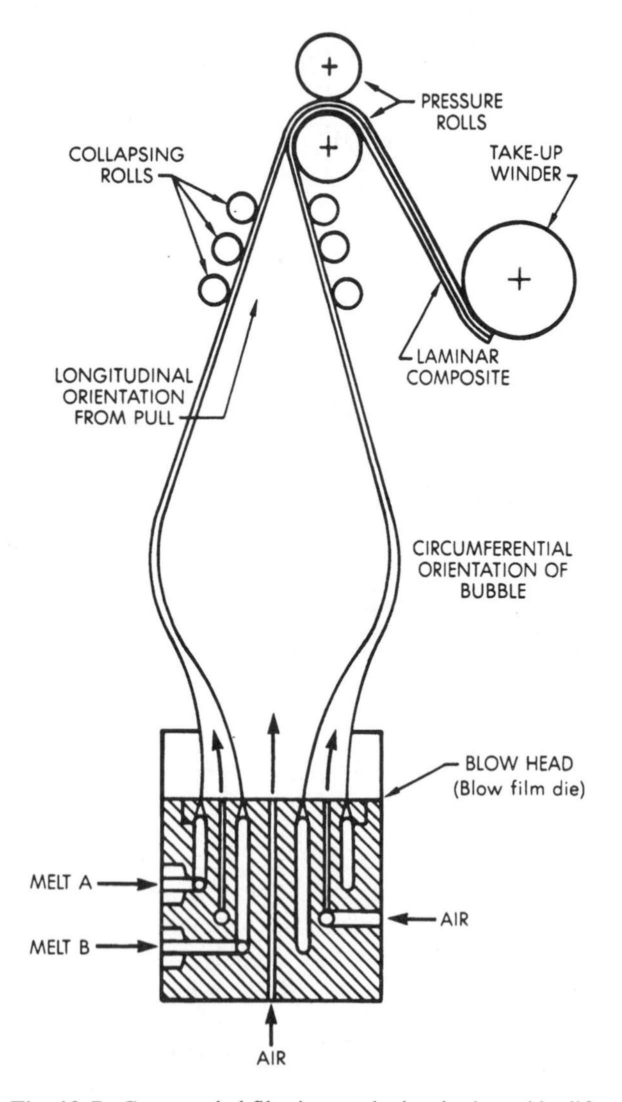

Fig. 12-7. Co-extruded film is stretched and oriented in different directions and then pressed together to form a composite laminar sheet.

Table 12-2. Selected Extruded Film Laminates and Applications

Laminate Material	Application
Paper-polyethylene-vinyl	Sealable pouches for dried milk, soups, etc.
Acetate-polyethylene	Tough, heat-sealable packing for nuts.
Foil-paper-polyethylene	Moisture barrier pouch for soup mixes, dry milk, etc.
Polycarbonate-polyethylene	Tough, puncture-resistant skin packages
Paper-polyethylene-foil-polyethylene	Strong heat seals for dehydrated soups
Paper-polyethylene-foil-vinyl	Heat-sealable pouches for instant coffee
Polyester-polyethylene	Touch, moistureproof boil-in bag pouches for foods
Cellulose-polyethylene-foil-polyethylene	Gas and moisture barrier for pouches of ketchup, mustard, jam, etc.
Acetate-foil-vinyl	Opaque, heat-sealing pouches for pharmaceuticals
Paper-actate	Glossy, scratch-resistant material for record covers, paperback books

Other products exemplifying this process are ID cards and drivers licenses. Because the pressure required to bond the film to the paper is rather low, these laminates are often called *low-pressure*.

In contrast, impregnated paper laminates are often called *high-pressure laminates*. Some laminates are processed at pressures exceeding 7000 kPa [1015 psi]. Many high-pressure laminates contain thermosetting resins—particularly urea, melamine, phenolic, polyester, and epoxy.

In the history of high-pressure laminates, J. P. Wright, the founder of the Continental Fiber Company, played a major role. In 1905, he produced one of the earliest phenolic laminates. This was five years before Baekeland had patented his ideas of using sheets impregnated with phenolic resins as laminates. Many high pressure industrial laminates were made from a layup of paper, cloth, asbestos, synthetic fiber, or fibrous glass using this basic patent. Today, there are over fifty standard industrial grades of laminates for electrical, chemical, and mechanical uses.

At first, high-pressure industrial laminates took the place of mica as a quality electrical insulation material. Around 1913, the Formica Corporation emerged, making high-pressure laminates to replace mica (*for mica*) in many electrical and mechanical uses. By 1930, the National Electrical Manufacturers Association (NEMA) saw the potential of decorative laminates, and in 1947, a separate NEMA section was established to work with government agencies and associations interested in setting up product standards for decorative laminates.

The basic process was to fuse layers of paper that had been impregnated with phenolic resin. Decorative layers and transparent exterior layers were also fused onto the layers of paper. To increase the production rate, many laminates were made at once by stacking a press. Figures 12-8 and 12-9 show this procedure.

Impregnating the paper prior to lamination was done by a number of methods. The most common are premix, dipping, coating, or spreading. Figure 12-10 shows these various methods.

After a drying period, the impregnated laminating stock is cut to the desired size and placed in multi-sandwich form between the metal plates (separation plates) of the press. Press platens are not smooth enough for the desired finish. These plates may be glossy, matte, or embossed. Metal foils are sometimes used between surface layers to produce a decorative finish. In decorative laminates, a printed pattern layer and a protective overlay sheet are superimposed on the base material. The prepared stock is subjected to heat and high pressure. The combined heat and pressure cause the resin to flow and the layers to compact into one polymerized mass. Polymerization may also be done by chemical or radiation sources. When the thermosetting resins have cured or the thermoplastic resins have cooled, the laminate is removed from the press.

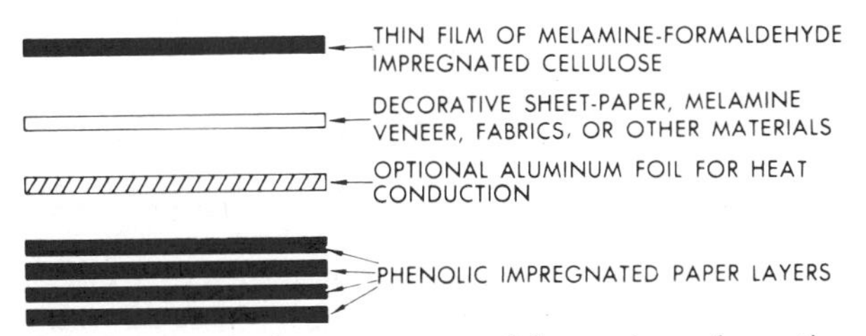

Fig. 12-8. Typical arrangement of layers in a decorative high-pressure laminate.

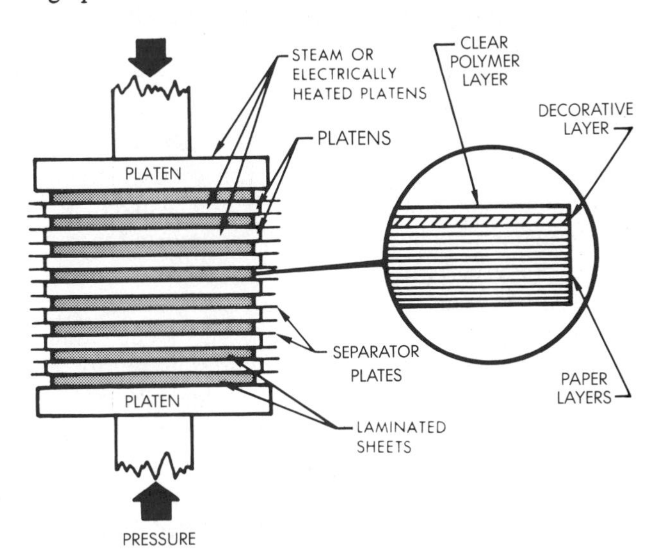

Fig. 12-9. Concept of multiple stacking of laminates in press.

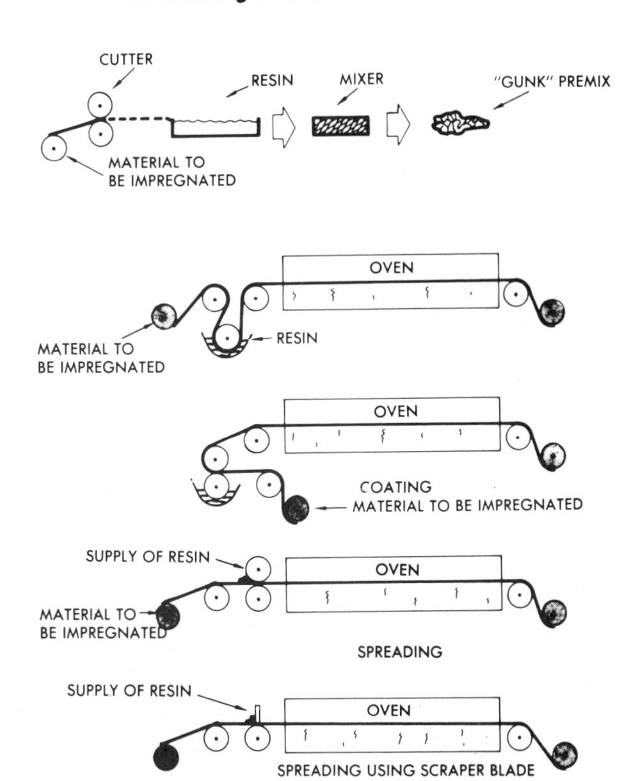

Fig. 12-10. Various impregnation methods.

Today, high-pressure laminates enjoy wide use. *Formica* and *Wilsonart* are brands of laminate often used on countertops. In addition to sheets of laminate, these materials appear as cams, pulleys, gears, fan blades, and printed circuit boards.

The major disadvantage of high-pressure lamination is slow production rates compared to high-speed injection molding. In order to speed production, some manufacturers have established continuous laminating processes. In *continuous laminating* (Fig. 12-11), fabrics or other reinforcements are saturated with resin and passed between two plastics film layers—cellophane, ethylene, or vinyl, for example.

The thickness of the laminated composite is controlled by the number of layers and by a set of squeeze rollers. The laminate is then drawn through a heating zone to speed polymerization. Corrugated awnings, skylights, and structural panels are products of continuous laminating.

Layers of Glass Cloth or Mat

Thermosetting resins are used in *hand layup laminates* (Figs. 12-12 and 12-13). These products are also called *contact moldings* or *open moldings.* After the mold (either male or female) is coated with a releasing agent,

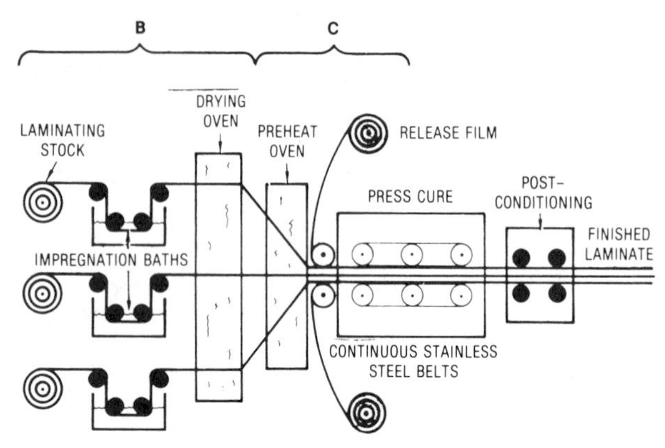

(A) Continuous lamination, in which the (B-stage) resin is changed into an infusible (C-stage) plastics.

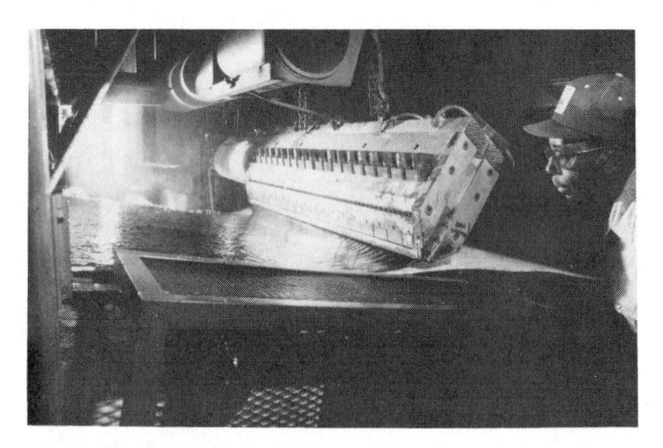

(B) Manufacturing a fiberglass reinforced thermoplastic composite sheet. (AZDELL, Inc.)

Fig. 12-11. Continuous lamination of thermosetting and thermoplastic matrix composite sheets.

(A) Hand layup or contact processing being used to apply fibrous glass to a titanium-honeycomb structure. (Bell Helicopter Co.)

(B) Contact mold for hand layup of honeycomb reinforcements for radome. (McMillan Radiation Labs, Inc.)

Fig. 12-12. Hand layup processing.

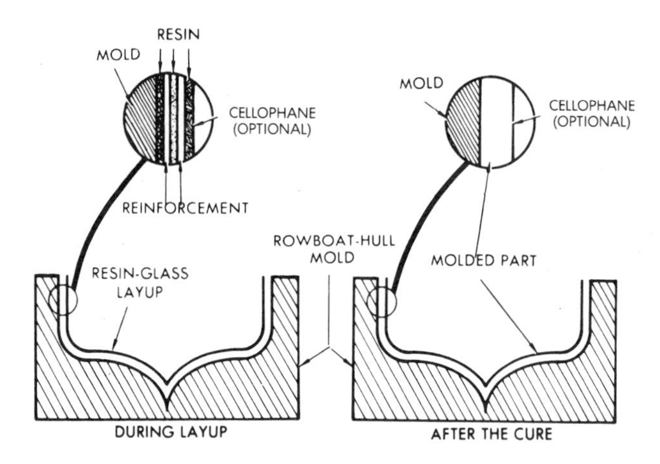

Fig. 12-13. In hand layup, reinforcing material in mat or fabric form is applied to the mold, then saturated with a selected thermosetting resin.

a layer of catalyzed resin is applied and allowed to polymerize to the gel (tacky) state.

This first layer is a specially formulated gel-coat resin used in industry to improve flexibility, blister resistance, surface finish or color, stain resistance, and weatherability. Gel coats based on neopentyl glycol, trimethylpentanediol glycol, and propylene glycol provide a major advantage as surface treatments for reinforced polyester products.

The gel coat forms a protective surface layer through which fibrous reinforcements do not penetrate. A prime cause of deterioration of fibrous reinforced plastics is penetration of water, which takes place when fibers protrude at the surface. Once the gel coat has partially set, reinforcement is applied. Then, more catalyzed resin is poured, brushed, or sprayed over the reinforcement. This sequence is repeated until the desired thickness is reached. In each layer, the mixture is worked into the mold shape with hand rollers, then the reinforced composite laminate is allowed to harden or cure (Fig. 12-14). External heating is sometimes used to speed polymerization (Fig. 12-15).

Hand layup and sprayed operations are often used alternately. In order to obtain a resin-rich, superior surface finish, a lightweight veil or surface mat is sometimes placed next to the gel layer. The coarser reinforcements are then placed over this layer. Some operations and designs use preforms, cloth, mat, and roving materials for directional or additional strength in selected areas of the part.

Among the major advantages of hand layup are low-cost tooling, minimal equipment required, and the ability to mold large components. As for disadvantages, the process is labor intensive and dependent upon the skill of the operator, while the production rate is low. Also the messy process exposes workers to hazardous chemicals.

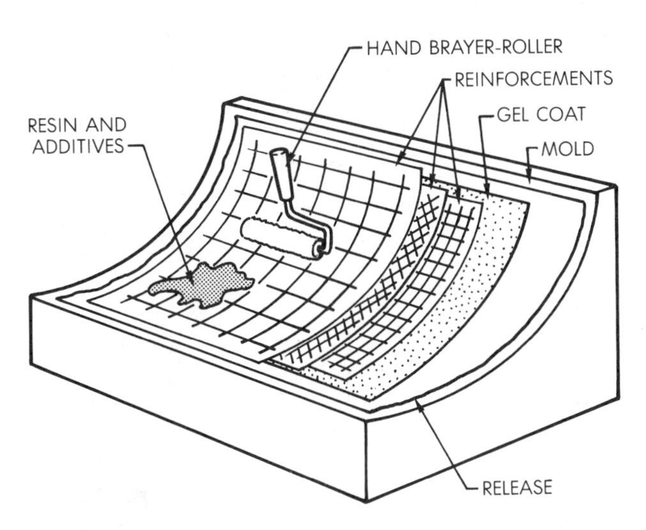

Fig. 12-14. Hand layup implies that little equipment is required. In all layup operations, it is important to bray out (remove) any bubbles trapped between the layers.

Fig. 12-15. Infrared ovens are used to cure resin in about 18 minutes. These glass-reinforced guitar bodies are produced by the hand layup method. (Fostoria Industries, Inc.)

Layers of Metal and Metal Honeycomb

Laminates of metal-face layers with lightweight cores are often called *sandwiches*. Core types include solid, corrugated paper, cellular plastics, and metallic or plastics honeycombs (Fig. 12-16). As a rule, honeycomb, waffle, and cellular sandwiches are isotropic, with excellent thermal, acoustical, and strength-to-weight ratios. The outer layers must be strong to carry the axial and inplane shear loading. Most of the tensile and com-

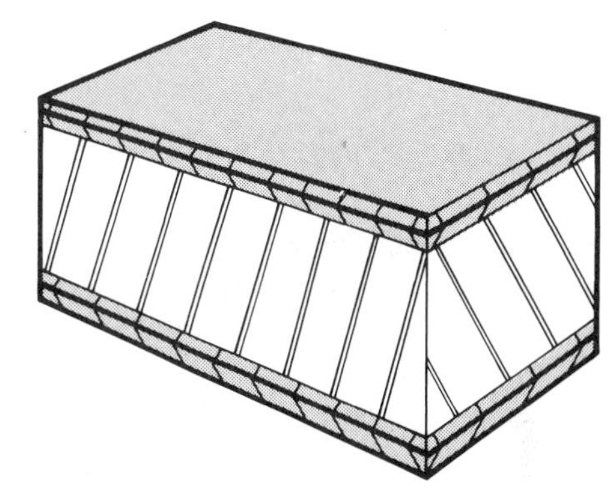

(A) Basic design.

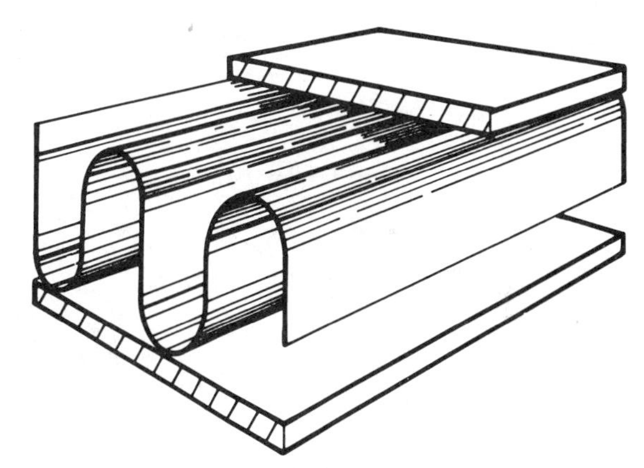

(B) Corrugated paper core.

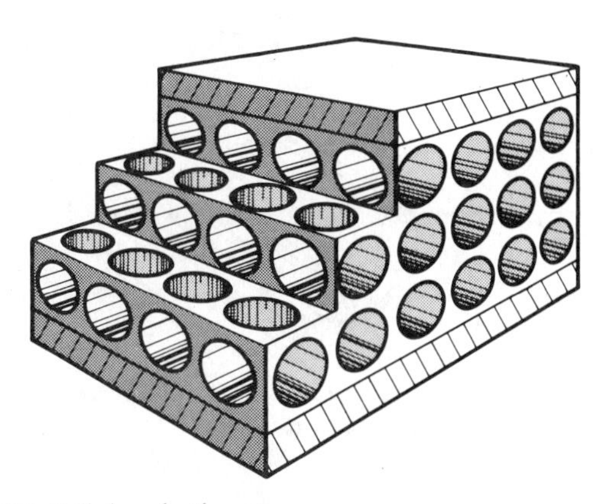

(C) Cellular plastics core.

Fig. 12-16. Various types of core construction.

(continued)

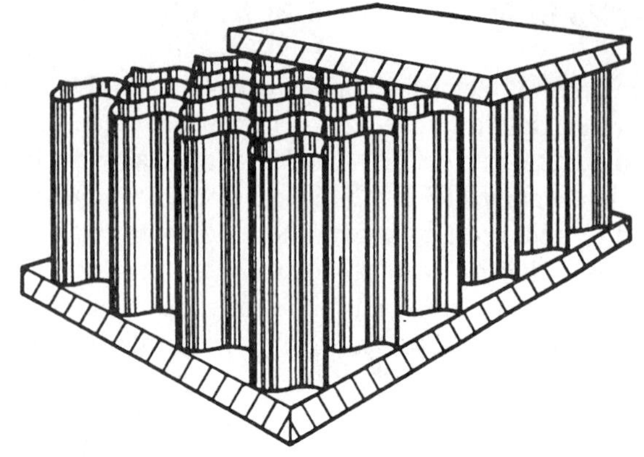

(D) Honeycomb core.

Fig. 12-16. Various types of core construction. *(Continued)*

pressive forces are transferred to these layers. The core material transfers the loads from one facing to the other.

All properties, including thermal and electrical, depend on the selection of facings, core, and bonding agent. Adhesive bonding is critical if shear and axial loads are to be transmitted to and from the core material. Polyimide, epoxy, and phenolics are commonly used. Resin-impregnated fiber matting, cloth, or paper are adhesive films that may be used in the core-to-face bond.

Honeycomb core materials made of resin-impregnated kraft paper, aluminum, glass-reinforced polymers, titanium, or other materials are among the strongest core structures for their mass. Aluminum is the most commonly used honeycomb core material. All honeycombs are anisotropic, and properties depend on the composition, cell size, and geometry. Two major methods of producing honeycomb core materials are illustrated in Figures 12-17 and 12-18.

A unique characteristic of the honeycomb manufactured by the expansion process is that it can be machined prior to expansion. Figure 12-19 shows unexpanded aluminum honeycomb. In this stage, it could be machined into the shape of an airfoil. After machining, it can be stretched or expanded as shown in Figure 12-20.

The properties of selected aluminum honeycomb materials are shown in Table 12-3. Properties of sev-

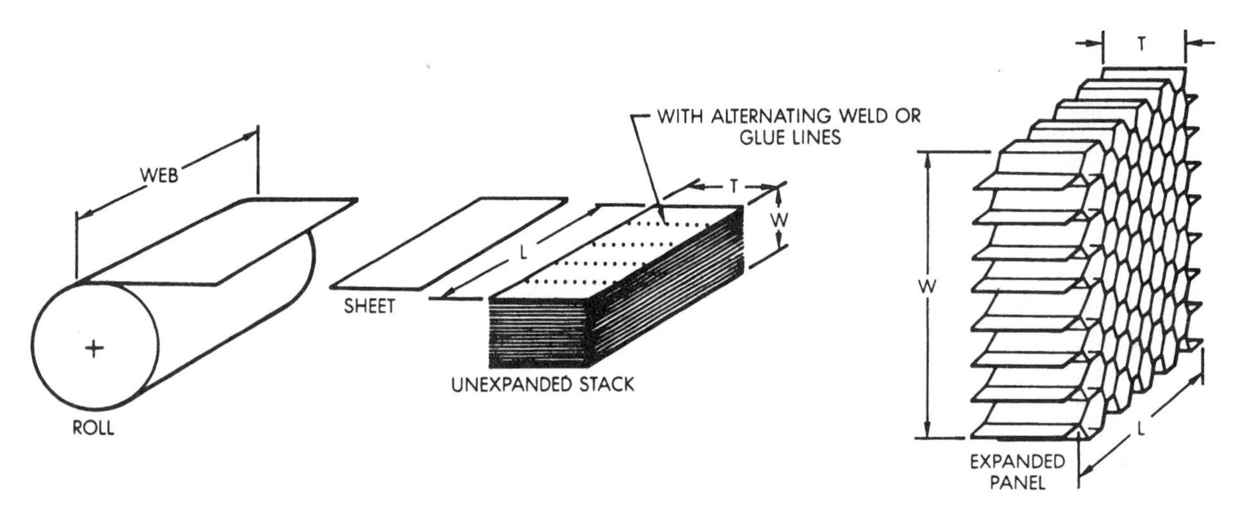

Fig. 12-17. Honeycomb manufacture by the expansion process.

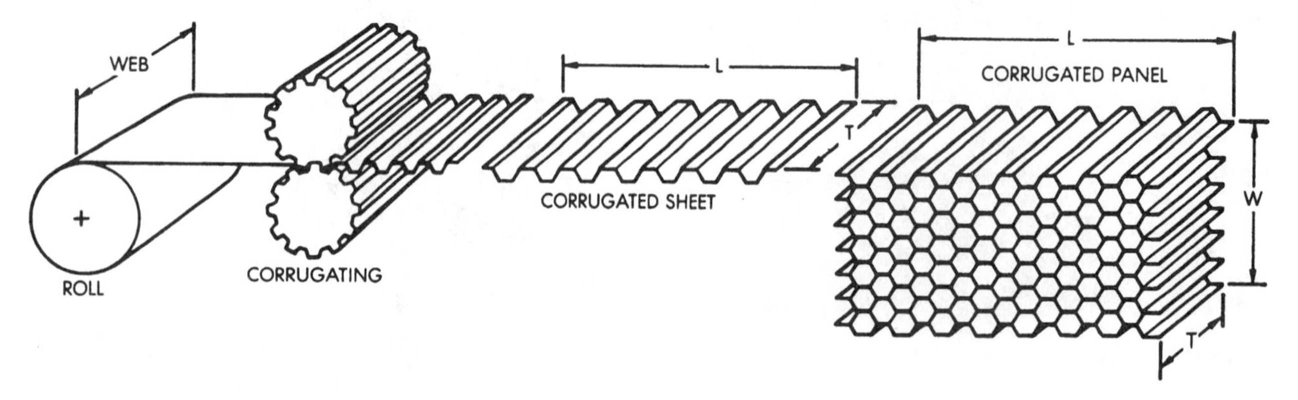

Fig. 12-18. Honeycomb manufacture by the corrugation process.

Fig. 12-19. Unexpanded aluminum honeycomb.

Fig. 12-20. Aluminum honeycomb at various stages of expansion.

eral glass-reinforced plastics honeycombs are shown in Table 12-4.

Layers of Metal and Foamed Plastics

Sandwich panels with foamed plastics as the core have wide application in the construction industry. Figure 12-21 shows metal-faced sandwiches with foamed-polyurethane cores. They are used for insulation and for structural components.

Other uses for cellular core sandwiches are refrigerator liners for truck boxes, railcars, food coolers, and exterior panels for mobile homes. Applications include doors and construction panels.

In a process called *foam reservoir molding* or *elastic reservoir molding,* open-celled polyurethane foam is impregnated with epoxy. The two skin layers are then pressed against the spongy core forcing some of the epoxy adhesive to adhere to the two face skins. The foam and matrix become a catacomb-like, skeletal structure.

(A) Installing panels.

(B) Sealing panels.

Fig. 12-21. Polyurethane-foamed roof panels require few nails.

Vocabulary

The following vocabulary words are found in this chapter. Look up the definition of any of these words you do not understand as they apply to plastics in the glossary, Appendix A.

Debond
Delamination
High-pressure laminates
Honeycomb
Impregnate
Interlaminar shear
Laminate
Laminated plastics
Lamination
Low-pressure laminates
Matrix
Sandwich construction

Table 12-3. Properties of 5056, 5052, and 2024 Hexagonal Aluminum Honeycomb

		Compressive			Plate Shear			
		Bare	Stabilized		"L" Direction		"W" Direction	
Honeycomb Cell-Material-Gauge	Nominal Density, kg/m^3	Strength, kPa	Strength, kPa	Modulus, MPa	Strength, kPa	Modulus, MPa	Strength, kPa	Modulus, MPa
5056 Hexagonal Aluminum Honeycomb:								
1/16–5056–0.0007	101	6894	7584	2275	4447	655	2551	262
1/16–5056–0.001	144	11721	12410	3447	6756	758	4136	344
1/8 –5056–0.0007	50	2344	2482	668	1723	310	1068	137
1/8 –5056–0.001	72	4343	4619	1275	2930	482	1758	262
5/32–5056–0.001	61	3275	3447	965	2310	393	1413	165
3/16–5056–0.001	50	2344	2482	669	1758	310	1069	138
1/4 –5056–0.001	37	1413	1448	400	1172	221	724	103
5056 Alloy Hexagonal Aluminum Honeycomb:								
1/16–5052–0.0007	101	5998	6274	1896	3516	621	2206	276
1/8 –5052–0.0007	50	1862	1999	517	1448	310	896	152
1/8 –5052–0.001	72	3585	3758	1034	2344	483	1517	214
5/32–5052–0.0007	42	1379	1482	379	1138	255	689	131
5/32–5052–0.001	61	2723	2827	758	1862	386	1207	182
3/16–5052–0.001	50	1862	1999	517	1448	310	896	152
3/16–5052–0.002	91	5309	5585	1517	3172	621	2068	265
1/4 –5052–0.001	37	1138	1207	310	965	221	586	113
1/4 –5052–0.004	127	9377	9791	2344	4826	896	3034	364
3/8 –5052–0.001	26	586	655	138	586	145	345	76

Table 12-4. Properties of Several Glass-Reinforced Plastics Honeycombs

	Compressive			Plate Shear			
	Bare	Stabilized		"L" Direction		"W" Direction	
Honeycomb Material-Cell-Density	Strength, kPa	Strength, kPa	Modulus, MPa	Strength, kPa	Modulus, MPa	Strength, kPa	Modulus, MPa
Glass-Reinforced Polyimide Honeycomb:							
HRH 327–3/16–4.0		3033	344	1930	199	896	68
HRH 327–3/16–6.0		5377	599	3171	310	1585	103
HRH 327–3/8 –4.0		3033	344	1930	199	1034	82
Glass-Reinforced Phenolic Honeycomb (Bias Weave Reinforcement):							
HFT–1/8–4.0	2688	3964	310	2068	220	1034	82
HFT–1/8–8.0	9997	11203	689	3964	331	2344	172
HFT–3/16–3.0	1896	2585	220	1378	165	689	62
Glass-Reinforced Polyester Honeycomb:							
HRP–3/16–4.0	3447	4137	393	1793	79	965	34
HRP–3/16–8.0	9653	11032	1131	4551	234	2758	103
HRP–1/4 –4.5	4344	4826	483	2068	97	1172	41
HRP–1/4 –6.5	7076	8136	827	3103	172	793	76
HRP–3/8 –4.5	4205	4757	448	2068	97	1172	41
HRP–3/8 –6.0	6205	6895	689	2758	155	1793	69

Questions

12-1. Two major disadvantages of high-pressure lamination are low __?__ rates and high pressures.

12-2. The process in which two or more layers of materials are bonded together is called __?__.

12-3. If there are unfavorable interlaminar stresses, what may occur? How can this be effectively prevented?

12-4. What are the major applications for high-pressure laminates?

12-5. How are extruders used to produce laminates? Are calenders used?

12-6. What properties are favorable for applications using honeycomb laminated components?

12-7. Name four honeycomb core materials and describe the merits of each in a particular application.

12-8. Define a laminated plastics and describe how several products may be formed.

12-9. Defend the selection of sandwich construction In house doors, airplane components and cargo containers.

12-10. Describe the process of continuous laminating and include the type of materials used and typical product applications.

Activities

Low-Pressure Thermoplastic Lamination

Introduction. Low-pressure, thermoplastic laminating is a process that bonds two sheets of thermoplastics around a selected paper or card item. Heat softens the plastics and pressure causes it to flow around the item. The edges of the plastics sheets bond together thermally.

Equipment. Lamination press, polished plates, blotter cushions, PVC or cellulose acetate sheets.

Procedure

12-1. Connect water-cooling hoses and adjust platen to 175 °C [347 °F].

12-2. Trim thermoplastic sheets 5 mm larger on all sides that the item to be laminated.

12-3. Assemble a laminating sandwich as follows:

1 top holding plate

2 top cushion blotters

1 polished plate

1 plastics sheet

1 item to be laminated

1 plastics sheet

2 cushion blotters

1 bottom holding plate

12-4. Place sandwich in press, close and apply about 37 MPa [5366 psi] to laminate card items or about 20 MPa [2901 psi] for paper items. Use slightly lower temperature and pressure for PVC sheets. (See Figure 12-22.)

12-5. To calculate the pressure needed for lamination, use 40 000 kg/m^2 as a minimum pressure.

12-6. Allow sandwich and plastics laminates to heat for 5 minutes. If multiple sandwiches are pressed, increase the heating time. (See Fig. 12-23)

12-7. When heating is complete, shut off electric power to heaters and slowly turn on water for cooling cycle.

CAUTION: Steam and return hoses are hot. Cool the platens until they reach 40 °C [104 °F]

12-8. Release the hydraulic pressure and remove the sandwich. Flex the sandwich to release the lam-

Fig. 12-22. Low-pressure laminated card about to be removed from sandwich.

Fig. 12-23. Sandwich assembly ready to be removed from compression press.

inate. Do not use a screwdriver to pry the sandwich apart. Any damage to the polished plate will be transferred to the next laminate.

12-9. Examine for bond, color bleeding, air bubbles, delamination, or surface damages.

High-Pressure Laminating

Introduction. In high-pressure laminating, the reinforcing substance in generally paper; although cloth, wood, or glass fabric may be impregnated with fusible B-stage resin. During the heat and pressure cycle, the impregnated sheets are fused together.

Equipment. Heated platen press, decorative sheet, melamine resin impregnated, kraft papers, phenolic resin impregnated, press plates.

Procedure

12-1. Connect water cooling hoses and adjust platen heater to 175 °C.

12-2. Trim sheets to 110 × 110 mm square. After laminating, sheets will be finished trimmed to 100 mm × 100 mm.

12-3. Assemble a sandwich as follows:

1 top holding plate

2 cushion blotters

1 overlay sheet

1 decorative sheet

4 or more layers of kraft paper

1 polished plate

2 cushion blotters

1 bottom holding plate

(See Figure 12-24.)

Fig. 12-24. Sandwich assembly for high-pressure lamination.

12-4. Place sandwich in press and apply about 200 MPa [29 101 psi] of pressure. In calculating pressure requirements, use 245 000 kg/m^2 [35.6 psi] as a minimum pressure for high-pressure lamination.

12-5. Heat sandwich for 10 minutes. Pressure will decay as sheets deform. Maintain pressure during the molding cycle. If multiple sandwiches are pressed, increase heating time.

12-6. After heating cycle, shut off electrical power to heaters and slowly turn on water for cooling cycle.

CAUTION: Steam and return hoses are hot. Cool platens until they reach 40 °C (104 °F)

12-7. Release hydraulic pressure and remove sandwich. Flex the sandwich to release the laminate. Do not use a screwdriver to pry apart.

12-8. Trim laminate and examine for delamination and surface defects.

High-Pressure Lamination

Introduction. If impregnated kraft paper and melamine impregnated decorative overlays are unavailable, the following exercise can provide experience with thermoset laminates.

Equipment. Heated platen press, pieces of smooth sheet metal (equal in size to the platens), phenolic resin in powder form, mold release, paper toweling.

Procedure

12-1. Cut several pieces of paper toweling to about the size of the sheet metal plates. Paper toweling is porous enough that the resin can flow in and through it. Less porous papers reduce resin flow and may tear rather that permit resin flow.

12-2. Lay down one layer of paper, then measure out about ½ cup of powdered phenolic resin. Cover the resin with another layer of paper, and place a top sheet metal plate on the paper.

12-3. Place the sandwich in a heated platen press, and exert pressure. The hydraulic pressure pointer will drop during the pressurization. This indicates flow of the resin. Depending on heat, pressure, and volume of resin/paper, the bounce of the hydraulic pressure will stop, indicating that cross-linking has occurred.

12-4. Release pressure and remove the sandwich. Remove the paper/resin materials. Figure 12-25 shows photos of laminates made with two, four, and six layers of paper.

12-5. Cut a strip out of the laminated paper, and form it into a dog bone for tensile testing. Figure 12-26 shows a strip cut with a band saw, and a prepared dog bone sample.

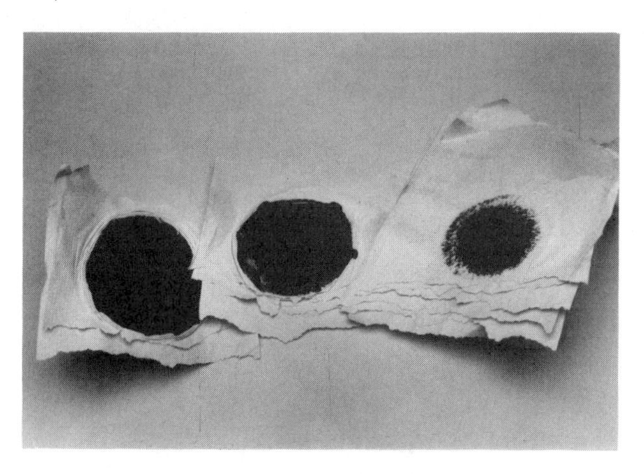

Fig. 12-25. Samples of two, four, and six-layer laminates.

Fig. 12-26. Laminate cut into a strip, then prepared as a dogbone.

12-6. Carefully measure the length, width, and thickness of the gage area. Tensile test. Because the cross-linked phenolic is rather hard, it may tend to slip in the jaws of a tensile tester. If this occurs, a piece of emery cloth folded over the grip area should eliminate slippage. The sample shown in Figure 12-26 failed at 37.1 MPa [5384 psi]

12-7. Make a disk of phenolic without any layers of paper, and test it. Does the paper increase the strength of the laminate? Does the paper increase the flexibility of the laminate?

12-8. Repeat the procedure with varying amounts of paper, or with other fibrous layers, such as cloth, glass cloth, glass mat, or Kevlar cloth.

12-9. To see if the layers of paper or other material affect the shrinkage of the laminate, scribe shallow lines on a sheet metal plate. Make sure that the spacings between the lines are known and accurate. Figure 12-27 shows the lines transferred to the laminate. These lines were spaced every 1 inch. By measuring the distance between the lines in a finished, cooled laminate, the material shrinkage can be calculated. (See the section in Chapter 6 on calculating mold shrinkage.)

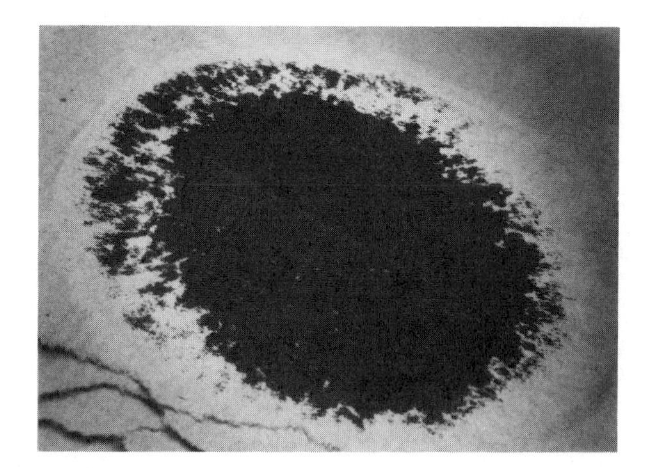

Fig. 12-27. Lines imprinted into phenolic laminate.

Chapter 13

Reinforcing Processes and Materials

Introduction

The term *reinforced plastics* is not very descriptive. It simply implies that an agent has been added to improve or *reinforce* the product. The SPE defines reinforced plastics as "a plastics composition in which reinforcements are embedded with strength properties greatly superior to those of the base resin." Specific terms such as *advanced, high-strength, engineered,* or *structural* composites came into use in the 1960s. With them, a stiffer, higher modulus material of exotic reinforcements in new matrices was used.

Today, *reinforced plastics* is used to describe several forms of composite materials, produced by any one of ten reinforcing processes. Someday, we may classify all laminating and reinforcing processes as *composite processing.*

Some composite-reinforcing techniques are variations of laminating. This is because the techniques may involve combining two or more different materials in layers. Other techniques are simply modifications of processing methods that produce a new material with specific or unique properties.

The following composite-reinforcing processes will be discussed in this chapter:

I. Matched die
   A. Bulk-moling compounds
   B. Sheet-molding compounds
II. Hand layup
III. Sprayup
IV. Rigidized vacuum forming
V. Cold-mold thermoforming
VI. Vacuum bag
VII. Pressure bag
VIII. Filament winding
IX. Centrifugal and blown-film reinforcing
X. Pultrusion
XI. Cold stamping/forming

In each process, the molds, dies, or rollers must be made with care to ensure proper release of the finished product. Film, wax, and silicone releasing agents are most often used on mold surfaces. Reinforced molding compounds should not be confused with laminates, although they occasionally are (Fig. 13-1).

In the past, only thermosetting plastics were commercially reinforced in large quantity; today, the demand for reinforced thermoplastics (RTP) is increasing. Because the thermoplastic materials may be processed in many different ways, many innovative uses have resulted.

Reinforced molding compounds may be molded by injection, matched die, transfer, compression, or extrusion methods to produce products with complex shapes and a broad range of physical properties. There is some difficulty in blow molding small, thin-walled items. Injection molding is the most common method of processing reinforced thermoplastics compounds. (See Injection, Extrusion, and Compression molding, in index.)

Short fibers of milled or chopped glass are most often used to reinforce molding compounds. See Table 13-1 for a list of properties of fibrous-glass reinforced plastics. Plastics fibers and exotic metallic and crystalline whiskers are used as well.

Table 13-1. Typical Properties of Fibrous-Glass Reinforced Plastics

Plastics	Relative Density	Tensile Strength, 1000 MPa	Compressive Strength, 1000 MPa	Thermal Expansion 10^{-4}/°C	Deflection Temperature (at 264 MPa),°C
Acetal	1.54–1.69	62–124	83–86	4.8–4.9	1062–1599
Epoxy	1.8–2.0	10–207	207–262	2.8–8.9	834–1599
Melamine	1.8–2.0	34–69	138–241	3.8–4.3	1406
Phenolic	1.75–1.95	34–69	117–179	2–5.1	1027–2178
Phenylene oxide	1.21–1.36	97–117	124–207	2.8–5.6	910–986
Polycarbonate	1.34–1.58	90–145	117–124	3.6–5.1	965–1000
Polyester (thermoplastic)	1.48–1.63	69–117	124–134	3.6	1379–1586
Polyester (thermosetting)	1.35–2.3	172–207	103–207	3.8–6.4	1406–1792
Polyethylene	1.09–1.28	48–76	34–41	4.3–6.9	800–876
Polypropylene	1.04–1.22	41–62	45–48	4.1–6.1	910–1027
Polystyrene	1.20–1.34	69–103	90–131	4.3–5.6	683–717
Polysulfone	1.31–1.47	76–117	131–145	4.3	1179–1220
Silicone	1.87	28–41	83–138	None	<3323

Matched Die

Bulk molding compound (BMC) and sheet molding compound (SMC) are the most common materials used in match die reinforcing.

Bulk-Molding Compounds

Bulk molding compounds (BMC) are a putty-like mixture of resin, catalysts, fillers, and short fiber reinforcements. BMC has many names. It is called *gunk, putty, dough, or slurry molding.* Dough molding compounds even have their own acronym, DMC. All these names refer to a premix of resins and reinforcements.

Bulk molding compounds are often formed into log or rope shapes to aid molding and handling operations. This fibrous putty may be extruded into H-beam

(A) A steering column lock housing made of injection molded, glass reinforced nylon instead of the traditional aluminum die-cast and machined housing.

(B) Fan and other parts for hair dryer made of reinforced plastics.

Fig. 13-1. Some examples of reinforced plastics parts. (Dow Chemical Co.)

(C) Reinforced parts and housing for a chain saw designed for rugged use.

Fig. 13-1. Some examples of reinforced plastics parts. (Dow Chemical Co.) *(Continued)*

or other profile shapes and automatically fed into the matched die.

BMCs are isotropic with fiber lengths usually less than .38 mm [0.015 inches]. The fiber lengths distinguish BMC from other molding compounds. Figure 13-2 shows a matched mold that converts BMC into a functional product.

Sheet Molding Compounds

Sheet molding compounds are leather-like mixtures of resins, catalysts, fillers, and reinforcements.

They are sometimes called the *flow mat* or *mold mat.* Since they are made in sheet form, fibers may be much longer than in BMC. A typical SMC incorporates about 30 percent random 25 mm [1 in.] chopped glass fibers, 25 percent resin, and 45 percent inorganic filler. SMC-25 indicates a 25 percent glass content.

In comparison to BMCs, SMCs provide greater glass loadings (70 percent glass) and lighter products. The longer fibers provide improved mechanical properties. SMCs include many specialized types, which have abbreviations such as UMC, TMC, LMC and XMC. Unidirectional molding compound (UMC) usually has about 30 percent of its continuous reinforcing fibers aligned in one direction. This provides greater

Fig. 13-2. Matched mold for making reinforced plastics chairs. (Cincinnati Milicron)

tensile strength in the direction of the fibers. Thick molding compounds (TMCs) are produced up to two inches thick. This thicker sheet allows greater variation in part wall thickness and a wider choice of reinforcements. TMCs are highly filled. Low-pressure molding compounds (LMCs) are SMC formulated to allow low-pressure molding techniques. High-strength molding compound (HMC) may contain more than 70 percent reinforcement for added strength and dimensional stability. Directionally reinforced molding compound (XMC) is a sheet containing about 75 percent directionally oriented continuous reinforcements.

SMCs are formed into final shape by the molding operation. An upper and lower carrier film (usually polyethylene) is used with resins. This film allows the SMCs to be stored neatly at hand and to be easily handled during processing. The carrier films are removed prior to the molding operation. (See Figure 13-3.)

SMCs require processing pressures in the range of 3.5 to 14 MPa [507 to 2030 psi]. Processing temperatures vary with product design and polymer formulation. SMCs are fed to molds and passed through pressing, curing, and demolding in one continuous cycle. This technique eliminates waiting through full cure cycle times at the press.

Major markets for BMC and SMC molded parts are in the transportation and appliance industry. Shower floors, heater housings, and appliance cases are made of BMC. As the name SMC implies, large parts such as automotive body panels, hoods, small boat hulls, furniture, and appliance components are made from this material in matched-die molds.

A variation of this process is *macerated* reinforced processing. Macerated parts are produced by chopping the reinforcing materials into pieces 0.2 to 10 mm [0.008 to 0.4 in.] long to be processed in the matched

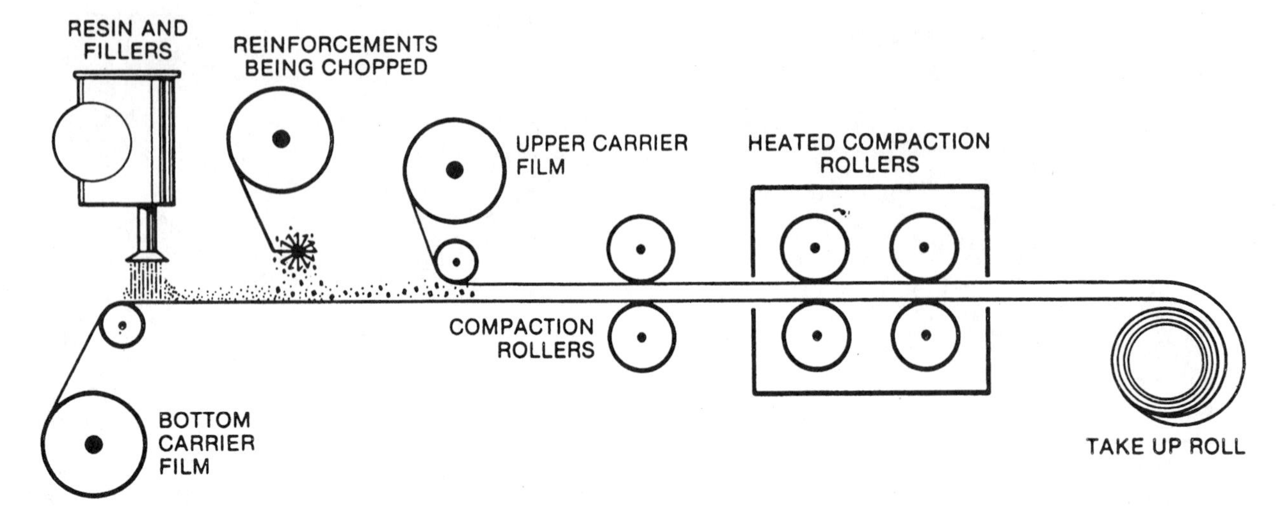

(A) Schematic of SMC.

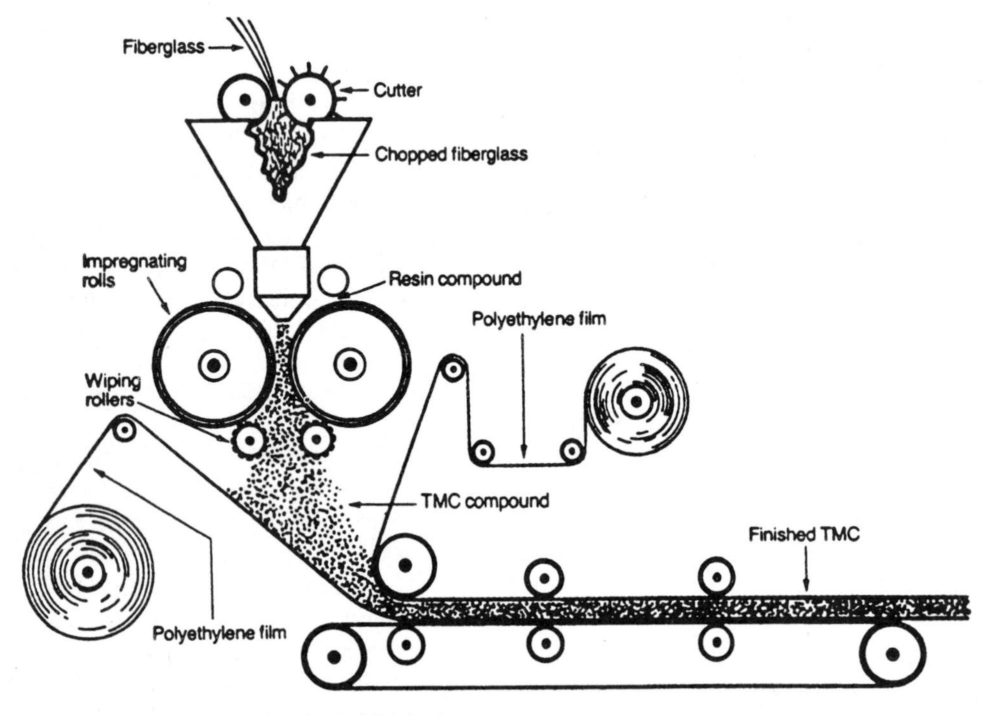

(B) Schematic of TMC (courtesy USS Chemicals Division).

Fig. 13-3. Production method for sheet molding compound.

molds. Reinforced resin products produced from matched-die molds are strong and may have a superb surface finish, both inside and out; however, mold and equipment costs are high. The following is a list of five advantages and disadvantages of matched-die processing.

Advantages of Matched Die

1. Both interior and exterior surfaces are finished.
2. Complex shapes (including ribs and thin details) are possible.
3. Minimum trimming of parts is needed.
4. Products have good mechanical properties, close part tolerances, and corrosion resistance.
5. Cost and reject rate are relatively low.

Disadvantages of Matched Die

1. Preform, BMC, TMC, XMC, and SMC require more equipment, handling, and storage.
2. Press guides must have good parallelism for close tolerances.
3. Molds and tooling are costly, compared to open molds.
4. Surfaces may be porous or wavy.
5. There are no transparent products.

Hand Layup or Contact Processing

Thermosetting resins are used in *hand layup moldings.* Because the reinforcement is generally a continuous layer, this process received description in Chapter 12 as a type of lamination; however, it can also be considered a reinforcing process. (See Chapter 12.)

Sprayup

In sprayup, catalyst, resin, and chopped roving may be sprayed simultaneously onto mold shapes (Fig. 13-4). While it is considered a variation of hand layup, this process can be accomplished by hand or machine. After the gel coat has been applied, sprayup of resin and chopped fibers begins. Careful roll out is important, to avoid damaging the gel coat. Roll out aids in *densifying* (eliminating air pockets and aiding in wetting action) the composite. Poor roll out can induce structural weakness by leaving air bubbles, dislocating the fibers, or causing poor *wet out* (coating of reinforcement). Heat may be applied to speed cure and to increase production. This low-cost method allows production of very complex shapes. Production rates are high compared to hand layup methods. Care must be taken to apply uniform layers of materials; otherwise, mechanical properties may not be consistent throughout the product. Highly contoured or stressed areas can be given additional thickness, or metallic tapping plates, stiffeners, or other reinforcing components can be placed in desired areas and over-sprayed.

Rigidized Vacuum Forming

In a process sometimes called *rigidized shell sprayup,* a thermoplastic sheet is thermoformed into the desired shape, eliminating the gel coat. PVC, PMMA, ABS, and PC are commonly used. This shell is reinforced (by spray or hand layup) on the back side to produce a strong composite bathtub, sink, bathtub-shower combination, small boat, exterior signs, car top carrier, or other similar products. This method is illustrated in Figure 13-5. The following list gives four advantages and three disadvantages of rigidized vacuum-forming.

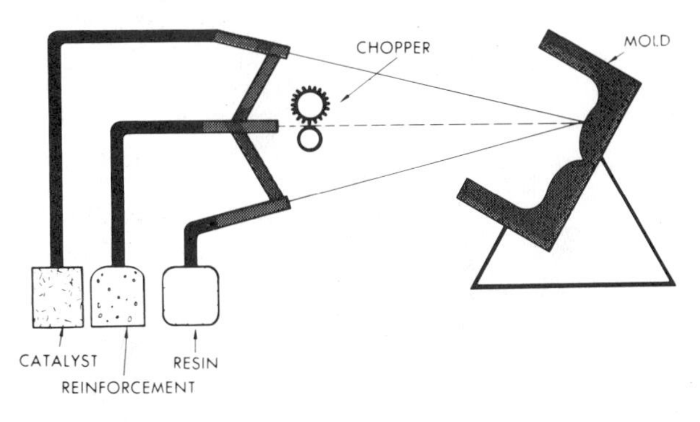

Fig. 13-4. The sprayup method can cover either simple or complex shapes easily, an advantage over the hand layup method.

Advantages of Rigidized Vacuum Forming

1. Thermoplastic skin gives smooth finish
2. Eliminates surface gel coat voids and gel time
3. Requires only a thermoforming mold
4. Output rates are faster than those obtained by sprayup methods

Disadvantages of Rigidized Vacuum Forming

1. Needs thermoforming equipment and sheet storage space
2. Costly materials (thermoplastic surface sheets)
3. Repair of damaged surface sheets is difficult

Cold-Mold Thermoforming

This process is like rigidized vacuum-forming, but the major differences are that two surfaces are finished, and that tolerances are more closely controlled. The process involves thermoforming a sheet, then reinforcing the back side by preform, mat or sprayup methods. Next, the composite is pressed between matched dies

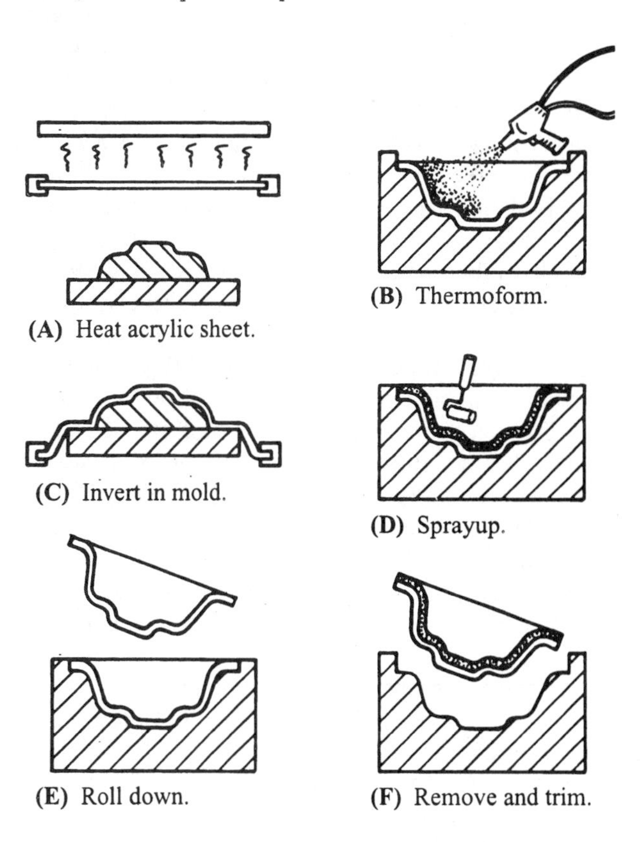

Fig. 13-5. The rigidized vacuum-forming process.

until cured. Polymerization is done by chemical means at room temperature. A major disadvantage is the additional die cost and the curing time in the mold.

Vacuum Bag

During vacuum-bag processing, a plastics film (usually polyvinyl alcohol, neoprene, polyethylene, or polyester) is placed over the layup. About 85 kPa of vacuum [25 in. of mercury] is drawn between the film and the mold (Fig. 13-6).

Vacuum is usually measured in millimeters of mercury drawn in a graduated tube, or in pascals of pressure. Vacuum in millimeters of mercury corresponding to 85 kPa can be calculated using this formula:

$$\frac{101 \text{ kPa}}{85 \text{kPa}} = \frac{760 \text{ mm}}{x}$$

or

$$x = 656 \text{ mm of mercury}$$

where

x = unknown (mm of mercury)

101 kPa = known atmospheric pressure,

760 mm = mm of mercury corresponding to 101 kPa of pressure.

The plastics film forces the reinforcing material against the mold surface, producing a high-density product free of air bubbles. Tooling for vacuum bag processing is costly when large pieces are made. Output is slow compared to the high-speed production rates for injection molding.

Both male and female tooling are used. If a smooth surface is required on the exterior of a boat hull, a female mold would be selected. A male mold would probably be selected for a sink. Because heat is required in many operations, ceramic or metal tooling is used. Infrared induction, dielectric, xenon flash, or beam radiation can be used to aid or speed cure.

The mold surface must be protected to allow removal of the finished composite. Plastic films, waxes, silicone resins, PE, PTFE, PVAI, polyester (Mylar), and polyamide films are used as releasing agents.

Popular *wet layup* resins are epoxies and polyesters. SMC, TMC, and other prepregs of polysulfone, polyimide, phenolics, diallyl phthalate, silicones or other resin systems may be used.

Reinforcements may include honeycomb materials, mats, fabrics, paper, foils, or other pre-impregnated forms.

Wet layup vacuum-bag processing is illustrated in Figure 13-7. After the tooling has been carefully protected with a releasing wax or film (depending on part geometry), a peel ply of finely woven polyester or polyamide fabric is carefully positioned. Sometimes a *sacrificial ply* (usually a fine, resin-impregnated fabric) is placed on the mold surface. The laminate layers are then placed in a specific design pattern on the peel ply. A second peel ply is placed on the laminated layers, followed by a release film or fabric. Dacron and Teflon are commonly used as release fabrics. Because perforations allow air and excess resin to escape, this

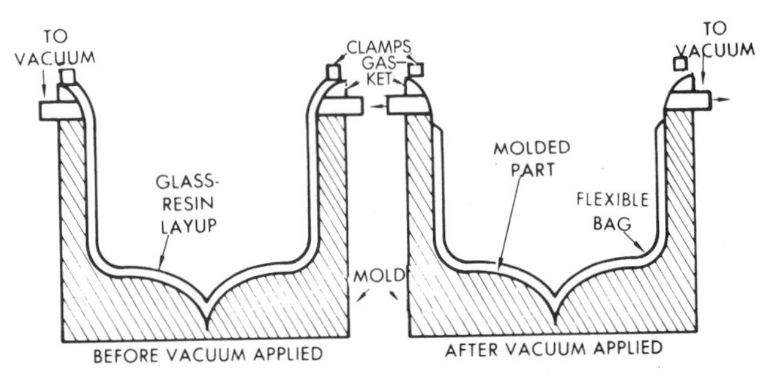

(A) Application of pressure to a layup results in improved strength and a better surface on the unfinished side of the product.

(B) Cargo carrier is vacuum-bag formed, using three layers of prepreg and one of dry glass cloth. (FMC Corp.)

(C) Cargo carrier installed on aircraft. (FMC Corp.)

Fig. 13-6. Vacuum-bag processing.

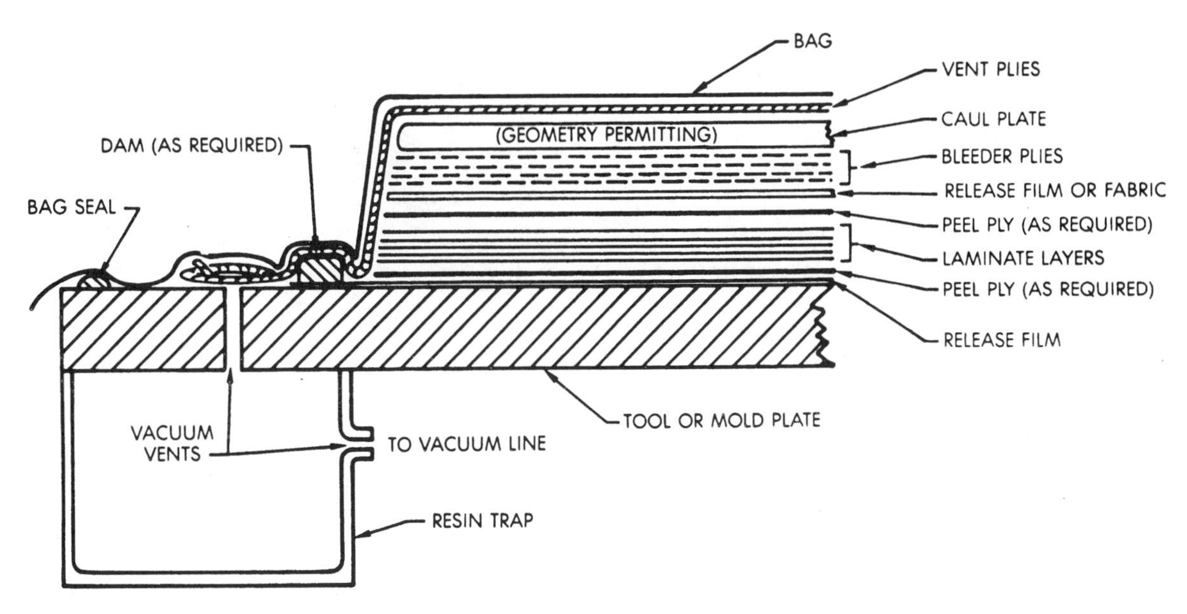

Fig. 13-7. Wet layup vacuum-bag processing.

layer is sometimes called the *breather ply*. Bleeder plies of cloth or mat are laid on the release fabric to collect air and resin that are forced in. On some composite compositions a caul plate is used to insure a smooth surface and minimize variations in temperature during the curing process. Several vent or breather plies are then laid so air can freely pass along the surface of the part inside the bag. The bag can be made of any flexible material that is airtight and won't dissolve in the matrix. Silicone rubber blanket, Neoprene, natural rubber, PE, PVAI, cellophane, or PA are commonly used. A vacuum of 25 inches of mercury, or about 12 psi of external pressure, is then drawn. To prevent excess liquid resin from being drawn into the vacuum lines, a resin trap is used. When additional density or difficult design requirements are needed, pressure-bag, rubber-plunger, rubber bag, autoclave, and hydroclave forming techniques are used.

Dry preimpregnated materials are usually more difficult to form into complex shapes. Additional pressure, plug assistance, and external heat sources are used to soften and to aid in shaping the composite against the tooling.

Pressure Bag

Pressure-bag processing is also costly and slow, but large, dense products with good finishes both inside and out are possible. Pressure-bag processing uses a rubber bag to force the laminating compound against the contours of the mold. About 5.1 psi [35 kPa] of pressure is applied to the bag during the heating and curing cycle (Fig. 13-8). Pressures seldom exceed 50.8 psi [350 kPa].

The mold and compounds may be placed in a steam or heated gas autoclave after layup. Autoclave pressures of 50.8 to 101.5 psi [350 to 700 kPa] will achieve greater glass loading and aid in air removal.

The term *hydroclave* implies that a hot fluid is used to press the plies against the mold. In all pressure designs, the tooling (including the flexible bag) must be able to withstand the molding pressures. Pressure-bag techniques to force the layup against the mold walls may be used for long hollow pipes, tubes, tanks or, other objects with parallel walls. At least one end of the object must be open to insert and remove the bag.

Three advantages and four disadvantages of vacuum and pressure-bag processing are found in the following list.

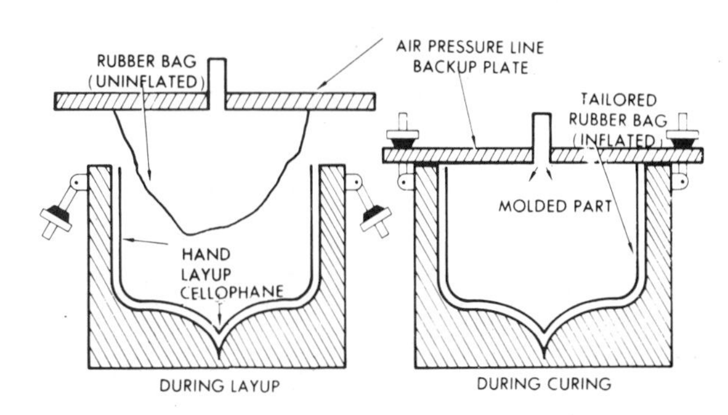

Fig. 13-8. Heat and an inflated rubber bag that applies pressure are used in the pressure-bag molding method.

Advantages of Vacuum and Pressure Bag

1. Greater glass loading and fewer voids than hand layup methods
2. Inside surface has better finish than hand layup methods
3. Better adhesion in composites

Disadvantages of Vacuum and Pressure Bag

1. More equipment needed than in hand layup methods
2. Inside surface finish not as good as matched die molding
3. Quality depends on skill of operator
4. Cycle times are long, limiting production with single mold

Filament Winding

Filament winding produces strong parts by winding continuous fibrous reinforcements on a mold.

Long continuous filaments are able to carry more load than random, short filaments. Over 80 percent of all filament winding is accomplished with E-glass roving. Higher modulus fibers of carbon, aramid, or Kevlar may be used. For some applications, boron, wire, beryllium, polyamides, polyimides, polysulfones, bisphenol, polyesters, and other polymers are also used. Specially designed winding machines may lay down these strands in a predetermined pattern to give maximum strength in the desired direction (Fig. 13-9).

During *wet winding,* excess resin matrix and entrapped air are forced (squeezed out from between strands). Filament-winding tension varies from 0.25 to 1 pound per end (a group of filaments). (See Figure

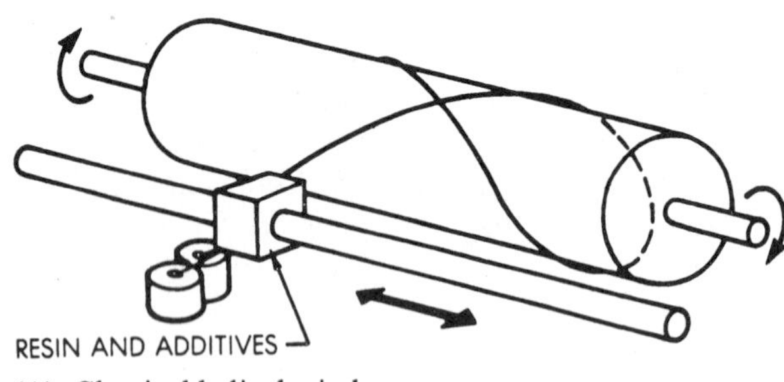

(A) Classical helical winder.

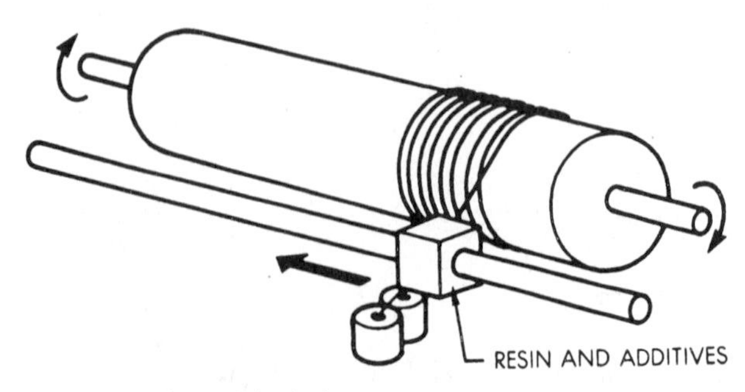

(B) Circumferential winder.

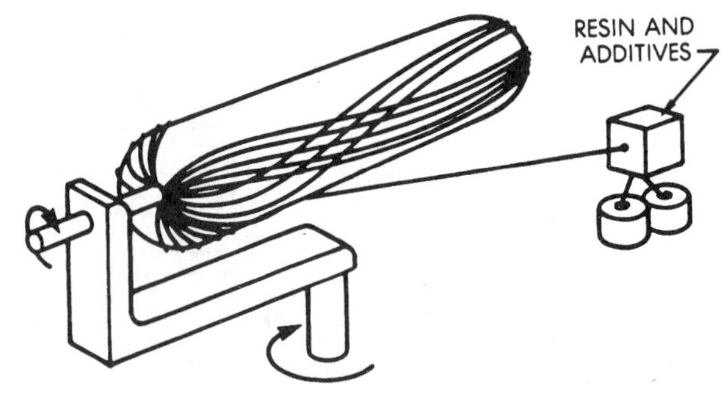

(C) Polar winder.

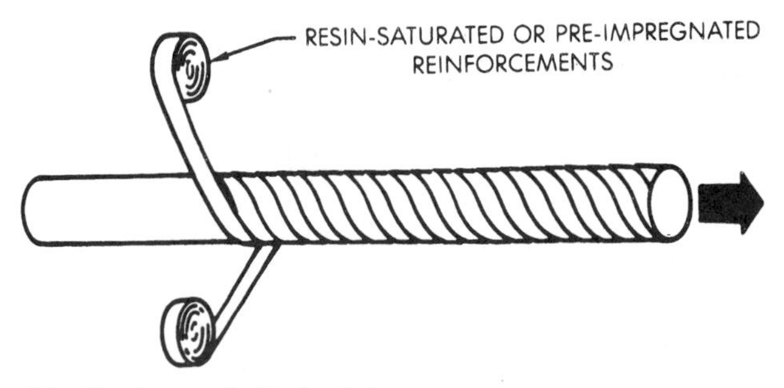

(D) Continuous helical winder.

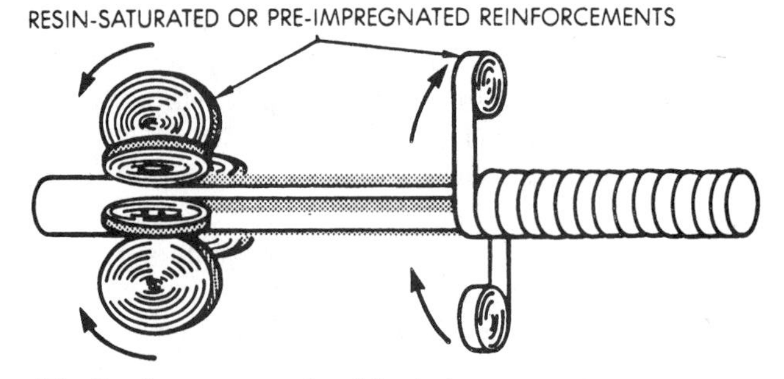

(E) Continuous normal-axial winder.

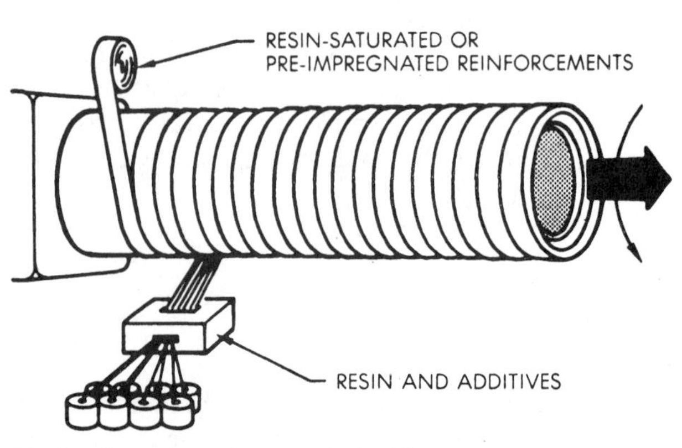

(F) Continuous rotating mandrel with wrap.

Fig. 13-9. Selected winding methods and designs.

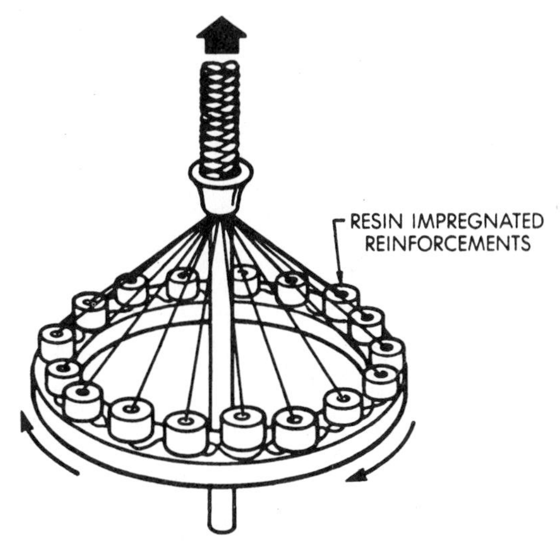

(G) Braid-wrap winder.

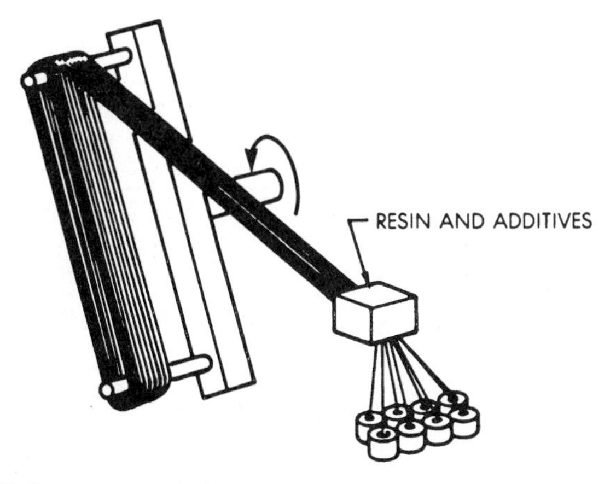

(H) Loop-wrap winder.

Fig. 13-9. Selected winding methods and designs. *(Continued)*

Fig. 13-10. Wet filament winding. (Owens-Corning Fiberglass Corp.)

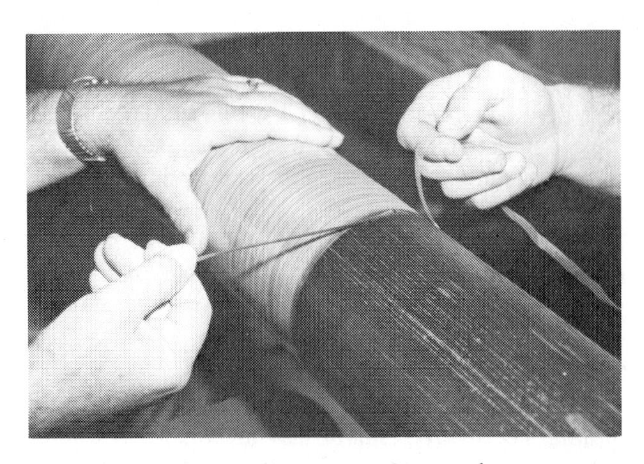

Fig. 13-11. Boron-epoxy prepreg tape, only $\frac{1}{8}$ inch wide, is wound by hand during production of a helicopter tail-rotor driveshaft. (Advanced Structures Division, TRE Corp.)

13-10 showing wet filament winding. Note restrictions to shape of filament wound parts.)

In *dry winding,* preimpregnated B-stage reinforcements help to insure consistency in resin-to-reinforcement content design. These preimpregnated reinforcements may be machine- or hand-wound on the tooling (Fig. 13-11). Curing may be accelerated by heated mandrels (tooling), ambient ovens, chemical hardeners, or other energy sources. Many cylindrical laminated forms are produced by this method. The collapsible mandrel must have the desired shape of the finished product. Soluble or low-temperature-melting mandrels may also be used for special complex shapes or sizes.

The advantage of filament winding is that it allows the designer to place the reinforcement in the areas subject to the greatest stress. Containers made by this process usually have a higher strength-to-mass ratio than those made by other methods. They may be produced at a lower cost in virtually any size. Figure 13-12 shows various winding patterns used for pressure vessels.

On many pressure vessels, the filament windings are not removed from a mandrel, but are overwrapped on thin metal or plastics containers.

Filament-wound applications include rocket engine cases, pressure vessels, underwater buoys, radomes, nose cones, storage tanks, pipes, automotive leaf springs, helicopter blades, spacecraft spars, fuselage, and other aerospace parts.

Centrifugal Reinforcing and Blown-Film Reinforcing

In centrifugal reinforcing, the resin and reinforced materials are formed against the mold surface as it rotates (Fig. 13-13). During this rotation resin is distributed uniformly through the reinforcement by centrifugal force. Heat is then applied to help polymerize the resin. Tanks and tubing may be produced in this manner.

Another specialty process involves reinforcing blown-film. In one proprietary blown-film process, a composite sheet is produced by filamnet reinforcing the inside of the hot blown film and pressing the fibrous layered film between pinch rollers. This concept is illustrated in Figure 13-14.

Pultrusion

In pultrusion, resin-soaked matting or rovings (along with other fillers) are pulled through a long die heated to between 120 and 150 °C [250 to 300 °F]. The product is shaped and the resin is polymerized as it is drawn through the die. Radiofrequency or microwave heating may also be used to speed production rates. The process appears to be like extrusion. In the extrusion process the homogeneous material is *pushed* through the die opening. In pultrusion, however, resin-soaked reinforcements are *pulled* through a heated die where the resin is cured (Fig. 13-15).

Pultrusion dies are generally 60 to 150 cm [24 to 60 in.] in length and heated to aid in the polymerization process. Cure must be carefully controlled to prevent cracking, delamination, incomplete cure, or sticking to the die surfaces.

Output varies from a few millimeters to over 3 m/min. The many resins in use include vinyl esters, polyesters, and epoxies. Fibrous glass is the most widely used reinforcement; although graphite, carbon, boron, polyester, and polyamide fibers may be used. Reinforcements can be positioned in the pultrusion product where needs for strength are the highest.

Hot-melt thermoplastic materials and reinforcements may also be used. Parallel orientation of reinforcements produces a strong composite in the direction of the fibers. Some operations may use SMC or wound preforms in combination with other continuous reinforcements to improve omnidirectional properties.

Siding, gutters, l-beams, fishing rods, automotive springs, frames, airfoils, hammer handles, skis, tent poles, golf shafts, ladders, tennis racquets, vaulting poles, and other profile shapes are some items produced by pultrusion.

Pulforming is a variation of pultrusion. As materials are pulled from the reinforcement creels and impregnated with resin and other compounds, a number of forming devices (molds) of various cross-sectional shapes form the composite part. In one method, rotary male and female dies are brought together on the pultruded material and the composite cured. Curved parts can be formed by forcing the pulform into a large circular female mold with a flexible steel belt. The

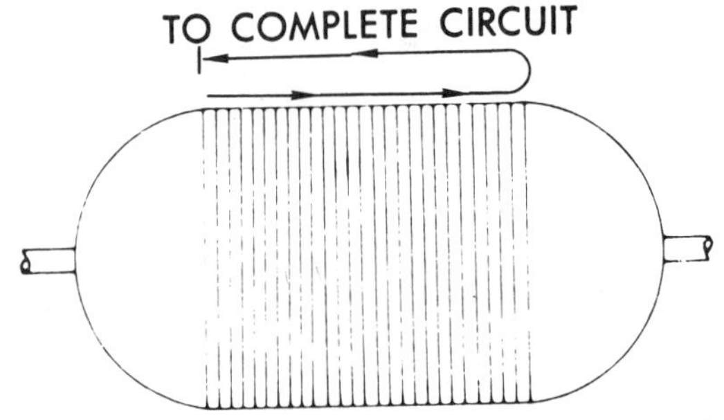

(A) Circular loop windings provide optimum girth or hoop strength in a filament wound structure.

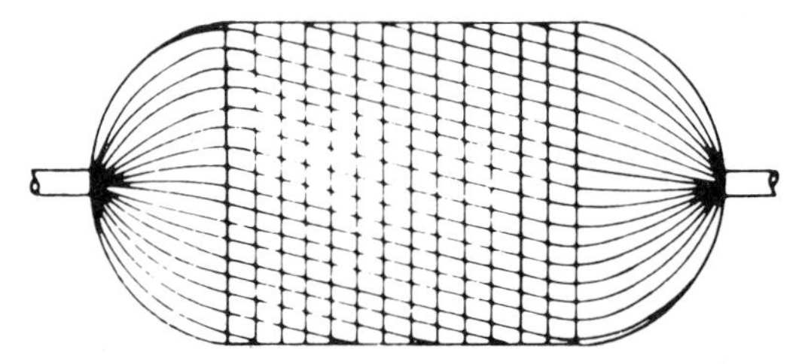

(B) Single circuit helical winding combined with circular loop windings provide high axial tensile strength.

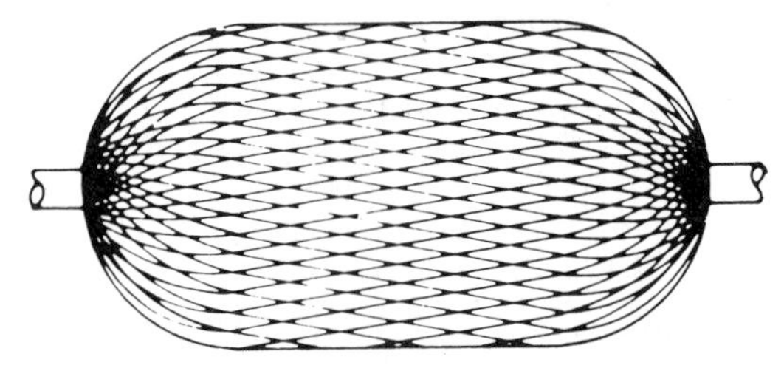

(C) Multiple circuit helical windings allow optimum use of the glass filament's strain characteristics, without the addition of loop windings.

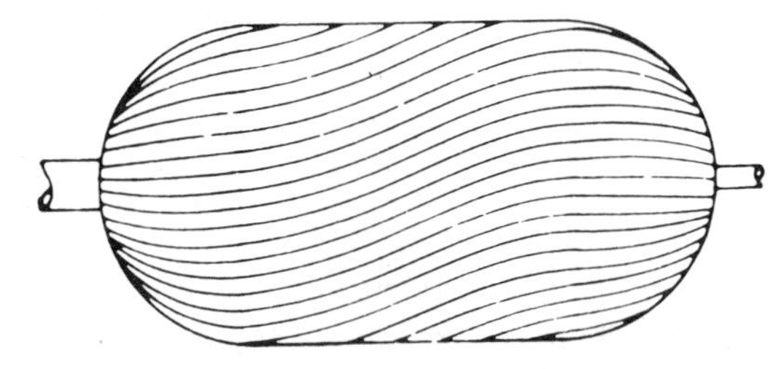

(D) Dual helical windings are used when openings at the ends of the structure are of different diameters.

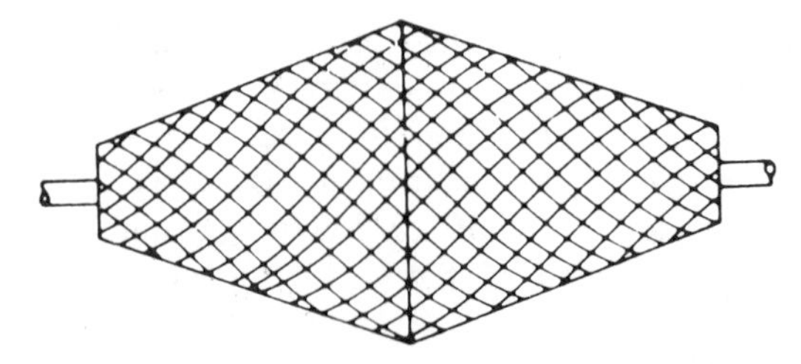

(E) Variable helical windings can produce odd-shaped struc tures.

Fig. 13-12. Advantages of various types of filament winding. (CIBA-GEIGY)

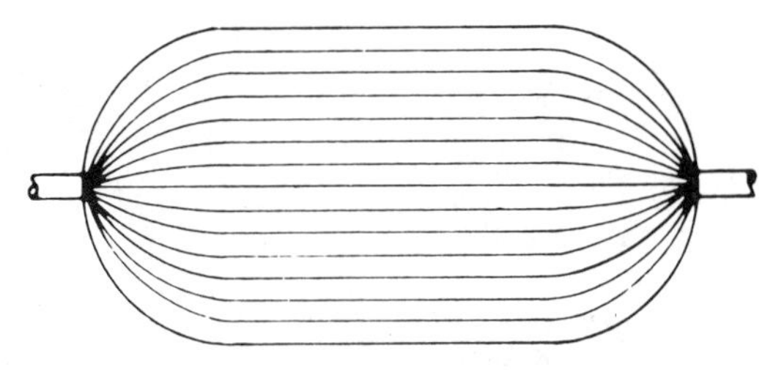

(F) Planar windings provide optimum longitudinal strength (with respect to the winding axis).

Fig. 13-12. Advantages of various types of filament winding. (CIBA-GEIGY) *(Continued)*

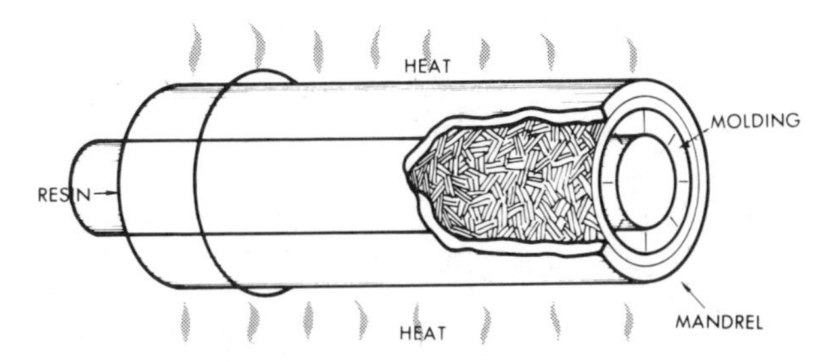

Fig. 13-13. In the centrifugal method, chopped reinforcement and resin are evenly distributed on the inner surface of a hollow mandrel. The assembly rotates inside an oven to provide heat for curing.

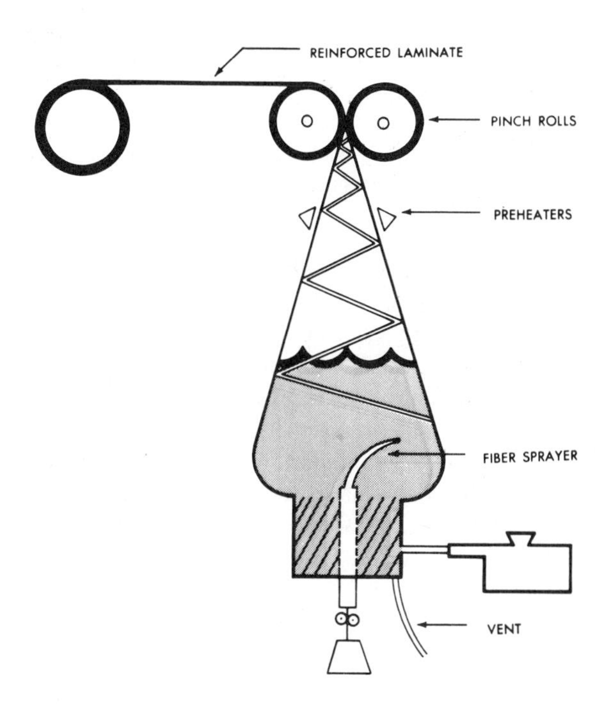

Fig. 13-14. Making filament-reinforced sheets by blown-film process.

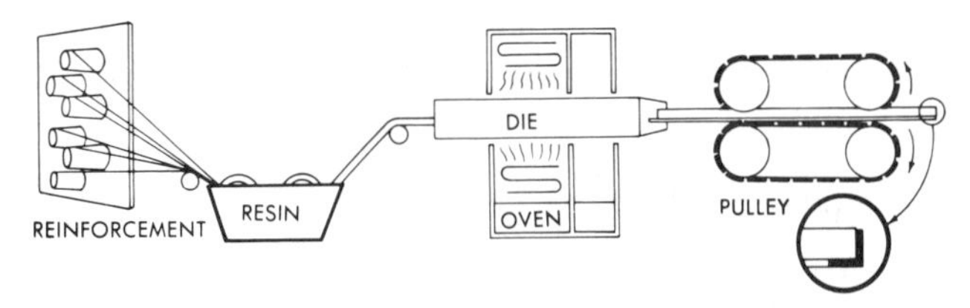

(A) Basic scheme of continuous pultrusion. (Morrison Molded Fiber Glass Co.)

(B) Structural supporting members of fiberglass-reinforced polyester for the operating floor of this chemical mixing plant were produced by pultrusion. (Morrison Molded Fiber-Glass Co.)

Fig. 13-15. Pultrusion method and a product application.

mold and belt are heated to speed cure in a continuous pulformed operation. See Figure 13-16. Applications include hammer handles, bows, curved springs, and other products that do not have a continuous cross-sectional shape.

Cold Stamping/Forming

Fibrous-glass-reinforced thermoplastics, available in sheet form, may be cold-formed much like metals (Fig. 13-17). Long reinforcements are used to improve the strength-to-mass ratio.

During the forming operation, the sheet is preheated to about 392 °F [200 °C] and then formed on normal metal stamping presses. It is possible to produce parts with complex designs and varying wall thickness with this method. Production rates may exceed 260 parts per hour. Uses include motor covers, fan guards, wheel covers, battery trays, lamp housings, seat backs, and many interior automotive trim panels.

According to one authority, stampable reinforced thermoplastic composite sheets could become a replacement for stamped steel from Detroit. Many of these sheets are paintable with a class-A finish out from the mold. Non-class-A automotive finish applications account for about 80 percent of the glass/polypropylene (PP) sheet demand. Polycarbonate/polybutylene terephthalate (PC/PBT), polyphenylene oxide/polybutylene terephthalate (PPO/PBT) and polyphenylene oxide/polyamide (PPO/PA) are alloys combined with modified glass mat or other special reinforcements in stampable, formable sheets.

(A) Glass rovings feeding through wet-out tank.

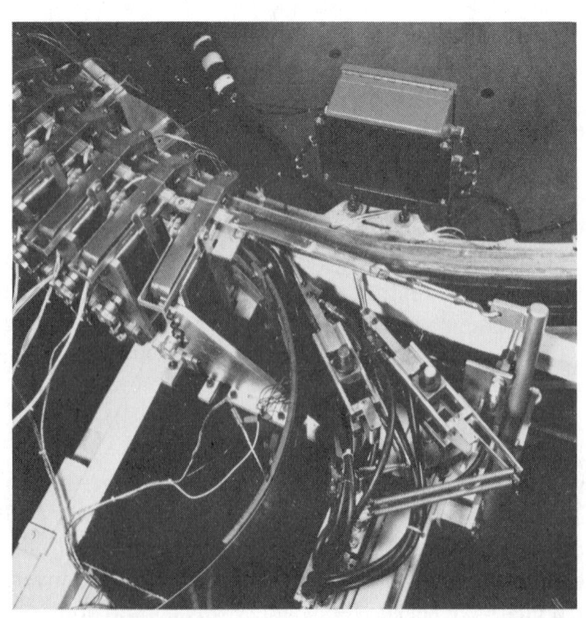

(B) Heated belt die closure on cure section.

(C) Spring stock exiting the die/belt section.

(D) Flying spring stock cut-off saw.

Fig. 13-16. Pulforming of a curved composite leaf spring. (Goldsworthy Engineering.)

Non-impregnated *commingled* blends of continuous thermoplastic filaments such as polyetheretherketone (PEEK) and polystyrene (PPS), with reinforcing filaments of carbon, glass aramid, or metals may be made into yarns, fabrics, or felts. Woven, braided, or knitted three-dimensional commingled preforms are then heated under pressure in the mold. The thermoplastic filaments melt and wet out the adjacent reinforcements. These forms are a versatile material for composites.

Vocabulary

The following vocabulary words are found in this chapter. Look up the definition of any of these words

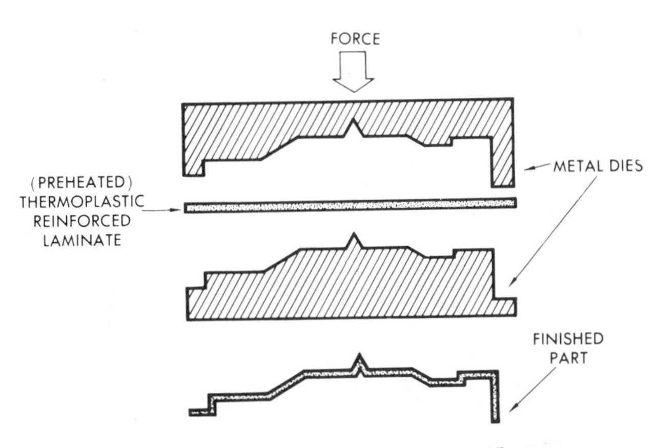

(A) Composite is preheated and formed on cold conventional metal-stamping dies and equipment.

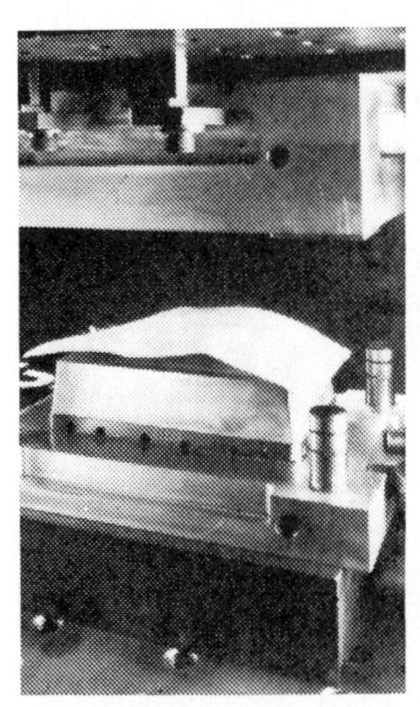

(B) Preheated sheet is formed between cooled matched metal dies. (G.R.T.L. Co.)

(C) Since the blank is smaller than the finished part, it flows out to the periphery of the dies without forming flash or trim. (G.R.T.L. Co.)

Fig. 13-17. Cold stamping.

you do not understand as they apply to plastics in the glossary, Appendix A.

Autoclave
Biaxial winding
Blanket
Bleeder cloth
Breather
Caul plate
Commingled
Fiber orientation
Filament winding
Gel coat
Mandrel
Polar winding
Pressure-bag molding
Pultrusion
Reinforced molding compound
Reinforced plastics
Resin rich area
Sprayup

Questions

13-1. Plastics with strength properties that increased by the addition of filler and reinforcing fibers to the base resin are called ___?___ .

13-2. The thin unreinforced layer of resin placed on the surface of a mold in the hand layup process is called a ___?___ .

13-3. The process of using reinforcements to improve selected properties of plastics parts is called ___?___ .

13-4. The two major disadvantages of high-pressure lamination are low ___?___ rates and high pressures.

13-5. In ___?___ molding, resin-soaked matting or rovings are pulled through a long heated die.

13-6. Name the processing technique that is the simplest for making a relatively strong vessel with a complex shape.

Chapter 14

Casting Processes and Materials

Introduction

This chapter focuses on plastics casting processes and the frequently used casting materials. *Casting* involves placing a plastics material in a mold and allowing it to harden. In contrast to molding and extrusion, casting does not require significant force to push the polymer into a mold cavity, but relies on atmospheric pressure to fill the mold.

To fill a mold using atmospheric pressure, the polymer must approach a liquid state. Many plastics simply do not become liquid enough to flow into molds, even at elevated temperatures. Many hot polymers have a viscosity similar to bread dough; consequently, acetal, PC, PP, and many other plastics are not casting materials. Monomers are typically more liquid than polymers, having viscosities similar to pancake syrup. Because of this property, monomers find considerable use as casting materials.

Casting includes a number of processes in which monomers, modified monomers, powders, or solvent solutions are poured into a mold, where they become a solid plastics mass. The transition from liquid to solid may be achieved by evaporation, chemical action, cooling, or external heat. After the cast material solidifies in the mold, the final product is removed from the mold, and finished.

Casting techniques may be placed into six distinct groups listed in the following chapter outline:

I. Simple casting
   A. Special types of simple castings
II. Film casting
III. Hot-melt casting
IV. Slush casting and static casting
   A. Slush casting
   B. Static casting
V. Rotational casting
   A. Centrifrugal casting
   B. Rotational casting
VI. Dip casting

Simple Casting

In simple casting, liquid resins or molten plastics are poured into molds and allowed to polymerize or cool. Molds may be made of wood, metal, plaster, selected plastics, selected elastomers, or glass. Silicones, for example, are often cast over patterns to make molds in which plastics or other materials may be cast.

Examples of products made by simple casting include: jewelry, billiard balls, cast sheets for windows, furniture parts, watch crystals, sunglass lenses, handles for tools, desk sets, knobs, table tops, sinks, and fancy buttons. Figure 14-1 shows the basic principle of simple casting.

Phenolic castings were part of the early developmemt of the plastics industry. Leo Baekeland introduced numerous articles cast of Bakelite. Today, the most important casting resins are polyester, epoxy, acrylic, polystyrene, silicones, epoxies, ethyl cellulose, cellulose acetate butyrate, and polyurethanes. Probably the most well known is polyester resin because it is used in crafts and hobby work.

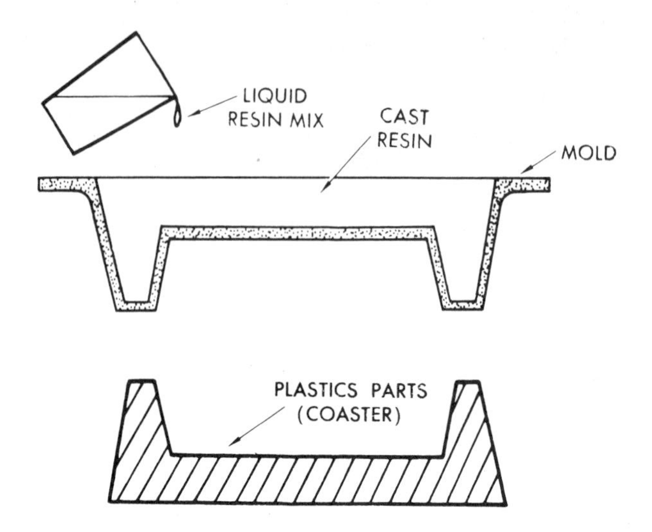

Fig. 14-1. Solid casting in an open, one-piece plastics mold.

Polyester casting resins may be unfilled or filled. To reduce the cost of unfilled polyester, casting resins are extended with water. Water-extended polyesters finds wide use to cast furniture and cabinet parts (Fig. 14-2).

Many polyester casting resins contain large amounts of fillers and reinforcements. For example, cultured marble is a product that contains marble dust or calcium carbonate (limestone) as the filler and polyester plastics as the binding material. It is used to produce lamp bases, table tops, exterior veneers, statues, and other marble-like products (Fig. 14-3).

Clear acrylic plastics are seen as cast rods, sheets, and tubing. Acrylic sheets are often produced by pouring a catalyzed monomer or partially polymerized resin between two parallel plates of glass (Fig. 14-4). The glass is sealed with a gasket material to prevent

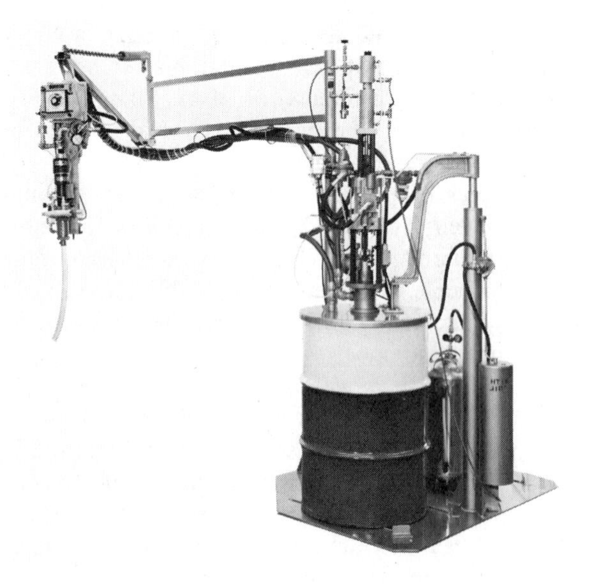

(A) This machine blends components for polyester products produced by simple casting. (Pyles Industries, Inc.)

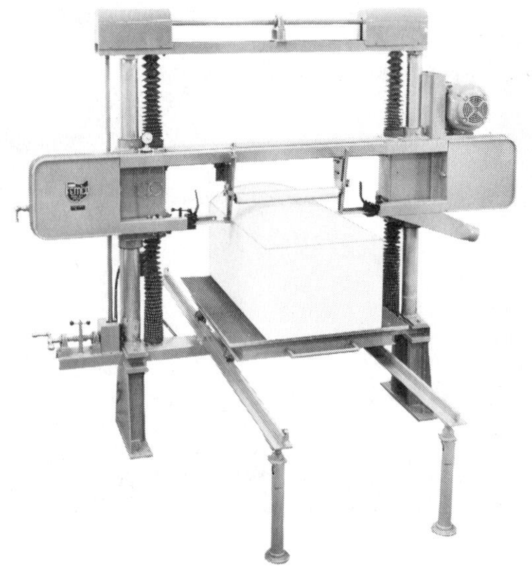

(B) A large block of cellular plastics, cast in an open mold, will be cut into slabs by this machine. (McNeil Akron Corp.)

(C) Water-extended polyester was used to make these picture and mirror frames by simple casting.

Fig. 14-2. Simple casting and cast products.

Fig. 14-3. This one-piece sink is a casting made with filled (marble dust) polyester. Acrylic and polyester resins are used as binders for many products.

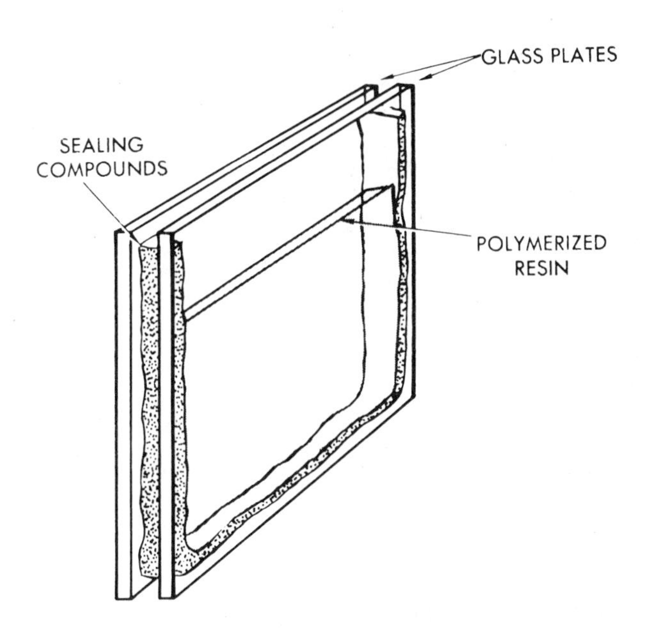

Fig. 14-4. Casting plastics sheets.

leakage and to help control the thickness of the cast sheet. After the resin has fully polymerized in an oven or autoclave, the acrylic sheet is separated from the glass plates and reheated to relieve stresses that occur during the casting process. The faces are covered with masking paper to protect the sheet during shipment, handling, and fabrication. Sheets may be purchased untrimmed, with the sealing material still sticking to the edges (Fig. 14-5).

Some materials are unsuitable for casting, but are needed in sheet form. In contrast to casting, a cutting process, called skiving, may be required. Sheets of cellulose nitrate and other plastics may be sliced (skived) from blocks that have been softened by solvents. After the residual solvents have evaporated, the skived piece is pressed between polished plates to improve the surface finish (Fig. 14-6).

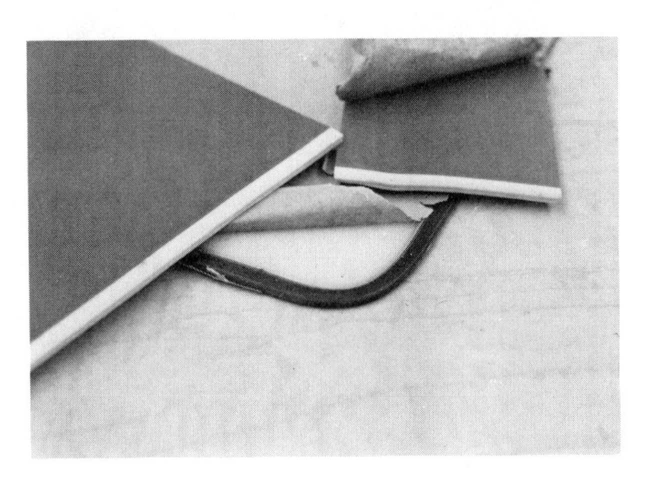

Fig. 14-5. Sealing materials still adhere to the edges of these untrimmed acrylic plastics sheets.

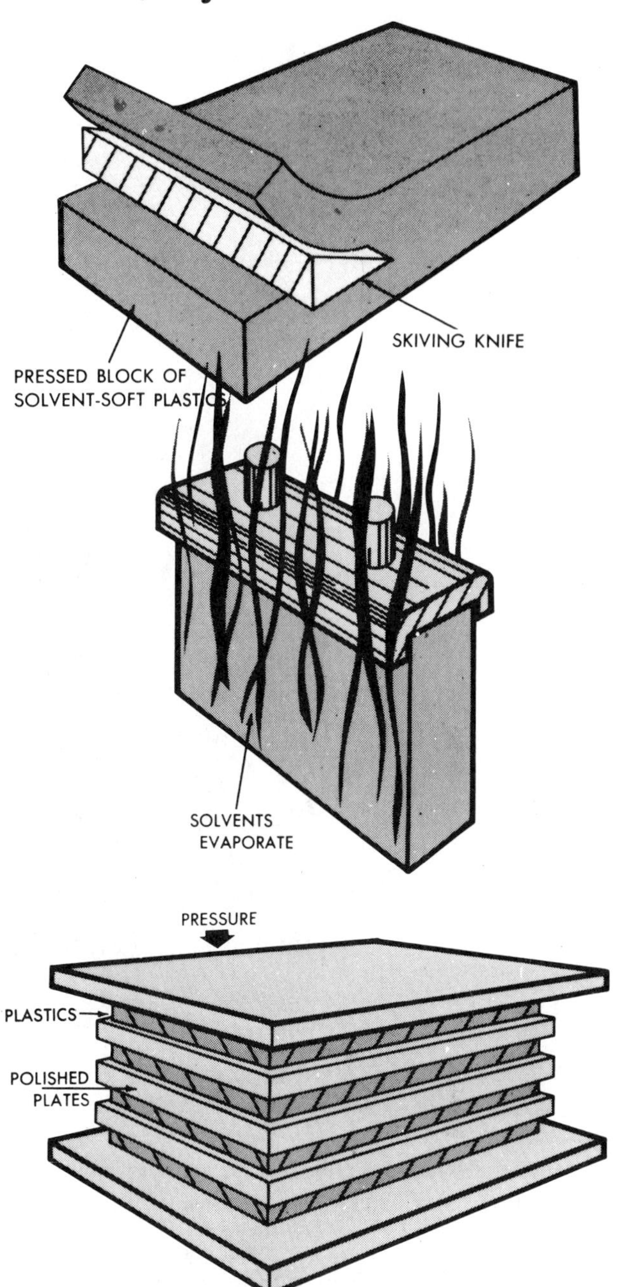

Fig. 14-6. Skiving sheets from a block of plastics.

Special Types of Simple Castings

In addition to the common simple castings, three special types of simple castings are popular. These are: embedment, potting, and encapsulation. Foams may be cast, but are discussed under the foaming processes in Chapter 16.

Embedments. Embedments encase objects completely with a transparent plastics. After polymerization, the casting is removed from the mold, and often polished (Fig. 14-7).

Objects are embedded for preservation, display, and study. In the biological sciences, animal and plant

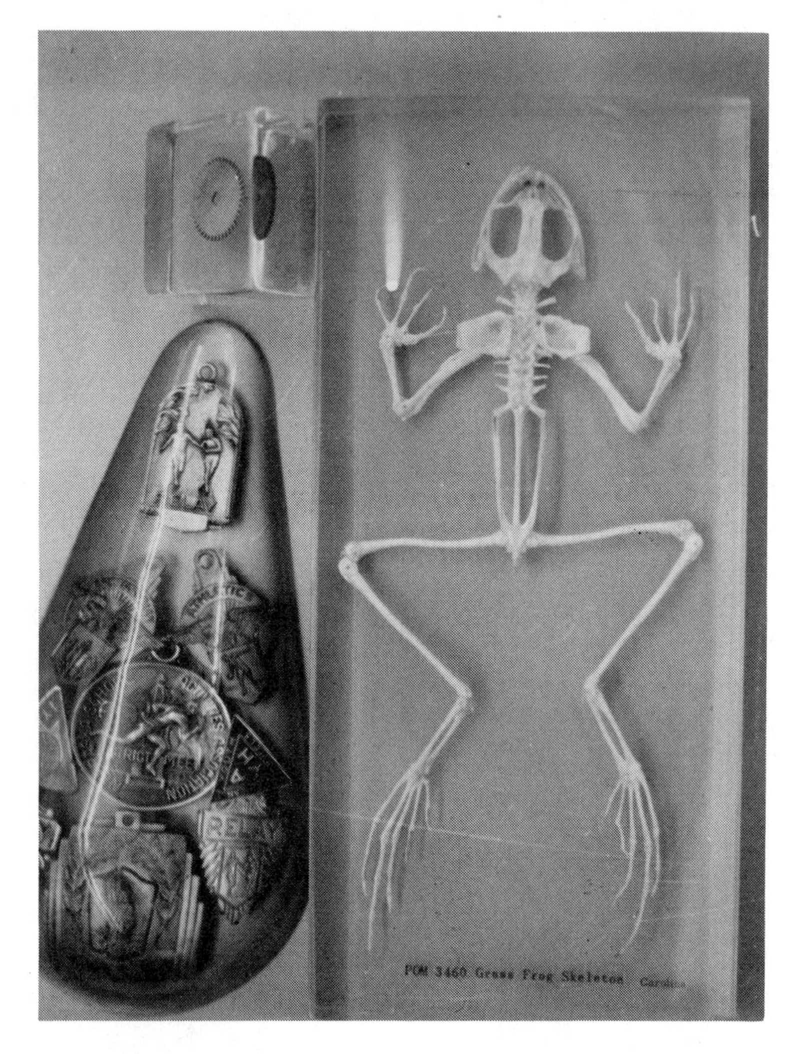

Fig. 14-7. Embedding in transparent polyester has wide use for objects such as classroom science samples.

specimens often are embedded to help preserve them. This allows safe handling of the most fragile sample.

Potting. Potting is used to protect electrical and electronic components from harmful environments. The potting process completely encases the desired components in plastics, and the mold becomes part of the product (Fig. 14-8). Vacuum, pressure, or centrifugal force are frequently applied to ensure that all voids are filled with resin.

Encapsulation. Encapsulation is similar to potting. Encapsulation is a solventless covering on electrical components (Fig. 14-9). This envelope of plastics does not fill all the voids. The process involves dipping the object in the casting resin. After potting, many components are encapsulated.

Film Casting

The water-soluble packaging for bleaches and detergent is an example of a cast film. Film casting involves

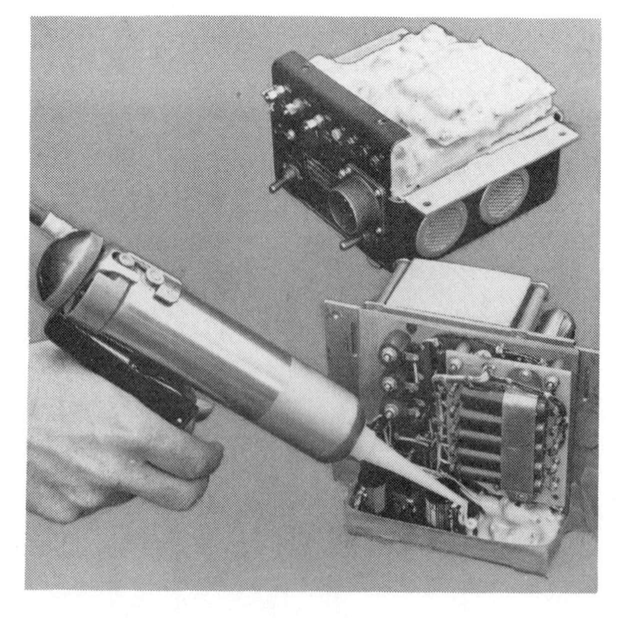

Fig. 14-8. Silicone elastomer used to pot an electronic unit. (Dow-Corning Corp.)

(A) Transformer being encapsulated with silicone elastomer.

(B) Electronic components are encapsulated (epoxy) into convenient modules.

Fig. 14-9. Examples of encapsulation. (Dow-Corning Corp.)

dissolving plastics granules or powder, along with plasticizers, colorants, or other additives, in a suitable solvent. The solvent solution of plastics is then poured onto a stainless steel belt. The solvents are evaporated by the addition of heat, and the film deposit is left on the moving belt. The film is stripped or removed and wound on a takeup roller (Fig. 14-10). This film may be cast as a coating or laminate directly on fabric, paper, or other substrates.

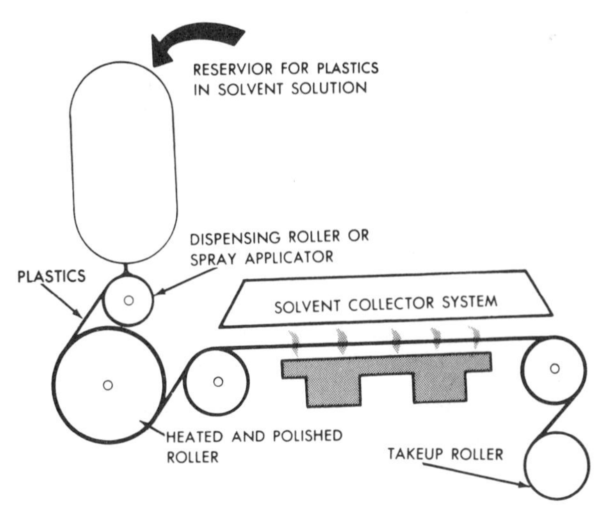

(A) Roller solvent casting.

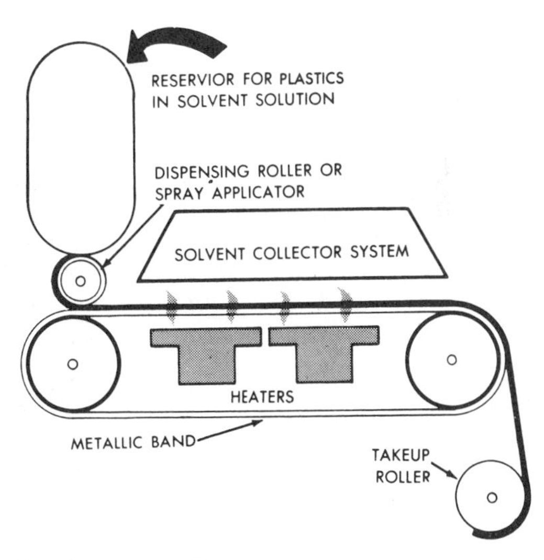

(B) Band solvent casting.

Fig. 14-10. Film casting.

Solvent casting of film offers the following three advantages over other heat-melt processes:

1. Additives for heat stabilization and for lubrication are not needed.
2. Films are uniform in thickness and optically clear.
3. No orientation or stress is possible with this method.

To be economically feasible, solvent casting of film requires a solvent recovery system. Plastics that may be solvent cast include: cellulose acetate, cellulose butyrate, cellulose propionate, ethyl cellulose, polyvinyl chloride, polymethyl methacrylate, polycarbonate, polyvinyl alcohol, and other copolymers. Casting liquid plastics latexes on Teflon-coated surfaces, rather than stainless steel, may also be used to produce special films.

Aqueous dispersions of polytetrafluoroethylene and polyvinyl fluoride are cast on heated belts at temperatures below their melting points. This method provides a handy way to make films and sheets of materials that are hard to process by other means. These films are used as nonstick coatings, gasket material, and sealing components for pipes and joints.

Hot-Melt Casting

Hot-melt plastics were used for casting as early as World War II. Today, hot-melt formulations may be based on ethyl cellulose, cellulose acetate butyrate, polyamide, butyl methacrylate, polyethylene, and other mixtures. The largest use is for strippable coatings and adhesives. Hot-melt resins may be used for making molds for casting other materials. Also, hot-melt resins are used in a casting process for potting and encapsulation (Fig. 14-11). Not all potting compounds are thermoplastic and hot-melting. Silicone is most often used for coating, sealing, and casting, but epoxy and polyester resins are also used for these purposes.

Electrical parts may be protected from hostile environments by being placed in molds and having hot resin poured over the components. When cool, the plastics provides protection for wires and vital parts. The encapsulated or potted components may then be placed with other assembles to produce the finished product. Some encapsulations and pottings are not cast in separate molds but are produced by pouring the molten compound directly over the components inside the case of the finished product. The insulation of parts in a radio chassis or a motor is a well-known example. If the components are cast in place and not removed from a mold shape, they must be classed as coatings.

Slush Casting and Static Casting

Slush casting and static castings share a type of process, but often do not share materials. Slush castings rely on liquid casting materials, while static castings usually start with powders.

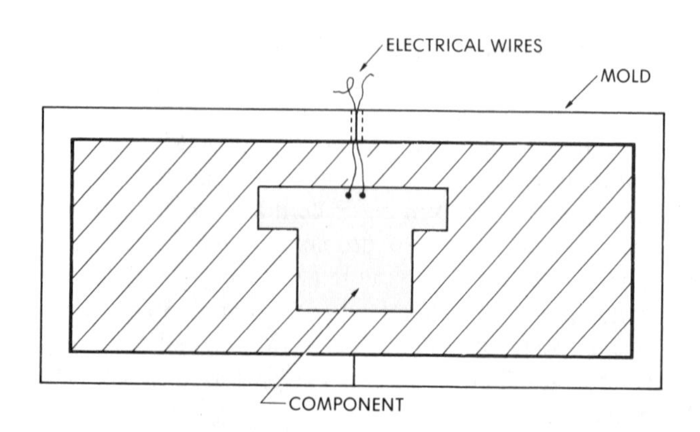

Fig. 14-11. Hot-melt encapsulation of an electronic component.

Slush Casting

The major materials for slush castings are *plastisols* and *organosols*. Plastisols are mixtures of only finely ground plastics and plasticizers (Fig. 14-12). Organosols are vinyl or polyamide dispersions in organic solvents and plasticizers (Fig. 14-13). Organosols may be 50 to 90 percent solids. The solids are tiny particles of

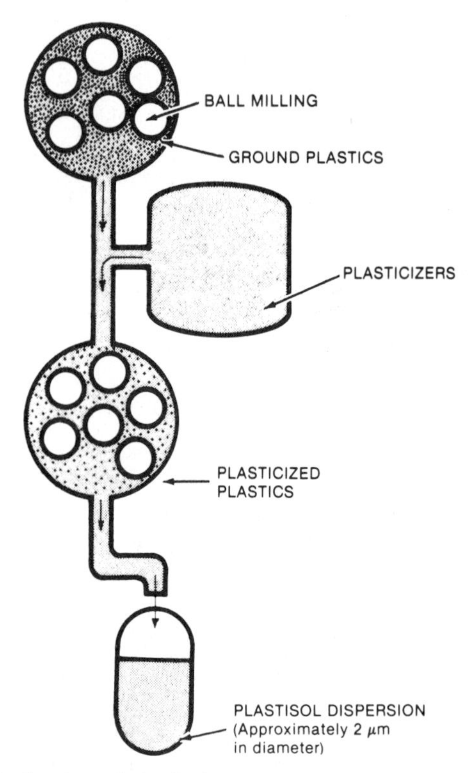

(A) Production of plastisols.

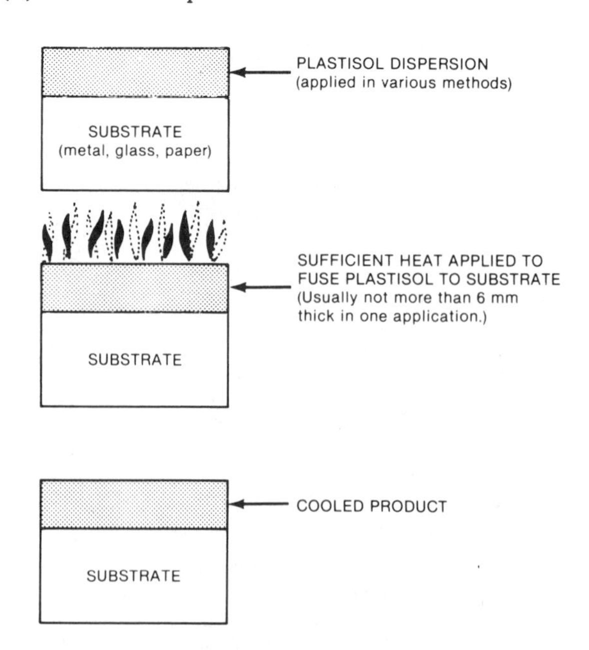

(B) Fusion of plastisols to substrate.

Fig. 14-12. Production and use of plastisols.

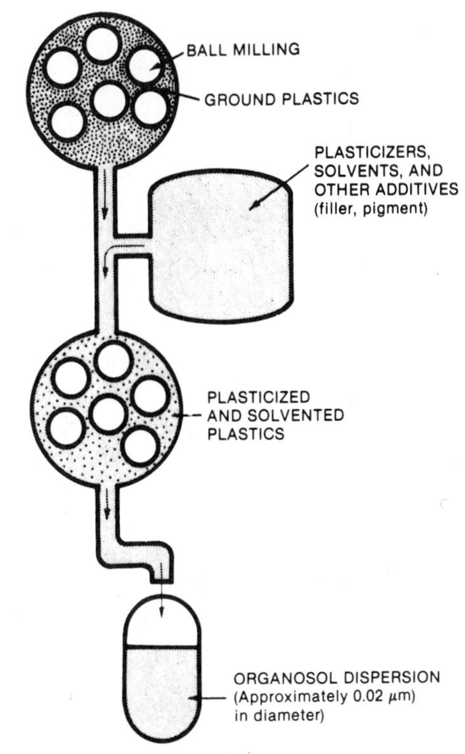

A) Production or organosols.

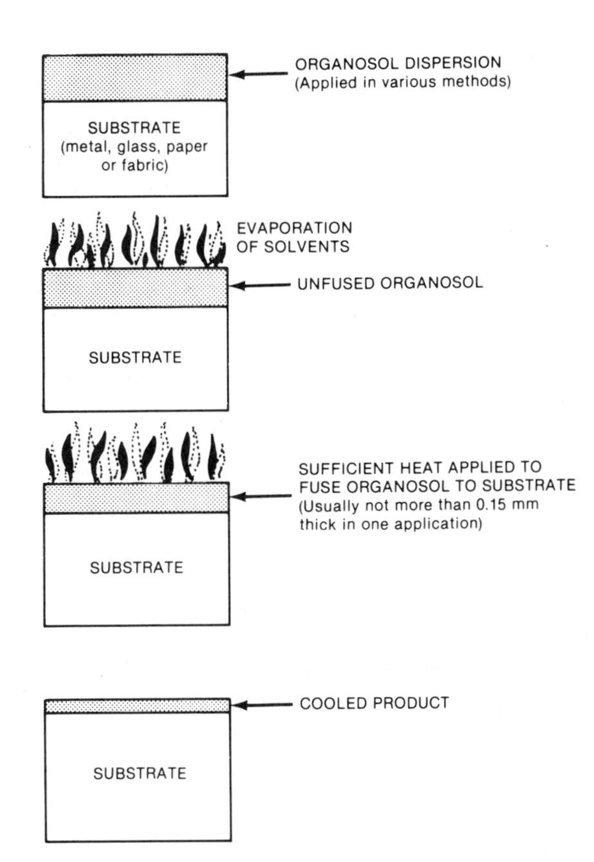

(B) Fusion of organosols on substrate.

Fig. 14-13. Production and use of organosols.

plastic, frequently ground PVC. Organosols contain both plasticizers and varying amounts of solvents. A plastisol may be converted to an organosol by the addition of selected solvents.

Slush-cast items are hollow but will have an opening like that found in doll parts, syringe bulbs, and special containers. Any design in the mold will be on the outside of the product.

Slush casting involves pouring dispersions of polyvinyl chloride or other plastics into a heated, hollow, open mold. As the material strikes the walls of the mold, it begins to solidify (Fig. 14-14). The wall thickness of the molded part increases as the temperature is increased or solution is left in the hot mold. When the desired wall thickness is reached, the excess material is poured from the mold and the mold is then placed in an oven until the plastics fuses together or evaporation of solvents is complete. After water cooling, the mold is opened and the product removed.

Commercial molds are usually made from aluminum since this metal allows rapid cycling and lower tooling costs. Ceramic, steel, plaster, or plastics molds may be used also. Vibrating, spinning, or the use of vacuum chambers may be necessary to drive out air bubbles in the plastisol product.

Organosols are cast and the solvents allowed to escape. The dry, unfused plastics is left on the substrate. Heat is then applied to fuse the plastics.

Static Casting

Thermoplastic powders are also used in a dry process sometimes called *static casting.* In static casting, a metal mold is filled with powdered plastics and placed in a hot oven (Fig. 14-15). As the heat penetrates the mold, the powder melts and fuses to the mold wall. When the desired wall thickness is obtained, the excess powder is removed from the mold. The mold is then returned to the oven until all powder particles have completely fused together.

Huge storage tanks and containers with heavy walls are examples of products made by this casting method. Cellular polystyrene or polyurethanes may be placed in the remaining space in manufacture of tough-skinned flotation devices.

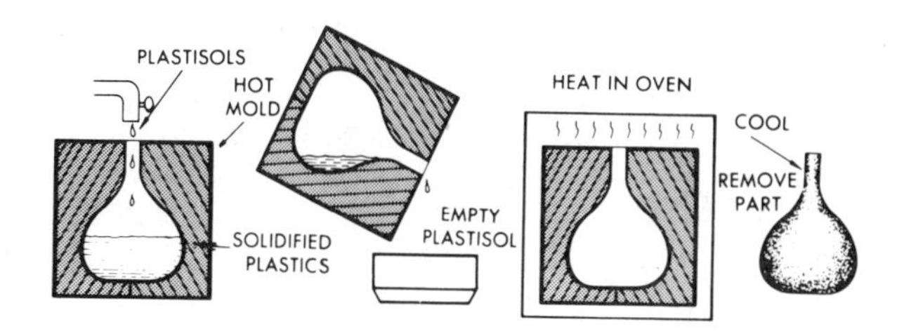

Fig. 14-14. Basic slush casting with plastisols.

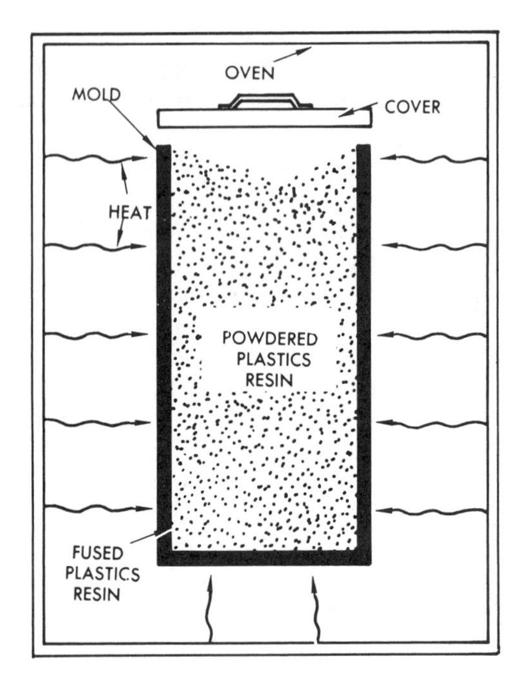

Fig. 14-15. Principle of slush casting.

In a related process known as *vibrational microlamination* (VIM), a combination of heat and vibration is used. Thin layers of homopolymer alternating with reinforced layers are used to produce huge storage tanks, hollow toys, syringe bulbs, or other containers (Fig. 14-16).

Rotational Casting

Rotational casting relies on the rotation of a mold to evenly distribute the casting material on the walls of the mold. The materials used are plastics powders, monomers, or dispersions. Rotational casting includes two basic categories, which stem from the numbers of axes of rotation. If a mold rotates only on one plane, the process is identified as *centrifugal casting*. If the mold moves on two planes of rotation, the process is *rotational casting*.

Centrifugal Casting

In centrigfugal casting, cylindrical shapes are common. Large pipes and tubes are typical products. In specialty processes, inflatable mandrels and reinforced materials may be placed next to the skin layer to produce a high-density composite with ribs or other geometric designs. In another operation, wet layup is placed on the moldwall. Centrifugal force causes the reinforcement and matrix to take the shape of the mold.

Rotational Casting

In rotational casting, plastics powders or dispersions are measured and placed in multi-piece aluminum molds. The mold is then placed in an oven and rotated in two planes (axes) at the same time (Fig. 14-17). This action spreads the material evenly on the walls of the hot mold. The plastics melts and fuses as it touches the hot mold surfaces, making a one-piece coating. The heating cycle is complete when all powders or dispersions have melted and fused together, but the mold continues to rotate as it enters a cooling chamber. Cast crystalline polymers are generally air cooled while amorphous polymers may be quickly cooled by water spray or bath. Finally, the cooled plastics product is removed.

By the programming of rotation speed, the wall thickness in different areas may be controlled. If it is desirable to have a thick wall section around the parting line of a ball, the minor axis can be programmed to turn at a faster speed than the minor axis. This places more powdered material against the hot mold in that area. Figure 14-18 shows how this is done.

Rotational casting may be used for hollow, completely closed objects such as balls, toys, containers, and industrial parts including armrests, sun-visors, fuel tanks, and floats. Figure 14-19 shows rotational casting equipment. Luggage may be cast as one piece which is then cut at the seam to form two perfectly fitting halves. Figure 14-20 shows some rotational-cast items.

Foam-filled and double-walled items, including true composites, can be produced. Short-fiber rein-

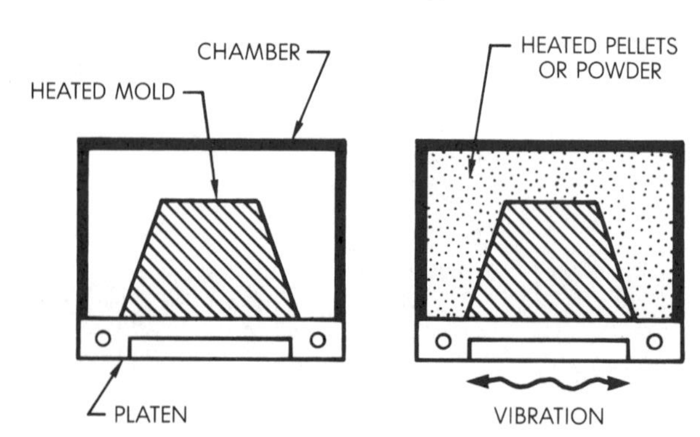

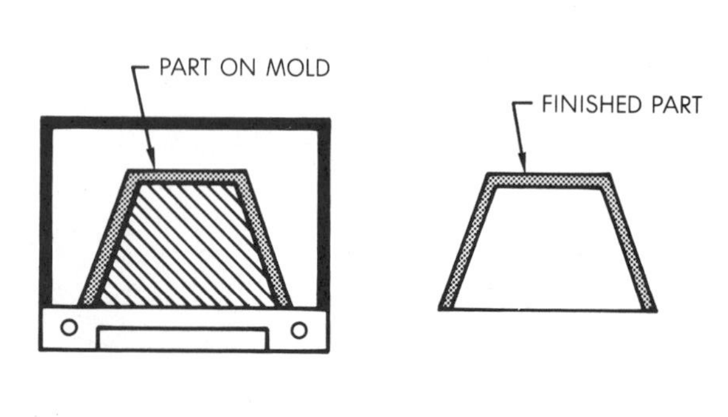

Fig. 14-16. In Vibrational microlamination (VIM), different formulations, types of plastics and/or reinforcements may be alternately used to produce a true laminate.

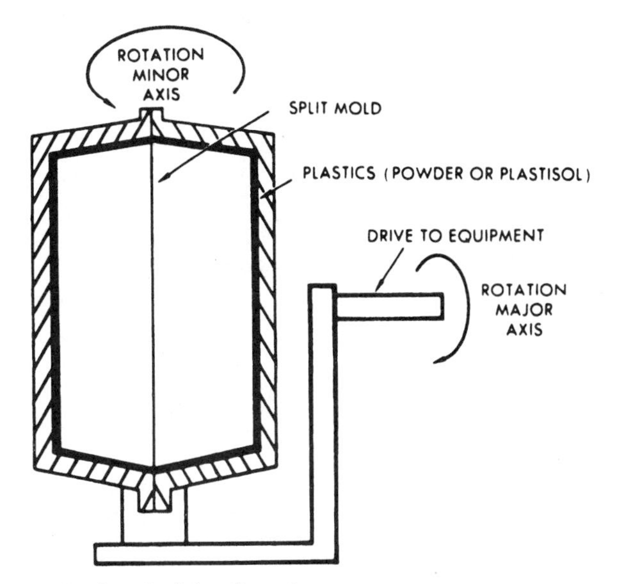

(A) Basic principle of rotation.

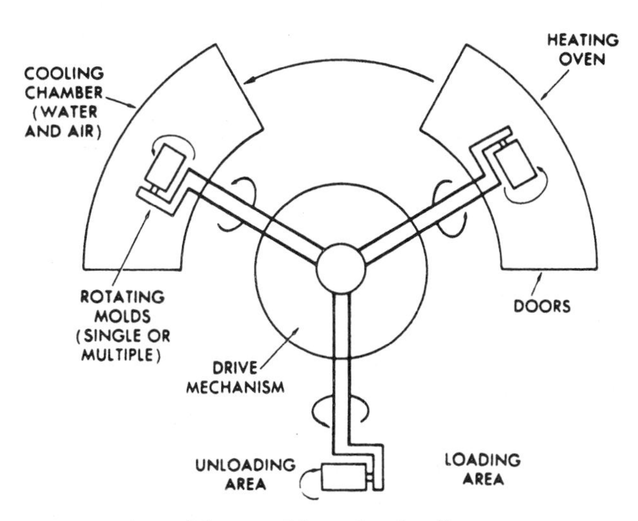

(B) Top view of three-mold rotational unit.

Fig. 14-17. Principle of rotational casting.

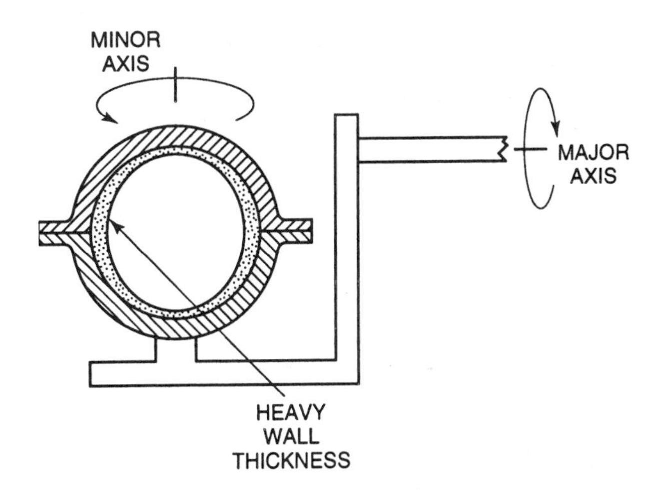

Fig. 14-18. Wall thickness in a ball can be varied by rotating the minor axis faster than the major axis.

(A) Large rotational-cast product removed from mold. (Plastics Design Forum)

(B) Rotational casting of large tank with spray water-cooling. (McNeil Femco Corp.)

(C) Rotational casting eight parts at once. (McNeil Femco Corp.)

Fig. 14-19. Rotational casting machines.

forcements are used, but care must be taken to prevent wicking of protruding reinforcements. Placing a homopolymer layer over the reinforced polymer may overcome this problem. In one operation, a solid outer skin layer is produced, followed by the release of a second charge of material from a *dump box* in the mold.

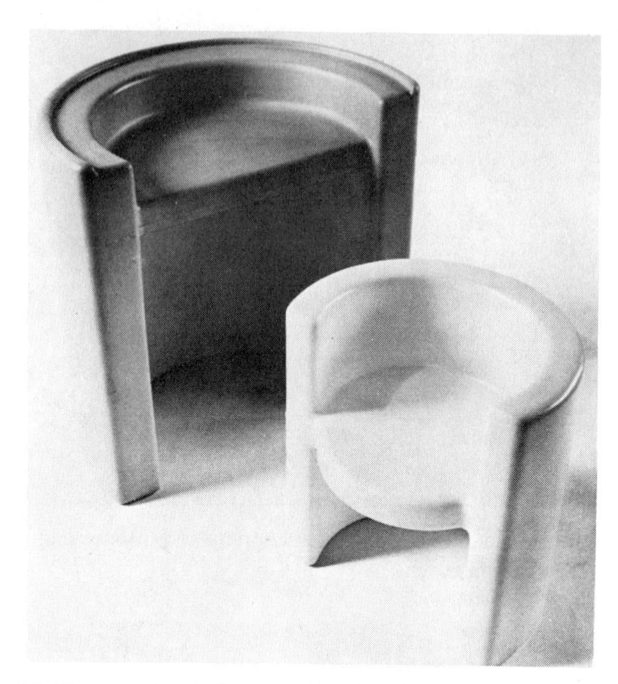

(A) Furniture made from rotational casting. (Design Forum)

(B) This horse was rotational cast from a three-piece mold. (McNeil Femco Corp.)

(C) Rotational-cast trash container. (Phillips Petroleum Co.)

Fig. 14-20. Rotational cast-products.

Dip Casting

Dip casting is a simple process in which a heated mold is dipped into liquid dispersions of plastics. The plastics melts and adheres to the hot metal surface. After removing the mold from the dispersion, it enters a curing oven to ensure proper fusion of the plastics particles. After curing, the plastics is peeled from the mold.

Dip casting should not be confused with *dip coating*. Coatings are not removed from the substrate. In dip casting, a preheated mandrel the shape and size of the *inside* of the product is lowered into a plastisol dispersion (Fig. 14-21). As the resin hits the hot mold surface, it begins to melt and fuse. The thickness of the piece continues to increase as it remains in the solution. If additional thickness is desired, the coated piece may be reheated and dipped again. After the desired thickness is obtained, the mold is removed from the oven and cooled and the part is then stripped from the mold. Thickness of the product is determined by the temperature of the mold and the time the hot mold is in the plastisol.

Several layers or alternate colors and formulations may be applied by alternately heating and dipping. Any design on the mold will appear on the inside of the product. Plastics gloves, overshoes, coin purses, spark plug covers, and toys are examples of dip-cast products. Listed below are six advantages and four disadvantages of casting processes.

Advantages of Casting

1. Cost of equipment, tooling, and molds are low.
2. It is not a complex forming method.
3. There are a wide number of treatment techniques.
4. Products have little or no stress.
5. Material costs are relatively low.
6. Rotational casting produces one-piece hollow objects.

Disadvantages of Casting

1. Output rate is low and cycle time is high.
2. Dimensional accuracy is only fair.
3. Moisture and air bubbles may be problematic.
4. Solvents and other additives may be dangerous.

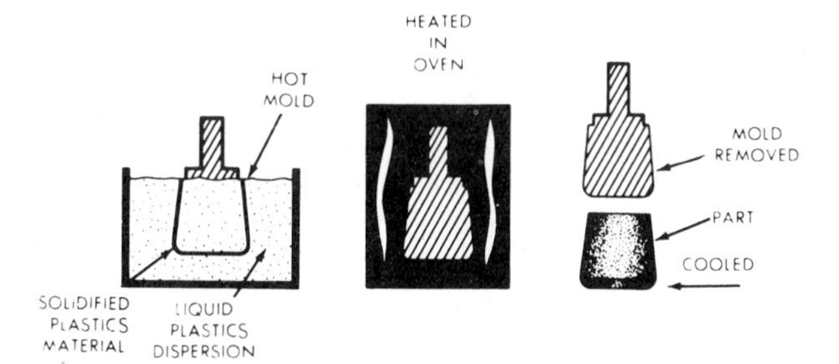

(A) Scheme of dip casting

(B) Dip-cast parts for recreational uses. (BF Goodrich Chemical Division.)

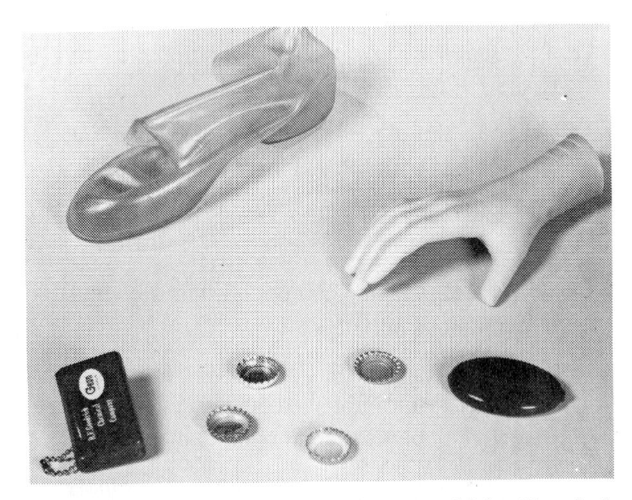

(C) Assorted dip-cast items. (BF Goodrich Chemical Division.)

Fig. 14-21. Dip casting and resulting products.

Vocabulary

The following vocabulary words are found in this chapter. Look up the definition of any of these words you do not understand as they apply to plastics in the glossary, Appendix A.

Casting
Dip casting
Embedment
Encapsulating
Hot melt
Plastisol
Potting
Organosol
Rotational casting
Slush casting
Vibrational microlamination

Questions

14-1. Identify the amount of pressure required for casting processes.

14-2. List three reasons for taking precautions when using polyester resins and catalyst.

14-3. The process of submerging a hot mold into a resin and removing the plastics from the molds is called ___?___.

14-4. A major disadvantage of rotational casting is the ___?___.

14-5. Name three methods that may drive out air bubbles in simple or plastisol castings.

14-6. What materials can be used as molds for dip casting?

14-7. Name the casting compounds that are vinyl dispersions in solvents.

14-8. Identify the plastics sheets that are often cast between two polished sheets of glass. Protective paper is then applied to the surface.

14-9. A process where an object is completely enclosed in transparent plastics is called ___?___.

14-10. Molds for static casting are normally made of ___?___.

14-11. What determines the wall thickness of dip castings?

14-12. Name three measures that would help to reduce the problem of air bubbles in simple castings.

14-13. Name a process similar to slush casting where dry thermoplastic powders are used.

14-14. Identify the major difference between dip coating and dip casting.

14-15. Hollow, one-piece objects may be made by the ___?___ casting process.

14-16. In order to completely fuse the powders or dispersions, a second heating cycle is necessary with ___?___ casting.

14-17. Because easily shaped tooling materials are used, casting molds are normally ___?___ expensive than molds for injection molding.

14-18. What two parameters determine the wall thickness of a rotational casting?

14-19. If a plastics is not easily processed by heat methods, what process may be used to make very thin films?

14-20. Plastisol products are popular because inexpensive methods and molds are used with ___?___ casting.

14-21. Name four reasons why castings processes are less costly than molding operations:

14-22. Name the common cause of pits or pockmarks in or on castings.

Activities

Rotational Casting

Introduction. In rotational casting, the mold receives a measured amount of plastics. The mold is closed, heated, and rotated on two-axes, producing hollow, one-piece products. Castings may include layers of differing materials.

Equipment. Rotational-casting equipment, rotational casting molds, mold release, powdered polyethylene, heat resistant gloves, bucket with water.

Procedure

14-1. Preheat molding oven to 200 °C [392 °F].

14-2. Clean molds and apply a light coat of mold release. Excessive mold release will damage the surface smoothness of the moldings.

14-3. Place a measured amount of polyethylene powder (precolored or add color pigment if desired) in one half of the mold. Clean all powder from mold lips to prevent flash. (See Figure 14-22.)

14-4. Close mold and place in rotational device.

CAUTION: Use protective gear. Oven is hot. (See Figure 14-23.)

14-5. Close oven door and set time for 10 minutes.

14-6. Set rotating mechanism at fastest setting for small mold, slower for larger molds.

14-7. Turn off electrical heaters and begin cooling cycle. Continue rotation becasue air stream aids cooling; otherwise, the hot plastics in the mold will run or sag. Air cool for 5 minutes.

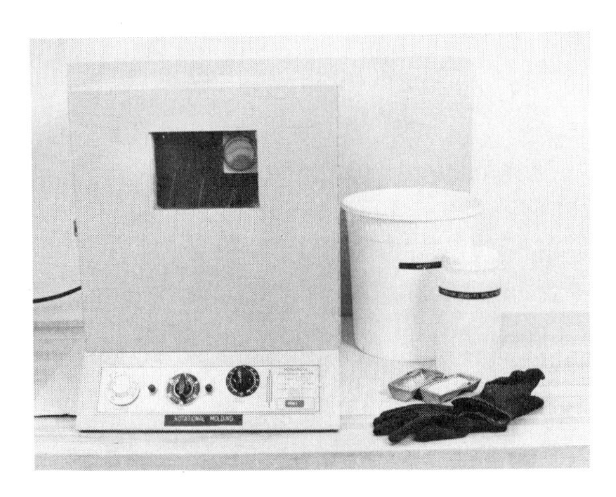

Fig. 14-22. Load polyethylene powder into mold cavity while oven is heating.

Fig. 14-23. Carefully place filled mold into hot oven.

Do not remove from oven until mold temperature is below 100 °C [212 °F].

14-8. Place mold in water to continue rapid cooling.

CAUTION: Mold is hot!

14-9. Remove mold from water and open mold. Do not use metal tools to remove parts because they may damage mold surfaces. (See Figure 14-24.)

14-10. Trim flash if necessary. Cut part in half and measure the wall thickness. Is the wall thickness uniform?

14-11. Try to predict the wall thickness that will result from a larger or smaller charge of powder. Make the molding to evaluate the prediction.

Slush Casting

Introduction. Slush-casting variations are used to produce hollow items. Many industrial products require automatic equipment, but some parts must be manually stripped from the mold. Plastisol products

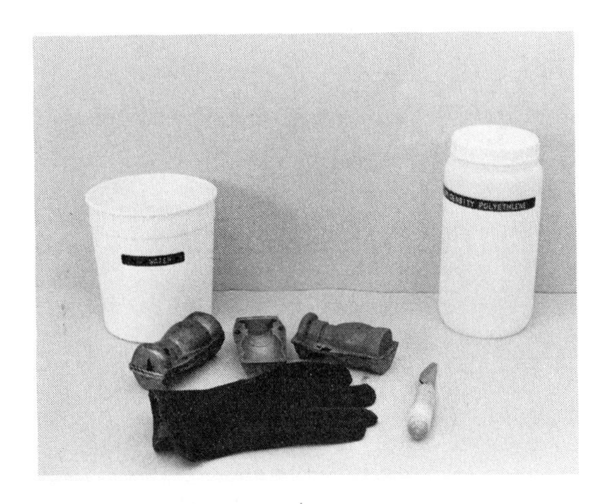

Fig. 14-24. After cooling cycle, remove cast object and trim.

range in texture from soft and supple to semirigid. This variation depends on the formulation of the plastisol.

Equipment. Heat-resistant gloves, oven, slush-casting mold, plastisol, mold release, container with water.

Procedure

14-1. Clean mold and apply light coat of mold release.

14-2. Preheat oven to 200 °C [392 °F].

14-3. Place mold in hot oven for 10 minutes. Use protective gear.

14-4. Remove hot mold from oven, place on heat-resistant surface, and quickly pour plastisol into mold. (See Figure 14-25.)

14-5. After 5 minutes, pour excess plastisol from hot mold. (See Figure 14-26.)

14-6. Set oven to 175 °C [347 °F].

Fig. 14-25. Fill hot mold with plastisol.

Fig. 14-26. After about 5 minutes, pour excess plastisol from hot mold.

14-7. Return mold to hot oven for 20 minutes. Steps 4 through 7 may be repeated if a heavier wall thickness is desired.

14-8. Remove hot mold from oven and quench in water. (See Figure 14-27.)

14-9. Remove part from mold. Do not use metal tools for part removal or mold surface may be scratched. (See Figure 14-28.) Figure 14-29 shows a slush-cast door stop.

14-10. Repeat entire process to find the effect of longer preheat times or greater preheat temperatures.

Polyester Casting

Introduction. An investigation of polyester casting resins can provide a demonstration of polymerization and an opportunity to observe the polymerization reaction. The catalyst (hardener) used to cure polyester resin is often methyl ethyl ketone peroxide (MEK).

Fig. 14-27. After plastisol has cured, quench object in water.

Fig. 14-28. Strip plastisol from mold.

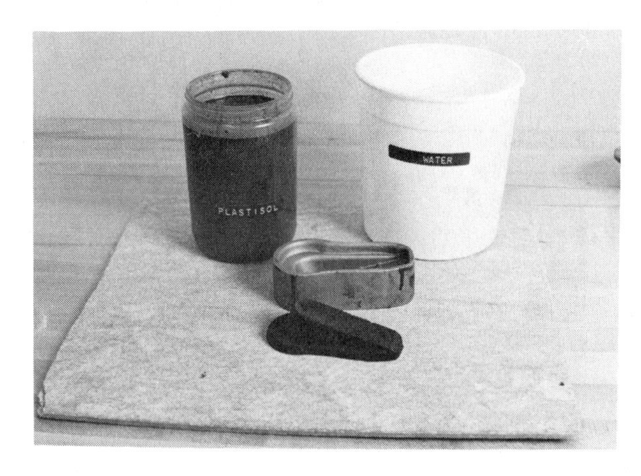

Fig. 14-29. Finished plastisol slush-cast item.

CAUTION: MEK is an organic peroxide and must be handled carefully.

MEK combines chemically with accelerators in the resin. That reaction releases chemicals called free radicals which cause the reactive (unsaturated) polyester molecules to begin to polymerize. To document the change from liquid resin to solid, polymerized polyester, complete the following activity.

Equipment. Paper cups, 4 ounce size, polyester casting resin and hardener, stir sticks, vinyl gloves, band saw, milling machine, rockwell hardness tester.

Safety Precautions. Wear vinyl gloves when handling liquid resin and catalyst. Work in a well ventilated area.

Procedure

14-1. Establish a volume for the samples. About half of a 4 ounce cup is convenient. Mark the cups to ensure that all samples are equivalent in volume.

14-2. Select steps of hardener concentration. Figure 14-30 shows cups with 4, 8, 12, 16, 20, and 24

Fig. 14-30. Polyester samples with various hardener concentrations

drops of hardener. After placing hardener in resin, stir thoroughly using stir sticks.

14-3. Allow 24 hours for polymerization.

14-4. Remove the paper cup from the hardened polyester. Figure 14-31 shows hardened samples. The sample with 4 drops did not harden.

14-5. Use a band saw to cut two sides to provide clamping surfaces. Also cut some material off the top and bottom to provide clean surfaces for machining. Figure 14-31 shows one sample ready for machining.

14-6. Mill the top and bottom flat. If these surfaces are not flat, hardness testing results may be erroneous. Figure 14-31 also shows a sample after machining.

14-7. Test the hardness of the samples with a Rockwell hardness tester. The Rockwell M scale should be appropriate.

14-8. Graph the results. If the steps of hardener concentration were too large or too small, repeat with smaller increments. Figure 14-32 shows the results of such a test. Notice that it includes data from samples with 10 and 14 drops of hardener. Notice also that at 8 drops, the material was too soft for reliable measurement. The samples with 10, 12, 14, and 16 drops shows the increasing completion of the polymerization reaction.

14-9. To gain additional information, select a concentration of hardener that yields fully hardened polyester.

14-10. Make one sample at the selected concentration and measure the changes in the temperature of the resin as it polymerizes. Figure 14-33 shows a set-up to gather the time/temperature data.

Fig. 14-31. One sample as removed from paper cup. One samples as cut with bandsaw, and one machined to create a flat surface.

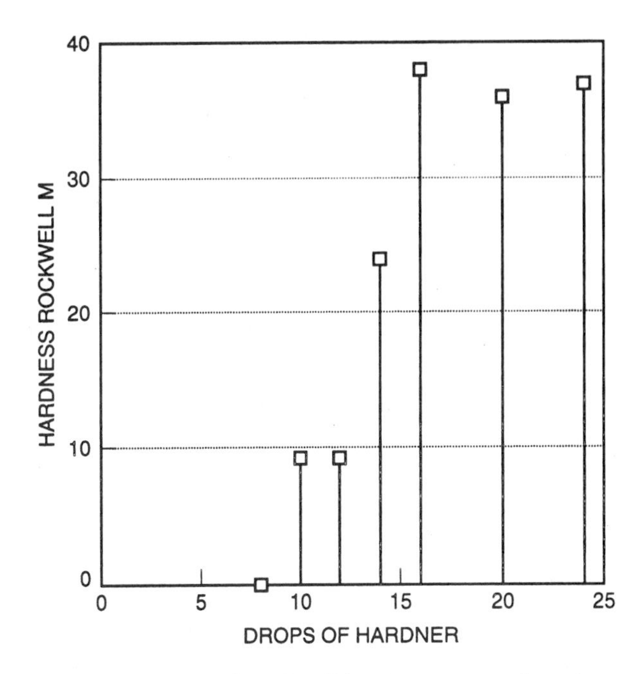

Fig. 14-32. Graph of Rockwell hardness test results.

Fig. 14-33. Set-up for measuring temperature of curing resin.

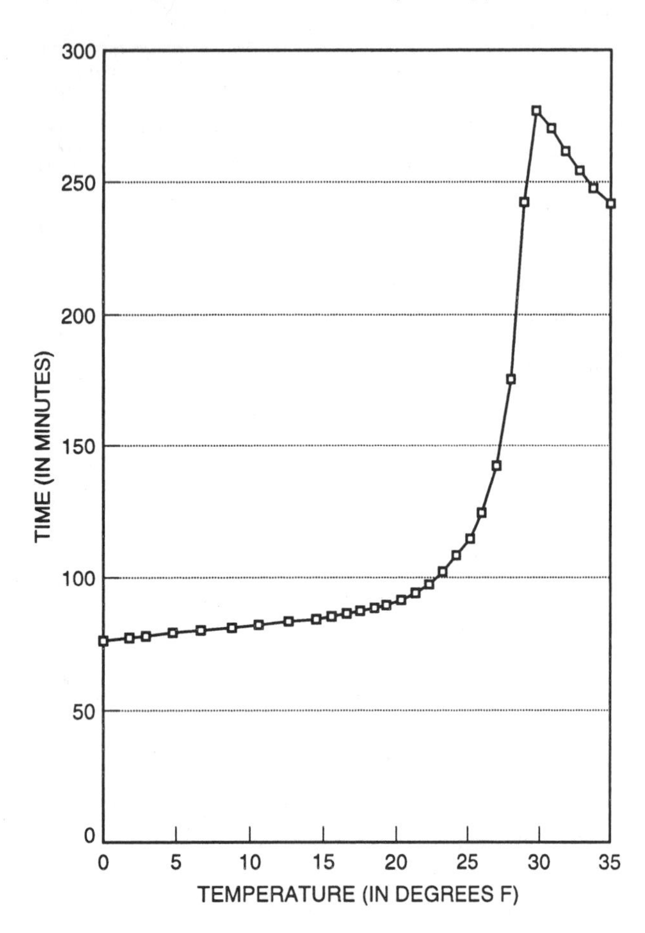

Fig. 14-34. Time/temperature graph of curing polyester resin.

14-11. Graph the results. Figure 14-34 shows a curve created by 16 drops of hardener. The highest temperature reached is called the peak exotherm.

14-12. Determine how changes in hardener concentration effect the time/temperature curve.

Chapter 15

Thermoforming

Introduction

Thermoforming is an ancient technique. Ancient Egyptians found that animal horns and tortoise shells could be heated and formed into a variety of vessels and shapes. When synthetic plastics became available, thermoforming was an early application. In the United States, John Hyatt thermoformed celluloid sheets over wooden cores for piano keys.

Today, thermoformed items surround us. They include: signs, light fixtures, ice-cube trays, ducts, drawers, instrument panels, tote trays, housewares, toys, refrigerator panels, transparent aircraft enclosures, and boat windshields (Fig. 15-1). The packaging industry relies heavily on thermoforming. Cookies, pills, and other products are commonly packaged by blister packaging. Single portions of butter, jellies, and other foods also appear in blister packs. Replacement parts and hardware are other examples of items that are sometimes skin packaged.

The materials used in thermoforming include most thermoplastics—except acetals, polyamides, and fluorocarbons. Usually, thermoforming sheets contain only one basic plastics; however, some thermoplastic composites are also thermoformed.

Thermoforming processes are possible because thermoplastic sheets can be softened and reshaped, and

(A) Sports car with body made of panels that were vacuum-formed, glued together, and painted. (U.S. Gypsum)

(B) Blister package with reclosable sliding door. (Celanese Plastic Materials Co.)

Fig. 15-1. Some articles fabricated by thermoforming processes.

(continued)

(C) Deep-draw, straight-walled containers are thermoformed. (Shell)

(D) Clear plastics frames for a display package are thermoformed on a continuous web from plastics sheet. (Celanese Plastic Materials Co.)

(E) Vacuum-formed clear covers protect and display bakery products.

Fig. 15-1. Some articles fabricated by thermoforming processes. *(Continued)*

the new shape is retained when the material is cooled. Since most thermoforming sheets were originally formed by sheet extrusion, considerable energy, time, and space can be saved by directly thermoforming sheets as they leave the extruder. However, many thermoforming industries change materials, colors, and textures frequently. They are not good candidates for immediate thermoforming of extruder sheets.

The force required to alter a sheet into a desired product can be mechanical, air, or vacuum pressure. In many cases, thermoforming requires a combination of two or three pressure sources.

Tooling can run from low-cost plaster molds to expensive water-cooled steel molds. The most common tooling material is cast aluminum. Wood, gypsum, hardboard, pressed wood, cast phenolic resins, filled or unfilled polyester or epoxy resins, sprayed metal, and steel also may be used for molds. Both male (plug) and female (cavity) molds are used. Molds require sufficient draft to assure stress-free parts removal.

Because tooling costs are usually low, parts with large surface areas may be produced economically. Prototypes and short runs are also practical. Although dimensional accuracy is good, thinning is a problem in some part designs.

In this chapter, thirteen basic thermoforming techniques are discussed. These are listed in the following chapter outline:

I. Straight vacuum forming
II. Drape forming
III. Matched-mold forming
IV. Pressure-bubble plug-assist vacuum forming
V. Plug-assist vacuum forming
VI. Plug-assist pressure forming
VII. Solid phase pressure forming (SPPF)
VIII. Vacuum snap-back forming
IX. Pressure-bubble vacuum snap-back forming
X. Trapped-sheet contact-pressure forming
XI. Air-slip forming

XII. Free forming
XIII. Mechanical forming

Some of the modern industrial thermoforming machines that perform these processes are shown in Figure 15-2.

(A) High-speed pressure/vacuum former operates from either roll stock or inline with an extruder.

(B) Rotary style unit used for large industrial components at a fairly high production rate.

(C) Twin-sheet thermoforming machine with separate, independent clamping frames.

Fig. 15-2. Modern industrial thermoforming machines. (Brown Machine Co.)

Straight Vacuum Forming

Vacuum forming is the most versatile and widely used thermoforming process. Vacuum equipment costs less than pressure or mechanical processing equipment.

In straight vacuum forming, a plastics sheet is clamped in a frame and heated. While the hot sheet is in a rubbery state, it is placed over a recessed mold cavity. The air is removed from this cavity by vacuum (Fig. 15-3) and atmospheric pressure (10 kPa) forces the hot sheet against the walls and contours of the mold. When the plastics has cooled, the formed part is removed, and final finishing and decorating may be

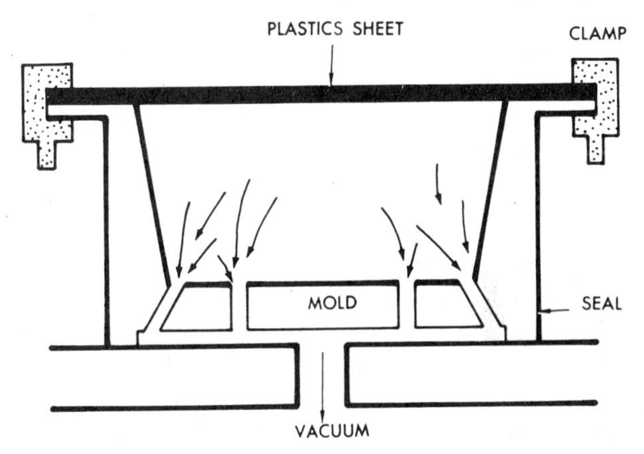

(A) A clamped and heated plastics sheet is forced down into the mold by atmospheric pressure after a vacuum is drawn in the mold. (Atlas Vac Machine Co.)

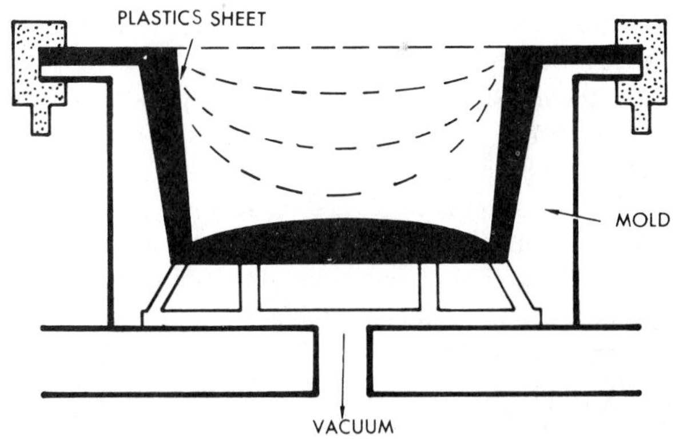

(B) Plastics sheet cools as it contacts the mold. (Atlas Vac Machine Co.)

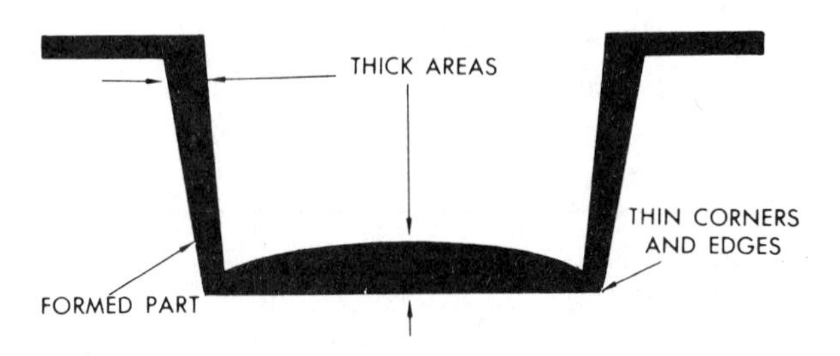

(C) Areas of the sheet that touched the mold last are the thinnest. (Atlas Vac Machine Co.)

Fig. 15-3. Straight vacuum forming.

(D) Note the frame that holds the heated thermoplastics sheet as it is drawn down over the mold. (Chemplex Co.)

Fig. 15-3. Straight vacuum forming. *(Continued)*

done, if necessary. Blowers or fans are used to speed cooling. One disadvantage of thermoforming is that formed pieces usually must be trimmed, and the scrap must be reprocessed.

Most vacuum systems have a surge tank to ensure a constant vacuum of 500 to 760 mm of mercury. Superior parts are formed by quickly applying the vacuum before any portion of the sheet has cooled. Slots are more desirable and efficient than holes in allowing the air to be drawn from the mold. Slots or holes should be smaller than 0.025 in. [0.65 mm] in diameter to avoid surface blemishes on the formed part. A hole or slot should be placed in all low or unconnected portions of the mold. If this is not done, air may be trapped under the hot sheet with no way to escape. Unless they are collapsible, molds should include a 2 to 7 degree (draft) angle for easy part removal.

Thinning at the edges and corners of a part is a disadvantage in using relatively deep recessed molds. As the sheet material is drawn into the mold, it stretches and thins. The regions that are stretched minimally remain thicker than regions that are stretched extensively. If preprinted flat sheets are formed, thinning must be kept in mind when trying to compensate for distortion during forming. Straight vacuum forming is limited to simple, shallow designs and thinning will occur in corners.

The *draw* or *draw ratio* of a recessed mold is the ratio of the maximum cavity depth to the minimum span across the top opening. For high-density polyethylene, the best results are achieved when this ratio does not exceed 0.7: 1. Thermoforming equipment and dies are relatively inexpensive.

When the plastics has cooled, it is removed for trimming or postprocessing, if needed. Markoff (marks from the mold) is on the *inside* of the product while such marks appear on the *outside* of the part in straight vacuum forming.

Drape Forming

Drape forming (incorrectly called mechanical forming) is similar to straight vacuum forming except that after the plastics is framed and heated, it is mechanically stretched over a male mold. A vacuum (actually, a pressure differential) is applied that pushes the hot plastics against all portions of the mold (Fig. 15-4). The sheet touching the mold remains close to its original thickness. Side walls are formed by the material draping between the top edges of the mold and the bottom seal area at the base.

It is possible to drape form items with a depth to diameter ratio of nearly 4:1. High draw ratios are possible with drape forming; however, this technique is also more complex. Male molds are easy to make and, as a rule, cost less than female ones, but male molds are more easily damaged.

Drape forming has also been used to form a hot plastics sheet over male or female molds by gravitational forces alone. Female molds are preferred for multicavity forming because there must be more spacing if male molds are selected.

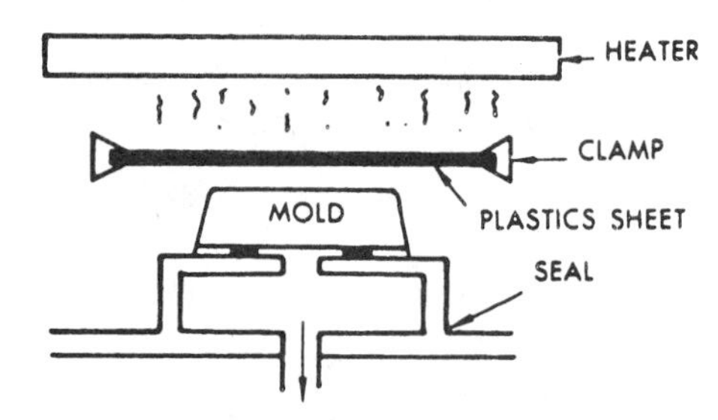

(A) Clamped heated plastics sheet may be pulled over the mold, or the mold may be forced into the sheet.

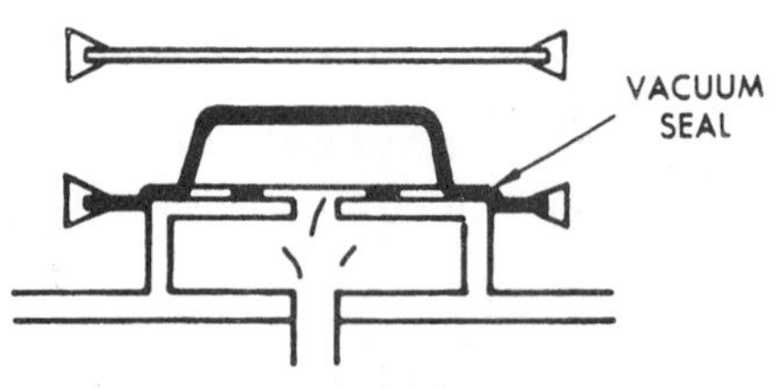

(B) Once the sheet has formed a seal around the mold, a vacuum is drawn to pull the plastics sheet tightly against the mold surface.

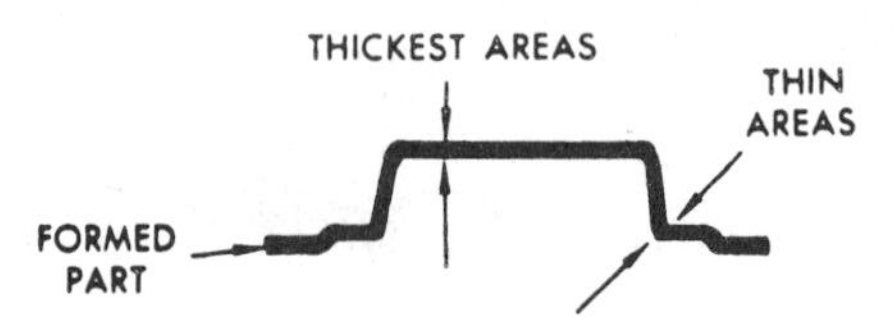

(C) Final wall thickness distribution in the molded part.

Fig. 15-4. Principle of drape forming plastics. (Atlas Vac Machine Co.)

Matched-Mold Forming

Matched-mold forming is similar to compression molding. A heated sheet is trapped and formed between male and female dies that may be made of wood, plaster, epoxy, or other materials (Fig. 15-5). Accurate parts with close tolerances may be quickly produced in costly water-cooled molds. Very good molded detail and dimensional accuracy can be obtained with water-cooled molds, including lettering and grained surfaces. There is mark-off on both sides of the finished product; therefore, mold dies *must* be protected from scratches or damage because such defects would be reproduced by the thermoplastic materials. A smooth-surface mold should not be used with polyolefins because air may be trapped between the hot plastics and a highly polished mold. Sandblasted mold surfaces are usually used for these materials.

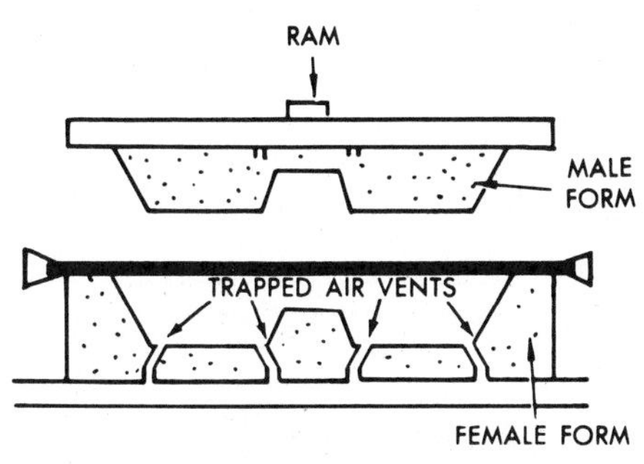

(A) The heated plastics sheet may be clamped over the female die, as shown, or draped over the mold form.

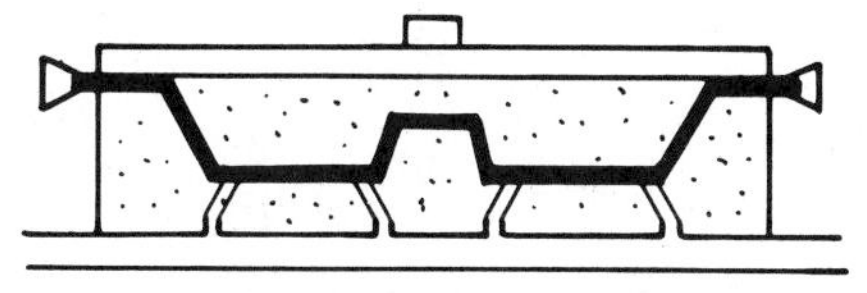

(B) Vents allow trapped air to escape as the mold closes and forms the part.

(C) Distribution of materials in the product depends on the shapes of the two dies.

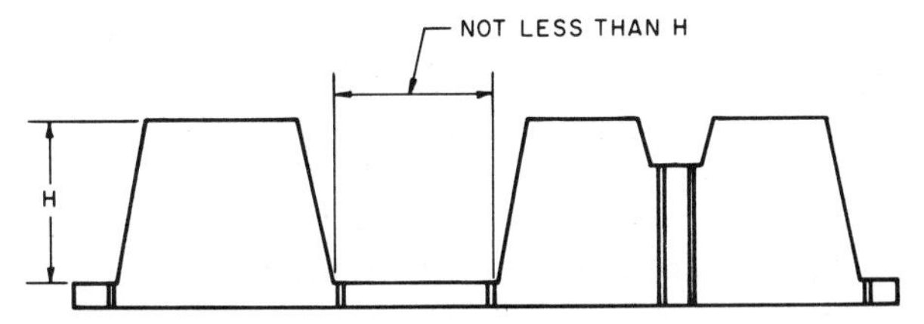

(D) Male mold forms must be spaced at a distance equal to or greater than their height or webbing may occur.

Fig. 15-5. Principle of matched-mold forming. (Atlas Vac Machine Co.)

Figure 15-6 shows a matched-mold operation in which the blank is smaller than the finished part. Under the rim, it flows out to the periphery without forming trim or flash. The total mold cycle is 10 to 20 seconds.

Pressure-Bubble Plug-Assist Vacuum Forming

For deep thermoforming, pressure-bubble plug-assist vacuum forming is an important process. By this process, it is possible to control the thickness of the formed article. The item may have uniform thickness or the thickness may be varied.

Once the sheet has been placed in the frame and heated, controlled air pressure creates a bubble (Fig. 15-7). This bubble stretches the material to a predetermined height, usually controlled by a photocell. The male plug assist is then lowered forcing the stretched stock down into the cavity. The male plug is normally heated to avoid chilling the plastics prematurely. The plug is made as large as possible so the plastics is stretched close to the final shape of the finished product. Plug penetration should be from 70 to 80 percent of the mold cavity depth. Air pressure is then applied from the plug side while at the same time a vacuum is drawn on the cavity to help form the hot sheet. For many products, vacuum alone is used to complete formation of the sheet. In Figure 15-7, both vacuum and pressure are applied during the forming process. The female mold must be vented to allow trapped air to escape from between the plastics and the mold.

Plug-Assist Vacuum Forming

To help prevent corner or periphery thinning of cup- or box-shaped articles, a plug assist is used to mechanically stretch and pull additional plastics stock into the female cavity (Fig. 15-8). The plug is normally heated

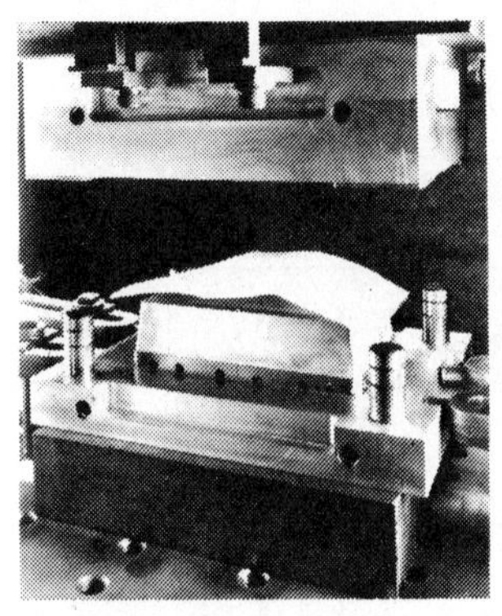

(A) Preheated sheet is placed between cooled matched metal dies.

(B) The finished formed part is larger than the original sheet.

Fig. 15-6. Matched-mold forming being done on a conventional metal-stamping press. (G.R.T.L. Co.)

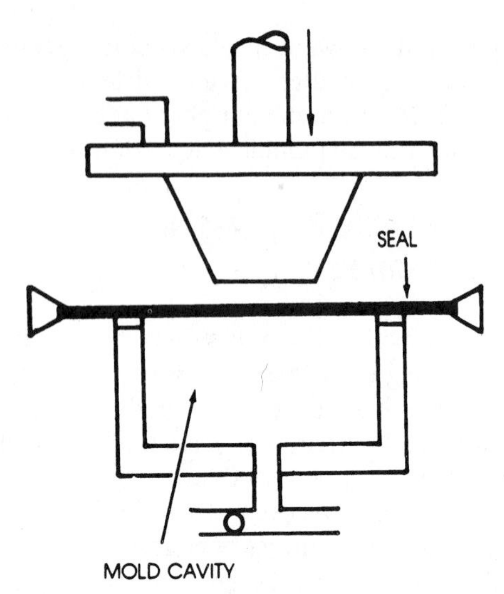

(A) The plastics sheet is heated and sealed across the mold cavity.

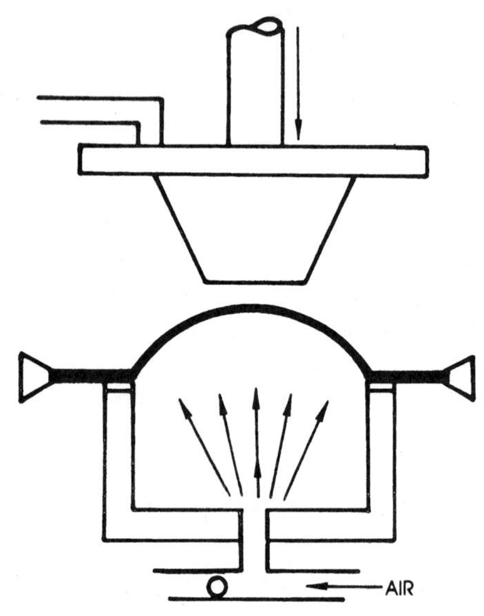

(B) Air is introduced, blowing the sheet upward into an evenly stretched bubble.

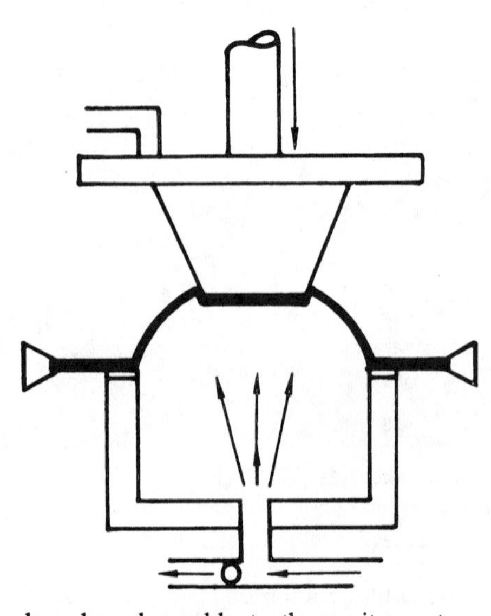

(C) A plug shaped roughly to the cavity contour presses downward into the bubble, forcing it into the mold.

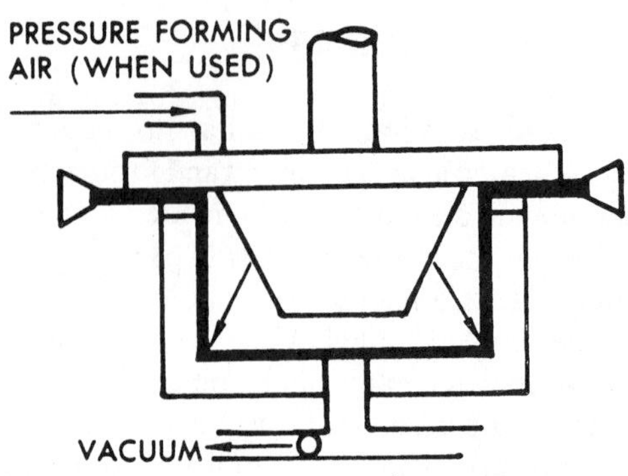

(D) When the plug reaches its lowest point, vacuum is drawn to pull the plastics against the mold walls. Air may be introduced from above to aid forming.

Fig. 15-7. Pressure-bubble plug-assist vacuum forming. (Atlas Vac Machine Co.)

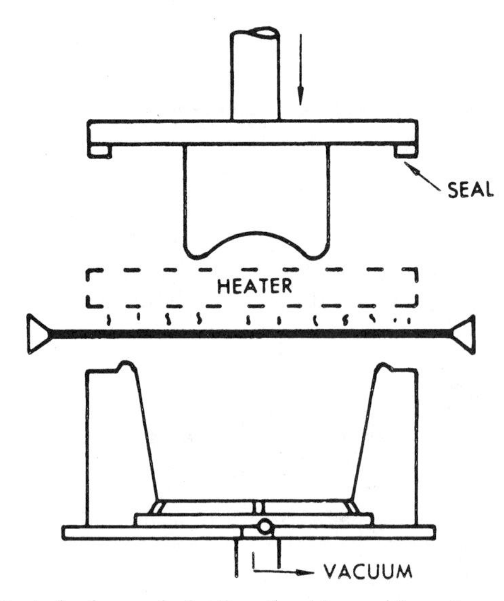

(A) Heated, clamped plastics sheet is positioned over mold cavity.

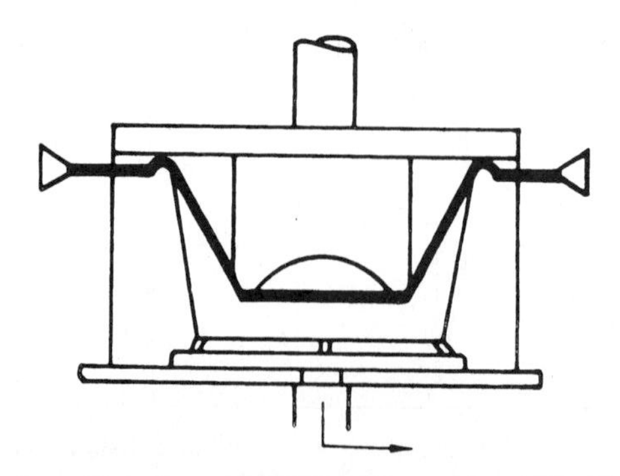

(B) A plug, shaped roughly like the mold cavity, plunges into the plastics sheet to prestretch it.

Fig. 15-8. Plug-assist vacuum forming. (Atlas Vac Machine Co.)

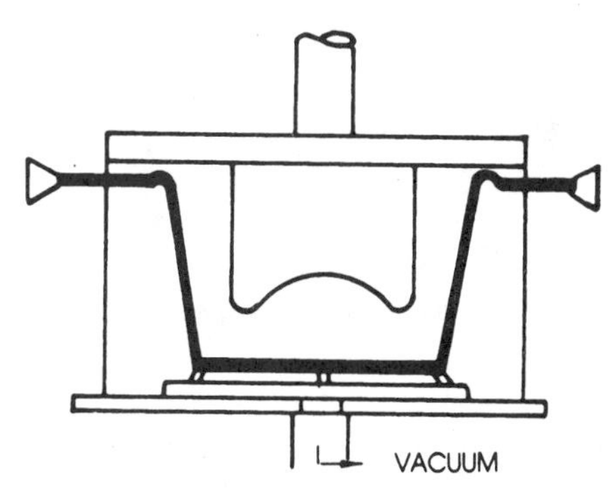

(C) When the plug reaches the limit of its travel, a vacuum is drawn in the mold cavity.

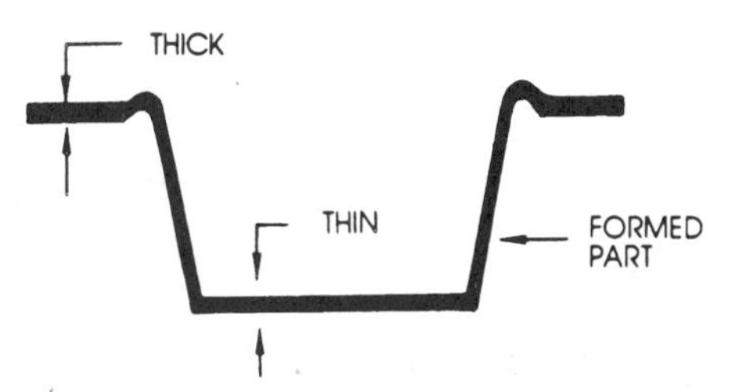

(D) Areas of the plug touching the sheet first form thickened areas due to chilling effect.

Fig. 15-8. Plug-assist vacuum forming. (Atlas Vac Machine Co.) *(Continued)*

to just below the forming temperature of the sheet stock. The plug should be from 10 to 20 percent smaller in length and width than the female mold. Once the plug has forced the hot sheet into the cavity, air is drawn from the mold, completing the formation of the part. The plug design or shape determines the wall thickness, as shown in cross-section in Figure 15-8D.

Plug-assist vacuum and pressure forming allows deep drawing, and permits shorter cooling cycles and better control of wall thickness; however, close temperature control is needed, and the equipment is more complex than straight vacuum forming (Fig. 15-9).

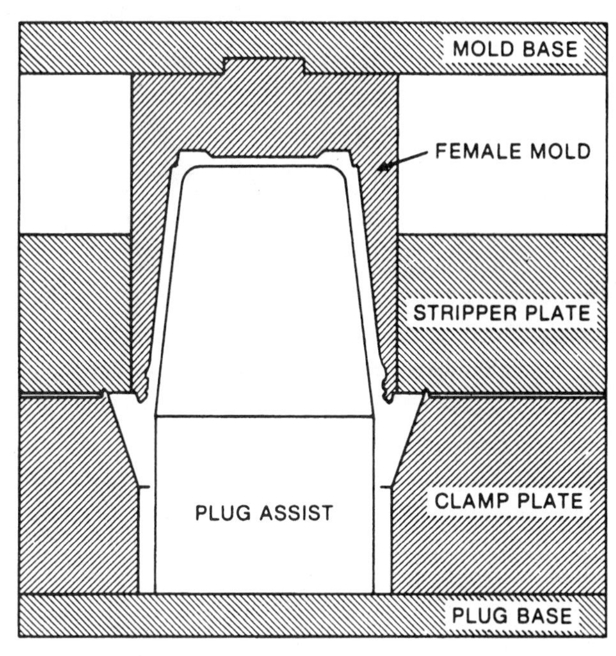

(A) Clamp layout.

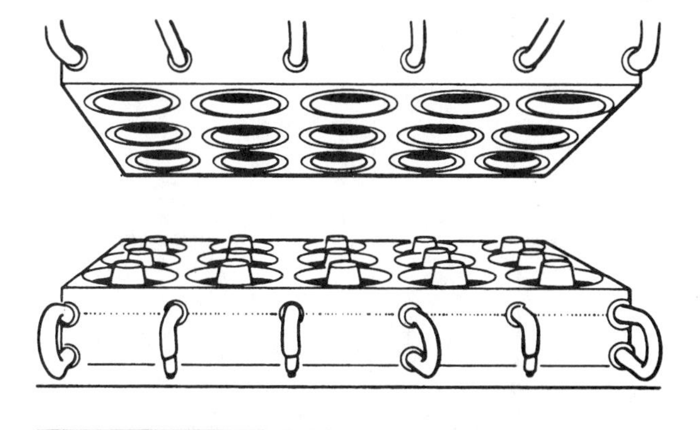

(B) RAM part clamps.

Fig. 15-9. Restricted-area molding (RAM), with individual part clamps built into the mold. This helps to control material draw and reduces draw ratio. (Brown Machine Co.)

Plug-Assist Pressure Forming

Plug-assist pressure forming is similar to plug-assist vacuum forming in that the plug forces the hot plastics into the female cavity. Air pressure applied from the plug forces the plastics sheet against the walls of the mold (Fig. 15-10).

Solid Phase Pressure Forming (SPPF)

A process called *solid phase pressure forming* (SPPF) is similar to plug-assist forming. The technique begins with a solid blank (extruded, compression molded, sintered powders) which is heated to just below its melting point. Polypropylene or other multilayered PP sheets are used. The blank is then pressed into a sheet form and transferred to the thermoforming press. A plug further stretches the hot material and air pressure forces the hot material against the mold sides (Fig. 15-11). The two (biaxial) stretching operations cause molecular orientation, which enhances the strength, toughness, and environmental-stress crack resistance of the thermoformed product.

Vacuum Snap-Back Forming

In vacuum snap-back forming, the hot plastics sheet is placed over a box and a vacuum is drawn, which causes

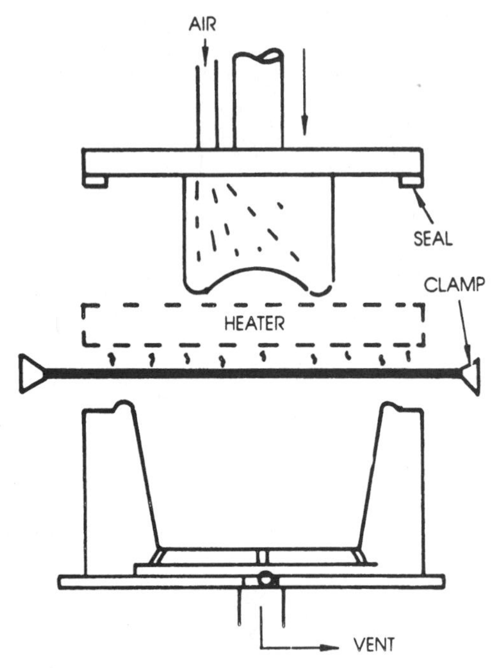

(A) Heated, clamped sheet is positioned over the mold cavity.

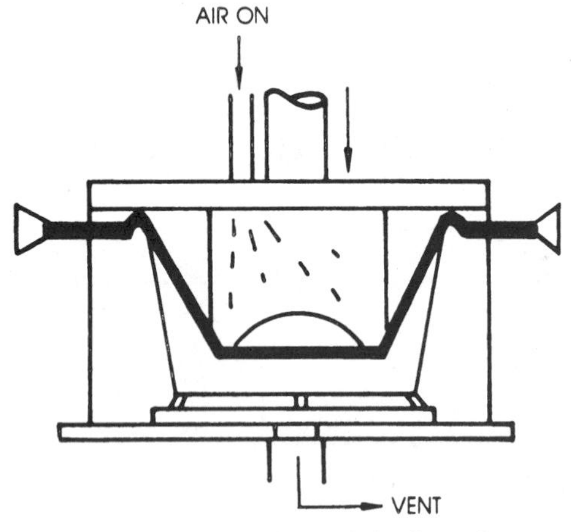

(B) As the plug touches the sheet, air is allowed to vent from beneath the sheet.

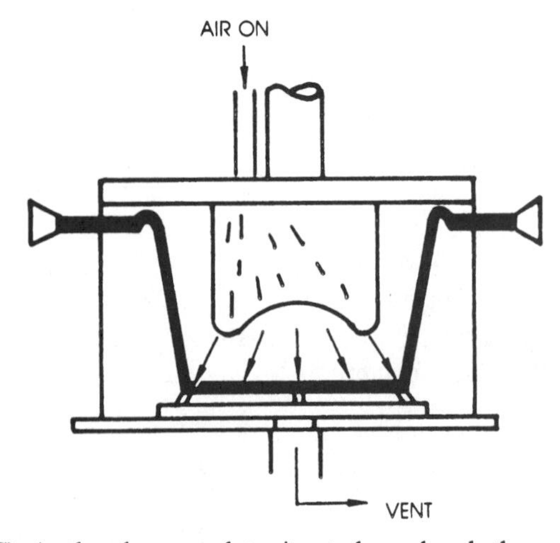

(C) As the plug completes its stroke and seals the mold, air pressure is applied from the plug side, forcing the plastics against the mold.

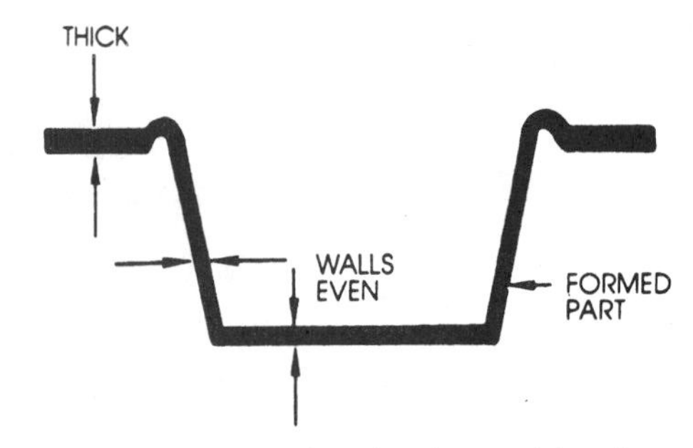

(D) Plug-assist pressure forming is capable of producing products with uniform wall thickness

Fig. 15-10. Plug-assist pressure forming. (Atlas Vac Machine Co.)

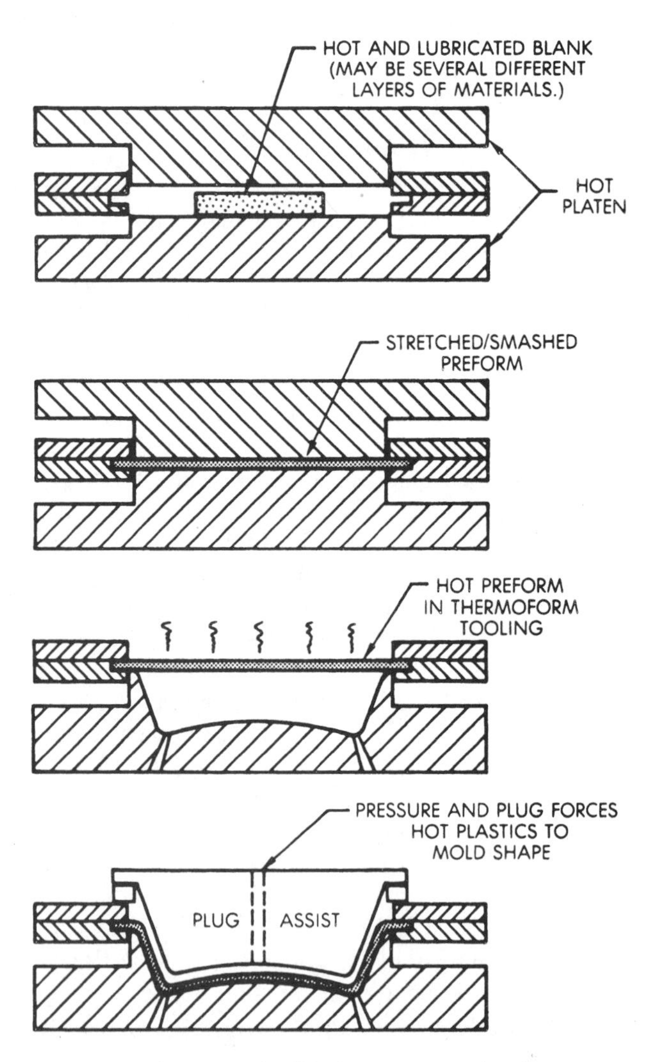

Fig. 15-11. Concept of solid phase pressure forming.

a bubble to be forced into the box (Fig. 15-12). A male mold is lowered and the vacuum in the box is released, causing the plastics to *snap back* around the male mold. A vacuum may also be drawn in the male mold to help pull the plastics into place. Vacuum snap-back forming allows complex parts with recesses to be formed.

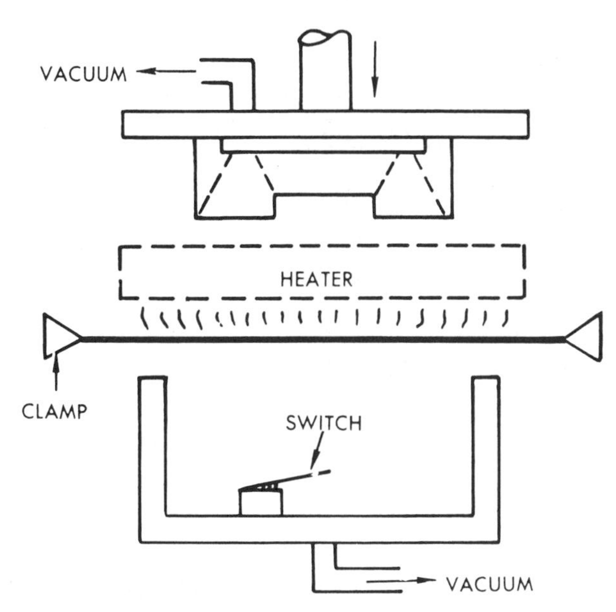

(A) Plastics sheet is heated and sealed over the top of the vacuum box. (Atlas Vac Machine Co.)

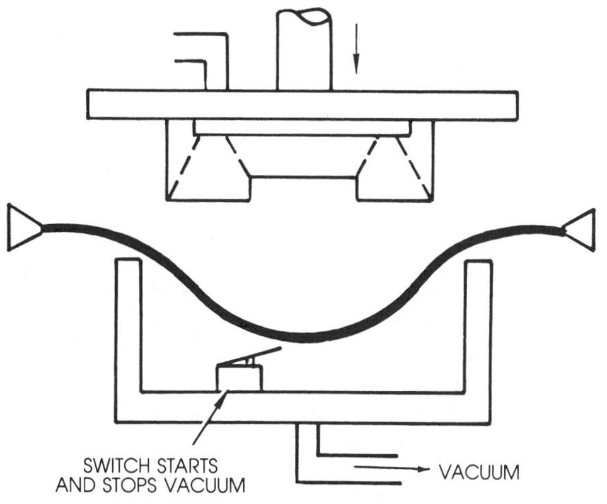

(B) Vacuum is drawn beneath the sheet, pulling it into a concave shape. (Atlas Vac Machine Co.)

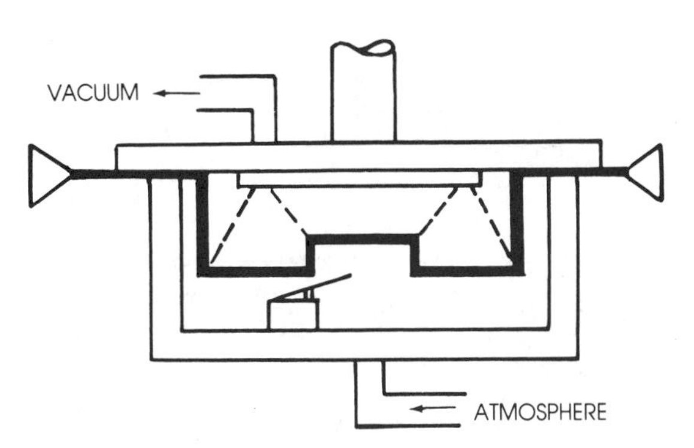

(C) The male plug is lowered and a vacuum drawn through it. At the same time, vacuum beneath the sheet is vented. (Atlas Vac Machine Co.)

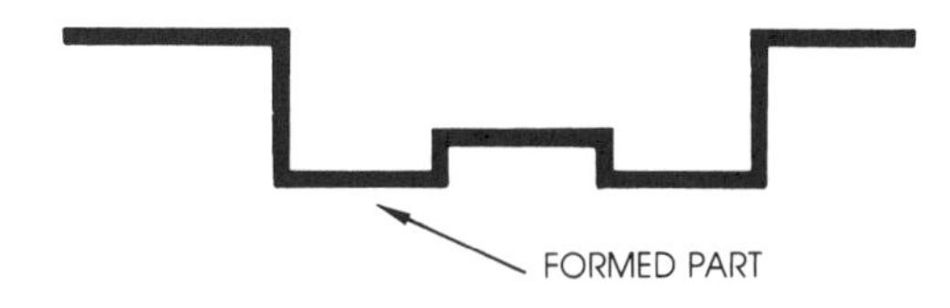

(D) External deep draws can be obtained with this process to form luggage, auto parts, and other items. (Atlas Vac Machine Co.)

(E) A complete auto body is vacuum-formed from 6.3 mm [1/4 in.] thick ABS-polycarbonate during a 20 minute cycle. (Borg-Warner)

Fig. 15-12. Vacuum snap-back forming. (Atlas Vac Machine Co.)

Pressure-Bubble Vacuum Snap-Back Forming

As the name implies, the sheet is heated and then stretched into a bubble shape by air pressure (Fig. 15-13). The sheet prestretches about 35 to 40 percent. The male mold is then lowered. A vacuum is applied to the male mold while air pressure is forced into the female cavity. This causes the hot sheet to snap back around the male mold. Markoff is on the male mold side.

Pressure-bubble vacuum snap-back forming allows deep drawing and the formation of complex parts, but the equipment is complex and costly.

Trapped-Sheet Contact-Heat Pressure Forming

This process is like straight vacuum forming except that air pressure and a vacuum assist may be used to force the hot plastics into a female mold. Figure 15-14 shows the steps of this process.

Air-Slip Forming

Air-slip forming is similar to snap-back forming except for the method of creating the stretch bubble. This concept is illustrated in Figure 15-15.

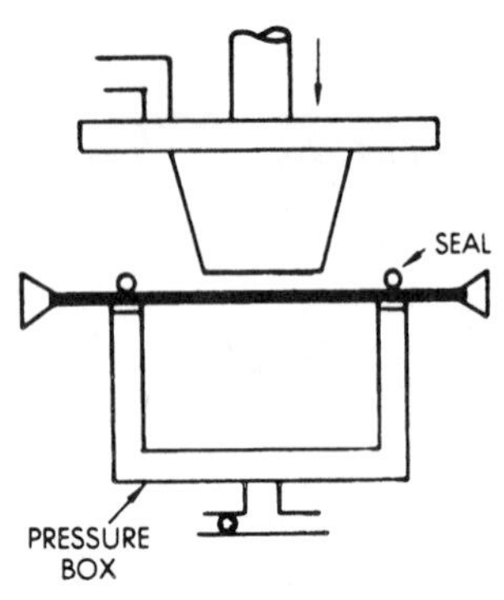

(A) Heated plastics sheet is clamped and sealed across a pressure box.

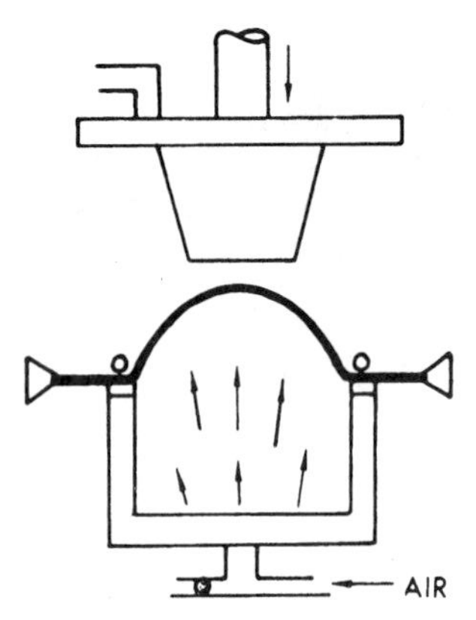

(B) Air pressure is introduced beneath the sheet, causing a large bubble to form.

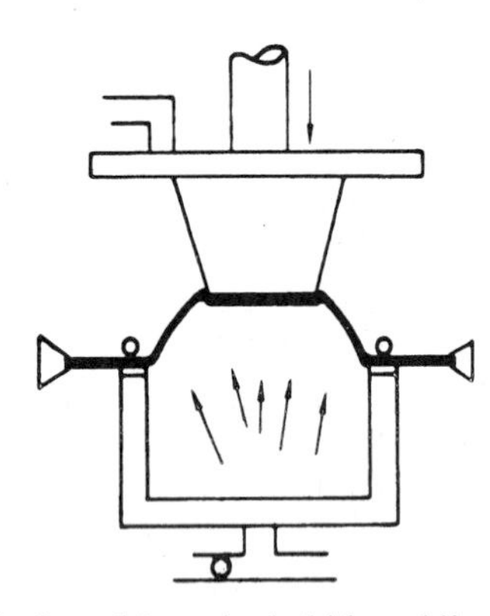

(C) A plug is forced into the bubble, while air pressure is maintained at a constant level.

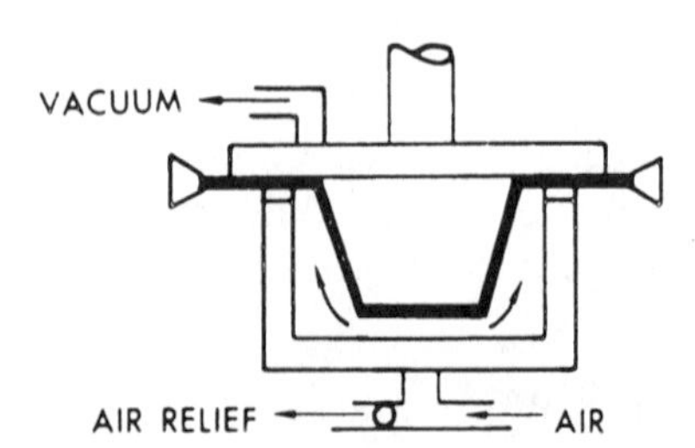

(D) Air pressure beneath the bubble and a vacuum at the plug side create a uniform draw.

Fig. 15-13. Pressure-bubble vacuum snap-back forming. (Atlas Vac Machine Co.)

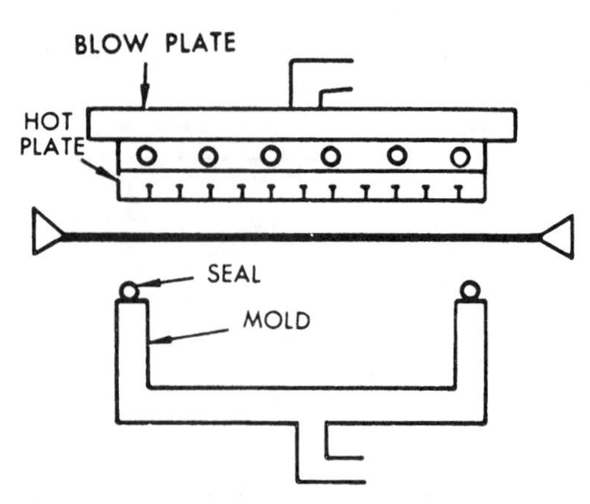

(A) A flat, porous plate allows air to be blown through its face.

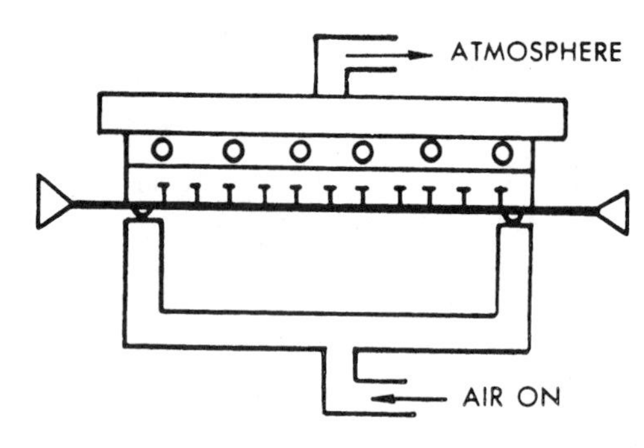

(B) Air pressure from below and a vacuum above force the sheet tightly against the heated plate.

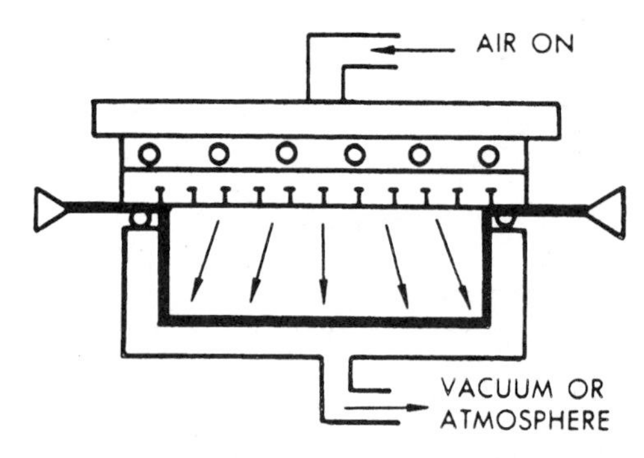

(C) Air is blown through the plate to force the plastics into the mold cavity.

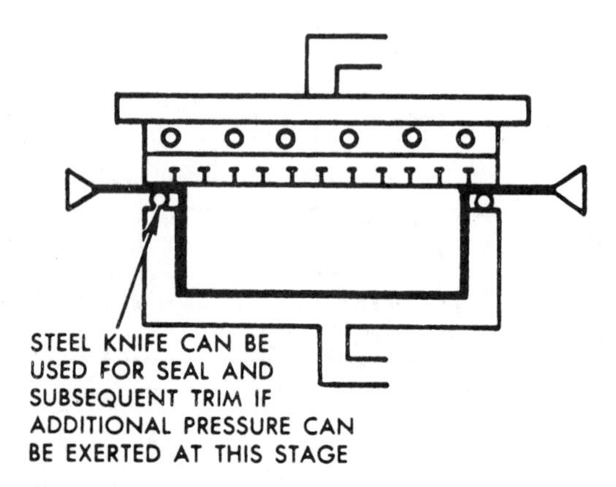

(D) After forming, additional pressure may be exerted.

Fig. 15-14. Trapped-sheet contact-heat pressure forming. (Atlas Vac Machine Co.)

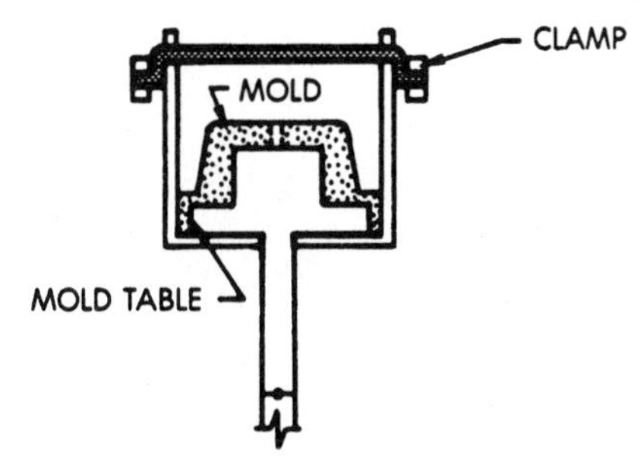

(A) Sheet is clamped to the top of a vertical walled chamber.

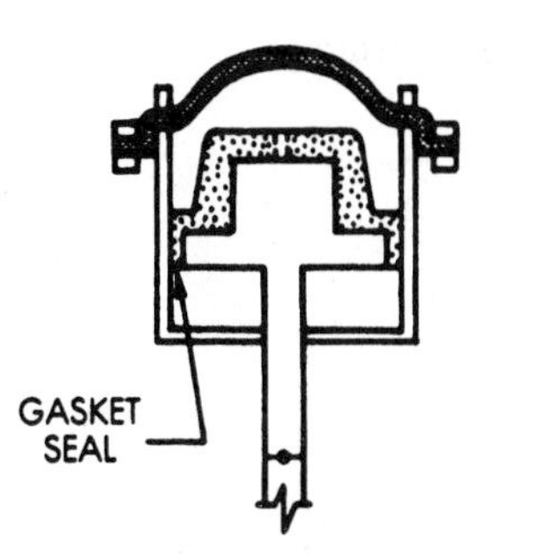

(B) Pre-bellow is acheived by a pressure buildup between the sheet and mold table.

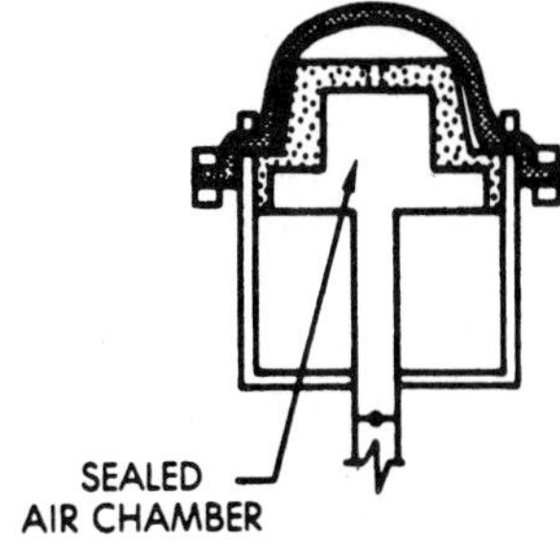

(C) Mold rises in chamber. Mold table is gasketed at edges of chamber wall.

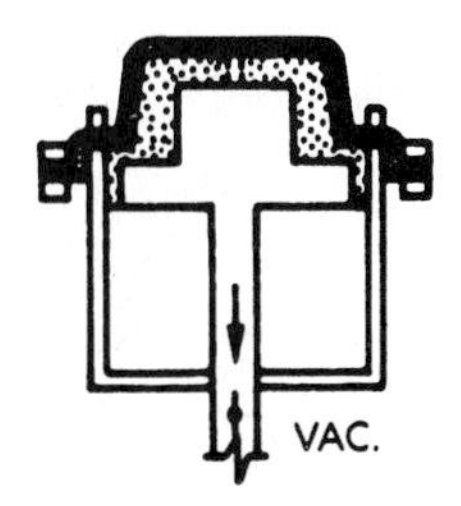

(D) Space between mold and sheet is evacuated as sheet is formed against mold by differential air pressure.

Fig. 15-15. Air-slip forming.

Free Forming

In free forming, air pressures of over 390 psi [2.7 MPa] may be used to blow a hot plastics sheet through the silhouette of a female mold (Fig. 15-16). The air pressure causes the sheet to form a smooth bubble-shaped article. A stop may be used to form special contours in the bubble. Skylight panels and aircraft canopies are well-known examples of this technique. Unless a stop is used, there is no markoff. Only air touches each side of the material. There will be markoff from clamping.

Mechanical Forming

In mechanical forming, no vacuum or air pressure is used to form the part. It is similar to matched molding; however, close-fitting matched male and female molds

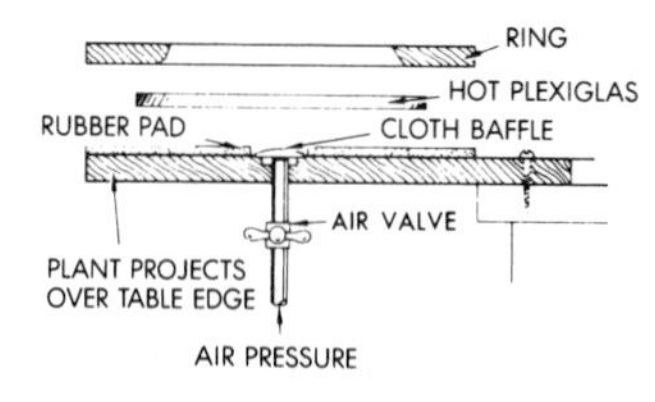

(A) Basic setup.

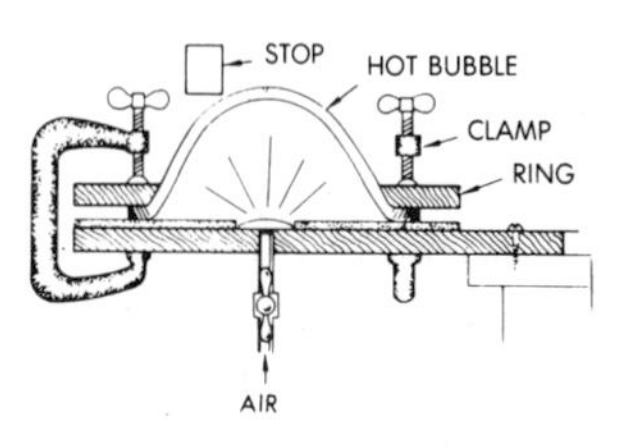

(B) Injection of air.

(C) Removing free-formed acrylic bubble-shaped product. (Rohm & Haas Co.)

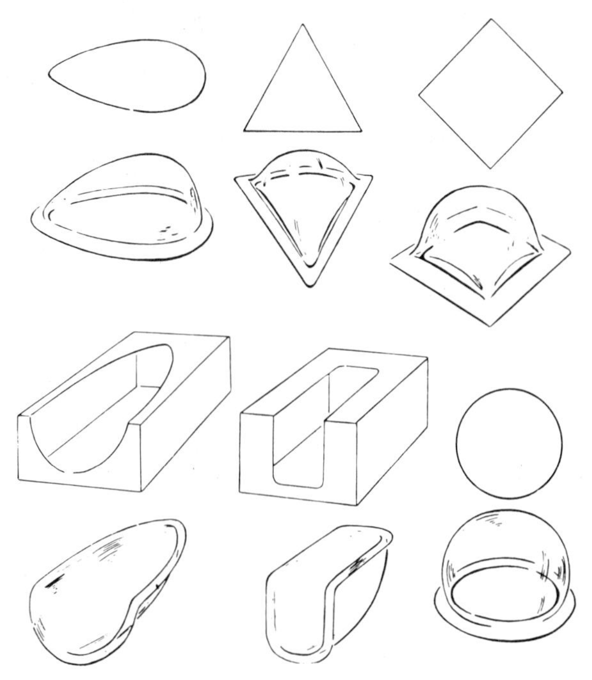

(D) Examples of free-form shapes that can be obtained with various openings. (Rohm & Haas Co.)

Fig. 15-16. Free forming of plastics bubbles.

are *not* used. Only the mechanical force of bending, stretching, or holding the hot sheet is used.

This process is sometimes classed as a fabrication or postforming operation. The forming process may make use of simple wooden forming jigs to give the desired shape using ovens, a strip heater, or heat guns for the heat source. Flat stock may be heated and wrapped around cylindrical shapes or stock may be heated in a narrow strip and bent at right angles. Tubes, rods, and other profile shapes may be mechanically formed (Fig. 15-17).

Plug-and-ring forming (Fig. 15-18) is sometimes classed as a separate forming process. No vacuum or air pressure is used, however, and it may be classed as a type of mechanical forming.

The process consists of a male mold shape and a similarly shaped female silhouette mold (not a matched mold). The hot plastics is forced through the *ring* (not necessarily a rough shape) of the female mold by the male. The cooling plastics take the shape of the male mold that it touches. Table 15-1 gives some common problems encountered in thermoforming plastics.

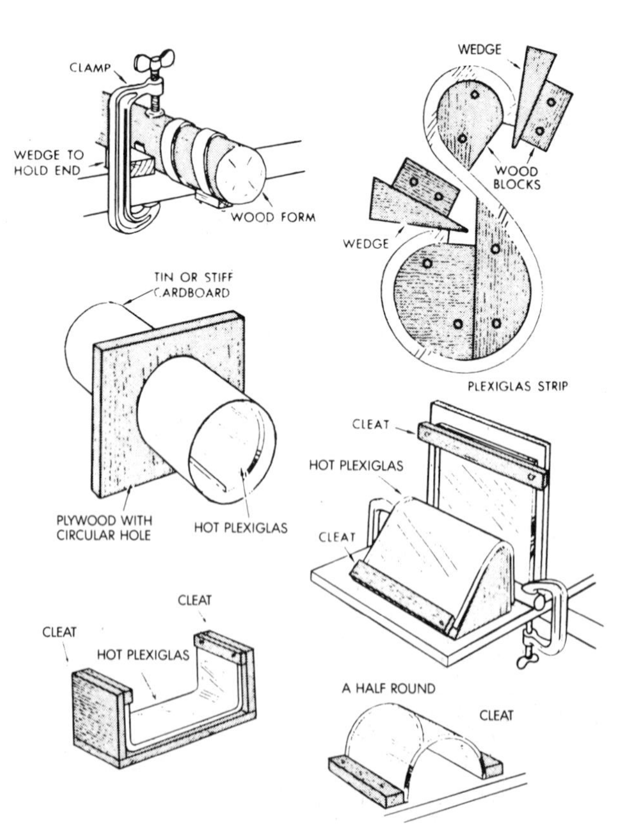

Fig. 15-17. Examples of mechanical forming. (Rohm & Haas Co.)

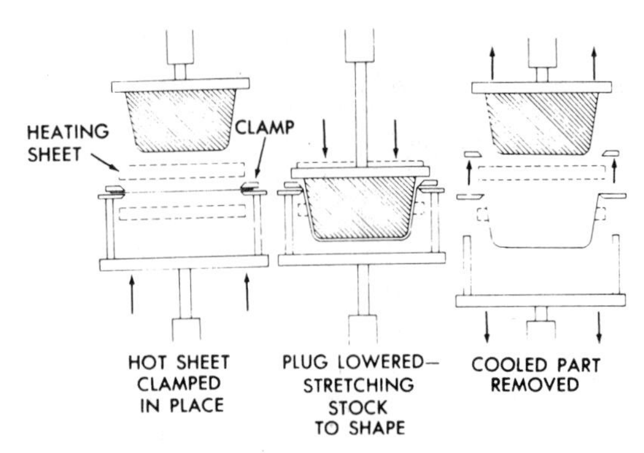

(A) Basic principle of plug-and-ring forming.

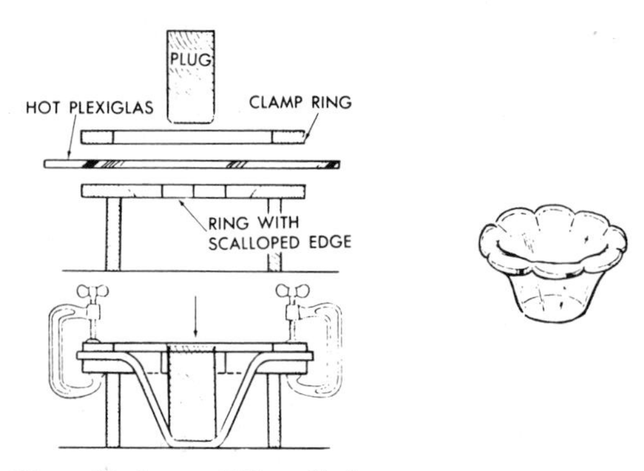

(B) Vase. (Rohm and Haas Co.)

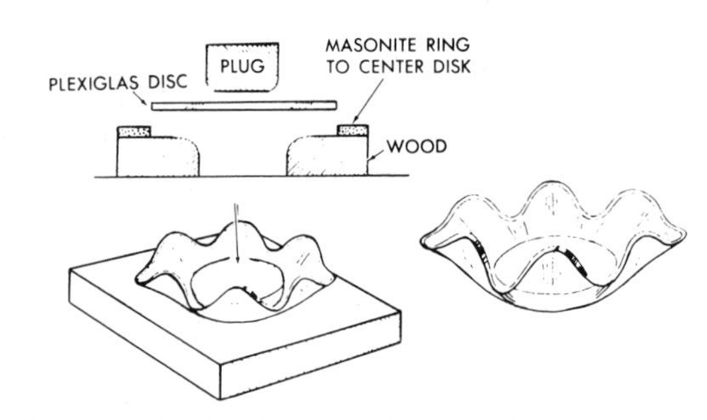

(C) Decorative bowl. (Rohm & Haas Co.)

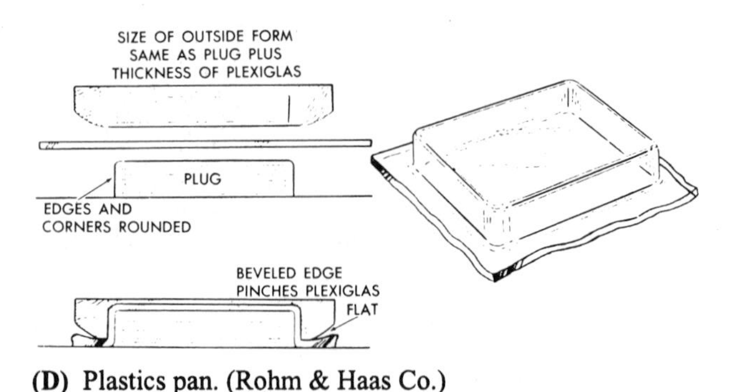

(D) Plastics pan. (Rohm & Haas Co.)

Fig. 15-18. Examples of plug-and-ring forming.

Table 15-1. Troubleshooting Thermoforming

Defect	Possible Remedy
Pinholes or ruptures	Vacuum holes too large, too much vacuum or uneven heating Attach baffles to the top clamping frame
Webbing or bridging	Sharp corners on deep draw, change design or mold layout Use mechanical drape or plug assists or add vacuum holes Check vacuum system and shorten heating cycle
Markoff	Slow draping action may trap air Clean mold or remove high surface gloss from mold Remove all tool marks or wood grain patterns from mold Mold may be chilling plastic sheet too quickly
Excessive post shrinkage	Rotate sheet in relation to mold Increase cooling time
Blisters or bubbles	Overheating sheet—lower heater temperature Ingredients of sheet formulation incorrect or hygroscopic
Sticking to mold	Smooth mold or increase taper and draft Use mechanical releasing tools, air pressure, or mold release Mold may be too warm or increase cooling cycle
Incompletely formed pieces	Lengthen heating cycle and increase vacuum Add vacuum holes
Distorted pieces	Poor mold design—check tapers and ribs Increase cooling cycle or cool molds Sheet removed too quickly while still hot
Change in color intensity	Use proper mold design and allow for thinning of piece Lengthen heating cycle and warm mold and assists Use heavier gauge sheet and add vacuum holes

Vocabulary

The following vocabulary words are found in this chapter. Look up the definition of any of these words you do not understand as they apply to plastics in the glossary, Appendix A.

Air-slip forming
Drape forming
Free forming
Matched-mold forming
Mechanical forming
Shrink wrapping
Snap-back forming
Thermoforming
Vacuum forming

Questions

15-1. Name three materials that vacuum-forming molds can be made from.

15-2. Identify the packaging that makes use of the product as the mold.

15-3. Identify the process that will produce products with the greatest detail.

15-4. Mechanical forming is the name incorrectly used for ___?___ forming.

15-5. The name of the unit or tool used to heat a small section of plastics so it may be bent at a sharp angle is ___?___.

15-6. The sides of a thermoforming mold are tapered to aid in removal of the part. This taper is called ___?___.

15-7. Name four typical thermoformed products.

15-8. Which thermoforming technique is used to form a part with a very deep draw?

15-9. A ___?___ or ___?___ is normally placed in all low or unconnected portions of the thermoforming mold.

15-10. The ___?___ of a female mold is the ratio of the maximum cavity depth to the minimum span across the top opening.

15-11. What is a major disadvantage of straight-vacuum forming using deep cavities?

15-12. In ___?___ forming no vacuum or air pressure is used to form the hot plastics sheet.

15-13. Name the term for marks left on the formed sheet if the mold is not smooth or clean.

15-14. If vacuum holes are too large, or if there is too much vacuum or uneven heating; ___?___ or ___?___ will occur.

15-15. Is a male or female mold used in free forming?

15-16. In plug-assisted methods, plug penetration would normally not exceed ___?___ percent of the mold cavity depth.

15-17. In ___?___ -and- ___?___ forming, a male and similarly shaped female silhouette mold shape the hot plastics.

15-18. Sharp corners on a deep draw may cause ___?___ or bridging.

15-19. Three advantages that metal thermoforming molds offer are ___?___.

15-20. What can we do with the scrap and trim that remain from thermoforming processes?

15-21. One of the most common tooling materials for thermoforming is ___?___.

15-22. Thermoforming is possible because thermoplastic sheets can be ___?___ and ___?___.

15-23. In vacuum forming, ___?___ forces the hot plastics sheet against the mold contours.

15-24. It is more desirable and efficient to have ___?___ rather than holes as passages for drawing the air from the mold.

15-25. Male molds are easy to make and generally cost less than ___?___ molds.

15-26. Accurate, close-tolerance parts with good detail may be made by ___?___ - ___?___ forming.

15-27. As a general rule, a smooth-surfaced mold should not be used when forming ___?___.

15-28. It is easier to control the thickness of deep thermoformed products by using the ___?___ or the pressure-bubble vacuum snap-back forming process.

15-29. Give two advantages of thermoforming.

15-30. Give two disadvantages of thermoforming.

15-31. Name a process where a heated thermoplastic sheet is pulled down into or over a mold surface.

15-32. Is a vacuum used in drape forming?

15-33. What is the major disadvantage of matched-mold thermoforming?

15-34. In vacuum forming, only ___?___ pressure is used.

15-35. The major difference between plug-assisted pressure forming and plug-assisted vacuum forming is that ___?___ pressures may be used with pressure forming.

15-36. Describe straight vacuum thermoforming.

15-37. Describe how product thickness is controlled in pressure-bubble plug-assist vacuum forming.

15-38. What determines product wall thickness in plug-assist pressure thermoforming?

15-39. What do the words *snap-back* refer to in vacuum snap-back forming?

15-40. Describe free forming.

15-41. Describe mechanical forming and plug-and-ring forming.

Activities

Free-Blow Thermoforming

Introduction. In free-blow thermoforming, a ring or yoke holds down a sheet of hot plastics. The ring must prevent air leakage, so that when air is forced below the sheet, it expands upward in a smooth bubble. Acrylic sheets are popular for free-blown products.

Equipment. A thermoforming machine, a ring or yoke, regulated air supply, personal safety equipment.

Procedure

15-1. Cut a sheet of plastics about 50 mm larger than the opening in the form. Figure 15-19 shows a laboratory thermoformer appropriate for this exercise. Figure 15-20 shows a typical ring for free blowing.

15-2. Set the air pressure to "0" on the regulator.

15-3. Mount the plastics sheet under ring and heat until formable. If forming acrylic, heat the sheet till it is between 150 °C [300 °F] and 190 °C [375 °F].

15-4. Accurately measuring the temperature of a thermoforming sheet with a pyrometer is difficult, because the sheet looses temperature very rapidly. If possible, acquire thermoforming temperature strips, as shown in Figure 15-21. The increments on these strips change color at pre-set temperatures. Figure 15-22 shows a piece of plastics that achieved a temperature of 143 °C. (290 °F)

15-5. When the thermoformer has been on long enough to stabilize in temperature, use the temperature strips to determine the oven time needed to achieve the desired temperature. Once the time is known, use this time and eliminate temperature measuring devices.

15-6. When the sheet is ready for forming, remove heat source and increase air pressure until the part blows to the desired height.

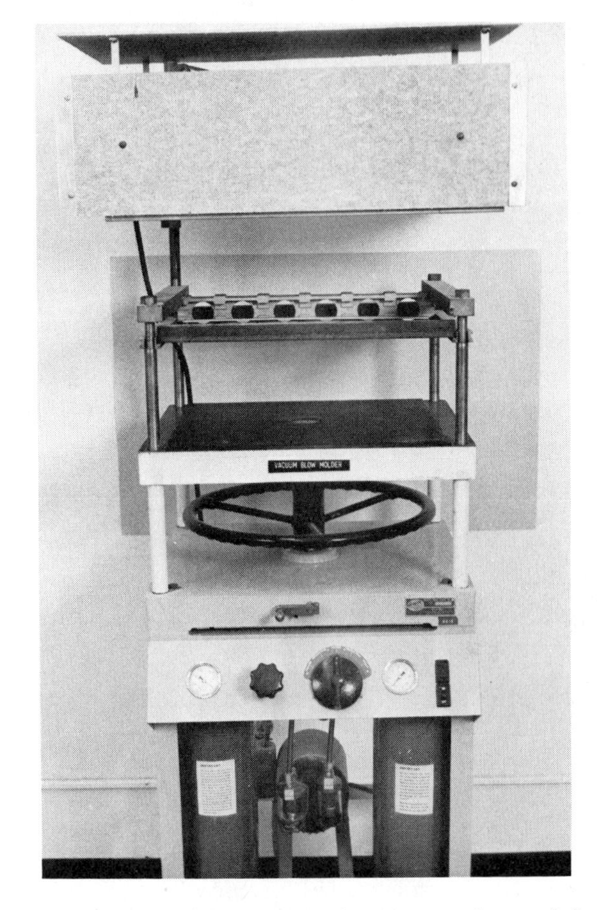

Fig. 15-19. This laboratory equipment may be used for free-blow thermoforming.

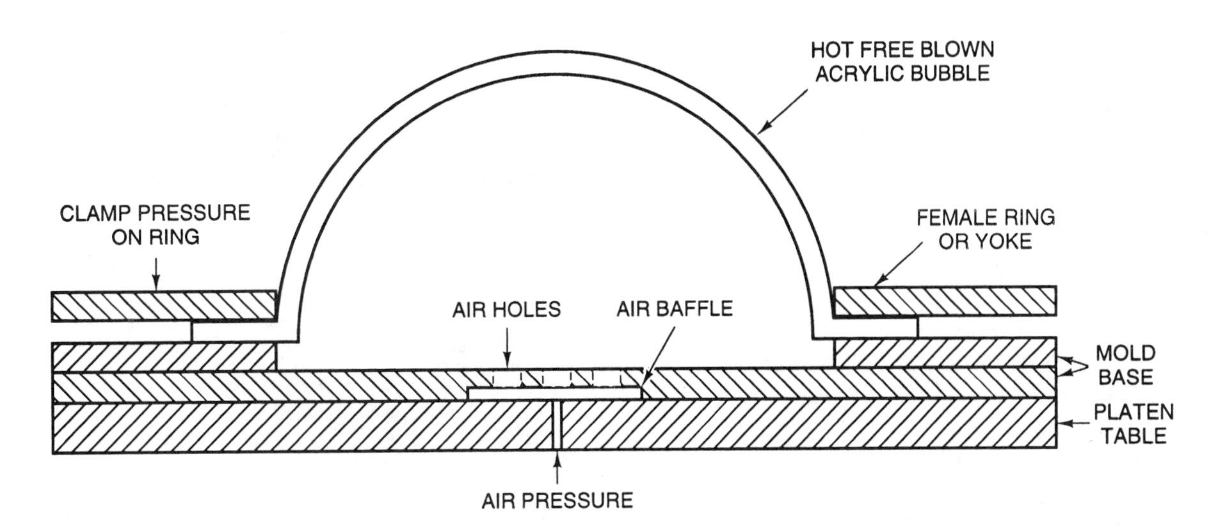

Fig. 15-20. Ring for free-blowing.

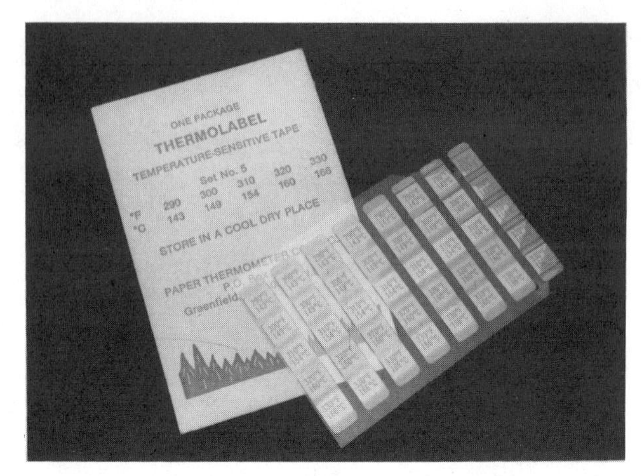

Fig. 15-21. Temperature-sensitive tapes come in varying temperature ranges. This sample covers the range 143–166 °C.

Fig. 15-22. This strip reached the temperature of 143 °C, but did not reach 149 °C.

15-7. Maintain pressure until part is cool, and remove from clamp.

15-8. Cut the part in half and measure the wall thickness. Is the wall uniform? How does the free blown wall compare to the wall in straight vacuum formed items?

Plastics Memory

Introduction. Thermoforming provides a good opportunity to observe plastics memory. The memory can lead to warp if a thermoformed item gets hot enough to permit stress relaxation.

Equipment. Thermoforming machine, thermoforming sheets, simple molds.

Procedure

15-1. If a laboratory thermoformer, similar to the one shown in Figure 15-19, is available, but an appropriate mold is not, refer to Figure 15-23 and 15-24 for an extremely simple mold. This mold is a piece of 2 by 6, with holes drilled, and self-adhesive form tape used to create both a seal and an opening to the vacuum holes. The aluminum disk permits thermoforming of shells for injection-molded Taber abrader disks. Other forms are equally useful in observing plastics memory.

15-2. After measuring the thickness of a sheet, heat it and form it around a mold, as shown in Figure 15-25.

15-3. Remove the sheet and let it cool. Then remount in thermoformer. Heat until the sheet pulls it flat, and then rapidly remove the heat source.

15-4. When the sheet is cool, measure it to see if any thinning remains. Figure 15-25 shows both a formed sheet and a sheet returned to flat by the action of memory.

15-5. Draw the same sheet further by using a thicker object. Figure 15-26 shows the use of a small C-clamp.

15-6. After removing the clamp, or other item, reheat to see if the sheet will flatten again. If no holes occurred during forming, the sheet should return completely.

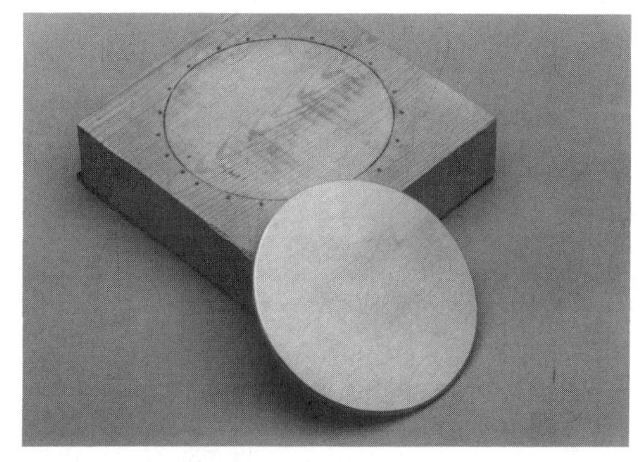

Fig. 15-23. Simple thermoforming mold for Taber abrader shells.

Fig. 15-24. Bottom of thermoforming mold, showing air holes and foam tape.

Fig. 15-25. Sheet thermoformed around aluminum disk, then flattened using plastics memory.

15-7. How many times will plastics memory work? How deep can the sheet draw and still retain its memory?

Straight Vacuum Forming and Plug-Assist Forming

Introduction. Localized thinning is a significant problem with straight vacuum forming. The purpose of this exercise is to compare straight vacuum forming to plug-assist forming.

Equipment. A thermoformer, a female mold, a plug for the selected mold, thermoforming sheets. The thermoformer used to demonstrate this activity is considerably larger than the one shown in Figure 15-19.

Procedure

15-1. If the thermoforming sheets are PS, no problems will arise with moisture; however, if the sheets are ABS, they may be wet and require drying. If not properly dried, these sheets may exhibit bubbles on the surface after forming.

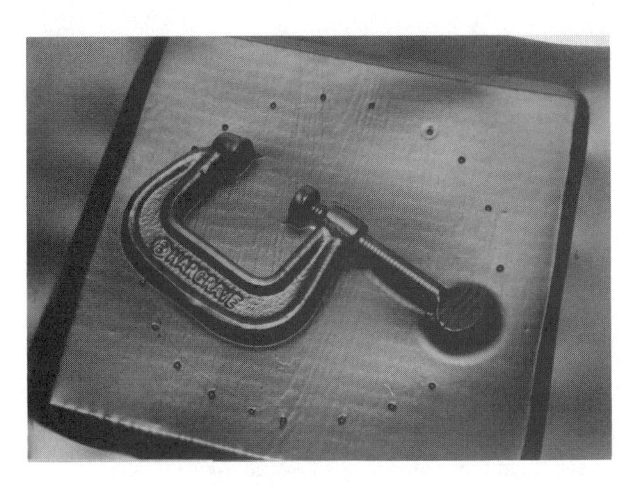

Fig. 15-26. Sheet formed around a small C-clamp.

15-2. Determine the time required to reach a chosen thermoforming temperature. If the sheets are PS, select a temperature between 130 °C and 180 °C. Grid the thermoforming sheets as shown in Figure 15-27. The grid shown is every 1 inch.

15-3. Locate the sheet in the clamp frame, making sure that the grid is parallel to the clamp. The mold used to demonstrate this exercise is seen in Figure 15-28.

15-4. Heat and form the sheet. Figure 15-29 shows straight vacuum forming. Notice how the sheet stretched. Section the formed sheet and measure the wall thickness at various places. If the elements have stretched in a regular manner, measure the stretched grid. Figure 15-30 shows a regularly stretched element. The original sheet was 0.0625 inches thick. The element stretched to 2.45 in.^{2} and, by calculation, should have a thickness of 0.0256 in. The measured thickness was 0.025 in.

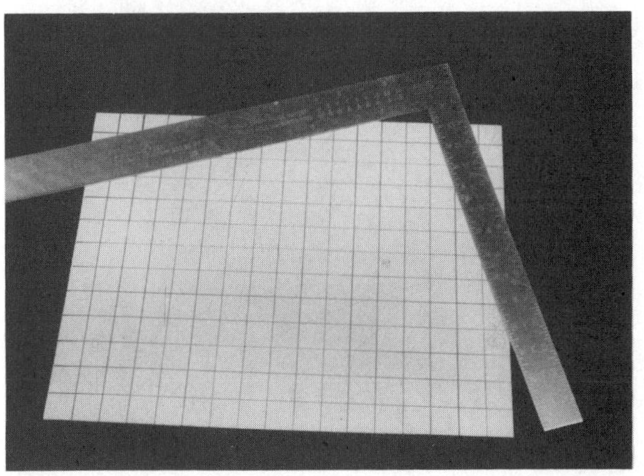

Fig. 15-27. Thermoforming sheet with 1 inch square grid pattern.

Fig. 15-28. Mold used to demonstrate thermoforming techniques.

15-5. If the elements are not uniform in stretch, as shown in Figure 15-31, it may be easier to measure the thickness than to calculate it.

15-6. Form a sheet using plug assist. The plug used for this demonstration is seen in Figure 15-28. The depth of plug penetration is critical. If the plug depth is too great, the area of the sheet in contact with the plug will stretch very little. That will force extensive stretch in neighboring elements, and may lead to greater localized thinning than straight vacuum forming. Figure 15-32 shows a draw with plug depth too great. If the plug depth is too shallow, there will be little change compared to straight vacuum forming. Figure 15-33 shows a draw in which the plug helped reduce localized thinning. Notice the change in the stretch patterns.

15-7. Adjust the plug to optimize the thickness at the corners of the part and calculate the maximum gain in wall thickness at the corners.

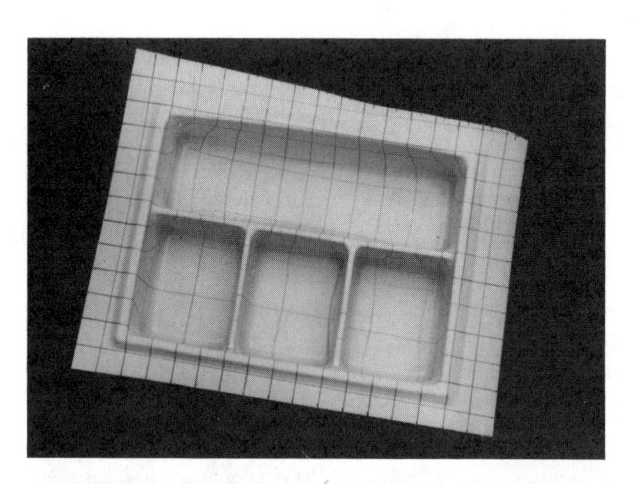

Fig. 15-29. Example of draw using straight thermoforming.

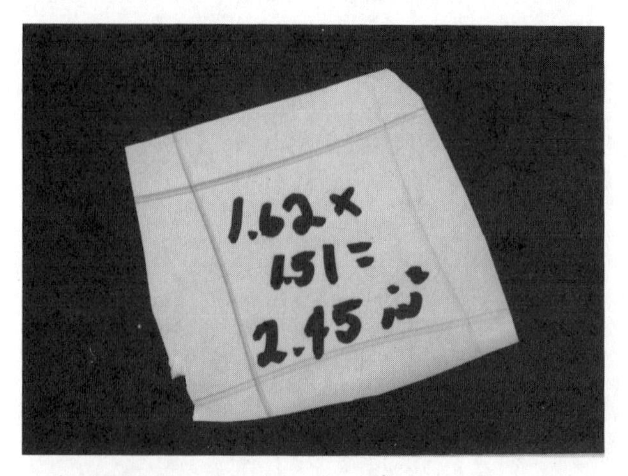

Fig. 15-30. This element was 1 in.^{2} before forming. Uniform biaxial stretch would result in a square. This element was almost uniform in stretch.

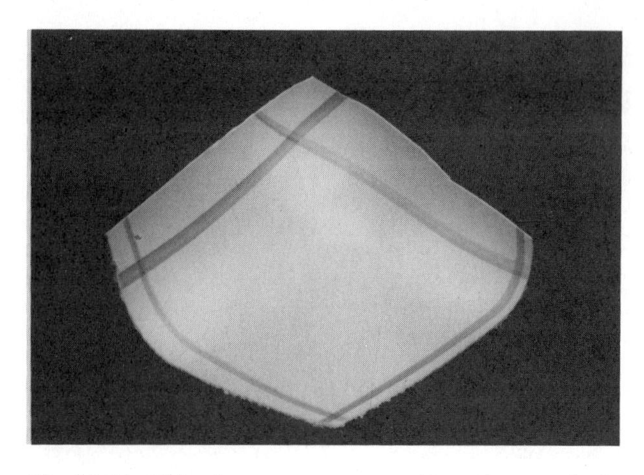

Fig. 15-31. This element was not uniform in stretch. Its thickness varies considerably.

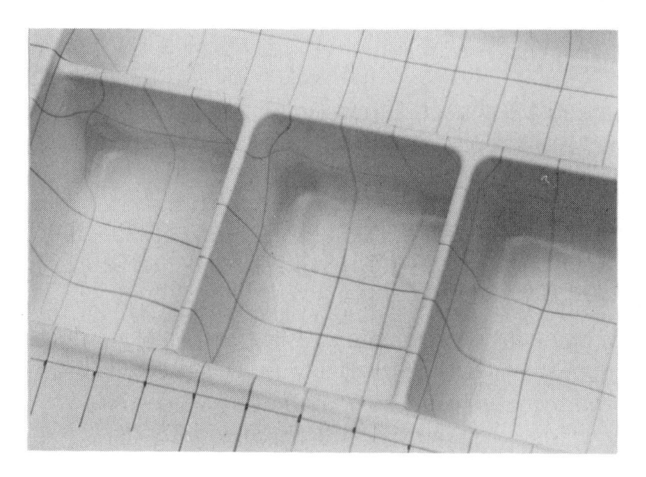

Fig. 15-32. Plug depth was too great and prevented the central elements from stretching, causing excessive thinning in neighboring elements.

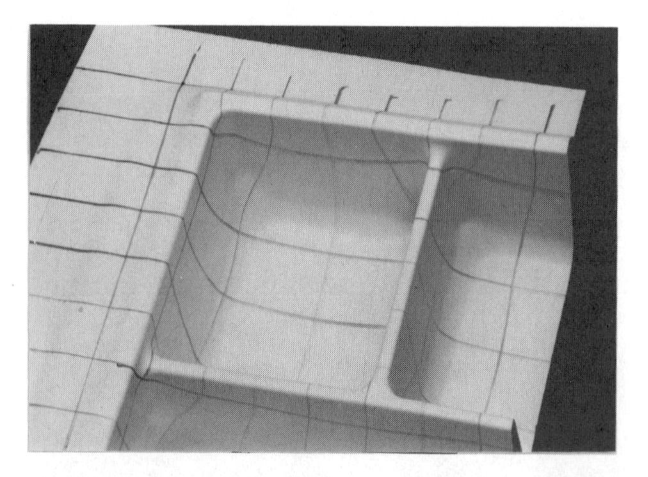

Fig. 15-33. This draw shown is an improvement compared to the straight thermoforming shown in Figure 15-29.

Chapter 16

Expansion Processes

Introduction

Methods of expanding plastics are described in this chapter. An expanded plastics is something like a sponge, bread, or whipped cream, since all are cellular in structure. Expanded plastics are sometimes called frothed, cellular, blown, foamed, or bubble plastics, and they may be classified by cell structure, density, type of plastics, or degree of flexibility including rigid, semirigid and flexible forms.

These low density cellular (from Latin *cellula*, meaning small cell or room) materials be classified as either closed-cell or open-cell. (See Figure 16-1.) If each cell is a discrete, separate cell, it is a closed-cell material. If the cells are interconnected, with openings between cells (sponge-like), the polymer is an open-celled material. These expanded (cellular) polymers may have densities ranging from that of the solid matrix to less than 0.56 lb/ft^3 [9 kg/m^3]. Nearly all thermoplastic and thermosetting plastics can be expanded. They may be made flame retardant. Table 16-1 lists selected properties of some expanded plastics.

Fig. 16-1. Examples of closed-cell PS (top) and open-cell PV (bottom) cellular plastics.

Resins are made into expanded plastics by six basic methods:

1. Thermal breakdown of a chemical blowing agent, freeing a gas in a plastics particle (a popular method).

Pentanes, hexanes, halocarbons, or mixtures of these materials are forced into the plastics particles under pressure. As the bead or granule of plastics is heated the polymer becomes soft, allowing the blowing agents to vaporize. This produces an expanded piece sometimes called the prepuff, preform, or pre-expanded bead. Cooling must be carefully controlled to prevent collapse of the cell or prepuff. Sudden cooling may create a partial, internal vacuum in the cell. This pre-expansion is facilitated by dry heat, radio-frequency radiation, steam, or boiling water. Pre-expanded materials must be used within a few days to prevent complete loss of all volatile expanding agents. They should be kept in a cool, airtight container until ready for molding. PS, SAN, PP, PVC, and PE are made cellular by this method.

Blowing agents vaporize quickly from polyethylene pellets; thus, their shelf life is very short. Therefore, most molders simply order pre-expanded *prepuffs*

Table 16-1. Selected Properties of Expanded Plastics

	Coefficient of Linear Expansion 10°/°C	Water Absorption, Vol %	Flammability, mm/min	Density Range kg/m³	Thermal Conductivity W/m* K	Max Service Temperature, °C	Compression Strength, kPa
Cellulose acetate	6.35	13–17	Slow burning	96–128	0.043	176	862–1 034
Epoxy							
Packed in place	38	1–2	Self-extinguishing	210–400	0.028–1.15	260	13–14 × 10^3
Foamed in place	102		Self-extinguishing	80–128	0.035	148	551.5–758
Phenolic							
Reactive type	5–10	15–50	Self-extinguishing	16–1280	0.036–6.48	121	172.3–419
Polyethylene	24.1	1.0	63.5	400–480	0.05–0.058	71	68.9–275.7
Polystyrene							
Extruded	11	0.1–0.5	Self-extinguishing	20–72	0.03–0.05	79	68.9–965
Expanded-beads	10.1	1.0	Self-extinguishing	16–160	0.03–0.039	85	68.9–1 375
Self-expanded-beads and others	10.1	0.01	Self-extinguishing	80–160	0.03	85	310.2–838
Polyvinylchloride							
Open cell			Self-extinguishing	48–169		50–107	
Closed cell			Self-extinguishing	64–400		50–107	
Silicone							
Premixed powder		2.1–3.2	Does not burn	192–256	0.043	343	689–2241
Liquid resin, rigid and semirigid		0.28	Self-extinguishing	56–72	0.04–0.43	343	55
Flexible			Self-extinguishing	112–144	0.045–0.052	315	
Urethane							
Rigid	1370		Self-extinguishing	32–640	0.016–0.024	148–176	172–210
Flexible	1650	10	Slow burning	22–320	0.032	107	

for final processing. Polyolefins are sometimes cross-linked by radiation to prevent cellular collapse before cooling. Prior to molding, the prepuffs are placed into holding tanks to diffuse air into them. Once pressurized, they are molded into a closed-cell product. Prepuffs are usually stabilized by thermal drying and annealing for a period of several hours. Pentane and butane are volatile organic compounds.

2. Dissolving in the resin a gas that expands at room temperature (a common method).

Nitrogen and other gases may be forced directly into the polymer melt. In injection or extruder equipment, special screw-shaft seals are used to prevent the escape of gas from the hot matrix. As the melt leaves the die or enters the mold cavity, the gas vaporizes and causes the polymer to expand. Once the matrix melt falls below the glass transition temperature, the expansion is stabilized.

3. Mixing in a liquid or solid component that vaporizes when heated in the melt (sometimes done to produce structural foams) (Fig. 16-2).

Granular, powder, or liquid blowing agents may be mixed and forced through the melt. They remain compressed in the mold until forced into the atmosphere. Rapid decompression (expansion) then occurs. PS, CA, PE, PP, ABS, and PVC are expanded by this method.

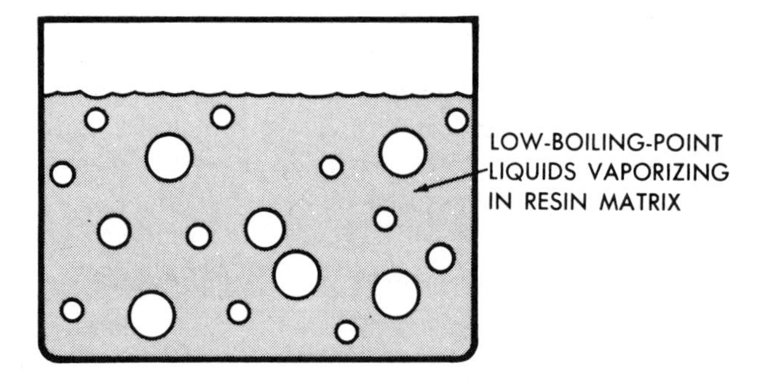

Fig. 16-2. Bubbles of gas are formed as chemicals change physical states.

4. Whipping air into the resin and then rapidly curing or cooling the resin (sometimes used in the production of carpet backing) (Fig. 16-3).

Vinyl esters, urea-formaldehyde (UF), phenolics, polyesters and some dispersion polymers are made cellular by mechanical methods.

5. Adding components that liberate gas within the resin by chemical reaction.

This is a favorite method for producing expanded materials from condensation polymers. Liquid resins, catalysts, and blowing agents are kept separate until molding. After mixing, a chemical reaction expands the matrix into a cured cellular material. As the matrix

Fig. 16-3. Mechanically frothed or foamed polyvinyl on thick flooring construction. Magnification 10×. (Firestone Plastics Co.)

expands, many cell walls rupture, forming a catacomb structure. Polyethers, polyurethane (PU), urea formaldehyde (UF), epoxy (EP), polyether polyurethane (EU), silicone (SI), isocyanurates, carbodiimide, and most elastomers use this expansion method. Polyurethanes and polystyrene account for more than 90 percent of all insulation in refrigerators, freezers, cryogenic tanks, and cellular construction insulation. Isocyanate is commonly expanded between layers of aluminum foil, felt, steel, wood, or gypsum facings. Flexible polyurethanes are open-celled, while rigid polyurethanes are closed-cell materials.

The EPA has expressed concern over the use of chlorofluorocarbons (CFCs), methylene chloride, pentane, and butane as blowing agents. To speed the reduction or elimination of CFCs, a special tax on their use began in 1994. Even pentane and butane may run afoul of clean air rules.

6. Volatizing moisture (steam left in resins by the heat generated by prior exothermic chemical reaction)

Water-blown systems may be used in some condensation polymers. In addition to the steam from the water condensate, carbon dioxide gas is liberated. Some systems combine water and halocarbon in the resin mixture.

Reinforced cellular materials may be produced with particulate and fibrous reinforcements dispersed in the polymer matrix. Fibers tend to orient themselves parallel to cell walls yielding improved rigidity. Reinforced epoxies have been mechanically mixed with pre-expanded beads (PS, PVC) and fully expanded and cured in a mold. Cellular materials are also used in the manufacture of sandwich composites. (See Sandwich and Continuous Reinforcing, Chapter 12.)

Syntactic plastics are sometimes included as a separate expanded-plastics group. Syntactic plastics are produced by blending microscopically small (0.03 mm) hollow balls of glass or plastics in a resin matrix binder (Fig. 16-4). This produces a light, closed-cell material. The putty-like mixture may be molded or applied by hand into spaces not easily reached by other means. The major uses of syntactic plastics include tooling, noise alleviation, thermal insulation and high-compression-strength flotation devices.

Cellular materials have been produced by placing glass spheres in the polymer matrix formulation before sintering. In a process called *leaching,* various salts or other polymers may be sintered together. A solvent solution dissolves the crystals or selected polymer to leave a porous matrix. Alternate layers of compatible polymers, reinforcements, and soluble crystal mixtures produce a true composite component.

Expanded plastics are used for insulation, packaging, cushioning, and flotation. Some act as acoustical, as well as thermal, insulation. Others are used as moisture barriers in construction. Expanded epoxy materials are used as light tooling fixtures and models. Expanded plastics may also be used as noncorroding, light, shock-absorbing materials for automobiles, aircraft, furniture, boats, and honeycomb structures. In the textile industry, expanded plastics are used as padding and insulation to give garments a special texture or feel.

During World War 11, the Dow Chemical Company introduced expanded polystyrene products in the United States and General Electric produced expanded phenolic products. The two main expanded plastics in use today are probably polystyrene and polyurethane. Polystyrene products are rigid, closed cellular structures (Fig. 16-5). Polyurethane products may be rigid or flexible. PU can also be either closed- or open-celled. Expanded products familiar to the consumer are ceiling tile, Christmas decorations, flotation materials, toys, package liners for fragile items, mattresses, pillows, carpet backing, sponges, and disposable containers (Fig. 16-6).

The following processes are explained in this chapter:

I. Molding
   A. Low-pressure processing
   B. High-pressure processing
   C. Other expanding processing
II. Casting
III. Expanding in place
IV. Spraying

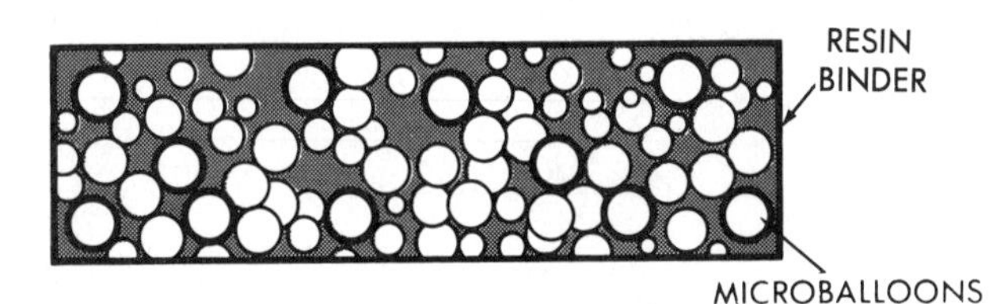

Fig. 16-4. Syntactic foam.

Fig. 16-5. Closed cellular polystyrene has excellent thermal insulating qualities. (Sinclair-Koppers Co.)

Molding

Various processes have been developed for molding expandable plastics, including injection molding, compression molding, extrusion molding, dri-electric molding (a high-frequency expanding method), steam chamber, or probe molding. Integral skinned cellular, or foamed, plastics are commonly cast or molded. A solid, dense skin of plastics is formed on the mold surface, which is heated to aid in forming this skin. During the skinning process, a cellular core is formed by forcing blowing agents or gas into the melt to cause the cellular structure (Fig. 16-7).

The term *structural foam* is used to include any cellular plastics with an integral skin. Its stiffness depends largely upon the skin thickness. There are many methods of making this integral skin; however, structural foam products are molded from the melt, by either a low pressure, or a high-pressure process (Fig. 16-8).

(A) Injection-molded polyethylene bowl.

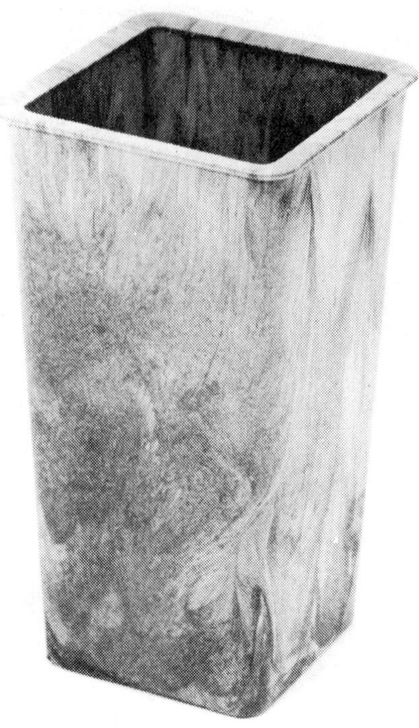

(B) Injection-molded wastebasket of polyethylene.

Fig. 16-6. Some molded products of expanded polyethylene. (Phillips Petroleum Co.)

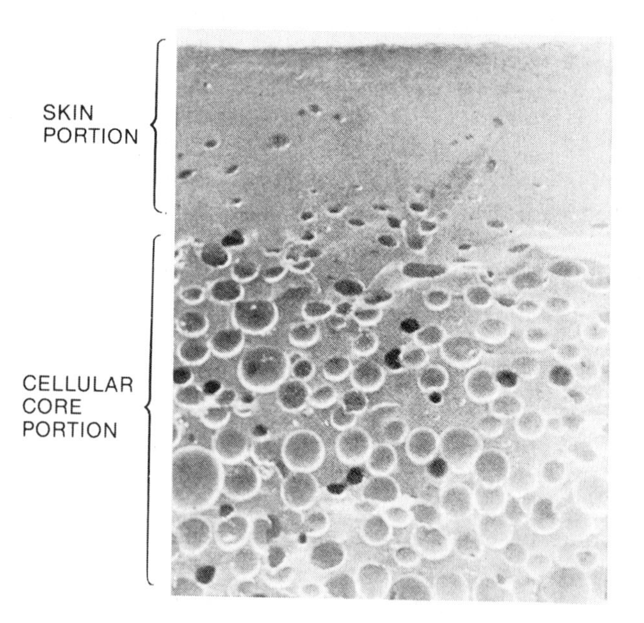

Fig. 16-7. In this photograph, the transition from solid skin to cellular core may be seen.

Low-Pressure Processing

Low-pressure processing is the simplest, most popular and most economical method for large parts. A mixture of molten plastics and gas is injected into molds, at a pressure of from 145 to 725 psi [1 to 5 MPa]. A skin is formed as the gas cells collapse against the sides of the

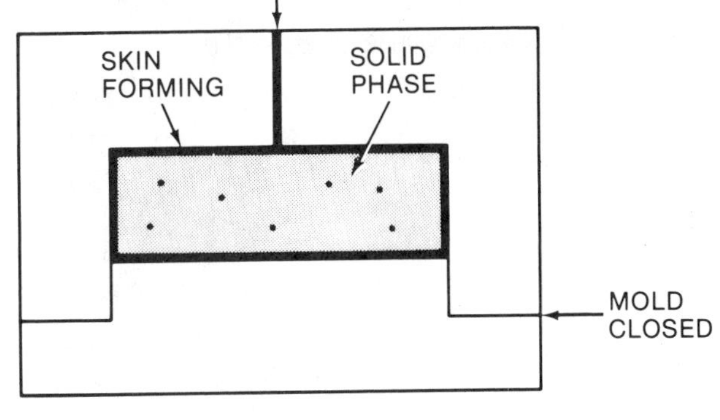

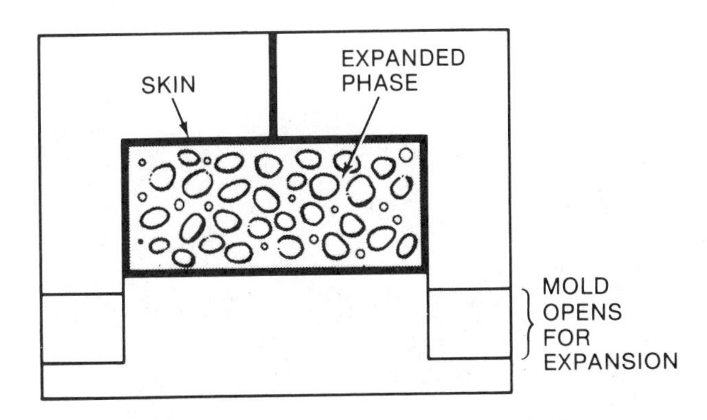

Fig. 16-8. Low- and high-pressure methods of expanding products with an integral skin.

mold. Skin thickness is controlled by the amount of melt forced into the mold, mold temperature, and pressure. The type of blowing agent or the amount of gas forced into the melt also helps determine the skin thickness and density of the part.

Low pressure techniques produce closed-cell parts that are nearly stress free. The collapse of the cells on the mold surface sometimes produces a surface swirl pattern on parts.

High-Pressure Processing

In high-pressure processing, the heated melt is forced into the mold with pressures of from 4351 to 20307 psi [30 to 140 MPa]. The mold is completely filled with the melt, thus allowing the melt to become solid against the mold or die surfaces. To allow for expansion, the mold cavity is increased by having the mold open slightly or by withdrawing cores. The melt expands within this increase in volume.

In one injection and extrusion method, the expanding agent is added directly to the hot mix. The pressure of the plunger or extruder does not allow the material to expand until it is forced into a mold (Fig. 16-9).

In other injection and extrusion methods, the blowing or expanding agent, colorants, and other additives are melted directly into the molten plasticjust

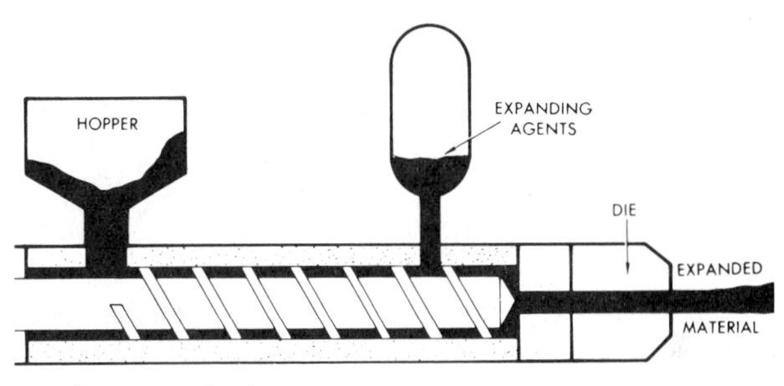

(A) Screw method.

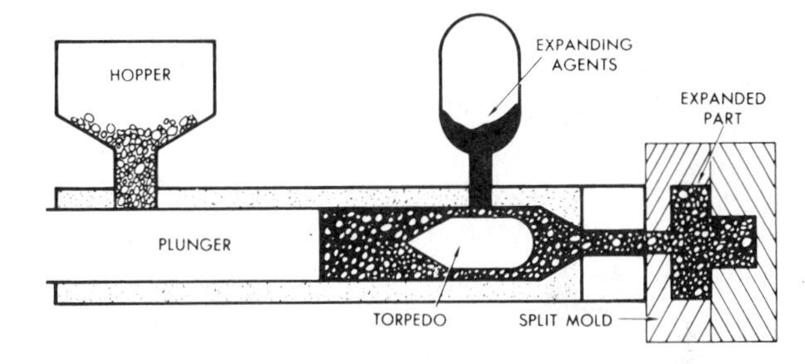

(B) Plunger method.

Fig. 16-9. Adding expanding agents directly to the hot mix.

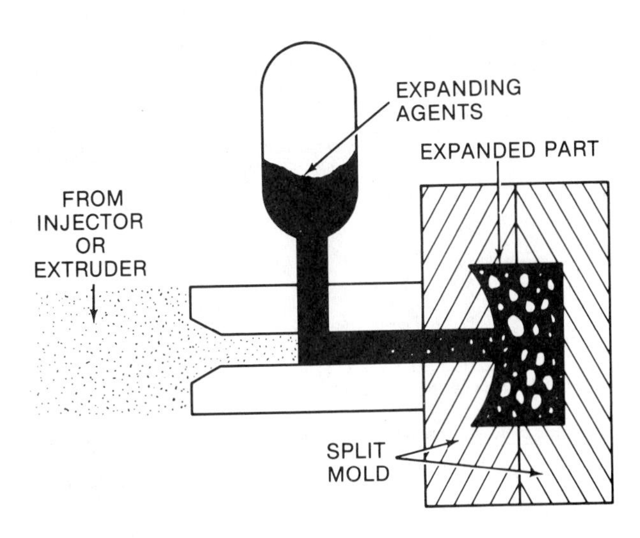

Fig. 16-10. A molding method in which the expanding agent is added to the molten plastics just before the plastics enters the mold.

before it enters the mold (Fig. 16-10). The expansion then takes place in the molding cavity.

In yet another process, the extruded material with expanding agents is fed into an accumulator (Fig. 16-11). When the predetermined charge is reached, a plunger forces the material into the mold cavity, where expansion takes place. (See Reaction Injected Molding in Chapter 10.)

Reaction injection molding (RIM), reinforced reaction injection molding (RRIM), and mat molding/RIM (MM/RIM) have grown rapidly after extensive use in the automotive industry and for numerous structural applications. Reinforcements greatly improve dimensional stability, impact strength, and modulus. In a modified resin transfer molding technique, long fibers or mats are placed in the mold and a mixture of reactive components are forced into the cavity. The resulting MM/RIM product is tough, light, and strong.

In Figure 16-12, an expanding automobile component is shown being formed by liquid reaction molding (LRM).

An expanded plastics with a skin (unexpanded layer) is produced by forcing the hot mixture around a fixed torpedo. The extruded shape is a hollow form the shape of the sizing die. The extrudate then expands, filling the hollow center, and the skin is formed by the cooling action of the sizing and cooling dies (Fig. 16-13). Structural profiles are produced by this method.

In another process, two plastics of either the same formulation or different families are injected one after the other into a mold. The first plastics does not contain expanding agents and is injected partly into the mold. The second plastics, containing the expanding agents, is then injected against the first plastics, forcing it against

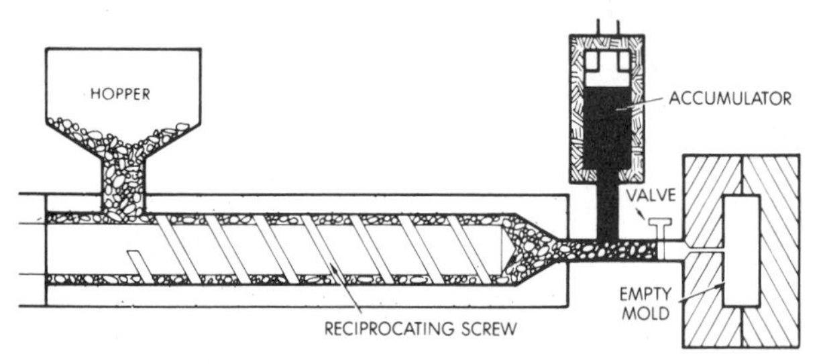

(A) Material enters accumulator.

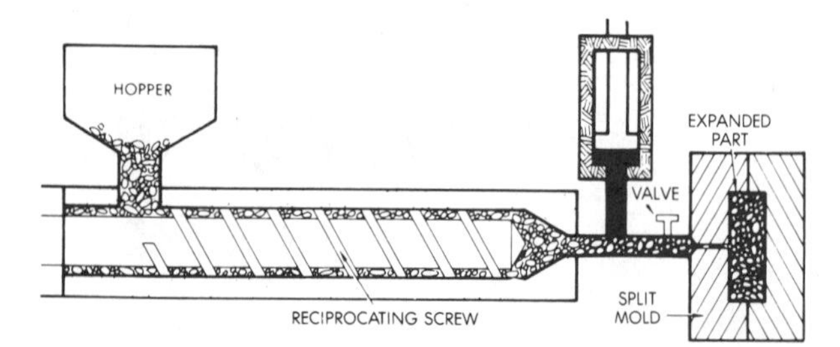

(B) Accumulator forces material into mold cavity.

Fig. 16-11. A molding method in which an accumulator is used.

(A) Machine and tooling for RIM process.

(B) Automobile part being removed from the mold.

Fig. 16-12. Reaction injection molding (RIM) also known as liquid reaction molding (LAM) is used to produce an automobile component with a skinned foam. (Cincinnati Milacron)

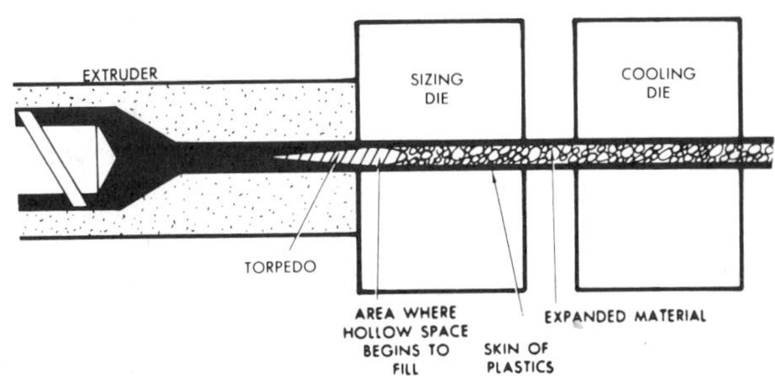

(A) A method of production.

(B) A vinyl foamed mat with a skin on both sides.

Fig. 16-13. Forming a skin on expanded plastics.

the edges of the mold and forming a shell around the expandable plastics. To close off the shell, more of the first resin is injected into the mold, fully encapsulating the second resin. The part made has an outer skin of one plastics and inner core of expanded plastics.

In a unique process called gas counter-pressure, gas is forced into an empty, sealed mold cavity. The melt is then forced into the cavity against the gas pressure. Expansion begins as the mold is vented.

Extruded polyvinyl materials may be expanded as they emerge from the die or may be stored for future expansion. They are used in the garment industry as single components or given cloth backing (Fig. 16-14).

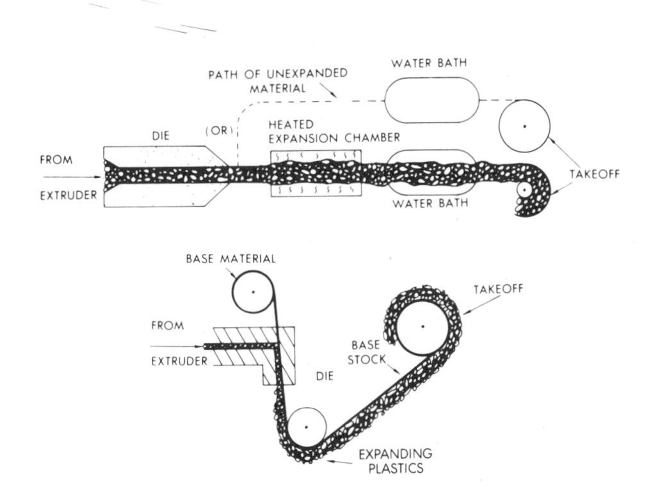

Fig. 16-14. Two methods of expanding a plastics as it emerges from the die.

Foamed materials are used as backing on car pets or other flooring, as shown in Figure 16-15.

In compression molding of expandable plastics, the resin formulation is extruded into the molding

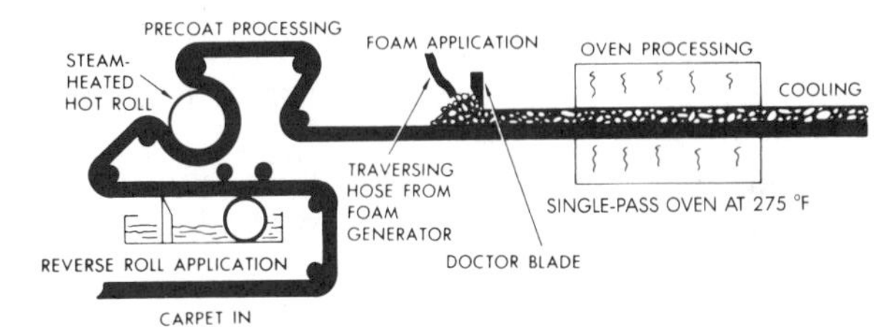

(A) Applying carpet backing. (Union Carbide Corp.)

(B) Tough artificial turf of polyamine will not pull loose and is widely used for athletic fields. (3M Co.)

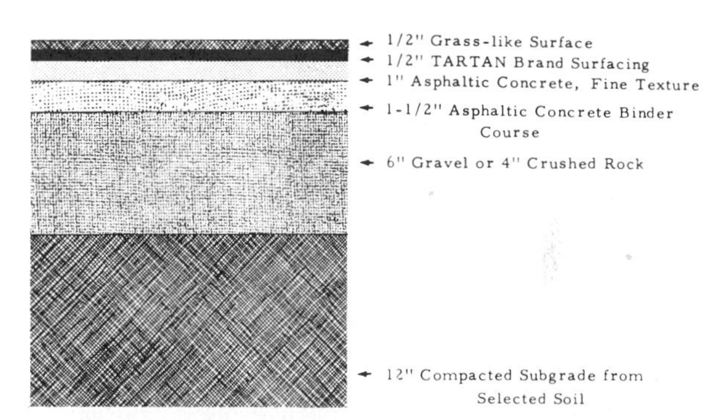

(C) Cross section of base used for artificial turf. (3M Co.)

(D) Workers spreading the cushioning layer that will be surfaced with the artificial grass material. (3M Co.)

Fig. 16-15. Foams are used in many flooring and athletic field applications.

chamber and the mold closed. The molten resin quickly expands, filling the mold cavity.

One of the largest markets is for extrude polystyrene logs, planks, and sheets. They are produced by extruding from a die the molten plastic containing the expanding agent. The expansion occurs rapidly at the die orifice. Rods, tubes, or other shapes may be produced in this manner.

Table 16-2 gives solutions to problems that may occur in expansion processes.

Other Expanding Processes

Not all expanded plastics make use of a hot melt method. Polystyrene is commonly produced in the form of small beads containing an expanding agent. These beads may be pre-expanded by heat or radiation and then placed in a mold cavity where they are heated again, often by steam, causing further expansion (Fig. 16-16). These beads expand up to forty times their original size (Fig. 16-17) and the expanding pressure packs the beads into a closed cellular structure. In Figure 16-16, a typical bead-expanding system is shown. Core box vents are used to allow the steam into the mold cavity, and these vents leave marks on the expanded product (Fig. 16-16C).

Table 16-2. Troubleshooting Expansion Processes

Defect	Possible Causes and Remedies
Mold not filled	Venting, short shot, trapped gas, increase pressure, increase amount of material, use fresh materials
Pits or holes in surface	Reduce amount of mold release—mold release interferring with blowing agent, increase mold temperature, not enough material in mold, melt temperature incorrect, concentration of blowing agent incorrect, polish die or mold surface
Distorted pieces	Poor mold design, increase mold time, increase cure cycle, allow thick sections to cool longer, increase mold strength
Sticking to mold	Increase mold release, select correct release, poor mold design, cool mold, polish mold surfaces
Parts too dense	Increase blowing agent or gas, use fresh pre-expanded beads, lower injection pressures, reduce melt
Part density varies	Mix compounds thoroughly, check screw design, increase mold temperature, increase melt temperature, increase dwell and cure time.

Thermal excitation of molecules by high-frequency radio energy is also used to expand beads. This is sometimes called dri-electric molding because it does not require steam lines, moisture and steam vents, or metallic molds. Inlays or decorative substrates of paper, fabric, or plastics may be molded in place by this method (Fig. 16-18). See Dielectric or High-Frequency Bonding in Chapter 18.

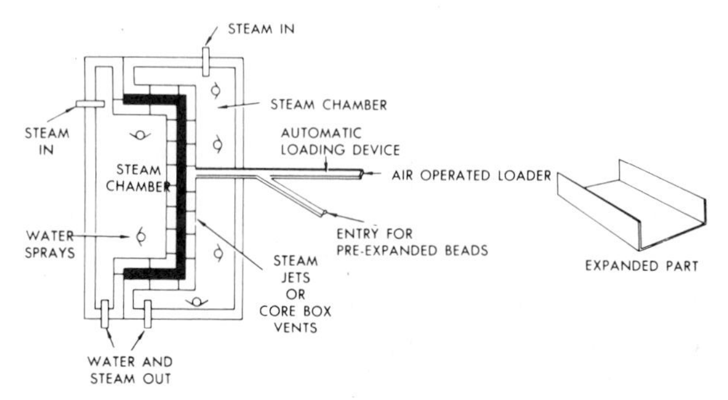

(A) Typical bead-expanding mold that uses steam.

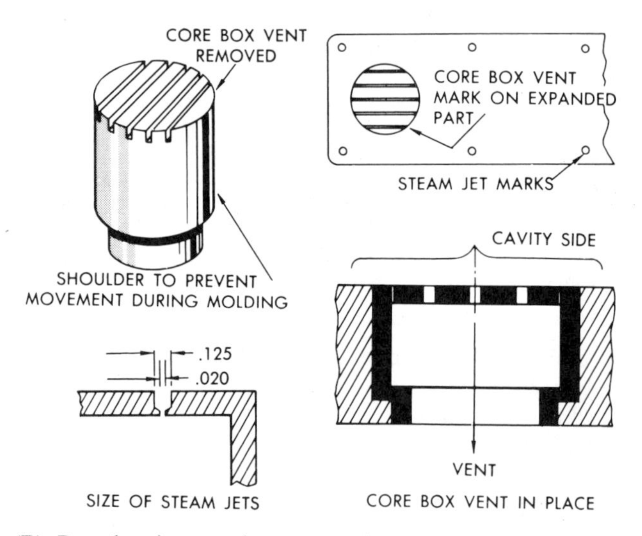

(B) Part showing core box vent and steam jet marks.

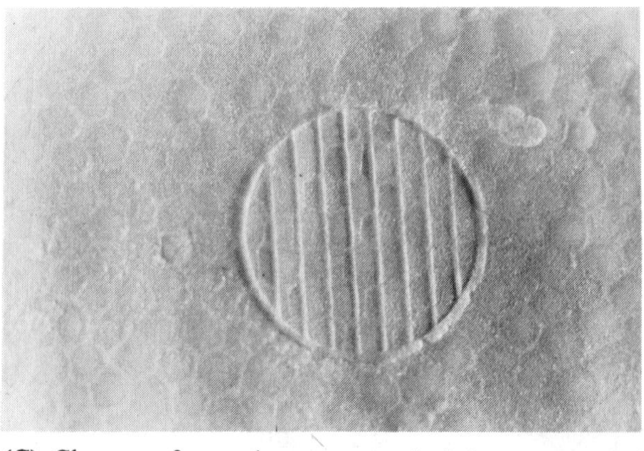

(C) Closeup of core box vent mark left on expanded product.

Fig. 16-16. Expanding plastics without using a hot-melt method.

(continued)

(D) Molding machine for producing expandable polysiyrene.

Fig. 16-16. Expanding plastics without using a hot-melt method. *(Continued)*

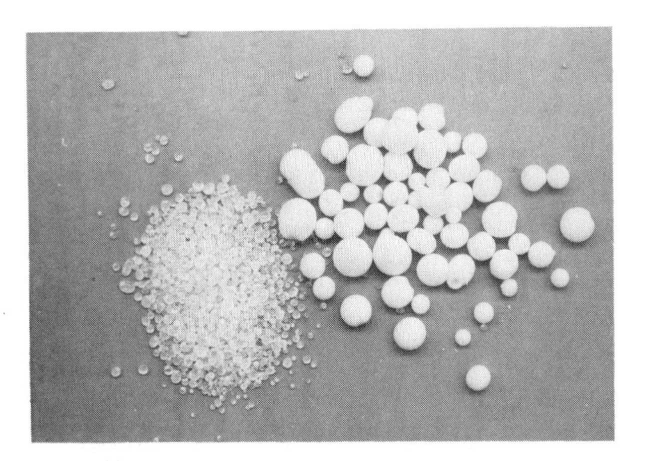

Fig. 16-17. Unexpanded (left) and expanded (right) beads of polystyrene.

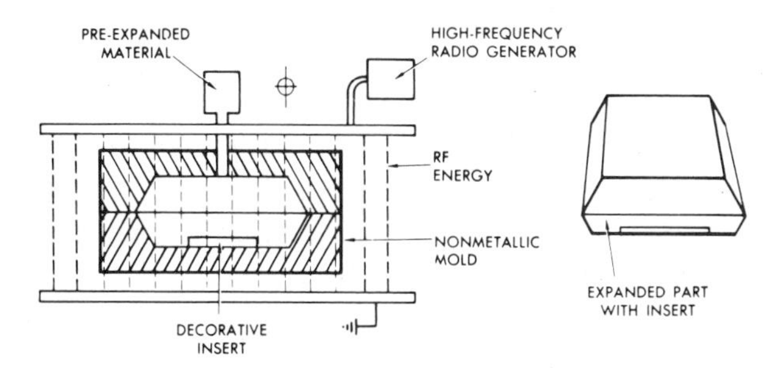

(A) Basic principle of expanding plastics using radio-frequency energy.

(B) Radiation may also be used to expand and cure plastics.

Fig. 16-18. Plastics may be expanded by radiation.

Insulated cups, ice chests, holiday decorations, novelty items, toys, and many flotation and thermal-insulating products are well known products of this method.

Casting

In casting expandable plastics materials, the resin mix containing catalysts and chemical expanding agents is placed in a mold where it expands into a cellular structure (Fig. 16-19). Polyurethanes, polyethers, urea-formaldehyde, polyvinyls, and phenolics are often cast-expanded. Flotation devices, sponges, mattresses, and safety cushioning materials are often cast. Large slabs or blocks of flexible polyurethanes are cast in open and closed molds. These slabs or blocks are cut into mattress stock or shredded for cushioning. Crash pads and pillow products may be cast in closed molds.

Slab stock is commonly produced by continuous production processes. Some are extruded, while many are simply cast or sprayed on a continuous belt. This stock is used as cores for sandwich or laminated composites.

Expanding-in-Place

Expanding-in-place is similar to casting except that the expanded plastics and the mold, together, become the finished product. Insulation in truck trailers, rail cars, and refrigerator doors; flotation material in boats; and coatings on fabrics are examples.

In this process, the resin, catalyst, expanding agents, and other ingredients are mixed and poured into the cavity (Fig. 16-20). The expansion takes place at room temperature, but the mixture may be heated for a greater expanding reaction. This method is said to be

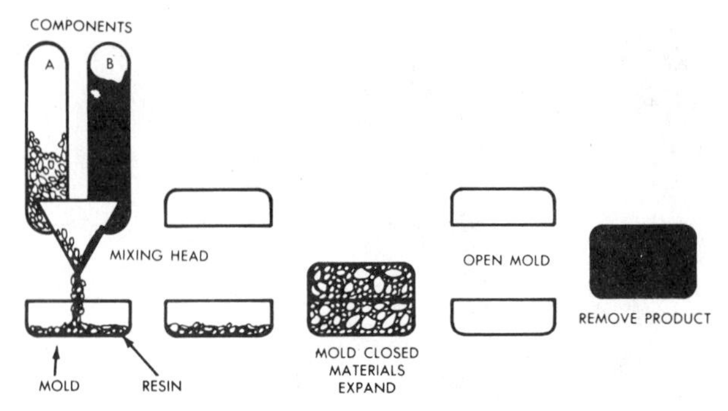

(A) Production scheme for casting plastics materials.

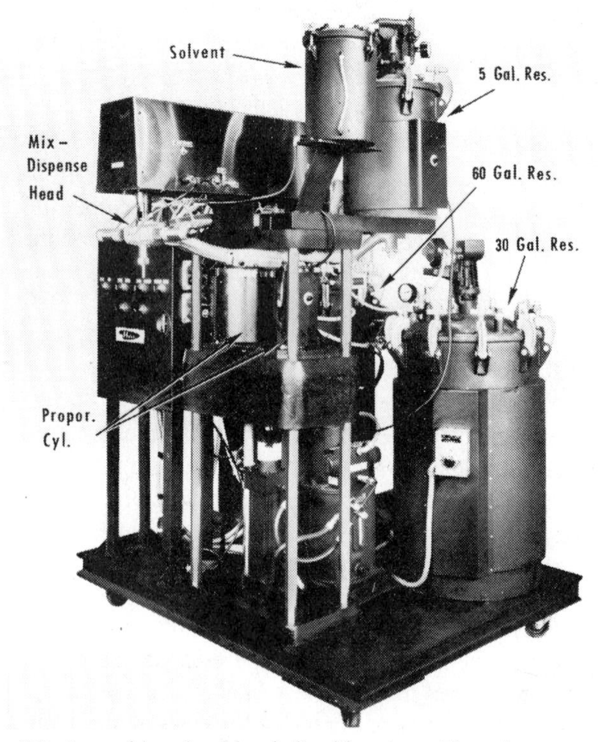

(B) A machine that blends liquid resins with curing agents to produce polyurethane foam. (Hull Corp.)

(C) Examples of shaped foam. (McNeil Akron Corp.)

Fig. 16-19. The production of cast-foamed plastics.

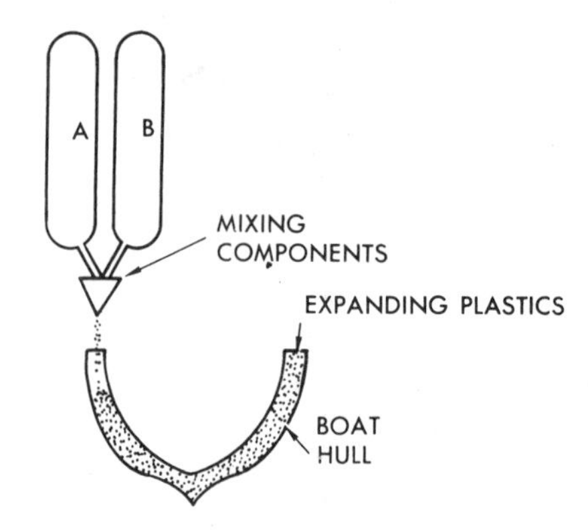

(A) Between inner and outer hulls of a boat.

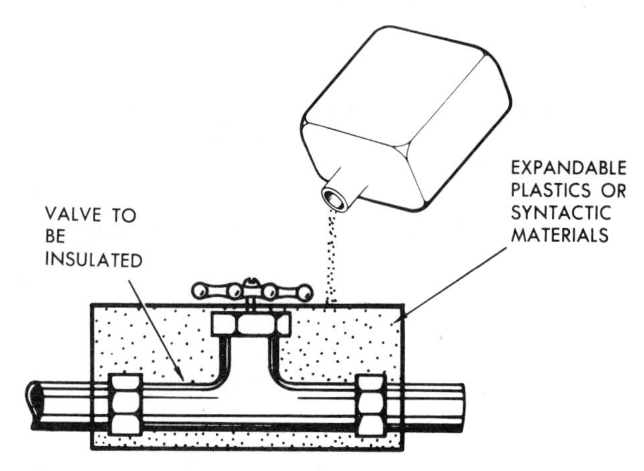

(B) Around a valve.

Fig. 16-20. Expanded-in-place forming.

done *in situ* (in place or position). Syntactic forms of plastics also may be placed in this category.

Spraying

A special spraying device is used to place expandable plastics on mold surfaces or on walls and roofs for insulation. Figure 16-21 shows two examples of this type of forming.

Five advantages and six disadvantages of expanded plastics are listed below.

Advantages of Expanded Plastics

1. Light, less costly products with low thermal conductivity
2. Wide range of formulations from rigid to flexible
3. Wide range of processing techniques
4. Less costly molds for low-pressure and casting methods, large products are possible
5. Parts may have high strength-to-mass ratios

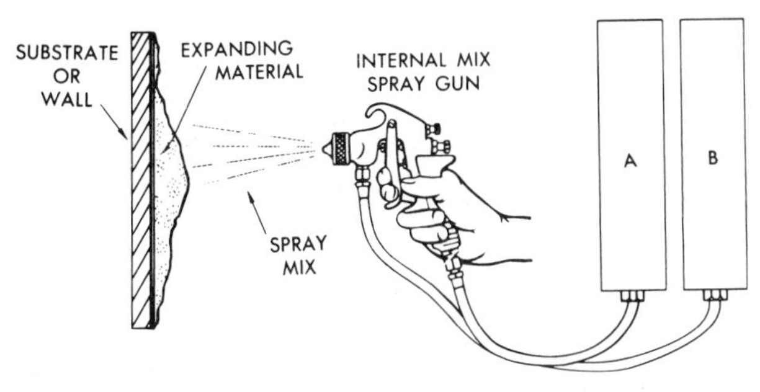

(A) Spraying on wall.

(B) Making a house shell.

Fig. 16-21. Some examples of spray forming.

Disadvantages of Expanded Plastics

1. Process slow, some need cure cycle
2. Special equipment needed for hot-melt methods
3. Tool and mold designs more costly for high-pressure method
4. Surface finish may be hard to control
5. Part size limited in high-pressure method
6. Some processes emit volatile gases or toxic fumes

Vocabulary

The following vocabulary words are found in this chapter. Look up the definition of any of these words you do not understand as they apply to plastics in the glossary, Appendix A.

Blow systems
Closed-cell
Expanded (foamed) plastics
Expand-in-place
Foaming agents
In-situ foaming
Leaching
Liquid reaction molding (LRM)
MM/RIM
Open-cell
Pre-expand
Prepuffs
Structural foam
Syntactic foam

Questions

16-1. Name three terms used to describe the expanding process.

16-2. The cellular structure of expanded plastics is ___?___ or ___?___.

16-3. Cellular plastics have ___?___ relative densities than solid plastics.

16-4. Plastics that are to be expanded in place are generally which of the two classes of plastics ___?___.

16-5. Name six methods of forming the cellular structure in the expanding processes.

16-6. Some expanded plastics may be thermoformed. They are commonly made of ___?___ (plastics).

16-7. List four broad product uses of expanded plastics.

16-8. What is used to expand polystyrene beads in the mold?

16-9. Cellular plastics with intregral skin are called ___?___.

16-10. What type of cell structure would a life vest have?

16-11. Supple expanded plastics are used chiefly for ___?___.

16-12. The two main expanded plastics in use today are probably ___?___ and ___?___.

16-13. Name four basic processes of foaming plastics.

16-14. What process would be selected to mold the front of an automobile?

16-15. Styroform is a ___?___ for polystyrene cellular plastics.

16-16. Which expanding process or processes would be used to produce each of the following products?

a. mattress

b. padded dashboard

c. egg carton

d. ice chest or cooler

e. insulation in home walls

16-17. As polystyrene beads are heated, the ___?___ agent causes the bead to swell.

16-18. Pre-expanded beads and unexpanded beads have a limited shelf life because they lose their ___?___ agent.

16-19. What does the term *in situ* mean?

16-20. As polystyrene beads expand and exert force against the walls of other beads, they form a ___?___ structure with no ruptured cells.

16-21. Name four items that may be produced by casting of expanded plastics.

16-22. Expanding in place is appropriate when the cellular material will not be ___?___.

16-23. In casting, the expanded plastics is always ___?___ from the mold.

16-24. Name the forming process in which the resin and expanding agent are atomized and forced out of a gun to strike a mold or substrate.

16-25. Where is a great amount of flexible polyurethane foam used?

16-26. It is the ___?___ that determines the physical properties of expanded plastics.

16-27. Name a major application for rigid polyurethane foams.

16-28. Many cellular profile shapes with a surface skin are produced by ___?___ methods.

16-29. Expanded polystyrene parts and molds must be ___?___ before removing the expanded part because latent heat in the center of the plastics part may continue to cause expansion.

16-30. Increasing foam density will ___?___ thermal conductivity but ___?___ nail-holding strength.

16-31. Small glass or plastics balls are sometimes used to make ___?___ plastics.

16-32. Name four methods of pre-expanding polystyrene beads.

16-33. In low pressure forming of structural framed products, pressures from 1 to ___?___ MPa are common.

16-34. If the product is not completely shaped or the mold not filled, ___?___ may be the cause.

16-35. On some expanded polystyrene products you can see marks left by the ___?___ used to allow steam into the mold cavity.

16-36. Two major disadvantages of high-pressure methods of expanding products are __?__ molds and __?__ part size.

16-37. Describe one way of producing an expanded plastics that has a skin.

16-38. Identify a use for syntatic foams and state why the foam would be advantageous for use in the application.

Activities

Blowing Agents

Introduction. One type of blowing agent consists of chemicals that decompose at processing temperatures and release gases that form tiny bubbles in a plastics. The amount of the blowing agent affects the size and number of cells created. A common agent of this type is based an azodicarbonamide. Such materials are often called azo-type blowing agents.

Equipment. Injection molder, HDPE, azo-type blowing agent, razor knife, baking powder, accurate scale, microscope.

Procedure.

16-1. A blowing agent needs to activate at the same temperature range required for processing the desired plastics. Acquire an azo-type blowing agent appropriate for HDPE. Figure 16-22 shows portions of three air shots of HDPE.

The top sample contains azo-type blowing agent, the lower two contain baking powder. The difference between the two samples blown with baking powder was the time in residence in the molding machine. The center sample is discolored compared to the azo material, but not as discolored as the lower sample, which darkened due to longer residence time.

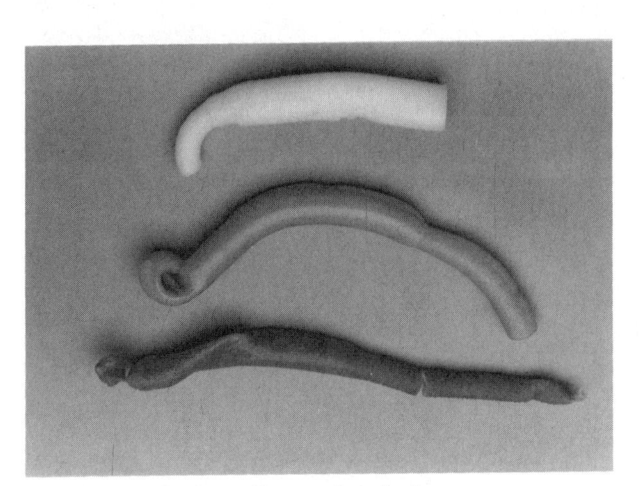

Fig. 16-22. Three small expanded air shots.

16-2. Mix the blowing agent and the basic plastics at a loading recommended by the manufacturer. If using baking powder in HDPE, use 3 percent as a starting point.

16-3. Make small air shots. Make sure that adequate ventilation is available to remove the fumes released from the hot plastics. Air shots will provide unhampered opportunity for the blowing agents to expand. Figure 16-23 is a section about 13 mm [0.5 in.] in diameter, cut from the top sample in Figure 16-22.

Don't confuse vacuum voids with cells caused by blowing agents. Figure 16-24 is the cross-section of a HDPE sprue. The hole was

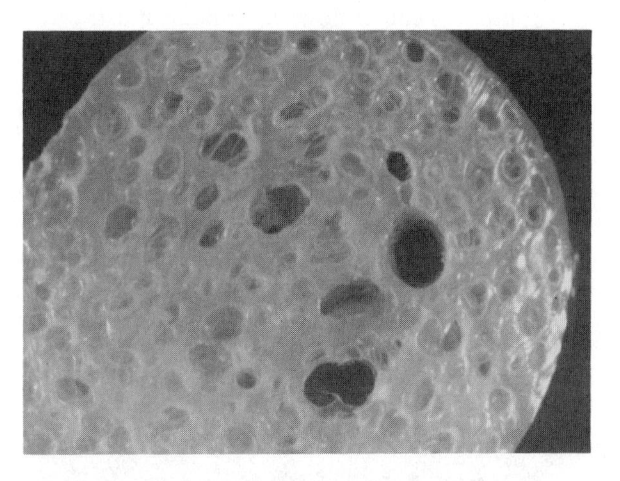

Fig. 16-23. Cross section of air shot showing cell structure.

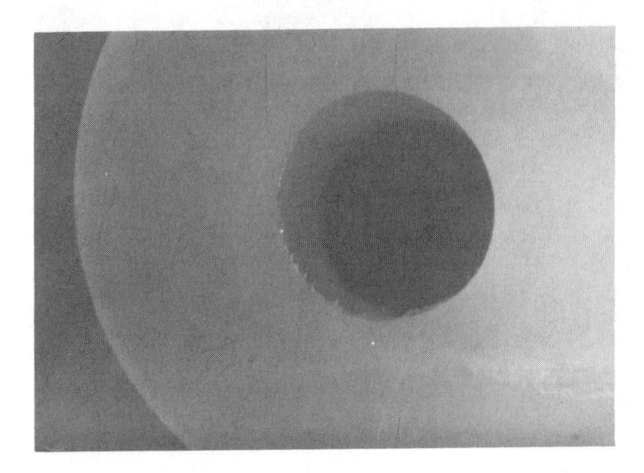

Fig. 16-24. A vacuum void in a sprue.

not an air bubble, but developed during cooling. When the forces caused by shrinkage overcome the strength of the outer surface of the part, the result is a sink mark. When the outer surface is stronger than the contraction forces, the result is often a vacuum void.

16-4. Measure the specific gravity of a portion of an air shot. Altering the percentage of blowing agent will affect the specific gravity. Visually inspect a slice of the air shot under a microscope. Natural HDPE is sufficiently translucent to permit good inspection in sections easily cut with a sharp razor knife.

16-5. Injection molding with conventional equipment does not yield uniform density products. Figure 16-25 shows parts made when the mold opened immediately after complete injection. The hottest regions, those in the center of the parts and near the gates, expanded more than neighboring regions.

16-6. In short shot moldings, the material at the end of flow has the opportunity to expand, because it is not pressurized, and has one open side in the mold. Figures 16-26A, 16-26B, and 16-26C show cross sections of a short shot, 16-26A near the end of flow, 16-26B about half way to the gate, and 16-26C near the gate. The differences in cell size demonstrate the decay in pressure as the flow gets further from the gate.

16-7. Inject full parts under various injection pressures. Measure the specific gravity of sections cut from the same region of the part. What range of specific gravities can be controlled with injection pressure?

16-8. Write a report summarizing your findings.

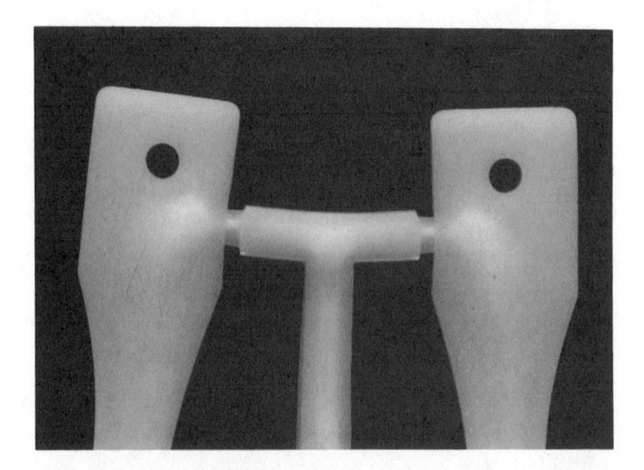

Fig. 16-25. Rapidly demolded part shows greatest expansion in hottest or thickest areas.

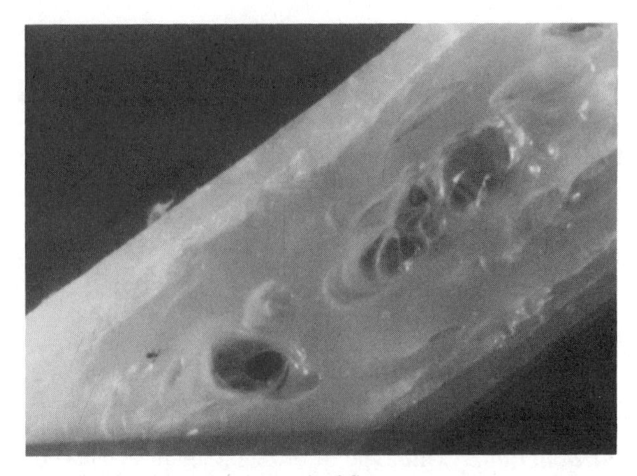

(A) Expansion near the end of flow.

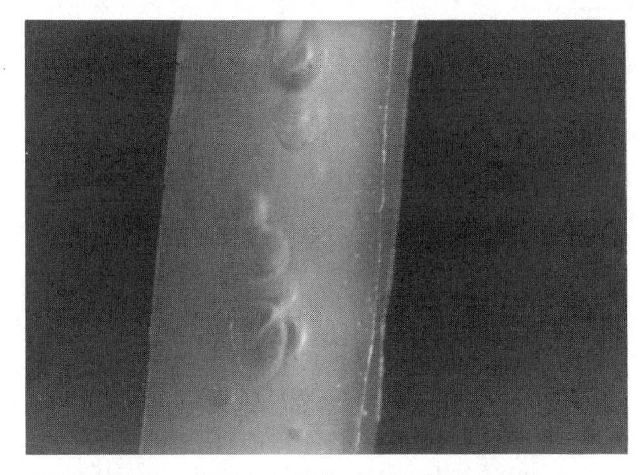

(B) Expansion near the middle of the flow length.

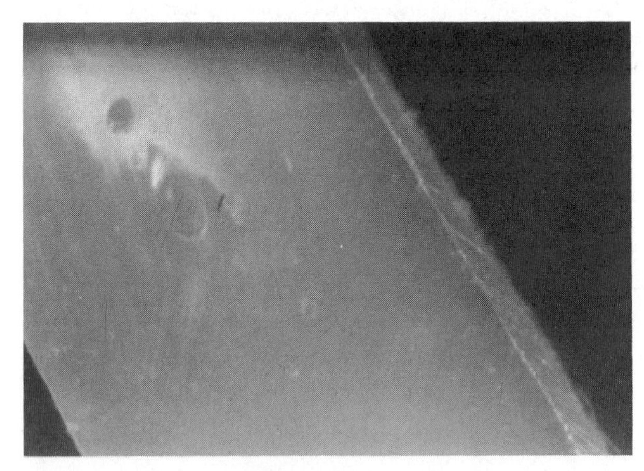

(C) Expansion near the gate.

Fig. 16-26. Expansion effected by location.

Microballoons

Introduction. Tiny glass spheres have been used for years to reduce weight in SMC and BMC parts. The glass spheres are so small that about 30 spheres can fit in the space occupied by a common grain of salt. However, attempts to use microballoons in injection-molded products were largely unsuccessful, because the forces needed in compounding and molding crushed the spheres. Recently, new high-strength microspheres have made applications in injection-molded products. One extrusion manufacturer claims survival rates of close to 90 percent.

Procedure

16-1. Acquire microballoons. Industrial grade microballoons are significantly different from the microballoons available in hobby shops. They are generally low-strength spheres, and will crush easily. If possible, acquire Scotchlite S60 glass bubbles. This product has a true density of 0.60 g/cc. Its bulk density is about 0.35 g/cc. The bulk density is lower, because it includes the voids between the particles in the calculation.

16-2. Examine the balloons under a microscope. Pay particular attention to the frequency of broken balloons.

16-3. Weigh out several batches of balloons and plastics, with incremental loadings, such as 1 percent, 2 percent, and 4 percent. Determine the specific gravity of the plastics, either experimentally or from manufacturers' material specifications.

16-4. Compound the microballoons into the plastics with an extruder.

16-5. Calculate the expected specific gravity of the extrudate, based on the loading of microballoons. Measure the specific gravity of the extrudate. Burn off the base material and examine the residual material under a microscope to see if the extrusion process broke any balloons. If it did break balloons, estimate the percent of breakage. Does the specific gravity data correspond to the breakage estimates?

16-6. Injection mold parts using the basic plastics and the compounded material. Measure the specific gravity of the parts containing microballoons. Burn off the plastics and examine the ash. Did the injection process cause breakage of the balloons?

16-7. Complete physical tests on parts with and without microballoons. Compare the weight reduction to the strength reduction.

16-8. Write up a report summarizing your findings.

Expanded Polystyrene Beads

Introduction. Cellular polystyrene plastics made from beads appear in many products. Small polystyrene beads containing a volatile gas expand when thermal energy (95 °C or 204 °F) vaporizes the gas. Under heat and pressure, the beads partially fuse together.

Equipment. PS beads, hot plate or stove-top burner, water container, mold, pigments, pot, pressure cooker, sieve, mold release.

Procedure

16-1. Boil about 2 L of water. (See Figure 16-27.)

16-2. Pour 50 ml fresh beads into boiling water. Stir until all beads rise to surface and expand to desired density. Too little pre-expansion will produce a dense, hard product. If beads are pre-expanded too much, they will not fill the mold properly during the expanding operation.

16-3. Using a sieve, remove pre-expanded beads from water to stop pre-expansion, as shown in Figure 16-28.

CAUTION: Water and beads are hot. It is best to use pre-expanded beads within 24 hours.

16-4. Make certain mold is clean. Apply a light coat of mold release.

Fig. 16-27. Make certain all materials are carefully prepared for expanding polystyrene beads.

Fig. 16-28. Pre-expanded beads being removed from water.

16-5. Fill mold half-full of pre-expanded beads. Powder pigments may be added and carefully mixed with beads at this time. Beads should be piled high. Remove beads from lips of mold. Add teaspoon of unexpanded beads for a harder, more dense product. Figure 16-29 shows the mold filling step.

16-6. Carefully assemble mold and shake mold to distribute beads evenly.

16-7. Place mold in pressure cooker containing at least 1 liter of water.

16-8. Place lid on cooker making certain lid is sealed and safety valve is functioning.

16-9. Allow pressure to increase to about 100 kPa [14.6 psi]. Hold pressure for 5 minutes. Do not allow pressure to exceed 140 kPa [20 psi]. (See Figure 16-30.)

16-10. Remove cooker from burner and slowly allow pressure to decease (about 5 minutes).

CAUTION: Beware of hot escaping steam.

16-11. After all steam has escaped, open cooker, remove mold, and place in cooling water.

CAUTION: Wear protective gloves and goggles. See Figures 16-31 and 16-32.

16-12. Open mold and remove expanded part. Compressed air may be used to aid removal of part. Trim flash. If beads were not fully expanded or fused together, increase pressure and/or cycle time. If product shrinks or has melted areas, reduce pressure, heating cycle, and/or pre-expansion (Fig. 16-33).

Optional Boiling Water Method

16-1. Prepare mold and pre-expanded beads.

16-2. Fill mold with pre-expanded beads and pack firmly.

Fig. 16-29. Fill mold half with pre-expanded beads.

Fig. 16-30. Place filled mold in cooker, put lid on cooker, and bring pressure up to 100 kPa. DO NOT EXCEED 140 kPa.

Fig. 16-31. Remove hot mold from cooker.

16-3. Place mold in boiling water for 40 to 45 minutes. Time will vary with the size of the mold, amount of pre-expansion and the age of the bead.

16-4. Cool in water.

16-5. Remove part.

Optional Dry Oven Method

16-1. Prepare mold and pre-expanded beads.

Fig. 16-32. Quench hot mold in water.

Fig. 16-33. Carefully remove expanded product and trim.

16-2. Fill 10 percent of the volume of the mold with water.

16-3. Fill mold with a mixture of pre-expanded and unexpanded beads. Use wet beads.

16-4. Place mold in 175 °C [347 °F] oven for about 10 minutes. Time will vary with the size of the mold, amount of pre-expansion and age of beads.

16-5. Cool in water.

16-6. Remove part.

Chapter 17

Coating Processes

Introduction

There are plastics coatings on cars, houses, machinery, and even fingernails. In this chapter, you will learn how coatings are applied. Many coatings are applied to a substrate to enhance the properties of the product by protecting, insulating, lubricating, or adding durable beauty. Coatings may have a combination of properties no other material can match such as flexibility, texture, color, and transparency.

For a process to be classed as *coating,* the plastics material must remain on the substrate. Dip casting and film casting products are not coatings, since the plastics is removed from the substrate or mold. Coating may be confused with other processes because similar equipment is used and variations in processing add to the confusion.

It is important to select coating materials that are reasonably close to the thermal expansion of the substrate to be coated. This becomes especially important if reinforcements are used. Some reinforced coating processes are probably better classified as modifications of laminating and reinforcing techniques. Reinforcements can help stabilize the coating matrix. Adhesion is one of the most critical factors in any coating operation. Some substrates must be properly prepared before coating. (See Corona discharge, Plasma, and Flame treatments in Chapter 19)

In extrusion processing, wire coating is a good example showing how more than one material may be put through an extrusion die. In extrusion or calendering of films, hot films are often placed on other substrates coating them. The use of liquid dispersions or solvent solutions is a casting method if the film is removed; it is a coating process if the film remains on the substrate.

There are nine broad (sometimes overlapping) techniques by which plastics are placed on substrates. These are listed in the following chapter outline:

I. Extrusion coating
II. Calender coating
III. Powder coating
  A. Fluidized-bed coating
  B. Electrostatic-bed coating
  C. Electrostatic powder gun coating
IV. Transfer coating
V. Knife or roller coating
VI. Dip coating
VII. Spray coating
VIII. Metal coating
  A. Adhesives
  B. Electroplating
  C. Vacuum metalizing
  D. Sputter coating
IX. Brush coating

Extrusion Coating

Single or coextrusion of hot melt may be placed on or around the substrate. Extrusion film coating is a technique in which a hot film of plastics is placed on a substrate and allowed to cool. For best adhesion, the hot film should strike the preheated and dried substrate before it reaches the nip of the pressure roller (Fig. 17-1). The chill roller is water-cooled to speed cooling of the hot film. It is usually chrome-plated for durability and high-gloss transfer, and it may be embossed to produce special textures on the film surface. The thickness of the film is controlled by the die orifice and by the surface speed of the chill roller. Because the substrate is moving faster than the hot extrudate as it comes out of the extruder die, the extrudate is drawn out to the desired thickness just before it reaches the nip of the pressure and chill rollers.

Several coating techniques result in a thin laminated composite. Because the primary objective of the coating is to provide protection, some materials are considered more effective for coating than others.

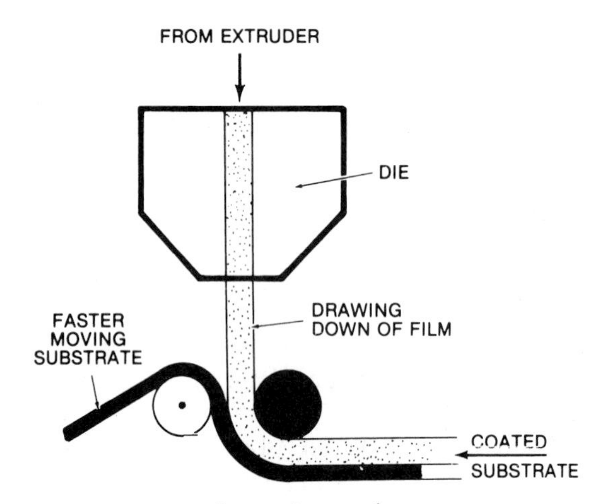

(A) Basic concept of extrusion coating.

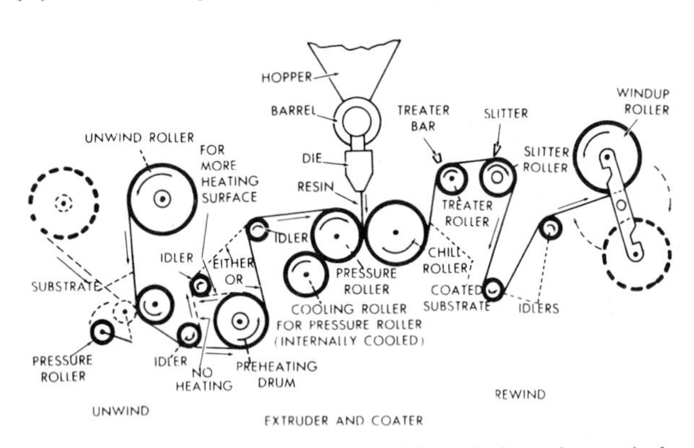

(B) Extrusion coating setup, with wind and unwind equipment. (USI)

Fig. 17-1. Extrusion film coating.

Polyolefins, EVA, PET, PVC, PA, and other polymers are commonly used in extrusion coating of various substrates to provide moisture, gas, and liquid barriers and heat-sealable surfaces. The polyethylene coating on paperboard milk cartons is a familiar example of liquid barrier and heat-sealable coatings.

Expanded plastics are also extruded onto various substrates. The substrate may also be drawn through the extrusion die, as in the coating of wire, cable, rods, and some textiles (Fig. 17-2). Five advantages and two disadvantages of extrusion coating are shown below.

Advantages of Extrusion Coating

1. Multilayer plastics may be placed on substrate
2. No solvents are needed
3. Thickness applied to substrate may be varied
4. Uniform coating thickness on wire and cable
5. Cellular coatings may be placed on substrate

Disadvantages of Extrusion Coating

1. Extrudates are hot melts
2. Equipment is expensive

Calender Coating

Calendered films may be used as a coating on many substrates, in a method similar to extrusion coating.

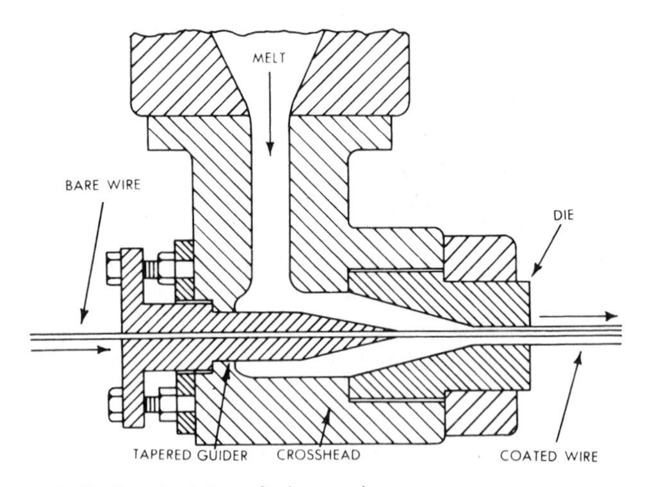

(A) Basic principles of wire coating.

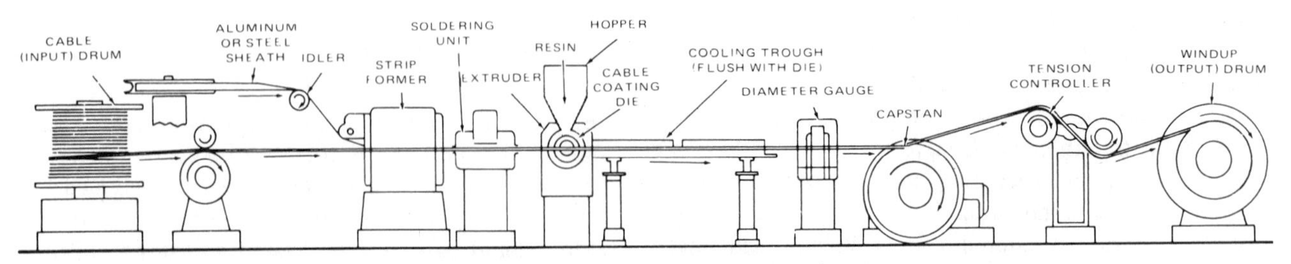

(B) Layout of cable-coating extrusion plant.

Fig. 17-2. Process and products of extrusion wire coating.

(C) An extruder coats cable with polyethylene plastics. (Western Electric Co.)

Fig. 17-2. Process and products of extrusion wire coating *(Continued)*

The hot film is squeezed onto the substrate by the pressure of the heated gauging rollers (Fig. 17-3)

Melt roll coating is a modification of calendering. In this process, the preheated substrate is pressed into the hot melt by a rubber-covered roller, or an embossing roll may be used. The coated material is cooled and placed on windup rolls (Fig. 17-4).

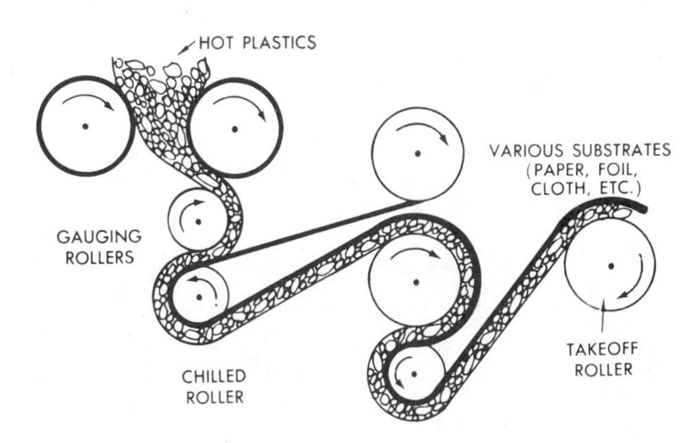

Fig. 17-3. Calender coating.

Pressure-sensitive and heat-reactive hot melts that are commonly used as adhesives may be coated on a substrate. Paper, plastics, and textiles are coated by this process (Fig. 17-5).

Coating on a paper substrate may add beauty, strength, scuff resistance, moisture, and soil resistance or provide a sealing system for making a package.

Five advantages and two disadvantages of calender coating are listed below.

Advantages of Calender Coating

1. High-speed continuous process
2. Precise thickness control
3. Pressure-sensitive and heat-reactive hot melts may be used
4. Coatings are stress-free
5. Short runs are relatively economical

Disadvantages of Calender Coating

1. Equipment cost is high
2. Additional equipment needed for flat stock

Powder Coating

Although ten techniques are known for applying plastics powder coatings, fluidized-bed, electrostatic-bed, and electrostatic powder gun techniques are the three major processes used today. The process of coating a substrate with a dry plastics powder is sometimes called *dry painting.*

PE, EP, PA, CAB (Cellulose acetate-butyrate), PP, PU, ACS (acrylonitrale-chlorinated polythylene-styrene), PVC, DAP (diallyl isophtholate resin), AN (acrylonitrile), and PMMA are made into powder (solventless) formulations for various powder coating techniques. After coating, some techniques require additional heating to assure complete fusion or cure.

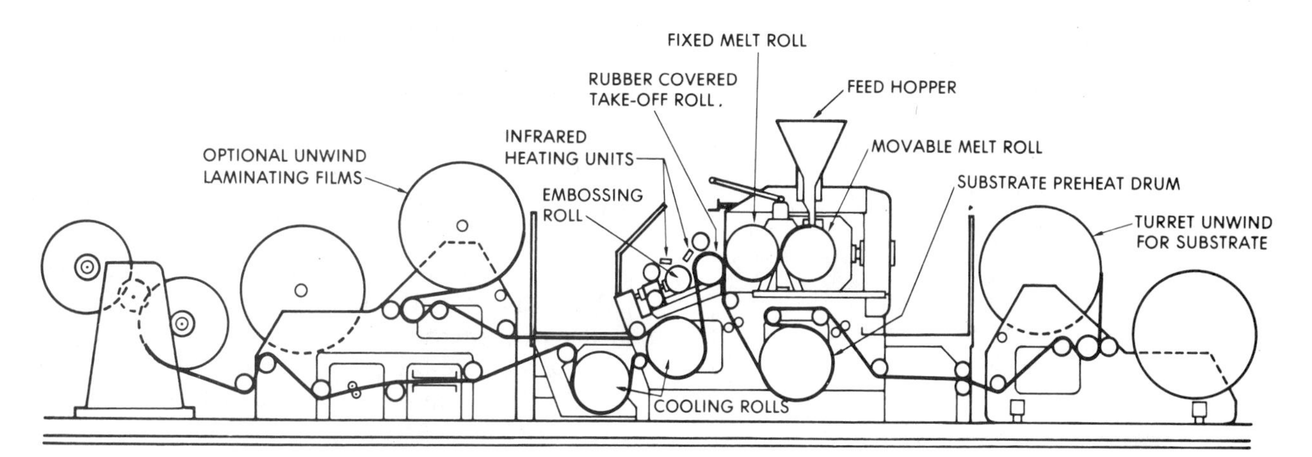

Fig. 17-4. Two-roll melt coater. (Zimmer Plastics, GMbH)

Fig. 17-5. Coating industrial fabrics with Zimmer coater. (Zimmer Plastics, GMbH)

Fluidized-Bed Coating

In fluidized-bed coating, a heated part is suspended in a tank of finely powdered plastics, usually a thermoplastic (Fig. 17-6A). The bottom of the tank has a porous base membrane to allow air (or inert gas) to atomize the powdered plastics into a cloud-like dust storm. Perhaps "fog cloud" would be more descriptive because the air velocity is carefully controlled. This air-solid phase looks and acts like a boiling liquid—hence the term fluidized bed.

When the powder hits the hot part, it melts and clings to the part surface. The part is then removed from the coating tank and placed in a heated oven where the heat fuses, or cures the powder coating. Part size is limited by the size of the fluidized tank. Epoxy, polyesters, polyethylene, polyamides, polyvinyls, cellulosics, fluoroplastics, polyurethanes, and acrylics are used in powder coating.

The fluidized-bed process originated in Germany in 1953 and has since grown into a useful plastics process in the United States.

In a variation of this process the fluidized powder is sprayed onto preheated parts in a separate chamber.

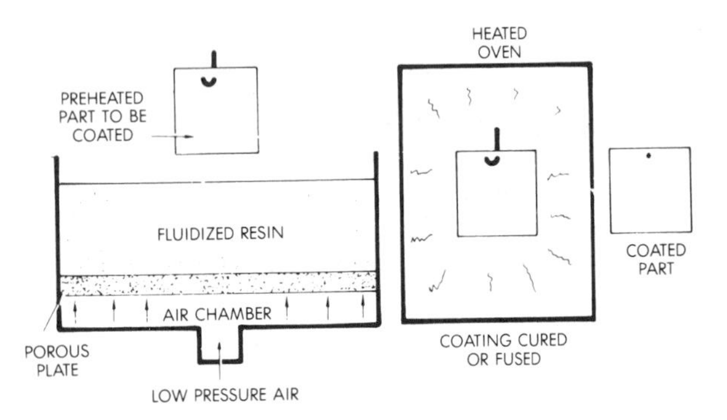

(A) Principle of operation (W.S. Rockwell Co.)

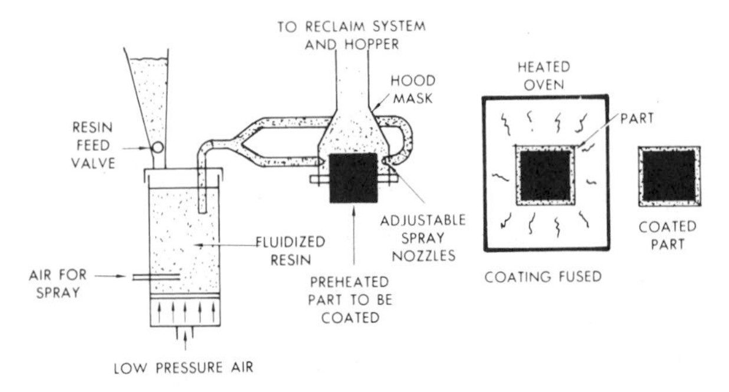

(B) Fluidized-bed spray coating technique. (W.S. Rockwell Co.)

(C) Large fluidized-bed coating operation. Metal dip preparation is shown at left. (Michigan OvenCo.)

(D) The parts on the conveyer are cleaned, heated, then coated by the fluidized-bed method. (Michigan Oven Co.)

(E) Transformer can tops, at right, are being given a primer application before being heated and coated by the fluidized-bed technique. (Michigan Oven Co.)

Fig. 17-6. Fluidized-bed coating process.

The overspray is collected and reused (Fig. 17-6B). (This process is sometimes called *fluidized-bed spray coating.*) The coating on the part is then fused in a heated oven.

The following list shows three advantages and six disadvantages of fluidized-bed coatings.

Advantages of Fluidized-Bed Coating

1. Thickness and uniformity
2. Thermoplastics and some thermosets may be used
3. No solvents needed

Disadvantages of Fluidized-Bed Coating

1. Substrate must be heated above plastics melt or fusion temperature
2. Primer may be needed
3. Thin coatings are hard to control
4. Continuous automation of line is difficult
5. Post cure is needed
6. Surface finish may be uneven (orange-peel)

Electrostatic-Bed Coating

In electrostatic bed-coating, a fine cloud of negatively charged plastics powders is sprayed and deposited on a positively charged object (Fig. 17-7).

Polarity may be reversed for some operations. Over 100000 volts with low (less than 100 mA) amperage are used to charge the particles as they are atomized by air or airless equipment. The electrostatic attraction causes the particles to cover all conductive surfaces of the substrate. These parts may or may not require preheating. If preheating is not used, the curing or fusing must take place before the plastics powder loses its charge. The curing is done in a heated oven, as shown in Figure 17-7. Thin foils, screens, pipes, parts for dishwashers, refrigerators, washing machines, cars, and marine and farm machines are electrostatic-bed coated. Five advantages and six disadvantages of electrostatic-bed coating are:

Advantages of Electrostatic-Bed Coating

1. Thin, even coats are easily applied
2. No preheating needed
3. Process is readily automated
4. Reduced overspray
5. Improved finish quality

Disadvantages of Electrostatic-Bed Coating

1. Thick coatings need preheating of substrate
2. Small openings or tight angles are hard to coat
3. Dust recovery system may be needed
4. Only ionic resins or plastics can be used
5. Post cure is generally needed
6. Substrates may require special preparation

Electrostatic Powder Gun Coating

The electrostatic powder gun process is similar to painting with a spray gun. In this process the dry plastics powder is given a negative electrical charge as it is sprayed on the grounded object to be coated (Fig. 17-8). Fusion or curing must take place in ovens before the powder particles lose their electrical charge, otherwise they will fall from the part. It is possible to coat complex shapes with this method. The fusing oven is the limiting factor in relation to size. Automobile manufacturers may replace liquid finishing processes with powder coating methods in the future. Hundreds of products are coated using this process, including outdoor fencing, chemical tanks, plating racks, and dishwasher, refrigerator, and washer parts.

The list below shows five advantages and seven disadvantages of electrostatic gun coating.

Advantages of Electrostatic Powder Gun Coating

1. Thin, even coats are easily applied
2. No preheating necessary

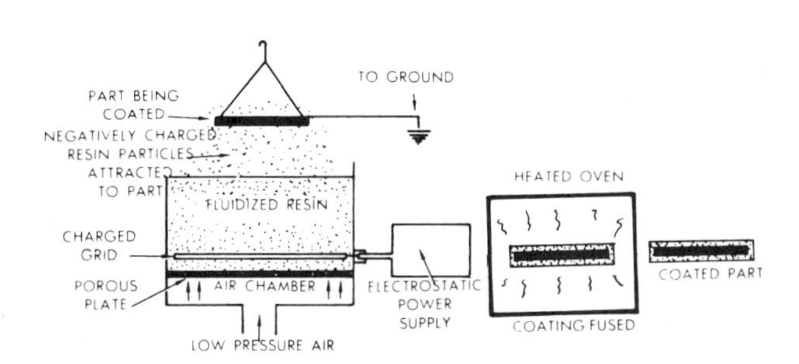

Fig. 17-7. Electrostatic-bed coating process. (W.S. Rockwell Co.)

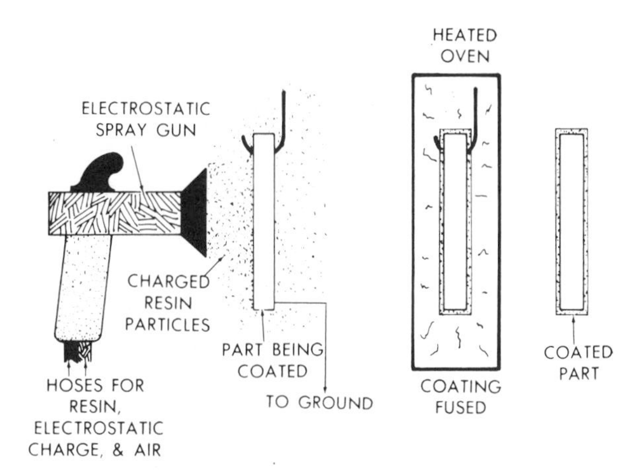

Fig. 17-8. Electrostatic powder gun coating process. (W.S. Rockwell Co.)

3. Process is readily automated
4. Short runs and coating of odd-shaped pieces are practical
5. Lower equipment cost than electrostatic-bed coating.

Disadvantages of Electrostatic Powder Gun Coating

1. Thick coatings need preheating of substrate
2. Small openings or tight angles are hard to coat
3. Dust recovery system may be needed
4. Only ionic resins or plastics can be used
5. Post cure is needed
6. High labor cost
7. Thickness harder to control

Transfer Coating

In transfer coating, a release paper is coated with plastics solution and dried in an oven. A second coat of plastics is then applied over the first coat, and a fabric layer is placed on this wet layer. The coated textile then passes through nip rollers and a drying oven. Finally, the release paper is stripped away from the coated fabric. This method produces a tough leather-like skin on the fabric (Fig. 17-9).

Polyurethanes and PVC are commonly used to coat fabrics in the manufacture of awnings, footwear, upholstery, and fashion apparel.

(A) The covers of these books are examples of a plastics coating on paper substrates.

Two advantages and one disadvantage of this process are:

Advantages of Transfer Coating

1. Multicoated and colored substrates possible
2. Wide choice of substrates may be coated

Disadvantage of Transfer Coating

1. Release paper and additional equipment needed

Knife or Roller Coating

Knife and roller coating methods are another means of spreading a dispersion or solvent mixture of plastics on a substrate. The curing or drying of the plastics coating may be done by heating ovens, evaporating systems, heated rollers, catalysts, or irradiation.

The knife method may involve a simple blade scraper or a narrow jet of air called an *air knife* (Fig. 17-10A). Both sides of the substrate may be coated by this method.

Coating may be done by a combination of rollers as shown in Figures 17-10B and 17-10F. Paper and fabric are often coated by this method. Listed below are four advantages and two disadvantages of knife or roller coating processes.

Advantages of Knife or Roller Coating

1. High-speed continuous process
2. Excellent thickness control
3. Plastisol coatings are stress- and strain-free
4. Thick coatings are possible

Disadvantages of Knife or Roller Coating

1. Equipment and setup time costly
2. Not justified for short runs

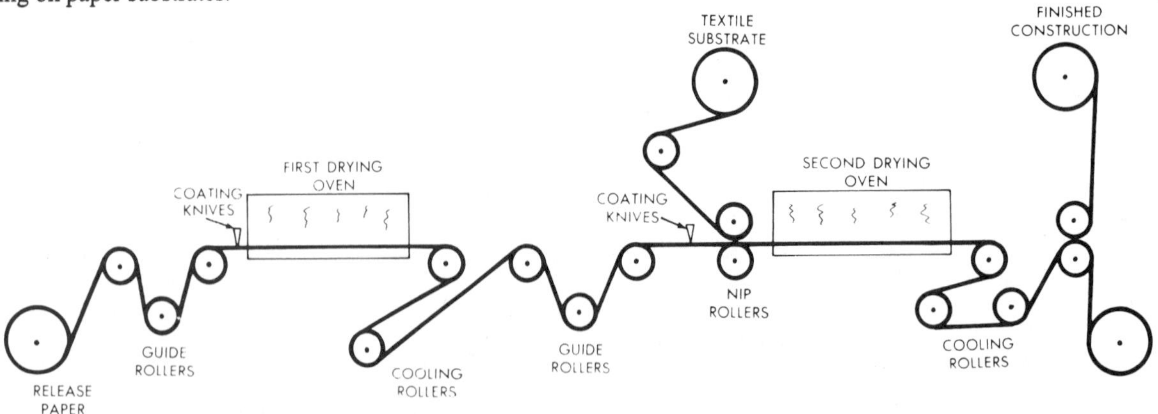

(B) Diagram of a transfer coating line.

Fig. 17-9. Transfer-coating process and products.

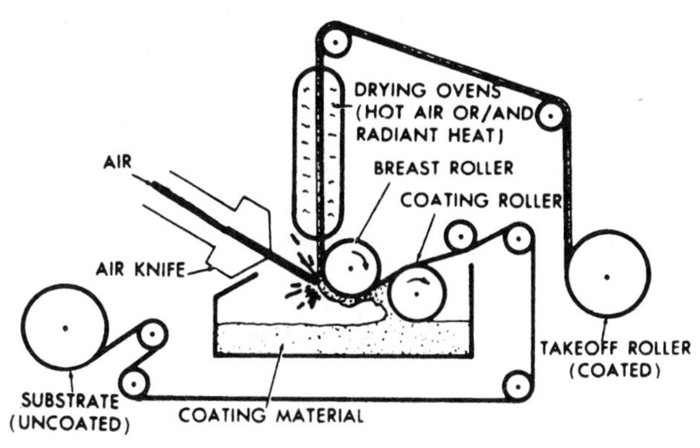

(A) Typical air-knife coating line.

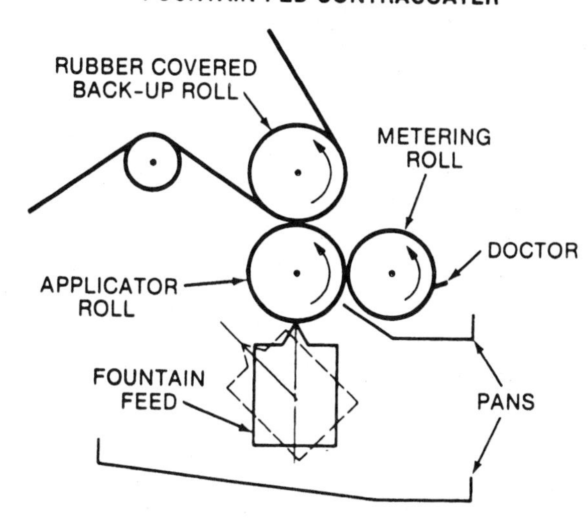

(B) Contracoater method. (Black-Clawson Co., Inc., Fulton Operations)

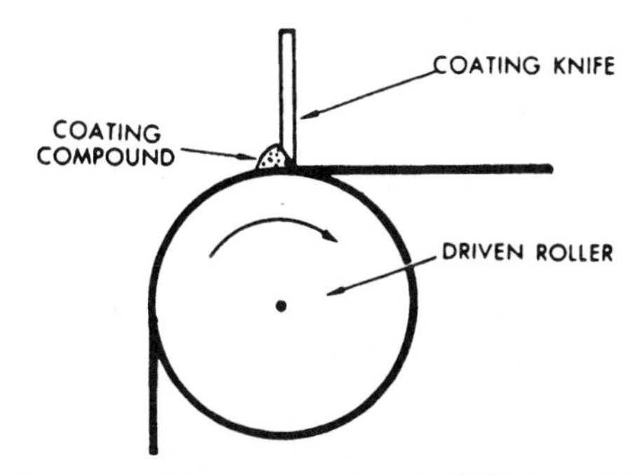

(C) Knife-over-roller coating head. (Waldron Division, Midland-Ross Corp.)

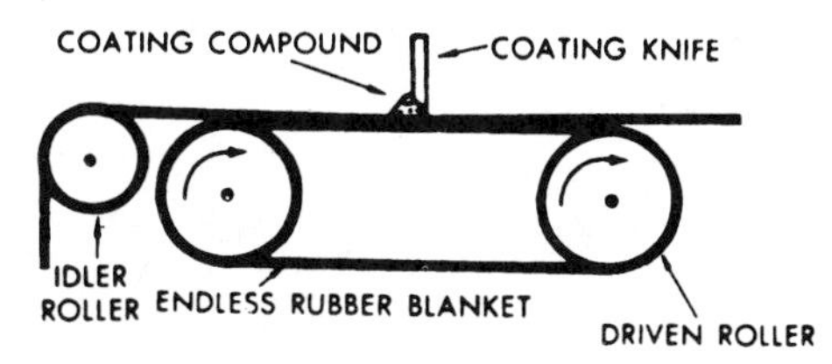

(D) Continuous blanket knife coater. (Waldron Division, Midland-Ross Corp.)

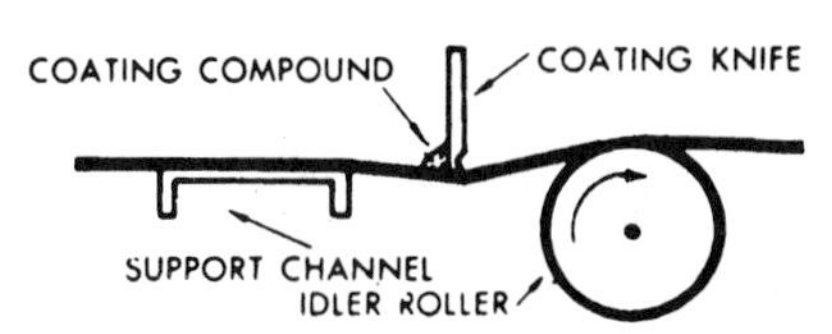

(E) Floating doctor knife. (Waldron Division, Midland-Ross Corp.)

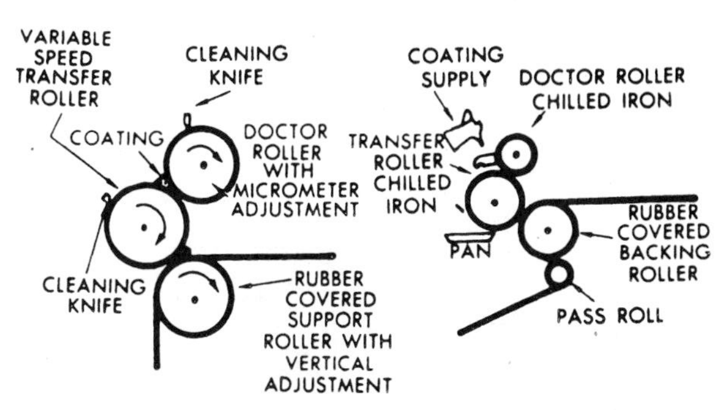

(F) Reverse roll coaters.

Fig. 17-10. Knife and roller coating processes.

Dip Coating

Dip coatings are applied by dipping a heated object in liquid dispersions or solvent mixtures of plastics. The most common plastics used is polyvinyl chloride. For dispersions, a heating cycle is required to fuse or cure the plastics on the coated object although some dip coatings may harden by simple evaporation of solvents. Generally, 10 minutes of heating is needed for each millimeter of coating thickness. Curing temperatures run from 350 to 375 °F [175 to 190 °C]. Tool handles and dish-drainer racks are the most common dip-coated products. Objects are limited by the size of the dipping tank (Fig. 17-11).

To ensure that replacement parts arrive in good condition, and so that they may be stored under varying conditions, strippable coatings are often used. They are placed on gears, guns, and other hardware. When machined or polished surfaces need protection during fabrication or other operations, strippable coatings may be applied. These coatings have good cohesion, but relatively poor adhesion; therefore they may be stripped or peeled from the part. Strippable coatings are sometimes used as masking films in electroplating or in applying paints (Fig. 17-12).

Wires, cables, woven cords, and tubing may be coated in a modified dipping process wherein these substrates pass through a supply of plastisol or organosol. Preheating the substrate and post-heating the product speeds fusion. The coat thickness is controlled in several ways. The strand may be passed through a die opening, fixing the size and shape of the coating. If no die is used, viscosity and temperature then decide the size and shape. Several passes through the plastisol

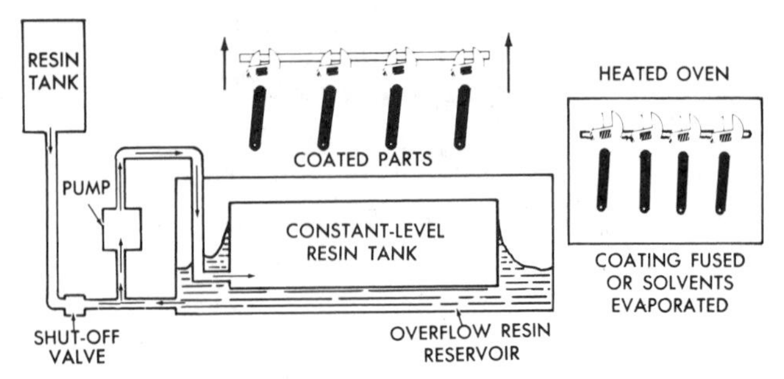

(A) Dip-coating technique.

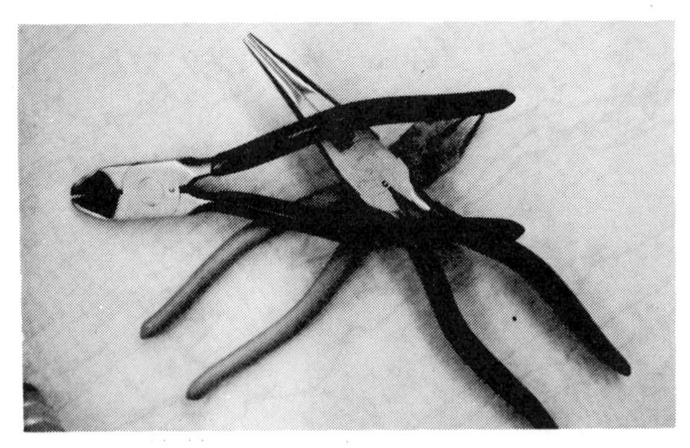

(B) Tool handles dip-coated with PVC.

Fig. 17-11. Dip-coating process and products.

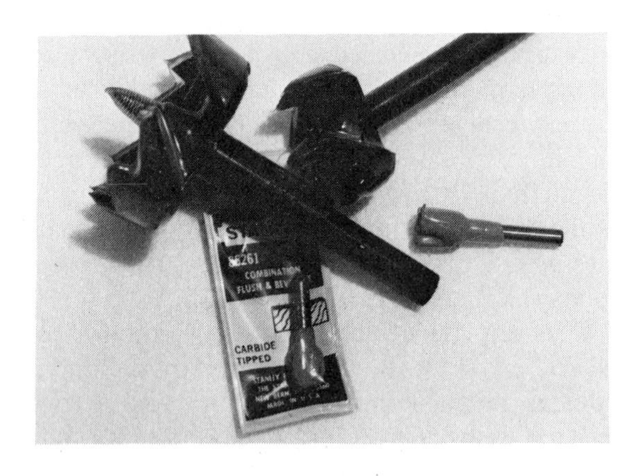

Fig. 17-12. Strippable coatings protect cutting tools.

or organosol will increase thickness. This process may be done in a vertical or horizontal position (Fig. 17-13). Two advantages and five disadvantages of dip coating are:

Advantages of Dip Coating

1. Light or heavy coatings may be applied on complex shapes
2. Relatively inexpensive equipment is used

Disadvantages of Dip Coating

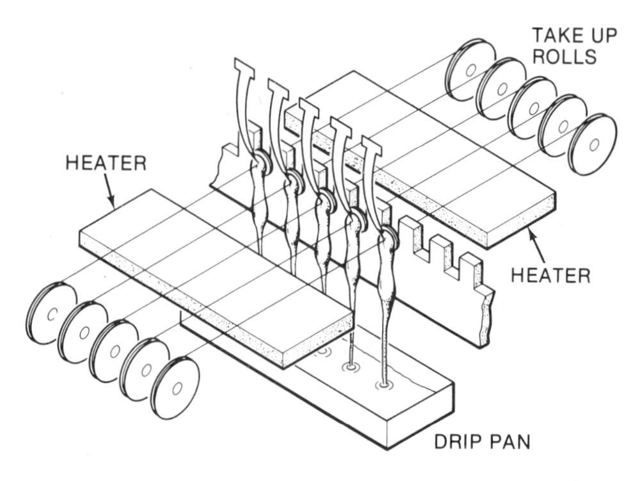

Fig. 17-13. This modified dipping process can apply a plastisol coating to wire, cable, woven cord, or tubing at very high speeds. (BF Goodrich Chemical Division)

1. Primers may be needed
2. Plastisols require preheated substrates and post-heating
3. Organosols require solvent recovery or exhaust
4. Pot life and viscosity must be controlled
5. Dip withdrawal rate must be controlled for even coating thickness

Spray Coating

In spray coating, dispersions, solvent solutions, or molten powders are atomized by the action of air (or inert gas) or the pressure of the solution itself (airless) and deposited on the substrate. Spray coating of furniture, houses, and vehicles with plastics paints or varnishes are examples. Dispersions of polyvinyl chloride (plastisol) have been spray coated on railroad cars (Fig. 17-14).

In a process sometimes called *flame coating,* finely ground powders are blown through a specially designed burner nozzle of a spray gun (Fig. 17-15). The powder is quickly melted as it passes through this gas or electrically heated nozzle. The hot molten plastics quickly cools and adheres to the substrate. This process is useful for items that are too large for other coating methods. Three advantages and disadvantages of spray coating are listed below.

Advantages of Spray Coating

1. Low equipment cost
2. Short runs are economical
3. Fast and adaptable to variation in size

Disadvantages of Spray Coating

1. Hard to control coating thickness
2. Labor costs may be high
3. Overspray and surface defects (runs and orange peel) may be a problem

(A) Coated parts for the electrical industry. (Quelcor, Inc.)

(B) Coated hood. (Michigan Chrome & Chemical Co.)

(C) Aerosol spray cans of plastics materials for coatings.

Fig. 17-14. Sprays and spray-coated products.

Metal Coating

Perhaps metal coating should not be classed as a basic process of the plastics industry; however, because many plastics are associated with this process, the following information is useful. In addition to use as a decorative finish, metal coatings may provide an elec-

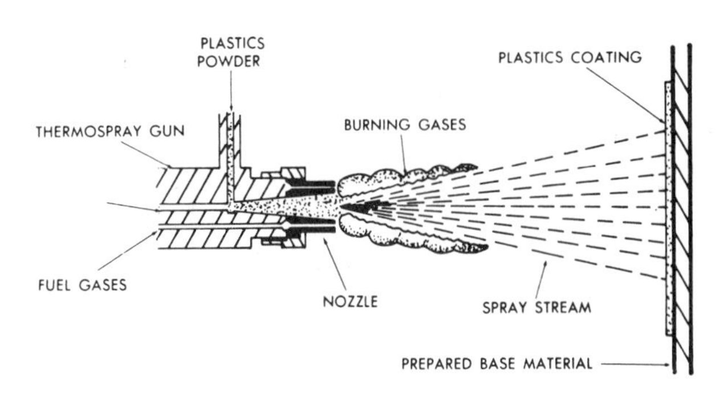

Fig. 17-15. Principle of flame coating.

trically conducting surface, a wear- and corrosion-resistant surface, or added heat deflection. The major methods for applying a metal coating on a substrate are with adhesives, through electroplating, through vacuum metalizing, and by sputter-coating techniques.

Adhesives

Adhesives are used to apply foils to many surfaces. The textile industry has used this method to adhere metal foils to special garment designs. Complex or irregular parts are difficult to coat, and polyethylene, fluoroplastics, and polyamides are difficult to adhere metals to.

Electroplating

Electroplating is done on many plastics. Both the resin and the mold design must be considered in producing a metal coating on plastics parts. Ribs, fins, slots, or indentations should be rounded or given tapers (Fig. 17-16). Phenolic, urea, acetal, ABS, polycarbonate, polyphenylene oxide, acrylics, and polysulfone are often plated.

An electrolysis preplating step is done by carefully cleaning the plastics part and etching the surface to ensure adhesion (Figs. 17-17, 17-18). The etched part

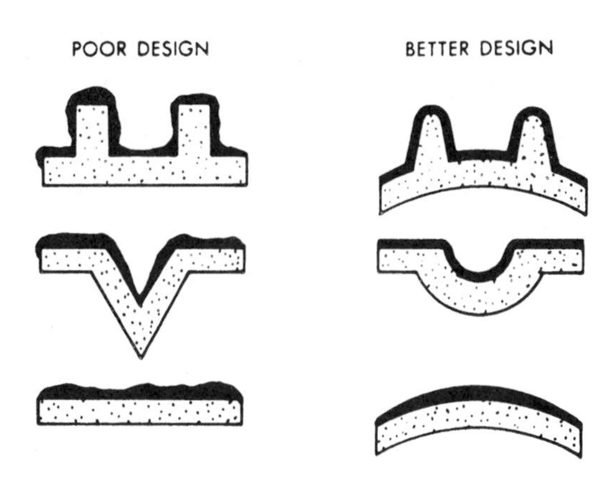

Fig. 17-16. When parts are to be plated, large-radius fillets and bends are desirable.

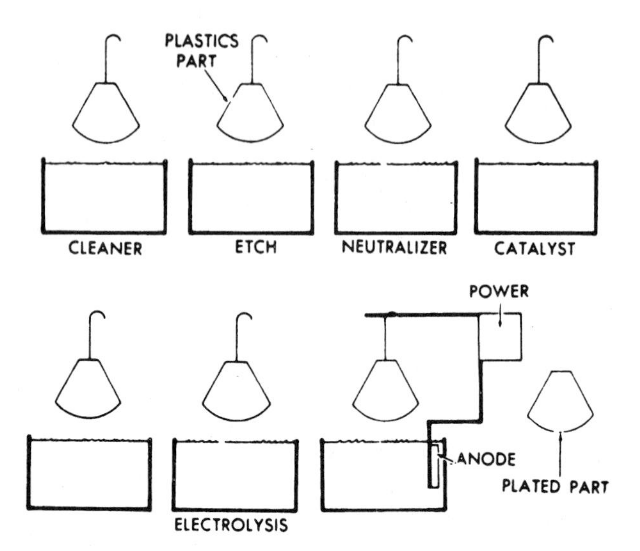

Fig. 17-17. Sequence of operations in electroplating plastics.

is again cleaned and the surfaces *seeded* with an inactive noble-metal catalyst. An accelerator is added to activate the noble metal, and the ionic solution of metal reacts autocatalytically in the electrolysis solution. Copper, silver, and nickel electrolysis solutions are used in preparing a deposit from 10 to 30 millionths of an inch [0.25 to 0.80 micrometres] thick. Once a conductive surface has been established, commercial plating solutions such as chrome, nickel, brass, gold, copper, and zinc, may be used. Most plated plastics acquire a chrome-like finish. The following list shows two advantages and seven disadvantages of electroplating on plastics.

Advantages of Electroplating

1. Mirror-like finishes
2. Good thickness control

Disadvantages of Electroplating

1. Holes and sharp angles are hard to plate
2. Some plastics are not easily plated
3. Plastics must be cleaned and etched before plating
4. Cycle time is long
5. High initial cost
6. Costly for short runs
7. Surface finish of plastic must be near perfect in smoothness

Vacuum Metalizing

In vacuum metalizing, plastics parts or films are thoroughly cleaned and given a base coat of lacquer to fill the surface defects and seal the pores of the plastics. Polyolefins and polyamides are chemically etched to ensure good adhesion. The plastics are then placed in a vacuum chamber, and small pieces or strips of the coating metal (chromium, gold, silver, zinc, or aluminum) are placed on special heating filaments. The chamber is sealed and the vacuum cycle started. When the desired vacuum is reached (0.5 micrometre Hg or about 0.07 Pa), the filaments are heated. The pieces of metal melt (by high voltage) and vaporize, coating everything the vapor touches in the chamber and condensing or solidifying on cooler surfaces (Fig. 17-19). Parts must be rotated for full coverage, because the vaporized metal travels in a straight path. Once the plating is done, the vacuum is released and the parts are removed. To help protect the plated surface from oxidization and abrasion, a lacquer coating is applied (Fig. 17-20). This finish is best suited for interior applications.

By alternately evaporating two or more metals, it is possible to create a chrome/copper/chrome or stain-

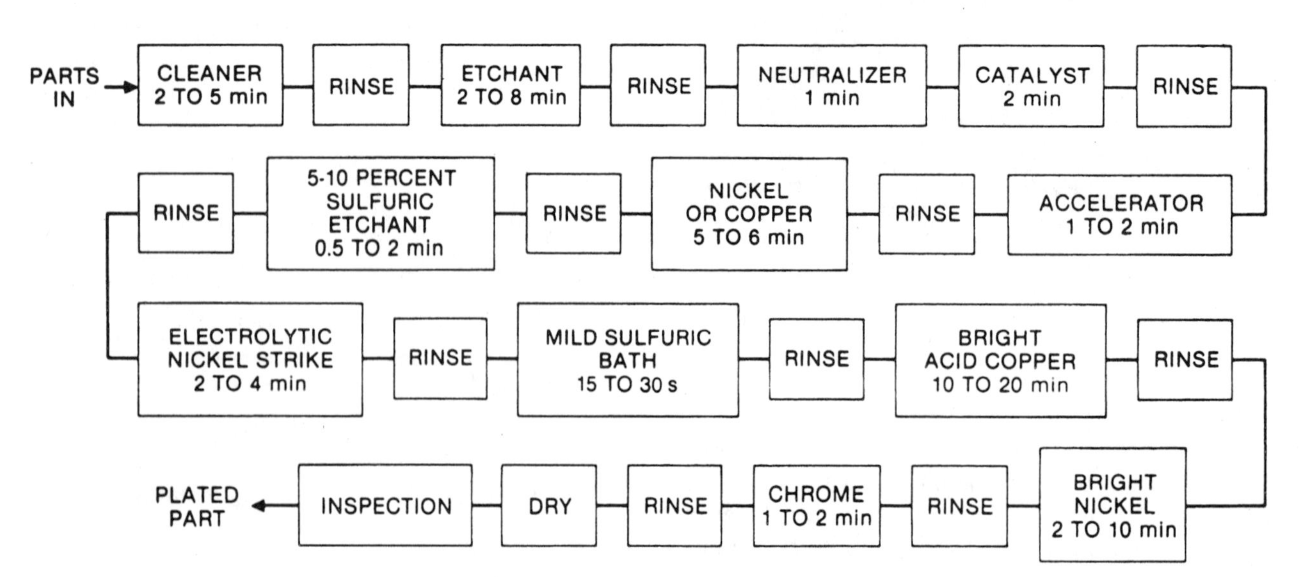

Fig. 17-18. Flow chart of typical plating process.

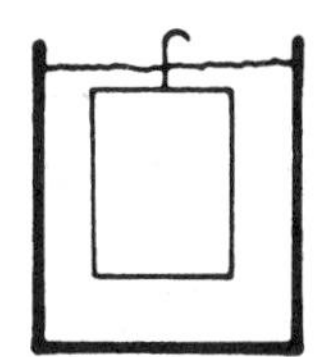

(A) Cleaning and etching part.

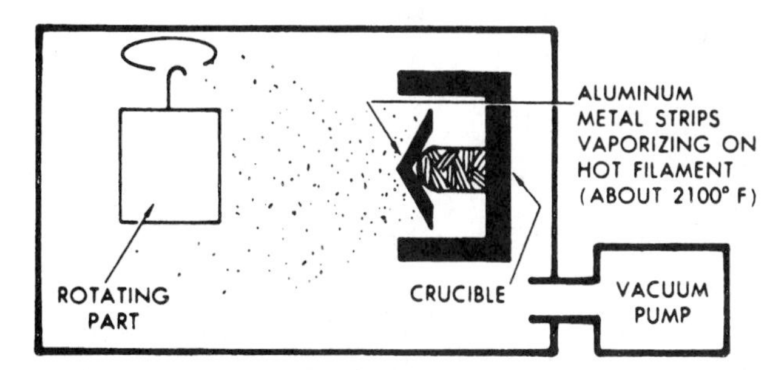

(B) Vaporizing aluminum to coat plastics.

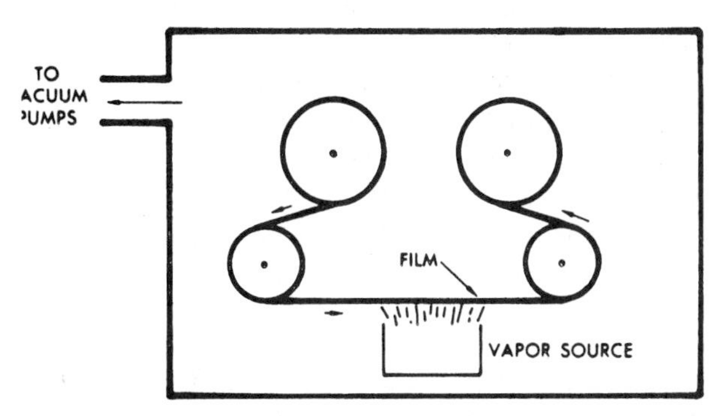

(C) Vacuum metallizing on plastics film.

(D) Type element for electric typewriter is metal-coated plastics.

Fig. 17-19. Vacuum metallizing of plastics parts.

less/copper/stainless laminate. The process is sometimes called *laminated vapor plating*. The list below shows four advantages and five disadvantages of the vacuum metalizing coating process.

Advantages of Vacuum-Metalizing Coating

1. Ultrathin, uniform coatings
2. Nearly all plastics may be used
3. Mirror-like finishes
4. No chemical processing

(A) Horns are loaded onto holding fixtures for placement into vacuum chamber. The fixture is rotated in the chamber during the metallizing process.

(B) A base coat of lacquer may be applied by dipping (shown here), spraying, or flow-coating, then baked in oven. The lacquer will smooth out small surface defects and provide an initial gloss.

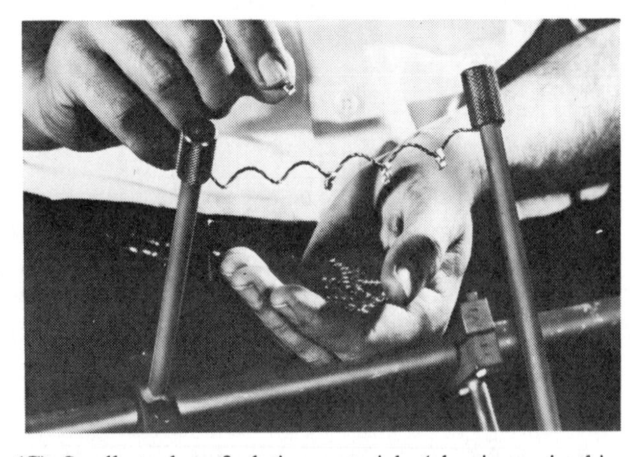

(C) Small staples of plating materials (aluminum, in this case) are placed on coils of stranded tungsten wire filaments. After the chamber is closed and proper vacuum drawn, the filaments are heated to incandescence.

Fig. 17-20. Technique for metallizing toy horns. (Pennwalt-Stokes Corp.)

(D) The aluminum melts, spreads in a thin layer over the elements, and vaporizes. Vaporization or flashing the filaments takes only 5 to 10 seconds,with a temperature of 1110 °C (2100 °F) attained in that time. The metallized products are then removed from the vacuum chamber and dipped in a protective topcoat lacquer. The transparent topcoat can be dyed, allowing a wide choice of colors.

Fig. 17-20. Technique for metallizing toy horns. (Pennwalt-Stokes Corp.) *(Continued)*

Disadvantages of Vacuum-Metalizing Coating

1. Plastics must be coated with lacquer for good results
2. Vacuum chamber limits part size and output rate
3. Scratches or flaws are exaggerated
4. High initial cost
5. Costly for short runs

Sputter Coating

Metals or refractories may be deposited by sputtering systems. Magnetron electronic equipment is used to spray the metal coating. Chromium atoms fall (sputter) on the plastics surface as argon gas strikes an electrode made of the coating metal. Typical thicknesses are 0.00019 to 0.00275 in. [0.005 mm to 0.07 mm]. A clear protective coating of PU, acrylate, or cellulosic is then applied as protection of the metallic coating. This process is used for coating knobs, films, light reflectors, car trim, and plumbing fixtures. Four advantages and three disadvantages of this process follow.

Advantages of Sputter Coating

1. Ultrathin coatings
2. Excellent adhesion
3. Automated line systems possible
4. Parts are electrically conductive

Disadvantatges of Sputter Coating

1. High technology and capital investment
2. Scratches or flaws are exaggerated
3. Protective coating needed over sputter layer

Brush Coating

Solvent and solventless coatings are often brushed onto a substrate by hand. Many paints and finishes are applied in this manner. Solventless finishes are two-component systems of resins and curing agents that are mixed and applied. Polyester, epoxy silicone, and some polyurethane resins are used in solventless formulations. Solvent-based coatings may need air drying or heating to cure.

Finish quality depends on the skill of the applicator and the type and viscosity of the material.

Protective coatings on large metal tanks are often applied by hand or spray-up methods. If placed underground, the tanks must be protected from corrosion and electrolytic action.

Familiar examples are coatings on houses, machinery, furniture, and fingernails. Two advantages and three disadvantages of brush coating processes are shown below.

Advantages of Brush Coating

1. Low equipment cost
2. Short runs and prototypes are not costly

Disadvantages of Brush Coating

1. High labor costs
2. Poor thickness control
3. Finish hard to control and reproduce

Vocabulary

The following vocabulary words are found this chapter. Look up the definition of any of the words you do not understand as they apply to plastics in the glossary, Appendix A.

Dip coating
Extrusion coating
Flame spraying
Fluidized bed
Knife coating
Laminated vapor plating
Vacuum metalizing

Questions

17-1. For a process to be classified as a coating the plastics must remain on the ___?___ .

17-2. The technique by which hot extruded plastics is pressed on a substrate without adhesive is called ___?___ .

17-3. The coating used to protect tools from rusting and from damage to cutting edges during shipment is called ___?___ .

17-4. Name two processes used to put a coating on wire.

17-5. Melt-roll coating is a modification of the ___?___ process.

17-6. What process is used to place coatings on tool handles?

17-7. What is the major advantage of electroplate coating?

17-8. What are two major advantages of brush coating?

17-9. Which coating process would be used to produce reflective window shade films?

17-10. How many minutes of heating time are re quired for each millimetre of dip-coating thickness.

17-11. Name four methods that may be used to spread dispersion or solvent mixtures on substrate.

17-12. The ideal wall thickness for a vinyl dipped coin purse is ___?___ .

17-13. The temperature used to cure plastisols is ___?___ .

17-14. The main element in curing a plastisol coating is ___?___ .

17-15. The main element in curing an organisol coating is ___?___ .

17-16. Name four methods or techniques that may be used to place a coating on a fabric.

17-17. Name the two major disadvantages of extrusion coating.

17-18. What causes the powder to atomize in fluidized bed processing?

17-19. The process of coating a substrate with a dry plastics powder is sometimes called ___?___ .

17-20. Name the major disadvantage of transfer coating.

17-21. In ___?___ coating processes, the plastics and the substrate are given opposite charges.

17-22. The most commonly used plastics for dip coating is ___?___ .

17-23. To speed fusion in dip coating, ___?___ and ___?___ the substrate are common practices.

17-24. A coating of ___?___ is sometimes applied to plated surfaces to minimize oxidization and abrasion.

17-25. Metals or refractories may be deposited by ___?___ coating systems.

17-26. Name four reasons for coating a substrate.

17-27. Name the three major processes of dry-powder coating.

17-28. Name three commonly transfer-coated materials.

17-29. A narrow jet of air used to spread or disperse resins or plastics on a substrate is called an ___?___ .

17-30. Name four products that are normally spray-coated.

17-31. Name the four major methods of metalizing a substrate.

17-32. Before vacuum metalizing, parts are given a ___?___ coat to minimize surface defects, provide a reflective surface and seal the substrate.

17-33. During the vacuum-metalizing cycle, the plastics parts must be ___?___ , because the vaporized metal travels in a line-of-sight path.

17-34. Textiles are coated for moisture and chemical resistance, while pots, pans and tools are coated for ___?___ and chemical resistance.

17-35. A manufacturer has the following products to be coated. Recommend coating techniques for each.

a. plastics grille to look like chrome

b. concrete wall slabs for construction

c. two-color jewelry piece (black and gold)

d. book covers

e. textile rainwear

17-36. Describe the fluidized-bed coating process.

17-37. In electrostatic-bed coating, how is the dry powder deposited on the part?

17-38. What is electrostatic powder gun coating? How does it differ from electrostatic-bed coating?

17-39. Briefly describe the electroplating process. Which plastics materials are well suited to this process?

17-40. How would you set up a simple process for coating tool handles with a resilient plastics?

17-41. List the coating processes requiring heat for curing. Also, list those that do not require heat for curing.

Activities

Fluidized-Bed Coating

Introduction. Fluidized-bed coating is a powder coating process that uses either thermoplastic or thermosetting materials. Figures 17-21, 17-22, and 17-23 show various types of fluidized-bed coaters. All three coaters contain a porous bottom that allows passage of air (or other gases). The air forces its way through the powder, causing the powder to float in the tank, somewhat like a fluid. The powder melts and sticks to the surface of pre-heated substrates, when they are dipped into the bed. A post-heating cycle will smooth the surface of thermoplastics and cause final curing of thermosets.

Equipment. Fluidized bed coater, polyethylene powder, caliper, oven, substrates for dipping, insulated gloves.

Fig. 17-21. Small commercial laboratory fluidizer.

Fig. 17-22. Small home-made laboratory fluidizer.

Fig. 17-23. Laboratory fluidizer with compressor.

Procedure

17-1. Make substrates for dipping. Figure 17-24 shows a steel substrate, cut from a strip 1/8 by 3/4 inches. These pieces are 1 3/4 inches long. Other size or shape substrates are also functional. Attach a short piece of wire to each substrate as shown. Measure the thickness of substrates.

17-2. Hang the substrates in an oven and heat to 350 °F.

17-3. Prepare fluidized bed by adding polyethylene powder and adjusting the air pressure to thoroughly fluidize the powder. Depending on the type of fluidizer, 20 to 34 kPa [2.9 to 5 psi] may be sufficient.

17-4. Remove one piece of metal from the oven, hold it by the wire and dip it quickly into the bed. Record the duration of time in the bed. Remove the piece and shake off excess powder.

17-5. Return the coated piece to the oven to smooth the surface by fusion of powder particles.

17-6. Remove from oven and cool.

17-7. Measure the thickness of the coated piece and calculate the thickness of the coating. Is the coating thicker at the bottom than at the top of the piece?

17-8. It is often possible to remove the coating by cutting along the edge of the metal and peeling off the coating. One item in Figure 17-24 shows a coating lifted off the substrate. Measuring the thickness on a peeled coating may increase accuracy of measurement.

17-9. Arbitrarily determine a desired thickness of coating. The coatings in Figure 17-24 were about 0.38 mm [0.015 in.] thick.

17-10. Determine a process that repeatedly yields a smooth, uniform coating of the desired thickness. The parameters of importance are:

- Preheat temperature
- Postheat temperature and time
- Dip time
- Depth substrate is dipped into the bed.
- Agitation during dipping
- Cleanness of substrate
- Air pressure

17-11. Write up a report summarizing the process which yields optimum coating.

Further investigations. To determine the effects of the substrate material, fabricate substrates from steel, aluminum, and copper or brass. Make sure they are dimensionally identical. Heat to the same temperatures and dip equally. Does the coating thickness change from one type of substrate to another?

Dip Coating

Introduction. A plastisol is a mixture of polyvinyl chloride powder and a plasticizer. When a preheated substrate is dipped into plastisol, the particles of PVC adhere to the substrate. Post-heating completes the fusion of the PVC particles.

Equipment. Vinyl dispersions, substrates for dipping, oven, calipers, tensile tester, insulated gloves.

Procedure

17-1. Fabricate substrates for dipping. The substrates shown in Figure 17-25 are aluminum, cut from a bar 0.2 × 1 inch. They are about 4 inches long, to fit in a quart can of vinyl dispersion.

17-2. Preheat the substrates to 200 °C (392 °F). Working rapidly, remove the substrate from the oven and hang in the dispersion. Figure 17-25 shows dip coating in progress. Record the dipping time. Remove from the dispersion and allow all drips of fall off.

17-3. Return to the oven and cure at 175 °C (347 °F).

17-4. When thoroughly fused, remove from oven and cool.

Fig. 17-24. Uncoated, coated, and peeled substrates.

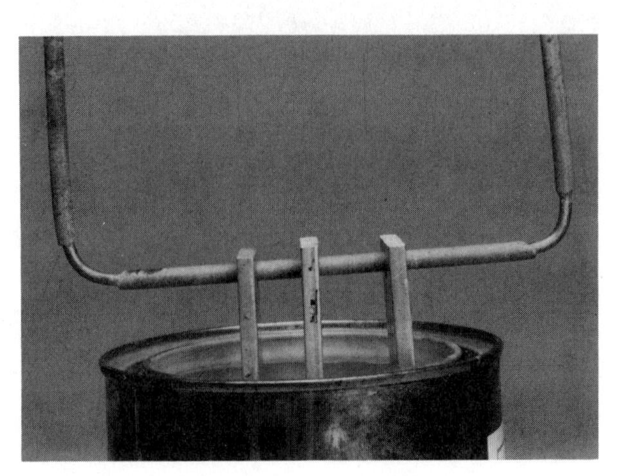

Fig. 17-25. Dipping into a quart of plastisol.

17-5. Slip coating off substrate or cut at edge of substrate.

17-6. Figure 17-26 shows a substrate as dipped, a piece of vinyl cut into a dog bone shape, and another piece after tensile testing. To cut the dog bone, use a cutter as shown in Figure 17-27. If a cutter is unavailable, use scissors to cut the shape. Tensile test the vinyl sample to determine the strength and percent elongation. Omitting the cutting into a dog bone shape will provide a strip of uniform width. This may lead to difficulties if the sample begins to slip out of the jaws of the tensile tester. The broken sample in Figure 17-26 had an ultimate tensile strength of 9.78 MPa [1417 PSI] and an elongation of 385 percent.

17-7. Experimentally determine a process that yields smooth, uniform, strong and elastic vinyl. The parameters of significance are:

- Preheat temperature
- Dip time
- Cure temperature
- Cure time
- Surface smoothness of the substrate

CAUTION: Overheating the plastisol will release toxic hydrogen chloride gas (HCl). Make sure the ventilation near the oven is adequate.

17-8. Write a report summarizing the process which yielded optimum vinyl.

Additional activities

a. To observe the effects of substrate smoothness, polish one surface of the substrate, and roughen the opposite side. What effect does a rough surface produce?

b. Investigate the effect of post-heating time. Produce several samples, incrementally increase the post-heat time, and test to determine the effects on the physical properties of the vinyl.

c. Do a research paper on latex rubber surgical gloves. Due to concern about AIDS, the use of latex gloves has increased. How do the manufacturers produce uniform thickness gloves, with no holes?

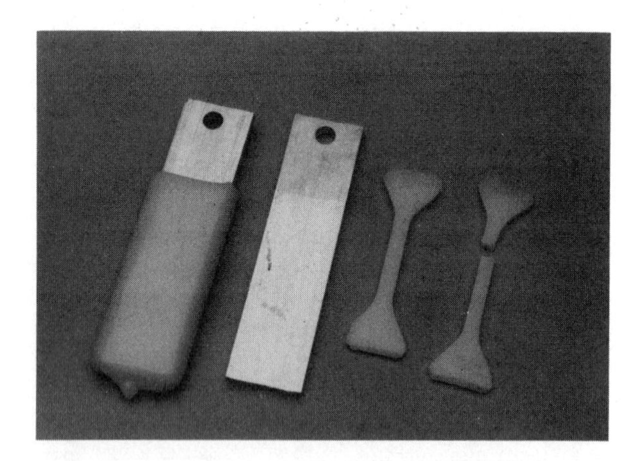

Fig. 17-26. Stages in testing of vinyl dip-coatings.

Fig. 17-27. Dog bone cutter showing cutting edges.

Chapter 18

Fabrication Processes and Materials

Introduction

As with wood, metal, and other materials, plastics components often require assembly or fabrication. This chapter will treat the major fabrication processes and the materials utilized in these processes. There are four broad methods by which plastics are joined. These are included in the following chapter outline:

I. Mechanical adhesion
   A. Thermoplastic resins
   B. Thermosetting resins
   C. Elastomeric types
II. Chemical adhesion
   A. Solvent bonding
   B. Frictional heating techniques
   C. Transferred-heat techniques
III. Mechanical fastening
IV. Friction fitting
   A. Press fitting
   B. Snap fitting
   C. Shrink fitting

Mechanical Adhesion

Adhesives are a broad class of substances that adhere materials together by a surface bond. If the adhesives hold parts together by interlocking the surfaces, they are mechanical or physical adhesives. Mechanical adhesives come in various forms; however, they must be in a liquid or semi-liquid state during the bonding operation. This ensures close contact with the *adherends* (surfaces being adhered). In mechanical adhesion, there is no flow of the adherends.

Animal and some natural plastics have been used as adhesives for centuries. Wax and shellac were once widely used to seal letters and important documents. Many ancient civilizations used pitch to seal cracks in boats and rafts. Archeologist have evidence that more than 30 centuries ago Egyptians used adhesives to attach gold leaf to wooden coffins and crypts.

At one time, the word *glue* referred to an adhesive obtained from hides, cartilage, bones, and other animal materials. Today, this term is synonymous with adhesive based on plastics and usually refers to bonding wood.

Mechanical adhesives generally fall into three basic categories: 1) thermoplastic resins, 2) thermosetting resins, and 3) elastomeric types. Table 18-1 lists a number of thermosetting and thermoplastic adhesives, and their available forms.

Thermoplastic Resins

Thermoplastic resins include adhesives based on acrylics, vinyls, cellulosics and hot-melt materials.

Acrylic adhesives. Acrylic adhesives range from flexible to hard materials. A popular form of acrylic adhesive is a *cyanoacrylate* adhesive. It is a rapid-setting adhesive that polymerizes when pressure is applied to the joint. The solvent *N,N-dimethylformamide* will thin this adhesive and clean unpolymerized excess; however, this adhesive does not cure by solvent evaporation, but by polymerization.

Table 18-1. Available Forms of Selected Plastics Adhesives

Plastic Adhesives	Available Forms
Thermosetting	
Casein	Po, F
Epoxy	Pa, D, F
Melamine formaldehyde	Po, F
Phenol formaldehyde	Po, F
Polyester	Po, F
Polyurethane	D, L. Po, F
Resorcinol formaldehyde	D. L, Po, F
Silicone	L, Po, Pa
Urea formaldehyde	D, Po, F
Thermoplastic	
Cellulose acetate	L, H. Po, F
Cellulose butyrate	L, Po, F
Cellulose, carboxymethyl	Po, L
Cellulose, ethyl	H, L
Cellulose, hydroxyethyl	Po, L
Cellulose, methyl	Po, L
Cellulose nitrate	Po, L
Polyamide	H, F
Polyethylene	H
Polymethyl methacrylate	L
Polystyrene	Po, H
Polyvinyl acetate	Pt, D, L
Polyvinyl alcohol	Po, D, L
Polyvinyl chloride	Pa, Po, L

Note: Po—Powder; F—Film; D—Dispersion; L—Liquid; Pa—Paste; H—Hot Melt; Pt—Permanently tacky

Vinyl adhesives. Vinyl adhesives include a variety of materials. *Polyvinyl alcohol* is a water-based adhesive used to bond paper, textiles, and leather. The interlayer of safety glass is *polyvinyl butyral* because it has excellent adhesion to glass. The electrical and insulation value of *polyvinyl formal* makes it ideal for wire enamels. *Polyvinyl acetal* excels as a bonding adhesive for metals.

One very popular modern adhesive is white glue, a polyvinyl acetate adhesive. This adhesive comes ready to use, in a fast-setting liquid form. This familiar adhesive is a dispersion of polyvinyl acetate in a solvent. Often the solvent is water, and as a result, it must be kept from freezing. Carpenters, artists, secretaries, and many other people make use of the adhesive properties of this material.

Cellulosic adhesives. Cellulosic adhesives are popular and are available in solvent, hot-melt, and dry-powder forms. Duco Cement is a general purpose cellulose nitrate adhesive. It is waterproof, clear, and will adhere to wood, metal, glass, paper, and many plastics. Cellulose acetates and butyrates are familiar cements for plastics models.

Hot-melt adhesives. Hot-melt materials are popular because they are easy to use, somewhat flexible, and obtain their highest adhesive qualities when cooled. Several thermoplastics are used as hot-melt adhesives, including polyethylene, polystyrene, and polyvinyl acetate.

Small sticks or rods of these plastics are heated in an electric gun. The hot plastics is forced from the gun nozzle onto the gluing surface. Figure 18-1 shows a leather strap being assembled using hot-melt plastics in an electrically heated applicator gun. The shoe industry is presently using this adhesive as an effective means of assembly of leather goods.

Small, rapidly assembled articles may be bonded with this method. Probably the most serious drawback to hot-melt adhesives is the difficulty in making large glue joints. The adhesive cools too quickly to ensure adhesion on large bonding surfaces.

Thermosetting Resins

Thermosetting resins gain their strength from polymerization reactions that generally occur after the resin covers the adherends. The polymerization generally occurs because of thermal reactions or catalysts. The major types include amino and phenolic resins, and epoxies.

Amino resins. Casein and urea formaldehyde (amino resins) are used in the woodworking industries. Some of the urea resins are sold in liquid forms for use in the manufacture of plywood, particle boards, and hardboards.

Shell-molding is an important process used by foundry workers in casting metal parts. Phenolic and amino resins are used to bond the sand mold together.

Phenolic resins. *Phenol-formaldehyde* (phenolic) resins are sold in liquid, powder, and film forms. The films, about 0.001 in. [0.025 mm] thick, are placed between the materials to be bonded. Moisture in the material or external steam causes the adhesive film to flow and liquify. The curing reaction takes place at temperatures in the 250 °F to 300 °F [120 °C to 150 °C] range. A reinforced film is often used. These films are usually thin, like tissue paper, and saturated with the

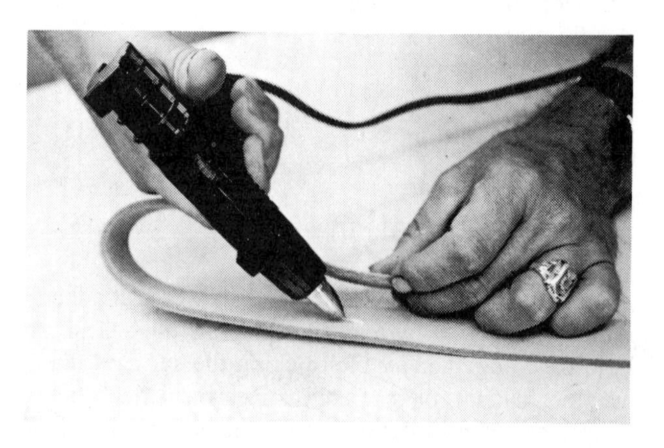

Fig. 18-1. This glue gun is used to apply hot-melt adhesives.

adhesive. They are used in the same manner as unreinforced films but are easier to handle and apply.

Large quantities of phenolic resin adhesives are used in the manufacture of exterior plywood and tempered or smooth-finished hardboard.

Resorcinol-formaldehyde resins are another phenolic-based adhesive that usually comes in a liquid form. It is mixed with a powdered catalyst at the time of use. This resin has the advantages of curing at room temperature and being water- and heat-resistant. High grades of exterior and marine plywoods are bonded with these adhesives. (Fig. 18-2)

High-frequency heating greatly speeds the curing or polymerization time of many plastics used as adhesives. The high-frequency field excites the molecules of the adhesive, causing heat and rapid polymerization. Wood joints are often assembled by this method, using resorcinol-formaldehyde adhesives.

Resin-bonded grinding wheels and sandpapers are made from abrasive grains and a plastics bonding agent. The grinding wheels are made from abrasive grains, powdered resin, and a liquid resin by a cold-molding process. Figure 18-3 shows some typical sandpapers and grinding wheels bonded with phenolic or other resins.

Epoxy resins. *Epoxy resin* adhesives are thermosetting plastics available in two-part paste components. Epoxy resin and either a powdered or a resinous catalyst are mixed to polymerize the resin. Heat is sometimes used to aid or speed this hardening process. Specially formulated one-part epoxies may be polymerized by the application of heat alone.

Epoxy adhesives have excellent adhesion to nearly all materials if the surfaces are properly prepared. The excellent adhesive properties of epoxies are used to mend broken china, bond copper to phenolic laminates in printed circuits, and to bond components in sandwich or skin-type structures. However, even epoxy adhesives have difficulty bonding polyethylene, silicones, and fluorocarbons.

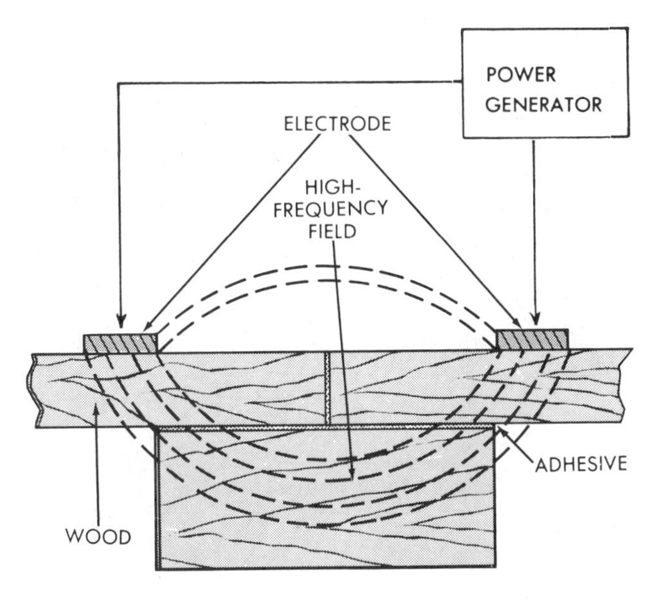

Fig. 18-2. High frequency (radio) waves may be used to heat adhesives.

Fig. 18-3. Many grinding wheels and sandpapers are resin-bonded.

Elastomeric Types

Elastomeric adhesives must bond effectively with the substrate to which they are applied. Their basic purpose is to keep moisture, air, or other agents out of cracks or small openings. To remain effective, these compounds must maintain adhesion and stretch or compress as the materials they contact expand or contract. For example, the compounds that seal glass windows to aluminum frames must withstand differential expansion, since aluminum expands about 2.5 times more than glass.

Elastomeric adhesives bear many names, including caulk, sealant, glazing compound, and putty. Some of the more common sealants include polysulfides, acrylics, polyurethanes, silicones, and both natural and synthetic rubber compounds.

If the openings and cracks are rather small, caulking and sealing materials are appropriate (Fig. 18-4). If the cracks or openings are rather large, the putties or patching compounds will contain a large amount of filler. They are also formulated to minimize shrinkage.

If glass is involved, the materials are usually called glazing compounds. Many glazing compounds and putties contain acrylic compounds. They are easily applied and will not crack, sag, or break down. Figure 18-5 shows automotive glazing applications, and Figure 18-6 includes window glazing.

If ceramics need sealing, special formulations of epoxy and silicone compounds find uses as sealants around lavatories and bathtubs.

The aircraft industry has developed many specialized adhesives to bond aluminum parts. The polymer polysulfide is a very effective sealant with a wide

Fig. 18-4. Many caulks, sealants, and glazing materials are available in cartridges for ease of application.

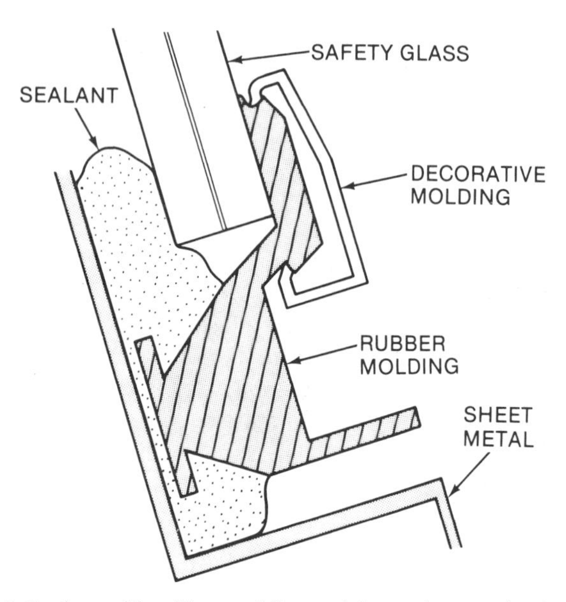

(A) Sealant with rubber molding and decorative metal strip.

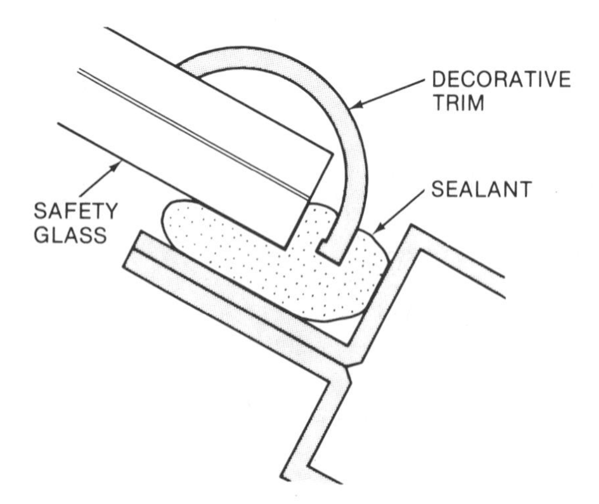

(B) Sealant with molding strip.

Fig. 18-5. Sealing methods used for automobile windshields.

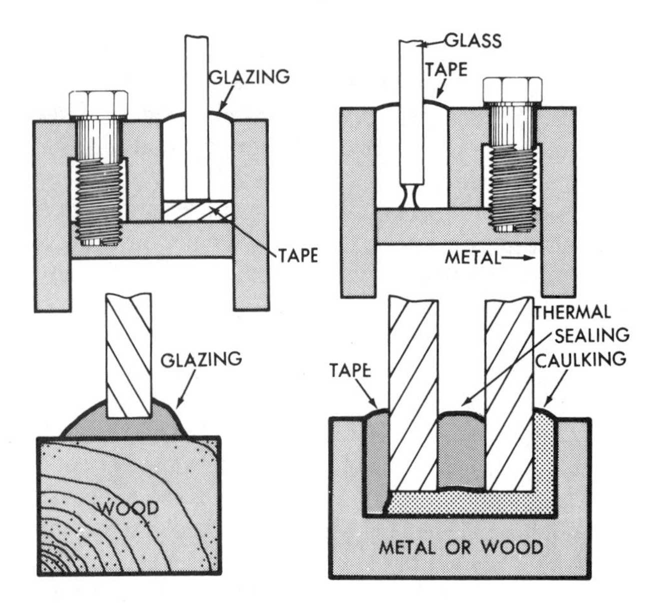

Fig. 18-6. Various sealing methods.

range of applications in the aircraft, electrical, and building industries (Fig. 18-7).

A modern and efficient method of applying sealants is by use of compressible tapes or extruder guns. Compressible tapes are rolls of sealant in ribbon form. They are used in the automotive industry for sealing metal joints and as adhesive sealants in window construction.

Extruder guns are convenient applicators that use disposable or refillable containers of sealing compound. These devices push the sealant out of a nozzle during application (Figs. 18-8 and 18-9).

Fig. 18-7. The aircraft and aerospace industries use polysulfide polymers in many sealing applications. (Thiokol Corporation, Chemical Division)

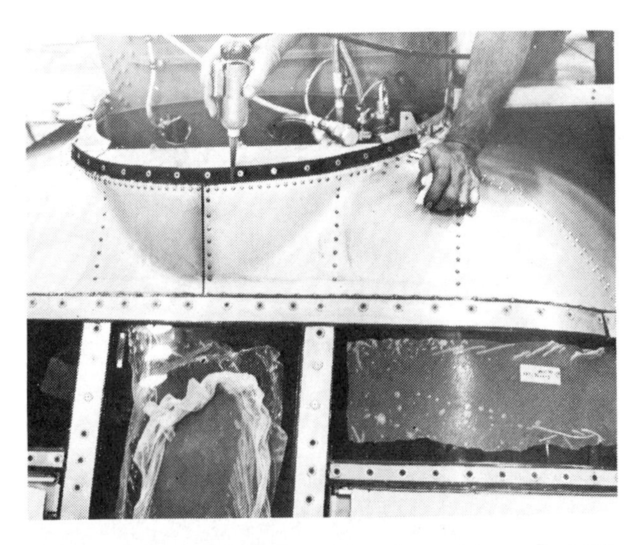

Fig. 18-8. A special applicator gun is used to apply LP® polysulfide base sealant for an aircraft motor housing. (Thiokol Corporation, Chemical Division)

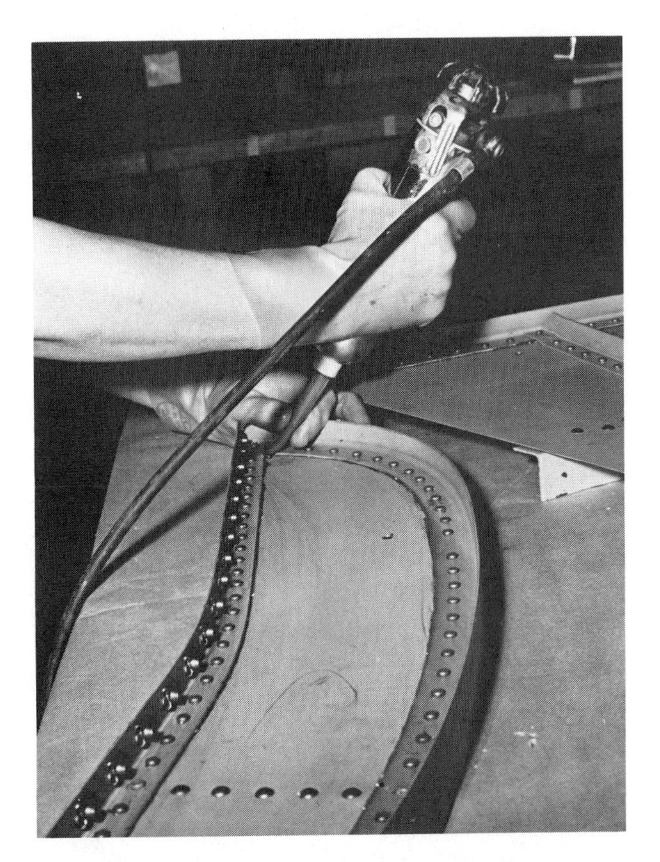

Fig. 18-9. Overlap angle joints, used in aerospace vehicles, are being sealed with LP® polysulfide base sealant. (Thiokol Corporation, Chemical Division)

Chemical Adhesion

Chemical or specific adhesion has been defined as adhesion between surfaces held together by valence forces of the same type as those that give rise to cohesion. The forces that hold the molecules of all materials together are referred to as *cohesive* forces. These forces include the strong primary valence bonds and the weaker secondary bonds.

In chemical adhesives, there is a strong valence attraction between the materials as the molecules flow together. During the welding of metals, for example, molten metal flows and there is a chemical cohesive action between the pieces.

It should be apparent that only by causing a softening or flow in the two materials can chemical bonding occur. If the surfaces are not caused to flow, only mechanical or physical forces hold the pieces together. In the cohesive bonding of metals, heat must be applied to cause the surface molecules to flow and intermingle. The cohesive bonding of plastics involves two major approaches; the use of solvents or heat. This type of bonding does not occur with thermoset materials. The heat-techniques cause melting or softening from friction between plastics parts or from heat transferred from hot metal.

Solvent Bonding

There are two basic forms of solvent based adhesives, *solvent cements* and *monomeric cements*. Solvent cements dissolve the surfaces of the plastics being joined. This forms strong intermolecular bonds as it evaporates. Monomeric cements are based on a monomer of at least one of the plastics to be joined. It is catalyzed so that a bond is produced by polymerization in the joint. Either cohesive or adhesive bonding will occur, depending on the chemical composition of the materials being joined.

Solvent cements and *dope cements* are two kinds of cements in common use. The first are solvents or blends of solvents that dissolve the material, then when the solvent evaporates, the items are fused together. Dope cements are sometimes called *laminating cements* or *solvent mixes*. They are composed of solvents and a small quantity of the plastics to be joined. This cement is a viscous (syrupy) material that leaves a thin film of the parent plastics on the joint when dried.

Solvents with low boiling points evaporate quickly; therefore, the joint must be positioned before all of the solvent evaporates (Table 18-2). An example is methylene chloride, with a boiling point of 104 °F [40 °C]. Solvent cements may be applied to the plastics joints by any of several methods mentioned below. Regardless of method, all joints should be clean and smooth. A V-joint is preferred for making butt joints by many manufacturers and fabricators (Fig. 18-10).

In the soaking method, joints may simply be soaked in a solvent until a soft surface is obtained. The pieces are then placed together at once under slight pressure until all solvents evaporate. If too much pres-

Table 18-2. Common Solvent Cements for Thermoplastics

		Boiling Point,	
Plastics	**Solvent**	**C°**	**F°**
ABS	Methyl ethyl ketone	40	[104]
	Methyl isobutyl ketone		
	Methylene chloride	40	[104]
Acrylic	Ethylene dichloride	84	[183]
	Methylene chloride	40	[104]
	Vinyl trichloride	87	[189]
Cellulose Plastics:			
Acetate	Chloroform	61	[142]
	Methylene dichloride	41	[106]
Butyrate, propionate	Ethylene dichloride	84	[183]
Ethyl acetate	Methyl ethyl ketone	80	[176]
Ethyl cellulose	Acetone	57	[135]
Polyamide	Aqueous phenol		
	Calcium chloride in alcohol		
Polycarbonate	Ethylene dichloride	41	[106]
	Methylene chloride	40	[104]
Polyphenylene oxide	Chloroform	61	[142]
	Ethylene dichloride	84	[183]
	Methylene chloride	40	[104]
	Toluene	110	[232]
Polysulfone	Methylene chloride	40	[104]
Polystyrene	Ethylene dichloride	84	[183]
	Methyl ethyl ketone	80	[176]
	Methylene chloride	40	[104]
	Toluene	110	[232]
Polyvinyl chloride and copolymers	Acetone	57	[135]
	Cyclohexane		
	Methyl ethyl ketone	80	[176]
	Tetrahydrofuran	65	[149]

sure is applied, the soft portion may be squeezed out of the joint, resulting in a poor bond.

Large surfaces may be dipped into, or sprayed with solvent cements. Cohesive bonds also may be made by allowing the solvent to flow into crack joints by capillary action. Small paint brushes and hypodermic syringes are handy cementing tools. Figure 18-11 shows a number of methods of cementing.

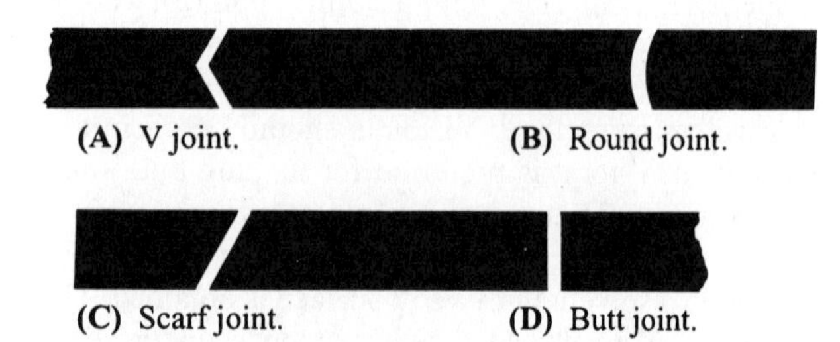

(A) V joint. (B) Round joint.

(C) Scarf joint. (D) Butt joint.

Fig. 18-10. Types of joints.

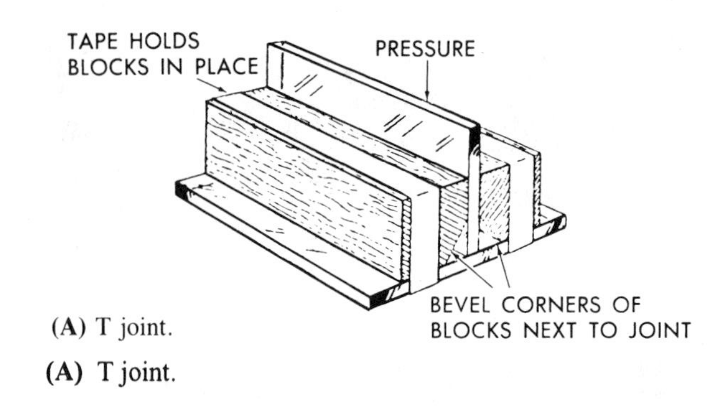

(A) T joint.

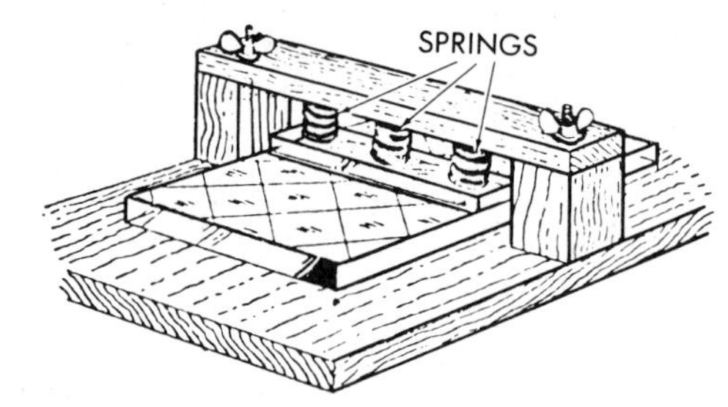

(B) Cementing a rib on a sheet

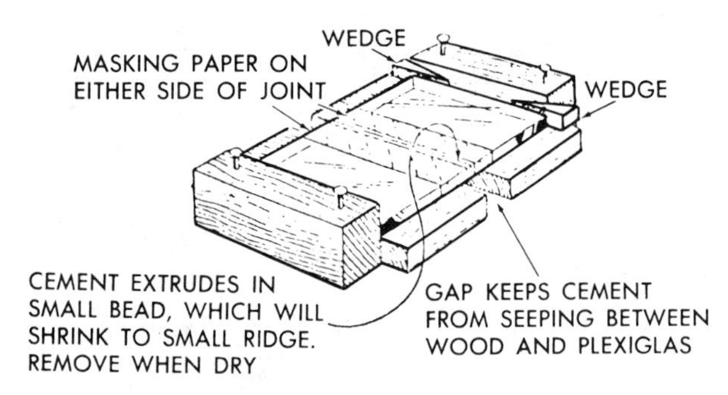

(C) Butt joint rig.

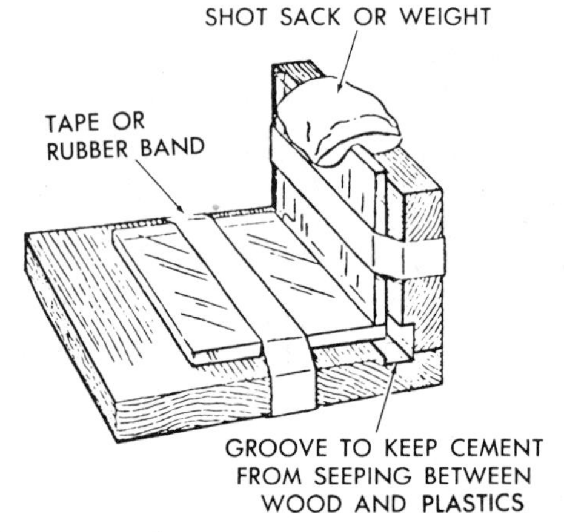

(D) Corner cementing.

Fig. 18-11. Various cementing methods. (Cadillac Plastics Co.)

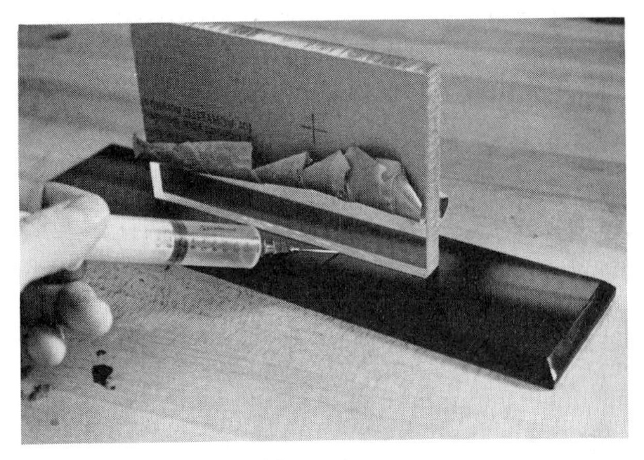

(E) Applying cement with a syringe

Fig. 18-11. Various cementing methods. (Cadillac Plastics Co.) *(Continued)*

Frictional Heating Techniques

The major techniques of chemical adhesion which use frictional heat are spin welding, dielectric bonding, and ultrasonic bonding.

Spin welding (bonding). *Spin welding* (bonding) is a friction method of joining circular thermoplastic parts. Frictional heat causes a cohesive melt when one or both parts are rotated against each other. Depending on the diameter and the material, joints must spin at 6 m/s [20 ft/s] with less than 138 kPa [20 psi] of contact pressure. When melting takes place, the spinning is stopped and the melt solidifies under pressure.

Joints may also be spin welded by rapidly rotating a filler rod on the joint. A heavy rod of the parent material is rotated at 5000 rpm and moved along the joints as it melts. (See Figure 18-12B.) The plastics weld looks like an arc weld on metal.

Vibration bonding, a variation of spin bonding, is a method by which non-circular parts may be bonded. Vibration frequencies are from 90 to 120 Hz and joint pressures range from about 1300 to 1800 kPa [200 to 250 psi].

Nearly any melt-processable, thermoplastic polymer (even dissimilar polymers with compatible melt temperatures) may be assembled into bottles, tubes, and other containers.

Dielectric or high-frequency bonding. *Dielectric bonding* is used to join plastics films, fabrics, and foams. Only plastics that have a high dielectric loss characteristic (dissipation factor) may be joined by this method. Cellulose acetate, ABS, polyvinyl chloride epoxy, polyether, polyester, polyamide, and polyurethane, have sufficiently high dissipation factors to allow dielectric sealing. Polyethylene, polystyrene,

(A) Plastics rod spin welded to plastics sheet.

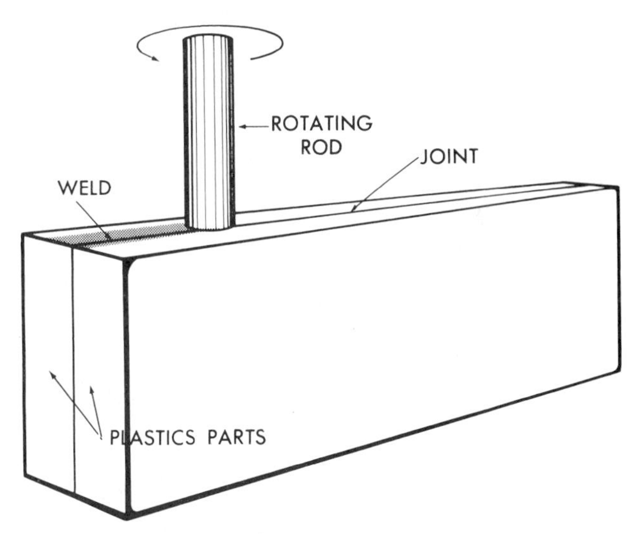

(B) A method of spin welding joints.

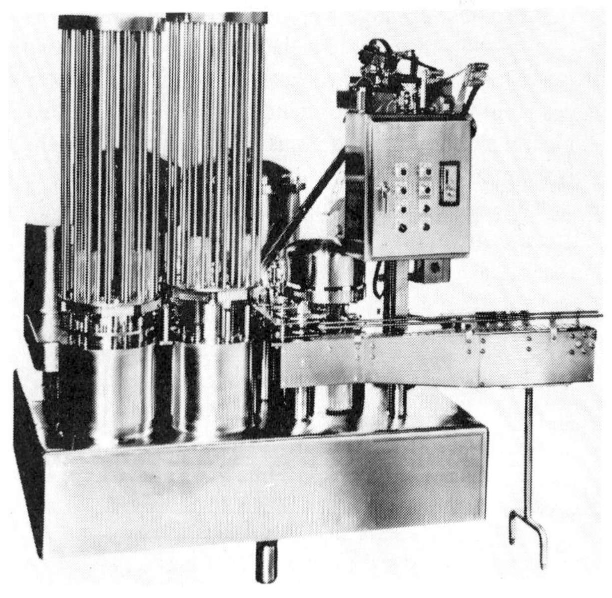

(C) This unit spin welds, fills, and caps preformed thermoplastic container halves. (Brown Machine Co.)

Fig. 18-12. Principle of spin welding (bonding).

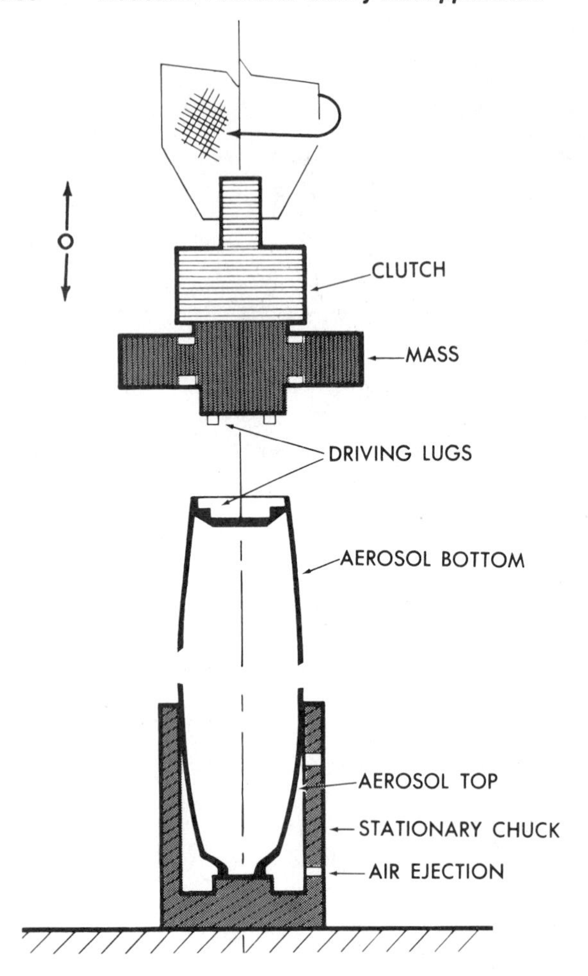

(D) Spin welding aerosol bottle halves. (DuPont)

Fig. 18-12. Principle of spin welding (bonding).

and fluoroplastics have very low dissipation factors and cannot be heat sealed electronically. The actual fusion is caused by high-frequency (radio-frequency) waves from transmitters or generators available in several kilowatt sizes. In the areas of the parts where the high-frequency waves are directed, molecules try to realign themselves with the oscillations (Fig. 18-13). This rapid molecular movement causes frictional heat and the areas become molten.

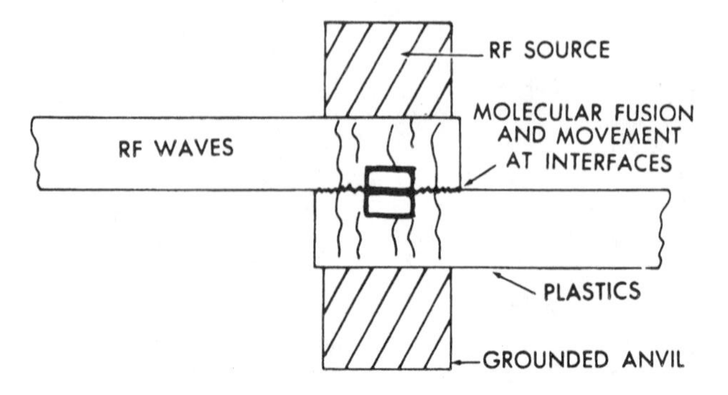

Fig. 18-13. Dielectric heat sealing with radio-frequency waves.

The Federal Communications Commission regulates the use of high-frequency energy. The generated signals are similar to those produced by TV and FM transmitters and operate at frequencies between 20 and 40 MHz.

Ultrasonic bonding. Ultrasonic energy is used to vibrate plastics mechanically. High-frequency mechanical vibrations in the range of 20 to 40 kHz are directed to the plastics part by a tool called a horn. (See Figure 18-14A.) An electronic transducer converts 60 Hz energy to the 20 to 40 kHz frequencies (Fig. 18-15). The high frequency causes the plastics molecules to vibrate, making sufficient frictional heat to melt the thermoplastic.

Ultrasonic techniques are used to activate adhesives to a molten state, to spot weld, and to sew or stitch films and fabrics together without the need for needles and thread. Simple joints of the type shown in Figure 18-16 may be welded in 0.2 to 0.5 seconds.

Staking is a term used to describe the ultrasonic or heated-tool formation of a locking head on a plastics stud that is like forming a head on a metal rivet. (See Figure 18-14E.) Plastics parts with studs may be assembled by this technique.

Many adhesives may be melted and caused to become cured by ultrasonic vibrations (Fig. 18-14G). Ultrasonic systems may be used to cut thermoplastic fabrics and degate parts from runner systems.

Spot bonding is a process similar to metal spot-welding used on plastics up to 0.25 in. [6 mm] in thickness using specially designed horns and high-power equipment. Vibrations from the horn penetrate the first sheet and nearly half the second; then the molten material flows into the space between the sheets. Films and fabrics are stitched in a similar fashion.

Insertion bonding uses ultrasonic means to place metallic inserts into plastics parts (Fig. 18-14F). The horn is used to hold the insert and direct the high-frequency vibrations into an undersized hole. As the plastics melts, the pressure of the horn forces the insert into the hole. Upon cooling, the plastics reforms itself around the insert.

Transferred-Heat Techniques

Chemical adhesion may also involve transferring heat into plastics parts from hot metals or gases. Common techniques include hot gas bonding, heated tool welding, impulse bonding, and electromagnetic bonding.

Hot-gas welding (bonding). Hot-gas welding consists of directing a heated gas (usually nitrogen) at temperatures of 400 to 800 °F [200 to 425 °C] onto the joints to be melted together. The temperature of the flameless hot-gas torch is controlled by regulating the gas flow or the heating source. Electric heating ele-

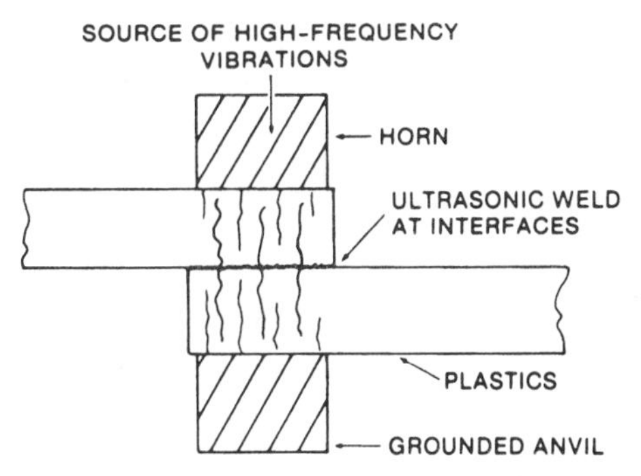

(A) Energy directors molded into parts eliminate movement of molten plastics to edges of joint.

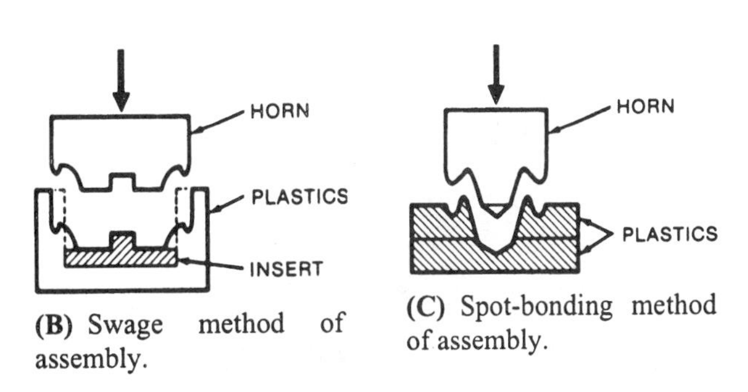

(B) Swage method of assembly.

(C) Spot-bonding method of assembly.

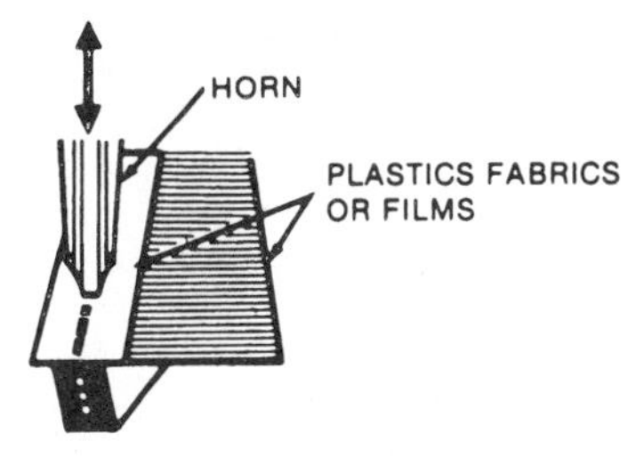

(D) Stitch method of bonding

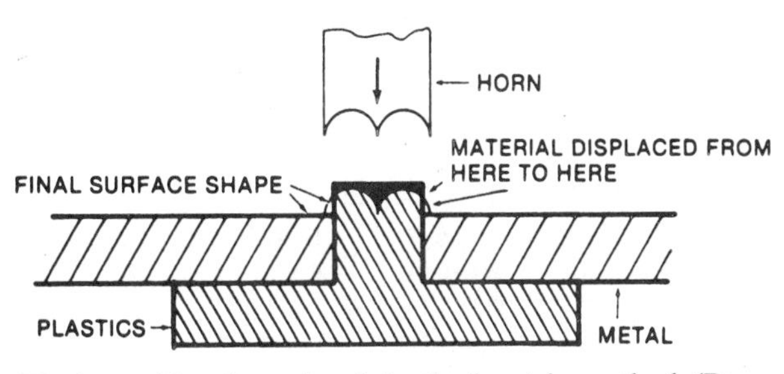

(E) Assembly of metal and plastics by stake method. (Branson Sonic Co.)

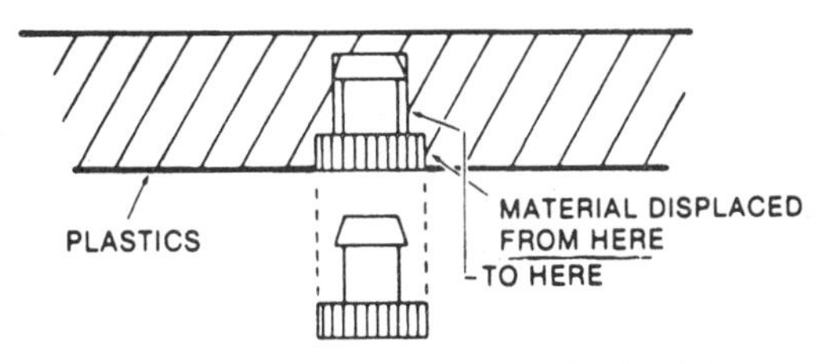

(F) Inserting a plastics stud. (Branson Sonic Co.)

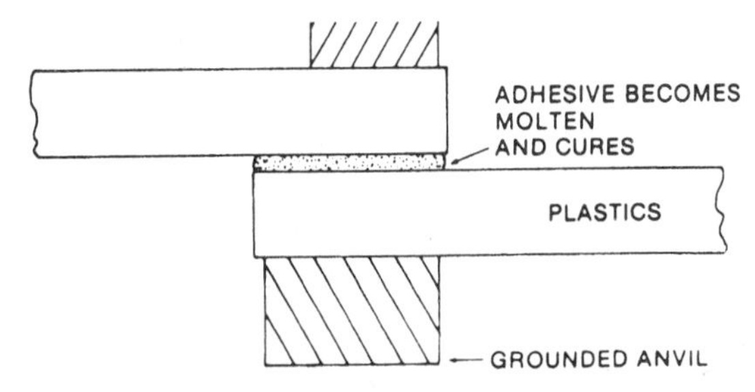

(G) Ultrasonic vibrations melt and cure adhesives.

Fig. 18-14. Ultrasonic bonding methods.

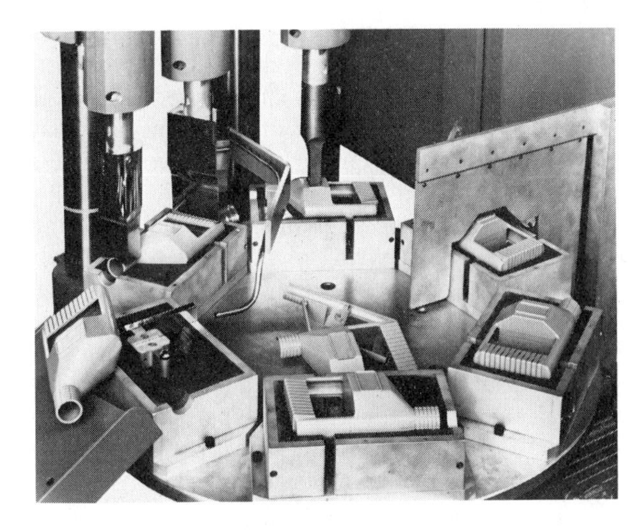

Fig. 18-15. Typical tooling for a rotary indexing ultrasonic assembly system. Three horns were used to bond a difficult plastics part. (Sonics and Materials, Inc.)

Fig. 18-16. A traversing ultrasonic head is used to seal polyester film. (Sonics and Materials, Inc.)

ments are preferred with a nitrogen or air pressure of 2 to 4 psi [14 to 28 kPa]. This process is similar to open-flame welding of metals (Figure 18-17). Filler rods or materials like the parent plastics are used to build up the welded area. Welds may exceed 85 percent of the tensile strength of the parent material (Table 18-3). As in any welding technique, the joint area must be properly cleaned and prepared and butt joints should be beveled to 60°.

Heated-tool welding (bonding). As shown in Figure 18-18, heated-tool bonding, or fusion welding, is a method in which like materials are heated and the joints brought together while in the molten stage. The melted areas are then allowed to cool under pressure (Fig. 18-18A). Electric strip heaters, hot plates, soldering irons, or special heating tools are used to melt the plastics surfaces. The surfaces of the heating tools may be coated with Teflon, but the use of lubricants or other materials to prevent sticking of the plastics to the hot metal is not advised because these materials contaminate and weaken the weld.

Pipes and pipe fittings may be joined by heated-tool bonding, and one of the most common uses of heat joining is the fusing of films (Fig. 18-18B). Not all thermoplastics can be heat sealed, but those that cannot can be coated with a layer of plastics that can be heat sealed. Electrically heated rollers, jaws, plates, or metal bands are used to melt and fuse film layers.

Impulse bonding. Impulse bonding may be thought of as a heated-tool method without using continuously heated tools. An impulse of electricity controlled to the proper amount is used to heat the tools (Fig. 18-19). Plastics films 0.01 in. [0.25 mm] thick are held under pressure as the tool is quickly heated and cooled.

Electromagnetic bonding. Induction bonding is an electromagnetic technique for bonding thermoplastics,

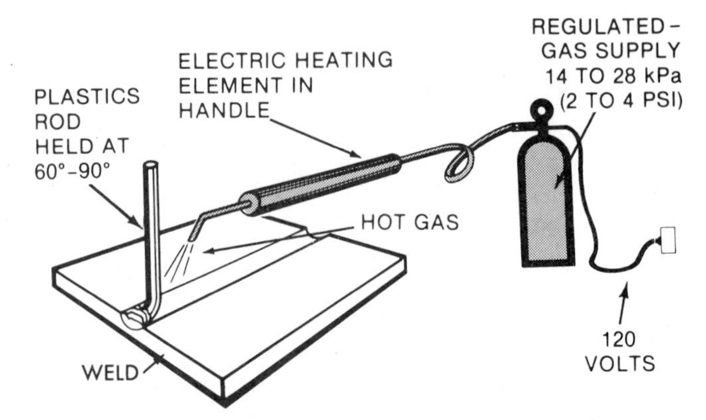

(A) Principle of hot-gas welding.

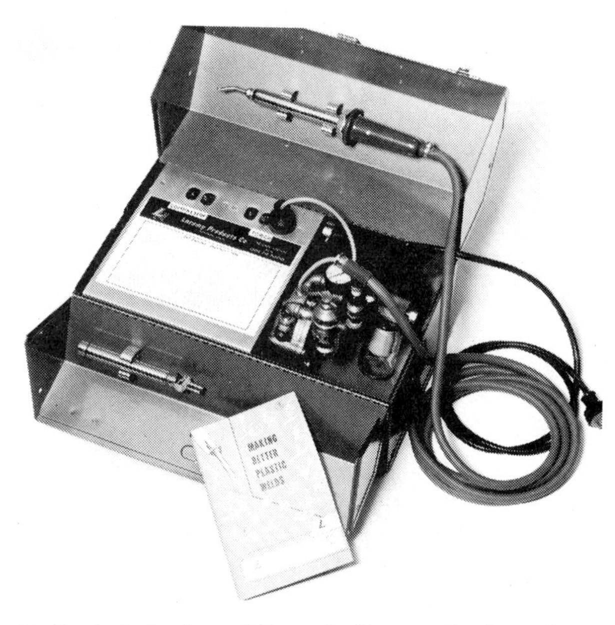

(B) Typical plastics welding unit. (Laramy Products Co.)

Fig. 18-17. Hot-gas welding of plastics.

Table 18-3. Weldability of Selected Plastics

Material	Bond Strength, %	Spot Weld	Staking and Inserting	Swaging	Hot-Gas Welding	Heated-Tool Bonding	Friction Bonding	Dielectric-Bonding
ABS	95–100	E	E	F	E	G	E	—
Acetal	65–70	G	E	P	G	G	G	G
Acrylics	95–100	G	E	P	E	F	G	G
Butyrates	90–100	G	G–F	G	P	G	G	E
Cellulosics	90–100	G	G–F	G	P	G	G	E
Phenoxy	90–100	G	E	G	G	G	G	G
Polyamide	90–100	E	E	F–P	G	F	G	G
Polycarbonate	95–100	E	E	G–F	E	G	G	G
Polyethylene	90–100	E	E	G	G–P	E	G	—
Polyimide	80-90	F	G	P	G	G	G	G
Polyphenylene	95–100	E	G	F–P	G	G	G	G
Polypropylene	90–100	E	E	G	G–P	G	G	—
Polysulfone	95–100	E	E	F	G	F	G	G
Polystyrene	95–100	E	E	F	E	G	E	—
Vinyls	40–100	G	G–F	F	F–P	E	E	E

Note: E—excellent, G—good, F—fair, P—poor.

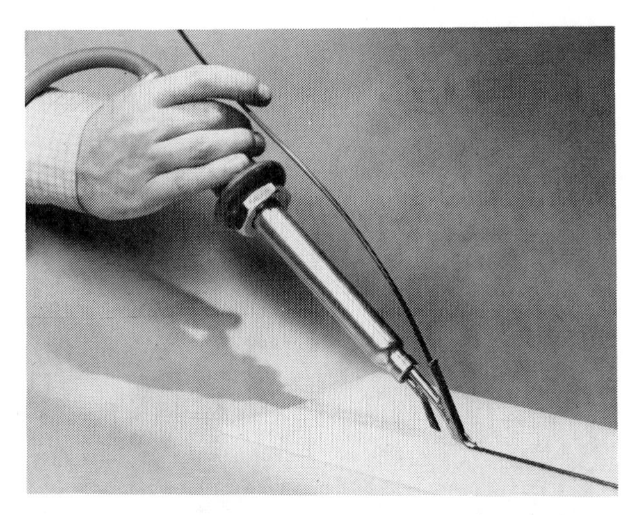

(C) Filler rod and welding tip. (Laramy Products Co.)

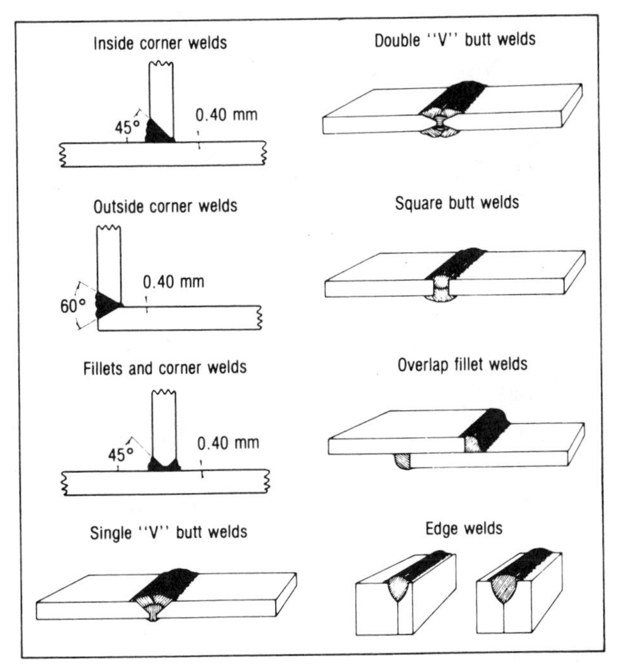

(D) Types of joints produced by hot-gas welding of thermoplastics. (*Modern Plastics Magazine*)

Fig. 18-17. Hot gas welding of plastics. *(Continued)*

although they cannot be heated directly by induction. The heat is produced by an induction generator with a power range from 1 to 5 kW and frequency range from 4 to 47 MHz. Metal powders (iron oxide, steel, ferrites, graphite) or inserts must be placed at the plastics joints. Then the metallic materials become hot when excited by the high-frequency induction source, melting the surrounding plastics. The metallic inserts or powders must remain in the final weld. Only a few seconds and slight pressure are needed for this rapid method of assembly (Fig. 18-20).

The induction coil must be as close to the joint as possible for rapid bonds. Non-metallic tooling must be used for alignment.

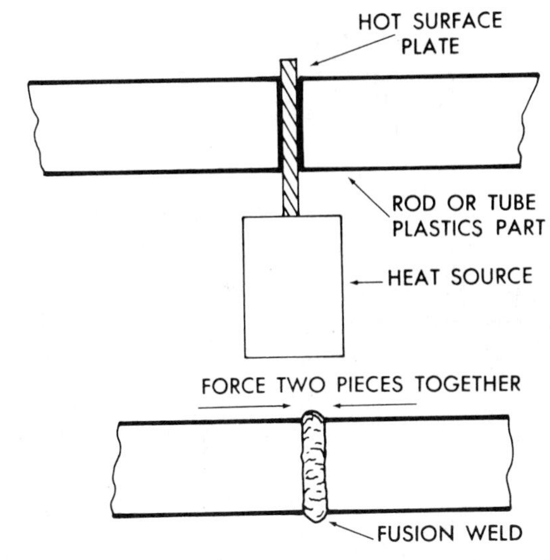

(A) Heated-tool bonding of plastics parts by fusion welding.

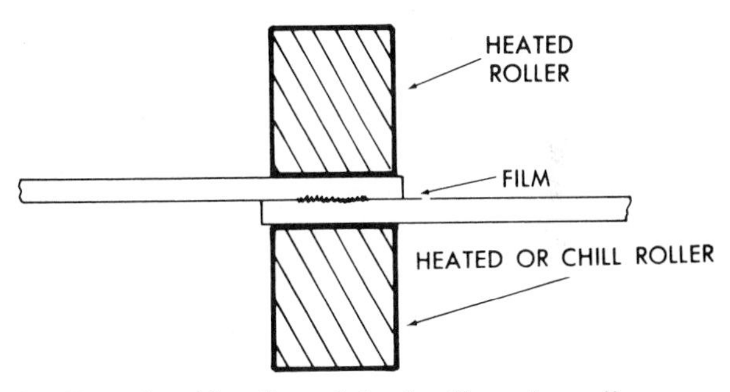

(B) Heated-tool bonding of plastics film using rollers.

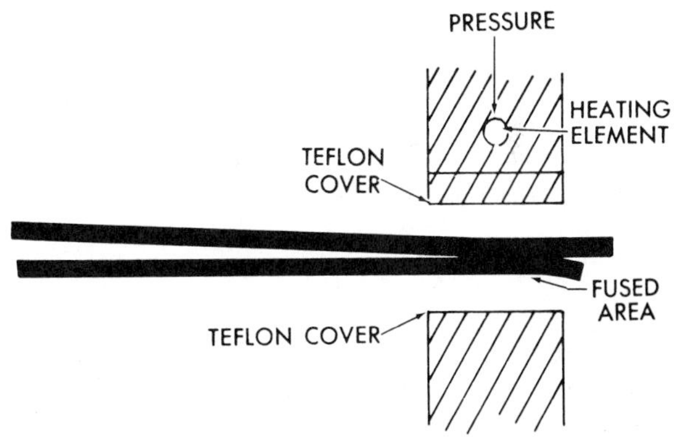

(C) Heated-tool bonding of plastics film using press.

Fig. 18-18. Heated-tool bonding of plastics.

Mechanical Fastening

There is a wide choice of mechanical fasteners for use with plastics. Self-threading screws are used if the fastener is not to be removed very often, but when frequent disassembly is required, threaded metal inserts are placed in the plastics.

In Figure 18-21, several types of metal inserts are shown. These may be molded in, or placed in the part after molding.

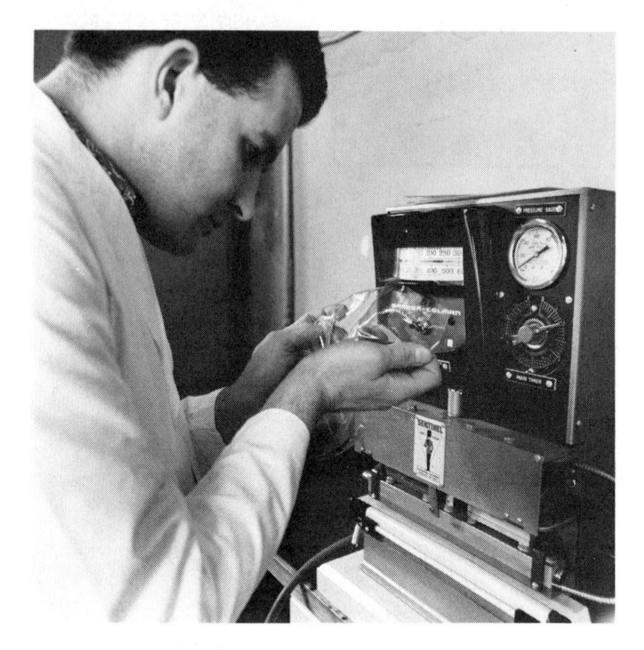

Fig. 18-19. An impulse-bonding machine being used to bond film.

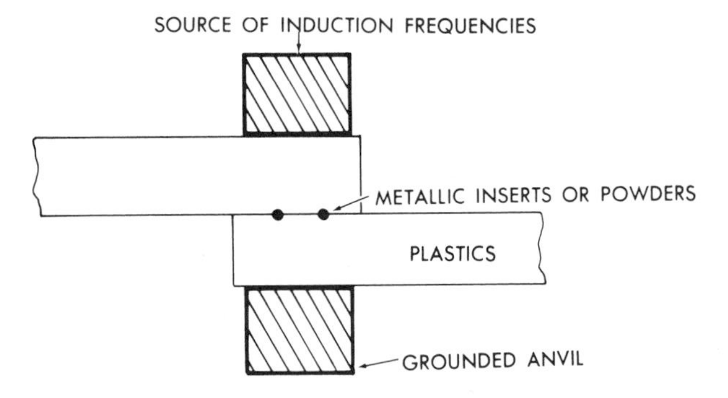

Fig. 18-20. Electromagnetic technique of induction welding.

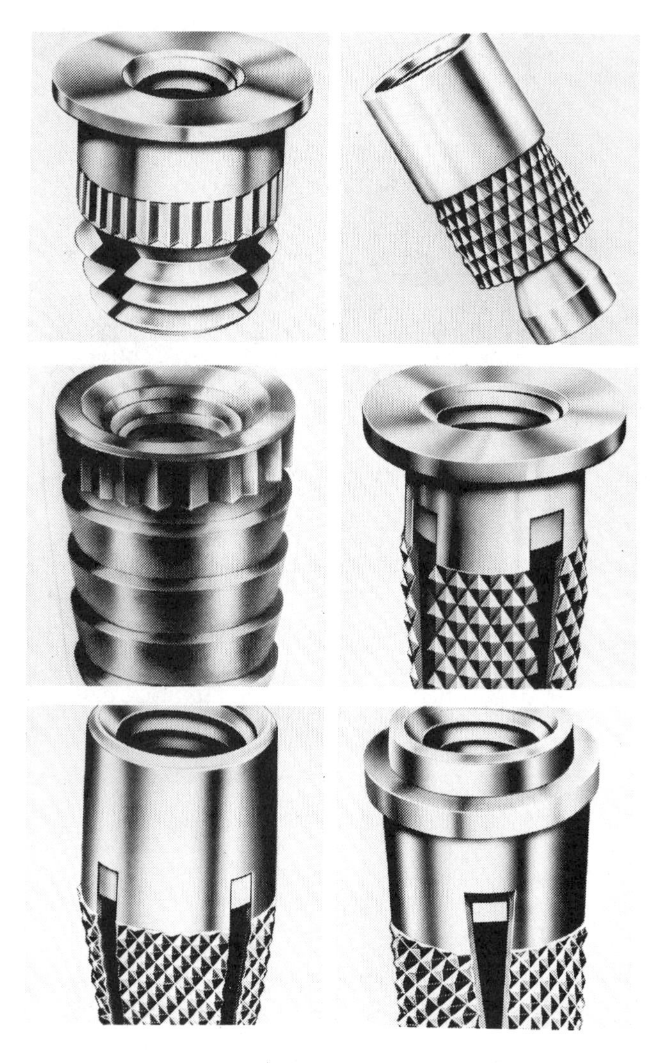

Fig. 18-21. Inserts designed for use in plastics are installed quickly and easily after molding. (Heli-Coil Products)

Screws and metallic or plastics rivets provide permanent assembly. Standard nuts, bolts, and machine screws, made in both metals and plastics, are used in common assembly methods (Fig. 18-22).

The spring clips and nuts shown in Figure 18-23 are low-cost, rapid mechanical fasteners. Hinges, knobs, catches, dowels, and other devices are also used for assembly of plastics.

Friction Fitting

Friction fitting is a term used to describe a number of pressure-tight joints of permanent or temporary assemblies. The most common are press fits, snap fits, and shrink fits. These techniques join like or unlike materials without any mechanical fasteners.

Press Fitting

Press fitting may be used to insert plastics or metallic parts into other plastics components. The parts may be joined while the plastics parts are still warm. When a shaft is press-fitted into a bearing or sleeve, the outside diameter—as well as the inside diameter—may be expanded (Fig. 18-24).

The distinction between a press fit and a snap fit is the undercut and amount of force required for assembly.

Snap Fitting

Snap fitting is a means of assembly in which parts are snapped into place. The plastics is simply forced over a lip or into an undercut retaining ring. Some examples are the simple locks or catches on plastics boxes or the covers on many parts including automotive dome

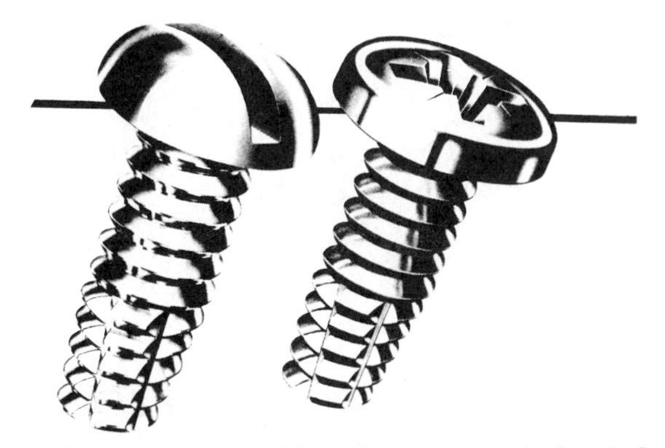

(A) Thread-cutting (self-threading) screws designed for hard plastics. (Parker-Kalon Fasteners Co.)

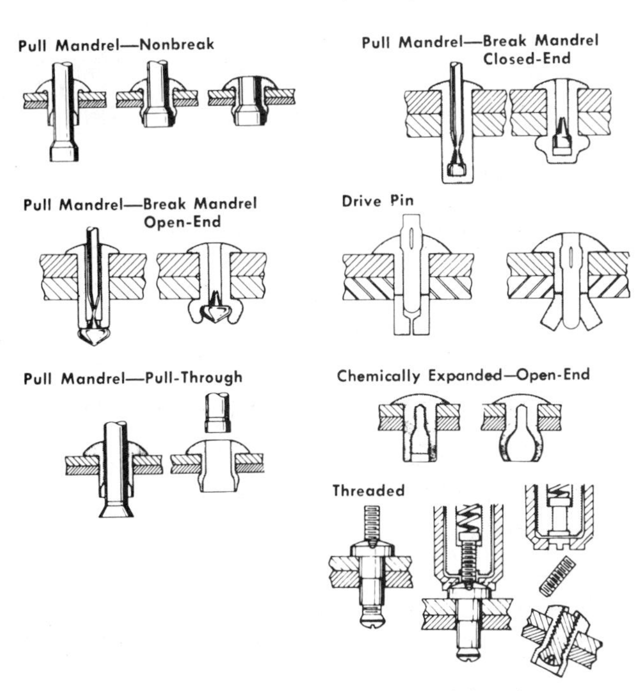

(B) Blind rivets. (Fastener Division, USM Corp.)

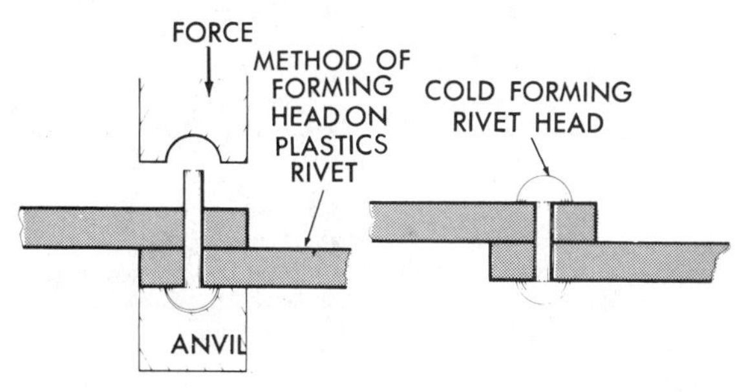

(C) Cold-forming of plastics rivets. Heads may be formed by mechanical or explosive means.

Fig. 18-22. Selected methods of assembly.

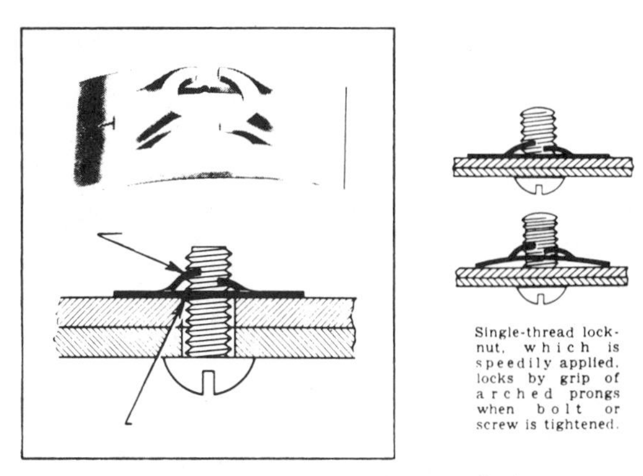

(A) Single-thread locknut. (Eaton Corporation)

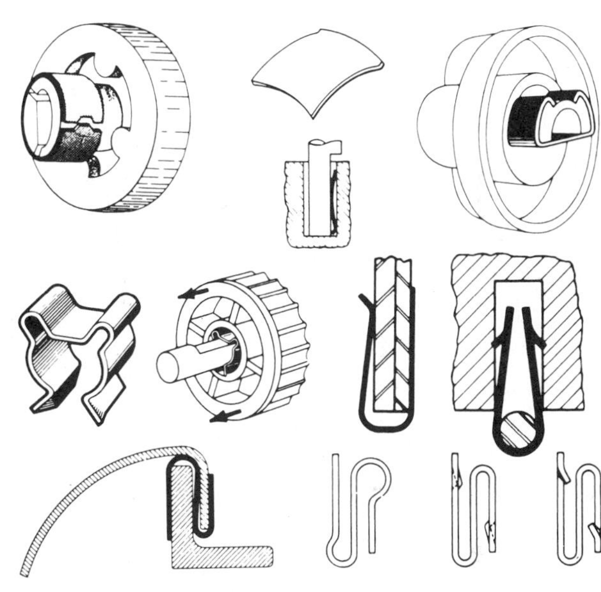

(B) Mechanical fasteners and devises for assembly of plastics parts.

Fig. 18-23. Inexpensive mechanical fasteners.

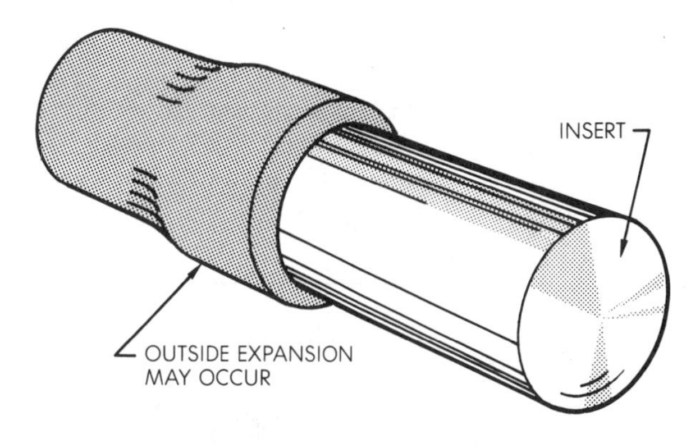

Fig. 18-24. Press-fitting, showing possible expansion of outside diameter of plastics part.

lenses, flashbulb covers, and instrument panels (Fig. 18-25).

In both press and snap fits, one of two parts is made smaller so that the two cannot be fit together without force. The intentional differences in the two part dimensions are called the *allowances.* A negative allowance is called *interference*, a characteristic necessary for a tight fit. Allowances made for unintentional variations in the dimensions are called *tolerances.* The maximum and the minimum dimensions are *limits.* These limits define the tolerance.

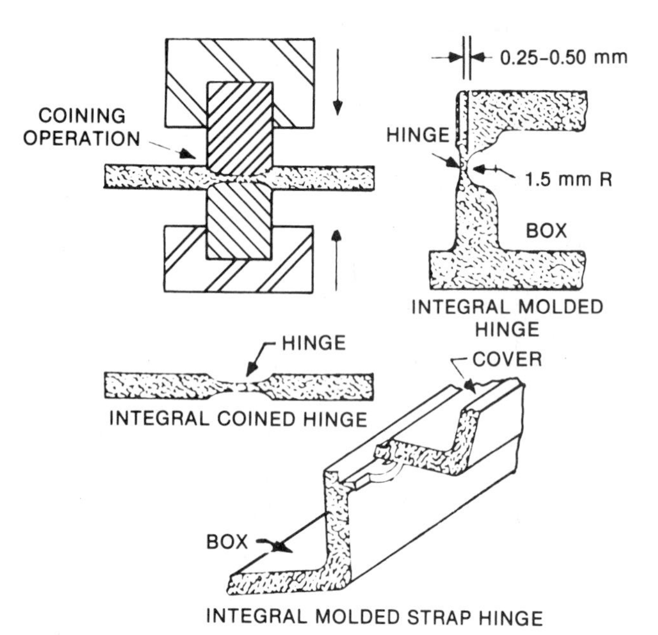

(A) Examples of integral molded and coined hinges.

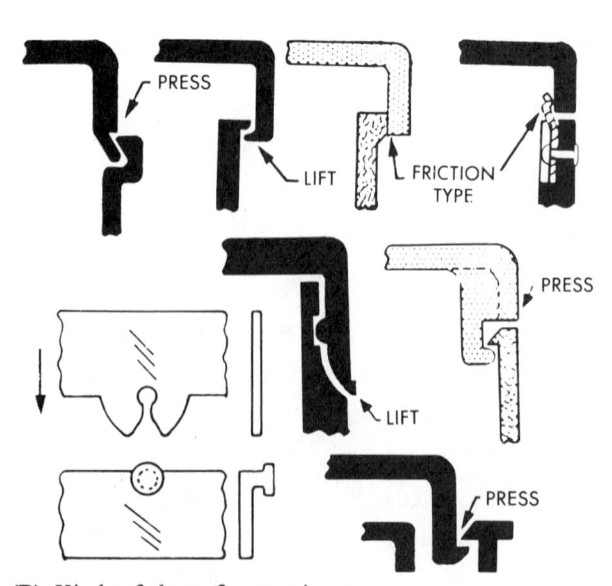

(B) Kinds of clasps for containers.

Shrink Fitting

Shrink fitting refers to placing inserts in the plastics just after molding and allowing the plastics to cool (Fig. 18-26). It also refers to the process of placing plastics parts over substrates where they are heated until the plastics shrinks to its original shape (Fig. 18-27).

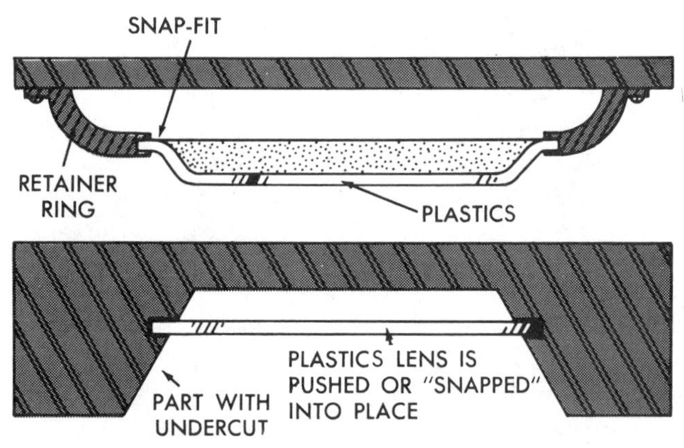

(C) Two examples of snap fitting.

Fig. 18-25. Assembly methods using snap-fitting and integral-hinge techniques.

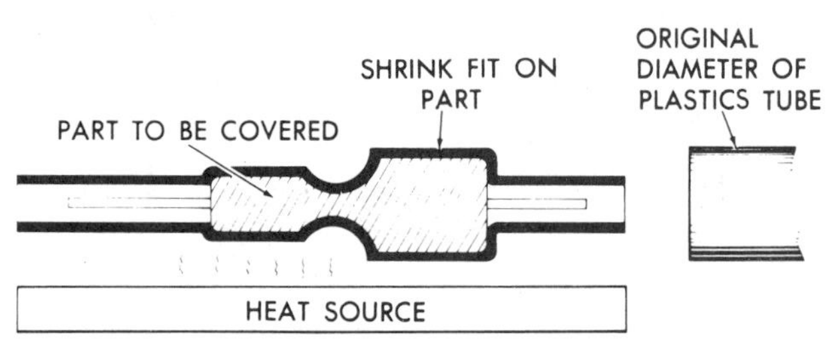

(A) Shrink-fit into plastics tube over electronic part. Note size of tube before heating.

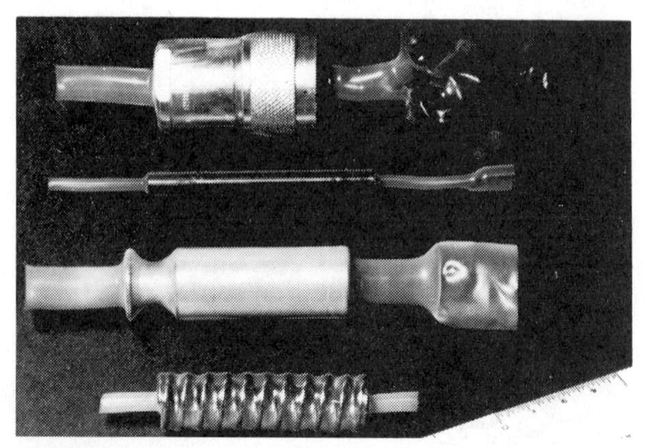

(B) Shrink-fitting plastics over parts for protection or for anti-stick properties. (Chemplast, Inc.)

Fig. 18-26. Shrink fitting technique and products.

Fig. 18-27. A shrink-fitting application demonstrating the strength of this packaging. (BASF)

Vocabulary

The following vocabulary words are found in this chapter. Look up the definition of any of these words you do not understand as they apply to plastics in the glossary, Appendix A.

Adhesive
Adhesion
Allowances
Cement
Cement bonding
Cohesion
Electromagnetic bonding
Glue
Heated-tool bonding
Hot-gas weld bonding
Hot-melt
Induction bonding
Interference
Limits
Mechanical fastening
Tolerances
Shrink fit
Spin bonding or welding
Ultrasonic bonding

Questions

18-1. Identify the type of adhesion created when materials are joined together and there is an intermingling of molecules.

18-2. Name a bonding process similar to spot welding of metals.

18-3. A __?__ dissipation factor is essential for dielectric or high-frequency bonding.

18-4. The bonding process that uses high-frequency vibration is called __?__.

18-5. The percent of bond strength for hot-gas weld bonding of polyethylene is __?__.

18-6. Name the bonding that is a method of frictionally joining circular thermoplastics together.

18-7. List three common solvents for solvent bonding acrylic plastics.

18-8. Which solvent from Table 15-2 will evaporate most rapidly?

18-9. Cements composed of solvents and a small quantity of the plastics to be joined are called __?__.

18-10. Name the term used to describe the ultrasonic forming of a locking head on plastics studs.

18-11. Insertion is a technique of which bonding method __?__.

18-12. Which bonding method would be used in the meat packaging department of a grocery supermarket?

18-13. Name two plastics often bonded by hot-gas welding.

18-14. Name the bonding process in which the plastics film is quickly heated and cooled by the die.

18-15. List two major advantages of impulse bonding.

18-16. Name the four basic ways plastics are joined.

18-17. Solvents or blends of solvents that melt selected plastics joints together are known as __?__ cements.

18-18. Into which of the four basic joining methods does spin welding fit?

18-19. In __?__, thermoplastic materials are softened by a jet of hot gas.

18-20. Identify the bonding method usually performed on films and fabrics.

18-21. Thermoplastics with __?__ factors such as ABS can be sealed by dielectric joining.

18-22. In __?__, high-frequency causes molecules to move rapidly, thus melting the plastics.

18-23. Thermosetting adhesives are __?__ in most solvents once they have cured.

18-24. In solvent cementing, evaporation of the __?__ may cause stress cracks.

18-25. Joint preparation for welding thermoplastics resembles that for __?__.

18-26. A variety of __?__ may be used to join thermoplastics, including knives, soldering irons, strip heaters and hot plates.

18-27. A special type of nut serving the function of a tapped hole is called an __?__.

18-28. Since __?__ materials are too brittle to be deformed by a self-tapping screw, a thread-cutter type must be used.

18-29. Name a gas that is used for hot gas welding.

Activities

Solvent Bonding

Introduction. Solvent bonds are readily produced, but often not easily evaluated. Tensile testing of solvent bonds often results in failure of the substrate rather than the bond. Shear testing can be successful if the samples are carefully prepared. To maintain uniform bond area, a fixture is recommended.

Equipment. Sheet acrylic, 6 mm [¼ in.] thick, acetone, eyedropper, universal testing equipment, band or table saw.

Procedure

18-1. Make or acquire an aluminum fixture similar to the one shown in Figure 18-28. This fixture is machined for strips of ¼ inch thick acrylic. This figure also shows three strips of acrylic in the fixture. These strips are 25 mm [1 in.] wide and about 400 mm [16 in.] long.

18-2. After cutting 1 inch strips of acrylic, deburr all edges and position bottom piece on fixture. Apply acetone to surface and lay second strip

Fig. 18-28. Aluminum fixture aligning three strips of acrylic for bond test samples.

into fixture. Then apply acetone to the top of the second strip, place third strip in place. Make sure that the strips are firmly against fixture and place a weight on the top strip.

18-3. Allow 24 hours for the solvent to evaporate from the bond area. Cut the bonded strips into pieces, about 25 mm [1 in.] long. (See Figure 18-29.) Measure carefully to determine the bonded area.

18-4. Set-up tester for compression test.

18-5. These pieces may shatter during testing. To protect operators and observers, place a cage around the sample. A one quart can, with the bottom removed, can provide needed protection. To hold the can or cage around the sample, hang it from the upper compression plate or place it on supports.

18-6. The sample shown in Figure 18-29 had a bonded area of 1.5 in^2, 0.75 in^2 per side. That much area requires considerable force to cause failure. Be sure that the load cell in the universal tester has large enough capacity.

Fig. 18-29. Bonded sample cut to size for testing.

18-7. Run the tester at a low speed, about 5 mm/minute [0.2 in./minute]. If the "legs" of the sample of equal in height and the bonded areas are equal, the bonds should fail simultaneously. Examine the broken pieces. If the test was successful, both bonds should fail, yielding three unfractured pieces. The bonds in this sample failed at 2400 psi.

Additional activities

a. Compare the strength of solvent bonds to joints made using various mechanical adhesives.

b. Some adhesives involve both cohesive forces and mechanical adhesion. Contact cement is applied to both surfaces and allowed to stiffen. When the two surfaces are pressed together, the resulting bond includes cohesive bonds between the two surfaces of contact cement, and mechanical adhesion between the cement and the adherends. Careful examination should indicate whether the failure was mechanical or within the cement.

c. Compare solvent and adhesive bonds to joints made using double sided adhesive tape.

Spin Welding

In spin welding, frictional heat softens two surfaces, allowing some molecular intertangling. When the heated areas cool, the bond gains considerable strength.

Equipment. Sheet acrylic material, acrylic rod or tube, 12 mm [0.5 in.] diameter or less, drill press, alignment fixtures, tensile tester.

Procedure

18-1. Cut acrylic sheet into squares 25 to 30 mm square.

18-2. Cut acrylic rod or tube into lengths, about 60 to 75 mm long. The rods need to be long enough to grip in the chuck of a drill press and to fit in the grips of the tensile tester. Figure 18-30 shows materials ready for spin welding.

18-3. Mount the rod in the chuck of the drill press, turn on the drill, and press the spinning rod against the acrylic square. When the heat is sufficient to soften the acrylic surfaces, hold downward pressure on the rod and turn off the press. Hold until the bond begins to cool.

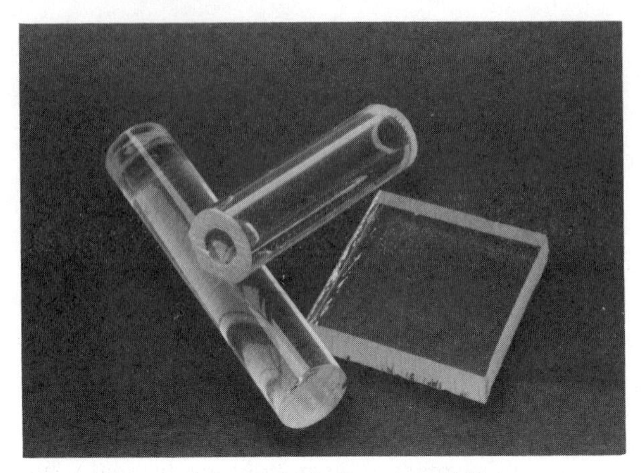

Fig. 18-30. Sample materials for spin welding.

18-4. In order to tensile test the bonds, the rods must be positioned directly opposite one another. To locate the second rod, make a simple fixture that contains a hole slightly larger in diameter than the rods. Position the fixture using a piece of rod in the chuck. Then load the square and rod (as seen in Figure 18-31) onto the fixture, and bond the second rod.

18-5. Tensile test the resulting sample. It may be necessary to put V-type grips into the tensile tester jaws. This sample seen in Figure 18-31 failed at 684 psi.

18-6. Determine the combination of RPMs, pressure, and time that result in the highest strengths.

Fig. 18-31. Fixture for aligning second rod.

Impulse Bonding

Introduction. Heat joining or sealing creates a thermal bond between two or more layers of thermoplastic film. Impulse bonding uses tools that are hot only during the bonding cycle. Impulse bonding refers to the quick heating and cooling of the bonding element. Silicone rubber, PTFE, and other antistick agents are used to prevent the heated plastics from sticking to the heating element or tool. This type of bonding is common in the packaging industry because it is often used on materials that will not stay bonded unless they are held together while cooling.

Equipment. Impulse bonder, sandwich bags.

Procedure

18-1. Carefully examine a sandwich bag. Was is originally made from sheet material or from a blown tube?

18-2. Cut off the bonded seams.

18-3. Rebond the seams and adjust the dwell and current of the bonder to achieve a good seal and a clean cut. Cut-wires are usually coated with PTFE. If the coating is cracked or missing, the cut will be adversely affected.

18-4. The cut-wire should run down the center of the bonded zone. If the cut is too close to one edge of the bonded zone, it may be incomplete.

18-5. If the nonstick coating above the heating element is damaged, the bond and/or cut will be poor.

Hot-Gas Welding

Introduction. Hot-gas welding usually implies joining thermoplastic materials that are greater than 1 mm in thickness. The hot gas melts the plastics so that they fuse. Hot air is appropriate for some materials. For other materials, nitrogen gas is essential to prevent oxidation and achieve strong welds. Table 18-4 indicates the temperature, gas, and weld angle for selected plastics.

There are four important variables to consider when making hot gas welds.

1. Gas temperature
2. Gas pressure
3. Filler rod and torch angle
4. Feed speed

Equipment. Materials to weld, approximately 2 x 20 x 100 mm, rod to match selected plastics, hot-gas welder.

Table 18-4. Welding Data for Thermoplastics

	Welding Temperature, °C	Welding Gas	Butt-Weld Strength, %	Weld Angle, degrees
ABS	175–200	Nitrogen	50–85	60
Acrylic	315–345	Air	75–85	90
Chlorinated Polyether	315–345	Air	65–90	90
Fluorocarbon	285–345	Air	85–90	90
Polycarbonate	315–345	Nitrogen	65–85	90
Polyethylene	285–315	Nitrogen	50–80	60
Polypropylene	285–315	Nitrogen	65–90	60
Polystyrene	175–400	Air	50–80	60
PVC	260–285	Air	75–90	90

Note: Welding temperature is measured 6 mm from welding tip.

Procedure

18-1. Protect table with temperature-resistant covering.

18-2. Choose plastics and determine parameters from Table 18-4.

18-3. Regulate gas supply to about 25 kPa [3.6 psi]. The volume of gas going through the heating element determines the welding temperature. *Do not* plug in heating unit until gas is flowing through the torch.

18-4. Check gas temperature.

18-5. Hold welding tip about 6 mm [¼ in.] away from thermometer to determine temperature of gas.

18-6. Direct about 60 percent of heat on plastics pieces and 40 percent on filler rod. Once the plastics are molten, push the filler rod into the weld joint with a light pressure.

18-7. At the end of the weld, continue to hold a light pressure on the filler rod until it has cooled.

18-8. Cut off filler rod.

18-9. Make several welds before turning off welder.

18-10. Tensile test the welds. Compare results with the strengths listed in Table 18-4.

Molded-In Threaded Inserts

Introduction. Threaded inserts enjoy wide use in plastics parts. Inserts of varying lengths, diameters and thread pitch are available from a number of manufacturers. The holding strength of inserts depends on the surrounding plastics, the temperature of the insert during molding, and several injection molding process conditions. To systematically examine the effects of varying process conditions or differing materials, it is necessary to prepare molding and testing equipment.

Equipment. Injection molding machine, mold to position and hold inserts, testing fixture, selected plastics, threaded inserts to match mold.

Procedure

18-1. Prepare a simple injection mold. A significant problem is removing the part from the mold. One solution is to use an ejector pin as shown in Figure 18-32. The pin protruding from the end of the ejector sleeve has a diameter equal to the minor diameter of the threaded brass insert. It also has a length equal to the insert, so that no plastic can flash over the end or into the threads. During ejection, the ejector presses on the insert and forces the molding out of the mold. Figure 18-33 shows an insert on the pin in the mold cavity. Notice the damage on the mold face caused when an insert fell off the pin during mold closing (Figures 18-32 and 18-33).

18-2. The complete part is seen in Figure 18-34. Testing the insert requires a fixture to hold it in a testing machine. Figure 18-35 shows a holding fixture. The hole in the fixture is large enough

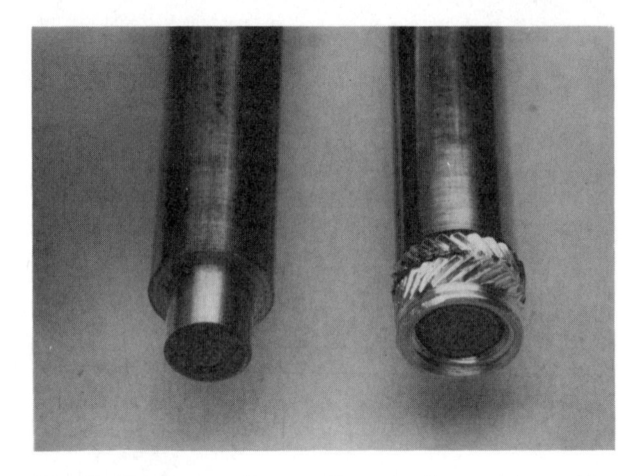

Fig. 18-32. Ejector sleeve with protruding rod to position threaded insert.

Fig. 18-33. Insert in mold on ejector pin.

Fig. 18-35. Holding fixture for testing threaded inserts.

to allow the insert to be pulled through without binding on the fixture. Figure 18-36 shows the part held in the fixture, ready for testing. The threaded ron will go into the upper jaw, and the fixture into the lower jaw of the tester.

18-3. Determine molding conditions or materials of interest, and mold a group of parts.

18-4. Wait 40 hours for the parts to stabilize before testing.

18-5. Calculate means and standard deviations to determine the relative magnitude of the effects.

Fig. 18-34. One trimmed sample.

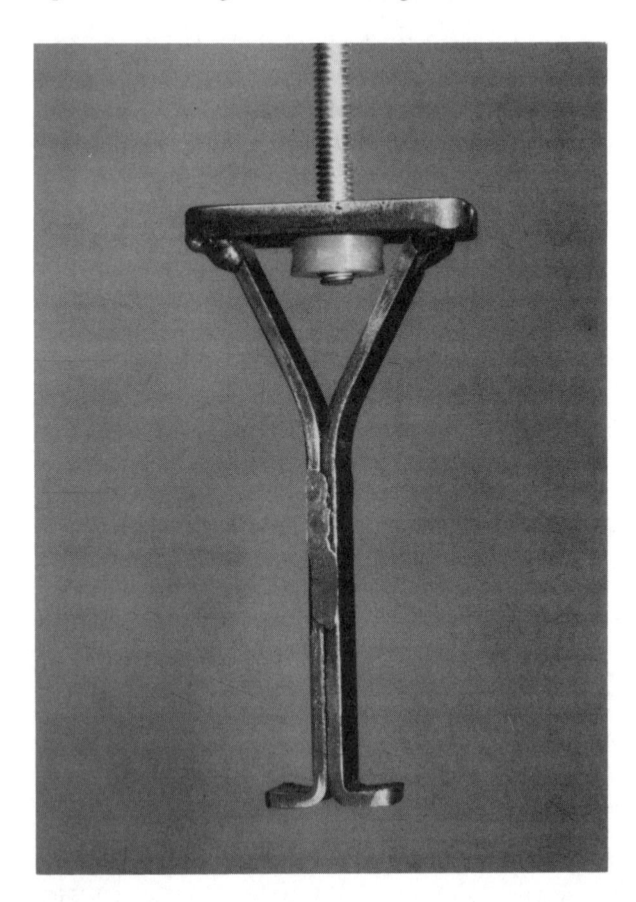

Fig. 18-36. Sample positioned in fixture, ready for loading into tensile tester.

Chapter 19

Decoration Processes

Introduction

In this chapter, you will learn that plastics may be decorated for some of the same reasons that cloth, ceramics, metals, and other materials are decorated, and that many of the processes are similar.

A number of decorating processes are used to produce plastics parts. Decorating may be done during molding, directly afterward, or before final assembly and packaging. The least costly way to produce decorative designs on the product is to include the desired design in the cavity of the mold. These designs may be textures, raised or depressed contours, or informative messages such as trademarks, patent numbers, symbols, letters, numbers, or directions.

Embossing or *texturing* of hot melts may be called a type of rotary molding. The majority of thermoplastics sheets or films are embossed against a composition roll or matched male and female rolls. The desired pattern will be retained if there is a proper balance between the pressure of the embossing roll, heat input, and subsequent cooling. Some polyvinyls and polyurethanes are embossed with textured casting paper. After the polymer has cooled or cured the paper is removed, leaving behind the pattern of the release paper.

In decorating plastics items, in the mold or out, surface treatment and cleanliness are of chief importance. Not only must the molds remain clean and mark-free, but the molded items must be properly prepared to ensure good decorating results. *Blushing* is the result of applying coatings over items that have not been properly dried to eliminate surface moisture. *Crazing* is due to solvent cutting along lines of stress in the molded plastics. The fine cracks (crazing) may be on or under the surface or extend through a layer of the plastics material.

A change in mold design may be needed to produce a stress-free molding, which will eliminate the problem.

Prior to decoration, the surface of the plastics must be cleaned. All traces of mold release, internal plastics lubricants, and plasticizers must be removed. Plastics parts become electrostatically charged, attracting dust and disrupting the even flow of the coating. Solvent or electronic static eliminators may be used to clean and prepare plastics articles for decorating.

Polyolefins, polyacetals, and polyamides must be treated by one of the methods described below to ensure satisfactory adhesion of the decorating media.

Flame treatment consists of passing the part through a hot oxiding flame of 2012 to 5072 °F [1100 to 2800 °C]. This momentary flame exposure does not cause distortion of the plastics but makes the surface receptive to decorating methods.

Chemical treatment consists of submerging the part (or portions of the part) in an acid bath. On polyacetals and polymethylpentene polymers, the bath etches the surface making the surface receptive to decorating. For many thermoplastics, solvent vapors or baths may be used for the etching treatment.

Corona discharge is a process in which the surface of the plastics is oxidized by an electron discharge (corona). The part or film is oxidized when passed between two discharging electrodes.

Plasma treating subjects plastics to an electrical discharge in a closed vacuum chamber. Atoms on the surface of the plastics are physically changed and rearranged, making excellent adhesion possible.

The nine most widely used decorating treatments for plastics are included in the following chapter outline:

I. Coloring
II. Painting
  A. Spray painting
  B. Electrostatic spraying
  C. Dip painting
  D. Screen painting
  E. Fill-in marking
  F. Roller coating
III. Hot-leaf stamping
IV. Plating
V. Engraving
VI. Printing
VII. In-mold decorating
VIII. Heat-transfer decorating
IX. Miscellaneous decorating methods

Coloring

The way to color plastics is to blend the least costly pigments into the base resin. Color matching may be a problem, therefore, successive batches of the same color of plastics may vary slightly. Most producers of colored resins and plastics encourage the use of stock or standard colors. Plastics parts of an assembly may be produced in different plant locations and at different times; therefore, it becomes necessary that color standards be carefully considered. Plasticizers, fillers, and the molding process may affect the final product color.

Colorants in the form of dry powders, paste concentrates, organic chemicals, and metallic flakes are usually blended with a given resin mix. Banbury, two-roll, and continuous mixers are used to disperse thoroughly the pigments in the resin. The colored resin may then be cast or extruded. Water- and chemical-solvent dyes have been used with success on many plastics. The procedure consists of dipping the parts in the dye bath and air drying. Three advantages and disadvantages of coloring decorating processes are given below.

Advantages of Coloring Plastics

1. Colored resin control is better in mass production.
2. Dyeing is less costly for short runs.
3. Surface dyeing is better for lenses.

Disadvantages of Coloring Plastics

1. Some colors are hard to produce and match.
2. There may be color migration and varying coloration in pieces with uneven thickness.
3. Pigment mixing in resin is more costly for short runs.

Painting

Painting plastics is a popular, low-cost way to decorate parts and provide flexibility in product color design. Transparent, clear, or colored plastics may be painted on the back surface for a striking contrast, variety, or appearance, an effect not possible by other methods. The six painting methods used in decorating plastics include:

1. Spray painting
2. Electrostatic spraying
3. Dip painting
4. Screen painting
5. Fill-in marking
6. Roller coating

The solvents or curing systems used in the paint must be chosen and controlled with care. As a rule, thermosetting plastics are less subject to swelling, etching, crazing, and deterioration from solvents. Temperature may be a limiting factor for the curing or baking of paints on many plastics. Radiation curing is one method used to cure coatings on plastics (See Chapter 20, Radiation Processes.)

Spray Painting

The most versatile and most often used method of decorating all sizes of plastics articles is spray painting. It is a less costly, rapid method of applying coatings. The spray guns may use air pressure (or hydraulic pressure of the paint itself) to atomize the paint.

Masking—needed when areas of the part are not to be painted—may be done with paper-backed masking tape or durable, form-fitting metal masks. Polyvinyl alcohol masks may be sprayed over areas and later removed by stripping or the use of solvents. Electroformed metal masks are preferred since they conform to the contour of the item and are durable. Four basic types of electroformed masks are shown in Figure 19–1.

Electrostatic Spraying

In electrostatic painting, the plastics surface must be treated to take an electrical charge; then the surface passes through atomized paint that has an opposite charge. The paint may be atomized by air, hydraulic pressure, or centrifugal force (Fig. 19–2). Nearly 95 percent of the atomized paint is attracted to the charged surface, thus this is a highly efficient way of applying paint. Narrow recesses are hard to coat, however, and

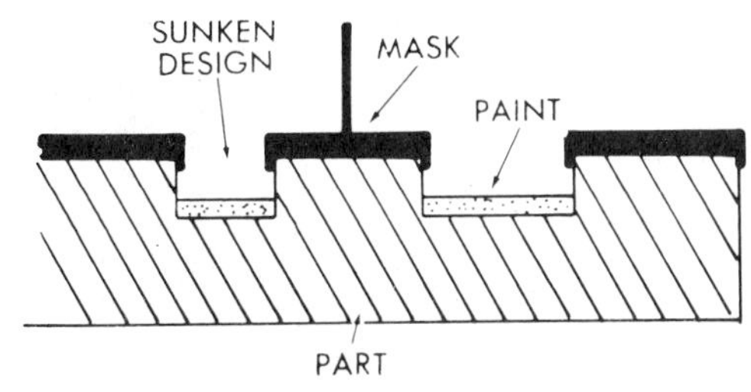

(A) Lip mask on sunken design.

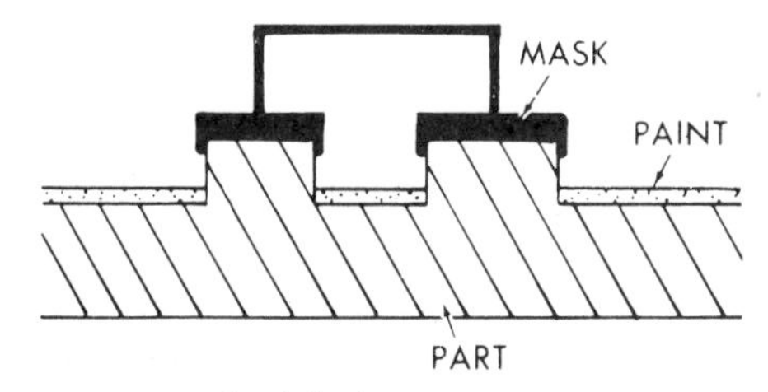

(B) Cap mask or raised design.

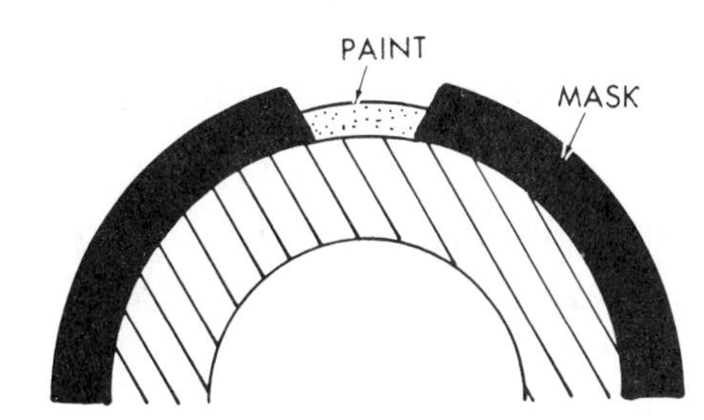

(C) Surface cutout mask.

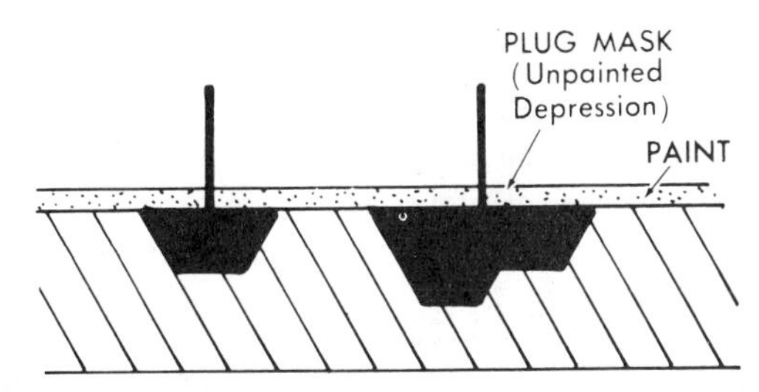

(D) Plug mask for unpainted depressions.

Fig. 19-1. Basic types of electroformed masks.

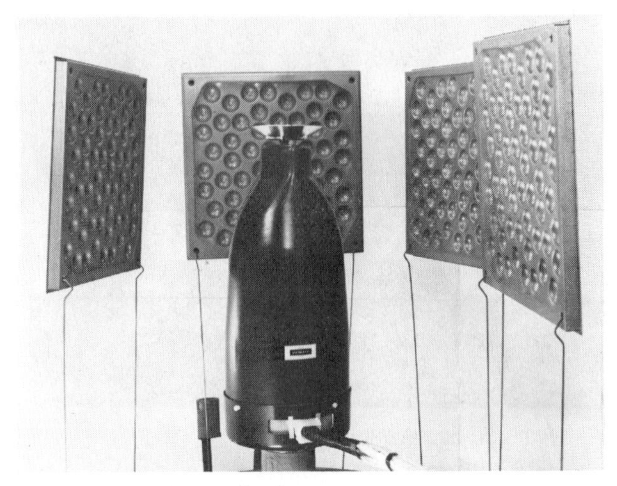
(A) Electrostatic atomization.

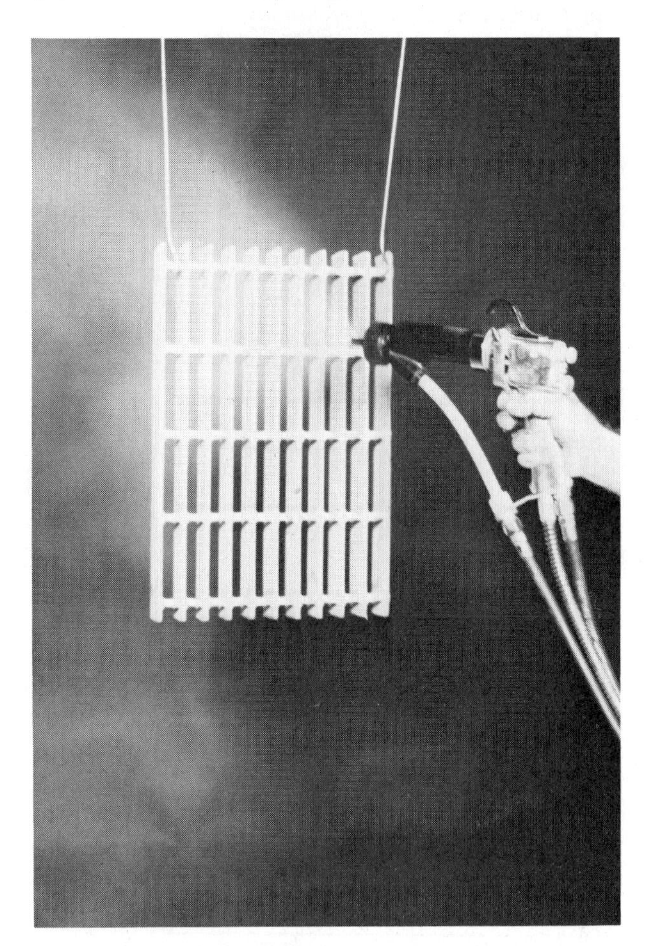
(B) Compressed air atomization.

Fig. 19-2. Methods of atomizing paint in electrostatic painting. (Ransburg Corp.)

metal masks are not practical. If dry plastics powders are used, the coated substrate must be placed in an oven to fuse the powder to the product. There are no solvents released during application or curing; however, the product must be able to withstand curing temperatures.

Dip Painting

Dip painting (Fig. 19–3) is useful when a single color or a base color is needed. A uniform coating may be applied if the part is withdrawn from the paint very slowly. Enough time must be allowed for drainage. Excess paint may be removed by spinning the part, hand wiping, or electrostatic methods.

Screen Painting

Screen painting is a versatile and attractive method of decorating plastics items. It consists of forcing a special ink or paint through the small openings of a stenciled screen onto the product surface. The process is sometimes referred to as *silk screen* painting, since

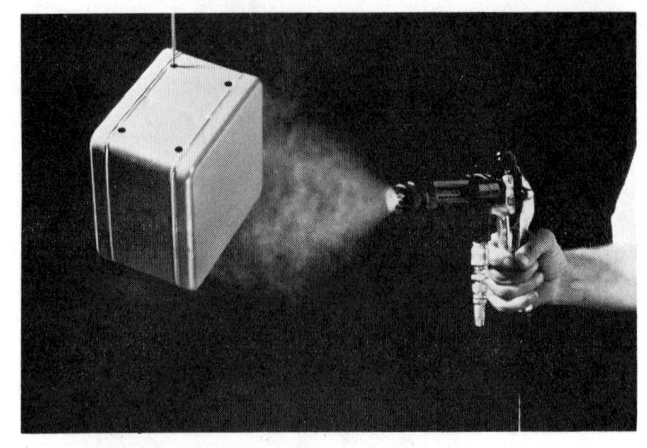

(C) Hydraulic atomization.

Fig. 19-2. Methods of atomizing paint in electrostatic painting. (Ransburg Corp.) *(Continued)*

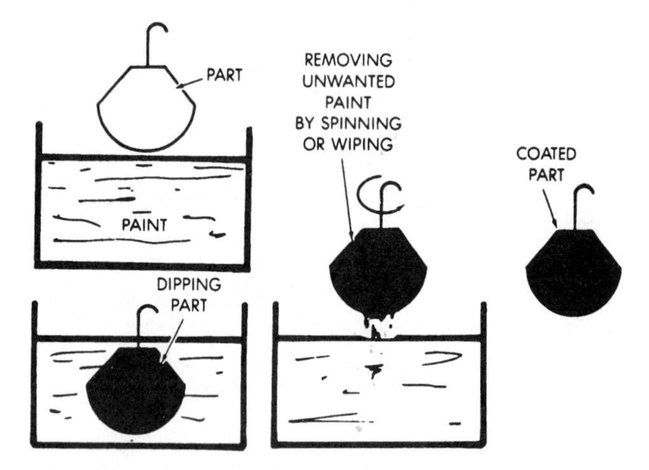

Fig. 19-3. Dip painting

early screens were made of silk. Screens may be made of metal mesh or finely woven polyamide, polyester, or other plastics. A simple screen stencil is prepared by blocking out the areas where no paint is wanted. For intricate designs or lettering, photographic stencils are applied to the screen. When exposed and immersed in a developer bath, the exposed areas wash away. It is through these openings that paint will be applied to the plastics surface beneath the screen.

Fill-In Marking

In the fill-in marking process, paint is placed in low or indented portions of the article (Fig. 19–4). Letters, figures, or designs on a plastics part are produced as depressions in the molded part. These recesses are filled by spraying or wiping paint into the depressions. To ensure a sharp image, the depression should be deep and narrow; otherwise, if the depression or design is too wide, the buffing or wiping action may remove the paint. Excess paint around the design may be removed by wiping or buffing operations.

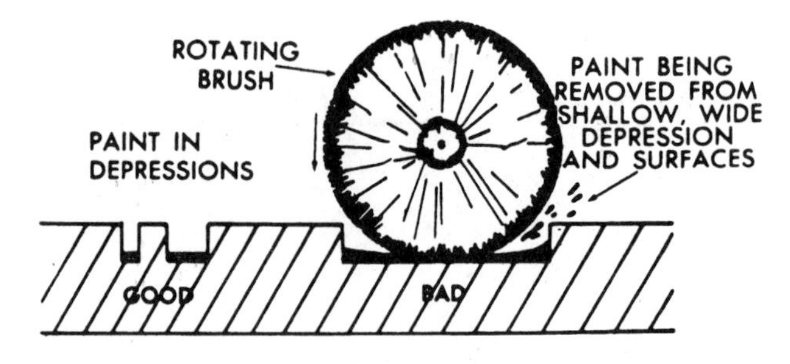

Fig. 19-4. Fill-in method of painting.

Roller Coating

Raised portions, letters, figures, or other designs may be painted by passing a coating roller over them (Fig. 19–5). In some cases, masking out portions of the article may be required. If edges and corners are sharp and highly raised, good coating details will be obtained. Roller coating may be automated or small runs may be done by hand with a brayer (hand roller).

Three advantages and six disadvantages of using painting decorating processes follow.

Advantages of Painting

1. Several inexpensive methods are possible.
2. Pretreatment of most plastics is not needed.
3. Variation of methods and designs may hide imperfections.

Disadvantages of Painting

1. Some plastics are solvent-sensitive.
2. Hand methods have higher labor costs.
3. Paint reduces cold impact resistance.
4. Fisheye blemishes may occur from having used silicone or other releases.
5. Solvents may be a health hazard.
6. Oven drying may be a problem with some thermoplastics.

Hot-Leaf Stamping

Hot-leaf stamping is sometimes called *roll-leaf stamping* or simply *hot stamping.* It offers a simple, economical method for producing a durable decoration on

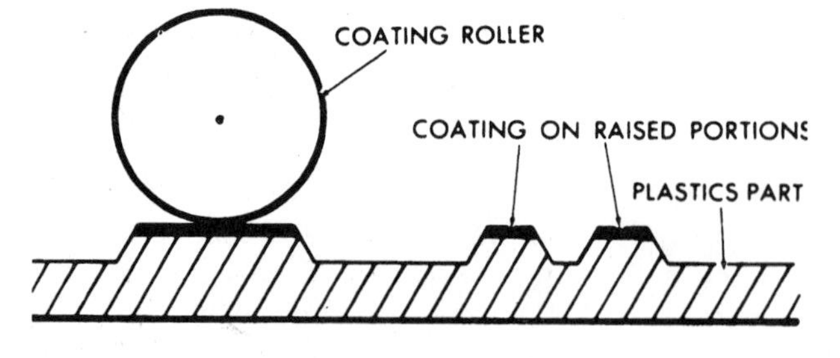

Fig. 19-5. Roller coating of raised portions.

plastics. Letters, designs, trademarks, or messages may be hot-leaf stamped. The process involves a film of metal or paint on a thin carrier (usually in roll form) and a hot-stamping die. The hot die strikes the surface of the plastics part through the carrier. The paint or metallic film is fused into the impression made by the stamp, providing a durable, clear decoration. The hot-stamping dies may be made of machine-engraved or chemically etched metal. Some are made of heat-resistant, flexible silicone (Fig. 19–6). For textured, uneven, or large surfaces, silicone dies may be preferred. Roller dies are used to transfer the design to large areas.

Hot-leaf stamping can be done on all thermoplastics and some thermosets, although thermosetting materials are not easily hot-stamped because of the high heat

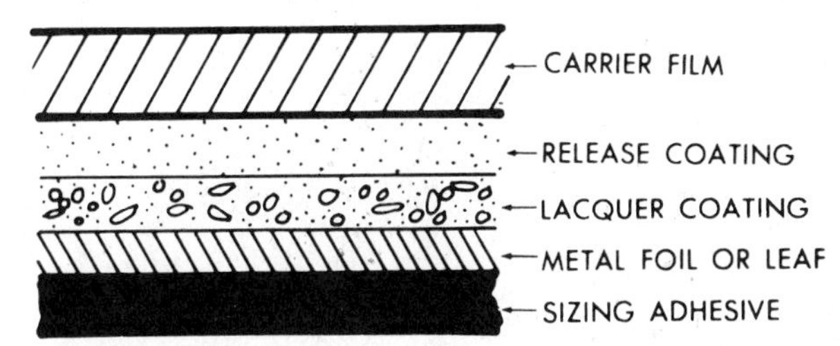

(A) Cross sectional diagram of a typical metallized hot-stamping foil.

(B) Textured silicone hot-stamping rolls produce continuous designs on flat products. (Gladen Division, Hayes-Albion Corp.)

(C) A ring hot-stamped by the roll shown in (B) above. (Gladen Division, Hayes-Albion Corp.)

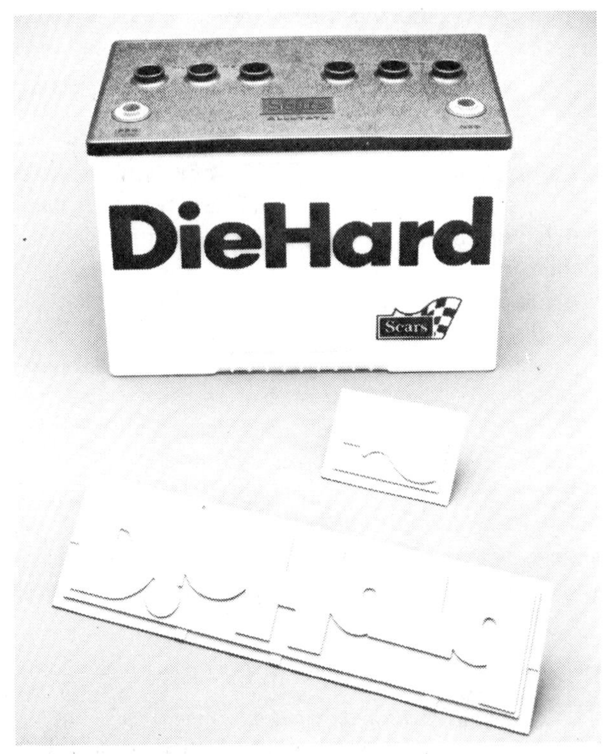

(D) Example of a product hot-stamped with a silicone die. (Gladen Division, Hayes-Albion Corp.)

(E) Molded silicone rubber die used to stamp glass. The glass bottom of the stein was coated with silk screening epoxy and forced-air dried before stamping. (Gladen Division, Hayes-Albion Corp.)

Fig. 19-6. Hot stamping and examples of products.

and pressure required. On thermosets, the process is similar to branding. Melamines are never hot-stamped. Urea-based resins are rarely decorated by this method.

Genuine gold, silver, or other metal foils (leaf), as well as paint pigments, may be placed on plastics. A typical bright metalized hot-stamping foil is shown in

Figure 19–6A. Because these foils and pigments are dry, they are easy to handle, they may be placed over painted surfaces, no masking is needed, and the process may be automatic or done by hand. The carrier film, which supports the decorative coatings until they are pressed on the plastics, is made of cellophane, acetate, or polyester. A thin layer of heat-sensitive material is placed on the carrier as a releasing agent. A lacquer coating is applied over the releasing layer to provide protection for the metal foil. If a paint is to be used instead of a metal foil, the lacquer and pigmented colors are combined into one layer. The bottom layer functions as a heat- and pressure-sensitive, hot-melt adhesive. Heat and pressure must have time to penetrate (dwell) the various layers and to bring the adhesive to a liquid state. Before the carrier film is stripped away, a short cooling time is allowed, permitting the adhesive to solidify. Four advantages and two disadvantages of the hot-leaf stamping decorating process are given below.

Advantages of Hot-Leaf Stamping

1. It is a high-speed, automated operation.
2. Foils or other patterns may hide flaws or gate marks.
3. No solvents are used.
4. Patterns may be changed in short runs.

Disadvantages of Hot-Leaf Stamping

1. Foils and release patterns are relatively costly.
2. Secondary functions and equipment are costly.

Figure 19–7A shows a hot-stamping press that applies multiple color decorations to a molded plastics

(B) Hot-stamping on four sides of a polyethylene beverage case. (Howmet Corp.)

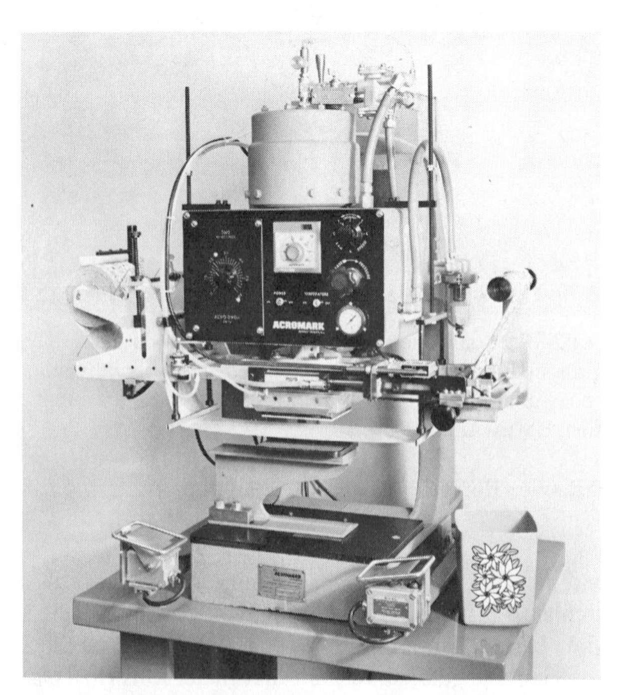

(A) Hot-stamping machine that applies multiple colors. (The Acromark Co.)

(C) Machine for hot-stamping squeezable plastics tubes. (The Acromark Co.)

Fig. 19-7. Hot stamping machines.

(D) Decorative hot stamping of a plastic drinking tumbler. (The Acromark Co.)

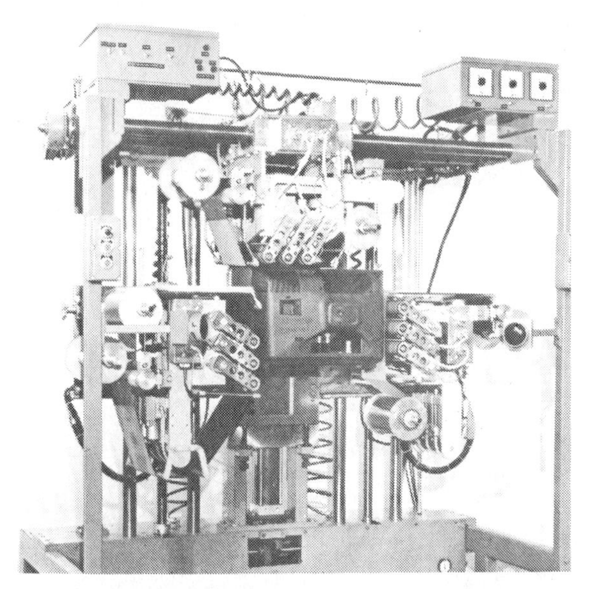

(E) Roll-on method of hot stamping being used to apply wood grain finish to a television cabinet. (Howmet Corp.)

Fig. 19-7. Hot stamping machines. *(Continued)*

canister in one operation. The designs are preprinted on the carrier and then are transferred and fused to the part with use of heat, pressure, and dwell. This process is dry, as is the case with all hot-stamping transfers; therefore, newly decorated parts can be handled, assembled, or packaged. This particular press setup can also use regular transfer dies and hot-stamping foil. Flat and shaped decorating areas can be accommodated and, with a special attachment, the complete circumference of cylindrical parts can be marked by this machine. Figure 19–7C shows a machine for hot-stamping a squeezable plastics tube with a highly decorative design. The tooling consists of a rotary dial table assembly that permits continuous operation of the press, requiring only the loading of the part to the nests. Ejection after marking is automatic.

Plating

Plating and vacuum metalizing have been discussed under the topic of metal coatings in Chapter 17. There are many functional applications of coating plastics with metal, but decorative applications outnumber functional ones. Metalized foils for dielectrics, electronics items such as semiconductors, and resistors are functional applications, as are flexible mirrors and plating for corrosion resistance. The mirror-like finish on automotive items, appliances, jewelry, and toy parts are examples of decorative applications. Four advantages and five disadvantages of plating are listed below.

Advantages of Plating

1. Metallic finish has a mirror-like quality.
2. Many plastics parts need little or no polishing before plating.
3. Electroplate thickness ranges from 0.00038 to 0.025 mm.
4. Plating is more durable than metalizing.

Disadvantages of Plating

1. Mold finish and design must be considered.
2. Not all plastics are easily plated.
3. Setup is expensive and includes many steps.
4. There are many variables to control for proper adhesion, performance, and finish.
5. Plating is more expensive than metalizing.

Engraving

Engraving is seldom used on a production scale, however, it does provide a durable means of marking and decorating plastics and is often used in engraved tool and die work. Pantographic engraving machines may be automatic or manual and are often used to engrave laminated nametags, door signs, directories, and equipment, and to place identifying names and marks on bowling balls, golf clubs, and other items. Laminated engraving sheets contain two or more layers of colored plastics. Engraving cuts through the top layer, exposing the contrasting second-color layer (Fig. 19–8).

Printing

There are over eleven distinct methods, and many combinations of these methods for printing on plastics.

Letterpress is a method in which raised, rigid printing plates are inked and pressed against the plastics part. The raised portion of the plate transfers the image.

(A) Engraving laminated plastics.

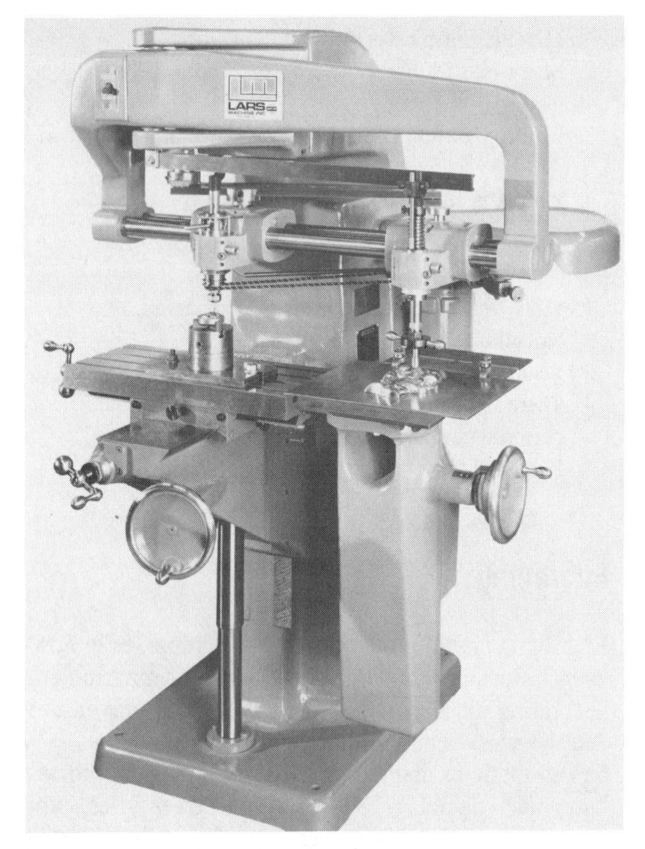

(B) Three-dimensional engraving. (Lars Corp.)

Fig. 19-8. Pantographic engraving machines.

Letterflex is similar to letterpress, except that flexible printing plates are used. Flexible plates may transfer their designs to irregular surfaces (Fig. 19–9A).

Flexographic printing is like letterflex, except that a liquid ink rather than a paste ink is used. The plate is often a rotary type, transferring inks that set or dry rapidly by solvent evaporation.

Dry offset is a method in which a raised, rigid printing plate transfers a paste ink image onto a special roller called an offset blanket. This roller then places the ink

(A) A one-color direct-printing letterflex press. The rubber plate allows printing on irregular surfaces.

(B) This machine will print one or more colors by either dry offset or letterflex methods.

Fig. 19-9. Letterflex and dry offset printing presses. (Apex Machine Co.)

image on the plastics part, thus the name *offset*. If multicolor printing is required, a series of offset heads can be used to apply different colors to the blanket roller. The multicolored image is then transferred (offset) to the plastics part in a single printing step (Fig. 19–9B).

Offset lithography is similar to dry offset, except that the impression on the printing plate is not raised or sunken. The process is based on the principle that oil and water do not mix. The image or message to be printed is placed on the plate by a photographic-chemical process. Images may be placed directly on the plate by special *grease* typewriter ribbons or pencils. The greasy or treated images will be receptive to the type of ink used. Those areas not treated will be receptive to water, but will repel ink. A water roller must first pass over the offset plate. Then, the ink roller will deposit ink on the receptive areas. The image is transferred from the printing plate to a rubber offset cylinder (blanket roller) that places the image on the plastics part.

Rotogravure or *intaglio* printing involves an image that is depressed or sunken into the printing plate. Ink is applied to the entire surface of the plate and a device called a doctor blade is used to scrape the plate and remove all excess ink. The ink left in the sunken areas is transferred directly to the product.

Silk screen printing is a process in which ink or paint is forced through a fine metallic or fabric screen onto the product. A rubber squeegee is used to force paint through the screen. The screen is blank or blocked off in areas where no ink is wanted.

Stenciling is similar to silk screen printing except that the open areas (those to be printed) do not have a connecting mesh. Stencils may be positive or negative. In positive stencil printing, the image is open and spray or rollers transfer the ink through these open areas onto the product. In negative stencil printing, the image is blocked out and the background is inked, leaving no ink in the stencil area. Stencil printing may be considered a masking operation.

Electrostatic printing has been adapted to several well-known printing techniques. In the process, dry inks are attracted to the areas to be printed by a difference in electrical potential. There is not direct contact between the printing plate or screen and the product. There are several methods by which a screen is made conductive in the image areas and nonconductive in other areas. Dry, charged particles are held in these open areas until discharged toward an oppositely charged back plate. The object to be printed is placed between the screen and the back plate. When the ink is discharged, it strikes the substrate surface. A fixing agent is then applied to provide a permanent image. The image is faithfully reproduced, regardless of the surface configuration of the substrate. Images can be printed on the yolk of an uncooked egg or similar products by this method. Edible inks are used to identify, decorate, and supply messages on fruits and vegetables.

Heat-transfer printing is used as a decorating process and as an important printing method. The process is similar to hot-leaf stamping in that a carrier film (or paper) supports the release layer and the ink image. The thermoplastic ink is heated and transferred to the product by a heated rubber roll.

Hot-leaf stamping is the process of transferring a colorant or a decorative material from a dry carrier film to a product by heat and pressure. It is sometimes used as a printing method.

In-Mold Decorating

During in-mold decorating, an overlay or coated film called a *foil* becomes part of the molded product. Both thermosetting and thermoplastic materials may receive decorative images through this process.

With thermosetting products, the film may be a clear cellulose sheet covered with a partially cured resin-like molding material. The in-mold overlay is placed in the mold cavity while the thermosetting material is only partially cured. The molding cycle then continues and the decoration becomes an integral part of the product. The bond between the image and the plastics substrate depends on the rear layer on the foil. This layer consists of material that will adhere to the type of thermosetting material selected.

With thermosetting products, an overlay may be held in place by cutting, so that it fits snugly in the cavity. Electrostatic methods also are used to hold foils in proper location.

With thermoplastic products, the film is usually polyester. In injection molding, the foil is placed in the mold before it closes prior to injection. As the melt flows into the cavity, the overlay bonds fully with the plastics substrate. When production runs are long, it is often desirable to automate the loading of the foil. Foil winding machines hold a fresh roll of foil on one side of the mold, and a take-up roll on the other side. When the mold opens and the decorated part is ejected, the foil winder takes up the used foil and locates fresh material.

As with hot-stamping foils, in-mold foils contain many layers. The bottom layer adheres to the selected thermoplastic material. The decoration, which often contains several colors or textures, is protected by a transparent topcoat or wear coat. The wear coat determines the abrasion and scratch resistance of a foil. A heat-activated release coat causes the separation of the decorating from the carrier film.

Blow-molded parts are often decorated in the mold. The ink or paint image is placed on a carrier film or paper. As the hot plastics expands and fills the mold cavity, the image transfers from the carrier to the molded part. The advantages and disadvantages of in-mold decorating are shown below.

Advantages of In-Mold Decorating

1. Full color images, halftones, or combinations may be used.
2. Strong bonding may be achieved.
3. Designs and short runs are economical.
4. Efficiency may reach well over 90 percent.

Disadvantages of In-Mold Decorating

1. Costs of foil, labor for hand loading, or automated loading machines are high.
2. Mold design must minimize washing and turbulence.
3. Mold design must cause foil to separate cleanly at edges, to minimize clean-up.

Heat-Transfer Decorating

In heat-transfer decorating, the image is transferred from a carrier film onto the plastics part. The structure of heat-transfer decorating stock is shown in Figure 19–10A. The preheated carrier stock is transferred to the product by a heated rubber roller (Fig. 19–10C.

A decorating or printing process that resembles a combination of engraving and offset printing is the *Tampo-Print.* In this process, a flexible transfer pad picks up the impression from the inked engraving plate (Fig. 19–11A) and transfers it to the item to be printed

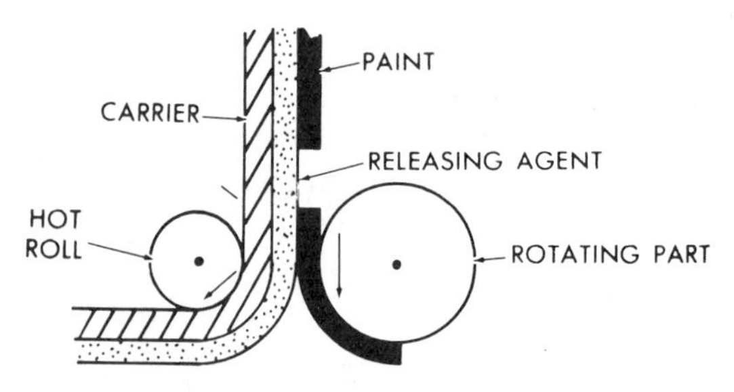

(C) Roller method of transferring design.

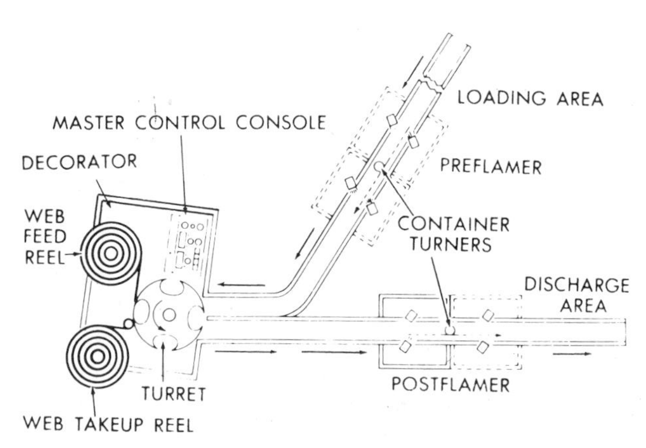

(D) Scheme for heat transfer of decorations to containers. (Therimage® Products Group, Dennison Mfg. Co.)

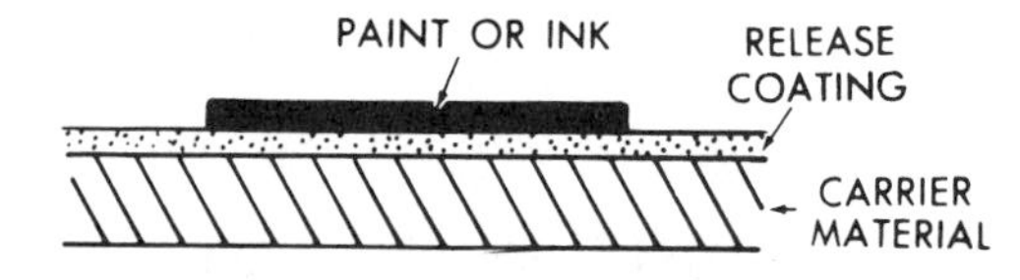

(A) Structure of heat-transfer decorating stock.

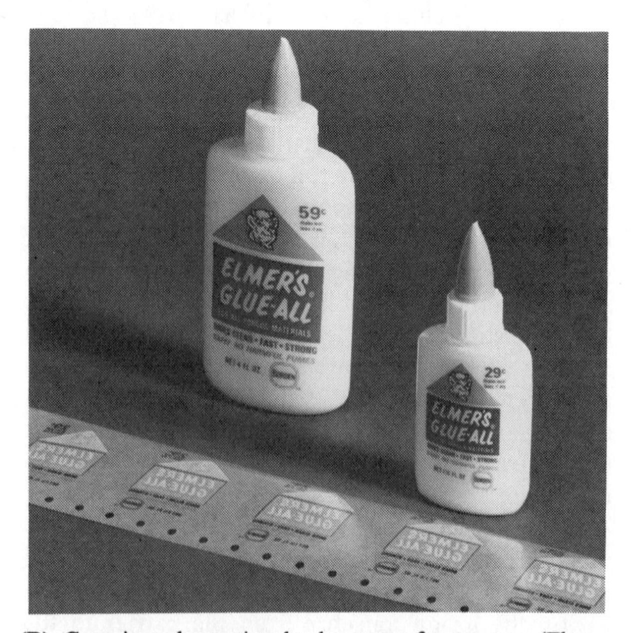

(B) Container decoration by heat-transfer process. (Therimage® Products Group, Dennison Mfg. Co.)

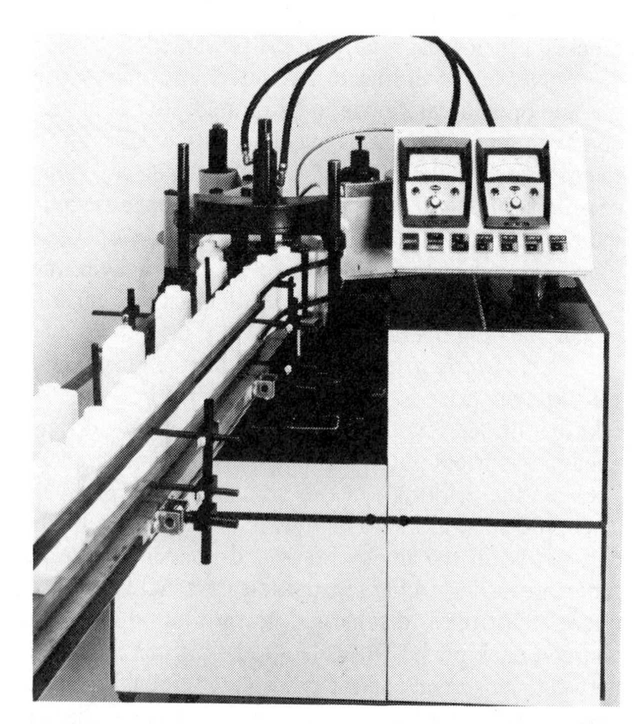

(E) Heat transfer of decorations, following the scheme shown in (D). (Therimage® Products Group, Dennsion Mfg. Co.)

Fig. 19-10. Heat-transfer decorating and examples of decorated products.

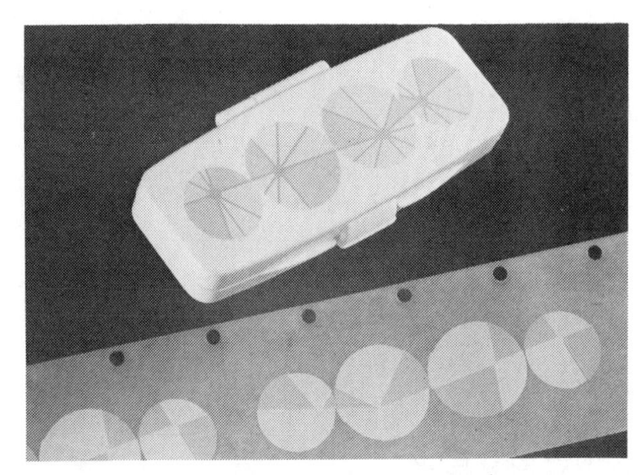

(F) Flat silicone rubber pad, 3 mm thick, [$^1/_8$ in] used to apply transfer to cosmetics container. (Gladen Division, Hayes-Albion Corp.)

(G) Various containers decorated by heat transfer process. (Therimage® Products Group, Dennison Mfg. Co.)

(H) Examples of heat-transfer decorations. Note rolls of decoration on carrier film. (Color-Dec Inc.)

Fig. 19-10. Heat transfer decorating and examples of decorated products. *(Continued)*

(Fig. 19–11B). The entire ink supply carried by the transfer pad is deposited on the part, leaving the pad clean. The flexible pad adapts to rough and uneven surfaces while maintaining absolute reproduction sharpness. Printing heads of various shapes can accommodate a wide variety of objects and textures, and multicolor wet-on-wet printing, including halftones, can be accomplished. Nearly any type of printing ink or paint may be used by this simple process. Depending on the type of product, up to 20000 parts per hour can be automatically decorated. Two advantages and disadvantages of the heat-transfer decorating process are listed below.

Advantages of Heat-Transfer Decorating

1. Similar to hot-leaf stamping, except multicolored designs are possible
2. Many heat-transfer systems are available

Disadvantages of Heat-Transfer Decorating

1. Carrier film and designs are costly
2. Secondary operation and equipment are needed

Miscellaneous Decorating Methods

There are many other decorating methods, including pressure-sensitive labels, decalcomanias, flocking, and decorative coatings or clads.

Pressure-sensitive labels are easy to apply. The designs or messages are usually printed on the adhesive-backed foil or film label and the labels are placed on the finished product by hand or mechanical means.

Decalcomanias, commonly known as decals, are a means of transferring a picture or design to plastics. They are generally a decorative film on a paper back-

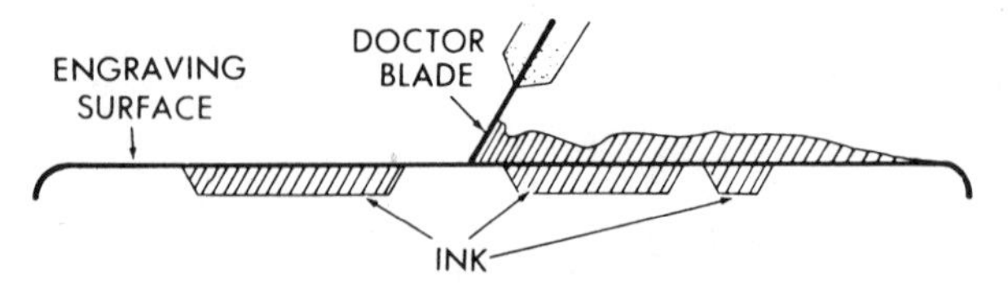

(A) Etched plate with ink held in recessed areas. Surface of the engraving is wiped clean by doctor blade.

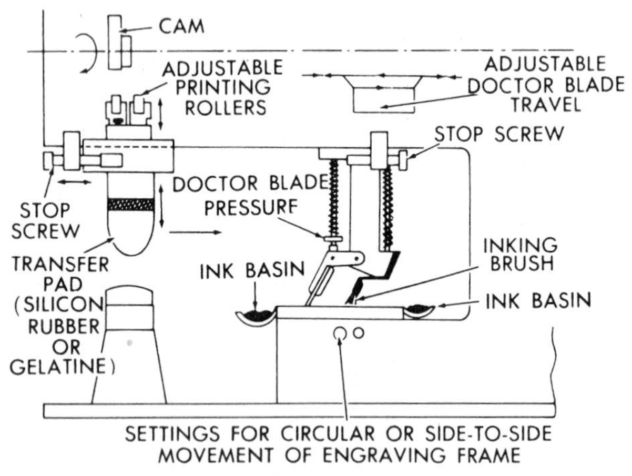

(B) Movement of the transfer pad and inking brush or squeegee.

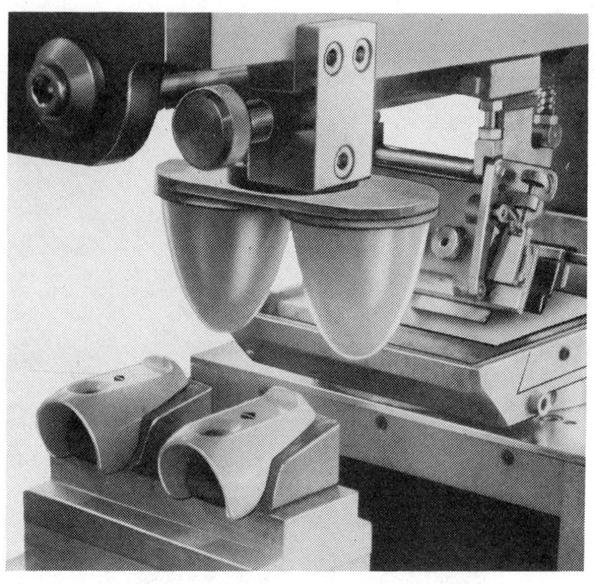

(D) Closeup of transfer pad and inking mechanism.

Fig. 19-11. The Tampo-Print process used a transfer pad to apply ink to products. (Dependable Machine Co.)

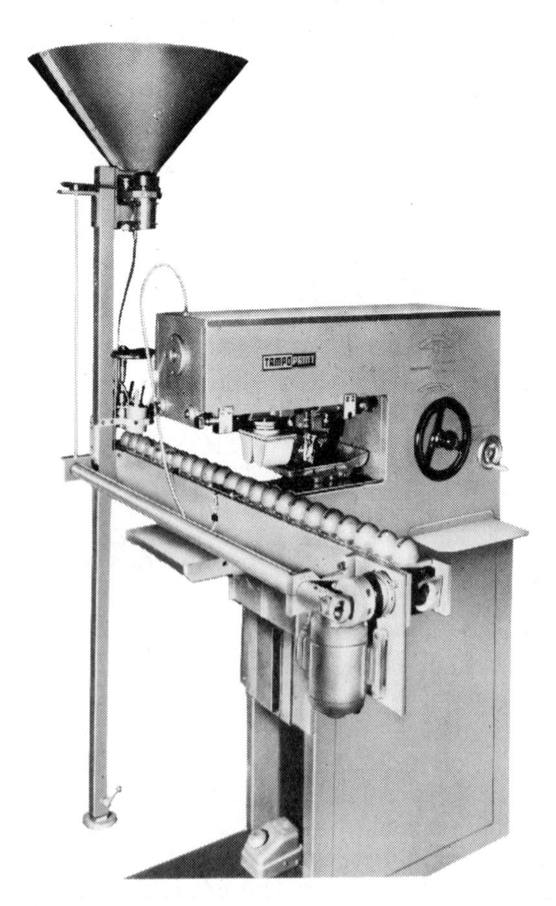

(C) Entire printing machine setup.

ing. The decalcomania is moistened in water and the adhesive-backed film is slipped off the paper and onto the plastics surface. This process is not widely used, mainly because decalcomanias are hard to place accurately and quickly on the plastics surface.

Flocking by mechanical or electrostatic means is an important method of placing a velvet-like finish on nearly any surface. The process consists of coating the product with an adhesive and placing plastics fibers on the adhesive areas. The velvet-like coatings or designs on wallpapers, toys, and furniture are examples.

There are a number of wood-graining decorative processes. Some are done by rolling engraved or etched wood-graining plates over a contrasting background color. This is actually a printing process adaptation. Some decorative laminates and clad coatings are also used to decorate substrates. Polyvinyl clad metal products are used in store fixtures, partitions, room dividers, furniture, automobiles, kitchen equipment, and bus interiors. These are durable as well as decorative.

Many foils and patterns are thermoformable, permitting the decoration of three-dimensional thermoformed parts. The thermoformed shell may be filled by casting or injection molding. Plastics thinning and pattern distortion must be controlled (Fig. 19–12).

Four advantages and two disadvantages of the Pressure sensitive decorating process are shown below.

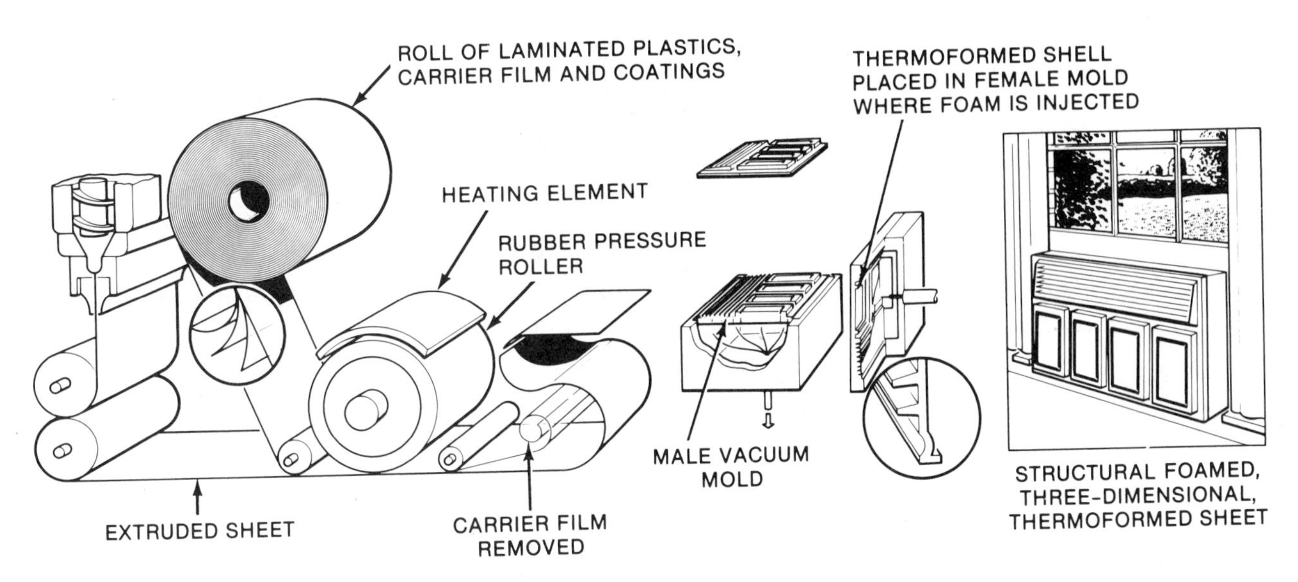

Fig. 19-12. Thermoforming a decorative cover for an appliance. (Dri Print Foils)

Advantages of Pressure Sensitive Decorating

1. Variable application rates (relatively high-speed to hand-dispensed)
2. Multi-colored patterns and designs
3. May be used on all plastics
4. Short runs and changes in patterns are economical

Disadvantages of Pressure Sensitive Decorating

1. Secondary operation and equipment are needed.
2. Surface labels may wear or be removed.

Vocabulary

The following vocabulary words are found in this chapter. Look up the definition of any of these words you do not understand as they apply to plastics in the glossary, Appendix A.

Corona discharge
Electrostatic printing
Engraving
Heat-transfer decorating
Hot-leaf stamping
In-mold decorating
Offset printing
Tampo-print

Questions

19-1. A ___?___ may be used to prevent silicone paint from depositing where it is not wanted.

19-2. Name two advantages of silicone hot-stamping dies.

19-3. Name the painting process in which surface wear does not easily remove the design.

19-4. What additive may lower the electrical resistance in a plastics?

19-5. Name the result of applying a coating over items that have not been properly dried of surface moisture.

19-6. The name of a well-known mixer used to blend plastics ingredients is ___?___.

19-7. Hot-leaf stamping is sometimes called roll-leaf stamping or simply ___?___.

19-8. Name three functional uses of plating.

19-9. Name three decorative uses of plating.

19-10. Electroplate thickness varies from ___?___ to ___?___ mm.

19-11. An important method of placing a velvet-like finish on a surface by mechanical or electrostatic means is called ___?___.

19-12. Name a process in which the image is transfelled from the carrier film to the product by stamping with rigid or supple shapes and with heat and pressure.

19-13. It may be less costly to incorporate the desired design in the ___?___, dies or rollers.

19-14. Most decorating processes require that the substrate be thoroughly ___?___ of mold release, lubricants and plasticizers.

19-15. The best and least costly way to color plastics products is to blend pigments into the basic ___?___.

19-16. Plasticizers, ___?___, and molding may affect color.

19-17. An inexpensive and popular method of painting plastics is ___?___.

19-18. A variety of shapes and different sizes of products may be rapidly painted by ___?___ methods.

19-19. Name the three ways of atomizing paint.

19-20. Excess paint may be removed in the dip-coating process by spinning the part, hand wiping, or by ___?___ methods.

19-21. The process of forcing special inks or paint through small openings of a stenciled screen into a product surface is called ___?___ or ___?___.

19-22. In ___?___, the image depression should be deep and narrow for sharp detail.

19-23. Raised portions, letters, figures, or other designs are easily decorated by ___?___ methods.

19-24. Name five methods of printing on plastics products.

19-25. The pattern or decoration becomes part of the plastics article as it is fused by heat and pressure of the plastics material during ___?___ decorating.

19-26. Because of the use of a flexible printing pad, the ___?___ process is especially useful for decorating irregular surfaces.

19-27. A metal term used to indicate that two or more layers of plastics (or metals) are pressed together under pressure is ___?___.

19-28. Labels are adhered to the substrate by ___?___, while decals are ___?___ activated.

19-29. Name four methods for pretreating polyethylene to receive paint or ink.

Activities

In-Mold Decoration

Introduction. In-mold decoration uses the heat of hot plastics to adhere decorations or messages to the product. Its major advantage is the elimination of secondary printing or decorating operations.

Equipment. Injection molder, in-mold foil, or hot-stamping foil, test-bar mold, Scotch 3M transparent tape, number 600.

19-1. If possible, acquire in-mold foil. Foil for ABS is popular and available. If in-mold foil is not on hand, use hot-stamping foil as a substitute. A major difference between in-mold foil and hot-stamping foil is the thickness of the polyester carrier film. In-mold foils have much thicker carriers than hot-stamping foils. Figure 19-13 shows an ABS disk covered with wood-grain

Fig. 19-13. In-mold foil on a Taber abrader disk.

in-mold material. This disk permits abrasion resistance testing in a Taber abrader.

19-2. If the injection mold used for in-mold decoration has a flat side, tape the foil to that side. Make sure that the melt will contact the back side of the foil. Should the melt contact the "wrong" side of the foil, no adhesion will occur.

19-3. If the melt forces the foil into a cavity, the resulting stretch may exceed the strength of the carrier film, particularly when using hot-stamping foil, and cause a rip. The rip may allow the melt to run on the "wrong" side of the foil. Figure 19-14 shows a example of this problem. Notice that the rip began at an edge of the part. To avoid ripping the foil, use a mold with one flat half. Even on the flat half, hot-stamping foil will stretch during injection. Figure 19-15 shows small wrinkles in the foil near the edge of the part. These would not occur if the carrier were thicker.

19-4. Inject at a range of temperatures. If melt temperature is low, the adhesion will be poor. If

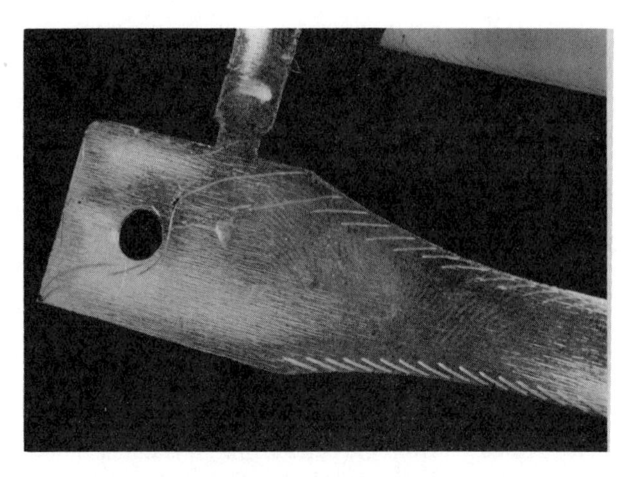

Fig. 19-15. Foil wrinkles near edge of part

melt temperature is appropriate, adhesion will be good. If melt is too hot, the foil may stretch and distort the design, loss of gloss may occur, and near the gate, a spot of foil may be completely removed by *burn through.* The problems of distortion will be exaggerated when using hot-stamping foil.

19-5. Test the adhesion of the foil using the tape test. First, crosshatch the sample as shown in Figure 19-16. Then adhere the tape firmly, taking care to remove all air bubbles under the tape by rubbing firmly with the back of a thumb nail. Pull off the tape at a 90° angle to the substrate. Check to see if the adhesion failed. If the tape pulled off pieces of the foil, the adhesion was inadequate. The number 600 tape, manufactured by 3M under the brand name Scotch is a tape used for foil adhesion testing.

19-6. Use of foils with various compatibilities will show significant differences. Figure 19-16 also

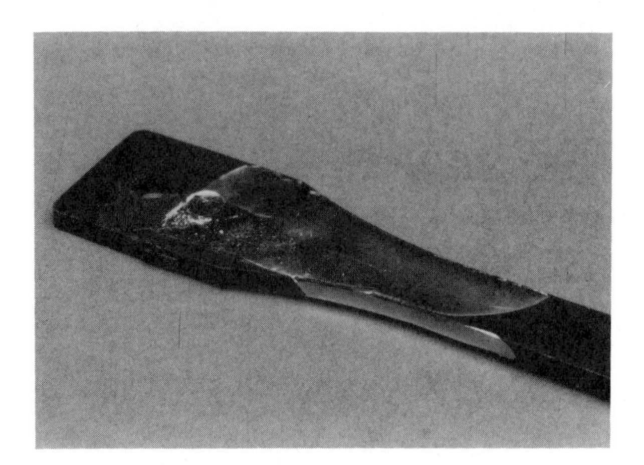

Fig. 19-14. This is an example of melt flow on the "wrong" side of the foil. The foil ripped due to stretching by the injected melt.

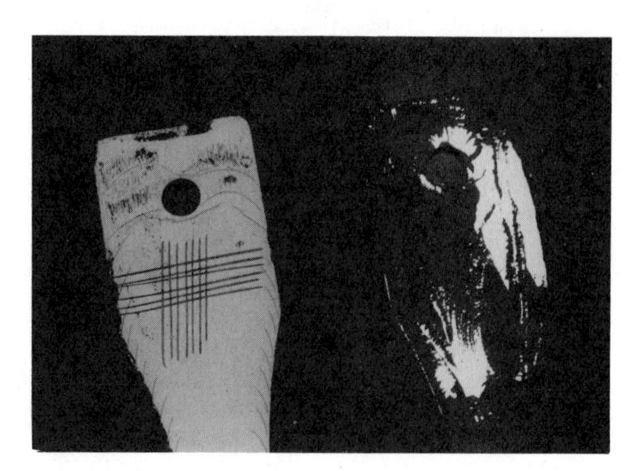

Fig. 19-16. One sample is crosshatched and ready for tape test. The other shows that styrenic foil did not adhere to PP part.

shows a hot-foil stamping foil compatible with styrenics used on polypropylene. Almost none of the foil adhered.

19-7. Write a report summarizing your findings.

Hot Stamping

Introduction. Hot-leaf stamping is a decorating process in which metal foils or pigments are placed on carrier films. The word *leaf* is used because thin metal foils of gold are called gold leaf. A heated die is used to press against the foil and plastics, and the foil or pigment is released from the carrier film. It is the heat and pressure of the process that bonds the decoration to the substrate.

Hot-foil stamping is a popular process for decorating plastics. The foil must be compatible with the plastics substrate to be decorated. Foils compatible with styrenics, such as PS, HIPS (high-impact polystyrene), and ABS are very common. Another type of foil adheres to olefins. A mismatch of foil type and plastics substrate will dramatically reduce adhesion.

Equipment. Hot-foil stamper, foil for styrenics, foil for olefins, stamping die or flat silicone rubber stamping plate.

19-1. Stamp words or letters onto various plastics with both styrenic foil and olefinic foil. Test the adhesion using the eraser on a pencil. If the adhesion is high, the force required to rub off the decoration will also be high.

19-2. Stamp using a flat silicone rubber die onto a flat piece of plastics material. This will yield a surface large enough for tape testing the adhesion of the foil. Crosshatch the surface and check the adhesion with a tape test. Both the time stamping and the temperature of the die will influence adhesion.

19-3. Stamp onto polyethylene or polypropylene. These materials often resist adhesion. To increase adhesion, flame treat the surface of the substrate. A bunsen burner or propane torch can provide the flame. Do not melt the substrate or char the surface. Vary the time of contact with the flame and test the effects on adhesion.

Chapter 20

Radiation Processes

Introduction

Energy savings, low pollution effects, and economic advantages are offered by radiation processing. You will learn how this technology offers new product and manufacturing ideas.

Radiation processing is a growing area of technology in which ionizing and nonionizing systems are used to change and improve the physical properties of materials or components. Radiation processes may find acceptance along with chemical and thermodynamic processes. The outline for this chapter is:

I. Radiation methods
II. Radiation sources
  A. Ionizing radiation
  B. Nonionizing radiation
  C. Radiation safety
III. Irradiation of polymers
  A. Damage by radiation
  B. Improvements by radiation
  C. Polymerization by radiation
  D. Grafting by radiation
  E. Advantages of radiation
  F. Applications

Radiation Methods

The term *radiation* can refer to energy carried by either waves or particles. The carrier of wave energy is called the *photon.* In radiant energy, the photon is wave-like when in motion. It is particle-like when absorbed or emitted by an atom or molecule.

The ordinary electric light bulb with an element temperature of 4172 °F [2300 °C] emits radiation waves that are visible. The sun with a surface temperature of 10832 °F [6000 °C] emits both visible and invisible radiation. Human eyes can see radiation with wavelength as short as 400 μm [15.7×10^{-6} in.) and as long as 700 μm [27.5×10^{-6} in.] *Ultraviolet* radiations are waves of energy that can bum or tan the exposed parts of the human body yet are invisible to the human eye (Fig. 20-1). Ultraviolet radiation has wavelengths shorter than 15.7×10^{-6} in. [400 μm]. Photon wavelengths are measured in micrometres. Radiation from the sun, burning fuels, or radioactive elements are considered natural sources of radiation. A few of the more important radioactive elements that occur naturally are uranium, radium, thorium, and actinium. These radioactive materials emit photons of energy and/or particles as the nuclei disintegrate and decrease in mass. The earth contains small traces of radioactive materials while the sun is intensely radioactive.

Radiation may be produced by nuclear reactors, by accelerators, or from natural or artificial radioisotopes. The most important source of controlled radiation is artificial radioisotopes. Scientists have brought the use and control of induced radiation to a point where it may serve human needs.

When the number of *protons* in the nucleus of an atom changes, a different element is formed. If the number of *neutrons* in the nucleus is changed, a new element is not formed; only the mass of the element is different. Different forms (masses) of the same ele-

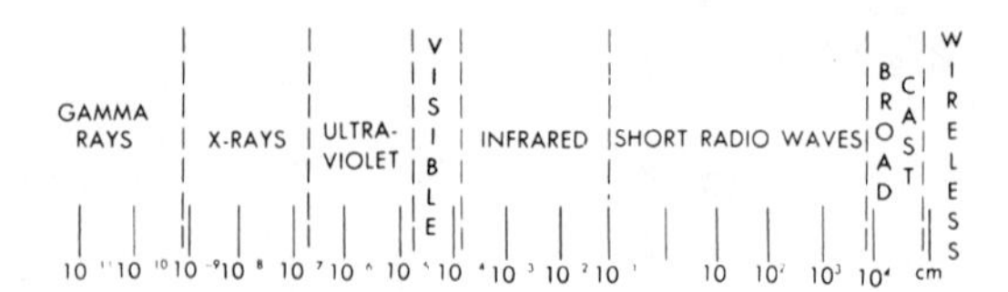

(A) Wavelengths of radiation

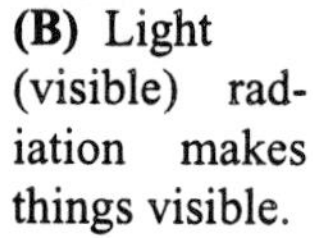

(B) Light (visible) radiation makes things visible.

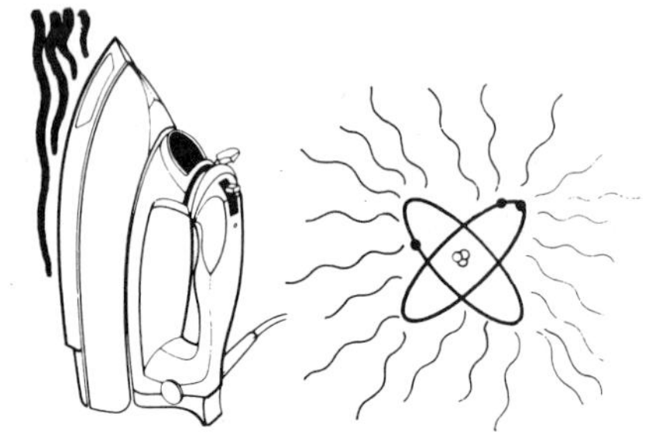

(C) Heat (infrared) radiation can be felt.

(D) Radioactive radiation can't be seen or felt.

Fig. 20-1. Types of radiation.

ment are called *isotopes*. Most elements have several isotopes. For example the simple element hydrogen occurs as three distinct isotopes (Fig. 20-2). Most hydrogen atoms have a mass number (the number of protons and neutrons) of 1, meaning they have no neutrons. A very small number of naturally occuring hydrogen atoms have one neutron and one proton and thus have a mass number of 2. Only when hydrogen has two neutrons and one proton (a mass number of 3) is it radioactive.

In 1900, the German physicist Max Planck advanced the idea that photons are bundles or packets of electromagnetic energy. This energy is either absorbed or emitted by atoms or molecules. The unit of energy carried by a single photon is called a *quantum* (*quanta* is the plural). Photons of energy radiation may be classified into two basic groups, electrically neutral and charged radiation.

Alpha particles are heavy, slow-moving masses with a double positive charge (two protons and two neutrons). When alpha particles strike other atoms, their double positive charge removes one or more electrons, leaving the atom or molecule in a dissociated or ionized state. Ionization, you will remember from Chapter 2, is the process of changing uncharged atoms or molecules into ions. Atoms in the ionized state have either a positive or a negative charge.

Electrons ejected from the nuclei of atoms at a very high speed and high energy are called *beta particles.* When a neutron disintegrates, it becomes a proton and an electron. The proton often stays in the nucleus while the electron is emitted as a beta particle. Beta particles are electrons with a negative charge. Because beta particles have only 0.000 544 times the mass of a proton, they move much faster and have greater penetrating power than alpha particles.

Most of the energy of alpha and beta particles is lost when they interact with electrons from other atoms. As the charged particles pass through matter, they lose or transfer all their excess energy to the nuclei or orbital electrons of the atoms they encounter. Since beta particles are negative, they can push or repel electrons leaving the atom with a positive charge, or beta particles may become attached to the atom giving it a negative charge.

Gamma radiations are short, very high-frequency electromagnetic waves with no electrical charge. *Gamma rays* and *X-rays* are alike except for origin and penetrating ability. Gamma photons can penetrate even the most dense materials. More than one metre of concrete is required to stop the radiation effect of gamma rays (Fig. 20-3).

The energy of gamma photons is absorbed or lost in matter in three major ways (Fig. 20-4):

1. Energy may be lost or transferred to an electron it strikes, forcing the electron out of orbit.
2. The gamma photon may strike an orbiting electron a glancing blow using only a part of its energy while the rest of the energy continues in a new direction.

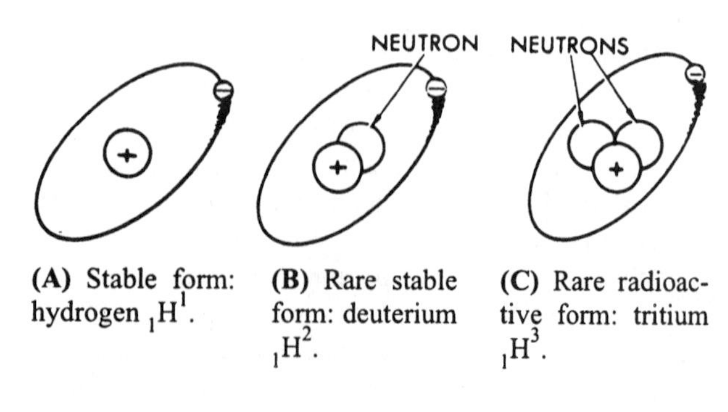

(A) Stable form: hydrogen $_1H^1$.

(B) Rare stable form: deuterium $_1H^2$.

(C) Rare radioactive form: tritium $_1H^3$.

Fig. 20-2. Isotopes of hydrogen.

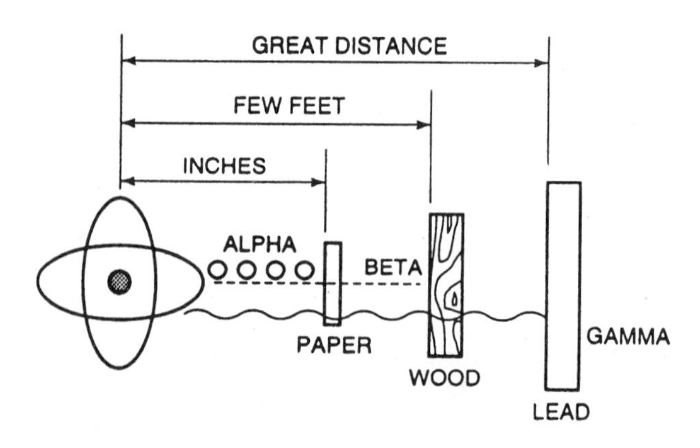

Fig. 20-3. Three types of radiation emitted by unstable atoms or radioisotopes: *alpha* (stopped by paper, *beta* (stopped by wood), and *gamma* (stopped by lead).

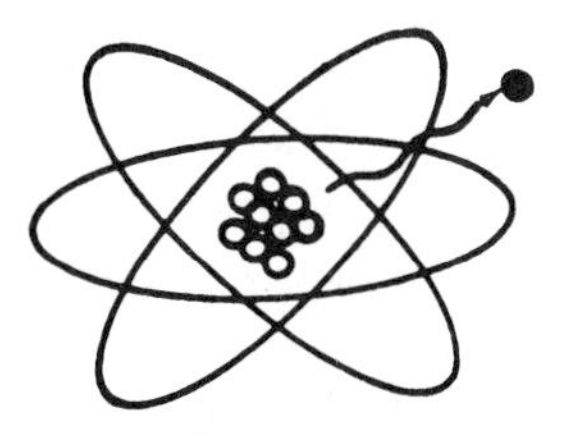

(A) Radioisotope with photon of energy being emitted.

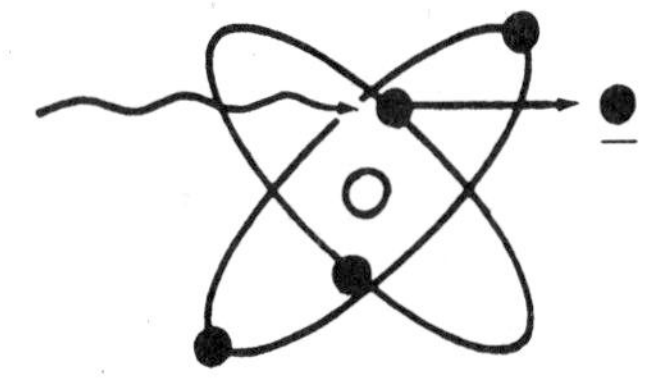

(B) Gamma energy being completely absorbed by forcing an electron from orbit and transferring energy.

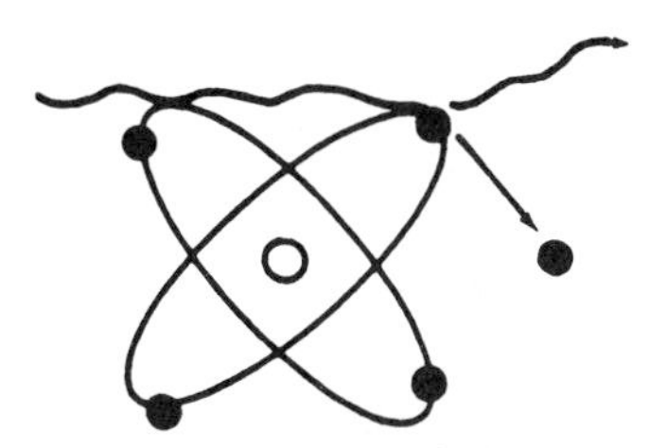

(C) Part of energy continues in new direction and part is used in ejecting electron from orbit.

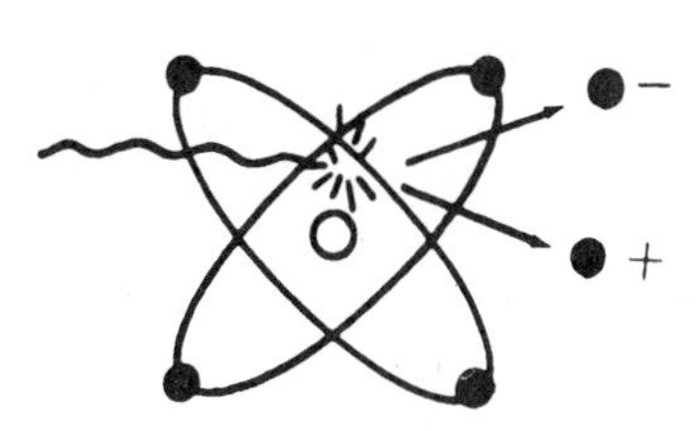

(D) Gamma radiation is annihilated. Electron and positron are created and share energy.

Fig. 20-4. Interaction of gamma radiation with matter.

3. The gamma ray or photon is annihilated when it passes near the powerful electrical field of a nucleus.

In the last method of gamma ray energy loss, the powerful electrical field of an atomic nucleus breaks the gamma photon into two particles of opposite charge, an electron and a positron. The positron quickly loses its energy by colliding with orbital electrons. The net effect of gamma radiation is like the effect of alpha and beta radiation. Electrons are knocked from orbit causing ionization and excitation effects in materials.

Neutrons are uncharged particles that may collide with atomic nuclei resulting in alpha and gamma radiation as energy is transferred or lost.

Radiation Sources

There are two basic types of radiation sources, those that produce ionizing radiation, and those that produce non-ionizing radiation.

Ionizing Radiation

Cobalt-60, strontium-90, and cesium-137 are three commercially available *radioisotope* sources that produce ionizing radiation. They are used because of their availability, useful characteristics, reasonably long half-life, and reasonable cost. Another source of ionizing radiation is used, or burnt, uranium slugs from fission reactor waste.

Gamma radiation, while very penetrating, is not a major source of ionizing radiation for several reasons. It is slow and may require several hours for treatment; its isotope sources are hard to control; the source cannot be turned off; experienced, well-trained workers are needed.

Electron beam accelerators are the primary ionizing source for radiation processing. Irradiation processing implies that a controlled or directed treatment of energy is being used on the polymer.

Nonionizing Radiation

Electron accelerators, such as Van de Graaff generators, cyclotrons, synchrotrons, and resonant transformers, may be used for producing nonionizing radiation.

Electrons from machines are less penetrating than radioisotope radiation; however, they may be easily controlled and turned off when not required. These machines are capable of delivering 200 kW of power. The dose rating may be expressed in a unit called a *gray* (Gy). One gray indicates a dose absorption of one joule of energy in one kilogram of plastics (1 J/kg). As a rule, 1 kW of power is required to deliver a dose of 793 lb [10 kGy to 360 kg] of plastics per hour.

Electron radiation penetrates only a few millimetres, into the plastics, but it can irradiate products at a very fast rate. Ultraviolet, infrared, induction, dielectric, and microwave are the most familiar sources of nonionizing radiation. They are generally used to speed up processing by heating, drying, and curing.

Ultraviolet radiation sources such as plasma arcs, tungsten filaments, and carbon arcs produce radiation with enough penetrating power for film and surface treatment of plastics. Prepregs that cure by exposure to sunlight have been produced.

Infrared radiation is often used in thermoforming, film extrusion, orientation, embossing, coating, laminating, drying, and curing processes.

Induction sources (electromagnetic energy) have been used to produce welds, preheat metal-filled plastics, and cure selected adhesives.

Dielectric sources (radio-frequency energy) have been used to preheat plastics, cure resins, expand polystyrene beads, melt or heat seal plastics, and dry coatings.

Microwave sources are used to speed curing and to heat, melt, and dry compounds.

Radiation Safety

All forms of natural radiation must be considered harmful to polymers and dangerous to work with, since they are not easily controlled; however, radiation processes are time-proven production methods used by a wide variety of industries today.

A disadvantage frequently discussed with radiation processing is safety, since an element of emotionalism often is associated with the word *radiation.*

Stray electrons and rays, high voltages, and ozone exposure are all potential hazards of radiation processing. However, safety in radiation processing is possible with an understanding of the hazards involved and with the adherence to a sound safety program. Safety standards have been published and maximum safe exposure levels for different kinds of radiation have been set by several governmental agencies.

For any company planning radiation processing, the wise manager will hire a qualified consultant to plan, design and implement a radiation safety program for the plant and its personnel.

Irradiation of Polymers

The transfer of energy from the radiation source to the material assists in breaking bonds and thus is used for rearranging atoms into new structures. The many changes in covalent substances directly affect important physical properties. The effects of radiation on plastics may be divided into four categories:

1. Damage by radiation
2. Improvements by radiation
3. Polymerization by radiation
4. Grafting by radiation

Damage by Radiation

Breaking covalent bonds by nuclear radiation is called scission. This separation of the carbon-to-carbon bonds may lower the molecular mass of the polymer. Figure 20-5 shows that irradiation of polytetrafluoroethylene causes the long linear plastics to break into short segments. As a result of this breaking, the plastics loses strength.

Degraditive symptoms include cracking, crazing, discoloration, hardening, embrittlement, softening, and other undesirable physical properties associated with molecular mass, molecular mass distribution, branching, crystallinity, and cross-linking.

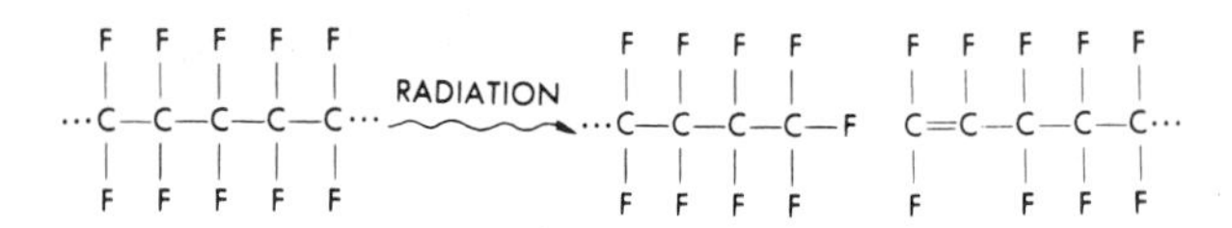

Fig. 20-5. Degradation by irradiation.

With controlled irradiation, polyethylene becomes a cross-linked, insoluble, and non-melting material. Improvements may include increased heat resistance and form stability at elevated temperatures, reduction in cold flow, stress cracking, and thermal cracking.

Effects of radiation on selected polymers are shown in Table 20-1.

The separation of the carbon-to-carbon bonds may also form free radicals, and this may lead to cross-linking, branching, polymerization, or the formation of gaseous by-products. In Figure 20-6A, radiation is shown as the cause of polymerization and cross-linking of hydrocarbon radicals. Figure 20-6B shows that a gaseous product is formed in the irradiation process. The radical (R) may be H, F, Cl, etc., as the gaseous product. The irradiation may result in atoms being knocked from the solid material, and this disassociation or displacement of an atom results in a defect in the basic structure of the polymer (Fig. 20-7). These vacancies in crystalline structures and other molecular changes result in changes in the mechanical, chemical, and electrical properties of polymers.

Cross-linkage in elastomers may be considered a form of degradation. For example, natural and synthetic rubbers become hard and brittle with more cross-linkage or branching (Figs. 20-8 and 20-9).

Mineral-, glass-, and asbestos-filled phenolics, epoxies, polyurethane, polystyrene, polyester, silicones, and furan plastics have superior radiation resis-

Table 20-1. Effects of Radiation on Selected Polymers

Polymer	Radiation Resistance	Radiation Dose for Significant Damage (Mrads)
ABS	Good	100
EP	Excellent	100–10,000
FEP	Fair	20
PC	Good	100+
PCTFE	Fair	10–20
PE	Good	100
PFV, PFV_2, PETFE, PECTFE	Good	100
PI	Excellent	100–10,000
PMMA	Fair	5
Polyesters (aromatic)	Good	100
Polyesters (unsaturated)	Good	1,000
Polymethylpentene	Good	30–50
PP	Fair	10
PS	Excellent	1,000
PSO	Excellent	1,000
PTFE	Poor	1
PU	Excellent	1,000+
PVC	Good	50–100
UF	Good	500

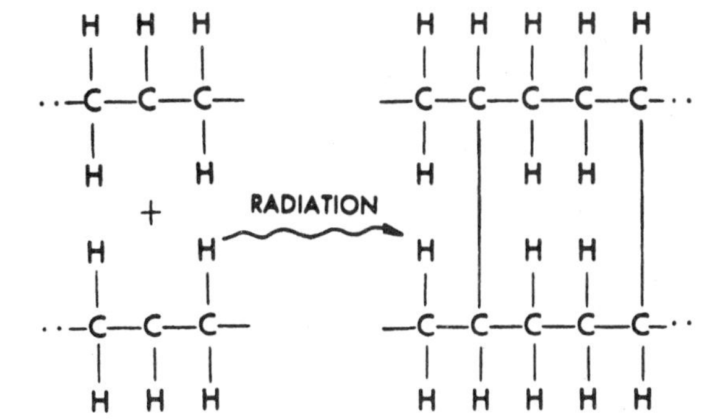

(A) Recombination leading to polymerization or cross-linking of hydrocarbon radicals.

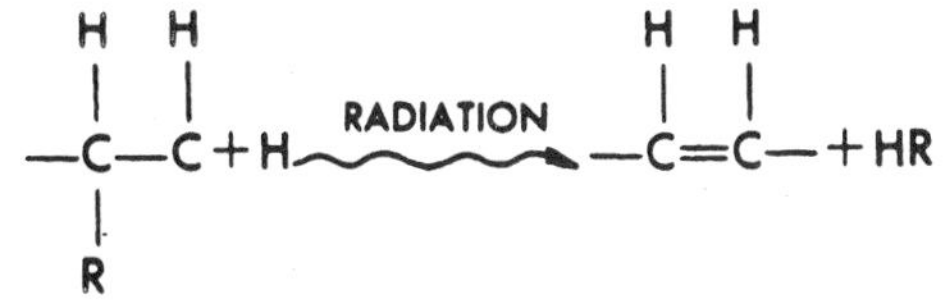

(B) Gaseous product formed by radiation.

Fig. 20-6. Formation of free radicals through irradiation.

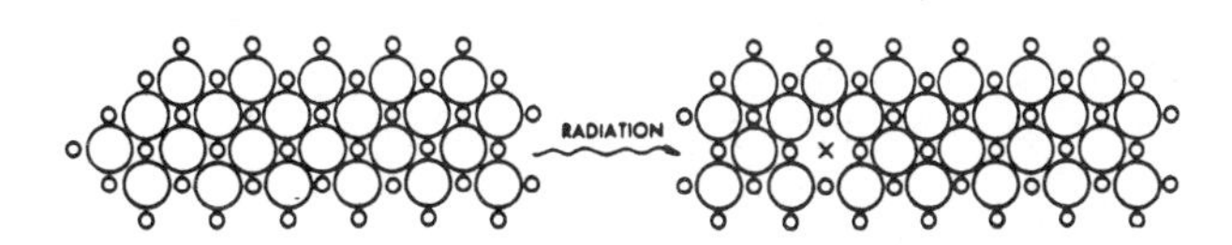

Fig. 20-7. Linear structure of plastics with missing atom. The vacancy in the crystalline structure is a potential site for radical attachment.

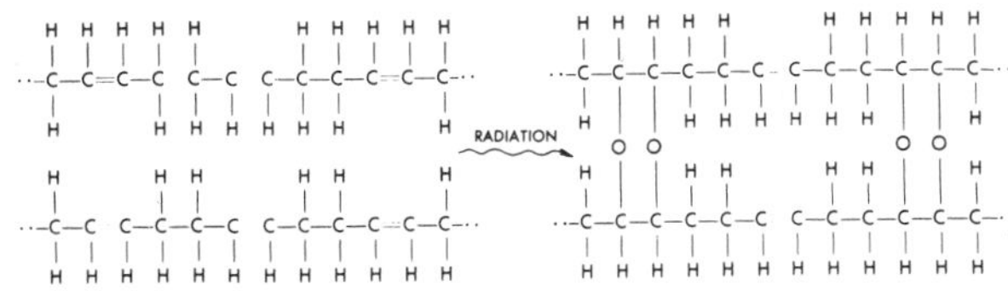

Fig. 20-8. Oxidation (radical attachment of oxygen) of polybutadiene. This cross-linking results in a rapid aging effect with loss of elastic strain.

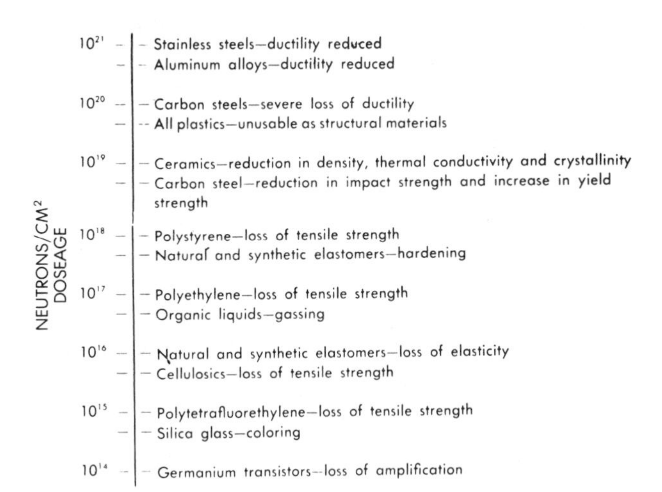

Fig. 20-9. Changes in materials properties caused by radiation. Controlled use of radiation can be beneficial.

tance. Unfilled methyl methacrylate, vinylidene chloride, polyesters, cellulosics, polyamides and polytetrafluoroethylene have poor radiation resistance. These plastics become brittle and their desirable optical properties are affected by discoloring and crazing. Fillers and chemical additives may help to absorb much radiation energy, while heavy pigmentation of the plastics may stop deep penetration of damaging radiation.

Improvements by Radiation

While some polymers are damaged by radiation, others may actually benefit from controlled amounts. Cross-linkage, grafting, and branching of thermoplastic materials may produce many of the desirable physical properties of the thermosetting plastics.

Polyethylene is one plastics that benefits from controlled, limited irradiation. Such radiation causes existing bonds to be broken and a rearrangement of the atoms into a branched structure. Branching of the PE chain elevates the softening temperature to above that of boiling water. (Excessive radiation may reverse the effect, however, by rupturing main links in the chains). Effects of radiation on selected polymers are shown in Table 20-1.

Radiation processing. Radiation processing today is most often done with electron machines or radioisotope sources such as cobalt-60. This radiation may increase molecular mass by linking molecules of some polymers together, or it may decrease molecular mass by degrading others. It is this cross-linking and degradation that accounts for most of the property changes in plastics.

The ability of radiation to begin ionization and free-radical formation may prove superior to the ability of other agents, such as heat or chemicals.

The main industrial disadvantage of radiation-induced chemical reaction is high cost. With radiation processing integrated directly into processing lines, the cost of radiation systems has been decreasing and it may soon be competitive with chemical processing for some uses.

Ultraviolet treatment may improve such surface characteristics as weather resistance, hardening, penetration, and neutralization of static electricity.

Cross-linking of wire insulation, elastomers, and other plastics parts improves stress-cracking, abrasion, chemical, and deformation resistance.

Polymerization by Radiation

During the dissociation of a covalent bond by irradiation, a free radical fragment is formed. This radical is available at once for recombinations. The same nuclear energy forces that cause depolymerization of plastics may begin cross-linkage and polymerization of monomer resins (Fig. 20-10).

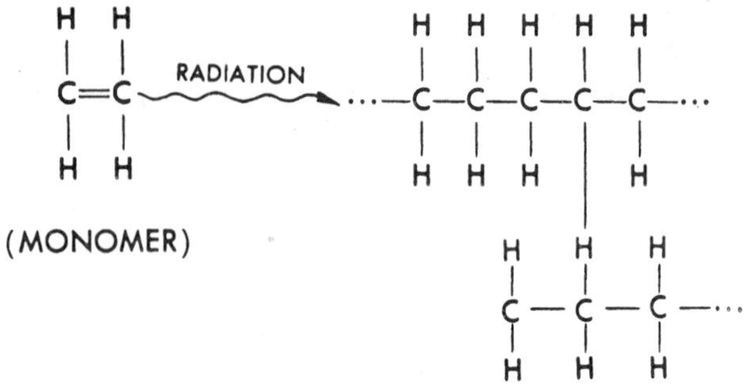

Fig. 20-10. Branching of polyethylene.

Polymerization and cross-linking are used to cure polymer coating, adhesives, or monomer layers. Typical doses (Mrads) for cross-linkable polymers range from 20-30 for PE, 5-8 for PVC, 8-16 for PVDF, 10-15 for EVA, and 6-10 for ECTFE.

Grafting by Radiation

When a given kind of monomer is polymerized and another kind of monomer is polymerized onto the primary backbone chain, a graft copolymer results. By irradiating a polymer and adding a different monomer and irradiating again, a graft copolymer is formed. The schematic structure of a graft copolymer is shown in Figure 20-11. The recombining or structuring of two different monomer units (A and B) often yields unique properties. It is possible that graft copolymers with highly specific properties could be combined for optimum product applications. Irradiation may produce a grafting reaction on a thin surface zone, or one conducted homogeneously throughout thick sections of a polymer.

Advantages of Radiation

Radiation processing may have numerous broad advantages to offset the main disadvantage of high cost.

The first advantage is that reactions can be initiated at lower temperatures than in chemical processing. A second advantage is good penetration, which allows the reaction to occur inside ordinary equipment at a uniform rate. Although gamma radiation from cobalt-60 sources can penetrate more than 12 in. [300 mm], the treatment rate is slow and exposure times are long. Electron radiation sources may react very rapidly with materials less than 0.39 in. [10 mm] thick. For these reasons, over 90 percent of irradiated products are processed by high-energy electron sources (Table 20-2).

A third advantage is that monomers can be polymerized without chemical catalysts, accelerators, and other components that may leave impurities in the polymer. A fourth advantage is that radiation-induced reactions are little affected by the presence of pigments, fillers, antioxidants, and other ingredients in the resin or polymer. A fifth advantage is that cross-linking and grafting may be done on previously shaped parts such as films, tubing, coating, moldings, and other products. Coatings in monomeric form may be applied with radiation processing, thus doing away with solvents and the collection or recovery systems for the solvents. Finally, the sixth advantage is that mixing and storing of chemicals used in chemical processing may be eliminated. (Fig. 20-12).

AAAAAAAAAAAAAAAAAAAAAAAAAAAAAA
BBBBBBBBBBB

Fig. 20-11. In graft polymerization, a monomer of one type (B) is grafted onto a polymer of a different type (A). Because graft copolymers contain long sequences of two different monomer units some unique properties result.

Applications

In addition to the processing advantages just mentioned, radiation processing may provide other marketable features not possible by other means (Fig. 20-13).

Graft and homopolymerization of various monomers on paper and fabrics improves bulkiness, resilience, acid resistance, and tensile strength. Irradiation of some cellulosic textiles has aided in the development of "dura-press" fabrics. Grafting selected monomers to polyurethane foam, natural fibers, and plastic textiles improves weather resistance, and eases ironing, bonding, dyeing, and printing. Small radiation dosages, which degrade the surface of some plastics, improve ink adhesion to their surface.

Impregnation of monomers in wood, paper, concrete, and certain composites has increased their hardness, strength, and dimensional stability after irradiation. For example, the hardness of pine has been increased 700 percent by this method. Novolacs and resols are soluble and fusible low-molecular-mass resins used in the production of prepregs (reinforcement-impregnated resins) and impregs (resin-impregnated materials). The term *A-stage* is used to refer to novolac and resol resins. Wood, fabric, glass fibers, and paper may be saturated with A-stage resins while under a high vacuum. This supersaturated prepreg or impreg may then be exposed to cobalt radiation causing the thermosetting, A-stage material to pass through a rubbery stage referred to as the *B-stage.* Further reaction leads to a rigid, insoluble, infusible, hard product. This last stage of polymerization is known as the *C-stage.* The terms A-, B-, and C-stage resins are also used to describe analogous states in other thermosetting resins. (See Phenolics in Appendix F.)

A commercially available *shrinkable* polyethylene film is often used for wrapping food items. This irradi-

Table 20-2. Industrial Applications for Electron Beam Processing

Product	Product Improvements and Process Advantages	Process
Wire and cable insulation, plastic insulating tubing, plastic packaging film	Shrinkability; impact strength; cut-through, heat, solvent, stress-cracking resistance; low dielectric losses.	Crosslinking, vulcanization
Foamed polyethylene	Compression and tensile strength; reduced elongation.	Crosslinking, vulcanization
Natural and synthetic rubber	High-temperature stability; abrasion resistance; cold vulcanization; elimination of vulcanizing agents.	Crosslinking, vulcanization
Adhesives: Pressure Sensitive Flock Laminate	Increased bonding; chemical, chipping, abrasion, weathering resistance; elimination of solvent;	Curing, polymerization
Coatings, paints, and inks on: Woods Metals Plastics	100% convertibility of coating; high-speed cure, flexibility in handling techniques; low energy consumption; room-temperature cure; no limitation on colors.	Curing, polymerization
Wood and organic impregnates	Mar, scratch, abrasion, warping, swelling, weathering resistance; dimensional stability, surface uniformity; upgrading of softwoods.	Curing, polymerization
Cellulose	Enhanced chemical combination	Depolymerization
Textiles and textile fibers	Soil-release; crease, shrink, weathering resistance; improved dyeability; static dissipation; thermal stability.	Grafting
Film and paper	Surface adhesion; improved wettability.	Grafting
Medical disposables	Cold sterilization of packages and supplies.	Irridiation
Packages and containers	Reduction or elimination of residual monomer.	Polymerization
Polymer	Controlled degradation or modification of melt index.	Irridiation, Depolymerization, Crosslinking.

ated film is cross-linked by radiation for increased strength. The film can be stretched more than 200 percent and is usually sold prestressed. When heated to 180 °F [82 °C] or higher, the film attempts to shrink back to its original dimensions, thus making a tight package.

(A) When the coating is passed under the electron beam, free radicals are produced by ionization. These free radicals start a rapid build-up of long-chain molecules that become the cured resin. This curing mechanism does not require heat or catalysts.

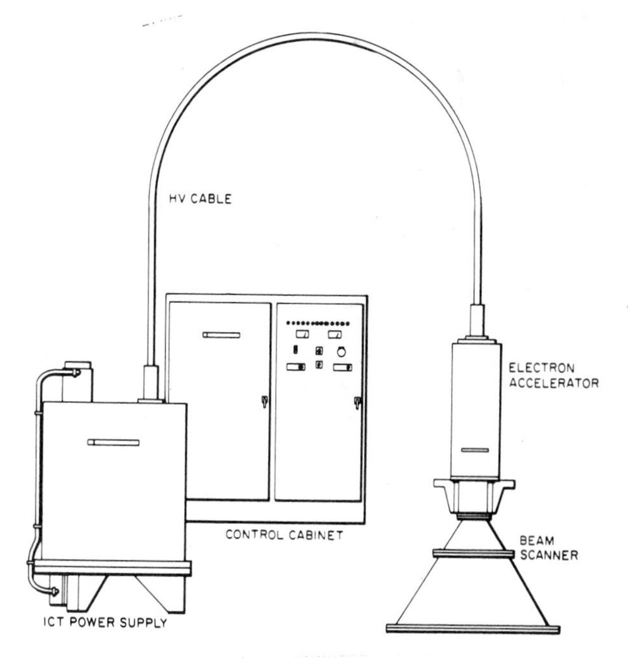

(B) Diagram of the major components of the electron-beam processing system.

Fig. 20-12. Principles of electron-beam radiation processing. (High Voltage Engineering Corp.)

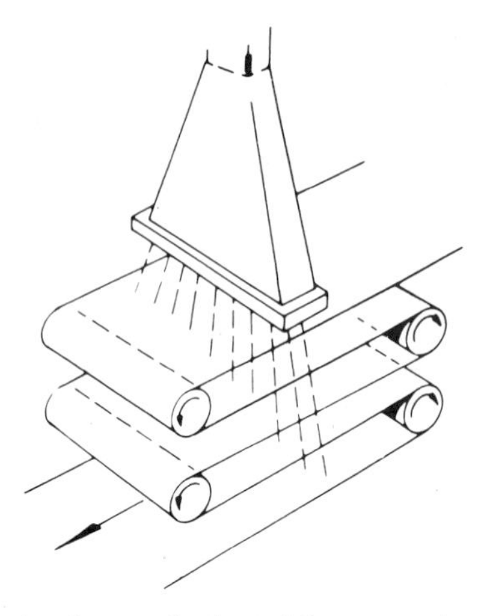

(C) Festooning is a method used for processing continuous flexible sheets or webs of material.

Fig. 20-12. Principles of electron-beam radiation processing. (High Voltage Engineering Corp.) *(Continued)*

Fig. 20-13. The polyethylene container in the center was exposed to controlled radiation to improve heat resistance. Containers at left and right were not treated. They lost shape at 350 °F [175 °C].

Radiation is also being used as a heatless sterilization system for packaged food and surgical supplies.

Radioisotopes are used in many measuring applications. Monomer resins, paints, or other coatings can be measured for thickness without contacting or marking the material's surface, as can extruded, or blown, films. Using this measuring method may reduce raw material consumption, reduce or eliminate scrap, ensure more uniform thickness, and speed up output (Fig. 20-14).

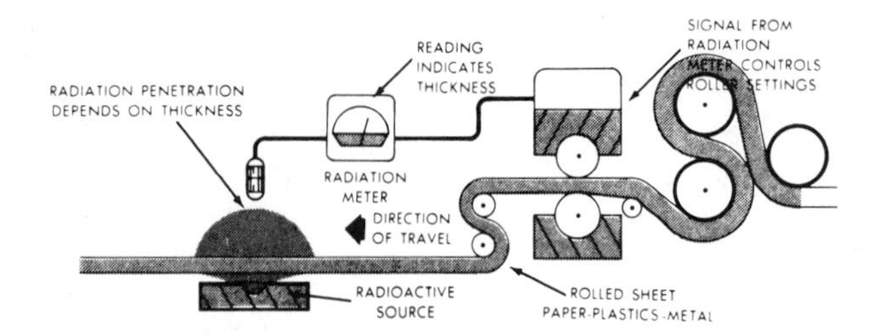

Fig. 20-14. Radioisotopes are used to continuously gauge thickness of a material without being in physical contact with it.

Four advantages of radiation processing and three of its disadvantages are listed below.

Advantages of Radiation Processing

1. It improves many important plastics properties.
2. Many non-ionizing radiation processes speed production by heating or initiating polymerization.
3. No physical contact is needed.
4. Machine sources may be controlled with ease and require less shielding.

Disadvantages of Radiation Processing

1. Gamma radiation equipment is somewhat costly; some is specialized.
2. It demands careful handling and trained personnel (especially ionizing radiation).
3. There is potential danger to operator due to ionizing radiation and radioisotopes.

Vocabulary

The following vocabulary words are found in this chapter. Look up the definition of any of these words you do not understand as they apply to plastics in the glossary, Appendix A.

Alpha particle
Beta particle
Gamma ray
Gray
Irradiation
Isotope
Photon
Radiation

Questions

20-1. Name the term that refers to bombardment of plastics with a variety of subatomic particles. It may be done to polymerize and change physical properties of plastics.

20-2. The most important source of controlled radiation is ___?___ .

20-3. When one or more different kinds of monomer are attached to the primary backbone of the polymer chain, a ___?___ results.

20-4. Name five plastics that have poor radiation resistance.

20-5. Name two additives for plastics that may help stop penetration of damaging radiation.

20-6. The carrier of wave energy is called ___?___ .

20-7. Electrons traveling at very high speed and with high energy are called ___?___ particles.

20-8. Different forms of the same element with different atomic masses are called ___?___ .

20-9. Breaking of covalent bonds by nuclear radiation is called ___?___ .

20-10. The two types of radiation systems or sources are ___?___ and ___?___ .

20-11. Controlled amounts of irradiation may cause ___?___ of bonds for free radical formation and cross-linkage.

20-12. Uncontrolled irradiation may break bonds, lowering ___?___ and ___?___ .

20-13. Name four adverse effects of irradiation.

20-14. Name three possible sources of ionizing radiation for irradiating plastics.

20-15. Name four possible sources of nonionizing radiation for irradiating plastics.

20-16. The ___?___ is the term often used to describe the dose given in irradiating polymers.

20-17. Radiation sources must be carefully handled only by trained ___?___ .

20-18. Accumulated dosages or exposure to ___?___ and ___?___ radiation may cause permanent cell damage.

20-19. Name four major advantages of irradiation processing.

20-20. Energy sources that may be used to preheat molding compounds, heat-seal films, polymerize resins and expand polystyrene beads are called ___?___ .

20-21. Radiations that sunburn or tan the human body are ___?___ rays.

20-22. Identify the particles that are heavy, slow-moving masses with a double positive charge.

20-23. Burnt ___?___ slugs from reactors or fission waste may be a source of radiation.

20-24. As a rule, all forms of natural radiation must be considered ___?___ to polymers.

20-25. Over ___?___ percent of irradiated products are processed by high-energy electron sources.

20-26. For the following products or applications, would you select nonionizing or ionizing processing?

a. polymerization of selected resins

b. surface treatment of films

c. drying plastics granules or preforms

Chapter 21

Design Considerations

Introduction

This chapter summarizes basic rules for designing products. Because of the diversity of materials, processes, and product uses, designing with plastics demands more experience than designing with other materials. The information presented herein should serve as a fundamental guide and a useful starting point in understanding the complexity of designing plastics products. See Appendix H for additional sources of information.

There are many sources for studying specific design problems and some of them are included in the discussions of individual materials and processes.

In the early years of their development, plastics were chosen mostly as a substitute for other materials. Some of those early products were very successful because of the consideration and thought given to the choice of materials. On the other hand, some of these products failed because the designers did not know enough about the properties of the plastics used or were motivated by costs rather than the practical use of the material. The products simply could not stand up to daily wear and tear. As the plastics industry has grown, so has the designers' knowledge of the properties of plastics. Because plastics have combinations of properties that no other materials possess, i.e., strength, lightness, flexibility, and transparency (Table 21-1), plastics are now chosen as primary materials rather than as substitutes.

The design considerations for polymer composites are more complex than those for homopolymers. Most composites vary with time under load, rate of loading, small changes in temperature, matrix composition, material form, reinforcement configuration, and fabrication method. They may be designed to be isotropic, quasi-isotropic, or anisotropic depending upon design requirements.

Three industrial tools that have revolutionized the design process, and became a fact of life in the 1980s are: computer assisted design (CAD), computer assisted manufacturing (CAM), and computer assisted moldmaking (CAMM). Most manual drafting and hand calculation will be eliminated resulting in designers, fabricators, materials manufacturers, and toolmakers making fewer errors in part designs, material selection, and tooling configurations. Figure 21-1 shows a

Fig. 21-1. Computer graphics (CAD-CAM) systems are designed to automate and integrate the many phases of the product development cycle. (Applicon Inc.)

Table 21-1. Plastics vs. Metals

Properties of Plastics Which May Be ...

Favorable

1. Light
2. Better chemical and moisture resistance
3. Better resistance to shock and vibration
4. Transparent or translucent
5. Tend to absorb vibration and sound
6. Higher abrasion and wear resistance
7. Self-lubricating
8. Often easier to fabricate
9. Can have integral color
10. Cost trend is downward. Today's composite plastics price is approximately 11% lower than five years ago. However. the long-established, high-volume plastics—phenolics, styrenes, vinyls, for example—appear to have reached a price plateau, and prices change only when demand is out of phase with supply
11. Often cost less per finished part
12. Consolidation of parts

Unfavorable

1. Lower strength
2. Much higher thermal expansion
3. More susceptible to creep, cold flow, and deformation under load
4. Lower heat resistance—both to thermal degradation and heat distortion
5. More subject to embrittlement at low temperature
6. Softer
7. Less ductile
8. Change dimensions through absorption of moisture or solvents
9. Flammable
10. Some varieties degraded by ultraviolet radiation
11. Most cost more (per cubic millimetre) than competing metals. Nearly all cost more per kilogram.

Either Favorable or Unfavorable

1. Flexible—even rigid varieties more resilient than metals
2. Electrical nonconductors
3. Thermal insulators
4. Formed through the application of heat and pressure

Exceptions

1. Some reinforced plastics (glass-reinforced epoxies, polyesters, and phenolics) are nearly as rigid and strong (particularly in relation to mass) as most steels, May be even more dimensionally stable.
2. Some oriented films and sheets (oriented polyesters) have greater strength-to-mass ratios than cold rolled steels
3. Some plastics are now cheaper than competing metals (Nylons vs. brass, acetal vs. zinc, acrylic vs, stainless steel)
4. Some plastics are tougher at low than at normal temperatures (acrylic has no known brittle point)
5. Many plastics-metal combinations extend the range of useful applications of both (metal-vinyl laminates, leaded vinyls, metallized polyesters, and copper-filled TFE)
6. Plastics and metal components may be combined to produce a desired balance of properties (plastics parts with molded-in, threaded-metal inserts; gears with cast-iron hubs and Nylon teeth; gear trains with alternate steel and phenolic gears; rotating bearings with metal shaft and housing and Nylon or TFE bearing liner)
7. Metallic fillers in plastics make them electrically or thermally conductive or magnetic

Source: Machine Design: Plastics Reference Issue

designer using a CAD/CAM system to improve productivity and smooth the path from design to production.

A designer is now able to use a computer in the design, engineering, and manufacture of all plastics products. During interactive process design, the designer uses the computer to rapidly draw and make changes that improve the appearance and function of the part. The graphic model appearing on the CRT screen may be rotated and viewed from different angles. Certain sections or complex shapes can be further detailed. The most important benefit of computer use in engineering is the improvement in design, labor productivity, market share, capital productivity, innovation, quality, and profitability. With computer-assisted engineering (CAE) the designer can use finite-element and other design analysis, system modeling, and simulated structural testing. CAM is useful for programming, robotic interfacing, quality control, and other operations associated with the manufacture of the product. CAD/CAE/CAM systems can evaluate and review such high-cost elements as material handling, tooling, tool maintenance, raw material costs, and scrap losses before production begins.

In computer-integrated manufacturing (CIM) all design, engineering, and manufacturing processes are intermixed to allow designers, engineers, technicians, accountants, and others access to the same database. Because these and business operations are integrated, there is a saving in terms of reduced down time, labor costs, just-in-time (JIT) manufacturing or zero-inventory, quick changeover for batch manufacturing, and rapid design changes. (See Tooling and Moldmaking in Chapter 22)

In most designs, a compromise must be made between highest performance, good appearance, efficient production, and lowest cost. Unfortunately, human needs are often given less importance than factors of cost, processes, or materials used. There are three (sometimes overlapping) major design considerations included in the following chapter outline:

I. Material Considerations
   A. Environment
   B. Electrical characteristics
   C. Chemical characteristics
   D. Mechanical factors

E. Economics
II. Design considerations
A. Appearance
B. Design limitations
III. Production considerations
A. Processes of manufacture
B. Material shrinkage
C. Tolerances
D. Mold design
E. Performance testing

Material Considerations

Materials must be selected with the right properties to meet design, economic, and service conditions. In the past, the design was commonly changed to compensate for the material limitations.

Caution must be exercized when using information obtained from data sheets or the manufacturer about the matrix performance, Much of this data is based upon laboratory-controlled evaluation. It is also difficult to compare proprietary data from several different suppliers. This does not imply, however, that this data cannot be utilized in the screening of candidate materials.

Customers often use *specifications* in a document to state all requirements to be satisfied by the proposed product, materials, or other standards. There are several different types of *standards,* including *physical* standards as kept by the National Bureau of Standards (NBS); *regulatory* standards such as those from the EPA; *voluntary* standards recommended by technical societies, producers, trade associations, or other groups such as Underwriters Laboratories (UL); mil (military) and *public* standards promoted by professional organizations such as the ASTM.

Metrication and internationalization of standards can reduce costs associated with materials, production, inventory, design, testing, engineering, documentation, and quality control. There is substantial evidence that metrication and standardization lowers costs. We must change our attitude about international measurement and standards if we are to significantly lower costs and increase our international trade.

With the computer, systematic methods of screening and selection of materials are easier. Computer models can predict and anticipate most of the ways a material can fail. In some computer models each property is assigned a value according to importance. Then each property that the part is expected to tolerate is entered. The computer will select the best combination of materials and processes.

Plastics materials must be chosen with care, keeping the final product use in mind. The properties of plastics depend more on temperature than do other materials. Plastics are more sensitive to changes in environment; therefore, many families of plastics may be limited in use. There is no one material that will possess all the qualities desired; however, undesirable characteristics may be compensated for in the product design.

The final material choice for a product is based on the most favorable balance of design, fabrication, and total cost or selling price of the finished item. Either simple or complicated designs may need fewer processing and fabrication operations using plastics, which may combine with the characteristics of the plastics material to make plastics cost competitive with other materials for specific parts.

Environment

When designing a plastics product, the physical, chemical, and thermal environments are very important considerations. The useful temperature range of most plastics seldom exceeds 392 °F [200 °C]. Many plastics parts exposed to radiant and ultraviolet energy soon suffer surface breakdown, become brittle, and lose mechanical strength. For products operating above 450 °F [230 °C], fluorocarbons, silicones, polyimides, and filled plastics must be used. The exotic environments of outer space and the human body are becoming commonplace for plastics materials. The insulation and ablative materials used for space vehicles, the artery reinforcements, monofilament sutures, heart regulators, and valves are only a small portion of these new uses.

Some plastics retain their properties at cryogenic (extremely low) temperatures. Containers, self-lubricating bearings, and flexible tubing must function properly in below-zero temperatures. The cold, hostile environments of space and earth are only two examples. Any time that refrigeration and food packaging is considered or where taste and odors are a problem, plastics may be chosen. The United States Food and Drug Administration lists acceptable plastics for packaging of foods.

The Child Protection and Toy Safety Acts of 1969 and 1976 govern the manufacture and distribution of children's toys. Should toys present an electrical, mechanical, toxic, or thermal hazard, they may be banned from sale.

In addition to extremes of temperature, humidity, radiation, abrasives, and other environmental factors, the designer must consider fire resistance. There are no fully fire resistant plastics.

Polyimide-boron fiber composites have high temperature resistance and high strength. Polyimide-graphite fiber composites can compete with metals in strength and achieve a significant weight saving at service temperatures up to 600 °F. Boron powder is sometimes added to the matrix to help stabilize the char that forms in thermal oxidation. Other flame-resistant additives or an ablative matrix may also be used. The danger from open flames is the most serious objection to the use of plastics in fabrics and architectural structures.

Remember, thermal degradation and cross-linking are not reversible phenomena. Glass transition, melting, and crystallization of most matrices are reversible.

Moisture may cause deterioration and weaken the reinforcement and matrix bond in composites. Any holes, exposed edges, or machined areas on composite designs should be given a protective coating to prevent moisture infiltration or wicking.

Electrical Characteristics

All plastics have useful electrical insulation characteristics. The selection of plastics is usually based on mechanical, thermal, and chemical properties; however, much of the pioneering in plastics was for electrical uses. The electrical insulating problems of high altitude, space, undersea, and underground environments are solved by the use of plastics. All-weather radar and underwater sonar would not be possible without the use of plastics. They are used to insulate, coat, and protect electronic components.

Particulate composites using carbon, graphite, metal, or metal-coated reinforcements provide EMI shielding for many products.

Chemical Characteristics

The chemical and electrical natures of plastics are closely related because of molecular makeup. There is no general rule for chemical resistance. Plastics must be tested in the chemical environment of their actual use. Fluorocarbons, chlorinated polyethers and polyolefins are among the most chemical-resistant materials. Some plastics react as semipermeable membranes. They allow selected chemicals or gases to pass while blocking others. The permeability of polyethylene plastics is an asset in packaging fresh fruits and meats. Silicones and other plastics allow oxygen and gases to pass through a thin membrane while at the same time they stop water molecules and many chemical ions. Selective filtration of minerals from water may be done with semipermeable plastics membranes.

Mechanical Factors

Material consideration includes the mechanical factors of fatigue, tensile, flexural, impact, and compressive strengths, hardness, damping, cold flow, thermal expansion, and dimensional stability. All of these properties were discussed in Chapter 6. Products that need dimensional stability call for careful choice of materials, although fillers will improve the dimensional stability of all plastics. A factor sometimes used for evaluation and selection is strength-to-mass ratio; the ratio of tensile strength to the density of the material. Plastics can surpass steels in strength-to-mass ratio.

Example:

Divide the tensile strength of the material by its density.

$$\text{Selected plastics} — \frac{0.70\text{ Gpa}}{2\text{g/cm}^3} = 0.350$$

$$\text{Selected steel} — \frac{1.665\text{Gpa}}{7.7\text{ g/cm}^3} = 0.214$$

The type and orientation of reinforcements greatly influences the properties of composite products. Some specifications in critical designs will specify a *safety factor* (SF). A safety factor (sometimes called design factor) is defined as the ratio of the ultimate strength of the material to the allowable working stress.

$$\text{SF} = \frac{\text{Ultimate Strength}}{\text{Allowable Working Stress}}$$

The SF for a composite aircraft landing gear might be 10.0, while for an automobile spring it will be only 3.0. With accurate, reliable data some designs use safety factors of 1.5 to 2.0.

Economics

The final phase of material selection is economic considerations. It is best not to include material costs in the preliminary screening of candidate materials. Materials with marginal performance properties or expensive materials need not be eliminated at this point. Either may continue to be a possible candidate depending on the processing parameters, assembly, finishing, and service conditions. A polymer with minimal performance characteristics may not be the best choice if reliability and quality are important.

Cost is always a major factor in design considerations or material selection. Strength-to-mass ratio, or chemical, electrical, and moisture resistance, may overcome price disadvantage. Some plastics cost more per kilogram than metals or other materials, but plastics often cost less per finished part. The most meaningful comparison between different plastics is cost per cubic centimetre. The apparent density and bulk factors are important in cost analysis in any molding operation.

Apparent density, sometimes called *bulk density,* is the mass per unit volume of a material. It is calculated by placing the test sample in a graduated cylinder and taking measurements. The volume *(V)* of the sample is the product of its height *(H)* and cross-sectional area *(A)*. Thus $V = HA$.

$$\text{Apparent density} = \frac{W}{V} \text{ where}$$

V = volume in cubic centimetres occupied by the material in the measuring cylinder

H = height in centimetres of the material in the cylinder

A = cross-sectional area in square centimetres of the measuring cylinder

W = mass in grams of the material in the cylinder

Many plastics and polymer composite parts cost ten times more than steel. On a volume basis, some are lower in cost than metals.

Bulk factor is the ratio of the volume of loose molding powder to the volume of the same mass of resin after molding. Bulk factors may be calculated as follows:

$$\text{bulk factor} = \frac{D_2}{D_1}$$

where

D_2 = average density of the molded or formed specimen

D_1 = average apparent density of the plastics material prior to forming.

Economics must also include the method of production and design limitations of the product. One-piece seamless gasoline tanks may be rotationally cast or blow-molded. The latter process uses more costly equipment but can produce the products more quickly, thus reducing costs. Conversely, large storage tanks may be produced at less cost by rotational casting than by blow-molding.

Capital investment for new tooling, equipment, or physical space could result in consideration of different materials and/or processes. The labor-intensive operations often associated with wet or open molding cannot hope to compete with the automated facilities of some companies. The number of parts to be produced and the initial production costs may be the decisive factor.

The three overriding factors in plastics design are service, production, and cost. The use and performance of the part or product must be a concern in defining some design factors. For each design, there may be several production or process options. Costs often appear to override most other concerns of design and development, and cost is often based on the production method.

Volume of sales is very important. If a mold costs $10000 and only 10000 parts are to be made, the mold cost would be $1.00 per part. If 1000000 parts are to be made, the mold cost would be $0.01 per part.

Many composite components may be more cost-effective in the long term. The products may simply outlast metals in many applications. One piece composite components could reduce the number of molds used, tooling, and assembly time in the production of a boat hull, fuselage, or automobile floor pan. Light, corrosion-resistant composites or plastics components could lower energy costs of fuel for the lifetime of the transporation vehicle. Less energy is consumed (including the energy content of the raw material) in the production of polymer parts than for metal parts. Plastics are mostly petroleum derived and must continue to compete for depleting resources.

Design Considerations

When considering the overall design conditions, the intended application or function, environment, reliability requirements, and specifications must be reviewed. The database in computer systems may alert the designer that a design is outside the parameters of the material or process selected. See Appendix H for additional sources of information.

Appearance

The consumer is probably most aware of a product's physical appearance and utility. This includes the design, color, optical properties, and surface finish. Elements of design and appearance encompass several properties at once. Color, texture, shape, and material may influence consumer appeal. The smooth, graceful lines of Danish-style furniture with dark woods and satin finish are one example. Changing any one of these elements or properties would drastically change the design and appearance of the furniture.

A few outstanding characteristics of plastics are that they may be transparent or colored, as smooth as glass, or as supple and soft as fur. For many uses, plastics may be the only materials with the desired combination of properties to fulfill service needs.

To ensure proper design, there must be close cooperation between mold makers, manufacturers, processors, and fabricators.

Thought must be given to the design of the plastics part before it is molded in order to ensure that the best combination of mechanical, electrical, chemical, and thermal properties will be obtained.

Residual stresses develop as a result of forcing the material to conform to a mold shape. These stresses are locked in during cooling or curing and matrix shrinkage. They could cause warpage in flat surfaces. Warpage is somewhat proportional to the amount of shrinkage of the matrix and is generally the result of differential shrinkage.

There are no hard and fast rules to determine the most practical wall thickness of a molded part. Ribs, bosses, flanges, and beads are common methods of adding strength without increasing wall thickness.

Large flat areas should be slightly convex or crowned for greater strength and to prevent warpage from stress (Fig. 21-2).

In Table 21-2 the complexity of parts is shown for several processes. In molding, it is important that all areas of the mold cavity be filled easily to uniformly minimize most of the stress in molding the part. A uniform wall thickness in the design is important to prevent uneven shrinkage of thin and thick sections. If wall thickness is not uniform, the molded part may distort, warp, and have internal stresses or cracks. From 0.24 to 0.51 in. [6 to 13 mm] may be considered heavy wall thicknesses in molded parts.

In general, the ribs should have a width at the base equal to one-half the thickness of the adjacent wall. They should be no higher than three times the wall thickness. Boss designs should have an outside diameter equal to twice the inside diameter of the hole. They should be no higher than twice their diameter.

Plastics parts should have liberal fillets and rounds to increase strength, assist molten material flow, and reduce points of stress concentration. All radii should be generous, and the recommended minimum radius is 0.020 in. [0.50 mm]. Optimum design is obtained with a radius-to-thickness ratio of 1:6.

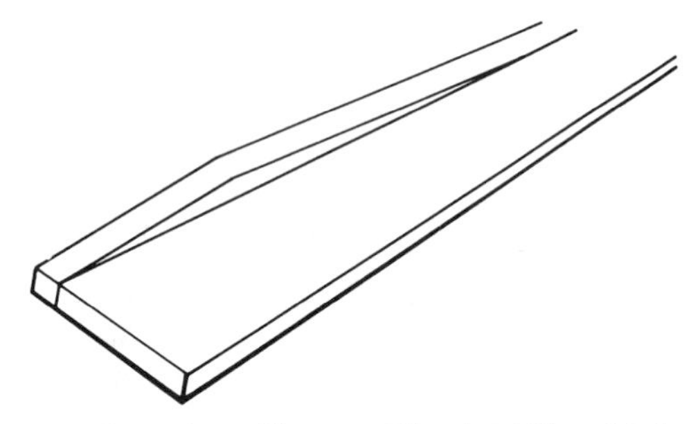

(A) Long, flat strips will warp. Ribs should be added, or the piece crowned in a convex shape.

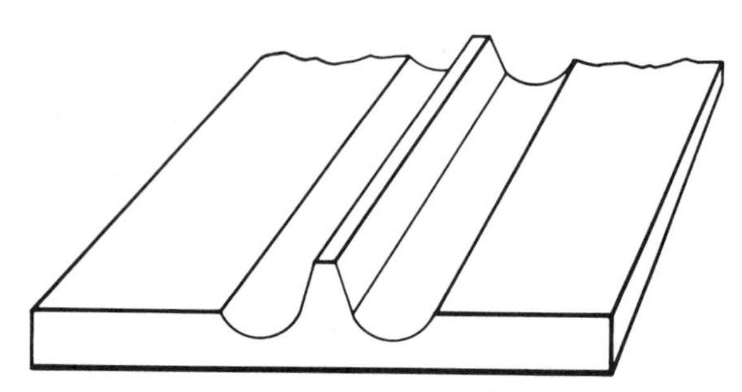

(B) Uneven sections will cause distortion, warpage, cracks, sinks, or other problems because of the difference in shrinkage from section to section.

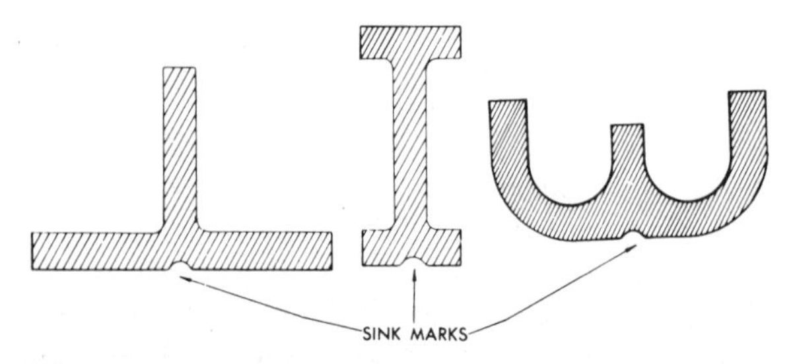

(C) Thickness of walls and ribs in thermoplastic part should be about 60 percent of the thickness of main walls This will reduce the possibility of sink marks.

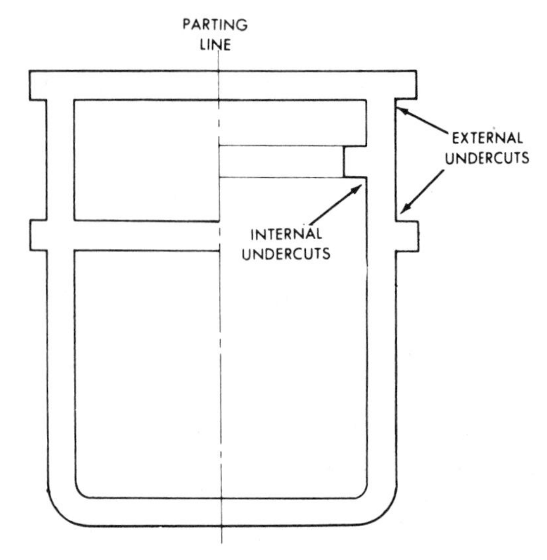

(D) A simple molding with internal and external undercuts.

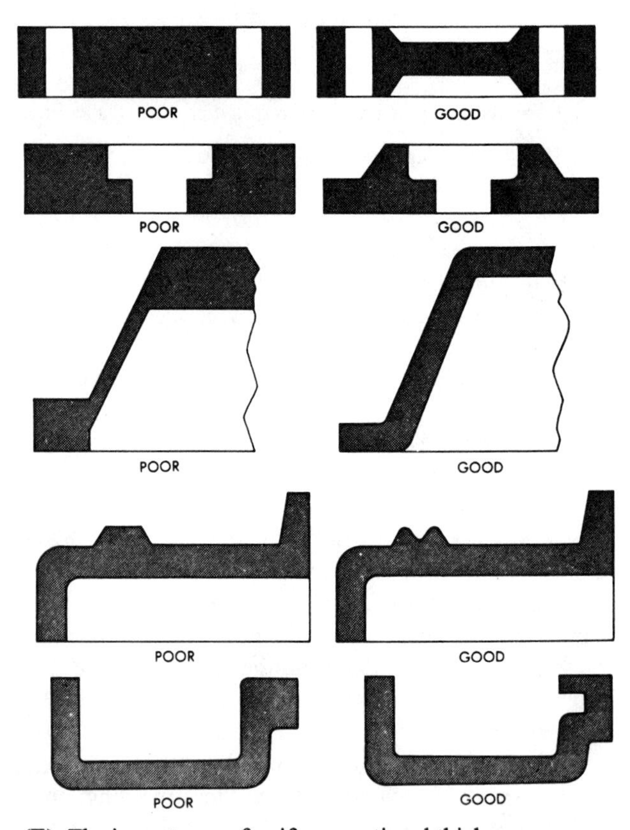

(E) The importance of uniform sectional thickness.

Fig. 21-2. Precautions to observe in production of plastics products.

Table 21-2. Plastics Processing Forms—Complexity of Part

Processing Form	Section Thickness, mm		Bosses	Undercuts	Inserts	Holes
	Max.	Min.				
Blow molding	> 6.35	0.254	Possible	Yes—but reduce production rate	Yes	Yes
Injection moldings	>25.4; normally 6.35	0.381	Yes	Possible—but undesirable; reduce production speed and increase cost	Yes—variety of threaded and non-threaded	Yes—both through and blind
Cut extrusions	12.7	0.254	Yes	Yes—no difficulty	Yes—no difficulty	Yes—in direction of extrusion only; 0.50-1.0 mm min.
Sheet moldings (Thermoforming)	76.2	0.00635	Yes	Yes—but reduce production rate	Yes	No
Slush moldings		0.508	Yes	Yes—flexibility of vinyl allows drastic undercuts	Yes	Yes
Compression moldings		0.889–3.175	Possible	Possible—but not recommended	Yes—but avoid long, slender, delicate inserts	Yes—both through and blind; but should be round, large, and at right angles to surface of part
Transfer moldings		0.889–3.175	Possible	Possible—but should be avoided: reduce production rate	Yes—delicate inserts may be used	Yes—should be round, large, and at right angles to surface of part
Reinforced plastics moldings	Bag: 25.4 matched die: 6.35	Bag: 2.54 matched die: 0.762	Possible	Bag: yes; matched die: no	Bag: yes. matched die: possible	Bag: only large holes; matched die: yes
Castings		3.175–4.762	Yes	Yes—but only with split and cored molds	Yes	Yes

Source: *Materials Selector,* Materials Engineering, Reinhold Publishing Corp., Subsidiary of Litton Publications, Inc., Division of Litton Industries.

Undercuts (internal or external) in parts should be avoided if possible. Undercuts usually increase tooling costs by requiring techniques for molding, part removal (which usually requires movable parts in the mold), and cooling jigs. A slight undercut could be tolerated in some products when using tough, elastic materials. The molded part can be successfully snapped or stripped out of the cavity while hot. Undercut dimensions should be less than 5 percent of the part diameter.

Decorating may be considered an important functional factor in plastics design. The product may include textures, instructions, labels, or letters, and it must be decorated in a manner that will not complicate removal from the mold (Fig. 21-3). Decorations should provide durable service to the consumer. Letters are commonly engraved, hobbed, or electrochemically etched into mold cavities (Fig. 21-4).

Design Limitations

Next to material selection, tooling and processing have a marked effect on the properties and quality of all plastics products.

Closely related to production is the design of the product, and ultimately, the design of the mold to produce the product. Output rates, parting lines, dimensional tolerances, undercuts, finish, and material shrinkage are among factors that must be kept in mind by the mold maker or tool designer. For example, undercuts and inserts slow output rates and require a more costly mold.

Fig. 21-3. Familiar examples of texture on plastics,which help to hide imperfections. (Mold-Tech. Rochlen Industries)

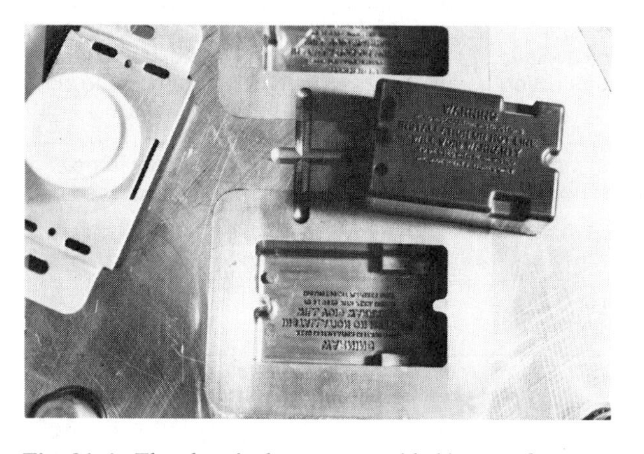

Fig. 21-4. The electrical part was molded in one of two cavities of the injection mold. Note engraved letters in the mold cavities.

The problem of material shrinkage is of equal importance to both the designer of molds and the designer of molded products. The loss of solvents, plasticizers, or moisture during molding, together with the chemical reaction of polymerization in some materials, results in shrinkage.

In the injection molding of crystalline materials shrinkage is affected by the rate of cooling.

Thicker sections, which take longer to cool, will experience greater shrinkage than adjacent thinner sections, which cool quickly.

The thermal contraction of the material must also be considered. The thermal expansion values for most plastics are relatively large. This is an asset in removal of molded products from mold cavities. If close tolerances are needed, material shrinkage and dimensional stability must be considered. Material shrinkage is sometimes used to ensure snug, or shrink, fits of metal inserts.

By using a computer model in a CAD system, it is possible to show the stress response of the part with a specific geometry, reinforcement content, reinforcement orientation, and molding orientation (flow). The computer model may require the use of ribs, contours, or other configurations to produce isotropic or anisotropic properties.

After the preliminary design is done, a physical prototype is often produced. This allows the design engineer and others to see and test a working prototype mold. Simulated performance and service tests may be performed with the prototype. If design or material errors were made, specifications for redesign can be made.

Both material and production considerations are important when designing products of plastics. The problems met in producing plastics products often demand selection of the production techniques before the material considerations are discussed.

Production Considerations

In any product design, the behavior of the material and cost are often reflected in molding, fabrication, and assembly techniques. The tooling design must consider material shrinkage, dimensional tolerance, mold design, inserts, decorations, knockout pins, parting lines, production rates, and other post-processing operations (Table 21-3).

Trimming, cutting, boring holes, or other fabricating or assembly techniques may slow production, lower performance properties, and increase costs.

The part shape, size, matrix formulation, and polymer form will often limit the means of production to one or two possibilities.

Processes of Manufacture

With new technology and materials, processing is often the decisive competitive factor. Today, there are fewer limitations in processing thermo-plastics and thermosetting materials than in years past. Processes that were unthinkable a few years ago have now become routine. Many thermosets are now injection molded or extruded. Some materials are processed in thermoplastic equipment and later cured. Polyethylene may be cross-linked after extrusion by chemical or radiation methods. Both thermoplastics and thermosets may be made cellular. Because of moldability, output

Table 21-3. Design Considerations

Plastics	Approximate Cost, ¢/in^3	Approximate Cost, ¢/cm^3	Linear Mold Shrinkage, mm/mm	Practical Dimensional Tolerances mm/mm (Single Cavity)			Taper Required, degrees		
				Fine	Standard	Coarse	Fine	Standard	Coarse
ABS	3–4	0.41	0.127–0.203	0.051	0.102	0.152	0.25	0.50	1.00
Acetal	7–8	0.46	0.508–0.635	0.102	0.152	0.229	0.50	0.75	1.00
Acrylic	4–5	0.27	0.025–0.102	0.076	0.127	0.178	0.25	0.75	1.25
Alkyd (filled)	5–6	0.33	0.102–0.203	0.051	0.102	0.127	0.25	0.50	1.00
Amino (filled)	3.5–5	0.27	0.279–0.305	0.051	0.076	0.102	0.125	0.5	1.00
Cellulosic	5–6	0.52	0.076–0.254	0.076	0.127	0.178	0.125	0.5	1.00
Chlorinated polyether	5–6	0.33	0.102–0.152	0.102	0.152	0.229	0.25	0.50	1.00
Epoxy			0.025–0.102	0.051	0.102	0.152	0.25	0.50	1.00
Fluoroplastic (CTFE)	240	14.7	0.254–0.381	0.051	0.076	0.127	0.25	0.50	1.00
Ionomer	4–5	0.27	0.076–0.508	0.076	0.102	0.152	0.50	1.00	2.00
Polyamide (6.6)	6–7	0.39	0.203–0.381	0.127	0.178	0.279	0.125	0.25	0.50
Phenolic (filled)	2.5–3	0.18	0.102–0.229	0.038	0.051	0.064	0.125	0.50	1.00
Phenylene oxide.	4.5–6	0.30	0.025–0.152	0.051	0.102	0.152	0.25	0.50	1.00
Polyallomer	2–2.5	0.14	0.254–0.508	0.051	0.102	0.152	0.25	0.50	1.00
Polycarbonate	7–8	0.46	0.127–0.178	0.076	0.152	0.203	0.25	0.50	1.00
Polyester (thermoplastic)	9–10	0.58	0.076–0.457	0.051	0.102	0.152	0.25	0.50	1.00
Polyethylene (high density)	1.5–2	0.15	0.508–1.270	0.076	0.127	0.178	0.50	0.75	1.50
Polypropylene	1–2	0.93	0.254–0.635	0.076	0.127	0.178	1.00	1.50	2.00
Polystyrene	2–2.5	0.14	0.025–0.152	0.051	0.102	0.152	0.25	0.50	1.00
Polysulfone	18–19	1.14	0.152–0.178	0.102	0.127	0.152	0.25	0.50	1.00
Polyurethane	7–8	0.46	0.254–0.508	0.051	0.102	0.152	0.25	0.50	1.00
Polyvinyl (PVC) (rigid)	35–40	2.26	0.025–0.127	0.051	0.102	0.152	0.25	0.50	1.00
Silicone (cast)	38–45	2.57	0.127–0.152	0.051	0.102	0.152	0.125	0.25	0.50

rates, and other material properties, seemingly costly materials become inexpensive products.

In Table 21-4, a comparison of processing and economic factors is shown.

Material Shrinkage

Irregularities in wall thickness may create internal stresses in the molded part. Thick sections cool more slowly than thin ones and may create *sink marks* as well as differential shrinkage in crystalline plastics. As a general rule, injection-molded crystalline plastics have high shrinkage while amorphous plastics shrink less.

Much pressure must be exerted to force the material through thin wall sections in the mold creating further problems because of material shrinkage. Polyethylenes, polyacetals, polyamides, polypropylenes, and some polyvinyls shrink between 0.020 and 0.030 in. [0.50 and 0.76 mm] after molding. Molds for these crystalline and other amorphous plastics must allow for material shrinkage.

Normally unfilled injection-molded plastics shrink more in the direction of flow as opposed to the axis transverse to flow. This is mainly caused by the orientation pattern developed by the flow direction from the gate or gates. The differential shrinkage results because oriented plastics normally have a higher shrinkage than non-oriented plastics. An exception is fiber-reinforced polymers.

Fiber-reinforced polymers will shrink more along the axis transverse to flow than along the axis of material flow. Typical shrinkage of fiber-reinforced polymers is about one-third to one-half that of non-reinforced polymers. The reason is that the fibers that are oriented in the direction of flow prevent the normal free shrinkage of the plastics or polymer.

Tolerances

Closely related to shrinkage is maintaining dimensional tolerances. Molding items with precision tolerances requires careful materials selection, and tooling costs are greater for precision molding. Dimensional tolerances of single-cavity molded articles may be held to ±0.002 in./in. [±0.05 mm/mm] or less with selected plastics. Errors in tooling, variations in shrinkage between multicavity pieces, and differences in temperature, loading, and pressure from cavity to cavity all increase the critical dimensional tolerances of multicavity molds. If, for example, the number of cavities is increased to 50, the closest practical tolerance may then be ±0.010 in./in. [±0.25 mm/mm.]

Table 21-4. Economic Factors Associated with Different Processes

Production Method	Economic Minimum	Production Rates	Equipment Cost	Tooling Cost
Autoclave	1–100	Low	Low	Low
Blow molding	1000–10000	High	Low	Low
Calendering (metres)	1000–10000	High	High	High
Casting processes	100–1000	Low-high	Low	Low
Coating processes	1–1000	High	Low-high	Low
Compression molding	1000–10000	High	Low	Low
Expanding processes	1000–10000	High	Low-high	Low-high
Extrusion (metres)	1000–10000	High	High	Low
Filament winding	1–100	Low	Low-high	Low
Injection molding	10000–100000	High	High	High
Laminating (continuous)	1000–10000	High	Low	High
Lay-up	1–100	Low	Low	Low
Machining	1–100	Low	Low	Low
Matched die	1000–10000	High	High	High
Mechanical forming	1–100	Low-high	Low	Low
Pressure-bag	1–100	Low	Low	Low
Pulforming (metres)	1000–10000	Low-high	Low	High
Pultrusion (metres)	1000–10000	Low-high	High	Low-high
Rotational casting	100–1000	Low	Low	Low
Spray-up	1–100	Low	Low	Low
Thermoforming	100–1000	High	Low	Low
Transfer molding	1000–10000	High	Low	High
Vacuum-bag	1–100	Low	Low	Low

Tolerance standards have been established by technical custom molders and by the *Standards Committee* of the *Society of the Plastics Industry, Inc.* These standards are to be used only as a guide since the individual plastics material and the design must be considered in determining dimensions.

There are three classes of dimensional tolerances for molded plastics parts. They are expressed as plus and minus allowable variations in inches per inch (in./in.) or millimetres per millimetre (mm/mm). *Fine* tolerance is the narrowest possible limit of variation possible under controlled production. *Standard* tolerance is the dimensional control that can be maintained under average conditions of manufacture. *Coarse* tolerance is acceptable on parts where accurate dimensions are not important or critical.

So that the part may be removed with ease from the molding cavity, draft should be provided (both inside and out). The degree of draft may vary according to molding process, depth of part, type of material, and wall thickness. A draft of 0.25 degree is sufficient for all shallow molded parts. For textured designs and cores, the draft angles should be increased.

If a part has a depth of 10 in. [250 mm] and 0.125 degree draft for a fine dimensional tolerance of 0.002 2 in. /in. [0.056 mm/mm], the total draft of the piece will be 0.22 in. [0.559 mm] per side (Fig. 21-5).

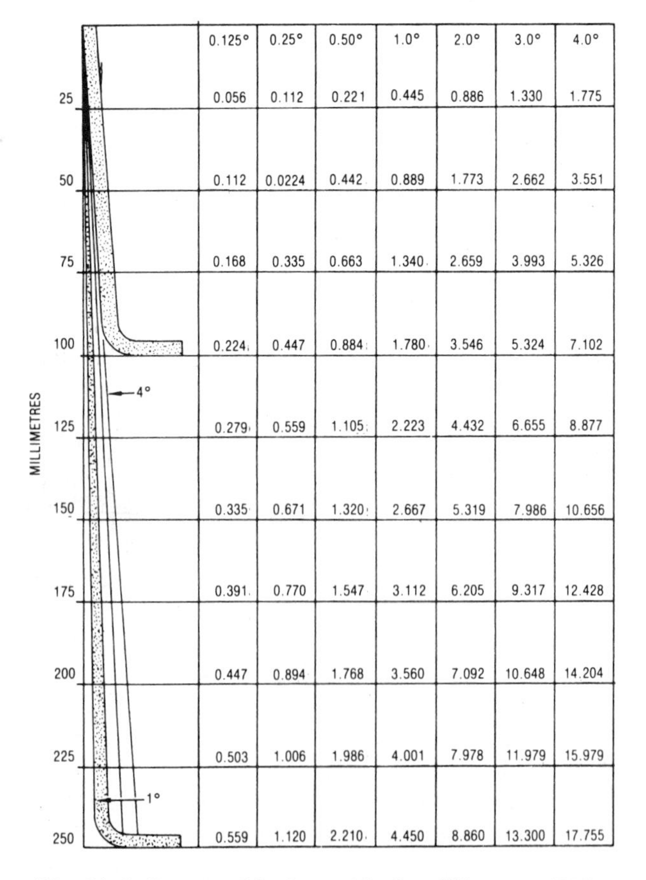

Fig. 21-5. Degree of draft per side, in millimeters. (SPI)

Mold Design

Mold design is an important factor in determining molding output. Because mold design is a complex subject, only a broad discussion is possible here. (See Chapter 22.) A design for a typical two-piece, two-cavity injection mold is shown in Figure 21-6. A three-plate mold is shown in Figure 21-7.

As the hot, molten material is forced from the nozzle into the mold, it flows through channels or passageways. The terms *sprue, runner,* and *gate* are used to designate these channels (Fig. 21-8).

The heavy tapered channel that connects the nozzle with the runners is called the *sprue.* In a single-cavity mold, the sprue feeds material directly through a gate into the mold cavity (Fig. 21-9). If the sprue feeds directly, it eliminates the need for a separate runner and gate. In most single cavity molds, there is no need for a runner unless the part is being gated in more than one place.

Runners are narrow channels that convey the molten plastics from the sprue to each cavity. In multicavity molds, the runner system should be designed so that

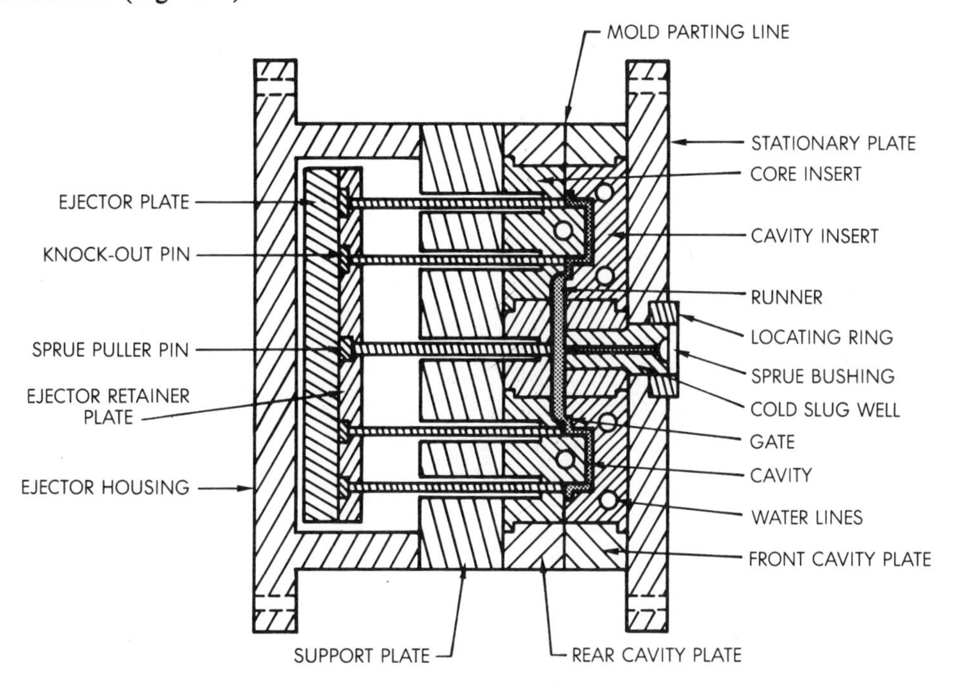

Fig. 21-6. Two-plate injection mold. (Gulf Oil Chemicals Co.)

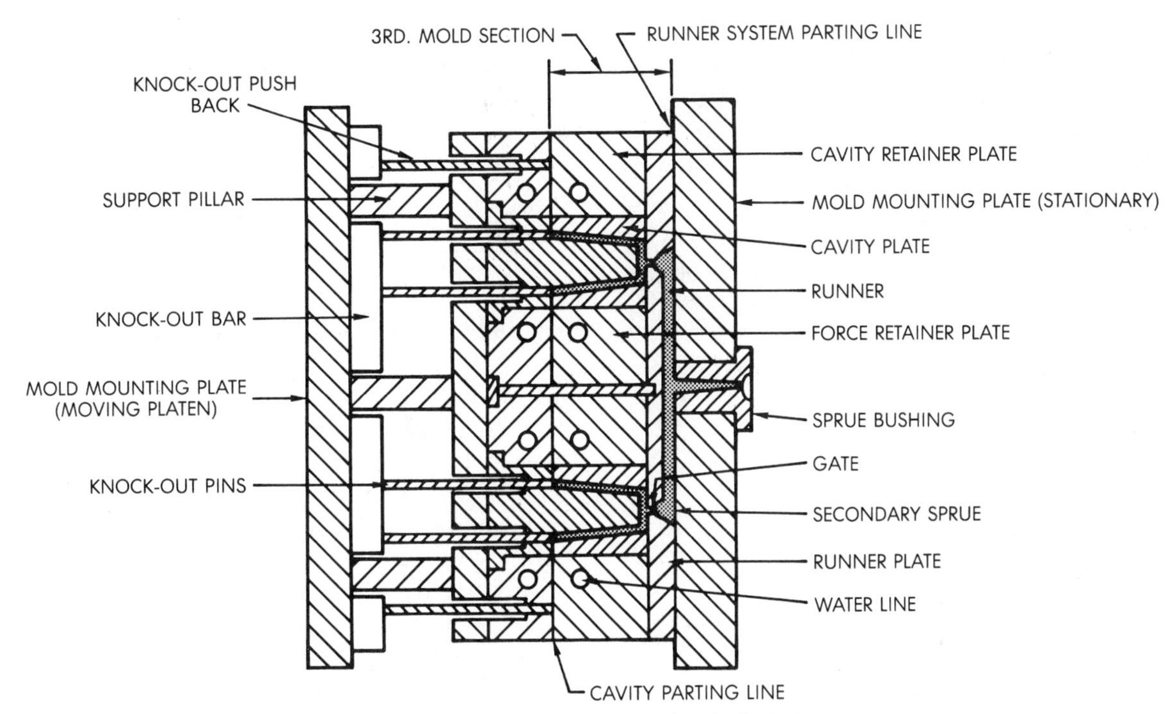

Fig. 21-7. Three-plate injection mold. (Gulf Oil Chemicals Co.)

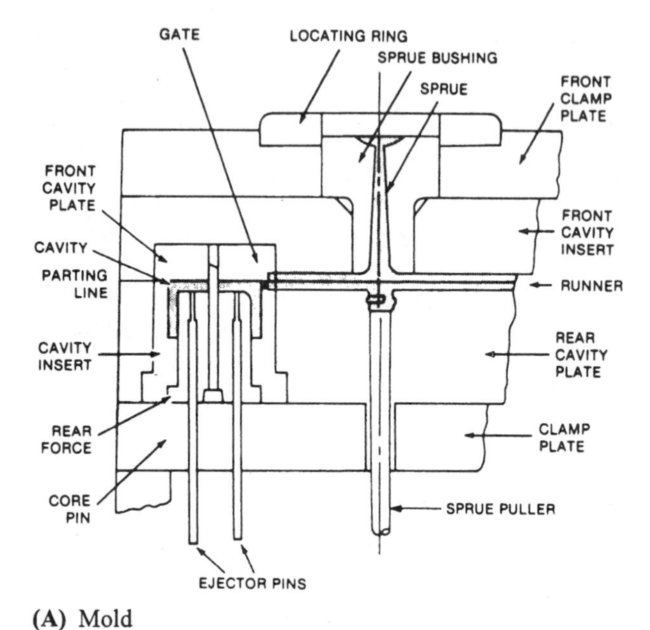

(A) Mold

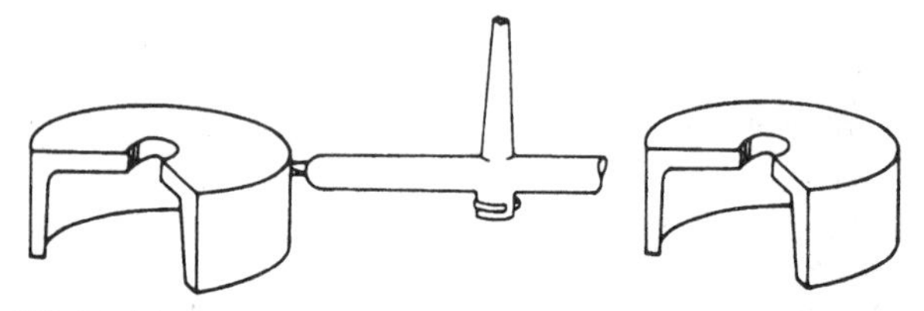

(B) Molded part, runner, gate, sprue, and typical part.

Fig. 21-8. Construction of an injection mold, and a molded part.

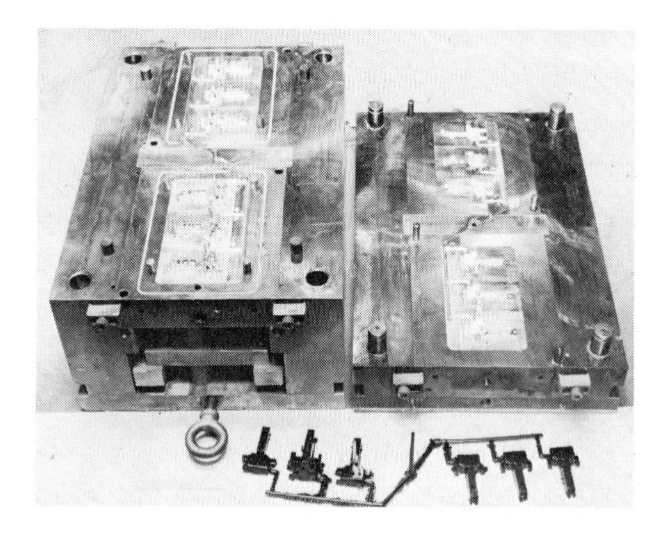

Fig. 21-9. Injection mold, with molded product. Note the flask, or fins, on runner and parts. (Hull Corp.)

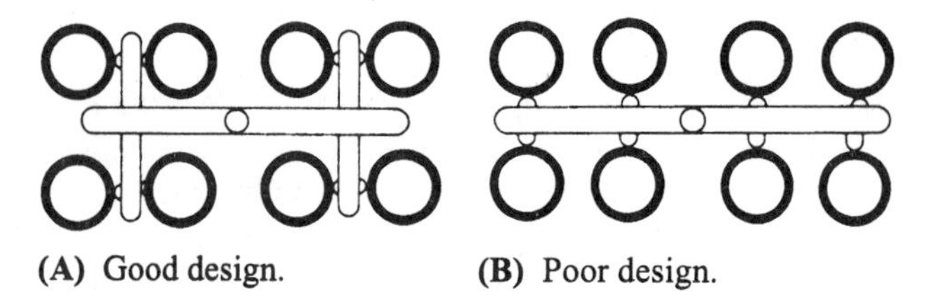

(A) Good design. (B) Poor design.

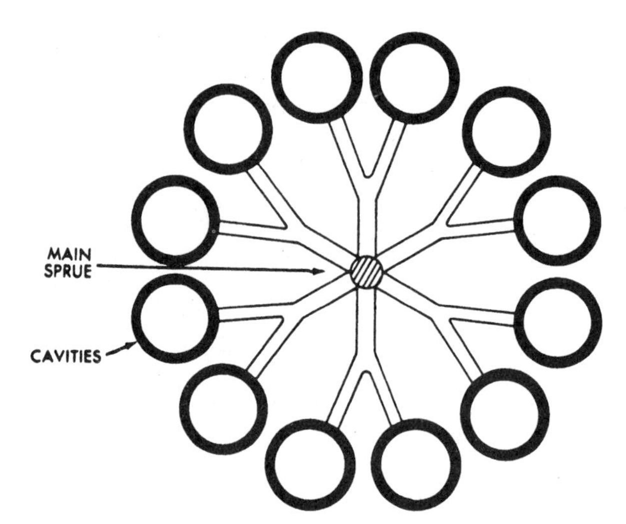

(C) Radial design.

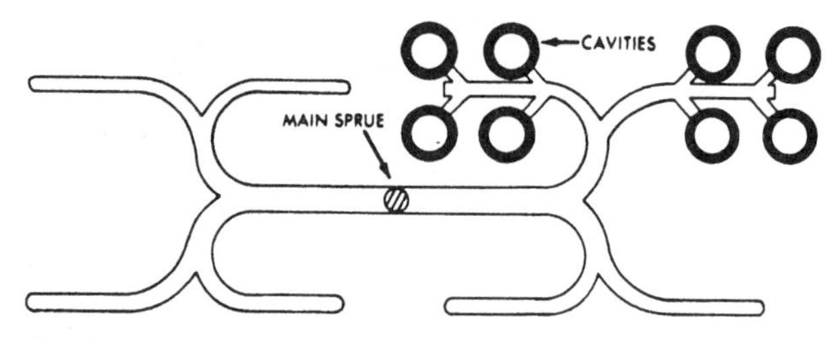

(D) Sweeping-curve design.

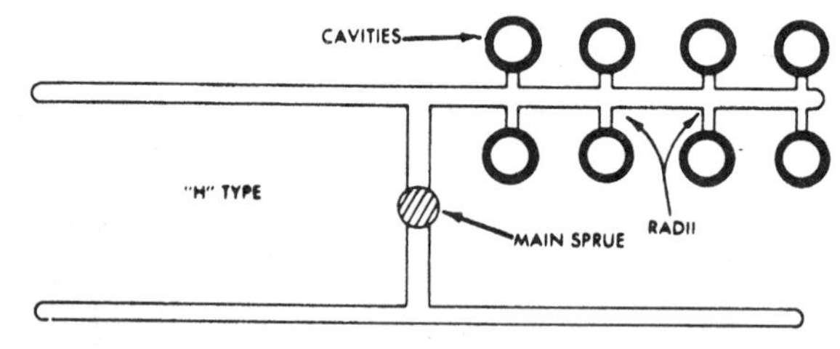

(E) H design.

Fig. 21-10. Some typical runner designs. (Du Pont)

all materials move the same distance from the sprue to each cavity (Fig. 21-10).

When a series of runners and gates is used, trapezoidal, half-round or full-round runners are machined into the mold. With trapezoidal and half-round runners and gates, only one half of the mold or die plate is machined (Fig. 21-11). Trapezoidal and half-round runners are easy to machine, but generally require more molding pressure. Round runners are advised for transfer molding if extruder-type plasticizers are used. Pinpoint (submarine) gating, as shown in Figure 21-12, may be used; however, this system also requires greater molding pressures.

A pinpoint gate is an opening of 0.030 in. [0.08 mm] or less through which the melt flows into a mold cavity. The submarine is a type of edge gate where the opening from the runner into the mold is located below

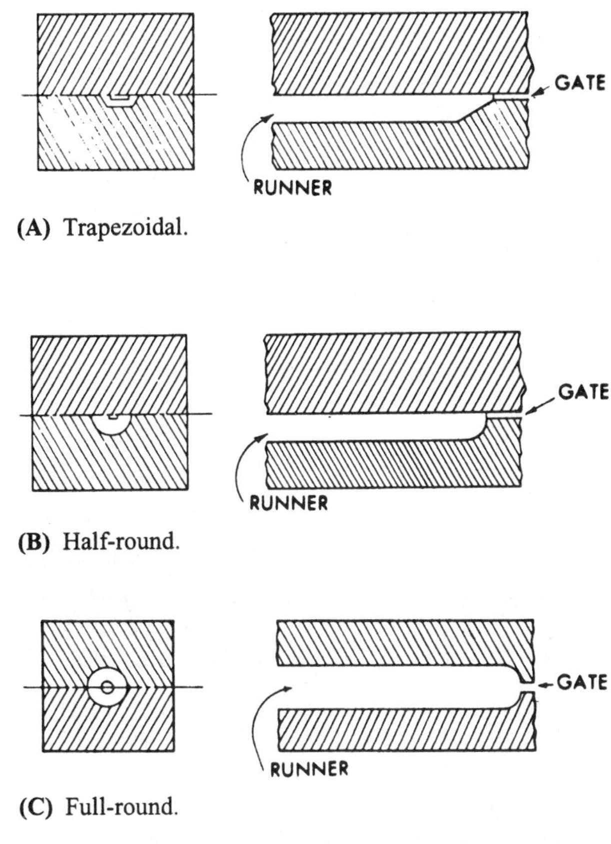

(A) Trapezoidal.

(B) Half-round.

(C) Full-round.

Fig. 21-11. Three basic runner systems used for transfer and injection molding.

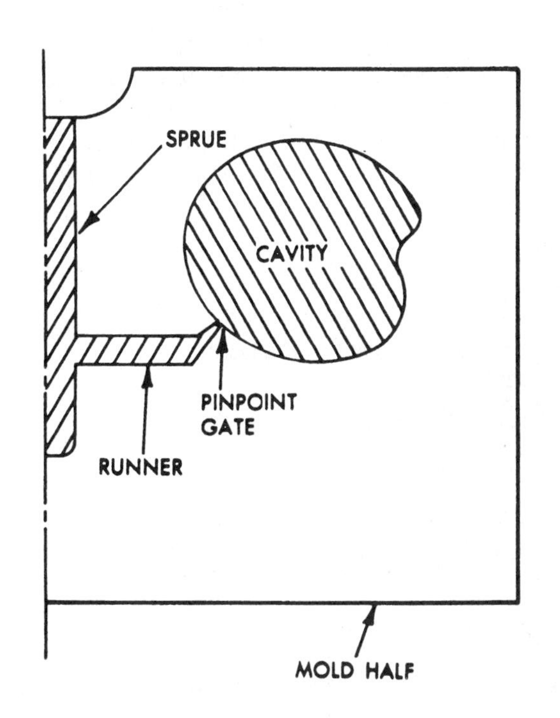

Fig. 21-12. Pinpoint (submarine) gate system, which eliminates gate break-off problems. It also reduces or eliminates finishing problems caused by large gate marks.

the parting line or mold surface. The part is broken from the runner system or ejection from the mold.

The cost of elaborate mold designs and high waste from culls, sprues, and flash are two major limitations of transfer molding.

In some molds, there is only one cavity; others have many. Regardless of the number of cavities the *gate* is the point of entry into each mold cavity. In multicavity molds, there is a gate entering each mold cavity. The gates may be any shape or size, however, they are usually small so as to leave as small a blemish as possible. Gates must allow a smooth flow of molten material into the cavity (Fig. 21-13). A small gate will help the finished item break away from the sprue and runners cleanly (Fig. 21-14).

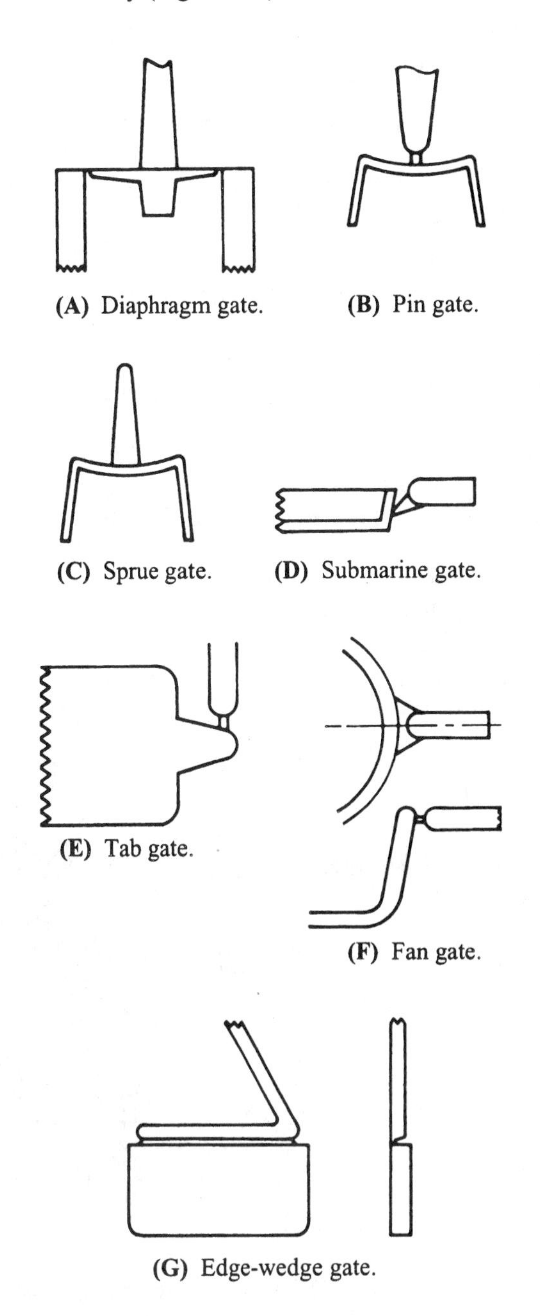

(A) Diaphragm gate. (B) Pin gate.

(C) Sprue gate. (D) Submarine gate.

(E) Tab gate.

(F) Fan gate.

(G) Edge-wedge gate.

Fig. 21-13. Some of the many possible gating systems.

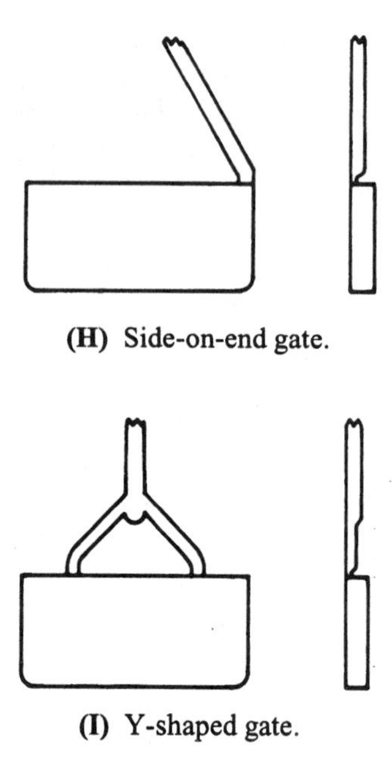

(H) Side-on-end gate.

(I) Y-shaped gate.

Fig. 21-13. Some of the many possible gating systems. *(Continued)*

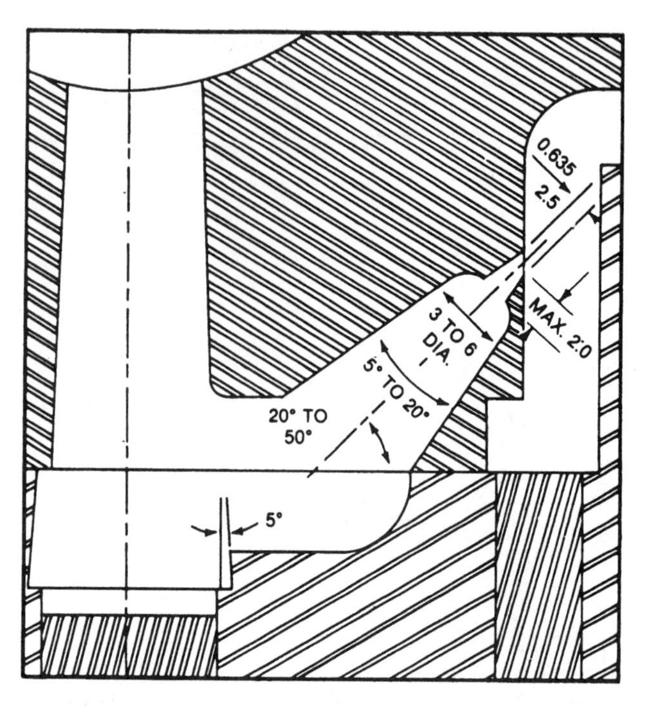

Fig. 21-14. A typical tunnel gate, which may be considered a variation of the pinpoint gate design.(Mobay Chemical Co.)

The sprues, gates, and runners are usually cooled and removed from the mold with the parts from each cycle. The sprues, runners and gates are removed from the parts, and then reground for molding, This reprocessing is costly and restricts the mass of molded articles per cycle of the injection-molding machine. In *hot runner molding,* the sprues and runners are kept hot by means of heating elements built into the mold. As the mold opens, the cured part pulls free from the still-molten hot runner system. On the next cycle, the hot material remaining in the sprue and runner is forced into the cavity (Fig. 21-15).

A similar system called *insulated runner molding* is used in molding polyethylene or other materials with low thermal transfer (Fig. 21-16). This system is also called *runnerless molding*. In this design, large runners are used. As the molten material is forced through these runners, it begins to solidify, forming a plastics lining that serves as insulation for the inner core of molten material. The hot inner core continues to flow through the tunnel-like runner to the mold cavity. A heated torpedo or probe may be inserted in each gate. This helps control freeze-up and drool.

There are many other mold designs that may include valve-gated molds, single or multiple molds (Fig. 21-17), unscrewable molds (for internal or external threads), cam and pin molds (for undercuts and cores), and multicolor, or multimaterial, molds. In

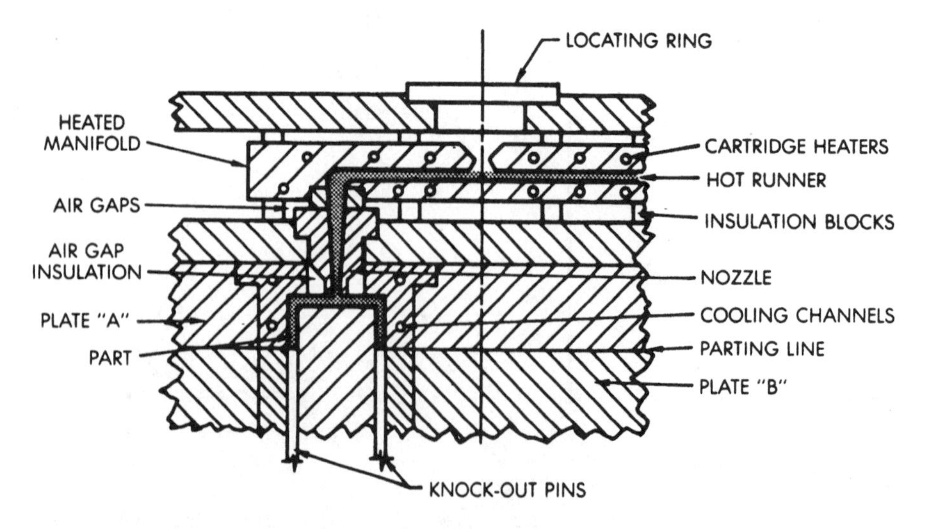

Fig. 21-15. Schematic drawing of hot runner mold.

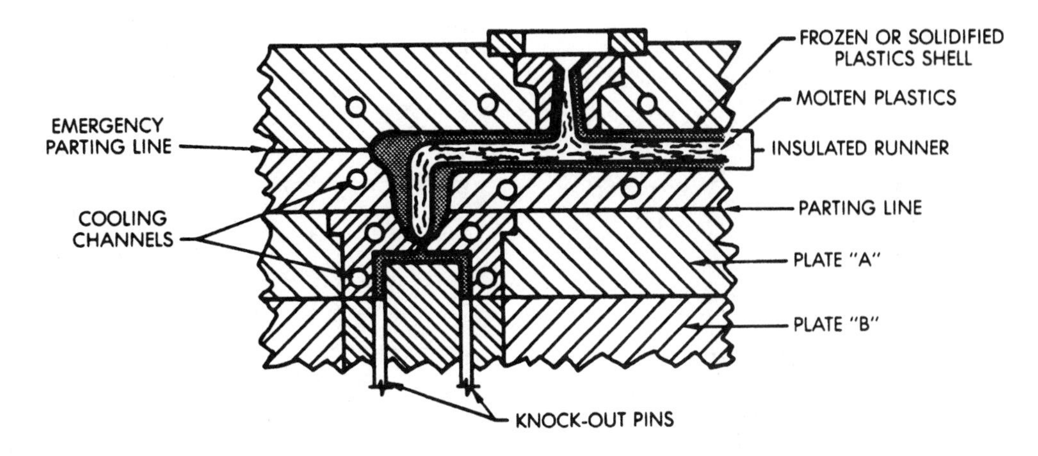

Fig. 21-16. Principle of insulated runner molding.

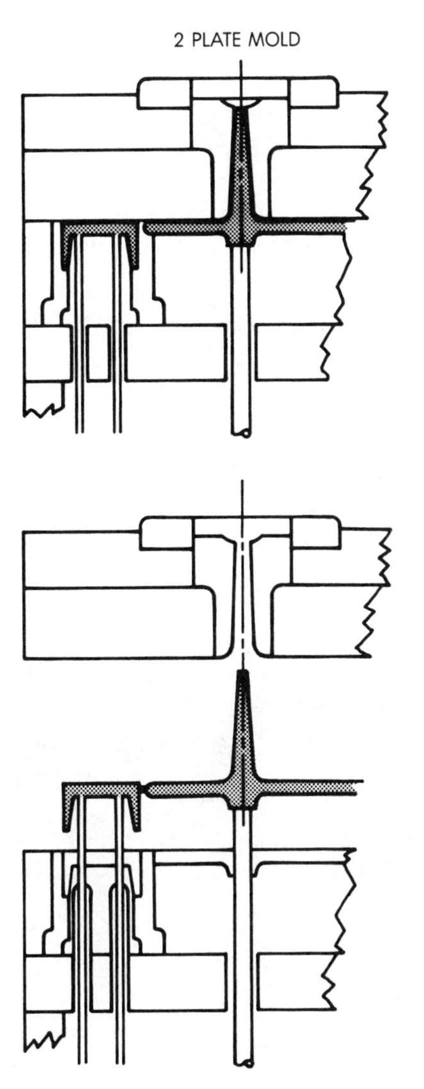

Fig. 21-17. Multiple mold design for injection molding. (Du Pont Co.)

Fig. 21-18. Calculator keys are an example of a multi-colored injection-molded product.

multicolor molds, a second color or material is injected around the first molding, leaving portions of the first molding exposed. Examples of multicolor products are special knobs, buttons and letter or number keys (Fig, 21-18).

Good draft, suitable gating, constant wall thickness, proper cooling, sufficient ejection, proper steels, and ample mold support are all important factors in mold design. (See Chapter 22, Tooling and Moldmaking).

Compression-mold design. There are three different types of compression-mold designs. Compression molds are usually produced of hardened steel that can withstand the great pressure and abrasive action of the hot plastics compound as it liquifies and flows into all parts of the mold cavity.

The *flash mold* is the least complex and most economical from the standpoint of original mold cost (Fig. 21-19). In this mold, excess material is forced out of the molding cavity to form a flash that becomes waste and must be removed from the molded part.

In positive mold design, vertical or horizontal flash provision is made for excess material in the cavity (Fig. 21-20). Vertical flash is easier to remove. The preforms must be measured with care if all parts are to be the same density and thickness. If too much material is loaded in the cavity, the mold may not fully close. In positive molds, little or no flash occurs. This design is used for molding laminated, heavily filled, or high-bulk materials.

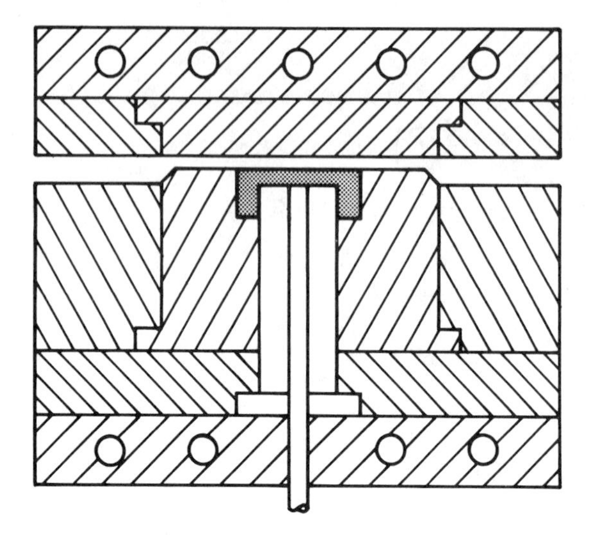

Fig. 21-19. Flash mold design.

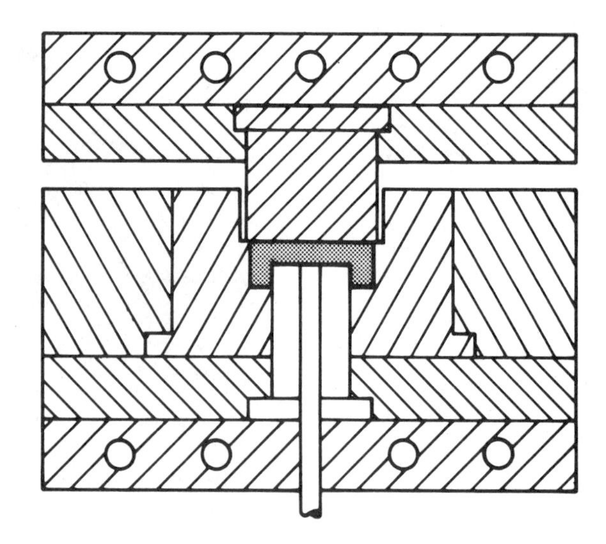

Fig. 21-21. Semipositive mold design.

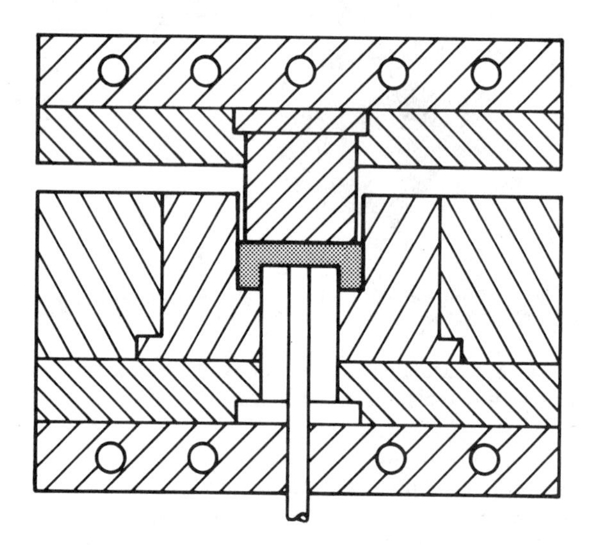

Fig. 21-20. Positive mold design.

With fully positive molds, gases freed during the chemical curing of thermosets may be trapped in the mold cavity. The mold may be opened briefly to allow gases to escape. This operation is sometimes called breathing.

Semipositive molds have horizontal and vertical flash waste (Fig. 21-21). This design is costly to make and maintain, but it is the most practical where many parts or long runs are needed. The design allows for some inaccuracy of charge by allowing flash thus giving a dense, uniform molded part. As the mold charge is compressed in the cavity, any excess material escapes. As the mold body continues to close, very little material is allowed to flash. When the mold fully closes, the telescoping male half is stopped by the *land.*

Blow-mold design. Construction of a *blow mold* is less costly. Aluminum, beryllium-copper, or steel are used as basic materials. Aluminum is one of the lowest cost blow-mold materials (Fig. 21-22). It is light and transfers heat rapidly. Beryllium-copper is harder and more wear resistant, but it is also more expensive. Steel is used at pinch-off points. If ferrous molds are used, they are plated to prevent rusting or pitting. (See Chapter 22.)

Parting lines. Parting lines are usually placed at the greatest radius of the molded part (Fig. 21-23). If the parting line isn't at the plane of greatest dimension, the mold must have movable parts or flexible molds or materials must be used. If the parting line cannot be placed on an inconspicuous edge or concealed, finishing is generally needed.

Fig. 21-22. Four containers are produced at one time from four parisons. Note aluminum tooling and cavity texture. (Uniloy Blowmolding Machinery Division, Hoover Universal.)

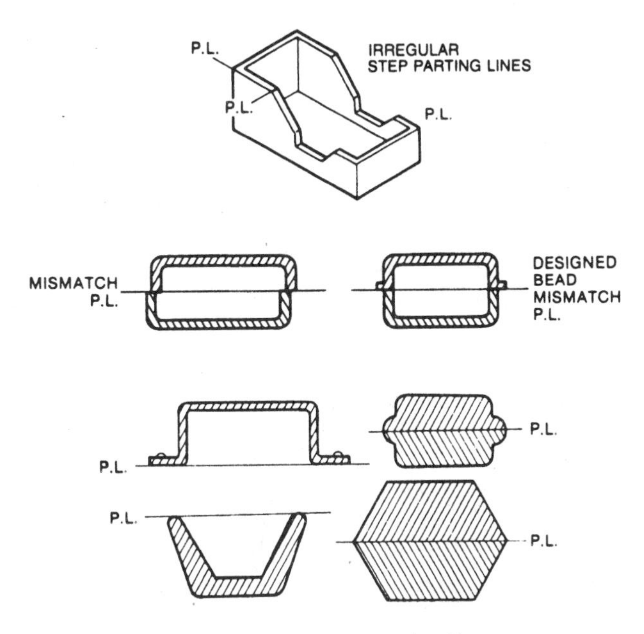

Fig. 21-23. Various locations for parting lines.

Ejector or knockout pins. Knockout or ejector pins push the hardened parts from the mold. They must touch the part in hidden or inconspicuous areas and should avoid contact with a flat surface, unless decorative designs can help conceal the marks. Pins should be made as large as possible for longest tool life. Knockout pins, blades or bushing shapes should not press against thin part areas.

The pins are usually attached to a master bar or pin plate. They are drawn flush with the surface of the mold by spring action. When the die is opened, a rod attached to the master bar or pin plate strikes a stationary stop, pushing all pins forward forcing the part from the cavity.

All molds need venting. Ejector pins may be altered to provide venting.

Inserts. Inserts and holes must be designed and placed with care in the molding cavity or part. A liberal draft should be given long pins and plugs. A general rule is to never have a hole depth more than four times the diameter of the pin or plug. Long pins are often made to meet halfway through a hole allowing twice the hole-depth limitation. Long pins are easily broken and bent by pressures of the flowing plastics.

The placement of molding gates in relation to holes is important. As the molten material is forced into the mold, it must flow around the pins which protrude into the mold cavity. These pins are withdrawn when the mold opens. If parts are to be assembled on these holes, the material should be made thicker by producing a boss (Fig. 21-24). The boss adds strength, preventing the material from cracking. The pins often restrict the flow of material and may cause flow marks, weld lines or possible cracking because of molding stress (Fig. 21-25). Flow lines and patterns are shown in Figure 21-26.

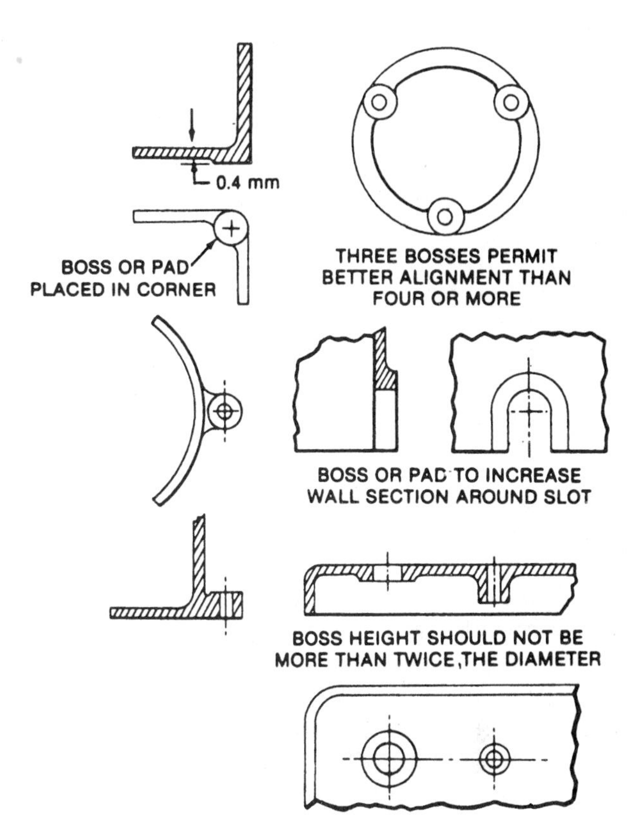

Fig. 21-24. Importance of boss design. (SPI)

When metal inserts are molded in place, the molten plastics material is forced around the insert. As the plastics cools, it shrinks around the metal insert, substantially contributing to the holding power of the insert. Inserts may be placed in thermoplastic parts by ultrasonic techniques. Before molding, inserts may be positioned automatically or by hand on small locating

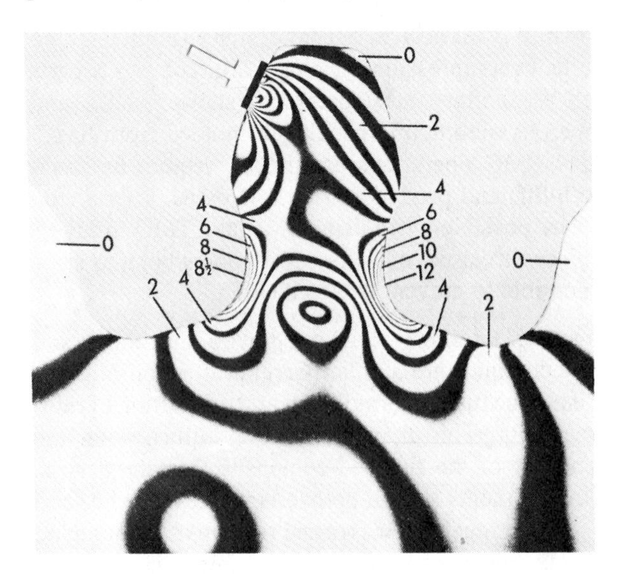

Fig. 21-25. Molding stress lines in this injection-molded part are visible under polarized light.

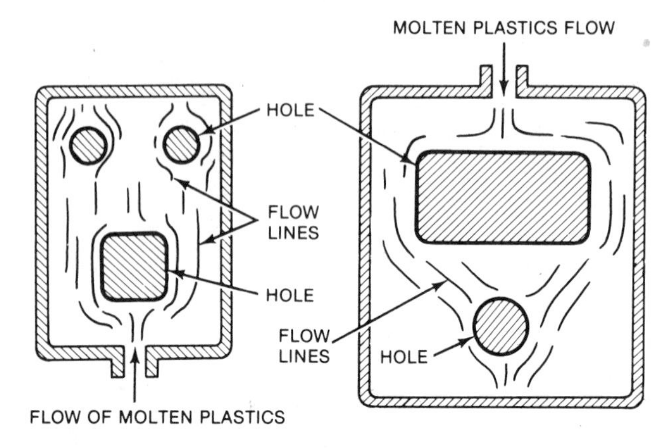

Fig. 21-26. Flow patterns must be kept in mind when designing molds. Flow lines may appear around holes and ribs, and opposite gates. (Vishay Intertechnology, Inc.)

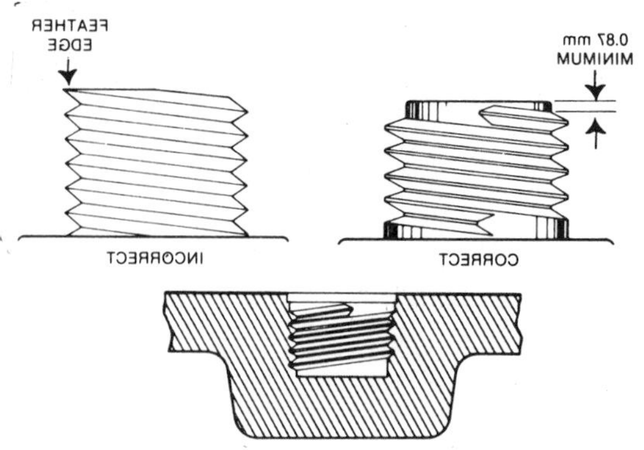

(A) Correct and incorrect threading.

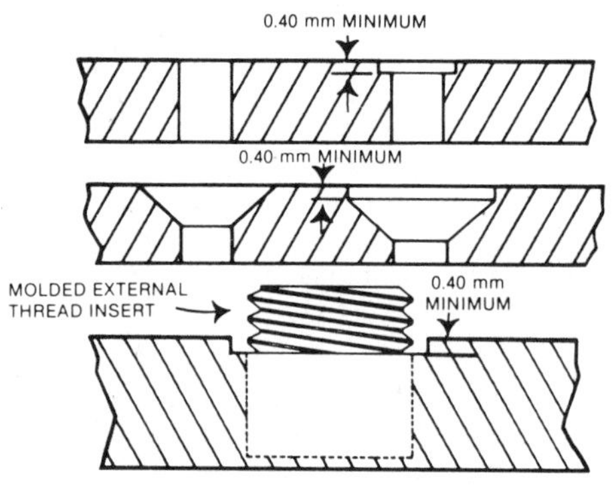

(B) Clearance of corepins.

Fig. 21-27. Putting threads and corepins in plastics

pins in the mold cavity. Enough material must be provided around all inserts to avoid cracking.

Pieces produced with pinpoint gating in the mold may not need any gate or sprue cutting. Many molds are designed so that gates are automatically sheared from the part as the press is opened.

Internal or external threads may be molded into plastics parts. Internal threaded parts may need an unscrewing mechanism for removal of the mold. A clearance of 0.03 in. [0.8 mm] should be provided at the end of all threads (Fig. 21-27A).

Metallic inserts molded into the part have greater strength. Generally, the ratio of wall thickness around the insert to the outer diameter of the insert should be slightly greater than one. Do not forget that materials have different expansion coefficients.

For blind core pin holes, a minimum 0.015 in. [0.04 mm] of clearance for screws, inserts, or other holding devices should be left (Fig. 21-27B).

Matched-mold design. The design parameters are similar to compression molding. Large composite parts such as sanitary tubs, bathroom shower stalls, and numerous automobile panels are molded from SMC. Many SMC operations are precut, require no mold pinch-off, and produce no flash. Bosses, inserts, and ribs are possible when using SMC and TMC. If layers or pieces are used, make the overlapping bond as large as possible to prevent stress cracking.

Open-mold design. Layup, sprayup, autoclave, and bag techniques are similar. Careful attention must be given to the matrix formulation and orientation of reinforcements. This may have more influence on the properties of the finished composite than the design. Reinforcements should be overlapped two inches and all joints staggered. Bosses and ribs are used for added strength but must be liberally tapered. Simple, integrated part designs with gradual changes in thickness are desired. To aid in part removal, blow-out holes (pneumatic) may be located in the bottom of the mold.

Pultrusion design. Bosses, holes, raised numbers, or textured surfaces are not possible with this continuous reinforcing process. Sharp corners or thickness transitions may result in resin rich zones with broken fibers.

Filament winding design. In this open-mold, continuous reinforcing process, the fibers are oriented to match the direction and magnitude of stresses. Computer-controlled placement of filaments is designed to compensate for angle, contour reinforcing, band width, equipment backlash, and other design considerations. Designs may call for permanent or removable mandrels (molds). (See Tooling in Chapter 22).

Laminar design. The principle design criterion is concerned with reinforcement orientation in each layer. A design close to optimum to resist all loads may be a laminate consisting of plies at 0°, ±45°, and 90°. There must be a –45° ply for every +45° ply to avoid distortion of the laminate. (See Figure 21-28.) The lamina should be oriented in the principle direction of anticipated stresses. (See Figure 21-29.) Fibers arranged in a random manner (isotropic), will have equal strength in all directions. The failure modes and deflection of composites are shown in Figure 21-30.

(A) All plies at 0°. Axial load results in stretching-shearing behavior.

(B) Two plies at ± (any angle). Opposing shear deformations in the plus and minus plies result in stretching-torsion interaction.

(C) A 0°/90° stacking. This arrangement bends under pure tension because the modulus-weighted centroid is not coincident with the geometric centroid, resulting in an offset load path.

(D) Another 0°/90° stacking. Because of different thermal expansion characteristics in each layer, this stacking deforms into a "saddle" when heated.

Fig. 21-28. Symmetry effects on deflection of composites. *(Machine Design)*

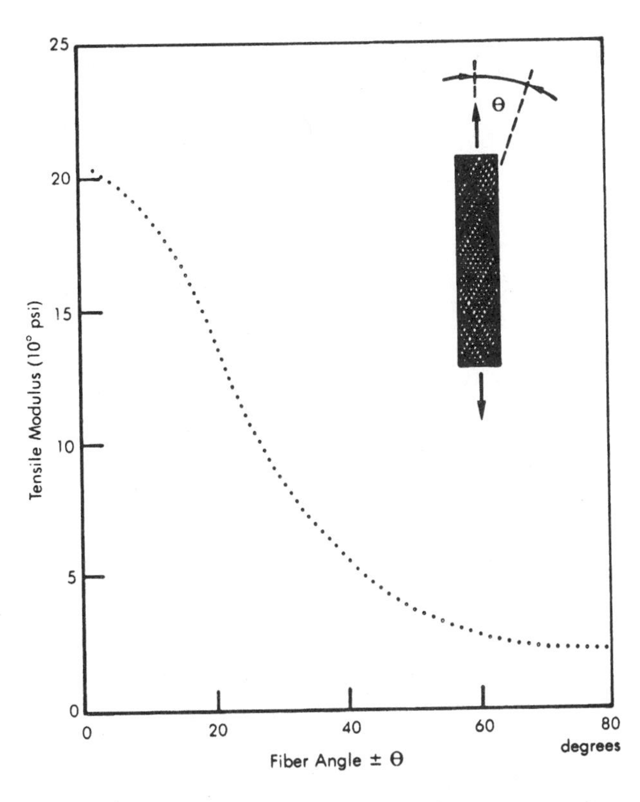

Fig. 21-29. Tensile modulus of carbon/epoxy composites drops steeply as the angle between the fibers and the direction of tensile load is increased. *(Machine Design)*

For sandwich laminates, the high-density facings must resist most of the applied and bending forces. The lightweight core must resist transverse tension, compression, shear, and buckling.

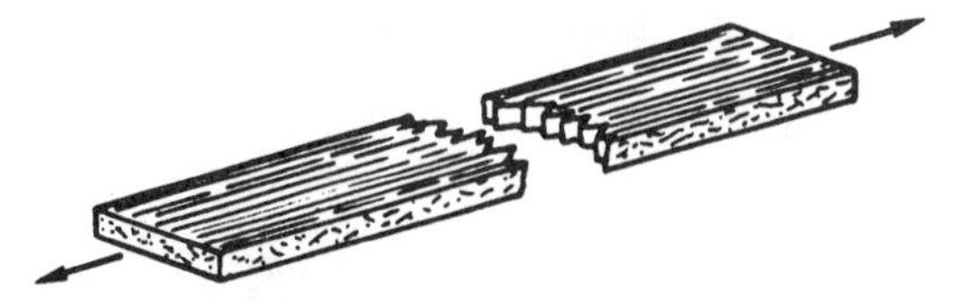

(A) Fiber tensile failure for all fibers in the direction of load (0°).

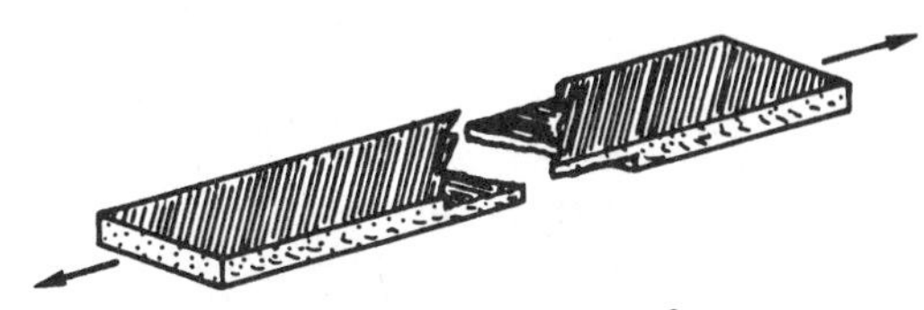

(B) Resin shear failure for fibers at ±45°.

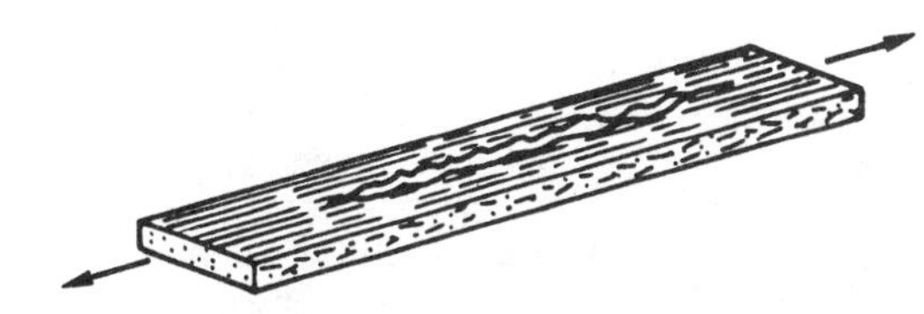

(C) Resin shear failure through the thickness, between fibers at 0°. Usually caused by poor fiber-to-resin adhesion.

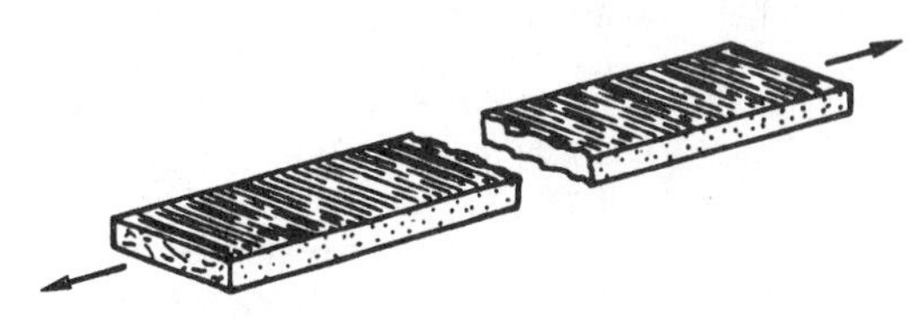

(D) Resin tensile failure between fibers at 90° to load.

Fig. 21-30. Failure modes of composites in tension. Tensional strength of a carbon/epoxy structural composite is always related to fiber direction. A simple tensile test shows strikingly different failure behavior in composites having different fiber orientation. In multidirectional composite, single-plies can fail with overall structural failure. Recognition of the various failure modes and knowing how composites fail are prerequisites to determining a fix. *(Machine Design)*

Table 21-5 shows advantages and limitations of various processes.

Performance Testing

The true test of any product is performance under actual service conditions. Tests can be used as indicators for product design, redesign, and reliable product service. The term *testing* implies that methods or procedures are employed to determine if parts meet the

Table 21-5. Plastics Processing Method—Processes, Advantages, Limitations

Processing Method	Process	Advantages	Limitations
Injection molding	Similar to die casting of metals. A thermoplastic molding compound is heated to plasticity in a cylinder at a controlled temperature and then forced under pressure through sprues, runners, and gates into a cool mold; the resin solidifies rapidly, the mold is opened, and the parts ejected; with certain modifications, thermosetting materials can be used for small parts	Extremely rapid production rate and hence low cost per part; little finishing required; excellent surface finish; good dimensional accuracy; ability to produce variety of relatively complex and intricate shapes	High tool and die costs; high scrap loss; limited to relatively small parts, not practical for small runs
Extrusion	Thermoplastic molding powder is fed through a hopper to a chamber where it is heated to plasticity and then driven, usually by a rotating screw, through a die having the desired cross section; extruded lengths are either used as is or cut into sections; with modifications, thermosetting materials can be used	Very low tool cost, material can be placed where needed. great variety of complex shapes possible; rapid production rate	Close tolerances difficult to achieve. openings must be in direction of extrusion; limited to shapes of uniform cross section (along length)
Thermoforming	VACUUM FORMING—Heat-softened sheet is placed over a male or female mold; air is evacuated from between sheet and mold, causing sheet to conform to contour of mold. There are many modifications, including vacuum snapback forming, plug-assist, drape forming, etc.	Simple procedure, inexpensive, good dimensional accuracy; ability to produce large parts with thin sections	Limited to parts of low profile
	BLOW OR PRESSURE FORMING—The reverse of vacuum forming in that positive air pressure rather than vacuum is applied to form sheet to mold contour	Ability to produce deep drawn parts, ability to use sheet too thick for vacuum forming, good dimensional accuracy; rapid production rate	Relatively expensive; molds must be highly polished
	MECHANICAL FORMING—Sheet metal equipment (presses, benders, rollers, creasers, etc.) forms heated sheet by mechanical means. Localized heating is used to bend angles; where several bends are required, heating elements are arranged in series	Ability to form heavy and/or tough materials: simple, inexpensive; rapid production rate	Limited to relatively simple shapes
Blow moldings	An extruded tube (parison) of heated plastics within the two halves of a female mold is expanded against the sides of the mold by air pressure; the most common method uses injection molding equipment with a special mold	Low tool and die cost; rapid production rate; ability to produce relatively complex hollow shapes in one piece	Limited to hollow or tubular parts; wall thickness difficult to control
Slush, rotational, dip castings	Powder (polyethylene) or liquid material (usually vinyl plastisol or organosol) is poured into a closed mold, the mold is heated to fuse a specified thickness of material adjacent to mold surface, excess material is poured out, and the semi-fused part placed in an oven for final curing. A variation, rotational molding, provides completely enclosed hollow parts	Low-cost molds, relatively high degree of complexity, little shrinkage	Relatively slow production rate, choice of materials limited

Table 21-5. Plastics Processing Method—Processes, Advantages, Limitations *(Continued)*

Processing Method	Process	Advantages	Limitations
Compression moldings	A partially polymerized thermosetting resin, usually pre-formed, is placed in a heated mold cavity; mold is closed, heat and pressure applied, and the material flows and fills mold cavity; heat completes polymerization and mold is opened to remove hardened part. Method is sometimes used for thermoplastics, e.g.. vinyl phonograph records; in this operation, the mold is cooled before it is opened.	Little waste of material and reduced finishing costs due to absence of sprues, runners, gates, etc.; large, bulk parts possible	Extremely intricate parts involving undercuts, side draws, small holes, delicate inserts, etc., not practical; extremely close tolerances difficult to achieve
Transfer moldings	Used primarily for thermosetting materials, this method differs from compression molding in that the plastic is 1) first heated to plasticity in a transfer chamber, and 2) fed, by means of a plunger, through sprues, runners and gates into a closed mold	Thin sections and delicate inserts are easily used; flow of material is more easily controlled than in compression molding, good dimensional accuracy; rapid production rate	Molds are more elaborate than compression molds, and hence more expensive; loss of material in cull and sprue; size of parts somewhat limited
Open-mold processing	CONTACT—The lay-up, which consists of a mixture of rein-forcement (usually glass cloth or fibers) and resin (usually thermosetting), is placed in mold by hand and allowed to harden without heat or pressure	Low cost; no limitations on size or shape of part	Parts are sometimes erratic in performance and appearance; limited to polyesters, epoxies and some phenolics
	AUTOCLAVE—The vacuum-bag setup is simply placed in an autoclave with hot air at pressures up to 1.38 MPa	Better quality moldings	Slow rate of production
	FILAMENT WOUND—Glass filaments, usually in the form of rovings, are saturated with resin and machine wound onto mandrels having the shape of desired finished part: finished part is cured at either room temperature or in an oven, depending on resin used and size of part	Provides precisely oriented reinforcing filaments; excellent strength-to-mass ratio; good uniformity	Limited to shapes of positive curvature, drilling or cutting reduces strength
	SPRAY MOLDING—Resin systems and chopped fibers are sprayed simultaneously from two guns against a mold: after spraying, layer is rolled flat with a hand roller. Either room temperature or oven cure	Low cost; relatively high production rate; high degree of complexity possible	Requires skilled workers; lack of reproducibility
Castings	Plastics material (usually themosetting except for the acrylics) is heated to a fluid mass, poured into mold (without pressure), cured, and removed from mold	Low mold cost, ability to produce large parts with thick sections; little finishing required; good surface finish	Limited to relatively simple shapes
Cold moldings	Method is similar to compression molding in that material is charged into a split, or open, mold; it differs in that it uses no heat—only pressure. After the part is removed from mold, it is placed in an oven to cure to final state	Because of special materials used, parts have excellent electrical insulating properties and resistance to moisture and heat; low cost; rapid production rate	Poor surface finish; poor dimensional accuracy; molds wear rapidly; relatively expensive finishing; materials must be mixed and used immediately
Bag molding	VACUUM BAG—Similar to contact except a flexible polyvinyl alcohol film is placed over layup and a vacuum drawn between film and mold (about 82 kPa)	Greater densification allows higher glass contents, resulting in higher strengths	Limited to polyesters, epoxies and some phenolics

(continued)

Table 21-5. Plastics Processing Method—Processes, Advantages, Limitations *(Continued)*

Processing Method	Process	Advantages	Limitations
	PRESSURE BAG—A variation of vacuum bag in which a rubber blanket (or bag) is placed against film and inflated to apply about 350 kPa	Allows greater glass contents	Limited to polyesters, epoxies and some phenolics
Matched-die molding	MATCHED DIE—A variation of conventional compression molding, this process uses two metal molds which have a close-fitting, telescoping area to seal (in the resin and trim the reinforcement; the reinforcement, usually mat or preform, is positioned in the mold, a premeasured quantity of resin is poured in, and the mold is closed and heated; pressures generally vary between 1.04 and 2.75 MPa	Rapid production rates; good quality and excellent reproducibility; excellent surface finish on both sides; elimination of trimming operations; high strength due to very high glass content	High mold and equipment costs; complexity of part is restricted; size of part limited

Source: Modified from Materials Selector, *Materials Engineering,* Penton/IPC a subsidiary of Pittway Corp.

required or specified properties. A pressure tank, rocket engine case, or pole vaulter's pole may be subjected to a critical pass/fail (proof) test. *Quality control* procedures must be used to determine if a product is being manufactured to specifications. It is primarily a technique used by management to achieve quality. *Inspection* ensures that manufacturing personnel check technique procedures, gauge readings, and detect flaws in processing of materials. Inspection is part of quality control. See Appendix G for additional sources of information.

Vocabulary

The following vocabulary words are found in this chapter. Look up the definition of any of these words you do not understand as they apply to plastics in the glossary, Appendix A.

Apparent density
Boss
Bulk factor
CAD
CAE
CAM
CAMM
CIM
Fillet
Flow line
Flow mark
Gate
Inspection
Orientation
Parameter
Parting lines
Quality control
Ribs
Runners
Safety factor
Specification
Sprue
Standards
Testing
Undercut

Questions

21-1. In ___?___ the sprues and runners are kept hot by means of heating elements built into the mold.

21-2. The ___?___ is the point of entry into the mold cavity.

21-3. Narrow channels that convey the molten plastics from the sprue to each cavity are called ___?___ .

21-4. Parting lines are normally placed at the ___?___ of the molded part.

21-5. The ___?___ is the opening in the mold where the product is formed.

21-6. The economic minimum number of pieces produced by hand lay-up is ___?___ .

21-7. Closely related to shrinkage is dimensional ___?___ .

21-8. The tapered channel connecting the nozzle and runners is called ___?___ .

21-9. In ___?___ compression mold designs no provision is made for placing excess material in the cavity.

21-10. Molded parts are pushed from the mold by ___?___ or ___?___ .

21-11. With ___?___ gating, pieces may not require any gate or sprue cutting. The dies are designed so that gates are sheared off automatically.

21-12. The three overriding requirements in plastics designing are:

21-13. If a mold costs $5000 and 10000 parts are to be made, the mold cost would be ___?___ per part.

21-14. In nearly every design, a compromise must be made between highest performance, attractive appearance, efficient production and ___?___ .

21-15. List four favorable properties possessed by most plastics:

21-16. List four unfavorable properties possessed by most plastics.

21-17. In addition to electrical, chemical, mechanical and economic considerations, name four additional requirements to consider before a product is made.

21-18. The most meaningful comparison in estimating the cost of a plastics product is cost per ___?___ .

21-19. Plastics are lighter per ___?___ than most materials.

21-20. Problems encountered in producing plastics products often require the selection of the ___?___ before material or ___?___ considerations.

21-21. There were many problems associated with early applications of plastics because designers forgot that the finished product must ___?___ as designed and desired.

21-22. Undercuts in parts usually increase ___?___ costs.

21-23. With relatively few exceptions, the ___?___ of plastics developed in the past have been by trial and error.

21-24. Plastics have replaced metals for many applications because of energy and ___?___ savings.

21-25. If wall thickness is not ___?___ the molded part may distort, warp or have internal stresses.

21-26. The mark on a molded piece resulting from the meeting of two or more flow fronts during the molding operation is called a ___?___ .

21-27. Wavey surface appearances caused by improper flow of the hot plastics into the mold cavity are known as ___?___ .

21-28. To help prevent flow marks, ___?___ or change gate location.

21-29. Name three methods of producing an internal thread in a plastics part:

Chapter 22

Tooling and Moldmaking

Introduction

In this chapter, tool-making processes, equipment, and methods for shaping plastics will be discussed. Not all of the information will apply to all molding processes, since some processes are very specialized.

Computer-assisted moldmaking (CAMM) is a result of the microprocessor technology and may improve productivity by more than 100 percent. Computer assisted design and manufacturing (CAD/CAM) systems are used to design and aid in machining molds. This equipment automatically adjusts cavity dimensions for different resins or special contours. Designing and machining information is stored in the system's memory for making multicavity molds or for the replacement of cores or cavities.

CAMM could allow the plastics industry to produce high quality configurations, close tolerances, and reliable designs at a competitive price. The industry of moldmaking and tooling is highly labor intensive. Here, computer systems are sophisticated tools that increase productivity. The drafting and design departments can utilize CAD systems to automatically dimension the drawing and allow for material shrinkage. The data base in CAMM programs can select the mold base and recommend the placement of ejector pins, sleeves, locating ring, return pins, pullers, or other details. While drafting time and changing tooling parameters represent a significant percentage of the total cost of the part, CAMM systems greatly reduce this time and facilitate modifications or changes in tooling parameters.

Most of the moldmaking industry is composed of custom shops that specialize in moldmaking or in offering a special service such as plating, polishing, heat-treatment, engraving, or machining molds.

Molding information is given in the discussion of each process; however, Chapter 21 should be reviewed for basic design considerations. You should also review the descriptions of the plastics families, because they contain information about properties and design that affect moldability.

The outline for this chapter is:

I. Planning
II. Tooling
   A. Tooling costs
III. Machine processing
   A. Chemical erosion

Planning

Most mold designs begin with sketches that allow the moldmaker to make decisions about layout and visualize how parts will be made. Final CAD drawings will contain notes, dimensions, and tolerances. Some critical dimensions may need tool tolerances as low as 9.8×10^{-7} in. [+0.0025 mm] for some parts. The design will also show any special requirements. Shrinkage tolerances, finish, engraving, plating, special materials, or other dimensioning factors will be noted.

CAM systems allow the user to determine cooling systems, finish, tool path, part geometry, feeds, and inherent equipment limitations (back lash, tool wear) before machining. After the program is verified it is stored for later use.

CAM systems can utilize the stored information to cut and form the molds. Nearly 80 percent of the moldmakers' time is devoted to setting up the machine tool. Only 20 percent is actually spent cutting the material. CAD/CAM systems substantially reduce setup, lead, and machining time. (See Fig. 22-1.)

Nearly every phase of the product development process, from concept to completion, can utilize CAMM to save time and reduce costs.

Tooling

Collectively, jigs, fixtures, molds, dies, gauges, clamping devices and inspection equipment will be referred to as *tooling*. The terms *jig* and *fixture* are often used interchangeably; both are devices used to locate and hold a workpiece in the correct position during machining, inspection, or assembly. A *jig* guides the tool during a manufacturing operation such as boring. A *fixture* does not have built-in tool guides. It is used primarily to hold the work securely during machining, cooling, and drying.

Part of the manufacturing cost of plastics is the cost of special fixtures. These are tools used to help measure or load a plastics charge in a molding machine, remove flash, remove molded parts, or hold parts for cooling. Some are used to aid machining. These include holding blocks, drill jigs, and punch dies.

Tooling Costs

Many factors affect tooling costs: the size of the production run, production technique, reinforcements, additives, fiber orientation, matrix, the complexity of design, the tolerance needed, the amount of mold maintenance, and machining.

Multiple-cavity molds, or those with inserts or special surface finishes, add to tooling costs. Designing a mold set to have interchangeable cavities may lower tooling costs because new cavities may be inserted to form other parts, thus extending the use of the original mold. For a small number of parts, a multiple-cavity mold may not be economical because tooling tolerances are much harder to maintain, and mold maintenance and machining are more costly for multiple-cavity molds.

Two remarks often quoted in the moldmaking industry are, "There is no such thing as a simple plastics part," and "The part is only as good as the mold that makes it." Each of these remarks show the importance of tooling and mold design. (See Figure 22-2.)

There are four broad types of tooling: 1) prototype, 2) temporary, 3) short run, and 4) production. These are shown in Table 22-1.

Table 22-1. General Types of Tooling

Tool Classification	Number of Parts	Tooling Materials
Prototype	1–10	Plaster, wood, reinforced plasters
Temporary	10–100	Faced plasters, reinforced plasters; faced plasters, backed metal-deposition, cast & machined soft metals
Short Run	100–1000	Soft metals, steel
Production	>1000	Steel, soft metals for some processes

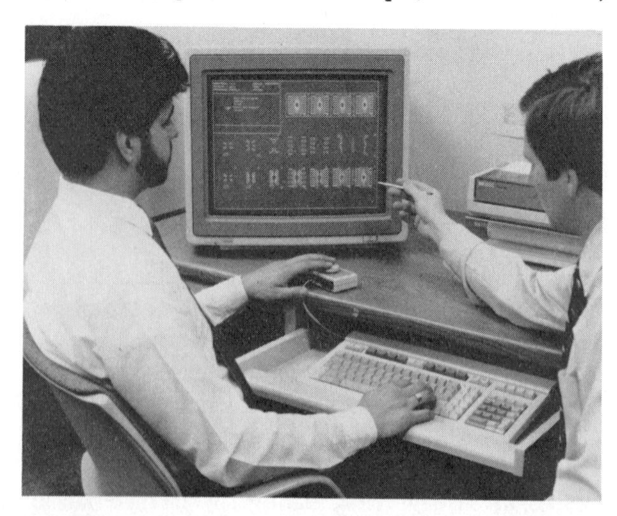

Fig. 22-1. With this sophisticated CAD-CAM installation, engineers can design, test, and refine even the most complex molds directly on the computer. The need for building costly prototypes is eliminated. (AGIE USA LTD)

Fig. 22-2. The importance of the tool and die personnel can not be understated. This tool and die maker puts finishing touches on a steel injection mold. (Bethlehem Steel.)

Types of materials used in tooling include gypsum plasters, plastics, wood, and metals.

Gypsum plasters. The United States Gypsum Company has developed a number of high-strength plaster materials. These materials have enough strength to produce prototype models, die models, transfer (take-off) tools, patterns, and die molds for forming plastics (Fig. 22-3).

Fibers, expanded metal, or other materials are often used to further strengthen the plaster tooling. Metal bases and frames provide secure mountings. In some techniques, templates (loft or templates) help shape the wet plaster. Even a pile of rocks or an expanded bladder (balloon) can be used to help form the general contour. Plaster is easily shaped by hand. Models may be produced by placing plaster over clay, wax, wooden, or wire frame shapes. Rocks, wax, and other model forms are generally removed and replaced by reinforced plastics support. Some plaster molds are designed to be used only once. Hollow designs, or some of the break-away designs require that the plaster mold be broken and washed away. Typical plaster molds are faced with metal deposition, polymer coating, or composite facing, to provide a durable surface for part removal. Metal coatings also improve thermal conductivity necessary in some molding techniques. Cooling coils may be cast in the tooling.

The trade names Ultracal, Hydrocal, and Hydrostone are found on plasters used for tooling. Vacuum-formed pattern molds are often made of low-cost plasters. Hydrostone has an average compressive strength of nearly 11 000 psi [76 MPa].

Plaster is an important master pattern material from which metal or polymer skin master molds are produced (Fig. 22-4).

Plastics. Polymer tooling (plastics and elastomers) are used to make master patterns, transfer tools, cores, boxes, templates, draw dies, jigs, fixtures, inspection tools, and prototypes. They are replacing wooden and plaster tooling (Fig. 22-5). Laminated, reinforced, and filled plastics are used mainly for making dies, jigs, and foundry patterns (Fig. 22-6).

Polymer molds are generally divided into two groups: those that are backed and those that are not. Foams and honeycomb sheets are often used to provide strong, lightweight support of tools.

The use of plastics in die fabrication is growing rapidly. Metal-filled and glass-reinforced phenolics, ureas, melamines, polyesters, epoxies, silicones, and polyurethanes are strong, light, and easy to machine. Alumina and steel are common fillers which offer improved thermal conductivity, machinability, strength, and extended service temperature and life. Many may be foam-filled. These materials are used in tooling in both the plastics and metals industries. Plastics tools have been used as bending dies, stretch-forming dies, and drop-hammer dies. Acetals, polycarbonates, high-density polyethylene, fluoroplastics,

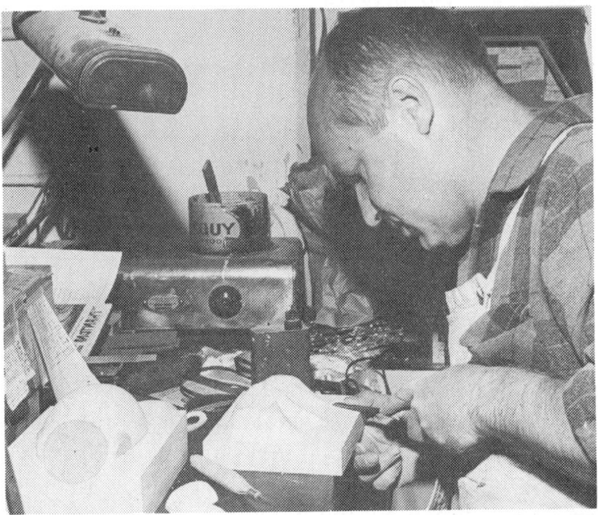

(A) This worker is making a gypsum plaster pattern for forming plastics. (Revell, Inc.)

(B) An epoxy resin master model of an aircraft pilot's enclosure after removal from the plaster mold. (U.S. Gypsum Co.)

Fig. 22-3. Gypsum plaster pattern and plaster mold.

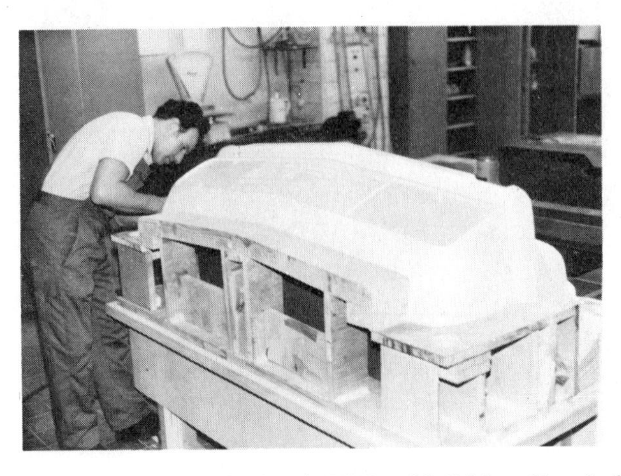

Fig. 22-4. Tooling for female RIM mold. (Mobay Chemical Co.)

(A) Basic shape and form (loft frame).

(B) Filling with syntactic foam.

(C) Machining to final shape and size.

Fig. 22-5. Light, strong tooling may be fabricated with honeycomb structures. Bonding and fill is done with syntactic foams of extrudable epoxy. (Ren Plastics, Inc.)

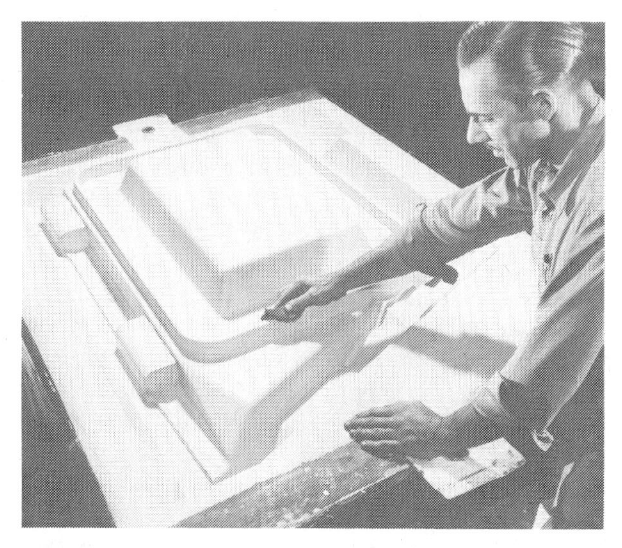

Fig. 22-6. A large foundry pattern being made by laminating glass cloth and epoxy resin. Plastics is replacing metal for pattern material because it is less expensive and because plastics patterns may be made faster. (U.S. Gypsum Co.)

and polyamides are also used as tooling. These materials find use as matched stamping dies, jigs, and fixtures. The acceptability of plastics tools is verified by their wide use in the aerospace, aircraft, and automotive industries.

Large epoxy-reinforced molds are popular for layup and sprayup techniques. These toolings are backed and supported by metal frames and bases (Fig. 22-7). In Figure 22-8, polymer tooling is used to form SMC. Polymer tooling must be able to withstand the prolonged exposure of curing temperatures.

Plastics tooling has several advantages over metal or wood tools. Plastics tools may be cast in inexpensive molds, they duplicate easily, and they allow frequent changes in design. Plastics tooling is also light in mass and corrosive resistant. Hot-melt compounds are replacing wood and steel in dies. hammers, mockups, prototypes, and other fixtures used by industry (Fig. 22-9).

Complex furniture parts are sometimes made from flexible polymer molds. Silicones are best known, but polysulfide and polyurethane elastomers are also used.

Unbacked, flexible molds are used in the furniture industry to faithfully reproduce wood grain designs (Fig. 22-10). The basic concept of flexible-plunger or elastomer press molding is shown in Figure 22-11.

Wood. Wooden tooling is used for prototypes and some short-run work (Fig. 22-12). It is also used for some thermoforming dies and for pattern work.

Metals. Although plastics may become the dominant material in the future, we cannot have plastics products without metals.

For prototype or short-run work, low-melting-point metals may be used. Zinc, lead, tin, bismuth, cad-

Fig. 22-7. Two layers of 10 oz. glass cloth and eight layers of 20 oz. cloth are alternately laid with a matrix of epoxy to produce this spa mold with intricate designs. (Ren Plastics)

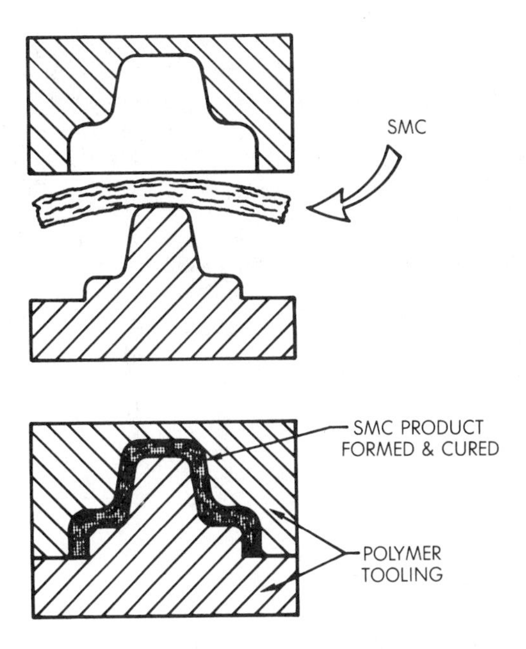

Fig. 22-8. Matched-die molding of SMC using polymer tooling.

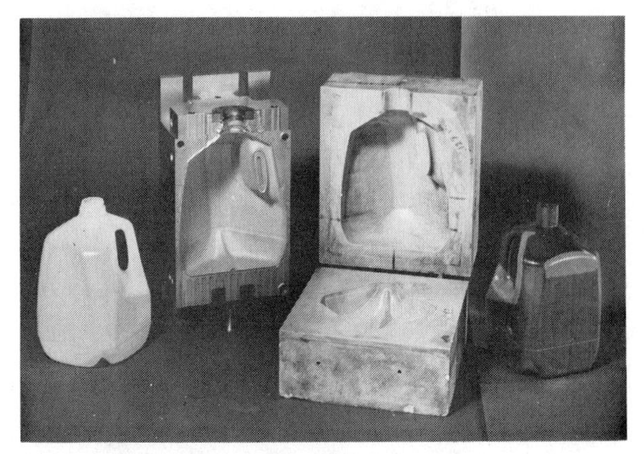

(A) Low-cost plastics pattern development.

(B) Duplicate milling of mold halves from plastics pattern on right.

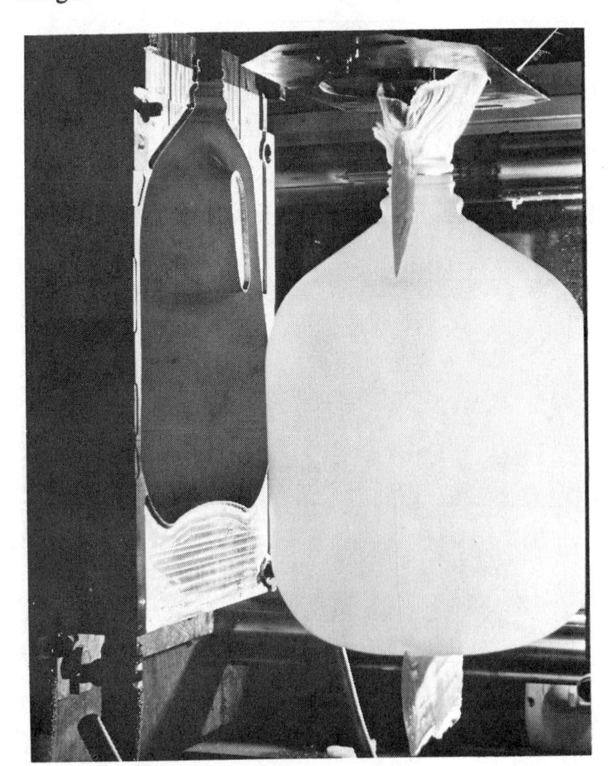

(C) Plastics product from metal-filled plastics mold. Note pinch-off and parting line. (Chemplex Co.)

Fig. 22-9. Low-cost molds made of metal-filled plastics.

Fig. 22-10. Woodgrain detail and miter joints are faithfully reproduced in this polyurethane furniture component. (The Upjohn Co.)

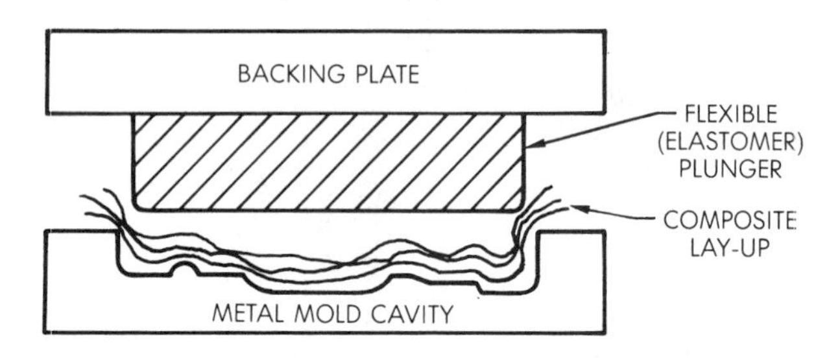

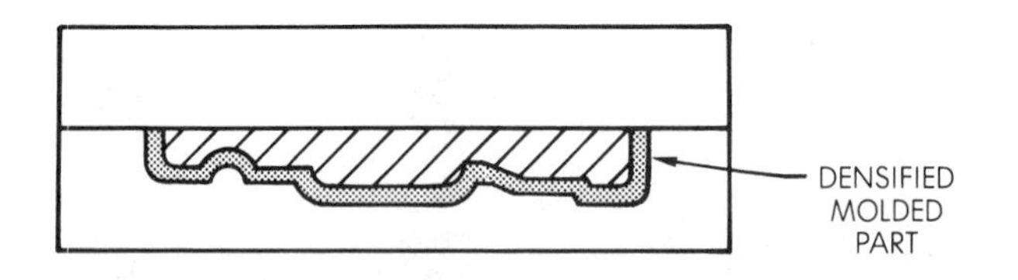

Fig. 22-11. Flexible-plunger or elastomeric press molding.

Fig. 22-12. A wooden pattern is made from photographs and drawings. It is double the size of the finished part. (Bethlehem Steel Co.)

mium, and aluminum are used in thermoforming dies, casting patterns, and duplicate models. Aluminum is a popular metal for many molding processes because it is light, easy to machine, and is a good thermal conductor.

Aluminum 7075 is a popular, high-grade, heat-treated alloy. It is sometimes annodized and plated to improve surface hardness and prolong tooling life. Aluminum (7075-T652) is used extensively in blow molding, bag molding, thermoforming, and prototypes. It is soft and lacks sufficient strength and hardness to prevent material wash in injection molds or pinch-off areas in blow molds.

Berylium-copper (C17200) is used in some injection molds and blow-molding processes (Fig. 22-13). This material can be cast, wrought, and drawn. It can

(A) Steps in producing a mold for simulated wood furniture components. From left: hand-carved wooden model, silicone impression, gypsum model, and beryllium-copper injection molds.

(B) The finished product.

Fig. 22-13. Beryllium-copper injection molds were used to produce this simulated wood furniture.

also be pressure cast using a master hob. Many cavities can be produced with one hob. The BeCu is generally supplied to the moldmaker tempered to 38-42 Rockwell C. Beryllium-copper will reproduce fine detail. Wood grain patterns are examples of the detail possible with BeCu molds.

Kirksite (Zamak, zinc alloy) molds, made of an alloy of aluminum and zinc, are used because they are low in cost. This metal will reproduce detail better than aluminum and also last longer. Because of the low pouring temperature of 800 °F [425 °C], it is possible to cast cooling lines in the mold. These alloys are used for short- and long-run blow-molding operations. Pinch-off edges must be protected by steel inserts against excessive stress concentrations.

Steel is vital to the plastics industry. Carbon, the basic element in plastics, is also an important ingredient of steel, but steel must also have additional alloying elements for moldmaking. Mold hardness can vary from 35 to 65 Rockwell C, depending upon needs (Fig. 22-14).

A popular, non-alloyed carbon steel used for machine bases, frames, and structural components is AISI 1020, 1025, 1030, 1040, and 1045.

For long production runs, various steel alloys may be used. In uses where high compressive strength and wear resistance at elevated temperatures are required, an AISI Type H21 steel may be used. The symbol H indicates hot-work tool steels.

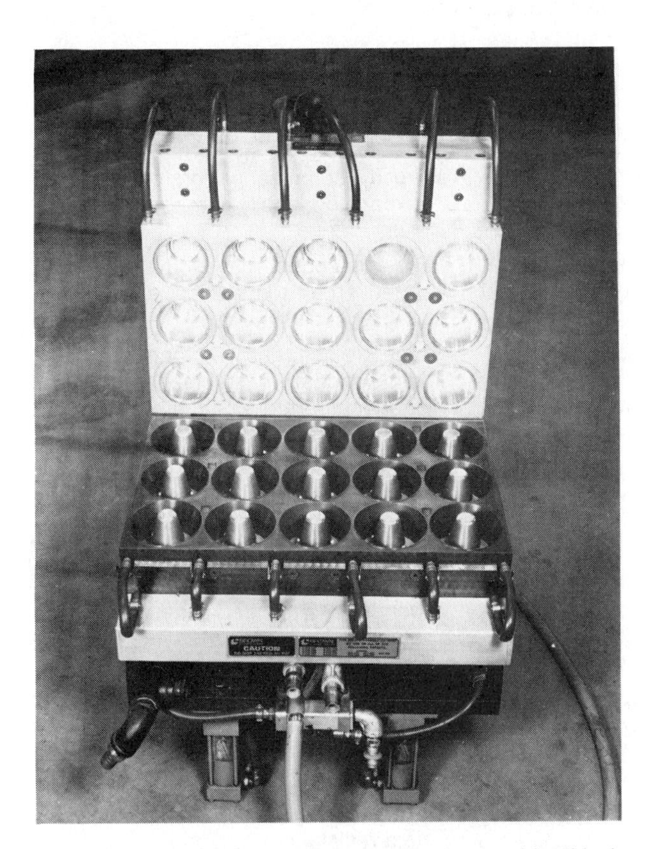

Fig. 22-14. A fifteen-cavity thermoforming mold. This is long-run, high production metal tooling. (Brown Machine Co.)

Type H21 steel analysis:
Carbon 0.35%
Manganese 0.25%
Silicon 0.50%
Chromium 3.25%
Tungsten 9.00%
Vanadium 0.40%

AISI Type WI is a high-quality, straight-carbon, water-hardened tool steel used in various tooling applications.

Type WI steel analysis:
Carbon 1.05%
Manganese 0.20%
Silicon 0.20%
Alloys None

Toolmakers have long wished to have a nondeforming die steel. One that combines the deep-hardening characteristics of air-hardened steels with the simplicity of low-temperature heat treatments possible in many oil-hardened steels is AISI Type A6. The symbol A is used for air-hardened steels.

Type A6 steel analysis:
Carbon 0.70%
Manganese 2.25%
Silicon 0.30%
Chromium 1.00%
Molybdenum 1.35%
Plus alloy sulfides

Popular, alloyed steels, including tool steels, are AISI type A2 and A6 for injection, transfer, compression, and master hob molds. Fully hardened steels such as D2 and D3 are also popular. The high-carbon and chromium D3 steels have good wear resistance. Mold cavities, back plates, knives, and die are made from the chromium-molybdenum steel 4140.

The symbol P means a precipitation-hardened steel. AISI Type P20 is usable for all types of injection mold cavities. It may be hardened to a core hardness of 38 Rockwell C.

Type P20 steel analysis:
Carbon 0.30%
Molybdenum 0.25%
Chromium 0.75%

Mold cavities, holding blocks, dies, and other tooling are made of P20 and P21 steels.

Stainless steel Type 420 and 440C may be used where corrosion resistance is needed or adverse atmospheric conditions exist. It can develop hardness of 45 to 50 Rockwell C.

Oil-hardened steels are sometimes selected for slides and bushings. Type 02 is most common.

Tungsten carbide is widely used for cutting taps because it has good wear and abrasion resistance. Mold cavities may be made from this ceramic type of material but it is brittle and harder to shape than tool steels.

Miscellaneous tooling. There are numerous miscellaneous and innovative tooling techniques. Wax and soluble salts are sometimes used. Mandrels, bladders, or other mold shapes have been made of air inflated elastomers. Even glass shapes are used as molds. After casting, forming, or filament winding, the glass is sometimes broken and removed. Concrete and ice have been used to produce unique tooling in some applications. The advantages and disadvantages of selected tooling materials are shown in Table 22-2.

Machine Processing

There are a number of processes used in making tools and dies from steels. Milling, turning, drilling, boring, grinding, hobbing, casting, planing, etching, electroforming, electrical-discharge machining, plating, welding, and heat-treating are only a few.

Toolmaking should be viewed as a tightly controlled process rather than a sequence of discrete tasks. Control of the toolmaking process may be done with the aid of the computer. CIM systems may control individual computer numerically controlled (CNC) machine tools. The manufacturing of tooling begins with planning and balancing the workload, specifications, and techniques with machine capabilities. Mistakes in the planning process can be costly and result in excessive production time and even tool failure. Machining of molds and tooling is sometimes separated into two broad areas: 1) initial machining which involves rough turning and milling, and 2) final machining which involves grinding, electrical-discharge machining (EDM), and polishing. These processes usually take place after hardening.

Milling, turning, drilling, boring, and grinding are cutting-tool processes. Shapers, planers, lathes, drilling machines, grinding machines, milling machines, and various pantograph duplicating machines are often used to cut metal in making molds and dies. Figure 22-15A shows steel being removed by a cutting tool in a vertical milling machine.

In Figure 22-15B, a specially modified vertical milling machine is used to cut and duplicate molds in steel from the master pattern. The tracing head on the

Table 22-2. Advantages and Disadvantages of Selected Tooling Materials

Tool Material	Advantages	Disadvantages
Aluminum	Low cost; good heat transfer, easily machined; corrosion resistant; lightweight; doesn't rust	Porosity; softness; galling; thermal expansion; easily damaged; limited runs
Copper alloyed (brass, bronzes, berylium)	Easily machined; good surface detail; high thermal conductivity; doesn't rust	Softness; copper may inhibit cure; attacked by some acids; easily damaged; limited runs
Miscellaneous (salts, inflatables, wax, ceramics)	Some are low cost, reusable; designed with undercuts; easily fabricated; light; hard; thermal conductors; ceramics are high temperature materials	Some are soft: easily damaged; dimensionally unstable; damaged by high temperature or chemicals; poor thermal conductors
Plaster	Low cost; easily shaped; good dimensional stability; doesn't rust	Porosity; softness; poor thermal conductivity: easily damaged; limited runs; limited thermal and strength range
Polymers (laminated, reinforced, filled)	Low cost; easily fabricated; thermal expansion similar to many composites; light-weight; doesn't rust; large designs economical; fewer parts	Limited design; poor thermal conductivity; limited dimensional stability; limited runs; limited thermal range
Steel	Most durable, high thermal resistance; strong and wear resistant; thermal conductivity	Most expensive tooling; machining more difficult; size limitations; many parts; rusts; heavy
Wood	Low cost. easily machined; lightweight; doesn't rust	Porosity; poor dimensional stability; soft; limited runs; poor thermal conductivity and resistance
Zinc alloyed (lead, tin)	Low cost; easily rnachined; good thermal conductivity; good detail; doesn't rust	Soft; easily damaged; limited runs; limited thermal and strength range

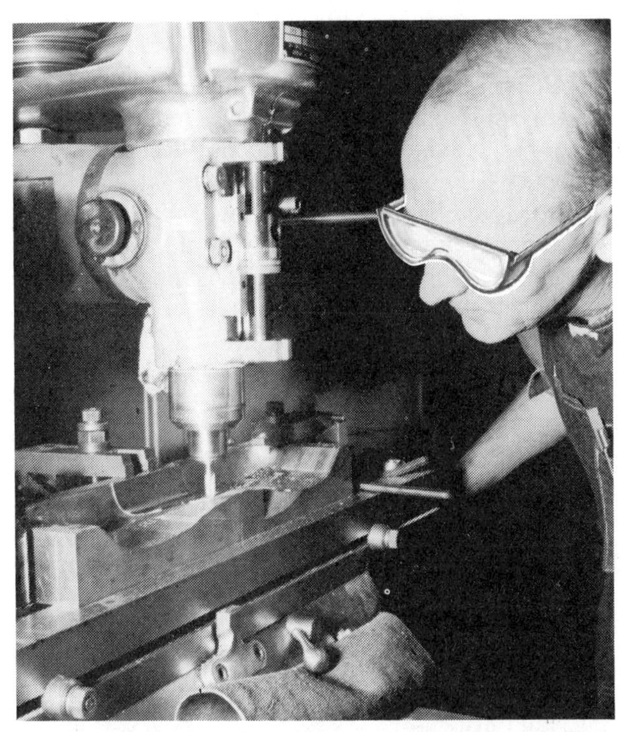

(A) Making a steel mold on a milling machine. (Revell, Inc.)

(B) Duplicating a steel mold from a pattern, at right. (Cincinnati Milacron)

(C) Pantograph machine with a plaster master model used to machine a metal piece. (U.S. Gypsum Co.)

Fig. 22-15. Making tools and dies from steel and plastics.

right controls movements of the work table and cutter spindle. This pattern is made of metal and the duplicator setup is machining a cavity for a blow mold. The workpiece ratio is 1:1.

Pantograph machines are similar to duplicating machines, except that they operate on variable ratios up to 20:1. In Figure 22-15C, a plaster master is shown much larger than the steel workpiece. The large ratio reduction allows the steel to be machined with very delicate detail by coordinated movements of the table and cutting tool.

Hobbing, etching, electroforming, and electrical discharge machining are *metal-displacement* processes. They are used in making molds where no cutting tools are involved.

Cold hobbing involves pushing a piece of very hard steel into a blank of unhardened steel (Fig. 22-16). The process is performed at room temperatures. Pressures vary from 200 to 400 $\times 10^3$, psi [1 380 MPa to 2760 MPa] depending on the hobbing metals and blanking material. Hobbing machines may need press capacity as high as 3 000 tons [2 722 tonnes].

In a proprietary procedure called CAVAFORM, a master hob can supply an unlimited number of impres-

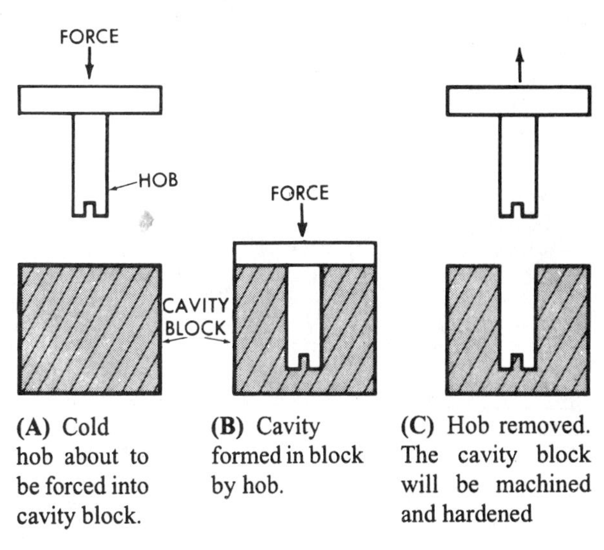

(A) Cold hob about to be forced into cavity block.

(B) Cavity formed in block by hob.

(C) Hob removed. The cavity block will be machined and hardened

Fig. 22-16. Diagram of metal displacement (hobbing) process.

sions distinguished by a remarkable fidelity to the size and finish of the hob. This cold-forming swaging procedure is accomplished in most steel in an annealed state. A cavity steel can be chosen to satisfy molding requirements with low heat treat distortion. It may also be vacuum-heat treated to minimize polishing after heat treatment. A major disadvantage is that this process generally requires a through hole at the bottom of the impression.

Hobs are often made from oil-hardened tool steels containing a high percentage of chromium. It may be economical to hob single die cavities, but hobbing is usually used for making large numbers of impressions for multicavity molds. Multicavity molds frequently are numbered to permit instant location of any molding troubles.

A slight draft must be provided to allow removal of the hob from the forming blank. The hob must be clean because even a pencil mark on the hob may be transferred to the cavity during the hobbing operation.

After hobbing, the blank is machined and hardened before placement in mold bases. In Figure 22-17A a finished hob (right) has formed the cavity (center) in the blank of steel. At left is the *force* or male portion of the compression mold. A finished, compression-molded part is shown in Figure 22-17B. It was molded in the finished mold cavity.

(A) Hob, at right, was pressed into steel block, at center.

(B) The finished, compression-molded part from the hobbed mold.

Fig. 22-17. Example of hobbing.

Electrical erosion or electrical-discharge machining (EDM) is a fairly slow method of removing metal, compared with mechanical methods. Steel is removed at about 0.016 in^3/min [4.37×10^{-4} mm^3/s]. The workplace may be hardened before the cavity is formed, which removes any problems from heat treatment after machining or forming. In the machining process, a master pattern is made of copper, zinc, or graphite. The pattern is then placed about 0.025 mm [0.000 98 in.] from the workpiece and both the workpiece and the master are submerged in a poor dielectric fluid, such as kerosene or light oil. Current is forced across the gap between the master and the workpiece, and each discharge removes minute amounts of substance from both. The loss of material from the tool master must be compensated for to obtain accurate cavities in the workpiece.

Most modern EDM machines now incorporate a multi-axis orbital movement in their spindle head that enables one electrode to be used for roughing, final sizing, and finishing.

For inexpensive tool masters made of carbon or zinc, the ratio of material removed from the workpiece to that removed from the tool may be more than 20:1. Accuracy may be within ± 0. 001 in. [±0.025 mm] with a finish cut of less than 30 microinches [0.007 62 mm).

Both wirecut and diesinking EDM techniques often eliminate the need for secondary finishing operations. Wire EDM will produce accurate, intricate cavities that are difficult or impossible to produce with conventional techniques. In Figure 22-18 a CNC controlled wire EDM cuts the final shape of these molding dies. The EDM principle is shown in Figure 22-19.

Chemical Erosion

In chemical erosion (etching) an acid or alkaline solution is used to create a depression or cavity. The pro-

Fig. 22-18. A CNC controlled wire EDM machine.

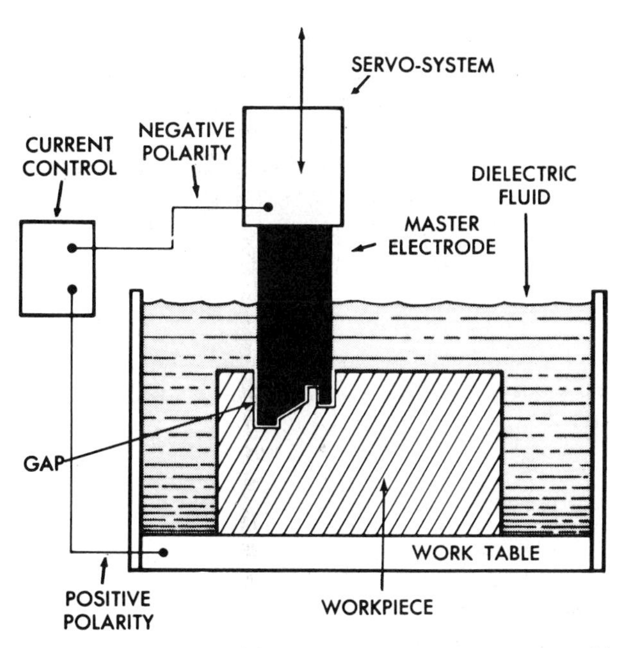

(A) Note the gap between the workpiece and master tool is uniform.

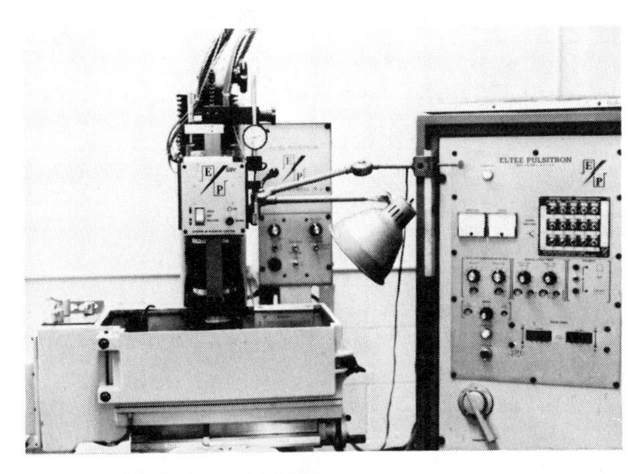

(B) A production model EDM.

(C) A laboratory size EDM.

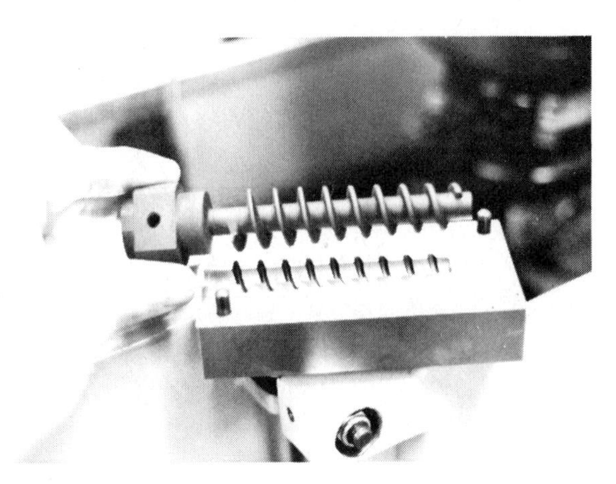

(D) Carbon master electrode used to make the die cavity.

Fig. 22-19. Electrical erosion (EDM) machining of a die.

cess usually involves the use of chemically resistant maskants such as wax, plastics-based paints, or films. The maskant is removed from those areas where the metal is to be chemically removed. Shallow cavities or designs are often reproduced with textures duplicating fabrics and leather. Photosensitive resistant materials are commonly used in the printing industry.

Allowance must be made to compensate for the effects of *etching radius,* or *etch factor.* As the etchant acts on the workpiece, it tends to undercut the maskant pattern. In deep cuts, the undercut may be serious. Figure 22-20 depicts the effects of the etch factor in chemical erosion.

Casting and *electroforming* are sometimes called *metal-deposition* processes. They involve depositing a metallic (or sometimes a ceramic or plastics) coating on a master form.

In Figure 22-21A a steel master mandrel is dipped into molten lead compounds until a coating is formed over it. The mandrel may then be removed and used again. Casting resins may be poured into the shell and removed when they are polymerized.

The hot casting of metals by the lost wax or sand processes, or by permanent metal molds may be used to produce precision molds. Molten metal may be poured over a hardened steel master to form a cavity, as shown in Figure 22-21B. This process is sometimes called *hot hobbing.* Molten metal is cast over a hob. Pressure is applied during the cooling.

Electroforming is an electroplating process. An accurate mandrel of plastics, glass, wax, or various metals is used as a master to electrically deposit the metallic ions from a chemical solution (Fig. 22-22). The molds are thin shelled and may have severe undercuts, but they usually have a highly polished finish (Fig. 22-23). The cavity may be strengthened by copper plating the back of the shell. Further strength may be provided by placing the die cavity into filled epoxy. The cavities may then be used for thermoforming, blow molding, or injection molding (Fig. 22-24).

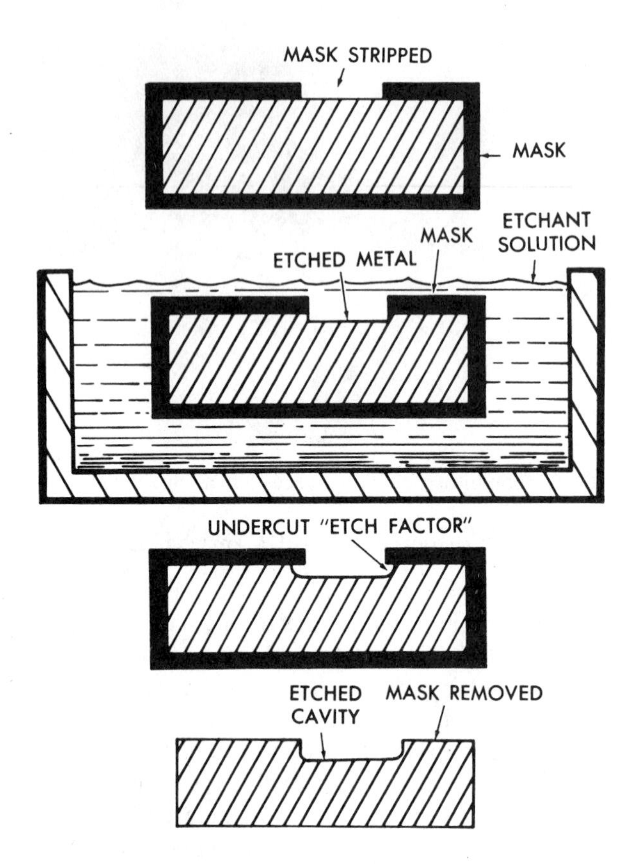

Fig. 22-20. Chemical erosion method of producing a die cavity.

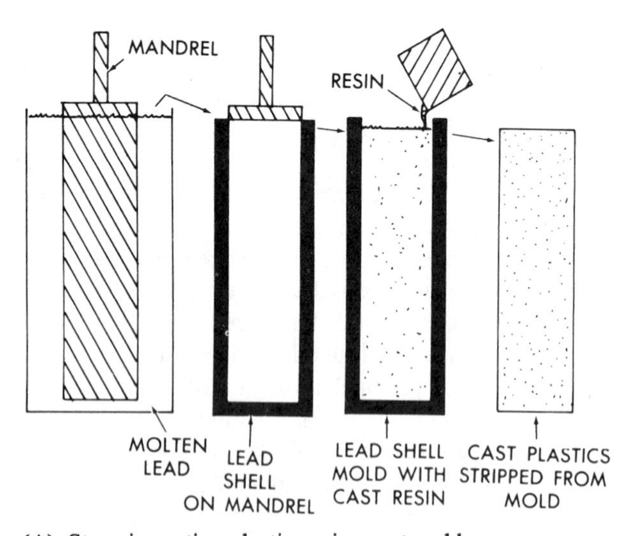

(A) Steps in casting plastics using cast molds.

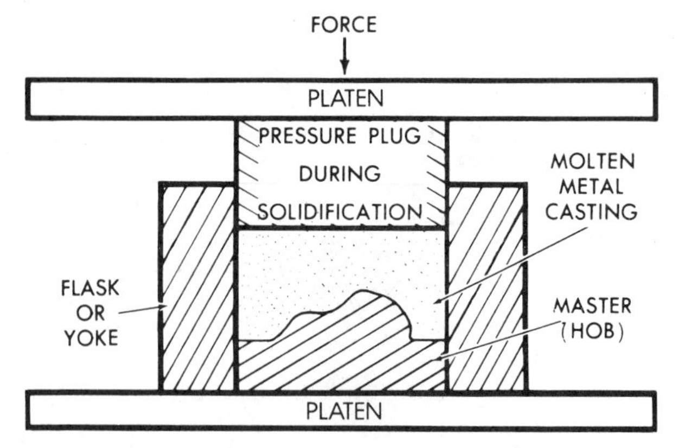

(B) Molten metal is cast on master. Pressure from plug results in a dense, sound casting.

Fig. 22-21. Casting of plastics and metal.

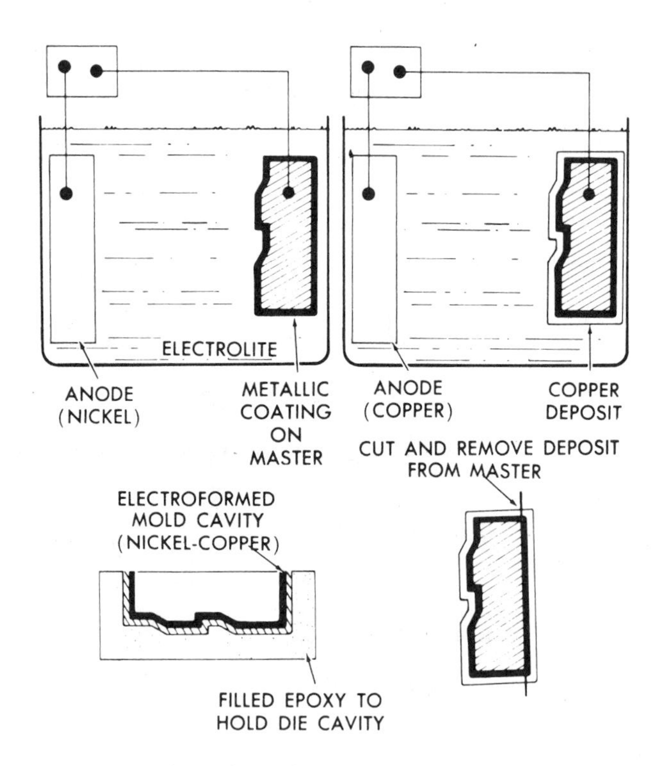

Fig. 22-22. Electroformed mold cavity. Copper coating and filled epoxy backing strengthen the nickel cavity.

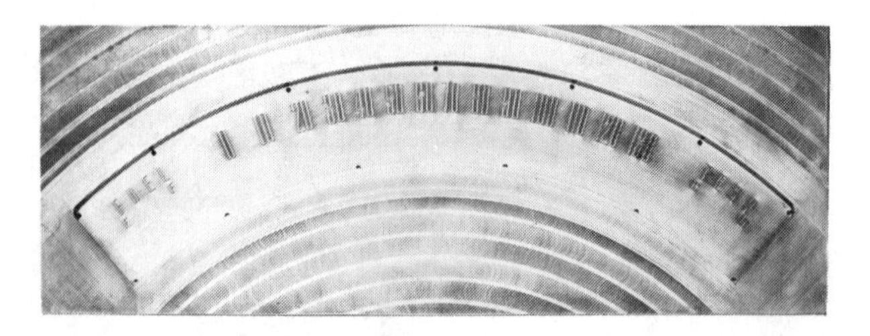

Fig. 22-23. An electroformed cavity showing raised detail from a polished background. The master is an engraved brass plate. (Electromold Corp.)

The major advantages of electroformed molds are their accurate reproduction of detail, zero porosity, zero shrinkage, and lower cost. The major disadvantages include design limitations, relative softness and difficulty with multiple cavities.

Sometimes it is desirable to have the molded piece cling slightly to one-half of the mold, depending on the

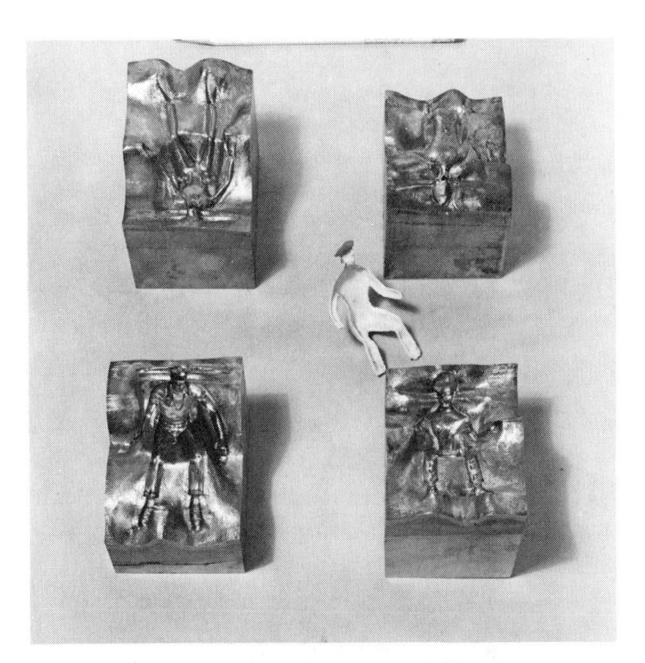

Fig. 22-24. A pair of cavities used to mold two halves of a hollow figure with a very elaborate match line. The cavities were filled with wax and the force plugs were electroformed against the cavity, producing perfect matched lines. Pencil shows relative scale. (Blectromold Corp.)

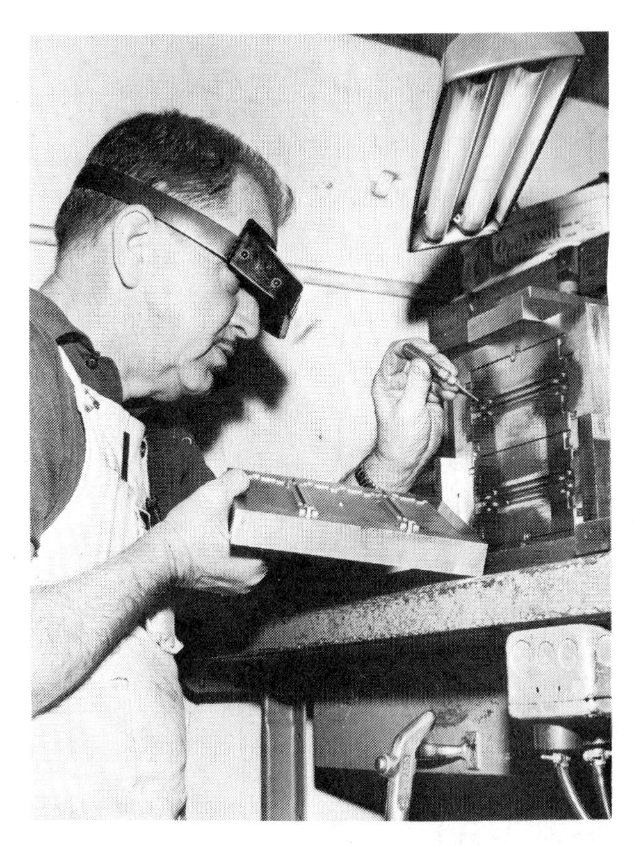

Fig. 22-25. Extreme care must be taken to prevent marring molds. (Revell, Inc.)

Fig. 22-26. This thermoforming cavity has an electroplated surface. (Electromold Corp.)

knockout mechanism. Excessive sticking may be caused by dents or undercuts in the mold or by a dirty cavity surface. In cleaning the dies and cavities a wax, lubricant, or silicone spray is often used. For stubborn spots, a wooden scraper or brass brush may be used. Never use steel scrapers when cleaning cavities because they may scratch or damage the polished finish of the cavity surfaces (Fig. 22-25).

There are a number of other metal-depositing methods that may be used for mold making. Flame spraying of metals and vacuum metallizing are two such methods. (See Chapter 17 Coating Processes.)

Electroplating, welding, and heat treatment are also used in making molds. Many finishing operations of molds are done by hand. A steel mold may be electroplated to protect the die cavity from corrosion and to provide the desired finish on the plastics product (Fig. 22-26).

Mold bases are important to the tool-maker. Bases hold the cavities in place and are made with enough thickness to provide heating and cooling for the cavities. They are made from steel, and are available in standard sizes. A standard mold base is shown in Figure 22-27. Bases may be purchased to accept most custom or proprietary molds (Fig. 22-28).

Alignment pins ensure proper matching of cavities when the mold base is brought together. If the assembly is not properly aligned and the parting lines are not in register, the mold cavities may have to be repositioned. If the molded part sticks in the die cavity, last-minute stoning, hand grinding, and polishing of the mold may be needed.

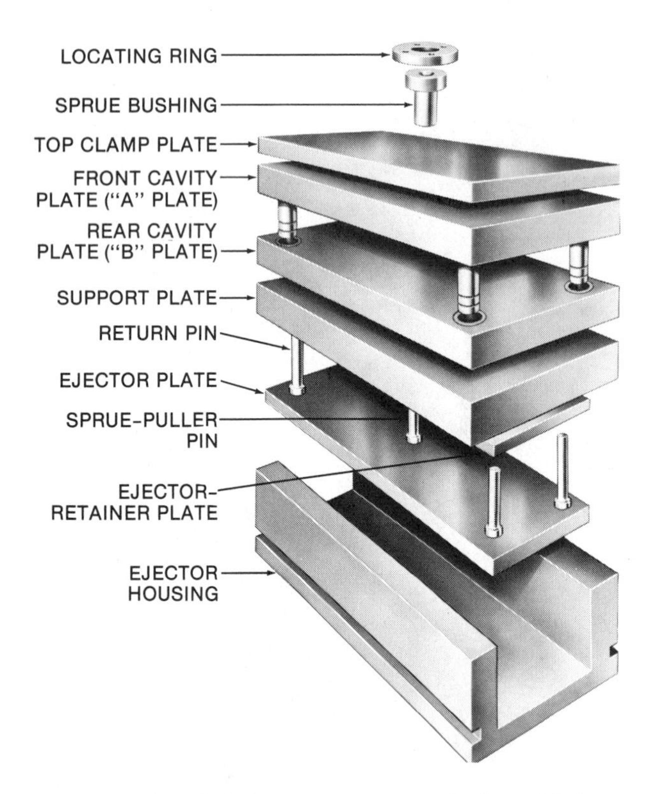

Fig. 22-27. Exploded view of a standard mold base, showing the parts. (D–M–E Co.)

(A) Standard cavity insert rounds showing bored holes in upper and lower cavity plated to receive inserts.

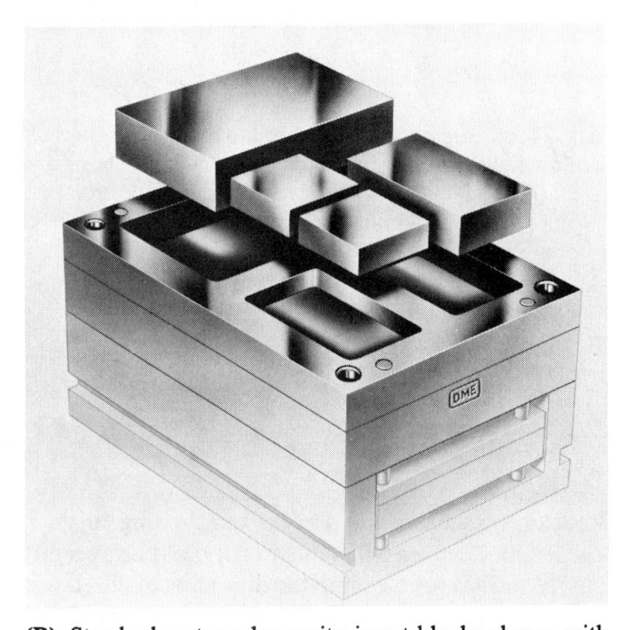

(B) Standard rectangular cavity insert blocks shown with pockets machined in mold base.

Fig. 22-28. Standard cavity insert blocks. (D–M–E Co.)

Vocabulary

The following vocabulary words are found in this chapter. Look up the definition of any of these words you do not understand as they apply to plastics in the glossary, Appendix A.

Air-hardening
Alignment pins
Chemical erosion
Deep-hardening
Die
EDM (Electrical-discharge machining)
Electroplating
Fixtures
Gypsum
Hobbing
Jig
Kirksite
Mold
Oil-hardened steel
Plastics tooling
Tools

Questions

22-1. Special fixtures, molds, dies etc., that enable a manufacturer to produce parts are called ___?___ .

22-2. Name an appliance for accurately guiding and locating tools during the operation involved in producing interchangeable parts.

22-3. Most mold designs begin from preliminary ___?___ .

22-4. A fundamental element in both plastics and steel is ___?___ .

22-5. Hobbing, etching, electroforming and electrical-discharge machining are generally classed as ___?___ processes.

22-6. An alloy of aluminum and zinc used for molds that has high thermal conductivity is known as ___?___ .

22-7. Much of the mold-making industry is composed of ___?___ shops.

22-8. Tooling ___?___ are more difficult to maintain in multiple cavity molds.

22-9. Tools for long run production work are usually made of ___?___ .

22-10. The trade names Ultracal, Hydrocal, and Hydrostone refer to ___?___ used for tooling.

22-11. Name the tooling material that has several advantages, including lightness, corrosion resistance, and low cost.

22-12. Devices that maintain proper alignment of the cavity as the mold closes are called ___?___ .

22-13. The act of shaping plastics or resins into finished products by heat and/or pressure is called ___?___ .

22-14. The tooling used to hold the cavities in place is called the ___?___ .

22-15. List four methods that may be used to produce a mold where no cutting tools are involved.

22-16. Because ___?___ are relatively inexpensive and have sufficient strength to produce some types of molds, they are used in prototype or experimental tooling.

22-17. Many metal tools and dies are being replaced by ___?___ because they are strong, light, and easy to machine.

22-18. A popular material for thermoforming and blow-molding molds is ___?___ because it is light, easy to machine, and has good thermal conductivity.

22-19. If a blow-molding mold requires fine mold detail, and if it must be easily machined but stronger than aluminum, ___?___ may be used.

22-20. For high strength, wear resistance, and high production runs, molds should be made of ___?___ .

22-21. A ___?___ machine is similar to and functions like a duplicator machine except that it is normally adjusted to operate at ratios as high as 20 to 1.

22-22. No cutting tools are used in machining operations classed as ___?___ processes.

22-23. The process where a piece of very hard steel is pushed into a blank of unhardened, unheated steel to form a mold cavity is called ___?___ .

22-24. An electrical means of removing metal by electrical erosion is called ___?___ .

22-25. In ___?___ or etching, an acid or alkaline solution is used to create a cavity.

22-26. Part of the manufacturing cost of plastics products is due to special ___?___ or tooling.

22-27. Gang or ___?___ molds and those with inserts require additional tooling costs.

22-28. Steel for molds may vary between 35 and 65 Rockwell C in ___?___ depending upon requirements.

22-29. Name the type of tooling you would select to produce:

a. ten thermoformed serving trays

b. ten thousand cast woodcut moldings for furniture doors

c. ten thousand molded polypropylene chair seats

d. ten million molded electrical switchplate covers

22-30. Where corrosion resistance is needed or adverse atmospheric conditions exist, ___?___ steel may be used for molds.

Chapter 23

Commercial Considerations

Introduction

In this chapter, you will learn that production personnel and management must work together for a successful plastics business. Financing, equipment, price quotations, plant locality, and other factors also must be considered.

Producing plastics parts is a competitive business. Material selection, processing techniques, production rates, and other variables must be considered in the selling price. (See Chapters 21 and 22.)

Resin manufacturers and custom molders are the best sources of information about the performance of a plastics material. The quantity and the compounding of the resin ingredients are important variables. When estimating or planning new items, confer with resin manufacturers and indicate all specifications. You should be able to answer the following questions:

1. How is the part to be used?
2. What grade of resin is to be used?
3. What physical requirements must the finished part meet?
4. What processing techniques will be used?
5. How many parts are to be made?
6. Will new capital investment for equipment be needed?
7. What are the specifications for reliability and quality of each part?
8. Will we be able to produce what the customer wants at a profit?
9. How and when will the customer pay for our services?

Price quotations may vary a great deal depending upon quantity. The price for polyethylene may exceed $0.75/lb in one pound bags, $0.48/lb in fifty pound bags, and $0.35/lb in a truck or train carload. This chapter includes the following topics:

I. Financing
II. Management and personnel
III. Plastics molding
IV. Auxiliary equipment
V. Molding temperature control
VI. Pneumatics and hydraulics
VII. Price quotations
VIII. Plant site
IX. Shipping

Financing

Strong interest and great prospects are not the only prerequisites to starting an industrial enterprise because without enough financing, no business firm can succeed and grow.

One of the major functions of management is to plan the capital structure of the business with care. In a proprietorship private capital or loans are used for financing, while in a corporation the sale of stock is used for capitalization.

Some equipment firms provide financing for purchases of their machinery through delayed or deferred payments, lease purchase plans, or direct financing. Insurance companies, commercial banks, mortgages, private lenders, and others are also sources of capital. The Small Business Investment Act of 1958 has helped many small firms. The Small Business Administration (a federal agency) has helped thousands of business to secure loans.

Management and Personnel

It has often been said, "a business is as strong or successful as its management operations." Many enterprises fail each year while others continue to struggle, barely surviving. Many of the problems of struggling and failing businesses may be associated with poor management. Management must coordinate the enterprise, regulating assets, personnel, and time to make a profit.

A major concern in the plastics industry is that the labor supply is not keeping pace with its growth rate. Employment opportunities for women are great since they make up nearly half the plastics industry workforce. The area of research in particular needs men and women to work with polymers, processes, and fabrication. Most plastics companies have need for professional personnel including executives, engineers, and supervisors; for technical personnel, including technicians and paraengineers; for skilled workers such as machinists, assistant technicians, and machinery set-up people; and for semiskilled and unskilled workers, who include material handlers, equipment operators, and packers.

Most of the unskilled or semiskilled personnel may be trained by the company; however, professional, technical, and skilled personnel must have college or technical school training.

The plastics industry will continue to compete for a limited supply of skilled designers, engineers, and moldmakers. The use of CAD, CAM, CAE, CAMM, and CIM systems might prove to be one way to meet the challenge of skilled personnel shortages and increase productivity.

Management must maintain good labor relations, which may include collective bargaining with labor unions. Successful relations between labor and management are part of the successful enterprise.

Plastics Molding

Much has been written about the general properties and forming processes of plastics. Molding plastics is difficult and often demands considerable experience to solve production problems. Technology in the plastics industry is constantly changing. Only basic information and precautions about molding plastics may be given here. (See also Chapter 21.)

The molding capacity of equipment may limit production output. Limitations include the available pressures of the press, amount of material it will mold, and physical size. Compression presses, for example, may vary in capacity from less than 5.5 to more than 1 653 tons (5 to 1 500 metric tons] of pressure. Extruder machines may plasticize less than 17 lb to more than 5.5 tons [8 to 5 000 kilograms] per hour. Injection machines may range from less than 0.70 oz to more than 44 lb (20 g to 20 kg] per cycle. The clamping pressures vary from less than 2.2 to more than 1 653 tons [2 to 1 500 metric tonnes]. It is common to run most equipment at 75 percent capacity, rather than maximum capacity (Table 23-1).

Many open-mold composite techniques are accomplished by hand. Placing layers of composite tape over specialized tools is slow. Handwork has the added disadvantages of possible incorrect tape orientation, process-induced voids, and/or porosity. One solu-

Table 23-1. Advantages and Disadvantages of Selected Manufacturing Methods

Manufacturing Method	Injection-Molding	Filament Winding	Blow-Molding	Pultrusion	Rotational Casting	Bag	Extrusion	Sprayup	Thermo-forming	Layup
Capital machine cost	high	low-high	low	high	low	low	high	low	low	low
Tool/mold costs	high	low-high	low	low-high	low	low	low	low	low	low
Material costs	high	high	high	high	low	high	high	high	high	high
Cycle times	low-high	high	low-high	high	high	high	low	high	high	high
Output rate	high	low	high	low	low	low	high	low	high	low
Dimensional accuracy	good	fair	fair	fair	fair	fair	fair	fair	poor	fair
Finishing stages	none	some	some	yes	some	some	yes	some	yes	some
Thickness variation	low	fair	fair	fair	low	low	low	low	high	low
Stress in molding	some	some	some	some	none	some	some	some	some	some
Can mold threads	yes	no	yes	no	yes	no	no	no	no	no
Can mold holes	yes	yes	yes	no	yes	yes	no	yes	no	yes
Open-ended components molded	yes	yes	yes	yes	yes	yes	yes	yes	yes	yes
Inserts molded-in	yes	yes	no	no	yes	yes	no	yes	no	yes
Waste material	none	some	some	none	none	some	some	some	some	some

tion to reduce manufacturing costs and assure consistent part quality is to utilize automated tape-laying equipment as part of the molding process. In Figure 23-1, course after course, layer upon layer, the part is formed by laminating the graphite/epoxy tape in a computerdesigned, cross-ply pattern. It is then cured in an autoclave under controlled heat and pressure. The resulting aircraft part has superb structural strength plus a weight advantage unmatched by any metal (Fig. 23-2).

Auxiliary Equipment

Plastics materials are poor conductors of heat. Some plastics are *hygroscopic,* that is, moisture-absorbing. For these reasons preheating auxiliary equipment may be needed to reduce moisture content and the polymerization or forming time. Often, hopper dryers are used on injection and extruders to remove moisture from the molding compounds and help assure consistent molding. Preheating of thermosets may be done by various thermal heating methods such as with infrared, sonic, or radio-frequency energy. Preheating may reduce cure and cycle time and prevent streaking, color segregation, molding stress, and part shrinkage. It also may allow for a more even flow of heavily filled molding compounds.

Fig. 23-2. Significant portions of the F-16, including the vertical fin and horizontal stabilizer, are being fabricated from composite graphite/epoxy materials. Future aircraft will utilize even larger amounts of composite materials. (General Dynamics)

Fig. 23-1. This automated tape-laying machine for laying exact courses of graphite/epoxy tape is an important step in the production of F-16 aircraft components. (General Dynamics)

Molders often must compound their own additives into resins or other compounds. Both hot and dry mixers may be needed to blend plasticizers, colorants, or other additives. Material silos or other conveyor systems may be required.

Microprocessors (CIM systems) can control many materials handling operations. Through centralized monitoring and control, all settings can be made and the status of all stations can be checked. These systems store molding parameters and actual batch data (formulas) for future use or to prevent the wrong materials from entering a process. Hoppers, loaders, and blenders are controlled to accurately weigh and meter ingredients for each machine. Most dryer suppliers agree that microprocessor control is the key in their technology. Microprocessors can accurately control energy use to create high performance, dust-free drying systems.

Preform equipment and loaders are important in compression molding and many composite systems.

Injection molding, blow molding, thermoforming, and other molding techniques may require the use of regrinders or granulators to grind sprues, runners, or other cutoff scrap pieces into usable molding materials. Some of these operations are done in-line. Processors look for noise and floorspace reduction, as well as energy efficiency and ease of maintenance when selecting granulators.

Annealing tanks are used with thermoplastic products to reduce sink marks and distortion. Large housings or similar large pieces are often placed over shrink blocks, or in jigs or dies, to help maintain correct dimensions and minimize warpage during cooling. Many automobile steering wheels once cracked after a length of time because of latent shrinkage of the molded part. Proper annealing and selection of materials have solved such problems.

Closed-loop processing with computer-controlled chillers and cooling towers is helping to reduce energy consumption and increase productivity.

The growth of the use of robotics has been phenomenal. The main reason for this growth is increased productivity. In addition, robots may remove human operators from hot, boring, and highly fatiguing jobs. The use of efficient and flexible robots can release humans for the creative and problem-solving tasks of the company (Fig. 23-3).

Although parts removal is the main application for robots, they could perform a variety of secondary operations that improve part quality and reduce labor costs. Robots may be used as parts-handling devices, transporting parts to secondary stations and packaging. This could eliminate nonvalue-added functions in the plant, such as assigning employees to package parts. Post-molding operations by robots may include assembling, gluing, sonic welding, decorating, or sprue snipping.

Some manufacturers utilize bar coding to keep track of material flow, inventory, and part production. Bar coding is used in flexible manufacturing systems (FMS). FMS consists of manufacturing cells that are equipped with a number of machining or molding operations or other automated equipment, all controlled by computer. As different coded parts come down the line, FMS or CIM systems determine which operations (assembly, decorate, finish) are to be performed.

(A) Robot completes broaching operation.

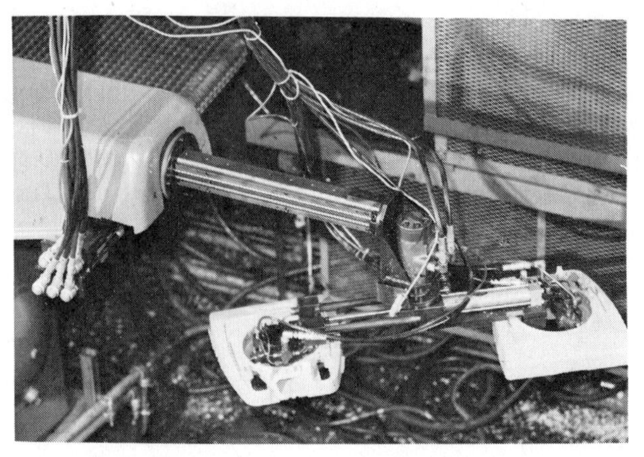

(B) Robot turns molded housings and places the parts on cooling conveyor.

(C) Robots, shown above removing a car bumper from an injection-molding press, do practically all of the material handling in the production process. Station-to-station movement is by automatically guided vehicles or an automated overhead electric monorail system. Painting also is done by robots. The bumpers, manufactured from a polycarbonate/polyester alloy called Xenoy, are as strong as steel, but lighter, less expensive, and easier to paint. They meet five-mile-an-hour crash standards and will not rust. (Ford Motor Co.)

Fig. 23-3. Use of robots in plastics products manufacturing. (Prab Robots, Inc.)

Molding Temperature Control

One of the most important factors in efficient molding is the control of temperature. The temperature control system may consist of four basic parts including the thermocouple, temperature controller, power output device, and heaters. A control system using these four parts is shown in Figure 23-4.

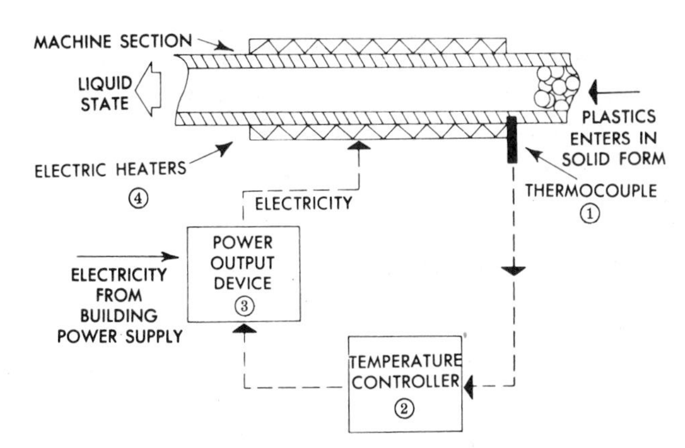

Fig. 23-4. The four basic parts of a temperature control system used in the plastics industry. (West Instruments, Gulton MCS Division)

The thermocouple is a device made of two unlike metals. Combinations of iron with constantan or copper with constantan are often used. When heat is applied to the junction of the two metals, electrons are freed, producing an electric current that is measured on a meter calibrated in degrees. Over 98 percent of all temperature sensors used by the plastics industry are thermocouples although resistance temperature detectors (resistance bulbs) have been used for a few installations. The West thermocouple system, used to minimize thermal variation, is shown in Figure 23-5.

The millivolt and the potentiometric controllers are two types of basic temperature controllers used widely. The potentiometric controller differs from the millivoltmeter in that the signal from the thermocouple is electronically compared to a setpoint temperature. Millivolt and potentiometric controllers may be designed to control the power to the heaters and hold them at a set temperature (stepless control), or they may be designed to turn off the power to the heaters when a set temperature is reached (proportional control). The power input device controls the power to the heaters. It usually consists of a mechanical relay or a solid-state control circuit. Instrumentation may also be used to control cooling cycles. Figure 23-6 shows an instrument panel for a complete operation.

Most modern machines are operated by hydraulic power and electrical energy. Vacuum, compressed air, hot water, and chilled water supplies may also be needed. For the most part, cold water is used to cool mold dies in order to reduce the molding cycle time.

It is important in injection molding that the cooling system is able to remove the total heatload generated in each cycle. The system must also be designed to insure that all molded sections, thick or thin, will cool at the same rate to minimize potential differential shrinkage.

The microprocessor has made major changes in all areas of plastics processing, including chillers and mold-temperature controllers. Microprocessor-based mold temperature controllers may keep molds or chillers within ±1 °F. Electromechanical and solid-state controllers usually have ±3 °F accuracy ratings.

Balances, scales, pyrometers, clocks, and various timing devices are important accessories. There are many manufacturers of machines and auxiliary equipment for the plastics industry.

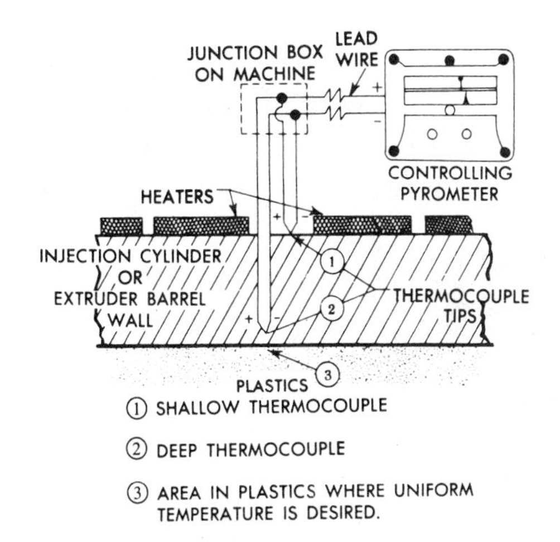

(B) Graph of temperatures.

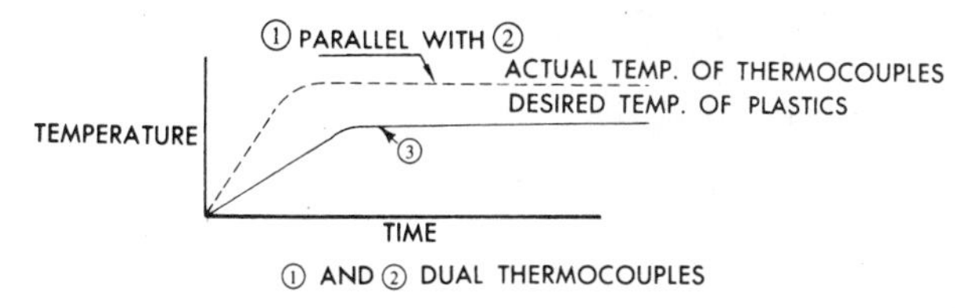

(A) Physical arrangement.

Fig. 23-5. The physical and electrical arrangements of dual thermocouples used to minimize temperature variations. (West Instruments, Gulton MCS Division)

Fig. 23-6. This instrument panel is used to control a large blow-molding operation. (Chemplex Co.)

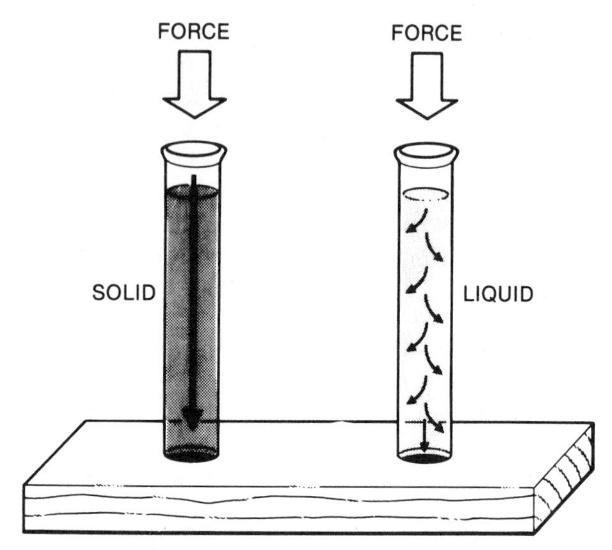

(A) Comparison of transmitting a force through a solid and a liquid.

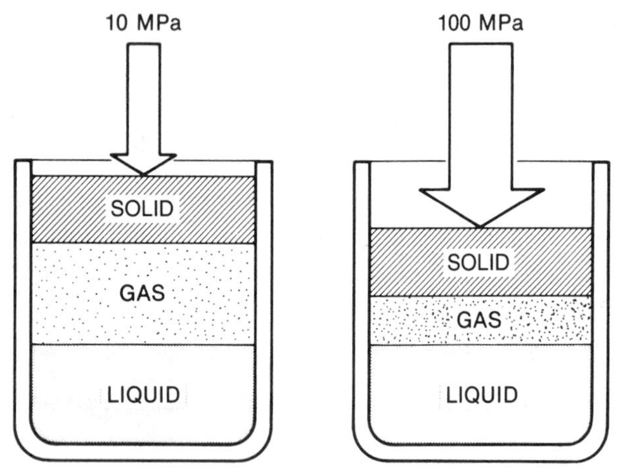

(B) Gas can be compressed, but a liquid resists compression.

Fig. 23-7. Force is transmitted differently by solids, liquids, and gases. A solid transmits force only in the direction of the applied force. Liquids (hydraulic systems) and gases (pneumatic systems) transmit force in all directions.

Pneumatics and Hydraulics

Pneumatic- and hydraulic-actuated accessories and equipment are important in plastics processing (Fig. 23-7).

Pneumatics are used to activate air-cylinders and provide compact, light, vibrationless power. Filters, air dryers, regulators, and lubricators are needed accessories for pneumatic systems.

Improper oil, or the presence of air or moisture, may cause noise in hydraulic lines. Chattering valves, pump wear, and high oil temperature may also cause noise problems. (See Chapter 4.)

Hydraulic power systems may be divided into four basic components:

1. *Pumps* that force the fluid through the system
2. *Motors or cylinders* that convert the fluid pressure into rotation or extension of a shaft
3. *Control valves* that regulate pressure and direction of fluid flow
4. *Auxiliary components,* which include piping, fittings, reservoirs, filters, heat exchangers, manifolds, lubricators, and instrumentation

Graphic symbols are used to show information about the fluid power system schematically. These symbols do not represent pressures, flow, or compound settings. Some devices have schematic diagrams printed on their faces (Fig. 23-8).

It is often hard to choose between a hydraulic and a pneumatic system for an application. As a general rule, when a great amount of force is needed, use hydraulics; when high speed or a rapid response is needed, use pneumatics.

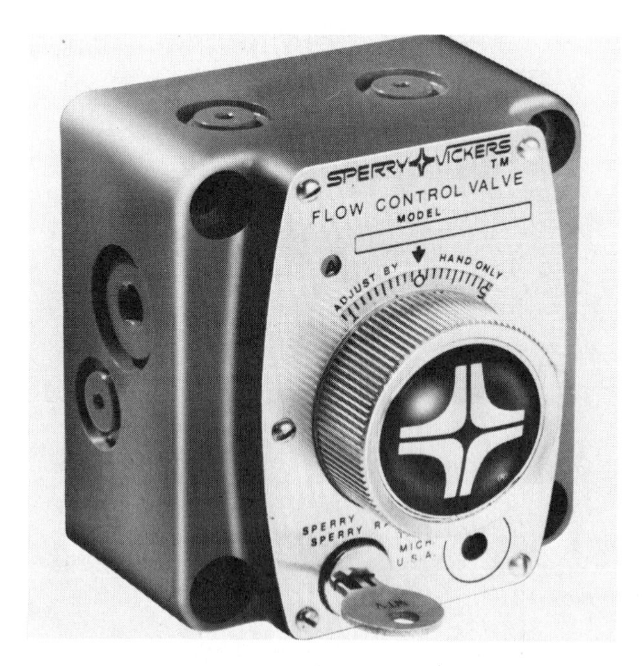

(A) Flow control valve compensates for fluid pressure and temperature variations.

Fig. 23-8. Fluid power control devices. (Sperry Vickers)

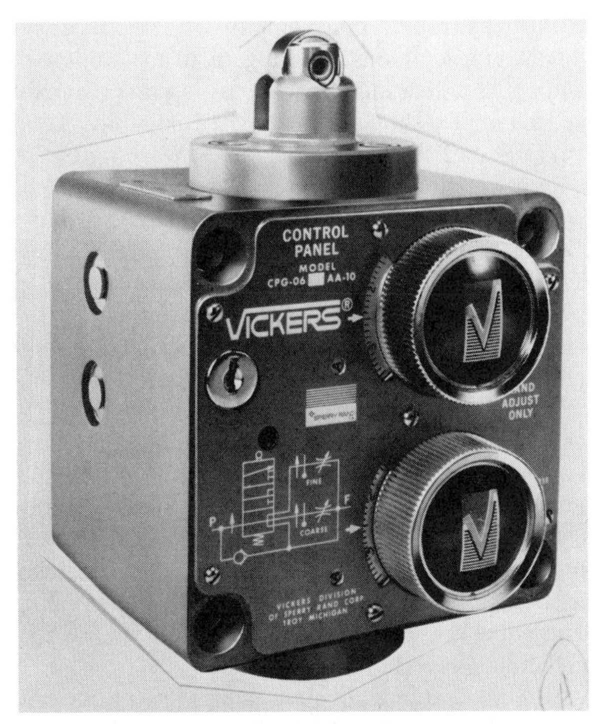

(B) Control panel with schematic on face.

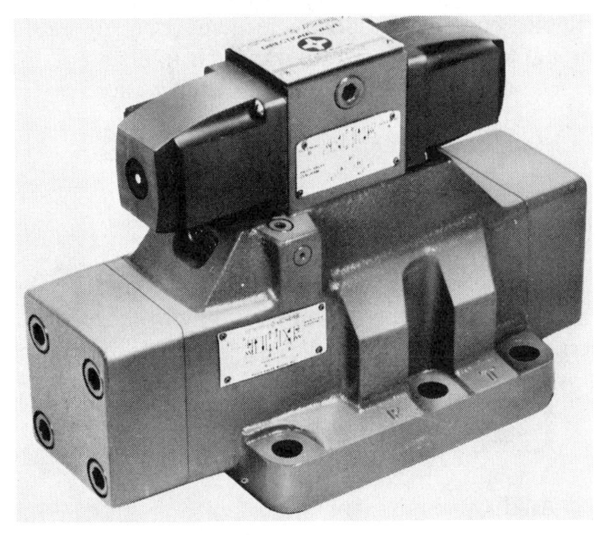

(C) Directional control valve.

Fig. 23-8. Fluid power control devices. (Sperry Vickers) *(Continued)*

(A) Male mold with part.

(B) Female mold cavities with part.

Fig. 23-9. Examples of male and female molds.

Price Quotations

Since all plastics parts are formed by processes utilizing molds or dies, it is only logical that much consideration be given to them.

The cornerstone of any plastics business is tooling. In a tough, competitive market, a business must produce molds faster, cheaper and better than the competition can if the business is to be profitable (Fig . 23-9).

Computer numerically controlled (CNC) machine tools and computer assisted designs and manufacturing (CAD/CAM) equipment help produce better, more accurate molds for a substantial savings in cost and time.

All price quotations for molded products should be based on the design of the mold and its general condition. In the long run, the less costly die is usually not the best buy. Compression, transfer, and injection molds are costly.

It is common for price quotations to be based on *custom molds*. These are molds owned by the customer and turned over to the molder to produce parts. The greatest danger in quoting prices is failure to take into consideration the condition of the custom mold. All quotations should be based on approval of the custom mold because repair or alteration of the mold may be necessary.

If the die is owned or made by the molder, it is called a *proprietary mold*. In quotations for products to be made with proprietary molds, the molder must make enough profit to amortize, or pay for, the mold. Often a part of the cost of making the mold is calculated into the price of each piece or each thousand pieces.

With custom or proprietary molds, corrosion is an enemy. Molds should be given a moisture-resistant, noncorrosive coating and water, air, or steam holes should be dried and given a coating of oil. The mold should then be placed in storage with all its accessories.

Large tooling (panels, boat hulls, sinks) require careful consideration of production space and adequate storage facilities. Both add to the cost of the product.

Plant Site

The plant site should be determined in relation to the nearness of raw material and potential market. Freight must be reflected in the cost of each plastics product. A quotation must reflect tax rates, labor conditions, and wages. If there is an anticipated increase in taxes or wages, the price quotation must include such an increase in production costs.

Many states and communities encourage new enterprises. They offer reduced taxes and adequate labor force. Labor relations and tax incentives may be major concerns in plant location.

Other considerations in site location are an available skilled-labor pool and proximity to educational institutions.

Shipping

Shipping plastics, resins, and chemicals falls under special governmental regulations. The United States Postal Service has ruled that any liquid that gives off flammable vapors at or below a temperature of 20 °F [–7 °C] is not mailable (Section 124.2d, Postal Manual). Poisonous materials are generally regarded as nonmailable by the provisions of 124.2d. Caustic or corrosive substances are prohibited in 124.22. The prohibitions mentioned are specified by the law in Section 1716 of Title 18, U. S. Code. Acids, alkalies, oxidizing materials, or highly flammable solids, highly flammable liquids, radioactive materials, or articles emitting objectionable odors are considered to be nonmailable as well.

Postmasters and other employees at post offices will not give opinions with regard to mailability of materials. To determine if material is properly mailable, the mailer should write to the Mailability Division, Office of the General Counsel, Washington, D. C., for instructions.

The Department of Transportation (DOT) regulates the interstate transportation of cellulose nitrate plastics by rail, highway, or water, and special packaging must be used when shipping this material.

The Interstate Commerce Commission (ICC) regulates the packing, marking, labeling, and transportation of dangerous or restricted materials. Standards for many shipping containers are specified and regulations also specify the use of labels. For example, flammable liquids require a red label, flammable solids a yellow label, and corrosive liquids a white label. There are also other labels for poisons and shipments of radioactive material. It is the responsibility of the shipping agency to check the label to make certain it is correct and that it is filled in by the shipper.

If there is any doubt about the shipment of flammable or plastics materials, it is best to check local ICC offices. Certain cities have ordinances that prohibit vehicles containing flammable items from operating through tunnels or via bridges. Such Interstate Commerce Commission regulations must always be observed.

Vocabulary

The following vocabulary words are found in this chapter. Look up the definition of any of these words you do not understand as they apply to plastics in the glossary, Appendix A.

Amortization
Annealing
Auxiliary equipment
Custom molds
Estimating
Flexible manufacturing systems (FMS)
Hydraulics
Hygroscopic
Pneumatics
Proprietary molds
Pyrometer

Questions

23-1. The greatest danger in quoting custom mold prices, is not taking into consideration the ___?___ of the custom mold.

23-2. Often a portion of the cost of making a mold is calculated into the production of each piece. This is called ___?___.

23-3. Molds made by the customer and used by the molder are called ___?___ molds.

23-4. The transmission of power through a controlled flow of liquids is known as ___?___.

23-5. Molds made and owned by the molder are called ___?___ molds.

23-6. The ___?___ is a device made of two dissimilar metals. Combinations of iron and constantan or copper and constantan are commonly used.

23-7. A ___?___ material tends to absorb moisture.

23-8. The ___?___ of the molding equipment will limit product size.

23-9. Name four pieces of auxiliary equipment that may be required for injection-molding operations.

23-10. Name four factors that may influence plant locality.

23-11. Name four factors that may influence the cost of each plastics product when estimating the planning of new items.

23-12. The Interstate Commerce Commission regulates the packing, marking, labeling and transportion of ___?___ or ___?___ materials.

23-13. Would you select a hydraulic or a pneumatic power system if great force is required?

23-14. When high speed and rapid response are required, would you select a hydraulic or a pneumatic power system?

23-15. The best sources of information about the performance of a plastics material are ___?___ manufacturers and ___?___ molders.

23-16. In a ___?___ enterprise, private capital or loans are used for financing.

23-17. The sale of stock shares is used for capitalization in a ___?___ enterprise.

23-18. Planning the capital structure of an enterprise is the major function of ___?___.

23-19. Successful relations between ___?___ and ___?___ are part of the successful enterprise.

23-20. Name the four basic components of hydraulic power systems.

23-21. The cornerstone of any plastics business is ___?___.

23-22. It is common to run or operate most molding equipment at ___?___ percent of maximum capacity.

23-23. Name the four general classifications of plastics personnel.

23-24. Approximately ___?___ percent of the plastics industry work force is female.

23-25. If ___?___ and CNC equipment is used, better, more accurate molds may be produced with a savings in cost and time.

23-26. In addition to increased productivity, ___?___ may be used to relieve human operators from hot, boring, and highly fatiguing jobs.

Appendix A

Glossary

A-stage—An early stage in the reaction of a thermosetting resin in which the material is fusible and is still soluble in certain liquids.

Ablative plastics—Plastics composites used as heat shields on aerospace vehicles. The intense heat encountered erodes and chars the top layers. The charred layer and some cooling effects from evaporation insulates the inner areas against further penetration of the intense heat.

Acetals (poly)—A polymer having the molecular structure of a linear acetal (polyformaldehyde), consisting of unbranched polyoxymethylene chains.

ACGIH—American Conference of Governmental Industrial Hygienists; this organization published guidelines and recommendations on the exposure limits to various chemicals.

Acrylic—A synthetic resin prepared from acrylic acid or from a derivative of acrylic acid.

Acrylonitrile—A monomer with the structure $CH_2{=}$ CHON. It is most useful in copolymers. Its copolymer with butadiene is nitrile rubber, and several copolymers with styrene exist that are tougher than polystyrene.

Acrylonitrile-butadiene-styrene (ABS)—Acrylonitrile and styrene liquids and butadiene gas are polymerized together in a variety of ratios to produce the family of ABS resins.

Acute Toxicity—The adverse effect on a human or animal body, with severe symptoms developing rapidly and quickly coming to a crisis. Examples include dizziness, nausea, skin rashes, inflammation, tearing of the eyes, unconsciousness, and even death.

Additional polymerization—Polymers formed by the combination of monomer molecules without the splitting-off of low molecular mass by-products, such as water.

Additive polymerization—A type of polymerization which adds one mer to another. It usually does not yield any by-products.

Adhesion—The state in which two surfaces are held together by interfacial forces that may consist of interlocking action (mechanical means).

Adhesive—A substance capable of holding materials together by surface attachment.

Air-assist forming—A method of thermoforming in which air flow or air pressure is used to partially preform the sheet just before the final pull-down into the mold using vacuum.

Air-hardening—Refers to steel that is cooled in air.

Air pollution—Filling the atmosphere with undesirable particles, some of which may be toxic.

Air slip forming—A thermoforming process in which air pressure is used to form a bubble and a vacuum is then used to form the hot plastics against the mold.

Alignment pins—Devices that maintain proper cavity alignment as a mold closes.

Aliphatic molecules—Organic compounds whose molecules do not have their carbon atoms arranged in a ring structure.

Alkanes—Hydrocarbons with the general formula C_nH_{2n+2}.

Alkenes—Hydrocarbons that have the general formula C_nH_{2n} and possess double covalent bonds.

Alkyd—Polyester resins made with some fatty acid as a modifier.

Alkynes—Hydrocarbons that have the general formula C_nH_{2n-2} and possess a triple bond between two carbon atoms.

Allowances—The intentional differences in the dimensions of two parts.

Allyl—A synthetic resin formed by the polymerization of chemical compounds containing the group $CH_2{=}CH{-}CH_2$. The principal commercial allyl resin is a casting material that yields allyl carbonate polymer.

Alternating copolymer—A copolymer that has a chemical structure in which the two types of monomers alternate in the polymer chain.

Amids—Organic compounds containing a—$CONH_2$ group, derived from organic acids.

Amines—Organic derivatives of ammonia (NH_3) obtained by substituting hydrocarbon radicals for one or more hydrogen atoms.

Amino—Chemical name showing the presence of an NH_2 or NH group. Also, materials with these groups.

Amorphous—A term which means not crystallized. Plastics that have an amorphous arrangement of molecular chains are often transparent.

Amortization—The gradual repayment of the cost of equipment such as molds. It may be done by contribution to a sinking fund at the time of each periodic interest payment.

Anisotropic—Exhibiting different properties when tested along axes in different directions.

Annealing—A process of holding a material at a temperature near, but below, its melting point, for a period of time to relieve internal stress without shape distortion.

ANSI—American National Standards Institute; a privately funded, voluntary membership organization that identifies industrial and public needs for national consensus standards and coordinates development of such standards. Many ANSI standards relate to safe design/performance of equipment, such as safety shoes, eyeglasses, smoke detectors, firepumps and household appliances; and safe practices or procedures, such as noise measurement, testing of fire extinguishers and flame arrestors, industrial lighting practices, and the use of abrasive wheels.

Antiblocking—Materials that prevent two plastics from undesired adhesion.

Antioxidant—A stabilizer that retards breakdown of the plastics by oxidation.

Antistatic—An additive that reduces static charges on a plastics surface.

Apparent density—The mass per unit volume of a material; includes the voids inherent in the material.

Aromatic hydrocarbons—Hydrocarbons derived from or characterized by the presence of unsaturated ring structures.

Ashing—The use of wet abrasives on wheels to sand and polish plastics.

Aspect ratio—The ratio of the length of a filler to its width.

ASTM—American Society for Testing Materials; voluntary membership organization with members from broad spectrum of individuals, agencies, and industries concerned with materials. As the world's largest source of voluntary consensus standards for materials, products, systems, and services, ASTM is a resource for sampling and testing methods, health and safety aspects of materials, safe performance guidelines, and effects of physical and biological agents and chemicals.

Atactic stereoisomerism—A random arrangement of molecular chains in a polymer.

Atomic mass (gram-atom)—Relative mass of an atom of any element, as compared with that of one atom of carbon taken as 12 g.

Atomic number (atomic mass number)—A number equal to the number of protons in the nucleus of an atom of the element.

Atoms—The smallest particles of an element that can combine with particles of other elements to produce the molecules of compounds. Atoms consist of a complex arrangement of electrons revolving about a positively charged nucleus containing particles called protons and neutrons.

Autoclave—Pressure vessel that can maintain temperature and pressure of a desired air or gas for the curing of organic-matrix composite materials.

Autoclave molding—A molding method in which, after final layup, an entire assembly is put into a steam or electrically heated autoclave at elevated pressure. Additional pressure achieves higher reinforcement loadings and improved removal of air.

Automatic cycle—The type of cycle of an injection-molding machine that requires no operator input.

Auxiliary equipment—Equipment needed to help control or form the product. Filters, vents, ovens, and takeup reels are examples.

Azo group—The group —N=N—, generally combined with two aromatic radicals. A whole class of dyestuffs is characterized by the presence of this group.

B-stage—An intermediate stage in the reaction of a thermosetting resin. In this stage, the material softens when heated and swells in contact with certain liquids, but it does not entirely fuse or dissolve. Resins in thermosetting molding compounds are usually in this stage.

Backbone—The main chain of a plastics molecule.

Back pressure—The viscous resistance of a material to continued flow when a mold is closed. In extrusion, resistance to the forward flow of molten material.

Baffle—A device used to restrict or divert the passage of fluid or gases through a pipeline or channel.

Bakelite—A phenolic thermosetting plastics invented by Leo Baekeland in 1907.

Banbury—A machine for compounding materials. The machine contains a pair of contrarotating rotors that masticate and blend the materials.

Barrel—The cylindrical housing in which the extruder screw rotates.

Barrel vent—An opening in a barrel wall to permit the escape of air and volatile matter from the material being processed.

Bell curve—A graphic representation of a standard, normal distribution.

Benzene—A clear, flammable liquid, C_6H_6. It is the most important aromatic chemical.

Biaxial winding—A type of winding, in filament winding, in which the helical band is laid in sequence, side by side, with no crossover fibers.

Bifunctional—A molecule with two active functional groups.

Binder—Resin that holds together filler or carriers in a pre-preg or molding compound. ***Carrier*** is the fabric portion of a composite. Fiberglass is the most commonly used fabric for composites.

Biodegradable—A material that chemically decomposes under the action of bio-organisms.

Blanket—Plies laid up in a complete assembly and placed on or in the mold all at one time (flexible-bag process); also, the form of bag in which the edges are sealed against the mold.

Blanking—The cutting of flat sheet stock to shape by striking it sharply with a punch while it is supported on a mating die. Punch presses are used.

Bleeder cloth—A nonstructural layer of material used in manufacture of composite parts to allow the escape of excess gas and resin during curing.

Block polymer—A polymer molecule that is made up of comparatively long sections that are of one chemical composition, those sections being separated from one another by segments of different chemical character.

Blow molding—A method of fabrication in which a parison is forced into the shape of the mold cavity by internal air pressure.

Blow systems—Any method or process used to cause polymers to expand or become cellular.

Blow-up ratio—In blow molding, the ratio of the mold-cavity diameter to the parison diameter. In blown film, the ratio of the final tube diameter to the original die diameter.

Boss—A protuberance on a part designed to add strength or to facilitate assembly.

Bottom ash—The residue of combustion that settles in the bottom of an incinerator.

Branching—Side chains attached to the main chain of the polymer. Side chains may be long or short.

Breaker plate—A perforated metal plate located between the end of the screw and the die head.

Breather—Porous material, such as a fabric or cilitate, for removal of air, moisture, and volatiles during curing.

Breathing—The opening and closing of a mold to allow gases to escape early in the molding cycle. Also called degassing.

Brittleness temperature—The temperature at which plastics and elastomers rupture by impact under certain conditions.

Buffing—An operation to provide a high luster to a surface. The operation, which is not intended to remove much material, usually follows polishing.

Bulk density—The mass per unit volume of a molding powder as determined in a reasonably large volume.

Bulk factor—The ratio of the volume of any given weight of loose plastics to the volume of the same weight of the material after molding or forming.

Bulk polymerization—The polymerization of a monomer without added solvents or water.

C-stage—The final stage in the reaction of a thermosetting resin, in which the material is relatively insoluble and infusible. Thermosetting resins, if fully cured plastics, are in this stage.

CAD—Computer-aided design; a computer system that aids or assists in the creation, modification, and display of a design. It is used to produce three dimensional designs and illustrations of the proposed part. The term **CAD** is also used to refer to computer-aided drafting.

CAE—Computer-aided engineering; a computer system that assists in the engineering or design cycle. It analyzes the design and calculates the performance predictions of service life and safety design factors.

Calendering—Process of forming a continuous sheet by squeezing the material between two or more parallel rolls to impart the desired finish or to ensure uniform thickness.

CAM—Computer-aided manufacturing; the utilization of computer systems in the management, control, and operations of the manufacturing facility through either direct or indirect computer interface with the physical and human resources of the company.

CAMM—Computer-assisted mold making; a computer system used to analyze mold temperatures and flows through finited element analysis or boundary element analysis. Orientation and placement of reinforcements in composites are also shown in some data base systems. This information is then used to compensate for matrix shortage in the design and machining of the mold.

Carbonaceous—Containing carbon.

Carcinogen—A substance or agent that can cause a growth of abnormal tissue or tumors in humans or animals. A material identified as an animal carcinogen does not necessarily cause cancer in humans. Examples of human carcinogens include coal tar, which can cause skin cancer, and vinyl chloride, which can cause liver cancer.

Casein—A protein material precipitated from skimmed milk by the action of either rennin or dilute acid. Rennet casein is made into plastics.

CAS number—(Chemical Abstracts Services registry number; these numbers unambiguously identify all known chemical substances.

Casting—The process of pouring a heated plastics or other fluid resin into a mold to solidify and take the shape of the mold by cooling, loss of solvent, or completing polymerization. No pressure is used. *Casting* should not be used as a synonym for *molding*.

Catalyst—A chemical substance added in minor quantity (as compared to the amounts of primary reactants) that markedly speed up the cure or polymerization of a compound. See also Initiator.

Caul Plate—A smooth metal plate used in contact with the layup during curing to transmit normal pressure and to provide a smooth surface to the finished part.

Ceiling—An exposure limit which should never be exceeded, even for short time durations.

Cellular or foamed—A sponge form. The sponge may be flexible or rigid, the cells closed or interconnected. The density of a cellular plastics may be anything from that of the solid parent resin down to 32 kg /m^3 [2 lbs/ft^3]. The terms *cellular, expanded,* and *foamed* plastics are used synonymously; however, *cellular* is most descriptive of the product.

Celluloid—A strong, elastic plastics made from nitrocellulose, camphor, and alcohol. Celluloid is used as a trade name for some plastics.

Cellulosics—A family of plastics with the polymeric carbohydrate cellulose as the main constituent.

Cement—To bond together, as to adhere with a liquid adhesive employing a solvent base of the synthetic elastomer or resin variety.

Centipoise—One one-hundredth of a poise, a unit of viscosity. Water at room temperature has a viscosity of about one centipoise.

Central tendency—The grouping of data points around one central point.

CERCLA—Comprehensive Environmental Response, Compensation, and Liability Act.

CFR—Code of Federal Regulations.

Chain growth polymerization—A type of polymerization in which the chains grow from initiation to completion almost instantaneously.

Char barrier—A layer of charred material that acts as an insulator to prevent a material from burning or continuing to burn.

Charging—In molding, the measurement or placing of the material in a mold. In polishing, the deposit of abrasive on a revolving wheel.

Chemical erosion—A chemical method of removing metal.

Chlorinated polyether—The polymer obtained from pentaerythritol by preparing a chlorinated oxethane and polymerizing it to a polyether by means of opening the ring structure.

Chronic toxicity—Adverse (chronic) effects resulting from repeated doses of or exposures to a substance over a relatively prolonged period of time. Ordinarily used to denote effects in experimental animals.

CIM—Computer-integrated manufacturing; the logical organization of individual engineering, production, and marketing, or other support functions into a computer-integrated system. Functional areas such as design, inventory control, physical distribution, cost accounting, planning, purchasing, etc. are integrated with direct materials management and shop floor management.

Clean Air Act—Federal act passed in 1970 that set levels of air pollution and caused the closure of many incinerators.

Closed-cell—Describing the condition of individual cells that make up cellular or foamed plastics when cells are not interconnected.

Coating—Placing a permanent layer of material on a substrate.

Cohesion—The propensity of a substance to adhere to itself; the internal attraction of molecular particles toward each other; the ability to resist partition from the mass.

Cold flow—See *Creep*.

Cold molding—A procedure in which a composition is shaped at room temperature and cured by subsequent baking.

Collodion—A thin, transparent film of dried pyroxylin.

Colorants—Dyes or pigments that impart color to plastics.

Combustible—A material that burns easily. Combustible liquids have a flash point of 38 °C or above.

Composite—A combination of two or more materials (generally a polymer matrix with reinforcements). The structural components of the composites are sometimes subdivided into: fibrous, flake, laminar, particulate, and skeletal.

Composting—Grinding trash into small pieces and mixing it with soil.

Compound—A substance composed of two or more elements joined together in definite proportions.

Compression mold—A mold that is open when the substance is introduced and that shapes the material by heat and by the pressure of closing.

Compression molding—A technique in which the molding compound is placed in an open mold cavity, the mold closed, and heat and pressure applied until the material has cured or cooled.

Compressive strength—The highest load sustained by a test sample in a compressive test, divided by the original area of the sample.

Condensation—A chemical reaction in which two or more molecules combine, with the separation of water or some other simple substance. If a polymer is formed, the condensation process is called *polycondensation.*

Condensation polymerization—Polymerization by chemical reaction that also produces a by-product.

Copolymerization—Addition polymerization involving more than one type of mer.

Corona discharge—A method of oxidizing a film of plastics to make it printable. Achieved by passing the film between electrodes and subjecting it to a high-voltage discharge.

Coumarone—A compound (C_8H_60) found in coal tar and polymerized with indene to form thermoplastic resins that are used in coatings and printing inks.

Coupling agents—Chemicals that promote the adhesion between reinforcements and the basic plastics material.

Covalent bonding—Atomic bonding by sharing electrons.

CPSC—Consumer Products Safety Commision; Federal agency with responsibility for regulating hazardous materials when they appear in consumer goods. For CPSC purposes, hazards are defined in the Hazardous Substances Act and the Poison Prevention Packaging Act of 1970.

Cracking—Thermal or catalytic decomposition of organic compounds to break down the high-boiling compounds into lower-boiling fractions.

Creep—The permanent deformation of a material resulting from prolonged application of a stress below the elastic limit. A plastics subjected to a load for a period of time tends to deform more than it would from the same load released immediately after application. The degree of the deformation is dependent on the load duration. Creep at room temperature is sometimes called *cold flow.*

Cross-linking—The tying together of adjacent polymer chains.

Crystallization—The process or state of molecular structure in some plastics that denotes uniformity and compactness of the molecular chains forming the polymer; usually attributed to the formation of solid crystals having a definite geometric form.

Curing agents—Chemicals that cause thermosetting plastics to cross-link or cure.

Custom molds—Molds owned by the customer and used by the molder.

Cyclic hydrocarbons—Cyclic or ring compounds. Benzene (C_6H_6) is one of the most important cyclic hydrocarbons.

Damping—Variations in properties resulting from dynamic loading conditions (vibrations). Damping provides a mechanism for dissipating energy without excessive temperature rise, prevents premature brittle fracture, and is important to fatigue performance.

Daylight opening—The clearance between platens when the press is fully opened. The opening must be large enough to allow a part to be ejected when the mold is in a fully opened position.

Debond—Area of separation with or between plies in a laminate, or within a bonded joint, caused by contamination; improper adhesion during processing, or damaging interlaminar stresses.

Deep-hardening—Refers to the depth of hardening possible in a piece of steel.

Degree of polymerization (DP)—Average number of structural units per average molecular mass. In most plastics, the DP must reach several thousand to achieve worthwhile physical properties.

Delamination—Debonding process primarily resulting from unfavorable interlaminar stresses; edge delamination, however, can be effectively prevented by a wrap-around reinforcement.

Denier—The mass in grams of 9 000 m (29 527 ft) of synthetic fiber in the form of a single continuous filament.

Density—Mass per unit volume of a substance, expressed in grams per cubic centimeter, or kilograms per cubic meter.

Depolymerization—A chemical reaction that breaks a polymer down into monomers or other short organic molecules.

Diamines—Compounds containing two amino groups.

Dibasic acid—An acid that has two replaceable hydrogen atoms.

Die—A forming piece used in shaping parts for quantity production.

Die cutting—Blanking or cutting shapes from sheet stock by striking sharply with a steel-rule die, which is a knife edge made in the shape of the item to be cut.

Dielectric strength—A measurement of the voltage required to break down or arc through a plastics material.

Difunctional—See Bifunctional

Dimensional stability—The ability of a plastics part to keep the precise shape in which it was molded, fabricated, or cast.

Dip casting—The process of submerging a hot mold into a resin. After cooling, the product is removed from the mold.

Dip coating—Applying a coating by dipping an item into a tank of melted resin or plastisol, then chilling. The object may be heated and powders used for the coating; the powders melt as they strike the hot object.

Distribution—A collection of values.

Drape forming—Method of forming thermoplastic sheet in a movable frame. The sheet is heated and draped over high points of a male mold. Vacuum is then pulled to complete the forming.

Drawing—The process of stretching a thermoplastic sheet, rod, film, or filaments to reduce cross-sectional area and change physical properties.

Dwell—A pause in the application of pressure to a mold, made just before the mold is completely closed. The pause allows gas to escape from the molding material.

EDM—See Electrical-discharge machining.

Elastic limit—The extent to which a material can be stretched or deformed before taking on a **permanent set**. Permanent set occurs when a material that has been stressed does not recover its original dimensions, as when a 12-inch piece of rubber that has been stretched becomes 13 inches long when relaxed.

Elastomer—A rubberlike substance that can be stretched to several times its original length, and which, on release of the stress, returns rapidly to almost its original length.

Electric dipole—A nonuniform charge distribution, in which one end of a molecule or ion is positive and the other negative.

Electrical-discharge machining (EDM)—A process in which a high-frequency intermittent electric spark is used to erode the workpiece.

Electromagnetic bonding—See Induction bonding.

Electron—A negatively charged particle that is present in every atom.

Electroplating—A method of applying metallic coatings to a substrate.

Electrostatic printing—The deposit of ink on a plastics surface where electrostatic potential is used to attract the dry ink through an open area defined by opaquing.

Elongation percent—The percentage a material elongated. It is usually measured at two points, one when a sample yields, and another when it breaks.

Elutriation—A process that separates contaminants and fines from a stream of chopped plastics materials by controlled up-drafts.

Embedment—Enclosing an object in an envelope of transparent plastics by immersing it in a casting resin allowing the resin to polymerize.

EMI—Electromagnetic Interference; refers to the use of conductive materials in composite to make composite conductive; thus, capable of protect (shielding) electronic devices from unwanted electrical interference (radiation, static electricity, lightning).

EMP—Electromagnetic Pulse; see EMI.

Emulsion polymerization—A process where monomers are polymerized by a water-soluble initiator while dispersed in a concentrated soap solution.

Encapsulating—Enclosing an item, usually an electronic component, in an envelope of plastics by immersing it in a casting resin and allowing the resin to solidify by polymerizing or cooling.

Engineering plastics—Those with good physical and mechanical properties, designed to meet a special need or use.

Engraving—The act of cutting figures, letters, or symbols into a surface. A plastics web is often printed or decorated by interposing a resilient offset roll between an engraved roll and the web.

EPA—U.S. Environmental Protection Agency; federal agency with environmental protection regulatory and enforcement authority. Administers Clean Air Act, Clean Water Act, FIFRA, RCRA, TSCA, other federal environmental laws.

Epoxy—Material based on ethylene oxide, its derivatives or homologs. Epoxy resins form straight-chain thermoplastics and thermosetting resins.

Ester—A compound formed by the replacement of the acidic hydrogen of an organic acid by a hydrocarbon radical; a compound of an organic acid and an alcohol formed with the elimination of water.

Esterification—The process of producing an ester by reaction of an acid with an alcohol and the elimination of water.

Estimating—The act of determining from statistical examples, experience, or other parameters, the cost of a product or service.

Exothermic—Evolving heat during reaction (cure).

Expanded (foamed) plastics—Plastics that are cellular or spongelike.

Expand-in-place—A process in which resin, catalyst, expanding agents and other ingredients are mixed and poured at the site where needed. Expansion takes place at room temperature.

Extrudate—The product delivered by an extruder.

Extrusion—The compacting and forcing of a plastics material through an orifice in more-or-less continuous fashion.

Extrusion coating—The resin coating placed on a substrate by extruding a thin film of molten resin and pressing it into or onto the substrate (or both), without use of adhesives.

FDA—The U.S. Food and Drug Administration; under the provisions of the Federal Food, Drug and Cosmetic Act, the FDA establishes requirements for the labeling of foods and drugs to protect consumers from misbranded, unwholesome, ineffective, and hazardous products. FDA also regulates materials for food contact service and the conditions under which such materials are approved.

Fatigue strength—The highest cyclic stress a material can withstand for a given number of cycles before failure occurs.

Feed—The distance the cutting tool moves into the work with each revolution.

Feedstocks—Refers to sources of raw material from which we make polymers.

Female mold—The indented half of a mold designed to receive the male half.

Fiber orientation—Fiber alignment in a nonwoven or a mat laminate where the majority of fibers are in the same direction, resulting in a higher strength in that direction.

Fibers—This terms usually refers to relatively short lengths of various materials with very small cross sections. Fibers can be made by chopping filaments.

Filament—A fiber characterized by extreme length, with little or no twist. A filament is usually produced without the spinning operation required for fibers.

Filament winding—A composite fabrication process that consists of winding a continuous reinforcing fiber (impregnated with resin) around a rotating and removable form (mandrel).

Filler—An inert substance added to a plastic to make it less costly. Fillers may improve physical properties. The percentage of filler used is usually small in contrast to reinforcements.

Fillet—A rounded filling of the internal angle between two surfaces of a plastic molding.

Fire resistant—A term for a substance that does not easily burn.

Fixture—A device used to support the work during processing or manufacture.

Flame retardant—A material that reduces the ability of a plastics to support combustion.

Flame spraying—Method of applying a plastics coating in which finely powdered plastics and suitable fluxes are projected through a cone of flame onto a surface.

Flammable—A material that burns readily. Flammable liquids have a flash point of 38 °C or above.

Flash—Extra plastics attached to a molding along the parting line. It must be removed to make a finished part.

Flash point—The temperature at which it gives off vapors sufficient to form an ignitable mixture with the air near the surface of the liquid.

Flexible manufacturing systems—A series of machines and associated workstations linked by a hierarchical common control, and providing for automatic production of a family of workpieces. A transportation system, for both the workpiece and the tooling, is as integral to an FMS as is computerized control.

Flexural modulus—The ratio, within the elastic limit, of the applied stress on a test sample in flexure to the matching strain in the outermost fibers of the sample.

Flexural strength (modulus of rupture)—The highest stress in the outer fiber of a sample in flexure at the moment of crack or break. In the case of plastics, is usually higher than tensile strength.

Flotation tank—A tank that promotes the separation of plastics or contaminants based on differing densities.

Flow line—Sometimes called a weld line. A mark on a molded piece made by the meeting of two flow fronts during molding.

Flow mark—Wavy surface appearance caused by improper flow of hot plastics into the mold cavity.

Fluidized bed—A method of coating heated items by immersion in a dense-phase fluidized bed of powdered resin. The objects are usually heated in an oven to provide a smooth coating.

Fluorescence—A property of a substance that causes it to produce light while it is being acted on by radiant energy such as ultraviolet light or X-rays.

Fluoroplastics—A group of plastics materials contain the element fluorene (F).

Fly ash—A type of ash, produced by incinerators, which is carried by the air. It must be removed before smokestack gases are exhausted into the atmosphere.

Foamed—See Cellular.

Foaming agents—Chemicals that generate inert gases on heating or chemical reaction, causing the resin to assume a cellular structure.

Formalin—A commercial 40 percent solution of formaldehyde in water.

Free forming—Air pressure is used to blow a heated sheet of plastics, the edges of which arc being held in a frame, until the desired shape or height is attained.

Frost line—In extrusion, a ring-shaped zone located at the point where the film reaches its final diameter.

Galalith—A plastics made by hardening casein with formaldehyde.

Gamma ray—Electromagnetic radiation originating in an atomic nucleus.

Gate—In injection and transfer molding, the orifice through which the melt enters the cavity. Sometimes the gate has the same cross section as the runner leading to it.

Gaylord—A large rectangular container with a volume of approximately one cubic yard. A gaylord of plastics pellets weighs about 1000 pounds.

Gel coat—A thin layer of resin that serves as the surface of the product. Reinforced layers may then be build up.

Glass transition—The change in amorphous or partially crystalline polymers from a viscous or rubbery condition to a hard and relatively brittle one (from hard to viscous condition).

Glass transition temperature (Tg)—A characteristic temperature at which glassy amorphous polymers become flexible or rubberlike because of the motion of molecular segments.

Glass, Type C—Chemically resistant glass fibers.

Glass, Type E—Electrical grade glass fibers.

Glue—Formerly, an adhesive prepared from animal hides, tendons, and other by-products by heating with water. In general use, the term is now synonymous with the term *adhesive.*

Gradient-density column—A means of conveniently measuring small plastics samples. The column is a glass gradient tube filled with a heterogeneous mixture of two or more liquids. The density of the mixture varies linearly or in other known fashion with the height. The specimen is placed, in the gradient tube and falls to a position of equilibrium that shows its density by comparison with positions of known standard samples.

Graft copolymer—A combination of two or more chains of constitutionally or configurationally different features, one of which serves as a backbone main chain, and at least one of which is bonded at some point(s) along the backbone and constitutes a side chain.

Gravity constant—9.807 meters per second squared is the acceleration caused on earth by gravity.

Gray (Gy)—The unit of measurement of the absorbed dose of ionizing radiation, defined as one joule per kilogram (1 Gy = 1 J/kg).

Gutta percha—A rubberlike product obtained from certain tropical trees.

Gypsum—Crystalline hydrated sulfate of calcium ($CaSO_4 \times 2H_20$), used for making plaster of Paris and Portland cement.

Halogens—The elements fluorine, chlorine, bromine, iodine, and astatine.

Hand layup—Method of positioning successive layers of reinforcement mat or web (which may or may not be preimpregnated with resin) on a mold or by hand. Resin is used to impregnate or coat the reinforcement, followed by curing of the resin to permanently fix the formed shape.

Hardness—The resistance of a material to compression, indentation, and scratching.

Haze—The cloudy or turbid appearance of an otherwise transparent sample, caused by the light scattered from within the sample or from its surfaces.

Heat-transfer decorating—A process in which the image is transferred from the carrier film to the product by stamping with rigid or flexible shapes, using heat and pressure.

Heated-tool bonding—A method of joining plastics by simultaneous application of heat and pressure to areas in contact. Heat may be applied by conduction or dielectrically.

High-pressure laminates—Laminates formed and cured at pressures higher than 7 000 kPa (1 015 psi).

Histogram—A vertical bar graph of the frequency of values in a distribution.

Hobbing—Forming cavities for multiple molds by forcing a hardened-steel shape, called a hob, into soft steel or beryllium-copper cavity blanks.

Hollow ground—A saw blade that has been specially ground so that the cutting teeth are the thickest portion to prevent binding in the kerf.

Homopolymer—A polymer consisting of like monomer structures.

Honeycomb—A manufactured product of metal, paper, or other materials that is resin-impregnated and has been formed into hexagonal-shaped cells. Used as a core material for sandwich or laminated construction.

Hopper dryer—A combination feeding and drying device for extrusion and injection molding of thermoplastics.

Hot-gas welding—A technique of joining thermoplastics materials in which the materials are softened by a jet of hot air from a welding torch and joined together at the softened points. Generally, a thin rod of the same material is used to fill and consolidate the gap.

Hot-leaf stamping—Decorating operation for marking plastics in which a metal leaf or paint is stamped with heated metal dies onto the face of the plastics. Ink compounds can also be used.

Hot melt—A general term referring to thermoplastic synthetic resins composed of 100 percent solids and used as adhesives at temperatures between 248 and 392 °F [120 °C and 200 °C].

Hydraulics—Referring to the branch of science that deals with liquids in motion; the transmission, control, or flow of energy by liquids.

Hydrocarbon—An organic compound containing only carbon and hydrogen and often occurring in petroleum, natural gas, coal, and bitumens.

Hydrocarbon plastics—Plastics based on resins made by polymerizing monomers composed solely of carbon and hydrogen.

Hydrolysis—A type of depolymerization that yields monomers by chemically attacking polymers.

Hygroscopic—Tending to absorb and retain moisture.

Impact modifier—A material that improves the impact resistance of plastics. Various types of rubber and elastomers are common.
Impact strength—The ability of a material to withstand shock loading.
Impingement—A method of mixing in which two or more materials collide.
Impregnate—To provide liquid penetration into a porous or fibrous material; the dipping or immersion of a fibrous substrate into a liquid resin. Generally, the porous material serves as a reinforcement for the plastic binder after curing.
Impregnation—The process of thoroughly soaking a material such as wood, paper, or fabric with synthetic resin, so that the resin gets within the body of the material.
Incineration—Burning of waste in a specially designed enclosed chamber.
Index of refraction (refractive index)—The ratio of the velocity of light in a vacuum to its velocity in a transparent sample. It is expressed as the ratio of the sine of the angle of incidence to the sine of the angle of refraction. The index of refraction of a substance usually varies with the wavelength of the refracted light.
Induction bonding—High-frequency electromagnetic fields are used to excite the molecules of metallic inserts placed in the plastics or in the interfaces, thus fusing the plastics. The inserts remain in the joint.
Industrial plastics—A plastics waste generated by various industrial sectors.
Inert (rare) gases—Gases that do not combine with other elements; helium, argon, neon, krypton, xenon, and radon (Group 8 of the periodic table).
Inhalation—Breathing a material into the lungs. Inhalation is the major route of exposure to toxic materials in the plastics processing industries.
Inhibitor—A substance that slows down a chemical reaction. Inhibitors sometimes used in certain monomers and resins to prolong storage life.
Initiation phase—The first of three steps in addition polymerization. Refers to producing a reactive state of the molecules, usually by some high-energy source catalysts, or radiation.
Initiator—An agent necessary to cause polymerization, especially in emulsion polymerization processes.
Injection blow molding—A blow-molding process in which the parison to be blown is formed by injection molding.
Injection molding—A molding procedure in which a heat-softened plastics is forced from a cylinder into a relatively cool cavity which gives the item the desired shape.
In-mold decorating—Making decorations or patterns on molded products by placing the pattern or image in the mold cavity before the actual molding cycle. The pattern becomes part of the plastics item as it is fused by heat and pressure.
In-situ foaming—The technique of depositing a foamable plastics into the place where foaming will take place.
Inspection—A term used to indicate that during manufacture of a part, personnel will conduct visual examinations of materials, placement of plies, gauge readings, etc.
Interference—The negative allowance used to assure a tight shrink or press fit.
Interlaminar shear—The shear strength at rupture in which the plane of fracture is located between the layers of reinforcement of a laminate.
Intermolecular forces—Secondary valence, or van der Waals, forces between different molecules.
Interpenetrating polymner network (IPN)—An entangled combination of two cross-linked polymers that are not bonded to each other.
Ion—An atom or group of atoms with a positive or negative electrical charge.
Ionic bonding—Atomic bonding by electrical attraction of unlike ions.
Ionomer—A polymer with ethylene as its major component, but containing both covalent and ionic bonds. The polymer exhibits very strong interchain ionic forces. These resins have high transparency, resilience, tenacity, and many of the characteristics of polyethylene.
IR—Infra-red; Infra-red analysis is a technique which uses infra-red light to identify the chemical make-up of plastics samples.
Irradiation—As applied to plastics, refers to bombardment with a variety of kinds of ionizing and nonionizing radiation. Irradiation has been used to begin polymerization and copolymerization of plastics, and in some cases, to bring about changes in the physical properties of a plastics.
Isocyanate resins—Resins synthesized from isocyanates and alcohols. Most uses are based on their combination with polyols.
Isomers—Molecules with the same chemical composition but different structures.
Isotatic stereoisomerism—A sequence of regularly spaced asymmetric atoms arranged in like configuration in a polymer chain.
Isotope—One of a group of nuclides having the same atomic number but differing atomic mass.
Isotropic—Properties of a material are equal in all directions.

Jig—An appliance for accurately guiding and locating tools during the making of interchangeable parts.

Kerf—The slit or notch made by a saw or cutting tool.

Kirksite—An alloy of aluminum and zinc used for molds. It has high thermal conductivity.

Knife coating—A method of coating a substrate by an adjustable knife or bar set at a suitable angle to the substrate.

Lac—A dark-red resinous substance deposited by scale insects on the twigs of trees; used in making shellac.

Lamellar—Sheetlike or platelike in shape. Referring to the aligned, looped molecular structure of crystalline polymers.

Laminar composite—Referring to a composite composed of layers of materials held together by the polymer matrix. They are divided into two classes: laminates and sandwiches.

Laminated plastics—A dense, tough solid produced by bonding together layers of sheet materials impregnated with a resin and curing them by application of heat, or heat and pressure.

Laminated vapor plating—The process of vacuum metalizing alternate layers of metal coatings on a polymer substrate.

Laminates—Two or more layers of material bonded together. The term usually applies to preformed layers joined by adhesives or by heat and pressure. The term also applies to composites of plastics films with other films, foil, and paper, even though they have been made by spread coating or by extrusion coating. A reinforced laminate usually refers to superimposed layers of resin-impregnated or resin-coated fabrics or fibrous reinforcements that have been bonded, especially by heat and pressure. When the bonding pressure is at least 7000 kPa [1015 psi) the product is called a high-pressure laminate. Products pressed at pressures under 7000 kPa are called low-pressure laminates. Products produced with little or no pressure, such as hand lay-ups, filament-wound structures, and spray-ups, are sometimes called contact-pressure laminates.

Laminating—The process of producing a composite laminate.

Lamination—The process of preparing a laminate; also, any layer in a laminate.

Laser cutting—A means of cutting materials by laser energy.

Latex—An emulsion of natural or synthetic resin particles dispersed in a watery medium.

LD—Lethal dose; a concentration of a substance being tested which will kill a test animal.

Leaching—Removing a soluble component from a polymer mix with solvents.

Limits—The maximum and minimum dimensions that define the tolerance.

Linear—Refers to a long straight-chain molecule, as contrasted with one having many side chains or branches.

Liquid Reaction Molding (LRM)—Same as reaction injection molding.

Low-pressure laminates—In general, laminates molded and cured at pressures ranging from 0.4 ksi (2.8 MPa) to contact pressure.

Lubrication bloom—An irregular, cloudy, greasy film on a plastics surface caused by excess lubricants.

Luminescence—Light emission by the radiation of photons after initial activation. Luminescent pigments are activated by ultraviolet radiation, producing very strong luminescence.

Macromolecules—The large (giant) molecules that make up the high polymers.

MM/RIM—Mat molding reaction-injection molding.

Mandrel—A form around which filament-wound and pultruded composite structures are shaped.

Mass—The quantity of matter; the physical amount of matter. When gravity acts on a mass of matter we say it has weight. See also weight.

Matched-mold forming—Forming hot sheets between matched male and female molds.

Matrix—The polymer material used to bind the reinforcements together in a composite.

Mean—The arithmetic average of the values in a distribution.

Mechanical fastening—Mechanical means of joining plastics with machine screws, self-tapping screws, drive screws, rivets, spring clips, clips, dowels, catches, or other devices.

Mechanical forming—Heated sheets of plastics are shaped or formed by hand or with the aid of jigs and fixtures. No mold is used.

Melt index—Also called MFR or melt flow rate; the amount of material in grams that is extruded through an orifice in 10 minutes under specified conditions.

Mer—The smallest repetitive unit in a polymer.

Methyl methacrylate—A colorless, volatile liquid derived from acetone cyanohydrin, methanol, and dilute sulphuric acid and used in production of acrylic resins.

m/s—metres per second.

Microballoons—Hollow glass spheres.

Mold—The cavity or matrix in which plastics are formed. Also, to shape plastics or resins into finished items by heat or heat and pressure.

Molding compounds—Plastics or resin materials in varying stages of formulation (powder, granular, or preform) comprising resin, filler, pigments, plasticizers, or other ingredients ready for use in the molding operation.

Molecular mass—The sum of the atomic mass of all atoms in a molecule. In high polymers, the molecular masses of individual molecules vary widely, so that they must be expressed as averages. Average molecular mass of polymers may be expressed as number-average molecular mass (Mn) or mass-average molecular weight (Mw). Molecular mass measurement methods include osmotic pressure, light scattering, solution pressure, solution viscosity, and sedimentation equilibrium.

Molecule—The smallest particle of a substance that can exist independently while retaining the chemical identity of the substance.

Monohydric—Containing one hydroxyl (OH) group in the molecule.

Monomers—A simple molecule capable of reacting with like or unlike molecules to form a polymer; the smallest repeating structure of a polymer, also called a mer.

MRF—Materials recovery facility; a facility for gathering, sorting, and baling recycled materials.

MSDS—Material Safety Data Sheet; a source of information about the health hazards caused by industrial chemicals.

MSW—Municipal solid waste; this term is used to describe the waste materials that are collected from homes and industries. MSW goes into landfills, unless recycling programs recover useful materials out of the waste stream.

n—An abbreviation that refers to the number of values in a set.

Negative rake—The angle of the face of a cutting tool ground so that the cutting end of a tool with positive rake is more blunt than that of a tool with no rake (rake angle equals zero).

NFPA—National Fire Prevention Association.

NIOSH—National Institute for Occupational Safety and Health of the Public Health Service, U.S. Department of Health and Human Services (DHHS); federal agency that recommends occupational exposure limits for various substances and assists OSHA and MSHA in occupational safety and health investigations and research.

Nitrocellulose (cellulose nitrate)—Material formed by the action of a mixture of sulphuric acid and nitric acid on cellulose. The cellulose nitrate used for Celluloid manufacture usually contains 10.8 to 11.1 percent nitrogen.

Normal distribution—A symmetrical distribution that has central tendency.

Novolac—A phenolic-aldehyde resin that remains permanently thermoplastic unless a source of methylene groups is added.

Nuisance plastics—Waste plastics that cannot he reprocessed under the existing technoeconomic conditions.

Number average molecular weight (M_n)—A type of molecular weight average based on the frequency of various length molecules in a distribution.

OSHA—Occupational Safety and Health Administration of the U.S. Department of Labor; federal agency with safety and health regulatory and enforcement authorities for most U.S. industry and business.

Offset printing—A printing technique in which ink is transferred from the printing plate to a roller. Subsequently, the roller transfers the ink to the object to be printed.

Offset yield—For materials that do not exhibit a clear yield point, an offset yield is calculated. It provides a yield point at a determined location on the stress/strain curve.

Oil-hardened steel—Steel that is cooled by an oil bath.

Open-celled—Referring to the interconnecting of cells in cellular or foamed plastics.

Open dumping—Placing trash or waste in open, uncontrolled areas on the land.

Open-grit sandpaper—Coarse sandpaper (number 80 or less).

Organosol—A dispersion usually of vinyl or polyamide, in a liquid phase containing one or more organic solvents.

Orientation—Plastics molecules can be oriented in one direction (uniaxially) or in two directions (biaxially.) Orientation is caused by flow or stretching and alters the physical properties of the material.

Oxygen index—A test for the minimum oxygen concentration in a mixture of oxygen and nitrogen that will, support a flame of a burning polymer.

Parameters—A term used loosely to denote a specified range of variables, characteristics, or properties relating to the subject being discussed; also, an arbitrary constant.

Parison—The hollow plastics tube from which a product is blow-molded.

Parkesine—An early plastics made by Alexander Parkes from collodion.

Particulate—Small particles with various shapes and sizes used to reinforce a polymer matrix.

Parting lines—Marks on a molding or casting where halves of the mold meet in closing.

PCR—Post consumer recycled; this refers to materials recycled from the post consumer waste stream. Containers for food and drinks fit in this category.

PEL—Permissible exposure limit; a measure used by OSHA to define the exposure level acceptable for an 8-hour day as part of a 40-hour week.

Pellets—One of the many formulations of molding compounds.

Periodic table—An arrangement of the elements in order of increasing atomic number, forming groups the members of which show similar physical and chemical properties.

Phenolic—A synthetic resin produced by the condensation of an aromatic alcohol with an aldehyde, particularly of phenol with formaldehyde.

Phenoxy—A high-molecular-mass thermoplastic polyester resin based on bisphenol A and epichlorohydrin.

Phosphorescence—Luminescence that lasts for a period after excitation.

Photodegradable—Materials that decompose due to the action of sunlight.

Photon—The least amount of electromagnetic energy that can exist at a given wavelength. A quantum of light energy is analogous to the electron.

Photosynthesis—Refers to the synthesis of chemicals with the aid of radiant energy from the light of the sun.

Picking line—A type of sorting technology in which operators select types of materials off a moving conveyor belt.

Pinch-off—A raised edge around the cavity in the mold that seals off the part and separates the excess material as the mold closes around the parison.

Plastic—An adjective, meaning pliable and capable of being shaped by pressure. Plastic is often incorrectly used as the generic word for the plastics industry and its products.

Plasticate—To make plastic. The plasticating capacity of an injection-molding machine is the maximum weight of material (PS) that it can prepare for injection in one hour.

Plasticizer—Chemical agent added to plastics to make them softer and more flexible.

Plastics—A noun; an organic substance, usually synthetic or semisynthetic, that can be formed into various shapes by heat and pressure and retain those shapes after heat and pressure have been removed. In its finished state, it is a rigid or flexible (but not elastic) solid containing a polymer of high molecular mass (weight).

Plastics alloy—Plastics made by physically mixing two or more polymers during the melt.

Plastics tooling—Tools, dies, jigs, or fixtures primarily for the metal-forming trades, constructed of plastics. (Usually laminates or casting materials).

Plastic strain—The strain permanently given to a material by stresses that exceed the elastic limit.

Platens—The mounting plates of a press to which the mode assembly is bolted.

Poise—A unit for measurements of viscosity.

Polar winding—A winding in which the filament path passes tangent to the polar opening at the other end. A one-circuit pattern is inherent in the system.

Polyacrylate—A thermoplastic resin made by the polymerization of an acrylic compound.

Polyallomer—Crystalline polymers produced from two or more olefin monomers.

Polyamide—A polymer in which the structural units are linked by amide or thioamide groupings.

Polyblend—Plastics that have been modified by the addition of an elastomer.

Polycarbonate—Polymers derived from the direct reaction between aromatic and aliphatic dihydroxy compounds with phosgene or by the ester exchange reaction with appropriate phosgene-derived precursors.

Polydispersity index—The ratio of the weight average molecular weight to the number average molecular weight.

Polyester—A resin formed by the reaction between a dibasic acid and a dihydroxy alcohol, both organic. Modification with multifunctional acids or acids and bases and some unsaturated reactants permits cross-linking to thermosetting resins. Polyesters modified with fatty acids are called alkyds.

Polyethylene—A thermoplastic material composed of polymers of ethylene. One of polyolefin family.

Polyimide—A group of resins made by reacting pyromellitic dianhydride with aromatic diamines. The polymer is characterized by rings of four carbon atoms tightly bound together.

Polymer—A compound with high molecular mass (weight), either natural or synthetic, whose structure can be represented by a repeated small unit (the mer). Some polymers are elastic and some are plastics.

Polymerization—The process of growing large molecules from small ones.

Polymerization Reaction—A chemical reaction in which the molecules of a monomer are linked together to form large molecules.

Polymethyl methacrylate—See Methyl methacrylate.

Polymethylpentene—An isotactically arranged aliphatic polyolefin of 4-methyl-pentene-1.

Polyolefin—A term used to indicate a family of polymers produced from hydrocarbons with double carbon-to-carbon bonds. Includes polyethylene, polypropylene, polymethylpentene.

Polyphenylene oxide—Currently made as a polyether of 2,6-dimethyl-phenol by an oxidative coupling process involving air or pure oxygen in the presence of a copper-arnine complex catalyst.

Polypropylene—A plastics material made by the polymerization of high-purity propylene gas in the presence of an organometallic catalyst at relatively low pressures and temperatures. One of the polyolefin family.

Polystyrene—A thermoplastic material produced by the polymerization of styrene (vinyl benzene).

Polysulfone—A thermoplastic consisting of benzene rings connected by a sulfone group (SO_2), an isopropylidene group, and an ether linkage.

Polyurethane—A family of resins produced by reacting diisocyanate with organic compounds containing two or more active hydrogens to form polymers having free isocyanate groups. These groups, under the influence of heat or certain catalysts, will react with each other, or with water, glycols, or other materials to form a thermoset.

Polyvinyls—A broad family of plastics derived from the vinyl group ($CH_2{=}H$—).

Pooled standard deviation—A standard deviation based on two or more standard deviations. It is used in graphic representations of distributions.

Pneumatics—A branch of science dealing with the mechanical properties of gases.

Postconsumer plastics waste—A plastics waste generated by a consumer.

Potting—An embedding process for parts, similar to encapsulating except that the object may be simply covered and not surrounded by an envelope of plastics. Normally considered a coating process.

ppm—Parts per million; a way of expressing tiny concentrations. In air, ppm is usually a volume/volume ratio; in water, a weight/volume ratio.

Pre-expanded—When polymer beads or granules are partially expanded prior to molding into cellular parts.

Preplasticator—The act of softening material before forcing it into the mold, another molding machine, or accumulator.

Preplasticizer—The act of adding a softening agent (plasticizer) before molding.

Prepuffs—The pre-expanded pieces of polymers used to make cellular polymer parts.

Pressure-bag molding—A process for molding reinforced plastics, in which a tailored flexible bag is placed over the contact layup on the mold, sealed, and clamped in place. Compressed air forces the bag against the part to apply pressure while the part cures.

Primary bonds—A strong association (interatomic attraction) between atoms.

Primary colors—The basic colors from which all others are made.

Primary recycling—The processing of scrap plastics into the same or similar types of product from which it has been generated, using standard plastics processing methods.

Promoter—A chemical, itself a feeble catalyst, that greatly speeds up the activity of a given catalyst.

Propagation phase—The second step in addition polymerization. Refers to rapid growth or addition of monomer units to the molecular chain.

Proportional limit—The greatest stress that a material can sustain without deviation from proportionality of stress and strain (Hooke's law); the point at which elastic strain becomes plastic strain. It is expressed in force per unit area (Pa).

Proprietary molds—Molds made and, owned by the molder.

Pultrusion—A continuous process for manufacturing composites with a constant cross-sectional shape. The process consists of pulling a fiber reinforcing material through a resin-impregnation bath and into a shaping die where the resin is subsequently cured.

Purging—Cleaning one color or type of material from the cylinder of a molding machine.

Pyrometer—A device used to measure thermal radiation.

Pyrolysis—Chemical decomposition of a substance by heat, and pressure used to change waste into usable compounds.

Pyroxylin—A moderately nitrated form of cellulose. It was used extensively in early photographic processes.

Quality control—A procedure to determine if a product is being manufactured to specifications; a technique of management for achieving quality. Inspection is part of that technique.

Quarternary recycling—The recovery of energy from waste plastics.

Radiation—See Irradiation.

Radical—A group of atoms of different elements that behave as a single atom in chemical reactions.

Random copolymer—A type of copolymer in which the two types of monomers have random arrangement along the length of the molecular chain.

Rare gases—See inert gases.

RCRA—Resource Conservation and Recovery Act; a federal act passed in 1976 which promoted reuse, reduction, incineration, and recycling of materials.

Reaction injection molding (RIM)—The molding process in which two or more liquid polymers are mixed by impingement-atomizing in a mixing chamber, then injected into a closed mold.

Recycling—Collection and reprocessing of waste materials.

Reinforced molding compound—A material reinforced with special fillers, fibers, or other materials to meet design needs.

Reinforced plastics—Plastics with strength increased by addition of filler and reinforcing fibers, fabrics, or mats to the base resin.

REL—Recommended exposure limit; A measure of acceptable exposure to hazardous chemicals.

Relative density—The density of any material divided by that of water at a standard temperature, usually 68 or 73 °F [20 or 23 °C]. Since water's density is nearly 1.00 g/cm^3, density in grams per cubic centimeter and relative density are numerically equal.

Rennin—An enzyme of gastric juice that causes the coagulation of milk.

Resin—Gum-like solid or semisolid substances that may be obtained from certain plants and trees or made from synthetic materials.

Resin-rich area—Localized area filled with resin and lacking reinforcing material.

Resin transfer molding (RTM)—The transfer of catalyzed resin into an enclosed mold in which the fiber reinforcing has been placed. Also called *resin-injection molding* and *liquid resin molding (LRM)*.

rpm—revolutions per minute.

r/s—revolutions per second.

Rib—A reinforcing member of a fabricated or molded part.

Rotational casting—A method used to make hollow objects from plastisols or powders. The mold is charged and rotated in one or more planes. The hot mold fuses the substance into a gel during rotation, covering all surfaces. The mold is then chilled and the product removed.

Roving—A bundle of untwisted strands, usually of fibrous glass.

RRIM—Reinforced reaction injection molding; A process that combines reaction injection molding with fibrous reinforcements, usually glass mats.

Runners—Channels through which plastics flow from the sprue to the gates of mold cavities.

SAE code J1344—A code for marking plastics to aid in their identification. This code should assist the recycling of automotive plastics.

Safety factor—The ratio of the ultimate strength of the material to the allowable working stress.

Sandwich—A class of laminar composites composed of a lightweight core material (honeycomb, foamed plastics, etc.) to which two thin, dense, high-strength faces or skins are adhered.

Sandwich construction—A structure consisting of relatively dense, high-strength facings bonded to a less dense, lower-strength intermediate material or core.

Sandwich panel—Panel consisting of two thin-face sheets bonded to a thick, lightweight, honeycomb or foam core.

Sanitary landfill—The controlled filling of lowlands or trenches with solid waste.

Saturated compounds—Organic compounds that do not contain double or triple bonds and thus cannot add elements or compounds.

Scleroscope—An instrument for measuring impact resilience by dropping a ram with a flattened-cone tip from a given height onto the sample and then noting the height of rebound.

Scrap plastics—Waste plastics that are capable of being reprocessed into commercially acceptable plastics products.

Secondary bonds—Forces of attraction, other than primary bonds, that cause many molecules to join.

Secondary color—A color obtained by mixing two or more primary colors.

Secondary recycling—The processing of scrap plastics into plastics products with less demanding properties.

Self-extinguishing—A somewhat loosely used term describing the ability of a material to cease burning once the source of flame has been removed.

Shell—A region about the nucleus of an atom in which electrons move; each electron shell corresponds to a definite energy level.

Shellac—A natural polymer; refined lac, a resin usually produced in thin, flaky layers or shells and used in varnish and insulating materials.

Shrink fit—A joining method in which an insert is put into a plastics part while that part is hot. Shrink fitting takes advantage of the fact that plastics expand when heated and shrink when cooled. The plastics is normally heated and the insert placed in an undersized hole. On cooling, the plastics shrinks around the insert.

Shrink wrapping—A technique of packaging in which the strains in a plastics film are released by raising the temperature of the film. This causes it to shrink over the package.

Silicosis—A disease of the lungs caused by the inhalation of silica dust.

Single covalent bond—The type of chemical bond found most frequently in plastics. It allows for rotation, thus permitting twisting and folding of molecular chains.

Sintering—Forming items from fusible powders. The process of holding the pressed powder at a temperature just below its melting point.

Slush casting—Resin in liquid or powder form is poured into a hot mold, where a viscous skin forms. The excess slush is drained off, the mold is cooled, and the casting removed.

Snap-back forming—A technique in which a plastics sheet is stretched to a bubble shape by vacuum or air pressure, a male mold is inserted into a bubble and the vacuum or air pressure is released allowing the plastics to snap-back over the mold.

Solid waste—Refuse that does not rot or decay, such as dirt, concrete, bricks, and many plastics.

Solubility parameter—A measure of the reactivity of plastics and organic solvents. A plastics with a lower solubility parameter than that of a selected solvent should dissolve in that solvent.

Solution polymerization—A process where inert solvents are used to cause monomer solutions to polymerize.

Solvent—A substance, usually a liquid, in which other substances are dissolved. The most common solvent is water.

Solvent resistance—The ability of a plastics material to withstand exposure to a solvent.

Specification—A statement of a set of requirements to be satisfied by a product, material, process, or system indicating (when appropriate) the procedure by which it may be determined whether the requirements are satisfied. Specifications may cite standards, be expressed in numerical terms, and include contractual agreements or requirements between the buyer and seller.

Specular gloss—The relative luminous reflectance factor of a plastics sample.

Spherulite—A rounded aggregate of radiating crystals with fibrous appearance. Spherulites are present in most crystalline plastics and may range in diameter from a few tenths of a micron to several millimeters.

Spinneret—A type of extrusion die with many tiny holes. A plastics melt is forced through the holes to make fine fibers and filaments.

Spin bonding or welding—A process of fusing two objects by forcing them together while one or both are spinning, until frictional heat melts the interface. Spinning is then stopped and pressure held until the parts are frozen together.

Sprayup—A general term covering several processes using a spray gun. In reinforced plastics, the term applies to the simultaneous spraying of resin and chopped reinforcing fibers onto the mold or mandrel.

Sprue—In the mold, the channel or channels through which the plastics is led to the mold cavity.

Stabilizer—An ingredient used in the formulation of some plastics (especially elastomers) to assist in holding the physical and chemical properties of the compounded materials at their initial values throughout the processing and service life of the material.

Standard—A document or an object for physical comparison to define nomenclature, concepts, processes, materials, dimensions, relationships, interfaces, or test methods.

Standard deviation—A measure of the *spread* of a distribution.

Standard normal distribution—A distribution with a mean of zero and a standard deviation of 1.0.

STEL—Short term exposure limit; a level of exposure to hazardous chemicals for 15 minutes.

Stepwise polymerization—A type of polymerization in which two mers combine to form chains two mers long. Then the two mer pieces combine to form pieces four mers long. The reaction continues in this manner until completed.

Stereoisomerism—The arrangement of molecular chains in a polymer. *Atactic* pertains to an arrangement that is more or less random. *Isotactic* pertains to a structure containing a sequence of regularly spaced asymmetric atoms arranged in like configuration in a polymer chain. *Syndiotactic* pertains to a polymer molecule in which groups of atoms that are not part of the primary backbone structure alternate regularly on opposite sides of the chain.

Stiffness—The capacity of a material to resist a bending force.

Strain—The ratio of the elongation to the gauge length of a test sample; that is, the change in length per unit of original length.

Strand—A bundle of filaments.

Strength-to-mass ratio—Pertaining to materials that are strong for their mass (weight). The strength/density value of a material.

Stress—The force producing, or tending to produce, deformation of a substance. Expressed as the ratio of applied load to the original cross-sectional area.

Structural foam—Cellular plastics with integral skin.

Substrate—A material onto which an adhesive or similar substance is applied.

Suspension polymerization—A process in which liquid monomers are polymerized as liquid droplets suspended in water.

Symmetrical distribution—A distribution that has similar frequencies of values above and below the mean.

Syndiotatic stereoisomer—A polymer molecule in which atoms that are not part of the primary structure alternate regularly on opposite sides of the chain.

Syntactic foam—Cellular resins or plastics with lowdensity fillers.

Synthetic—Materials produced by chemical means, rather than natural origin.

Tampo-Print—A process of transferring ink from an engraved ink-filled surface to a product surface by the use of a flexible printing (transfer) pad.

Technology—The science of efficient application of scientific knowledge.

Termination phase—The last of three steps in addition polymerization. Refers to ending molecular growth of polymers by adding chemicals.

Tertiary recycling—The recovery of chemicals from waste plastics.

Testing—A term that implies that methods or procedures are used to determine physical, mechanical, chemical, optical, electrical, or other properties of a part.

Tex—An ISO standard unit of linear density used as a measure of yarn count. One tex is the linear density of a fabric that has a mass of 1 g and a length of 1 km and is equal to 10^{-6} kg/m.

Thermal Expansion Resin Transfer Molding (TERTM)—A variation of the RTM process in which, after the resin is injected, heat causes the cellular core material to expand forcing reinforcements and matrix against the mold walls.

Thermoforming—Any process of forming thermoplastic sheet that consists of heating the sheet and pulling it down onto a mold surface.

Thermoplastic—(adj.) Capable of being repeatedly softened by heat and hardened by cooling. (n) A linear polymer that will repeatedly soften when heated and harden when cooled.

Thermoset or *Thermosetting*—A network polymer that will undergo or has undergone a chemical reaction by the action of heat, catalysts, ultraviolet light, etc., leading to a relatively infusible state.

Thixotropy—State of materials that are gel-like at rest but fluid when agitated. Liquids containing suspended solids are apt to be thixotropic.

TLV—Threshold limit value; a term used by ACGIH express thee airborne concentration of a material to which nearly all persons can be exposed day after day, without adverse effects. ACGIH exprsses TLV's in three ways: 1) *TLV-TWA*: the allowable time weighted average concentration for a normal 8-hour work day or 40-hour work week. 2) *TLV-STEL:* the short-term exposure limit, or maximum concentration for a continuous 15-minute exposure period (maximum of four such periods per day, with at least 60 minutes between exposure periods, and provided that the daily *TLV-TWA* is not exceeded). 3) *TLV-C:* the ceiling limit—the concentration that should not be exceeded even instantaneously.

Tools—Special fixtures, molds, dies, or other devices that enable a manufacturer to produce parts.

Tool steels—Steels used to make cutting tools and dies. Many of these steels have considerable quantities of alloying elements such as chromium, carbon, tungsten, molybdenum, and other elements. They form hard carbides that provide good wearing qualities but at the same time decrease machinability. Tool steels in the trade are classified for the most part by their applications, such as hot die, cold work die, high speed, shock resisting, mold, and special purpose steels.

Toughness—A term with a wide variety of meanings, no single mechanical definition being generally recognized. Represented by the energy required to break a material, equal to the area under the stress-strain curve.

Toxicity—The degree of danger posed by a substance to animal or plant life.

Toxic Substance—Any substance which can cause acute or chronic injury to the human body, or which is suspected of being able to cause diseases or injury under some conditions.

TPE—Thermoplastic elastomers; a group of materials that can be processed like plastics but have physical characteristics similar to rubber.

Trade name—A name given to a product to make it easy to recognize, spell, and pronounce. In the plastics industry, a trade name is used by the manufacturer to identify a particular resin or product.

Transfer molding—A method of molding plastics where the material is softened by heat and pressure in a transfer chamber, then forced by high pressure through the sprues, runners, and gates in a closed mold for final curing.

Tripoli—A silica abrasive.

Tumbling—Finishing operation for small plastics articles. Gates, flash, and fins are removed and surfaces are polished by rotating the parts in a barrel or on a belt together with wooden pegs, sawdust, and some polishing compounds.

Ultrasonic bonding—A method of joining using vibratory mechanical pressure at ultrasonic frequencies. Electrical energy is changed to ultrasonic vibrations through the use of either a magetostrictive or piezoelectric transducer. The ultrasonic vibrations generate frictional heat to melt the plastics allowing them to join.

Undercut—Having a protuberance or indention that impedes withdrawal from a two-piece rigid mold. Flexible materials can be ejected intact with slight undercuts.

USDA—U.S. Department of Agriculture; prior to 1971, USDA performed tests and issued approvals on respirators for use with pesticides. In 1971, the Bureau of Mines took over the pesticide respirator testing/approval functions/procedures later delegated to the Testing and Certification Branch (TCB) of NIOSH.

Vacuum forming—Method of sheet forming in which the edges of the plastics sheet are clamped in a stationary frame, and the plastics is heated and drawn down by a vacuum into a mold.

Vacuum metalizing—A process in which surfaces are thinly coated by exposing them to a metal vapor under vacuum.

Valence electrons—The electrons in the outermost shell of an atom.

Van der Waals' forces—Weak secondary interatomic attraction arising from internal dipole effects.

Vibrational microlamination (VIM)—A casting process where heated molds are vibrated in a bed of polymer pellets or powder.

Vicat softening point—The temperature at which a flat-ended needle of 1 mm^2 circular or square cross section will penetrate a thermoplastic specimen to a depth of 1 mm under a specified load using a uniform rate of temperature rise (Definition from ASTM D1525.)

Viscosity—A measure of the internal friction resulting when one layer of fluid is caused to move in relationship to another layer.

Vulcanize—The process of toughening natural rubber by compounding it with powdered sulfur.

Warp—The lengthwise direction of the weave in cloth or roving; also the dimensional distortion of a plastic object. See Weft.

Waste plastics—A plastics resin or product that must be reprocessed or disposed of.

Water pollution—Waste products discharged into rivers and waterways.

Weft—The transverse threads or fibers in a woven fabric; those fibers running perpendicular to the warp; also called *fill, filler, yarn, woof, pick.*

Weight—Weight is the force exerted by a mass as a result of gravity. In common usage, weight is used as a synonym for mass. The word "weight" should be avoided in technical practice, and replaced by gravitational force acting on the object, measured in newtons.

Weight average molecular weight—A type of molecular weight based on the contribution of each weight fraction to the total molecular weight of a sample.

Whiskers—Single crystals used as reinforcements.

Whiting—Calcium carbonate powder abrasive.

WTE—Waste-to-energy; a term to describe facilities that use municipal solid wastes as fuel in incinerators that produce electricity and steam.

$\bar{x}$—A symbol for the mean of a distribution. It is pronounced x-bar.

Yarn—Bundle of twisted strands.

Appendix B

Abbreviations for Selected Materials

Abbreviation	Polymer Term or Generic Name
ABS	Acrylonitrile-butadiene-styrene
ACS	Acrylonitrile-chlorinated polyethylene-styrene
AES	Acrylonitrile-ethylpropylene-styrene
AI	Amide-imide polymers
AMMA	Acrylonitrile-methyl-methacrylate
AN	Acrylonitrile
AP	Ethylene propylene
ASA	Acrylic-styrene-acrylonitrile
AU	Polyester polyurethane
BBP	Butyl benzyl phthalate
BFK	Boron fiber reinforced plastic
BMC	Bulk molding compounds
CA	Cellulose acetate
CAB	Cellulose acetate-butyrate
CAP	Cellulose acetate propionate
CAR	Carbon fiber
CF	Cresol-formaldehyde
CFRP	Carbon fiber reinforced plastics
CMC	Carboxymethyl cellulose
CN	Cellulose nitrate
CP	Cellulose propionate
CPE	Chlorinated polyethylene
CPET	Crystallized PET
CPVC	Chlorinated polyvinyl chloride
CS	Casein
CTFE	(Poly) Chlorotrifluoro-ethylene
DAIP	Diallyl isophthalate resin
DAP	Diallyl phthalate resin
DCHP	Dicyclohexyl phthalate
DCPD	(Poly) Dicyclopentadiene
DGEBA	Diglycidyl ether of bisphenol A (epoxy)
DMAL	Dimethylacetamide
DMC	Dough molding compound
DOPT	Di-2-ethylhexyl terephthalate
DP	Degree of polymerization

Abbreviation	Polymer Term or Generic Name
EC	Ethyl cellulose
ECTFE	Ethylene-chlorotrifluoroethylene
EEA	Ethylene-ethyl acrylate
EMA	Ethylene-methyl acrylate
EP	Epoxy
EPDM	Ethylene propylene diene rubber
EPE	Epoxy resin ester
EP-G-G	Prepreg of epoxy resin and glass fabric
EPM	Ethylene propylene copolymer
EPR	Ethylene and propylene copolymer
EPS	Expanded polystyrene
ETFE	Ethylene tetrafluoroethylene
EU	Polyether polyurethane
EVA	Ethylene vinyl acetate
EVOH	Ethylene vinyl alcohol
FEP	Fluorinated ethylene-propylene (Also PFEP)
FRP	Glass fiber reinforced polyester
FRTP	Fiberglass reinforced thermoplastics
GF	Glass fiber reinforced
GF-EP	Glass fiber reinforced epoxy resin
GR	Glass fiber reinforced
GRP	Glass reinforced plastics
HIPS	High-impact polystyrene
HMW-HDPE	High molecular weight-high density polyethylene
HVBME	High-vinyl modified epoxy
IPN	Interpenetrating polymer network
LCP	Liquid crystal polymers
LDPE	Low-density polyethylene
LIM	Liquid impingement molding
LLDPE	Linear low-density polyethylene
LRM	Liquid reaction molding
MA	Maleic anhydride
MBS	Methacrylate-butadiene-styrene

Abbreviation	Polymer Term or Generic Name
MDI	Methylene diisocyanate
MEKP	Methyl ethyl ketone peroxide
MF	Melamine formaldehyde
NBR	Acrylonitrile-butadiene rubber
OPP	Oriented polypropylene
OPVC	Oriented polyvinylchloride
OSA	Olefin-modified styrene-acrylonitrile
PA	Polyamide
PAA	Polyacrylic acid
PAI	Polyamide-imide
PAN	Polyacrylonitrile
PAPI	Polyrnethylene polyphenyl isocyanate
PB	Polybutylene
PBAN	Polybutadiene-acrylonitrile
PBS	Polybutadiene-styrene
PBT	Polybutylene terephthalate
PC	Polycarbanate
PCTFE	Polymonochlorotrifluoroethylene
PDAP	Polydiallyl phthalate
PE	Polyethylene
PEEK	Polyetheretherketone
PEI	Polyetherimide
PEO	Polyethylene oxide
PES	Polyether sulfone
PET	Polyethylene terephthalate
PETG	Glycol-modified PET
PF	Phenol-formaldehyde resin
PFA	Perfluoroalkoxy
PFEP	Polyfluoroethylenepropylene
PHEMA	Polyhydroxyethyl methacrylate
PI	Polyimide
PMCA	Polymethylchloroacrylate
PMMA	Polymethyl methacrylate
POM	Polyoxymethylene
PP	Polypropylene
PPC	Polyphthalate carbonate
PPE	Polyphenylene ether
PPO	Polyphenylene oxide
PPSO	Polyphenylsulfone
PPS	Polystyrene
PSO	Polysulfone
PTFE	Polytetrafluoroethylene
PTMT	Polytetramethylene terephthalate
PU	Polyurethane
PUR	Polyurethane rubber
PVAC	Polyvinyl acetate
PVAl	Polyvinyl alcohol
PVB	Polyvinyl butyral
PVC	Polyvinyl chloride
PVDC	Polyvinylidene chloride
PVDF	Polyvinylidene fluoride
PVF	Polyvinyl fluoride
SAN	Styrene-acrylonitrile
SBP	Styrene-butadiene plastics
SBR	Styrene-butadiene rubber
SI	Silicone
SMA	Styrene-maleic anhydride
SMC	Sheet molding compounds
SRP	Styrene-rubber plastics
TDI	Toluene diisocyanate
TFE	Polytetrafluoroethylene
TPE	Thermoplastic elastomer
TPU	Thermoplastic polyurethane
TPX	Polymethylpentene
UF	Urea-formaldehyde
UHMWPE	Ultra-high molecular weight polyethylene
UP	Urethane plastics
VCP	Vinyl chloride-propylene
VDC	Vinylidene chloride
VLDPE	Very low density polyethylene

Appendix C

Trade Names and Manufacturers

Trade Name	Polymer	Manufacturer
Abasfil	Reinforced ABS	AKZO Engineering
Absinol	ABS	Allied Resinous Products, Inc.
Abson	ABS resins and compounds	BF Goodrich Chemical Co.
Acelon	Cellulose acetate film	May & Baker, Ltd.
Acetophane	Cellulose acetate film	UCB-Sidac
Aclar	CTFE fluorohalocarbon films	Allied Chemical Corp.
Acralen	Ethylene-vinyl acetate polymer	Verona Dyestuffs Div. Verona Corp.
Acrilan	Acrylic (acrylontrile-vinyl chloride)	Monsanto Co.
Acroleaf	Hot stamping foil	Acromark Co.
Acrylaglas	Fiberglass reinforced styrene-acrylonitrile	Dart Industries, Inc.
Acrylicomb	Acrylic-sheet-faced honeycomb	Dimensional Plastics Corp.
Acrylite	Acrylic molding compounds; cast arylic sheets	Cyro
Acryloid	Acrylic modifiers for PVC; coating resins	Rohm & Haas Co.
Acrylux	Acrylic	Westlake Plastics Co.
Adell	Nylon 6/6	Adell
Aeroflex	Polyethylene extrusions	Anchor Plastics Co.
Aeron	Plastic-coated nylon	Flexfilm Products, Inc.
Aerotuf	Polypropylene extrusions	Anchor Plastics Co.
Afcolene	Polystyrene and SAN copolymers	Pechiney-Saint-Gobain
Afcoryl	ABS copolymers	Pechiney-Saint-Gobain
Akulon	Nylon 6 and 6/6	Schulman
Alathon	Polyethylene resins	E. I. du Pont de Nemours & Co.
Alfane	Thermosetting epoxy resin cement	Atlas Minerals & Chemicals Div., of ESB Inc.
Algoflon	PTFE	Ausimont
Alpha	Vinyl resins	Alpha Chemical and Plastics
Alpha-Clan	Reactive monomer	Marvon Div., Borg-Warner Corp.
Alphalux	PPO	Marbon Chemical Co.
Alsynite	Reinforced plastic panels	Reichhold Chemicals, Inc.
Amberlac	Modified alkyd resins	Rohm & Haas Co.
Amberol	Phenolic and maleic resins	Rohm & Haas Co.
Amer-Plate	PVC sheet material	Ameron Corrosion Control Div.
Ampol	Cellulose acetates	American Polymers, Inc.
Amres	Cellulose acetates	Pacific Resins & Chemicals, Inc.
Ancorex	ABS extrusions	Anchor Plastics Co.

Trade Name	Polymer	Manufacturer
Anvyl	Vinyl extrusions	Anchor Plastics Co.
APEC	Polycarbonate	Miles
Apogen	Epoxy resin series	Apogee Chemical, Inc.
Araclor	Polychlorinated polyphenyls	Monsanto Co.
Araldite	Epoxy resins and hardeners	CIBA Products Co.
Armorite	Vinyl coating	John L. Armitage & Co.
Arnel	Cellulose triacetate fiber	Celanese Corp.
Arochem	Modified phenolic resins	Ashland Chemical Co.
Arodure	Urea resins	Ashland Chemical Co.
Arofene	Phenolic resins	Ashland Chemical Co.
Aroplaz	Alkyd resins	Ashland Chemical Co.
Aroset	Acrylic resins	Ashland Chemical Co.
Arothane	Polyester resin	Ashland Chemical Co.
Artfoam	Rigid urethane foam	Strux Corp.
Arylon	Polyarl ether compounds	Uniroyal, Inc.
Arylon T	Polyaryl ether	Uniroyal, Inc.
Ascot	Coated spunbonded polyolefin sheet	Appleton Coated Paper Co.
Ashlene	Nylon 6/6	Ashley
Astralit	Vinyl copolymer sheets	Dynamit Nobel of America, Inc.
Astroturf	Nylon, polyethylene	Monsanto Co.
Atlac	Polyester resin	Atlas Chemical Industries, Inc.
Averam	Inorganic	FMC Corp.
Avisco	PVC films	FMC Corp.
Avistar	Polyester film	FMC Corp.
Avisun	Polypropylene	Avisun Corp.
Bakelite	Polyethylene, ethylene copolymers, epoxy, phenolic, polystyrene, phenoxy, ABS and vinyl resins and compounds	Union Carbide Corp.
Beetle	Urea molding compounds	American Cyanamid Co.
Betalux	TFE-filled acetal	Westlake Plastics Co.
Blanex	Crosslinked polyethylene compounds	Reichhold Chemicals, Inc.
Blapol	Polyethylene compounds and color concentrates	Reichhold Chemicals, Inc.
Blapol	Polyethylene molding and extrusion compounds	Blane Chemical Div., Reichhold Chemicals, Inc.
Blendex	ABS resin	Marbon Div., Borg-Warner Corp.
Bolta Flex	Vinyl sheeting and film	General Tire & Rubber Co., Chemical/ Plastics Div.
Bolta Thene	Rigid olefin sheets	General Tire & Rubber Co., Chemical/ Plastics Div.
Boltaron	ABS or PVC rigid plastic sheets	GenCorp
Boronal	Polyolefins with boron	Allied Resinous Products
Bostik	Epoxy and polyurethene adhesives	Bostik-Finch, Inc.
Bronco	Supported vinyl or pyroxylin	General Tire & Rubber Co., Chemical/ Plastics Div.
Budene	Polybutadiene	Goodyear Tire & Rubber Co., Chemical/ Plastics Div.
Butaprene	Styrene-butadiene latexes	Firestone Plastics Co. Div., Firestone Tire & Rubber
Cadco	Plastics rod, sheet, tubing, and film	Cadillac Plastic & Chemical Co.
Calibre	Polycarbonate	Dow
Capran	Nylon 6 film	Allied Chemical Corp.
Capran	Nylon films and sheet	Allied Chemical Corp.
Capron	Nylon/6/6	Allied Signal
Carbaglas	Fiberglass reinforced polycarbonate	Fiberfil Div., Dart Industries, Inc.
Carolux	Filled urethane foam, flexible	North Carolina Foam Industries, Inc.
Carstan	Urethane foam catalysts	Cincinnati Milacron Chemicals, Inc.
Castcar	Cast polyolefin films	Mobil Chemical Co.
Castethane	Castable molding urethane elastomer system	Upjohn Co., CPR Div.
Castomer	Urethane elastomer system	Baxenden Chemical Co.
Castomer	Urethane elastomer and coatings	Isocyanate Products Div., Witco Chemical Corp.
Celanar	Polyester film	Hoechst Celanese
Celanex	Thermoplastic polyester	Hoechst Celanese
Celcon	Acetal copolymer resins	Hoechst Celanese
Cellasto	Microcellular urethane elastomer parts	North American Urethanes, Inc.

Trade Name	Polymer	Manufacturer
Cellofoam	Polystyrene foam	US Mineral Products Co.
Cellonex	Cellulose acetate	Dynamit Nobel of America, Inc.
Celluliner	Resilient expanded polystyrene foam	Gilman Brothers Co.
Cellulite	Expanded polystyrene foam	Gilman Brothers Co.
Celpak	Rigid polyurethane foam	Dacar Chemical Products Co.
Celthane	Rigid polyurethane foam	Decar Chemical Products
Chem-o-sol	PVC plastisol	Chemical Products Corp.
Chem-o-thane	Polyurethane elastomer casting compounds	Chemical Products Corp.
Chemfluor	Fluorocarbon plastics	Chemplast, Inc.
Chemglaze	Polyurethane-based coating materials	Hughson Chemical Co., Div., Lord Corp.
Chemgrip	Epoxy adhesives for TFE	Chemplast, Inc.
Chevron PE	Polyethylene	Chevron
Cimglas	Fiber glass reinforced polyester moldings	Cincinnati Milacron, Molded Plastics Div.
Clocel	Rigid urethane foam system	Baxenden Chemical Co.
Clopane	PVC film and tubing	Clopay Corp.
Cloudfoam	Polyurethane foam	International Foam Div., Holiday Inns of America
Co-Rexyn	Polyester resins and gel coats; pigment pastes	Interplastic Corp., Commercial Resins Div.
Cobocell	Cellulose acetate butyrate tubing	Cobon Plastics Corp.
Coboflon	Teflon tubing	Cobon Plastics Corp.
Cobothane	Ethylene-vinyl acetate tubing	Cobon Plastics Corp.
Colorail	Polyvinyl chloride handrails	Blum, Julius & Co.
Colovin	Calendered vinyl sheeting	Columbus Coated Fabrics
Conathane	Polyurethane casting, potting, tooling and adhesive compound	Conap, Inc.
Conolite	Polyester laminate	Woodall industries, Inc.
Cordo	PVC foam and films	Ferro Corp., Composites Div.
Cordoflex	Polyvinylidene fluoride solutions, etc.	Ferro Corp., Composites Div.
Corlite	Reinforced foam	Snark Products, Inc.
Coror-Foam	Urethane foam systems	Cook Paint & Varnish
Coverlight HTV	Vinyl-coated nylon fabric	Reeves Brothers, Inc.
Creslan	Acrylic	American Cyanamid Co.
Crystic	Unsaturated polyester resins	Scott Bader Co.
Cumar	Coumarone-indene resins	Neville Chemical Co.
Curithane (Series)	Polyaniline polyamine; organo-mercury catalyst	Upjohn Co., Polymer Chemicals Div.
Curon	Polyurethane foam	Reeves Brothers, Inc.
Cycolac	ABS resins	Marbon Div., Borg-Warner Corp.
Cycolon	Synthetic resinous compositions	Marbon Div., Borg-Warner Corp.
Cycoloy	Alloys of synthetic polymers w/ABS resins	Marbon Div., Borg-Warner Corp.
Cycopac	ABS and nitrile barrier	Borg-Warner Chemicals
Cyovin	Self-extinguishing ABS graft-polymer blends	Marbon Div., Borg-Warner Corp.
Cyglas	Glass-filled polyester molding compound	American Cyanamid Co.
Cymel	Melamine molding compound	American Cyanamid Co.
Cyrex	Acrylic-PC alloy	Cyro
Dacovin	PVC compounds	Diamond Shamrock Chemical Co.
Dacron	Polyester	E. I. du Pont de Nernours & Co.
Dapon	Diallyl phthalate resin	FMC Corp., Organic Chemicals Div.
Daran	Polyvinyldene chloride emulsion coatings	W. R. Grace & Co., Polymers & Chemicals Div.
Daratak	Polyvinyl acetate homopolymer emulsions	W. R. Grace & Co., Polymers & Chemicals Div.
Darex	Styrene-butadiene latexes	W. R. Grace & Co., Polymers & Chemicals Div.
Davon	TFE resins & reinforced compounds	Davies Nitrate Co.
Delrin	Acetal resin	E. I. du Pont de Nemours & Co.
Densite	Molded flexible urethane foam	General Foam Div., Tenneco Chemical, Inc.
Derakane	Vinyl ester resins	Dow Chemical Co.
Dexon	Propylene-acrylic	Exxon Chemical USA
Diaron	Melamine resins	Reichhold Chemicals, Inc.
Dielux	Acetal	Westlake Plastics Co.

Trade Name	Polymer	Manufacturer
Dion-Iso	Isophthalic polyesters	Diamond Shamrock Chemical Co.
Dolphon	Epoxy resin & compounds polyester resins	John C. Dolph Co.
Dorvon	Molded polystyrene foam	Dow Chemical Co.
Dow Corning	Silicones	Dow Corning Corp.
Dri-Lite	Expanded polystyrene	Poly Foam, Inc.
Duco	Lacquers	E. I. du Pont de Nemours & Co.
Duracel	Lacquers for cellulose acetate & other plastics	Maas & Waldstein Co.
Duracon	Acetal copolymer	Polyplastics Co.
Duraflex	Polybutylene	Shell
Dural	Acrylic modified semirigid PVC	Alpha Chemical & Plastics Corp.
Duramac	Oil modified alkyds	Commercial Solvents Corp.
Durane	Polyurethane	Raffi & Swanson, Inc.
Duraplex	Alkyd resins	Rohm & Haas Co.
Durelene	PVC flexible tubing	Plastic Warehousing Corp.
Durethan	Nylon 6	Miles
Durethene	Polyethylene film	Sinclair-Koppers Co.
Durez	Phenolic and Alkyd resins	Chemical/ Occidental/ Plastics Corp.
Duron	Phenolic resins & molding compounds	Firestone Foam Products Co.
Dyal	Alkyd and styrenated-alkyd resins	Sherwin Williams Chemicals
Dyalon	Urethane elastomer material	Thombert, Inc.
Dyloam	Expanded polystyrene	W. R. Grace & Co.
Dylan	Polyethylene	ARCO/Polymers, Inc.
Dylel	ABS plastics	Sinclair-Koppers Co.
Dylene	Polystyrene resin and oriented sheet	ARCO
Dylite	Expandable polystyrene bead, extruded sheets, etc.	ARCO
E-Form	Epoxy molding compounds	Allied Products Corp.
Easy-Kote	Fluorocarbon release compound	Borco Chemicals, Inc.
Easypoxy	Epoxy adhesive kits	Conap, Inc.
Ebolan	TFE compounds	Chicago Gasket Co.
Eccosil	Silicone resins	Emerson & Cuming, Inc.
Ektar	Polyester, thermoplastic	Eastman Performance Plastics
El Rexene	Polyethylene, polypropylene, polystyrene and ABS resins	Dart Industries. Inc.
Elastolit	Urethane engineering thermoplastic	North American Urethanes, Inc.
Elastollyx	Urethane engineering thermoplastic	North American Urethanes, Inc.
Elastolur	Urethane coatings	BASF
Elastonate	Urethane isocyanate prepolymers	BASF
Elastonol	Urethane polyester polyols	North American Urethanes, Inc.
Elastopel	Urethane engineering thermoplastics	North American Urethanes, Inc.
Electroglas	Cast arylic	Glasflex Corp.
Elvace	Acetate-ethylene copolymers	E. I. du Pont de Nemours & Co.
Elvacet	Polyvinyl acetate emulsions	E. I. du Pont de Nemours & Co.
Elvacite	Acrylic resins	E. I. du Pont de Nemours & Co.
Elvamide	Nylon resins	E. I. du Pont de Nemours & Co.
Elvanol	Polyvinyl alcohols	E. I. du Pont de Nemours & Co.
Elvax	Vinyl resins; acid terpolymer resins	E. I. du Pont de Nemours & Co.
Empee	Polyethylene	Monmouth
Ensocote	PVC lacquer coating	Uniroyal, Inc.
Ensolex	Cellular plastic sheet material	Uniroyal, Inc.
Ensolite	Cellular plastic sheet material	Uniroyal, Inc.
Epi-Rez	Basic epoxy resins	Celanese Coatings Co.
Epi-Tex	Epoxy ester resins	Celanese Coatings Co.
Epikote	Epoxy resin	Shell Chemical Co.
Epocap	Two-part epoxy compounds	Hardman, Inc.
Epocast	Epoxies	Furane Plastics, Inc.
Epocrete	Two-part epoxy materials	Hardman, Inc.
Epocryl	Epoxy acrylate resin	Shell Chemical Co.
Epocure	Epoxy curing agents	Hardman, Inc.
Epolast	Two-part epoxy compounds	Hardman, Inc.
Epolite	Epoxy compounds	Hexcel Corp., Rezolin Div.
Epomarine	Two-part epoxy compounds	Hardman, Inc.
Epon	Epoxy resin; hardener	Shell Chemical Co.

Trade Name	Polymer	Manufacturer
Eponol	Linear polyether resin	Shell Chemical Co.
Eposet	Two-part epoxy compounds	Hardman, Inc.
Epotuf	Epoxy resins	Reichold Chemicals, Inc.
Escorene	LDPE and LLDPE	Exxon
Estane	Polyurethane resins and compounds	BF Goodrich Chemical Co.
Estron	Acetate	Eastman Kodak Co.
Ethafoam	Polyethylene foam	Dow Chemical Co.
Ethocel	Ethyl cellulose resin	Dow Chemical Co.
Ethofil	Fiber-glass reinforced polyethylene	AKZO Engineering
Ethoglas	Fiber-glass reinforced polyethylene	Fiberfil Div., Dart Industries, Inc.
Ethosar	Fiber-glass reinforced polyethylene	Fiberfil Div., Dart Industries, Inc.
Ethylux	Polyethylene	Westlake Plastics Co.
Evenglo	Polystyrene resins	Sinclair-Koppers Co.
Everflex	Polyvinyl acetate co-polymer emulsion	W. R. Grace & Co., Polymers & Chemicals Div.
Everlon	Urethane foam	Stauffer Chemical Co.
Excelite	Polyethylene tubing	Thermoplastic Processes
Exon	PVC resins, compounds and latexes	Firestone Tire & Rubber Co.
Extane	Polyurethane tubing	Pipe Line Service Co.
Extrel	Polyethylene & polypropylene films	Exxon Chemical, USA
Extren	Fiber-glass reinforced polyester shapes	Morrison Molded Fiber Glass Co.
Fabrikoid	Pyroxylin-coated fabric	Stauffer Chemical Co.
Facilon	Reinforced PVC fabrics	Sun Chemical Corp.
Fassgard	Vinyl coating on nylon	M. J. Fassler & Co.
Fasslon	Vinyl coating	M. J. Fassler & Co.
Felor	Nylon filaments	E. I. du Pont de Nemours & Co.
Fiber foam	Polyester-reinforced foam	Weeks Engineered Plastics
Fiberite	Melamine molding compound	ICI Fiberite
Fibro	Rayon	Courtaulds NA, Inc.
Fina	Polypropylene	Fina
Flexane	Urethanes	Devcon Corp.
Flexocel	Urethane foam systems	Baxenden Chem Co.
Floranier	Cellulose for esters	ITT Rayonier, Inc
Fluokem	Teflon spray	Bel-Art Products

Trade Name	Polymer	Manufacturer
Fluon	TFE resin	ICI American, Inc.
Fluorglas	PTFE coated and impregnated woven glass fabric, laminates, belting	Dodge Industries, Inc.
Fluorocord	Fluorocarbon material	Raybestos Manhattan
Fluorofilm	Cast Teflon films	Dilectrix Corp.
Fluoroglide	Dry-film lubricant of TFE	Chemplast, Inc.
Fluororay	Filled fluorocarbon	Raybestos Manhattan
Fluorored	Compounds of TFE	John L. Dore Co
Fluorosint	TFE-fluorocarbon base composition	Polymer Corp.
Foamthane	Rigid polyurethane foam	Pittsburgh Corning Corp.
Formadall	Polyester premix compound	Woodall Industries, Inc.
Formaldafil	Fiberglass reinforced acetal	AKZO Engineering
Formaldaglas	Fiberglass reinforced acetal	Fiberfil Div., Dart Industries, Inc.
Formaldasar	Fiberglass reinforced acetal	Fiberfil Div., Dart Industries, Inc.
Formica	High-pressure laminate	American Cyanamid Co.
Formrez	Urethane elastomer chemicals	Witco Chemical Corp., Organics Div.
Formvar	Polyvinyl formal resins	Monsanto Co.
Forticel	Cellulose propionate flake, resins	Hoechst Celanese
Fortiflex	Polyethylene resins	Solvay Polymers
Fortilene	Polypropylene	Solvay Polymers
Fortrel	Polyester	Fiber Industries, Inc.
Fosta-Net	Polystyrene foam extruded mesh	Foster Grant Co.
Fosta Tuf-Flex	High-impact polystyrene	Foster Grant Co.
Fostacryl	Thermoplastic polystyrene resins	Foster Grant Co.
Fostafoam	Expandable polystyrene beads	Foster Grant Co.
Fostalite	Light-stable polystyrene molding powder	Foster Grant Co.
Fostarene	Polystyrene molding powder	Foster Grant Co.
Futron	Polyethylene powder	Fusion Rubbermaid Co.
Gelva	Polyvinyl acetate	Monsanto Co.
Genal	Phenolic compounds	General Electric Co.
Genthane	Polyurethane rubber	General Tire & Rubber Co.
Gentro	Styrene butadiene rubber	General Tire & Rubber Co.

Trade Name	Polymer	Manufacturer
Geon	Vinyl resins, compounds latexes	BF Goodrich Chemical Co.
Gil-Fold	Polyethylene sheet	Gilman Brothers Co.
Glaskyd	Alkyd molding compound	American Cyanamid Co.
Glyptal	Alkyd resins	General Electric Co.
Gordon Superdense	Polystyrene in pellet form	Hammond Plastics, Inc.
Gordon Superflow	Polystyrene in granular or pellet form	Hammond Plastics, Inc.
Gracon	PVC compounds	W. R. Grace & Co.
GravoFLEX	ABS sheets	Hermes Plastics, Inc.
GravoPLY	Acrylic sheets	Hermes Plastics, Inc.
Grilamid	Nylon 12	EMS
Grilon	Nylon 6	EMS
Halon	TFE molding compounds	Allied Chemical Corp.
Haylar	CTFE	Ausimont
Haysite	Polyester laminates	Synthane-Taylor Corp.
Herculon	Olefin	Hercules, Inc.
Herox	Nylon filaments	E. I. du Pont de Nemours & Co.
Hetrofoam	Fire retardant urethane foam systems	Durez Div., Hooker Chemical Corp.
Hetron	Fire retardant polyester resins	Durez Div., Hooker Chemical Corp.
Hex-One	High-density polyethylene	Gulf Oil Co.
Hi-fax	Polyethylene	Hercules, Inc.
HiGlass	Glass reinforced PP	Himont
Hi-Styrolux	High-impact polystyrene	Westlake Plastics
Hostalen	Polyethylene	Hoechst Celanese.
Hostaflon	PTFE	Hoechst Celanese
Hydrepoxy	Water-based epoxies	Acme Chemicals Div., Allied Products Corp.
Hydro Foam	Expanded phenolformaldehyde	Smithers Co.
Hyflon	Perfluoroalkoxy	Ausimont
Hylar	Polyvinylidene fluoride	Solvay Polymers
Implex	Acrylic molding powder	Rohm & Haas Co.
Intamix	Rigid PVC compounds	Diamond Shamrock Chemical Co.
Interpol	Copolymeric resinous systems	Freeman Chemical Corp.
Irvinil	PVC resins and compounds	Great American Chem.
Isoderm	Urethane rigid and flexible intergral-skinning foam	Upjohn Co., CPR Div.
Isofoam	Urethane foam systems	Witco Chemical Corp.
Isonate	Diisocyanates and urethane systems	Upjohn Co., CPR Div.
Isoteraglas	Isocyanate elastomercoated Dacron glass fabric	Natvar Corp.
Isothane	Flexible polyurethane foams	Bernel Foam Products Co.
Jetfoam	Polyurethane foam	International Foam
K-Prene	Urethane cast material	Di-Acro Kaufman
Kalex	Two-part polyurethane elastomers	Hardman, Inc.
Kalspray	Rigid urethane foam system	Baxenden Chemical Co.
Kamax	Acrylic	Rohm & Haas
Kapton	Polyimide	E. I. du Pont de Nemours & Co.
Keltrol	Vinyl toluene copolymer	Spencer Kellogg
Ken U-Thane	Polyurethanes; urethane foam ingredients	Kenrich Petrochemicals, Inc.
Kencolor	Silicone/pigments dispersion	Kenrich Petrochemicals, Inc.
Kodacel	Cellulosic film and sheeting	Eastman Chemical Products, Inc.
Kodar	Copolyester thermoplastics	Eastman Chemical Products, Inc.
Kodel	Polyester	Eastman Kodak Co.
Kohinor	Vinyl resins and compounds	Pantasote Co.
Korad	Acrylic film	Rohm & Haas Co.
Koroseal	Vinyl films	BF Goodrich Chemical Co.
Kralastic	ABS high-impact resin	Uniroyal, Inc.
Kralon	High-impact styrene and ABS resins	Uniroyal, Inc,
Kraton	Styrene-butacliene polymers	Shell Chemical Co.
Krene	Plastic film and sheeting	Union Carbide Corp.
Krystal	PVC sheet	Allied Chemical Corp.
Krystaltite	PVC shrink films	Allied Chemical Corp.
Kydene	Acrylic/PVC powder	Rohm & Haas Co.
Kydex	Acrylic/PVC sheets	Rohm & Haas Co.
Kynar (Series)	Polyvinylidene fluoride	Elf Atochem North America
Lamabond	Reinforced polyethylene	Lamex, Columbian Carbon Co.
Lamar	Mylar vinyl laminate	Morgan Adhesives Co.
Laminac	Polyester resins	American Cyanamid Co.
Last-A-foam	Plastic foam	General Plastics Mfg.

Trade Name	Polymer	Manufacturer
Lexan	Polycarbonate resins, film, sheet	General Electric Co., Plastics Dept.
Lucite	Acrylic resins	E. I. du Pont de Nemours & Co.
Lumasite	Acrylic sheet	American Acrylic Corp.
Lustran	SAN and ABS molding and extrusion resins	Monsanto Co.
Lustrex	Polystyrene molding and extrusion resins	Monsanto Co.
Lycra	Spandex	E. I. du Pont de Nemours & Co.
Lytex	Epoxy	Quantum Composites
Macal	Cast vinyl film	Morgan Adhesives Co.
Maclin	Vinyl resins	Maclin
Makrolon	Polycarbonate	Miles
Marafoam	Polyurethane foam resin	Marblette Co.
Maraglas	Epoxy casting resin	Marblette Co.
Maranyl	Nylon 6/6	ICI Americas
Maraset	Epoxy resin	Marblette Co.
Marathane	Urethane Compounds	Allied Products Corp.
Maraweld	Epoxy resin	Marblette Co.
Marlex	Polyethylenes, polypropylenes, other polyolefin plastics	Phillips Petroleum
Marvinol	Vinyl resins and compounds	Uniroyal, Inc.
Meldin	Polyimicle and reinforced polyimide	Dixon Corp.
Merlon	Polycarbonate	Mobay Chemical Co.
Metallex	Cast acrylic sheets	Hermes Plastics, Inc.
Meticone	Silicone rubber dies and sheets	Hermes Plastics. Inc.
Metre-Set	Epoxy adhesives	Metachem Resins Corp.
Micarta	Thermosetting laminates	Westinghouse Electric Corp.
Micro-Matte	Extruded acrylic sheet with matte finish	Extrudaline, Inc.
Micropel	Nylon powders	Nypel, Inc.
Microsol	Vinyl plastisol	Michigan Chrome & Chemical Co.
Microthene	Powdered polyolefins	U.S. Industrial Chemicals Co.
Milmar	Polyester	Morgan Adhesives Co.
Mini-Vaps	Expanded polyethylene	Malge Co., Agile Div.
Minit Grip	Epoxy adhesives	High-Strength Plastics Corp.
Minit Man	Epoxy adhesive	Kristal Draft, Inc.
Mipoplast	Flexible PVC sheets	Dynamit Nobel of America, Inc.
Mirasol	Alkyd resins: epoxy ester	C. J. Osborn Chemicals, Inc.
Mirbane	Amino resin	Shows Highpolymer Co.
Mirrex	Calendered rigid PVC	Tenneco Chemicals, Inc., Tenneco Plastics Div.
Mista Foam	Urethane foam systems	M. R. Plastics & Coatings, Inc.
Mod-Epox	Epoxy resin modifier	Monsanto Co.
Molycor	Glass-fiber reinforced epoxy tubing	A. O. Smith, Inland, Inc.
Mondur	Isocyanates	Mobay Chemical Co.
Monocast	Direct Polymerized nylon	Polymer Corp.
Moplen	Isotactic Polypropylene	Montecatini Edison S.p.A.
Multrathane	Urethane elastomer chemical	Mobay Chemical Co.
Multron	Polyesters	Mobay Chemical Co.
Mylar	Polyester film	E.I. du Pont de Nemours & Co.
Napryl	Polypropylene	Pechiney-Saint-Gobain
Natene	High-density polyethylene	Pechiney-Saint-Gobain
Naugahyde	Vinyl coated fabrics	Uniroyal, Inc.
NeoCryl	Acrylic resins and resin emulsions	Polyvinyl Chemicals, Inc.
Neopolen	Expanded PE bead	BASF
NeoRez	Styrene emulsions and urethane solutions	Polyvinyl Chemicals. Inc.
NeoVac	PVA emulsions	Polyvinyl Chemicals, Inc.
Nestorite	Phenolic and urea-formaldehyde	James Ferguson & Sons
Nevillac	Modified coumarone-indene resin	Neville Chemical Co.
Nimbus	Polyurethane foam	General Tire & Rubber Co.
Nitrocol	Nitrocellulose base pigment dispersion	C. J. Osborn Chemicals, Inc.
Nob-Lock	PVC sheet material	Ameron Corrosion Control Div.
Nopcofoam	Urethane foam systems	Diamond Shamrock Chemical Co., Resinous Products Div.
Norchem	Low-density polyethylene resin	Northern Petrochemical Co.
Noryl	Modified polyphenylene oxide	General Electric Co., Plastics Dept.
Novacor	Polystyrene	Novacor
Nupol	Thermosetting acrylic resins	Freeman Chemical Corp.

Trade Name	Polymer	Manufacturer
NYCOA	Nylon 6/6	Nylon Corp of America
Nyglathane	Glass-filled polyurethane	Nypel, Inc.
Nylafil	Fiberglass reinforced nylon	AKZO Engineering
Nylaglas	Fiberglass reinforced nylon	AKZO Engineering
Nylasar	Fiberglass reinforced nylon	AKZO Engineering
Nylasint	Sintered nylon parts	Polymer Corp.
Nylatron	Filled nylons	Polymer Corp.
Nylo-Seal	Nylon 11 tubing	Imperial-Eastman Corp.
Nylux	Nylon	Westlake Plastics Co.
Nypelube	TFE-filled Nylons	Nypel, Inc.
Nyreg	Glass-reinforced Nylon molding compounds	Nypel, Inc.
Oasis	Expanded phenol-formaldehyde	Smithers Co.
Oilon Pv 80	Acetal-based resin sheets rods, tubing, profiles	Cadillac Plastic & Chemical Co.
Olefane	Polypropylene film	Amoco Chemicals Corp.
Olefil	Filled polypropylene resin	Amoco Chemicals Corp.
Oleflo	Polypropylene resin	Amoco Chemicals Corp.
Olemer	Copolymer polypropylene	Amoco Chemicals Corp.
Oletac	Amorphous polypropylene	Amoco Chemicals Corp.
Opalon	Flexible PVC materials	Monsanto Co.
Oppanol	Polyisobutylene	BASF Wyandotte Corp.
Orgalacqe	Epoxy and PVC powders	Aquitaine-Organico
Orgamide R	Nylon 6	Aquitaine-Organico
Orlon	Acrylic fiber	E. I. du Pont de Nemours & Co.
Oxy	PVC series	Occidental Chemical Co.
Oxyblend	PVC	Occidental Chemical Co.
Panda	Vinyl and urethane coated fabric	Pandel-Bradford Inc.
Papi	Polymethylene polyphenyliso-cyanate	Upjohn Co., Polymer Chemicals Div.
Paradene	Dark coumarone-indene resins	Neville Chemical Co.
Paraplex	Polyester resins and plasticizers	Rohm & Haas Co.
Paxon	Polyethylene	Allied Signal
Pelaspan	Expandable polystyrene	Dow Chemical Co.
Pelaspan-Pac	Expandable poly-styrene	Dow Chemical Co.
Pellethane	Thermoplastic urethane	Dow Plastics
Pellon Aire	Nonwoven textile	Pellon Corp.
Penton	Chlorinated polyether	Hercules, Inc.
PermaRex	Cast epoxy	Permali, Inc.
Permelite	Melamine molding compound	Melamine Plastics, Inc.
Petion	Glass filled PET	Miles
Petra	Polyester	Allied Signal
Pethrothene	Low-, medium-, and high-, density polyethylene	Quantum USI
Petrothene XL	Crosslinkable polyethylene	U.S. Industrial Chemical Co.
Phenoweld	Phenolic adhesive	Hardman, Inc.
Philjo	Polyolefin films	Phillips-Joana Co.
Philprene	Styrene-butadiene	Phillips Chemical Corp.
Piccoflex	Acrylontrile-styrene resins	Pennsylvania Industrial Chemical Corp.
Piccolastic	Polystryrene resins	Pennsylvania Industrial Chemical Corp.
Piccotex	Vinyl-toluene copolymer	Pennsylvania Industrial Chemical Corp.
Piccournaron	Coumarone-indene resins	Pennsylvania Industrial Chemical Corp.
Piccovar	Alkyl-aromatic: resins	Pennsylvania Industrial Chemical Corp.
Pienco	Polyester resins	Mol-Rex Div., American Petrochemical Corp.
Pinpoly	Reinforced polyurethane foam	Holiday Inns of America, Inc.
Plaskon	Plastic molding	Allied Chemical Corp.
Plastic Steel	Epoxy tooling and repair	Devcon Corp.
Plenco	Melamine and phenolic	Plastics Engineering
Pleogen	Polyester resins and gel coats; polyurethane systems	Mol-Rex Div., Whittake Corp.
Plexiglas	Acrylic sheets and molding powders	Rohm & Haas Co.
Plicose	Polyethylene film, sheeting, tubing, bags	Diamond Shamrock Corp.
Pliobond	Adhesive	Goodyear Tire & Rubber Co.
Pliolite	Styrene-butadiene resins	Goodyear Tire & Rubber Co.
Pliothene	Polyethylene-rubber blends	Ametek/ Westchester Plastics
Pliovic	PVC resins	Goodyear Tire & Rubber Co.

Trade Name	Polymer	Manufacturer
Pluracol	Polyethers	BASF Wyandotte Corp.
Pluragard	Urethane foams	BASF Wyandotte Corp.
Pluronic	Polyethers	BASF Wyandotte Corp.
Plyocite	Phenolic-impregnated overlays	Reichhold Chemicals, Inc.
Plyophen	Phenolic resins	Reichhold Chemicals, Inc.
Pocan	Polyester, thermoplastic	Miles
Polex	Oriented acrylic	Southwestern Plastics, Inc.
Pollopas	Urea-formaldehyde compounds	Dynamit Nobel of America, Inc.
Polvonite	Cellular plastic material in sheet form	Voplex Corp.
Polycarbafil	Glass reinforced PC	Akzo Engineering
Poly-Dap	Diallyl phthalate electrical molding compounds	U.S. Polymeric, Inc.
Poly-Eth	Low-density polyethylene	Gulf Oil Corp.
Poly-Eth-Hi-D	High-density polyethylene	Gulf Oil Corp.
Polycarbafil	Fiber-glass reinforced polycarbonate	Fiberfil Div., Dart Industries, Inc.
Polycure	Crosslinked polyethylene compounds	Crooke Color & Chemical Co.
Polyfoam	Polyurethane foam	General Tire & Rubber Co.
Polyfort	Reinforced Polypropylene	Schulman
Polyimidal	Thermoplastic polyimide	Raychem Corp.
Polylite	Polyester resins	Reichhold Chemicals, Inc.
Polymet	Plastic-filled sintered metal	Polymer Corp.
Polymul (series)	Polyethylene emulsions	Diamond Shamrock Chemical Co.
Polyteraglas	Polyester-coated Dacron-glass fabric	Natvar Corp.
Polywrap	Plastic film	Flex-O-Glass, Inc.
Poxy-Gard	Solventless epoxy compounds	Sterling, Div. Reichhold Chemicals, Inc.
PPO	Polyphenylene oxide	Reichhold Chemicals, Inc.
Pro-fax	Polypropylene	Hercules, Inc,
Profil	Fiber-glass reinforced polypropylene	AKZO Engineering
Proglas	Fiber-glass reinforced polypropylene	Fiberfil Div., Dart Industries, Inc.
Prohi	High-density polyethylene	Protective Lining Corp.
Propathene	Polypropylene polymers and compound	Imperial Chemical Ind., Ltd., Plastics Div.
Propylsar	Fiber-glass reinforced polypropylene	Fiberfil Div., Dart Industries, Inc.
Propylux	Polypropylene	Westlake Plastics Co.
Protectolite	Polyethylene film	Protective Lining Corp.
Protron	Ultrahigh-strength polyethylene	Protective Lining Corp.
Pulse	PC/ABS blend	Dow Plastics
Purilon	Rayon	FMC Corp.
Quelflam	Urethanes, low surface spread flame	Baxenden Chemical Co.
Radilon	Nylon 6/6	Polymers International
Rayflex	Rayon	FMC Corp.
Regalite	Press polished clear flexible PVC	Tenneco Advanced Materials, Inc.
REN-Shape	Epoxy material	Ren Plastics, Inc.
Ren-Thane	Urethane elastomers	Ren Plastics, Inc
Resiglas	Polyester resins, etc.	Kristal Draft, Inc.
Resimene	Melamine resins	Monsanto Co.
Resinoid	Phenolic molding compound	Resinoid
Resinol	Polyolefins	Allied Resinous Products, Inc.
Resinox	Phenolic resins	Monsanto Co.
Resorasa-bond	Resorcinol and phenol-resorcinol	Pacific Resins & Chemicals, Inc.
Restfoam	Urethane foam	Stauffer Chemical Co., Plastics Div.
Rexene	PE film	Rexene
Rexolene	Crosslinked polyolefin sheet	Brand-Rex Co.
Rexolite	Polystyrene rod and sheet stock	Brand-Rex Co.
Reynosol	Urethane, PVC	Hoover Ball & Bearing Co.
Rhodiod	Cellulose acetate sheet	M & B Plastics, Ltd.
Rhoplex	Acrylic emulsion	Rohm & Haas Co.
Richfoam	Urethane foam	E. R. Carpenter Co.
Rigidite	Modified acrylic and sheet polyester resins	American Cyanamid Co.
Rigidsol	Rigid plastisol	Watson-Standard Co.
Rimtec	Vinyl resins	Rimtec
Rogers	Epoxy molding compound	Rogers
Rolox	Two-part epoxy compounds	Hardman, Inc.
Royalex	Structural cellular thermoplastic sheet material	Uniroyal, Inc.

Trade Name	Polymer	Manufacturer
Royalite	Thermoplastic sheet material	Uniroyal Inc., Uniroyal Plastic Products
Roylar	Polyurethane elastoplastic	Uniroyal, Inc.
Rucoam	Vinyl film and sheeting	Hooker Chemical Corp.
Rucoblend	Vinyl compounds	Hooker Chemical Corp.
Rucon	Vinyl resins	Hooker Chemical Corp.
Rucothane	Polyurethanes	Hooker Chemical Corp.
Ryton	Polyphenylene Sulfide	Phillips Chemical Co.
Santolite	Aryl sulfonamide-formaldehyde resin	Monsanto Co.
Saran	Polyvinylidene chloride resin	Dow Chemical Co.
Satin Foam	Extruded polystyrene foam	Dow Chemical Co.
Scotchpak	Heat-sealable polyester film	3M Co.
Scotchpar	Polyester film	3M Co.
Selectrofoarn	Urethane foam systems and polyols	PPG Industries, Inc.
Selectron	Polymerizable synthetic resins; polyesters	PPG Industries, Inc.
Shareen	Nylon	Courtaulds North America, Inc.
Shuvin	Vinyl molding compounds	Blane Chemical Div., Reichhold Chemicals, Inc.
Silastic	Silicone rubber	Dow Corning Corp.
Sipon	Alkyl and Aryl resin	Alcolac, Inc.
Siponate	Alkyl and aryl sulfonates	Alcolac, Inc.
Skinwich	Urethane rigid and flexible integral-skinning foam	Upjohn Co.
Softlite	Ionorner foam	Gilman Brothers Co.
Solarflex	Chlorinated polyethylene	Pantasote Co.
Solef	Polyvinylidene fluoride	Solvay Polymers
Solithane	Urethane prepolymers	Thiokol Chemical Corp.
Sonite	Epoxy resin compound	Smooth-On, Inc.
Spandal	Rigid urethane laminates	Baxenclen Chemical Co.
Spandofoarn	Rigid urethane foam board and slab	Baxenden Chemical Co.
Spancloplast	Expanded polystyrene board and slab	Baxenden Chemical Co.
Spectran	Polyester	Monsanto Textiles Co.
Spenkel	Polyurethane resins	Spencer Kellogg Div., Textron Inc.
Starez	Polyvinyl acetate resin	Standard Brands Chemical Ind., Inc.
Structoform	Sheet molding compounds	Fiberite Corp.
Stryton	Nylon	Phillips Fibers Corp.
Stylafoam	Coated polystyrene sheet	Gilman Brothers Co.
Stypol	Polyesters	Freeman Chemical Corp., Div., H. H. Robertson Co.
Styrafil	Fiberglass reinforced polystyrene	AKZO Engineering
Styroflex	Biaxially oriented polystyrene film	Natvar Corp.
Styrofoam	Polystyrene foam	Dow Chemical Co.
Styrolux	Polystyrene	Westlake Plastics Co.
Styropor	Expanded polystyrene	BASF
Styron	Polystrene resin	Dow Chemical Co.
Styronol	Styrene	Allied Resinous Products, Inc.
Sulfasar	Fiberglass reinforced polysulfone	Fiberfil Div., Dart Industries, Inc.
Sulfil	Fiberglass reinforced polysulfone	AKZO Engineering
Sunlon	Polyamide resin	Sun Chemical Corp.
Super Aeroflex	Linear polyethylene	Anchor Plastic Co.
Super Coilife	Epoxy potting resin	Westinghouse Electric Corp.
Super Dylan	High-density polyethylene	Sinclair-Koppers Co.
Superflex	Grafted high-impact polystyrene	Gordon Chemical Co.
Superflow	Polystyrene	Gordon Chemical Co.
Sur-Flex	Ionomer film	Flex-O-Glass, Inc.
Surlyn	Ionomer resin	E. I. du Pont de Nemours & Co.
Syn-U-Tex	Urea-formaldehyde and melamine-formaldehyde	Celanese Resins Div., Celanese Coatings Co.
Syntex	Alkyd and polyurethane ester resins	Celanese Resins Div., Celanese Coatings Co.
Syretex	Styrenated alkyd resins	Celanese Resins Div., Celanese Coatings Co.
TanClad	Spray or dip plastisol	Tamite Industries, Inc.
Tedlar	PVF film	E. I. du Pont de Nemours & Co.
Tedur	Reinforced polyphenylene sulfide	Miles

Trade Name	Polymer	Manufacturer
Teflon	FEP and TFE fluorocarbon resins	E. I. du Pont de Nemours & Co.
Tefzel	PE-TFE	DuPont
TempRite	PVC for pipe	BF Goodrich
Tenite	PE and cellulosic compounds	Eastman Chemical Products, Inc.
Tenn Foam	Polyurethane foam	Morristown Foam Corp.
Tere-Cast	Polyester casting compounds	Sterling Div., Reichhold Chemicals, Inc.
Terucello	Carboxymethyl cellulose	Showa Highpolymer Company
Tetra-Phen	Phenolic-type resins	Georgia-Pacific Corp. Chemical Div.
Tetra-Ria	Amino-type resins	Georgia-Pacific Corp. Chemical Div.
Tetraloy	Filled TFE molding compounds	Whitford Chemical Corp.
Tetran	Polytetrafluoroethyl ene	Pennwalt Corp.
Texalon	Acetal and nylon	Texapol
Texin	Urethane elastomer molding compound	Miles
Textolite	Industrial laminates	General Electric Co., Laminated Products Dept.
Thermalux	Polysulfone	Westlake Plastics Co.
Thermasol	Vinyl plastisols and organosols	Lakeside Plastics International
Thermco	Expanded polystyrene	Holland Plastics Co.
Thermocomp AE	Reinforced ABS	Thermofil
Thermocomp	Reinforced Nylon	LNP
Thorane	Rigid polyurethane foam	Dow Chemical Co.
T-Lock	PVC sheet material	Amercoat Corp.
Torlon	Polyamide-imide	Amoco Performance Products
TPX	Polymethyl pentene	Mitsue Petrochemical Industries
Tran-Stay	Flat polyester film	Transilwrap Co.
Transil GA	Precoated acetate sheets	Transilwrap Co.
Tri-Foil	TFE-coated aluminum foil	Tri-Point Industries, Inc.
Trilon	Polytetrafluoro-ethylene	Dynamit Nobel of America, Inc.
Triocel	Acetate	Celanese Fibers Marketing Co.
Trolen (series)	Polyethylene and polypropylene sheets	Dynamit Nobel of America, Inc.
Trolitan (series)	Phenol-formalde-hyde com-pounds; boron	Dynamit Nobel of America, Inc.

Trade Name	Polymer	Manufacturer
Trolitrax	Industrial laminates	Dynamit Nobel of America, Inc.
Trosifol	Polyvinyl butyral film	Dynamit Nobel of America, Inc.
Tuffak	Polycarbonate	Rohm & Haas Co.
Tuftane	Polyurethane film and sheet	BF Goodrich Chemical Co.
Tybrene	Acrylonitrile-butadiene-styrene	Dow Chemical Co.
Tynex	Polyamide filaments	E. I. du Pont de Nemours & Co.
Tyril	Styrene-acrylonitrile resin	Dow Chemical Co.
Tyrilfoarn	Styrene-acrylonitrile foam	Dow Chemical Co.
Tyrin	Chlorinated polyethylene	Dow Chemical Co.
Udel	Sulfone polymers	Amoco Performance Products
U-Thane	Rigid insulation board stock urethane	Upjohn Co., CPR Div.
Uformite	Urea and melamine resins	Rohm and Haas Co.
Ultem	Polyetherimide	GE Plastics
Ultradur	Polyester, thermoplastic	BASF
Ultraform	Acetal	BASF
Ultramid	Polyamide 6; 6, 6;and 6, 10	BASF Wyandotte Corp.
Ultrapas	Melamine-formaldehyde compounds	Dynamit Nobel of America, Inc.
Ultrathene	Ethylene-vinyl acetate resins and copolymers	U.S. Industrial Chemicals Co.
Ultron	PVC film and sheet	Monsanto Co.
Unifoam	Polyurethane foam	William T. Burnett & Co.
Unipoxy	Epoxy resins, adhesives	Kristal Kraft, Inc.
Urafil	Fiberglass reinforced polyurethane	Fiberfil Div., Dart Industries, Inc.
Uraglas	Fiberglass reinforced polyurethane	Fiberfil Div., Dart Industries, Inc.
Uralite	Urethane compounds	Rezolin Div., Hexcel Corp.
Uramol	Urea-formaldehycle molding compounds	Gordon Chemicals Co.
Urapac	Rigid urethane systems	North American Urethanes, Inc.
Urapol	Urethane elastomeric coating	Poly Resins
Uvex	Cellulose acetate butyrate sheet	Eastman Chemical Products, Inc.
Valite	Phenolic molding compound	Valite

Trade Name	Polymer	Manufacturer
Valox	Thermoplastic polyester	General Electric Co.
Valsof	Polyethylene emulsions	Valchem Div., United Merchants & Mfrs., Inc.
Vandar	PBT alloy	Hoechst Celanese
Varcum	Phenolic resins	Reichhold Chemicals, Inc.
Varex	Polyester resins	McCloskey Varnish Co.
Varkyd	Alkyd and modified alkyd resins	McCloskey Varnish Co.
Varkyclane	Urethane vehicles	McCloskey Varnish Co.
Varsil	Silicone-coated fiber glass	New Jersey Wood Finishing Co.
V del	Polysulfone resins	Union Carbide Corp.
Vectra	Polypropylene fibers	Exxon Chemical USA
Velene	Styrene-foam laminate	Scott Paper Co., Foam Division
Velon	Film and sheeting	Firestone Plastics Co., Div., Firestone Tire & Rubber Co.
Versel	Thermoplastic polyester	Allied Chemical Corp.
Versi-Ply	Coextruded films	Pierson Industries, Inc.
Vestamid	Nylon 6/12	Huels America
Vibrathane	Polyurethane elastomer	Uniroyal, Inc.
Vibrin-Mat	Polyester-glass molding compound	Marco Chemical Div., W. R. Grace & Co.
Vibro-Flo	Epoxy and polyester coating powders	Armstrong Products Co.
Vinoflex	PVC resins	BASF Wyandotte Corp.
Vista	Vinyl resins	Vista
Vitel	Polyester resin	Goodyear Tire & Rubber Co., Chemical Div.
Vithane	Polyurethane resins	Goodyear Tire & Rubber Co., Chemical Div-
Vituf	Polyester resin	Goodyear Tire & Rubber Co., Chemical Div.
Volara	Closed-cell, low-density polyethylene foam	Voltek, Inc.
Volaron	Closed-cell, low density polyethylene foam	Voltek, Inc.
Volasta	Closed-cell, medium-density polyethylene foam	Voltek, Inc.
Voranol	Polyurethane resins	Dow Chemical Co.
Vult-Acet	Polyvinyl acetate latexes	General Latex & Chemical Corp.
Vultafoarn	Urethane foam systems	General Latex & Chemical Corp.
Vultathane	Urethane coatings	General Latex & Chemical Corp.
Vycron	Polyester	Beaunit Corp.
Vydyne	Nylon 6/6	Monsanto
Vygen	PVC resin	General Tire & Rubber Co., Chemical/ Plastics Div.
Vynaclor	Vinyl chloride emulsion coatings and binders	National Starch & Chemical Corp.
Vynaloy	Vinyl sheet	BF Goodrich Chemical Co.
Vyram	Rigid PVC materials	Monsanto Co.
Weldfast	Epoxy and Polyester adhesives	Fibercast Co.
Wellamid (series)	Polyamide 6 and 6,6 molding resins	Wellman, Inc., Plastics Div.
Well-A-Meld	Reinforced nylon resins	Wellman, Inc.
Westcoat	Strippable coatings	Western Coating Co.
Whirlclad	Plastic coatings	Polymer Corp.
Whitcon	Fluoroplastic lubricants	Whitford Chemical Corp.
Wicaloid	Styrene-butadiene emulsions	Wica Chemicals, Div. Ott Chemical Co.
Wicaset	Polyvinyl acetate emulsions	Wica Chemicals, Div, Ott Chemical Co.
Wilfex	Vinyl plastisols	Flexible Products Co.
Xenoy	PBT	GE Plastics
Xylon	Polyamide 6 and 6.6	Fiberfil Div., Dart Industries, Inc.
Zantrel	Rayon	American Enka Co.
Zefran	Acrylic, nylon polyester	Dow Badische Co.
Zelux	Polyethylene films	Union Carbide Corp., Chemicals & Plastics Div.
Zendel	Polyethylene films	Union Carbide Corp., Chemicals & Plastics Div.
Zerion	Copolymer of acrylic and styrene	Dow Chemical Co.
Zetafin	Ethylene copolymer resins	Dow Chemical Co.
Zytel	Nylon	E. I. du Pont de Nemours & Co.

Appendix D

Material Identification

Identifying Plastics

Plastics are complex materials. Identification is not easy because they can contain several polymers as well as other ingredients, including fillers. The insolubility of some plastics adds to the problem. Correct identification requires complex tools and techniques.

A student or a consumer may have to identify a plastics so that repairs or new parts can be made. Laboratory tests can help to identify ingredients of the unknown material. The methods shown in this appendix are meant for easy identification of basic polymer types. Identification methods that involve complex instruments are not covered.

More sophisticated methods include infrared spectroscopy analysis, which is the only accurate method of obtaining the quantitative identification of unknown polymers. Another highly complex and costly method is X-ray diffraction, which is used for identification of solid crystalline compounds.

Identification Methods

There are five broad identification methods for plastics:

1. Trade name
2. Appearance
3. Effects of heat
4. Effects of solvents
5. Relative density

Trade Names

The numerous trade names used today identify the product of a fabricator, a manufacturer, or a processor. Trade names may be associated with the product or with the plastics material. In either case, trade names can serve as a guide to identification of the plastics. See Appendix C for trade names of plastics materials on the commercial market.

If the trade name is known, the supplier or manufacturer may be the most reliable source of information about the kind of plastics, ingredients, additives, or physical properties. Batch and lot numbers may vary, but most of the essential information will be known by the supplier. Like gasoline brands, additives may vary in each family of plastics manufactured.

Appearance

Many physical or visual clues can be used to help identify plastics materials. Plastics in the raw, uncompounded, or pellet stage are harder to identify than finished products. Thermoplastics are generally produced in powder, granular, or pellet forms. Thermosetting materials are usually made in the form of powders, preforms, or resins.

The method of fabrication and the product application are good clues to identity of plastics. Thermoplastic materials are commonly extruded, injection-formed, calendered, blow-molded, and vacuum-molded. Polyethylene, polystyrene, and cellulosics are used extensively in the container and packaging industry. Harsh chemicals and solvents are likely to be stored in polyethylene containers. Polyethylene, polytetrafluoroethylene, polyacetals, and polyamides have a waxy feel not found in most polymers.

Thermosetting plastics are usually compression-molded, transfer-molded, cast, or laminated. Some thermosets are not reinforced. Others are reinforced or heavily filled. Some identifiable characteristics of plastics are given in Table D-1.

Effects of Heat

When plastics specimens are heated in test tubes, distinct odors of specific plastics may be identified. The actual burning of the sample in an open flame may provide further clues. Polystyrene and its copolymers burn with black (carbon) smoke. Polyethylene burns with a clear, blue flame and drips when molten. (See Table D-2.)

The actual melting point may provide further clues to identification. Thermosetting materials don't melt. Several thermoplastics melt at less than 195 °C [203 °F]. An electric soldering gun can be pressed on the surface of the plastics. If the material softens and the hot tip sinks into it, the material is a thermoplastic. If the material stays hard and merely chars, it is a thermoset.

The melting or softening point may be observed by placing a small piece of the unknown thermoplastic on an electrically heated platen or in an oven. The temperature must be carefully controlled and recorded. When the specimen is within a few degrees of the suspected melting point, the temperature should be increased at a rate of 1 °C/min.

A standard method of testing polymers is given in ASTM D-2117. For polymers that have no definite melting point (such as polyethylene, polystyrene, acrylics, and cellulosics) or for those that melt with a broad transition temperature, the Vicat softening point may be of some aid in identification. The Vicat softening point test is described in Chapter 6. Melting points of selected plastics are shown in Table D-2.

Two tests that rely on the effects of heat are the Beilstein test and the Lassaigne test.

The Beilstein test. The Beilstein test is a simple method of determining the presence of a halogen (chlorine, fluorine, bromine, and iodine). For this test, heat a clean copper wire in a Bunsen flame until it glows. Quickly touch the hot wire to the test sample, then return the wire to the flame. A green flame shows the presence of a halogen. Plastics containing chlorine are polychlorotrifluoroethylene, polyvinyl chloride, polyvinylidene chloride, and others. They give positive results to the halogen test. If the halogen test is negative, the polymer may be composed of only carbon, hydrogen, oxygen, or silicon.

The Lassaigne test. For further chemical analysis, the Lassaigne procedure of sodium fusion may be used.

CAUTION: While very useful, this test is dangerous because sodium is highly reactive. It should be performed with great care.

To conduct the Lassaigne test, place five grams of the sample in an ignition tube with 0.1 gram of sodium. Heat the tube until the sample decomposes, keeping the open end of the tube pointing away from you. While the tube is still red-hot, place it in distilled water and grind it with a mortar and pestle device. Filter out the carbon and glass fragments while the mixture is still hot. Divide the resulting filtrate into four equal portions. Use these portions to perform standard tests for nitrogen, chlorine, fluorine, and sulphur.

Effects of Solvents

Tests for the solubility or insolubility of plastics are easily performed identification methods. Except for polyolefins, acetals, polyamides, and fluoroplastics—thermoplastic materials can be considered soluble at room temperature. Thermosetting plastics may be considered solvent resistant.

CAUTION: When making solubility tests, remember that solutions may be flammable, may give off toxic fumes, may be absorbed through the skin, or all three. Appropriate safety precautions should be taken.

Before a solubility test can be made, a chemical solvent must be selected. To help identify polymers and solvents that may react molecularly with each other, solubility parameter numbers have been assigned to selected polymers and solvents (Table D-3).

A polymer should disslove in a solvent that has a similar or lower solubility parameter. This does not always happen, due to crystallization or hydrogen bonding.

When making solubility tests, use a ratio of one volume of plastics sample to twenty volumes of boiling or room temperature solvent. A water-cooled reflux condenser may be used to collect solvent or minimize solvent loss during heating. Table D-4 shows selected solvents with selected plastics.

Relative Density

The presence of fillers or other additives and the degree of polymerization (DP) can make identification of plastics by relative density tests difficult. The presence of fillers and additives can cause the relative density to differ greatly from that of the plastics itself. Polyolefins, ionomers, and low-density polystyrene

Table D-1. Identification of Selected Plastics

Plastics	Appearance	Applications
ABS	Styrene-like, tough, metal-like ring when struck, translucent	Appliance and tool housings, instrument panels, luggage, packing crates, sporting goods.
Acetal	Tough, hard, metal-like ring when struck, waxy feel, translucent	Aerosol stem valves, cigarette lighters, conveyor belts, plumbing, zippers
Acrylics	Brittle, hard, transparent	Models, glazing, wax
Alkyds	Hard, tough, brittle, usually bulk-filled, opaque	Electrical, paints
Allyl	Hard, filled, reinforced, transparent to opaque	Electrical, sealers
Aminos	Hard, brittle, opaque with some translucence	Appliance knobs, bottle caps, dials, handles
Cellullosics	Varies; tough, transparent	Explosives, fabrics, packaging, pharmaceuticals, handles, toys
Chlorinated Polyesters	Tough, translucent or opaque	Electrical, laboratory equipment, plumbing
Epoxies	Hard, mostly filled, reinforced, transparent	Adhesives, casting, finishes
Fluoroplastics	Tough, waxy feel, translucent	Anti-stick coatings, bearings, gaskets, seals, electrical
Ionomers	Tough, impact resistant, transparent	Containers, paper coatings, safety glasses, shields, toys
Nitrile barrier plastics	Tough, transparent, impact resistant	Packaging
Phenolics	Hard, brittle, filled, reinforced, transparent	Adhesives, billiard balls, handles, molding powders
Phenylene oxide	Tough, hard, often filled, reinforced, opaque	Appliance housings, consoles, electrical, respirators
Polyamides	Tough, waxy feel, translucent	Combs, door catches, castors, gears, valve seats
Polyaryl ether	Impact resistant, polycarbonate-like, translucent to opaque	Appliances, automobile paint, electrical
Polyaryl sulfone	Tough, stiff, opaque, polycarbonate-like	High-temperature uses in aerospace, industry, and consumer goods
Polycarbonate	Styrene-like, tough, metal-like ring when struck, translucent	Beverage dispensers, films, lenses, light fixtures, small appliances, windshields
Polyester, aromatic	Stiff, tough, opaque	Coatings, insulation, transistors
Polyester, thermoplastic	Hard, tough, opaque	Beverage bottles, packaging, photography, tapes, labels
Polyester, unsaturated	Hard, brittle, filled, transparent	Furniture, radar domes, sports equipment, tanks, trays
Polyolefins	Waxy feel, tough, soft, translucent	Carpeting, chairs, dishes, medical syringes, toys
Polyphenylene sulfide	Stiff, hard, opaque	Bearings, gears, coatings
Polystyrene	Brittle, white bend marks, metal-like ring when struck, transparent	Blister packages, bottle caps, dishes, lenses, transparent display boxes
Polysulfone	Rigid, polycarbonate-like, transparent to opaque	Aerospace, distributor caps, hospital equipment, shower heads
Silicones	Tough, hard. filled, reinforced, some flexible, opaque	Artificial organs, grease, inks, molds, polishes, Silly Putty, waterproofing
Urethanes	Tough castings, mostly foams, flexible, opaque	Bumpers, cushions, elastic thread, insulation, sponges, tires
Vinyls	Tough, some flexible, transparent	Balls, dolls, floor coverings, garden hoses, rainwear, wall tiles, wallpaper

Table D-2. Identification Test for Selected Plastics

Thermoplastic	Flame-Burn, Smoke & Flame Danger	Odor and Respiratory Danger	Melting Point, °C
ABS	Yellow flame, black smoke, drips, continues to burn	Rubber, sharp, biting	100
Acetal	Blue flame, no smoke, drips, melts, may burn, continues to burn	Formaldehyde	181
Acrylic	Blue flame, yellow top, white ash, black smoke, popping sound, spurts, continues to burn	Fruit, floral	105
Cellulose acetate	Yellow or yellow-orange-to-green flame, melts, drips, continues to burn, black smoke	Burnt sugar, acetic acid, burning paper	230
Cellulose acetate butyrate	Blue flame, yellow top, sparks, melts, drips, dripping may burn, continues to burn	Camphor, rancid butter	140
Ethyl cellulose	Pale yellow to blue-green flame with blue edge, melts, drips, drips burn	Burnt wood, burnt sugar	135
Fluoronated ethylene propylene	Melts, decomposes, poison gases formed	Slight acid or burned hair. DO NOT INHALE	275
Ionomer	Yellow flame with blue edge, continues to burn, some black smoke, melts, bubbles, drips burn	Paraffin	110
Phenoxy	Burns, no drips	Acid	93
Polyallomer	Yellow or yellow-orange flame, with blue edge, continues to burn, black smoke, melts clear, spurts, drips burn	Paraffin	120
Polyamides 6, 6	Blue flame, yellow tip, melts and drips, self-extinguishing, froths	Burned wool or hair	265
Polycarbonate	Decomposes, chars, self-extinguishing, dense black smoke, spurts orange flame	Characteristic, sweet, compare known sample	150
Polychlorotrifluoro-ethylene	Yellow flame, won't support combustion	Slight, acidic fumes. DO NOT INHALE	220
Polyethylene	Blue flame, yellow top, drippings may burn, transparent hot area, burns rapidly, continues to burn	Paraffin	110
Polyimides	Chars, brittle, blue flame		300
Polyphyenylene oxide	Yellow to yellow-orange flame, no drips, spurts, difficult to ignite, thick black smoke, decomposes	Paraffin, phenol	105
Polypropylene	Blue flame, drips, transparent hot area, burns slowly, trace of white smoke, melts, swells	Heavy, sweet, paraffin, burning asphalt	176
Polysulfones	Yellow or orange flame, black smoke, drips, self-extinguishing, sparkles, decomposes	Acid	200
Polystyrene	Yellow flame, dense smoke, clumps of carbon in air, drips, continues to burn, bubbles	Illuminating gas, sweet, marigold, floral	100
Polytetrafluoroethylene	Yellow flame, slightly green near base, won't support combustion, self-extinguishing, turns clear	None. DO NOT INHALE	327
Polyvinyl acetate	Yellow flame, black smoke, spurts, continues to burn, some soot, green on copper wire test	Vinegar, acetic acid	60
Polyvinyl alcohol	Yellow, smoky	Unpleasant, sweet	105
Polyvinyl chloride	Yellow flame, green at edges, black or gray smoke, chars, self-extinguishing, leaves an ash	Hydrochloric acid	75
Polyvinyl fluoride	Pale yellow	Acid	230
Polyvinylidene chloride	Yellow with green base, spurts green smoke, smoky, self extinguishing	Pungent	210
Casein	Yellow flame, burns with flame contact, chars	Burnt milk	

Table D-2. Identification Test for Selected Plastics *(Continued)*

Thermoplastic	Flame-Burn, Smoke & Flame Danger	Odor and Respiratory Danger	Melting Point, °C
Diallyl phthalate	Yellow flame, green-blue edge, self-extinguishing	Acid	
Epoxy	Yellow flame, some soot, spits black smoke, chars, continues to burn	Phenolic phenol, acid	
Melamine formaldehyde	Difficult to burn, self-extinguishing, swells, cracks, yellow flame, blue-green base, turns white	Fish-like, formaldehyde	
Phenolic	Cracks, deforms, difficult to burn, self extinguishing, yellow flame, little black smoke	Phenolic phenol	
Polyester	Yellow flame, blue edge, burns, ash and black beads, continues to burn, dense black smoke, no drips	Sweet, bitter-sweet, burning coal	
Polyurethane	Yellow with blue base, thick black smoke, spurts, may melt and drip, continues to burn	Acid	
Silicone	Low, bright yellow-white flame, white smoke, white ash, continues to burn	None	
Urea formaldehyde	Yellow flame with green-blue edge, self-extinguishing	Pancakes	

will float in water (which has a relative density of 1.00). For comparison of relative densities of various selected materials, see Table 6-7.

If a density gradient column is unavailable, a few solutions of known density can provide considerable data. Mix solutions of distilled water and calcium nitrate and measure them with technical-grade hydrometers until a desired relative density is obtained. For densities less than that of water (1.00) isopropyl alcohol may be used. Full-strength isopropyl alcohol has a relative density of 0.92. By adding small amounts of distilled water, this value may be raised.

If a plastics floats in a solution with a relative density of 0.94, it may be a medium- or low-density polyethylene plastics. If the sample floats in a solution of 0.92, it must be either low-density polyethylene or polypropylene. If the sample sinks in all solutions below a relative density of 2.00, the sample is a fluorocarbon plastics.

Table D-3. Solubility Parameters of Selected Solvents and Plastics

Solvent	Solubility Parameter
Water	23.4
Methyl alcohol	14.5
Ethyl alcohol	12.7
Isopropyl alcohol	11.5
Phenol	14.5
n-Butyl alcohol	11.4
Ethyl acetate	9.1
Chloroform	9.3
Trichloroethylene	9.3
Methylene chloride	9.7
Ethylene dichloride	9.8
Cyclohexanone	9.9
Acetone	10.0
Isopropyl acetate	8.4
Carbon tetrachloride	8.6
Toluene	9.0
Xylene	8.9
Methyl isopropyl ketone	8.4
Cyclohexane	8.2
Turpentine	8.1
Methyl amyl acetate	8.0
Methyl cyclohexane	7.8
Heptane	7.5
Plastics	**Solubility Parameter**
Polytetrafluoroethylene	6.2
Polyethylene	7.9–8.1
Polypropylene	7.9
Polystyrene	8 5–9 7
Polyvinyl acetate	9.4
Polymethyl methacrylate	9.0–9.5
Polyvinyl chloride	9.38–9.5
Bisphenol A polycarbonate	9.5
Polyvinylidene chloride	9.8
Polyethylene terephthalate	10.7
Cellulose nitrate	10.56–10.48
Cellulose acetate	11.35
Epoxide	11.0
Polyacetal	11.1
Polyamide 6, 6	13.6
Coumarone indene	8.0–10.6
Alkyd	7.0–11.2

Table D-4. Identification of Selected Plastics by Solvent Test Methods

Plastics	Acetone	Benzene	Furfuryl Alcohol	Toluene	Special Solvents
ABS	Insoluble	Partially soluble	Insoluble	Soluble	Ethylene dichloride
Acrylic	Soluble	Soluble	Partially soluble	Soluble	Ethylene dichloride
Cellulose acetate	Soluble	Partially soluble	Soluble	Partially soluble	Acetic acid
Cellulose acetate, butyrate	Soluble	Partially soluble	Soluble	Partially soluble	Ethyl acetate
Fluorocarbon	Insoluble (most)	Insoluble	Insoluble	Insoluble	Dimethyacetamide (not FEP-TFE)
Polyamide	Insoluble	Insoluble	Insoluble	Insoluble	Hot aqueous ethanol
Polycarbonate	Partially soluble	Partially soluble	Insoluble	Partially soluble	Hot benzene-toluene
Polyethylene	Insoluble	Insoluble	Insoluble	Insoluble	Hot benzene-toluene
Polypropylene	Insoluble	Insoluble	Insoluble	Insoluble	Hot benzene-toluene
Polystyrene	Soluble	Soluble	Partially soluble	Soluble	Methylene dichloride
Vinyl acetate	Soluble	Soluble	Insoluble	Soluble	Cyclohexanol
Vinyl chloride	Soluble	Insoluble	Insoluble	Partially soluble	Cyclohexanol

All of the solutions may be stored in clean containers and reused; however, the density of solutions should be checked during the testing procedure. Factors such as temperature and evaporation may radically change the relative density value.

The strength-to-mass ratio of a plastics can also help in identification. Plastics usually weigh considerably less in relation to volume than metals and most other materials. A reinforced structural foam plastics with a density of 35 lbs/ft^3 [550 kg/m^3] and a tensile strength of 100 x 10^6 psi [700 MPa] would have a strength-to-mass ratio of 550 kg/m^3 [700 MPa] = 1.272. Steel, with a tensile strength of 0.29 x 10^6 psi [2000 MPa] and a density 484 lbs/ft^3 [7 750 kg/m^3], would have a ratio of 0.258. These ratios are sometimes used in design criteria.

Appendix E

Thermoplastics

It is important that you become familiar with thermoplastic plastics. In this appendix look for the outstanding properties and applications of each plastics since plastics properties affect product design, processing, economics, and service.

This appendix treats individual groups of thermoplastics in alphabetical order:

- acetals (polyacetals)
- acrylics
- acrylic-styrene-acrylonitrile
- acrylonitrile-chlorinated polyethylene styrene
- cellulosics
- chlorinated polyethers
- coumarone-indene plastics
- ethylene acid
- ethylene-ethyl acrylate
- ethylene-methyl acrylate
- ethylene-vinyl acetate
- fluorplastics
- ionomers
- nitrile barrier plastics
- phenoxy
- polyallomers
- polyamides
- polyarylsulfone
- polyetheretherketone
- polyetherimide
- polycarbonates
- polyphenylene ether
- polymethylpentene
- polyolefins
- polyphenylene oxides
- polystyrene
- polysulfones
- polyvinyls
- styrene-maleic anhydride
- styrene-butadiene plastics
- thermoplastic polyesters
- thermoplastic polyimides.

Polyacetal Plastics (POM)

A highly reactive gas, formaldehyde (CH_2O), may be polymerized in a number of ways. Formaldehyde, or *methanal,* is the simplest of the aldehyde group of chemicals. The ending for the aldehydes is *al*, derived from the first syllable of aldehyde.

Simple polymers based on formaldehyde have been known since 1859. The first polyformaldehyde was put on the market by Du Pont in 1960. The polyformaldehyde (polyacetal) polymer is basically a linear, highly crystalline, long molecular structure. The term *acetal* refers to the oxygen atom that joins the repeating units of the polymer structure. Polyoxymethylene (POM) is the correct chemical term for this polymer. Acetal is a generic term.

A number of initiators or catalysts are used to polymerize the basic polyacetal resin including acids, bases, metallic compounds, cobalt, and nickel.

The polyacetal structure is:

$$H\text{-}O\text{-}(CH_2\text{-}O\text{-}CH_2\text{-}O)nH : R$$

OR

$$n\,\underset{H}{\overset{H}{C}}=0 \rightarrow \underset{O}{CH_2}\;\underset{O}{CH_2}\;\underset{O}{CH_2}\;\underset{O}{CH_2}\;\underset{O}{CH_2}\;\underset{O}{CH_2}\;R$$

R = Ether or Ester

The best known polyformaldehyde plastics is the oxymethylene linear structure with attached terminal groups. There are, however, many miscellaneous aldehyde-derived polymers.

Thermal and chemical resistance is increased when esters or ethers are attached as terminal groups. Both esters and ethers are relatively inert toward most chemical reagents. They are chemically compatible in organic chemical reactions.

H-O-H	R-O-R	$R\text{-}\overset{O}{\overset{\|}{C}}\text{-}OH$	$R\text{-}\overset{O}{\overset{\|}{C}}\text{-}OR$
Water	*Ether*	*Carboxylic Acid*	*Ester*

The formulas above show some of the structural relationships between water, ethers, carboxylic acids, and esters.

Close packing and short bond lengths are typical of polyacetal plastics, providing a hard, rigid, dimensionally stable material. Polyacetals have high resistance to organic chemicals and a wide temperature range (Fig. E-1).

Polyacetals are easy to fabricate, offer properties not found in metals, and are competitive in cost and performance with many nonferrous metals. They are similar to polyamides in many respects. Acetals provide superior fatigue endurance, creep resistance, stiffness, and water resistance. They are among the strongest and stiffest thermoplastics and may be filled for even greater strength, dimensional stability, abrasion resistance and improved electrical properties (Fig. E-2).

At room temperatures, polyacetals are resistant to most chemicals, stains, and organic solvents. These include tea, beet juice, oils, and household detergents. Hot coffee, however, will usually cause staining. Resistance to strong acids, strong alkalies, and oxidizing agents is poor. Copolymerization and filling improves the chemical resistance of the material.

Moisture and thermal resistance are characteristic of acetal polymers. For this reason, they are used for plumbing fixtures, pump impellers, conveyor belts, aerosol stem valves, and shower heads.

Acetals must be protected from prolonged exposure to ultraviolet light. Such exposure causes surface chalking, reduced molecular mass, and slow degradation. Painting, plating, and/or filling with carbon black or ultraviolet-absorbing chemicals will protect acetal products for outdoor use.

Acetals are available in pellet or powder form for processing in conventional injection molding, blow molding, and extrusion machines. Because of the highly crystalline structure of polyacetals, it is not possible to make optically transparent film. The operator must have adequate ventilation when processing polyacetal materials because upon degrading at high temperature, acetals release a toxic and potentially lethal gas.

Table E-1 gives some of the most important properties of acetal plastics while a list of six advantages and disadvantages follows.

Advantages of Polyacetals

1. High tensile strength with rigidity and toughness
2. Excellent dimensional stability
3. Glossy molded surfaces

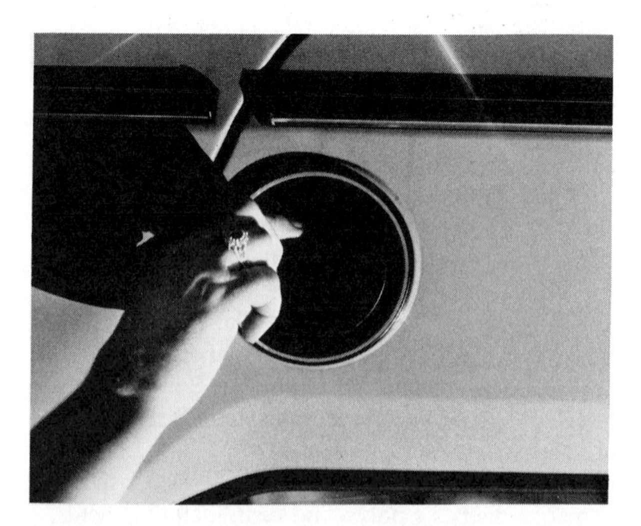

Fig. E-1. This automobile door handle made of polyacetal plastics will remain strong and keep its glossy finish even though it will be exposed to all kinds of weather and the ultra-violet rays of the sun. (Du Pont Co.)

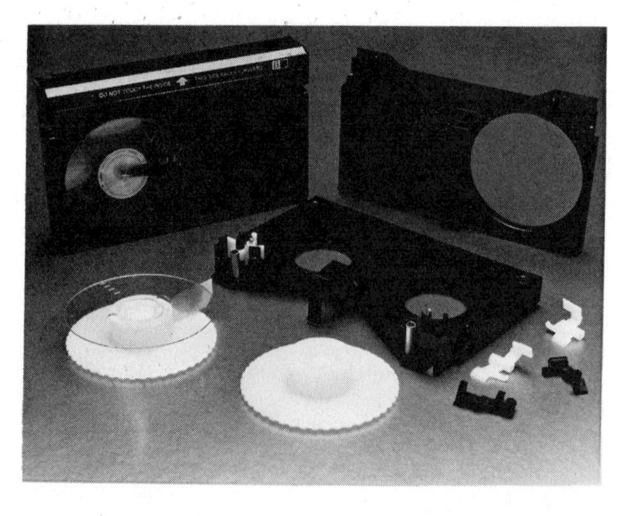

Fig. E-2. These high-stress videocassette parts are molded from acetal plastics. (Du Pont Co.)

Table E-1. Properties of Acetals

Property	Acetal (Homopolymer)	Acetal (20% Glass-Filled)
Molding qualities	Excellent	Good to Excellent
Relative density	1.42	1.56
Tensile strength, MPa	68.9	58.6–75.8
(psi)	(10 000)	(8 500–11 000)
Compressive strength (10% defl.), MPa	124	124
(psi)	(18 000)	(18 000)
Impact strength, Izod, J/mm	0.07 (Inj) 0.115 (Ext)	0.04
(ft lb/in)	1.4 (Inj) 2.3 (Ext)	(0.8)
Hardness, Rockwell	M94, R120	M75–M90
Thermal expansion, (10^{-4}/°C)	20.6	9–20.6
Resistance to heat, °C	90	85–105
(°F)	(195)	(185–220)
Dielectric strength, V/mm	14 960	22 835
Dielectric constant (at 60 Hz)	3.7	3.9
Dissipation factor (at 60 Hz)	3.7	3.9
Arc resistance, s	129	136
Water absorption (24 h), %	0.25	0.25–0.29
Burning rate, mm/min	Slow to 28	20–25.4
(in/min)	(Slow to 1.1)	(0.8–1.0)
Effect of sunlight	Chalks slightly	Chalks slightly
Effect of acids	Resists some	Resists some
Effect of alkalies	Resists some	Resists some
Effect of solvents	Excellent resistance	Excellent resistance
Machining qualities	Excellent	Good to fair
Optical qualities	Translucent to opaque	Opaque

4. Low static and coefficient of friction
5. Retention of electrical and mechanical properties to 248 °F [120 °C]
6. Low gas and vapor permeability

Disadvantages of Polyacetals

1. Poor resistance to acids and bases
2. Subject to UV degradation
3. Flammable
4. Unsuitable for contact with food
5. Difficult to bond
6. Toxic, releases fumes upon degradation

Acrylics

In 1901, Otto Rohm reported much of the research that later led to the commercial exploitation of acrylics. Dr. Rohm, pursuing his research in Germany, took an active part in the first commercial development of polyacrylates in 1927. By 1931, there was a Rohm and Haas Company plant operating in the United States. Most of these early materials were used as coatings or for aircraft components. For example, acrylics were used for windshields and bubble turrets on aircraft during World War II. From these early compounds, an extensive group of monomers has become available, and commercial applications of these polymers have grown steadily.

The term *acrylic* includes acrylic and methacrylic esters, acids, and other derivatives. The principal acid and ester monomers are shown in Table E-2. In order to avoid possible confusion, the basic acrylic formula is shown, with possible side groups R_1, and R_2, in Figure E-3.

There are many monomer possibilities and preparation methods. The most important is the commercial preparation of methyl methacrylate from acetone cyanohydrin. These homomonomers and comonomers may be polymerized by various commercial methods, including bulk, solution, emulsion, suspension, and granulation polymerization. In all cases, an organic peroxide catalyst is used to start polymerization. Many

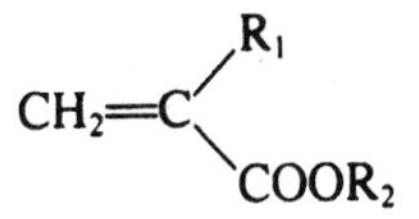

$$CH_2{=}C(R_1)COOR_2$$

(A) Basic acrylic formula.

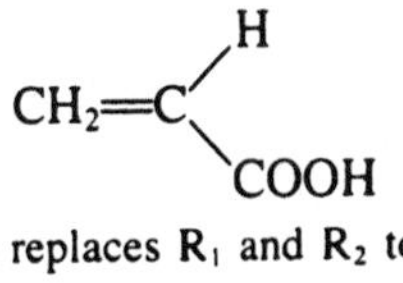

$$CH_2{=}C(H)COOH$$

(B) Hydrogen replaces R_1 and R_2 to produce acrylic acid.

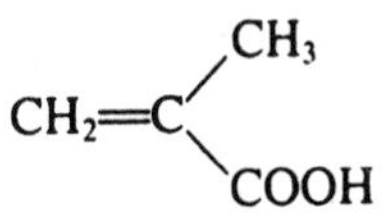

$$CH_2{=}C(CH_3)COOH$$

(C) Methyl group replaces R_1 to produce methacrylic acid.

Fig. E-3. Acrylic formula, with two possible radical replacements.

Table E-2. Principal Acid and Ester Monomers

Acrylic acid	Methyl acrylate	Ethyl acrylate	*n*-Butyl acrylate	Isobutyl acrylate	2-Ethylhexyl acrylate
$CH_2 = CHCOOH$	$CH_2 = CHCOOCH_3$	$CH_2 = CHCOOC_2H_5$	$CH_2 = CHCOOC_4H_9$	$CH_2 = CHCOOCH_2CH(CH_3)_2$	$CH_2 = CHCOOCH_2CH(C_2H_5)C_4H_9$

Methacrylic acid	Methyl methacrylate	Ethyl methacrylate	n-Butyl methacrylate	Lauryl methacrylate
$CH_2 = CCOOH$	$CH_2 = CCOOCH_3$	$CH_2 = CCOOC_2H_5$	$CH_2 = CCOOC_4H_9$	$CH_2 = CCOO(CH_2)_nCH_3$
|	|	|	|	|
CH_3	CH_3	CH_3	CH_3	CH_3

Stearyl methacrylate	2-Hydroxyethyl methacrylate	Hydroxypropyl methacrylate	2-Dimethylaminoethyl methacrylate	2-t-Butylaminoethyl methacrylate
$CH_2 = CCOO(CH_2)_6CH_3$	$CH_2 = CCOOCH_2CH_2OH$	$CH_2 = CCOO(C_3H_6)OH_4$	$CH_2 = CCOOCH_2CH_2N(CH_3)_2$	$CH_2 = CCOOCH_2CH_2NHC(CH_3)_3$
|	|	|	|	|
CH_3	CH_3	CH_3	CH_3	CH_3

of the molding powders are made by emulsion methods. Bulk polymerization is used for casting sheets and profile shapes.

The versatility of acrylic monomers in processing, copolymerization, and ultimate, or final state, properties has contributed to their wide use. Table E-3 gives some of the basic properties of acrylics.

Polymethyl methacrylate is an atactic, amorphous, transparent thermoplastic. Because of its high transparency (about 92 percent), it is used for many optical applications (Fig. E-4A). It is a good electrical insulator for low frequencies, and has very good resistance to weathering (Fig. E-4B). Outdoor advertising signs are a familiar use of acrylics.

Polymethyl methacrylate is a standard material for automobile taillight lenses and covers (Fig. E-4C.) This material is used for aircraft windshields and cockpit covers, and for bubble bodies on helicopters.

Polymethyl methacrylate may be produced by any usual thermoplastic process. It may be fabricated by solvent cementing. Cast and extruded sheets and profile shapes are popular forms. Sheet forms are widely

Table E-3. Properties of Acrylics

Property	Methyl Methacrylate (Molding)	Acrylic-PVC Copolymer (Molding)	ABS (High Impact)
Molding qualities	Excellent		Good to excellent
Relative density	1.17–1.20	1.30	1.01–1.04
Tensile strength, MPa	48–76	38	30–53
(psi)	(7 000–11 000)	(5 500)	(4 500–7 500)
Compressive strength, MPa	83–125	43	30–55
(psi)	(12 000–18 000)	(6 200)	(4 500–8 000)
Impact strength, Izod J/mm	0.015–0.025	0.75	0.25–0.4*
(ft lb/in)	(0.3–0.5)	(15)	(5.0–8.0)†
Hardness, Rockwell	M85–M105	R104	R75–R105
Thermal expansion, 10^{-4}/°C	12–23	12–29	24–33
Resistance to heat, °C	60–94	60–98	60–98
(°F)	(140–200)	(140–210)	(140–210)
Dielectric strength. V/mm	15 800–20 000	15 800	13 800–18 000
Dielectric constant (at 60 Hz)	3.3–3.9	4	2.4–5.0
Dissipation factor (at 60 Hz)	0.04–0.06	0.04	0.003–0.008
Arc resistance, s	No track	25	50–85
Water absorption (24 h), %	0.1–0.4	0.13	0.20–0.45
Burning rate, mm/min	Slow 0.5–30	Nonburning	Slow to self-extinguishing
(in/min)	(Slow 0.6–1.2)		
Effect of sunlight	Nil	Nil	Yellows
Effect of acids	Attacked by strong oxidizing acids	Slight	Attacked by strong oxidizing acids
Effect of alkalies	Attacked	None	None
Effect of solvents	Soluble in ketones, esters, and aromatic and chlorinated hydrocarbons	Attacked by ketones esters, and aromatic and chlorinated hydrocarbons	Soluble in ketones and esters
Machining qualities	Good to excellent	Excellent	Excellent
Optical qualities	Transparent to opaque	Opaque	Translucent

Notes: *At 23 °C, 3x 12 mmL bar
†At 73 °F, 1/8 x ½ in bar

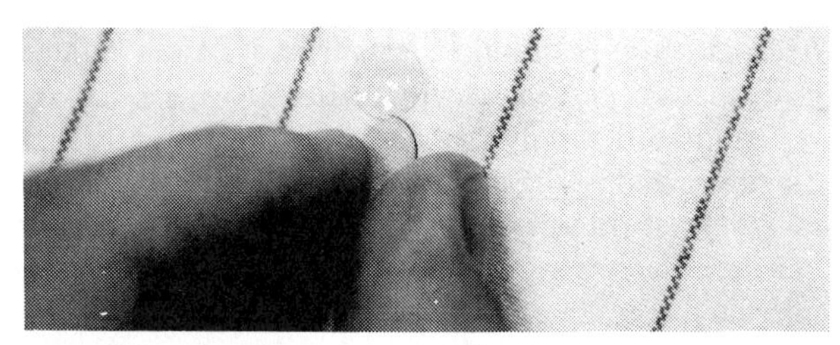

(A) Contact lenses.

(B) Panels on buildings.

(C) Tailgate lenses.

(D) Wall paints.

Fig. E-4. Acrylics have many useful properties.

used for room dividers and skylight domes, and as a substitute for glass in windows. There has been wide use of these plastics in the paint industry in the form of emulsions (Fig. E-4D). Emulsion acrylics are popular as a clear, hard, and glossy "wax" coating for floors. Acrylic-based adhesives are available with a wide range of uses and properties. These adhesives are transparent and available in solvent-based (air-drying), hot-melt, or pressure-sensitive forms. Figure E-5 shows an acrylic sealant being applied directly to an oily aluminum window frame under water.

Fig. E-5. An acrylic sealant being applied under water. (Cabot Corp.)

Glass-reinforced sheets are used to produce sanitary ware, vanities, tubs, and counters. Protective liquid coatings, known as gel coats, may be used with reinforcements as cover stock. Heavily filled and reinforced resins formulated to cross-link into a thermosetting matrix are used to produce marble-like bathroom fixtures and furniture.

Well-known trade names for polymethyl methacrylate are Plexiglas, Lucite, and Acrylite. From the following lists you can see there are more advantages (11) of acrylics than disadvantages (5).

Advantages of Acrylics

1. Wide range of colors
2. Outstanding optical clarity
3. Slow-burning, releasing little or no smoke
4. Excellent weatherability and ultraviolet resistance
5. Ease of fabrication
6. Excellent electrical properties
7. Unaffected by food and human tissues
8. Rigidity with good impact strength
9. High gloss and good *feel*
10. Excellent dimensional stability and low mold shrinkage
11. Stretch forming improves biaxial toughness

Disadvantages of Acrylics

1. Poor solvent resistance
2. Possibility of stress cracking
3. Combustibility
4. Limited continuous service temperature of 93 °C [200 °F]
5. Inflexibility

Polyacrylates

Polyacrylates are transparent, resistant to chemicals and weather, and have a low softening point. Applications include films, adhesives, and surface coatings for paper and textiles. They are usually copolymer compositions. Polyethyl acrylate may be cross-linked to form thermosetting elastomers. Polyacrylate monomers are used as plasticizers for other vinyl polymers.

Acrylic esters may be obtained from the reaction of ethylene cyanohydrin with sulphuric acid and an alcohol.

$$HO \cdot CH_2 \cdot CH_2 \cdot CN \xrightarrow[H_2SO_4]{ROH} CH_2 : CH \cdot CO \cdot O \cdot R$$

Acrylic and polyvinyl chloride (PMMA/PVC) may be alloyed to produce a tough, durable sheet, easily thermoformed into signs, aircraft trays, and public service seating. (Fig. E-6)

Polyacrylonitrile and Polyrmethacrylonitrile

The elastomers and fibers produced from polyacrylonitrile and polymethacrylonitrile materials were merely laboratory curiosities before World War II. Since that time there has been a rapid expansion of the use of acrylonitrile often as the main ingredient in acrylic fibers. These polymers are copolymerized and stretched to orient the molecular chain. Orlon and Dynel, classified as *modacrylic fibers,* contain less than 85 percent acrylonitrile. Modacrylic fibers contain at least 35 percent acrylonitrile units. *Acrylic fibers* such as Acrilan, Creslan, and Zefran contain more than 85 percent acrylonitrile.

Fig. E-6. This continuous cast, thermoformed acrylic sheet offers superior impact strength and sunlight resistance. (United States Steel)

Unmodified polyacrylonitrile is only slightly thermoplastic and is difficult to mold because its hydrogen bonds resist flow. Copolymers of styrene, ethyl acrylates, methacrylates, and other monomers are extruded into the amorphous fiber form. At this point, however, the fiber is too weak to be of value. It is then stretched to produce a greater degree of crystallization. Tensile strength is greatly increased as a result of this molecular orientation. (See Nitrile barrier plastics.)

Monomers of acrylonitrile and methacrylonitrile are shown below.

$CH_2{=}CHCN$ — *Acrylonitrile*

$CH_2{=}C(CH_3){-}CN$ — *Methacrylonitrile*

Acrylic-Styrene-Acrylonitrile (ASA)

The terpolymer of acrylic-styrene-acrylonitrile may vary in the percentages of each component to enhance or tailor a specific property. The excellent surface gloss makes it attractive as a cover layer for coextrusion over ABS, PC, or PVC. Exterior applications include signs, downspouts, siding, recreational equipment, camper tops, ATV bodies, and gutters.

This material is also blended with PC (ASA/ PC) and alloyed with PVC (ASA/PVC) and PMMA (ASA/ PMMA) to enhance specific properties. ASA/PMMA alloys have outstanding weatherability, gloss, and toughness. They are used to produce spas and hot tubs. (See Alloy and Blend)

Acrylonitrile-Butadiene-Styrene (ABS)

ABS polymers are opaque thermoplastic resins formed by the polymerization of acrylonitrilebutadiene-styrene monomers. Because they possess such a diverse combination of properties, many experts classify them as a family of plastics. They are, however, terpolymers ("ter" meaning three) of three monomers and *not* a distinct family. Their development resulted from research efforts on synthetic rubber during and after World War II. The proportions of the three ingredients may vary, which accounts for the great number of possible properties.

The three ingredients are shown below. Acrylonitrile is also known as vinyl cyanide and acrylic nitrile.

$CH_2{=}CHCN$ — *Acrylonitrile*

$CH_2{=}CH{-}CH{=}CH_2$ — *Butadiene*

$C_6H_5{-}CH{=}CH_2$ — *Styrene*

Below is a representative structure for acrylonitrile-butadiene-styrene.

```
H   H   H            H       H   H   H             H
|   |   |            |       |   |   |             |
C — C — C ————————— C ————— C — C — C ——————————— C ——
|   |   |            |       |   |   |             |
H  [C6H5] C≡N   H — C — H    H [C6H5] C≡N     H — C — H
                     |                             |
                     C — H                         C — H
                     |                             |
                     C — H                         C — H
                     |                             |
                 H — C — H                     H — C — H
                     |                             |
                     H                             H
```

Acrylonitrile-Butadiene-Styrene

Graft polymerization techniques are commonly used to make various grades of this material.

The resins are hygroscopic (moisture-absorbing). Predrying before molding is advisable. ABS materials can be produced on all thermoplastic processing machines.

ABS materials are characterized by resistance to chemicals, heat, and impact. They are used for appliance housings, light luggage, camera bodies, pipe, power tool housings, automotive trim, battery cases, tool boxes, packing crates, radio cases, cabinets, and various furniture components. They may be electroplates and used in automotive, appliance, and housewares applications (Fig. E-7).

Table E-3 lists some of the properties of ABS materials and a reference list of nine advantages and five disadvantages follows.

Advantages of ABS

1. Ease of fabrication and coloring
2. High impact resistance with toughness and rigidity
3. Good electrical properties
4. Excellent adhesion by metal coatings
5. Fairly good weather resistance and high gloss
6. Ease of forming by conventional thermoplastic methods
7. Good chemical resistance
8. Light weight
9. Very low moisture absorption

Fig. E-7. The bodies of these electric powered vehicles used by the U.S. Postal Services are made of ABS thermoplastic. (Borg-Warner Chemicals, Inc.)

Disadvantages of ABS

1. Poor solvent resistance
2. Subject to attack by organic materials of low molecular mass
3. Low dielectric strength
4. Only low elongations available
5. Low continuous service temperature

Acrylonitrile-Chlorinated Polyethylene-Styrene (ACS)

Because of the chlorine content, this terpolymer surpasses ABS in flame-retardant properties, weatherability, and service temperatures. Applications include office machine housing, appliance cases, and electrical connectors.

ABS/PA blends have excellent chemical and temperature resistance for underhood automobile components. ABS/PC alloys fill the price and performance gap between polycarbonate and ABS. Typical applications include typewriter housing, headlight rings, institutional food trays, and appliance housings. ABS/PVC alloys are used for air conditioner fans, grills, luggage shells, and computer housings because of their outstanding impact strength, toughness, and cost. ABS/EVA alloys have good impact and stress crack resistance. The elastomer content in ABS/EPDM improves low-temperature impact and modulus.

Cellulosics

Cellulose ($C_6H_{10}O_5$) is the material that composes the framework, or cell walls, of all plants, It is our oldest, most familiar, most useful industrial raw material because cellulose is abundant everywhere in one form or another. Plants are also a very inexpensive raw material. From them, we produce shelter, clothing, and food. Cereal straws and grass are composed of nearly 40 percent cellulose, wood is about 50 percent cellulose, and cotton may be nearly 98 percent cellulose. Wood and cotton are major industrial sources of this material. Long-chained molecules of repeating glucose units are referred to as "chemically modified natural plastics."

The chemical structure of cellulose is shown in Figure E-8. Each cellulose molecule contains three hydroxyl (OH) groups at which different groups may attach to form various cellulosic plastics. Cellulose can undergo reaction at the ether linkage between the units.

The term *cellulosics* refers to plastics derived from cellulose, a family that consists of many separate and

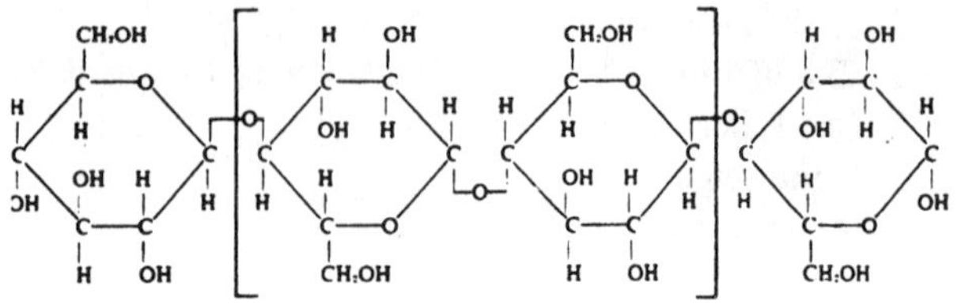

Fig. E-8. Chemical structure of cellulose.

distinct types of plastics. The relationship of cellulose to many plastics and applications is shown in Figure E-9.

There are three large groups of cellulosic plastics. *Regenerated cellulose* is first chemically changed to a soluble material then reconverted by chemical means into its original substance. *Cellulose esters* are formed when various acids react with the hydroxyl (OH) groups of the cellulose. *Cellulose ethers* are compounds derived by the alkylation of cellulose.

Regenerated Cellulose

Regenerated cellulose products are cellophane, viscose rayon, and cuprammonium rayon (no longer of commercial importance).

In its natural form, cellulose is insoluble and incapable of flow by melting. Even in the powdered form, it retains a fibrous structure.

There is evidence that in 1857 cellulose was found to be dissolvable in an ammoniacal solution of copper oxide. By 1897, Germany was commercially producing a fibrous yarn by spinning this solution into an acid or alkaline coagulating bath. Any remaining copper ions were removed by additional acid baths. This expensive process was called the *cuprammonium* process (for copper and ammonia), and the fiber was known as *cuprammonium rayon.* Newer synthetic fibers with equally desirable properties are less costly to produce, therefore cuprammonium rayon has lost its, popularity.

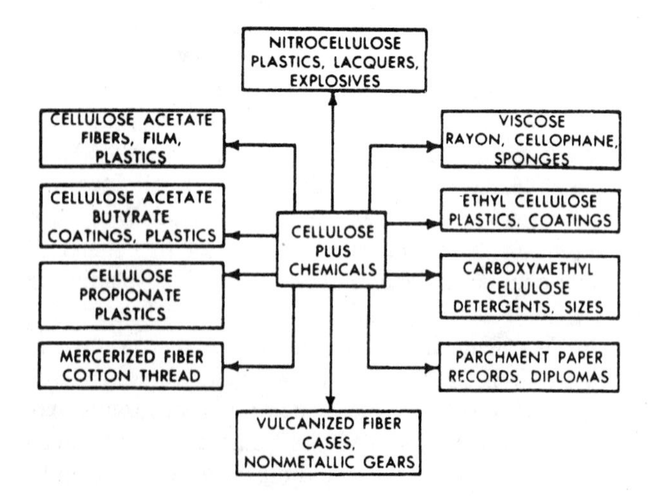

Fig. E-9. The cellulosic plastics. (Du Pont Co.)

In 1892, C. F. Cross and E. J. Bevan of England produced a different cellulose fiber. They treated alkali cellulose (cellulose treated with caustic soda) with carbon disulfide to form a xanthate. Cellulose xanthate is soluble in water to give a viscous solution (hence the name) called *viscose.* Viscose was then extruded through spineret openings into a coagulating solution of sulfuric acid and sodium sulfate. The regenerated fiber is called *viscose rayon.* Rayon has become an accepted generic name for fibers composed of regenerated cellulose. It is still found as a clothing fabric and has some use as tire cord.

Production patents were granted to J. F. Brandenberger of France in 1908 for an extruded, regenerated cellulose film called *cellophane.* Like viscose rayon, the xanthate solution is regenerated by coagulation in an acid bath. After the cellophane is dried, it is usually given a water resistant coating of ethyl acetate or cellulose nitrate. Cellophane films, coated and uncoated, are used for packaging food products and pharmaceuticals.

Cellulose Esters

Among the esters of cellulose are cellulose nitrate, cellulose acetate, cellulose butyrate, and cellulose acetate propionate. In this group of plastics, acids are used to react with the hydroxyl (OH) groups to form esters.

Professor Bracconot of France first carried out the nitration of cellulose in 1832. The discovery that mixing nitric and sulfuric acids with cotton will produce nitrocellulose (or cellulose nitrate) was made by C. F. Schonbein of England. This material was a useful military explosive but it found little commercial value as a plastics. Figure E-10 shows the nitration of cellulose.

At the London International Exhibition in 1862, Alexander Parks of England was awarded a bronze medal for his new plastics material called *Parkesine.* It was composed of cellulose nitrate with a plasticizer of castor oil.

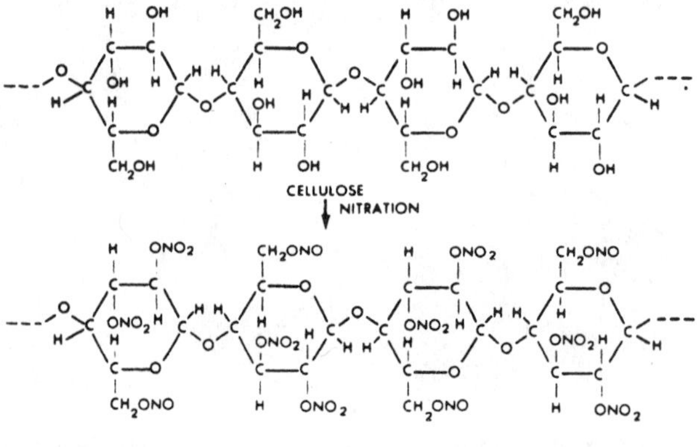

Fig. E-10. Nitration of cellulose used to produce nitrocellulose.

In the United States, John Wesley Hyatt created the same material while seeking a substitute for ivory billiard balls. His experiments followed the earlier work of the American chemist Maynard. Maynard had dissolved cellulose nitrate in ethyl alcohol and ether to form a product used as a bandage for wounds. He gave this solution the name *Collodion*. When the Collodion solution was spread over a wound, the solvent evaporated to leave a thin, protective film. One story of Hyatt's discovery of the material that became known as *Celluloid* tells of his accidently spilling camphor on some pyroxylin (cellulose nitrate) sheets and noticing the improved properties. Another version tells that he had been treating a wound covered by Collodion with a camphor solution and discovered a change in the cellulose nitrate product.

In 1870, Hyatt and his brother patented the process of treating cellulose nitrate with camphor and by 1872, Celluloid became a commercial success. Products made entirely of nitrocellulose were highly explosive, became brittle, and suffered greatly from shrinkage. The use of camphor as a plasticizer eliminated many of these disadvantages. Celluloid is made from pyroxylin (nitrated cellulose), camphor, and alcohol. It is highly combustible but not explosive.

Cellulose Nitrate (CN)

Cellulose nitrate was once used in photographic film, bicycle parts, toys, knife handles, and table-tennis balls (Fig. E-11). Today, it is seldom used because of the difficulty in processing and its high flammability. Cellulose nitrate cannot be injection or compression molded. It is usually extruded or cast into large blocks from which sheets are sliced. Films are made by continuous casting of a cellulose solution on a smooth surface. As the solvents evaporate, the film is removed from the surface and placed on drying rollers. Sheets and films may be vacuum-processed. Table-tennis balls and a few novelty items still may be made of cellulose nitrate plastics. Cellulose nitrate esters are found in lacquers for metal and wood finishes and are common ingredients for aerosol paints and fingernail polishes.

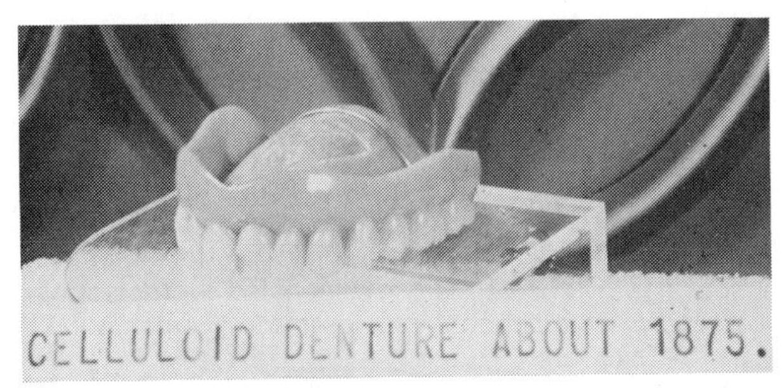

(A) Dental application.

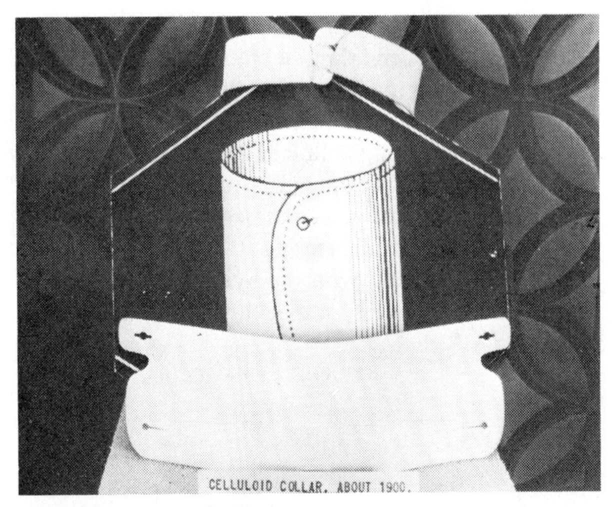

(B) Use in men's clothing.

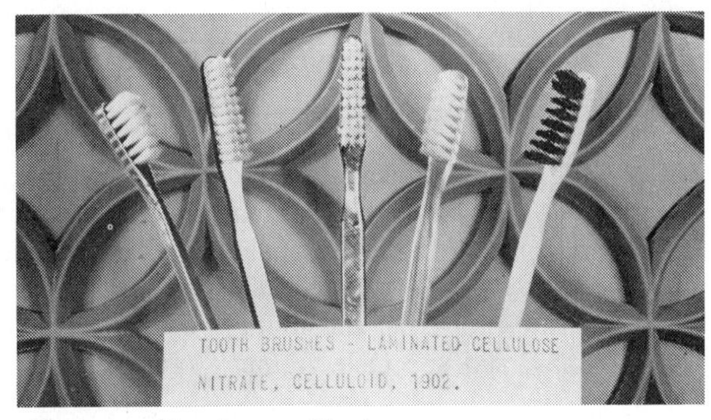

(C) Personal hygiene application.

(D) Film base, which led to growth of photography as a popular hobby.

Fig. E-11. Some early uses of Celluloid. (Celanese Plastic Materials Co.)

Cellulose Acetate (CA)

Cellulose acetate is the most useful of the cellulosic plastics. During World War I, the British secured the aid of Henry and Camille Dreyfus of Switzerland to start large-scale production of cellulose acetate. Cellulose acetate provided a fire-retardant lacquer for the fabric-covered airplanes then in use. By 1929, commercial grades of molding powder, fibers, sheets, and tubes were being made in the United States.

Basic methods for making this material resemble those used in making cellulose nitrate. Acetylation of cellulose is carried out in a mixture of acetic acid and acetic anhydride, using sulfuric acid as a catalyst. The acetate or acetyl group or radical (CH_3CO) is the source of chemical reaction with the hydroxyl (OH) groups. The structure of cellulose triacetate is as follows:

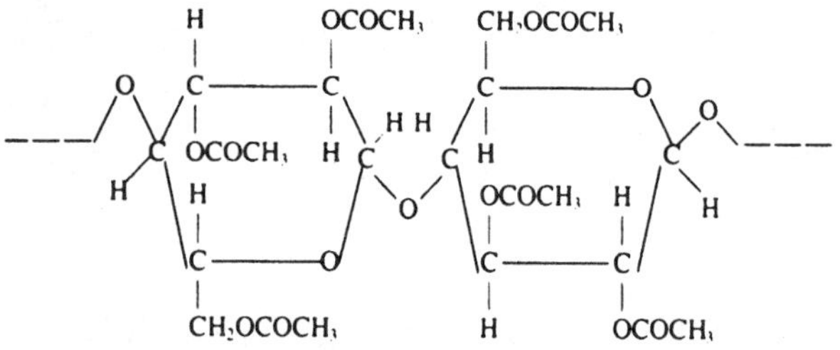

H OCOCH₃ CH₂OCOCH₃
OCOCH₃ H H H H
H OCOCH₃ H H
CH₂OCOCH₃ H OCOCH₃

Cellulose Triactate

Cellulose acetates exhibit poor heat, electrical, weathering, and chemical resistance. They are fairly inexpensive and may be transparent or colored. Their main uses are as films and sheets in the packaging and display industries. They are fabricated by nearly all the thermoplastic processes and molded into brush handles, combs, and spectacle frames. Vacuum-formed display containers for hardware or food products are common and films that permit the passage of moisture and gases are used in commercial packaging of fruits and vegetables. Coated films are used in magnetic recording tapes and photographic film. Cellulose acetate plastics are made into fibers for use in textiles. They are also employed as lacquers in the coating industry. Table E-4 gives some of the properties of cellulose acetate.

Cellulose Acetate Butyrate (CAB)

Cellulose acetate butyrate was developed in the mid-1930s by the Hercules Powder Company and Eastman Chemical. This material is produced by reacting cellulose with a mixture of sulfuric and acetic acids. *Esterification* is completed when the cellulose is reacted with butyric acid and acetic anhydride. The reaction is much like the making of cellulose acetate, except that butyric acid is also used. The product that results has acetyl groups (CH_3CO) and butyl groups ($CH_3CH_2CH_2CH$) in the repeating cellulose unit. This product has improved dimensional stability, weathers well, and is more chemical and moisture resistant than cellulose acetate.

Cellulose acetate butyrate is used for tabulator keys of office machines, automobile parts, tool handles, display signs, strippable coatings, steering wheels,

Table E-4. Properties of Cellulosics

Property	Ethyl Cellulose (Molding)	Cellulose Acetate (Molding)	Cellulose Propionate (Molding)
Molding qualities	Excellent	Excellent	Excellent
Relative density	1.09–1.17	1.22–1.34	1.17–1.24
Tensile strength, MPa	13.8–41.4	13–62	14–53.8
(psi)	(2 000–8 000)	(1 900–9 000)	(2 000–22 000)
Compressive strength, MPa	69–241	14–248	165.5–152
(psi)	(10 000–35 000)	(2 000–36 000)	(2 400–22 000)
Impact strength, Izod, J/mm	0.1–0.43	0.02–0 26	0.025–0.58
(psi)	(2.0–8.5)	(0.4–5.2)	(0.05–11.5)
Hardness, Rockwell	R50–R115	R34–R125	R10–R122
Thermal expansion, 10^{-4}/°C	25–50	20–46	28–43
Resistance to heat, °C	46–85	60–105	68–105
(°F)	(115–185)	(140–220)	(155–220)
Dielectric strength, V/mm	13 800–19 685	9 840–23 620	11 810–17 715
Dielectric constant (at 60 Hz)	3.0–4.2	3.5–7.5	3.7–4.3
Dissipation factor (at 60 Hz)	0.005–0.020	0.01–0.06	0.01–0.04
Arc resistance, s	60–80	50–310	175–190
Water absorption (24 h), %	0.8–1.8	1.7–6.5	1.2–2.8
Burning rate, mm/min	Slow	Slow to self-extinguishing	Slow 25–33
(in/min)			(1.0–1.3)
Effect of sunlight	Slight	Slight	Slight
Effect of acids	Decomposes	Decomposes	Decomposes
Effect of alkalies	Slight	Decomposes	Decomposes
Effect of solvents	Soluble in ketones, esters, chlorinated hydrocarbons, aromatic hydrocarbons	Soluble in ketones, esters, chlorinated hydrocarbons, aromatic hydrocarbons	Soluble in ketones, esters, chlorinated hydrocarbons, aromatic hydrocarbons
Machining qualities	Good	Excellent	Excellent
Optical qualities	Transparent to opaque	Transparent to opaque	Transparent to opaque

tubes, pipes, and packaging components. Probably the most familiar application of this material is in screwdriver handles. (Fig. E-12)

Cellulose Acetate Propionate

Cellulose acetate propionate (also called simply cellulose propionate) was developed by the Celanese Plastics Company in 1931. It found little use until materials shortages developed during World War II. It is made like other acetates with the addition of propionic acid (CH_3CH_2COOH) in place of acetic anhydride. Its general properties are similar to those of cellulose acetate butyrate. However, it exhibits superior heat resistance and lower moisture absorption. The main uses of cellulose acetate propionate include pens, automotive parts, brush handles, steering wheels, toys, novelties, and film packaging displays. Below is a list of nine advantages and four disadvantages of cellulosic esters.

Advantages of Cellulosic Esters

1. Forms glossy moldings by thermoplastic methods
2. Exceptional clarity (butyrates and propionates)
3. Toughness, even at low temperatures
4. Excellent colorability
5. Nonpetrochemical base
6. Wide range of processing characteristics
7. Resists stress cracking
8. Outstanding weatherability (butyrates)
9. Slow burning (except cellulose nitrate)

Disadvantages of Cellulosic Esters

1. Poor solvent resistance
2. Poor resistance to alkaline materials
3. Relatively low compressive strength
4. Flammability

Cellulose Ethers

Among the ethers of cellulose are ethyl cellulose, methyl cellulose, hydroxymethyl cellulose, carboxymethyl cellulose, and benzyl cellulose. Their manufacture generally involves the preparation of alkali cellulose and other reactants as shown in Figure E-13.

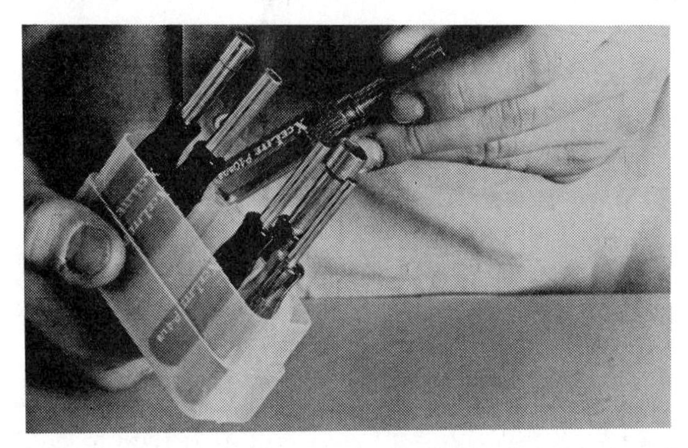

Fig. E-12. Cellulose butyrate and cellulose acetate are used for the handles of these tools. (Eastman Chemical Products, Inc.)

Ethyl cellulose (EC) is the most important of the cellulose ethers and is the only one used as a plastics material. Basic research and patents were established by Dreyfus in 1912. By 1934, the Hercules Powder Company offered commercial grades in the United States.

Alkali cellulose (soda cellulose) is treated with ethyl chloride to form ethyl cellulose. The radical substitution (*etherification*) of this ethoxy may cause the final product to vary over a wide range of properties. In etherification the hydrogen atoms of the hydroxyl groups are replaced by ethyl (C_2H_5) groups (Fig. E-14). This cellulose plastic is tough, flexible, and moisture resistant. Table E-5 gives some of the properties of ethyl cellulose.

Ethyl cellulose is used in football helmets, flashlight cases, furniture trim, cosmetic packages, tool handles, and blister packages. It has been used for protective coatings on bowling pins and in the formulations of paint, varnish, and lacquers. Ethyl cellulose is a common ingredient in hair sprays. It is often used as a hot melt for strippable coatings. These coatings protect metal parts against corrosion and marring during shipment and storage.

Methyl cellulose is prepared like ethyl cellulose, using methyl chloride or methyl sulphate instead of ethyl chloride. In etherification, the hydrogen of the OH group is replaced by methyl (CH_3) groups:

$$R(ONa)_{3n} + CH_3Cl \longrightarrow R(OCH)_{3n}$$

Methyl cellulose

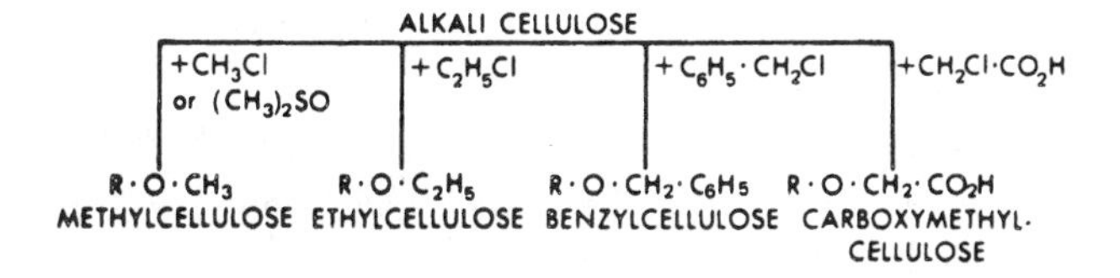

Fig. E-13. Cellulose plastics from alkali cellulose and other reactants. The R in these formulas represents the cellulose skeleton.

Fig. E-14. Ethyl cellulose (fully ethylated.)

Table E-5. Properties of Chlorinated Polyether

Property	Chlorinated Polyether
Molding qualities	Excellent
Relative density	1.4
Tensile strength, MPa	41
(psi)	(6 000)
Impact strength, Izod, J/mm	0.002
(ft lb/in)	(0.4)
Hardness, Rockwell	R100
Thermal expansion, 10^{-4}/°C	20
Resistance to heat, °C	143
(°F)	(290)
Dielectric strength, V/mm	16 000
Dielectric constant (at 60 Hz)	3.1
Dissipation factor (at 60 Hz)	0.01
Water absorption (24 h), %	0.01
Burning rate, mm/min	Self-extinguishing
Effect of sunlight	Slight
Effect of acids	Attacked by oxidizing acids
Effect of alkalies	None
Effect of solvents	Resistant
Machining qualities	Excellent
Optical qualities	Translucent to opaque

Methyl cellulose finds a wide number of uses. It is water soluble and edible. It is used as a thickening emulsifier in cosmetics and adhesives, and is a well-known wallpaper adhesive and fabric-sizing material. It is useful for thickening and emulsifying water-based paints, salad dressings, ice cream, cake mixes, pie filling, crackers, and other products. In pharmaceuticals, it is used to coat pills and as contact lens solutions (Fig. E-15).

Hydroxyethyl cellulose is produced by reacting the alkali cellulose with ethylene oxide. It has many of the same applications as methyl cellulose. In the schematic equation below, R represents the cellulose skeleton.

$$R(ONa)_{3n} + \overset{O}{\overbrace{CH_2CH_2}} \rightarrow R(OCH_2CH_2OH)_{3n}$$

Hydroxyethyl cellulose

At one time, *benzyl cellulose* was used for moldings and extrusions. In the United States, however, it is unable to compete with other polymers.

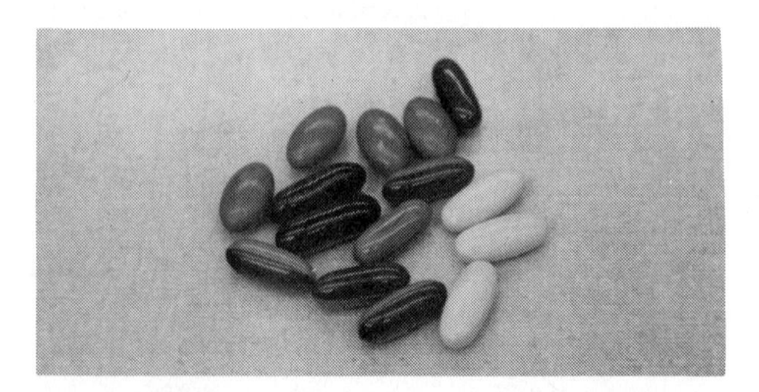

Fig. E-15. Methyl cellulose is used as a coating for pharmaceuticals.

Carboxymethyl cellulose (sometimes called sodium carboxymethyl cellulose) is made from alkali cellulose and sodium chloracetate. Like methyl cellulose, it is water soluble and used as a sizing, gum, or emulsifying agent. Carboxymethyl cellulose may be found in foods, pharmaceuticals, and coatings. It is a first rate water-soluble suspending agent for lotions, jelly bases, ointments, toothpastes, paints, and soaps. It is used to coat pills, paper, and textiles.

$$R(ONa)_{3n} + ClCH_2\ COONa \rightarrow R(OCH_2COONa)_{3n} + NaCl$$

Sodium Carboxymethyl Cellulose

Chlorinated Polyethers

In 1959, the Hercules Powder Company introduced *chlorinated polyethers* with the trade name Penton. In 1972, however, Hercules discontinued production and sale. This thermoplastic material was produced by chlorinating pentaerythritol. The resulting dichloromethyl oxycyclobutane was polymerized into a crystalline, linear product. This material was over 45 percent chlorine by mass (Fig. E-16).

Chlorinated polyethers may be processed on thermoplastic equipment. These materials are high-performance, high-priced plastics (Table E-5). They have been coated on metal substrates by fluidized-bed, flame-spraying, or solvent processes. Molded parts possess high strength, heat resistance, excellent electrical and chemical resistance, and low water absorption. Although the high price restricted their wider use, they did find use as coatings for valves, pumps, and meters. Molded parts included chemical meter components, pipe lining, valves, laboratory equipment, and electrical insulation. At degrading temperatures, lethal chlorine gas is released, however.

To date, there are no plans for another producer to make this plastics with its distinct chemical nature.

$$OHCH_2-C(CH_2OH)_2-CH_2OH \xrightarrow{\text{CHLORINATION}} ClH_2C-C(CH_2Cl)_2-CHOH \rightarrow$$

(PENTAERYTHRITOL)

$$\rightarrow ClH_2C-C(CH_2Cl)(CH_2-O)-CH_2 \rightarrow \left[-H_2C-C(CH_2Cl)_2-CH_2-O-\right]_n$$

DICHLOROMETHYL OXYCYCLOBUTANE — CHLORINATED POLYETHER PENTON

Fig. E-16. Production of Penthol, a chlorinated polyether.

Coumarone-Indene Plastics

Coumarone and indene are obtained from the fractionation of coal tar, but are seldom separated. These inexpensive products resemble styrene in chemical structure. Coumarone and indene may be polymerized by the ionic catalytic action of sulfuric acid. A wide range of products, from sticky resins to brittle plastics, may be obtained by varying the coumarone-indene ratio or by copolymerization with other polymers.

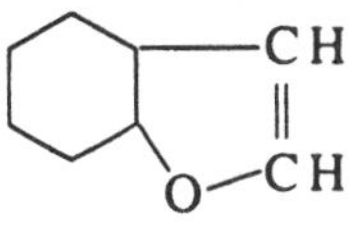

Coumarone 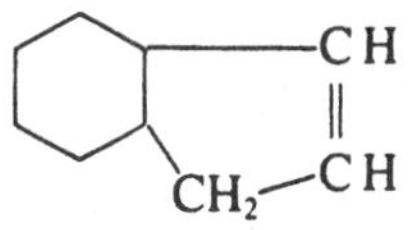 *Indene* 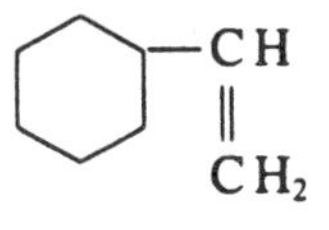 *Styrene*

Although available since long before World War II, they are not available as molding compounds and have found only limited uses. They are used as binders, modifiers, or extenders for other polymers and compounds. The largest quantities of coumarone and indene are used in the making of paints or varnishes and as binders in flooring tiles and mats.

The traits of these compounds vary greatly. Coumarone-indene copolymers are good electrical insulators. They are soluble in hydrocarbons, ketones, and esters. They range from light to dark in color and are inexpensive to make. They are true thermoplastic materials with a softening point of 100 to 120 °F [<35 °C to >50 °C]. Applications include printing inks, coatings for paper, adhesives, encapsulation compounds, some battery boxes, brake linings, caulking compounds, chewing gum, concrete-curing compounds, and emulsion binders.

Fluoroplastics

Elements in the seventh column of the Periodic Table are closely related. These elements (fluorine, chlorine, bromine, iodine, and astatine) are called *halogens* from the Greek word that means "salt-producing." Chlorine is found in common table salt. All of these elements are electronegative (they can attract and hold valence electrons) having only seven electrons in their outermost shell. Fluorine and chlorine are gases not found in a pure or free state. Fluorine is the most reactive element known. Large quantities of fluorine are needed for processes connected with nuclear energy technology, such as isolation of uranium metal.

Compounds containing fluorine are commonly called fluorocarbons although in the strictest definition, *fluorocarbon* should be used to refer to compounds containing only fluorine and carbon.

The French chemist Moissan isolated pure fluorine in 1886, but it remained a laboratory curiosity until 1930. In 1931, the trade name Freon was announced. This fluorocarbon is a compound of carbon, chlorine, and fluorine, for example, CCl_2F_2. Freons are used extensively as refrigerants. Fluorocarbons are also used in the formation of polymeric materials.

In 1938, the first polyfluorocarbon was discovered by accident in the Du Pont research laboratories. It was discovered that tetrafluoroethylene gas formed an insoluble, waxy, white powder when stored in steel cylinders. As a result of this chance discovery, a number of fluorocarbon polymers have been developed.

The term *fluoroplastic* is used to describe alkene-like structures that have some or all of the hydrogen atoms replaced by fluorine.

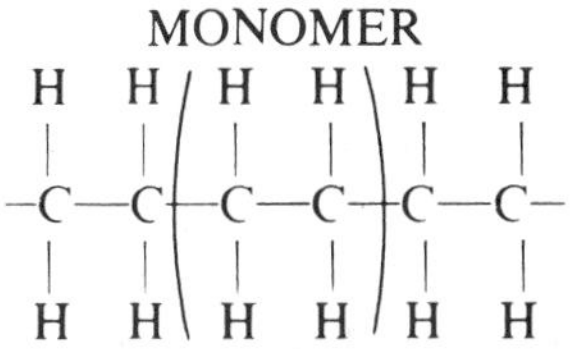

MONOMER MONOMER

Polyethylene 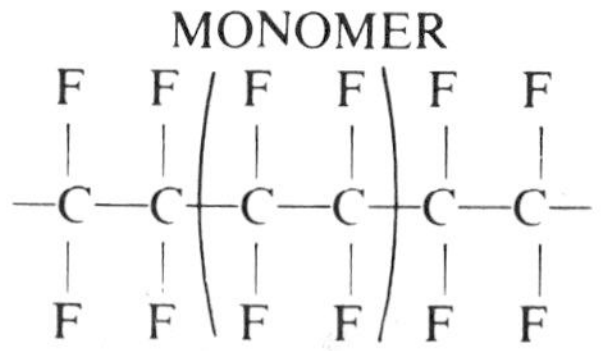 *Polytetrafluoroethylene*

It is the presence of the fluorine atoms that provides the unique properties characteristic of the fluoroplastic family. These properties are directly related to the high carbon-to-fluorine bonding energy and to the high electronegativity of the fluorine atoms. Thermal stability and solvent, electrical, and chemical resistance are weakened if fluorine (F) atoms are replaced with hydrogen (H) or chlorine (Cl) atoms. The C—H and C—Cl bonds are weaker than C—F bonds and are more vulnerable to chemical attack and thermal decomposition.

The major fluoroplastics are shown in Figure E-17. There are only two types of fluorocarbon plastics: polytetrafluoroethylene (PTFE) and polytetrafluoropropylene (FEP or fluorinated ethylene propylene). The others must be considered copolymers or fluorine-containing polymers.

Polytetrafluoroethylene (FTFE)

Polytetrafluoroethylene $(CF_2{=}CF_2)_n$ accounts for nearly 90 percent (by volume) of the fluorinated plastics. The monomer tetrafluoroethylene is obtained by pyrolysis of chlorodifluoromethane. Tetrafluoroethylene is polymerized in the presence of water and a peroxide catalyst under high pressure. PTFE is a highly crystalline, waxy thermoplastic material with a service temperature of –450 °F to +550 °F [–268 °C to +288 °C]. The high bonding strength and compact interlocking of fluorine atoms about the carbon backbone prevent PTFE processing by the usual thermoplastic methods. At present, it cannot be plasticized to aid pro-

```
   F  F            F  F
   |  |            |  |
 —C——C—          —C——C—
   |  |            |  |
  Cl  F            F  F
```

(A) Polychlorotrifluoroethylene. **(B)** Polytetrafluoroethylene.

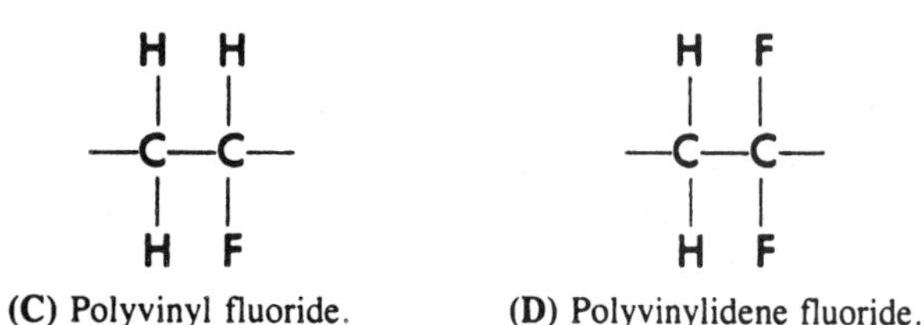

(C) Polyvinyl fluoride. **(D)** Polyvinylidene fluoride.

```
      F       F
      |       |
      C═══════C
      |       |
   F—C—F      F
      |
      F
```

(E) Polyhexafluoropropylene.

Fig. E-17. Monomers of fluoroplastics and fluorine-containing polymers.

(A) Cookware and utensils coated with Teflon-II.

(B) Tools coated with Teflon-S.

Fig. E-18. Some applications of Teflon. (Chemplast, Inc.)

cessing. Most of the material is made into preforms and sintered.

Sintering is a special fabricating technique used for metals and plastics. The powdered material is pressed into a mold at a temperature just below its melting or degradation point until the particles are fused (sintered) together. The mass as a whole does not melt in this process. Sintered moldings may be machined. Special formulations may be extruded in the form of rods, tubes, and fibers by using organic dispersions of the polymer. These are later vaporized as the product is sintered. Coagulated suspensoids may be used in much the same way. Presintered grades of this material may be extruded through extremely long compacting and sintering zones of special dies. Many films, tapes, and coatings are cast, dipped, or sprayed from PTFE dispersions by a drying and sintering process. Films and tapes may also be cut or sliced from sheet stock.

Teflon is a familiar trade name for homopolymers and copolymers of polytetrafluoroethylene. Its antistick property (low coefficient of friction) makes it a very useful coating. Teflon is applied to many metallic substrates, including cookware (Fig. E-18). There is no known solvent for these materials. They may be chemically etched and adhesive-bonded with contact or epoxy adhesives. Films may be heat-sealed together, but not to other materials.

Fluorocarbons have greater mass than hydrocarbons, because fluorine has an atomic mass of 18.9984 and hydrogen only 1.00797. Fluoroplastics are thus heavier than other plastics. Relative densities range from 2.0 to 2.3.

PTFE requires special fabricating techniques. Its chemical inertness, unique weathering resistance, excellent electrical insulation characteristics, high heat resistance, low coefficient of friction, and nonadhesive properties have led to numerous uses. Parts coated with PTFE have such a low coefficient of friction they release and slide easily, and require no lubrication. Saw blades, cookware, utensils, snow shovels, bakery equipment, and bearings are common applications. Aerosol spray dispersions of micron-sized particles of polytetrafluoroethylene are used as a lubricant and antistick agent for metal, glass, or plastics substrates, Many profile shapes are used for chemical, mechanical, and electrical applications (Fig E-19). Shrinkable tubing is used to cover rollers, springs, glass, and electrical parts. Skived or extruded tapes and films may be used for seals, packing, and gasket materials. Bridges, pipes, tunnels, and buildings may rest on slip joints,

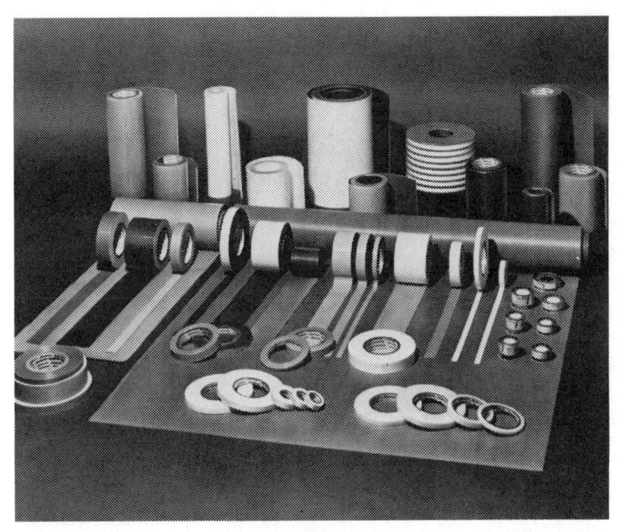

(A) A variety of tapes for many uses.

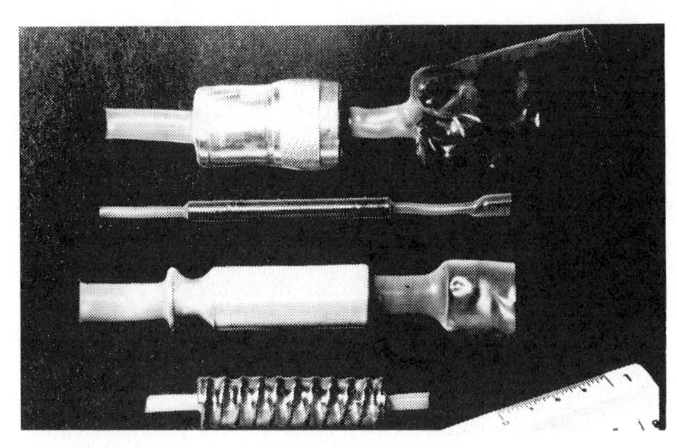

(B) Shrinkable protective tubing.

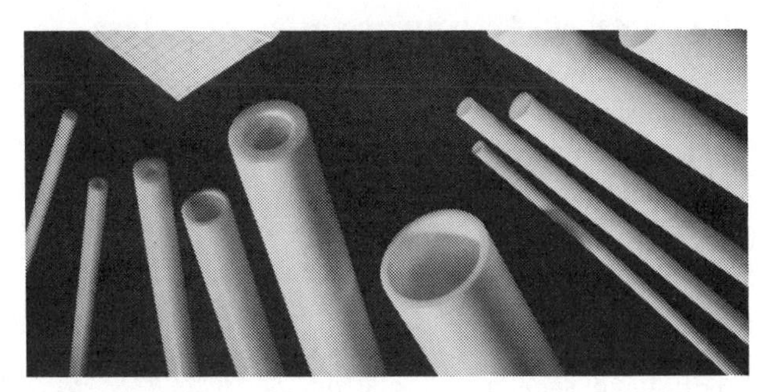

(C) Various rods and tubes.

Fig. E-19. Various PTFE applications. (Chemplast, Inc.)

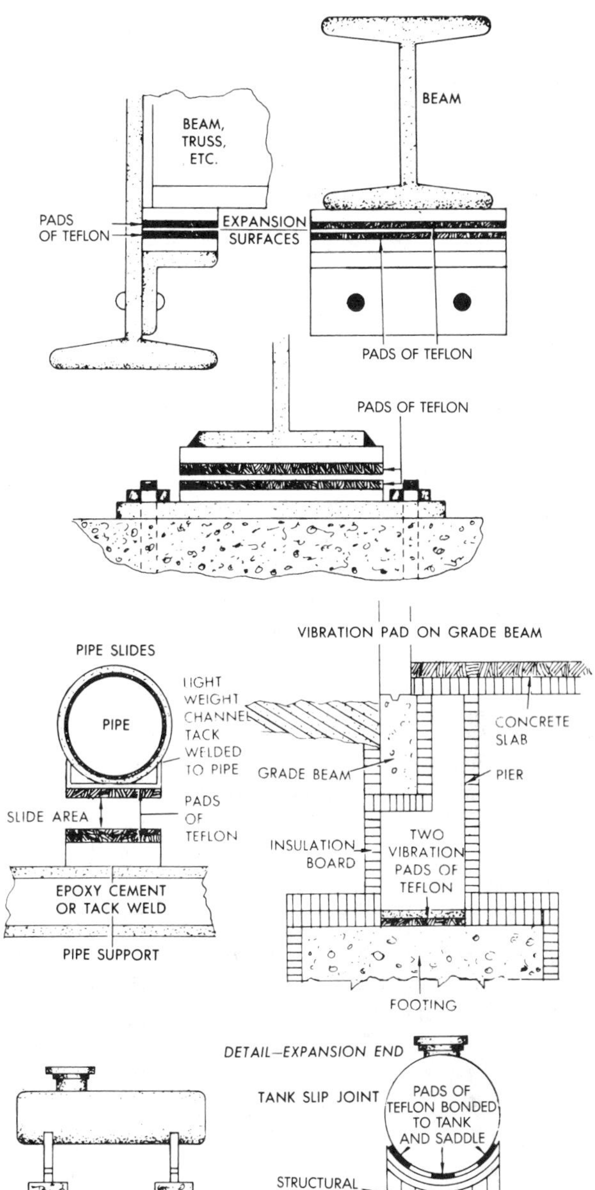

Fig. E-20. Teflon pads have many construction uses. (Du Pont Co.)

expansion plates, or bushing or bearing pads of polytetrafluoroethylene (Fig. E-20).

The excellent electrical insulating and low dissipation factors of polytetrafluoroethylene make it useful for wire and cable insulation, coaxial wire spacers, laminates for printed circuits, and many other electrical applications. Six advantages and eight disadvantages of PTFE are shown below.

Advantages of Polytetrafluoroethylene (PTFE)

1. Nonflammable
2. Outstanding chemical and solvent resistance
3. Excellent weatherability
4. Low friction coefficient (antistick property)
5. Wide thermal service range
6. Very good electrical properties

Disadvantages of Polytetrafluoroethylene

1. Not processable by common thermoplastic methods
2. Toxic in thermal degradation
3. Subject to creep
4. Permeable

5. Requires high processing temperatures
6. Low strength
7. High density
8. Comparatively high cost

Polyfluoroethylenepropylene (PFEP) or (FEP)

In 1965, DuPont announced another Teflon fluoroplastic wholly composed of fluorine and carbon atoms. Polyfluoroethylenepropylene (PFEP or FEP) is made by copolymerizing tetrafluoroethylene with hexafluoropropylene (Fig.E-21).

The partial disruption of the polymer chain by the propylene-like groups $CF_3CF{=}CF_2$ reduces the melting point and viscosity of FEP resins. Polyfluoroethylenepropylene may be processed with normal thermoplastic methods. This reduces production costs of items previously molded of PTFE. Because of pendant CF_3 groups this copolymer is less crystalline, more processable, and transparent in films up to 0.25 mm [0.01 in.] thick.

Commercial PFEP plastics possess properties similar to those of PTFE. They are chemically inert, have very good electrical insulation properties, and a somewhat greater impact strength. Service temperatures may exceed 400 °F [205 °C]. Polyfluoroethylenepropylene plastics are used extensively by the military and aircraft and aerospace industries for electrical insulation and high reliability at cryogenic temperatures. They are used for lining chutes, pipes, and tubes, and coating objects where a low coefficient of friction or nonadhesive characteristics are required. PFEP is molded into parts including gaskets, gears, impellers, printed circuits, pipes, fittings, valves, expansion plates, bearings, and other profile shapes (Fig. E-22).

As early as 1933, both Germany and the United States were making a fluoroplastic material in connection with atomic bomb research. It was used in handling uranium fluoride, a compound of uranium with fluorine.

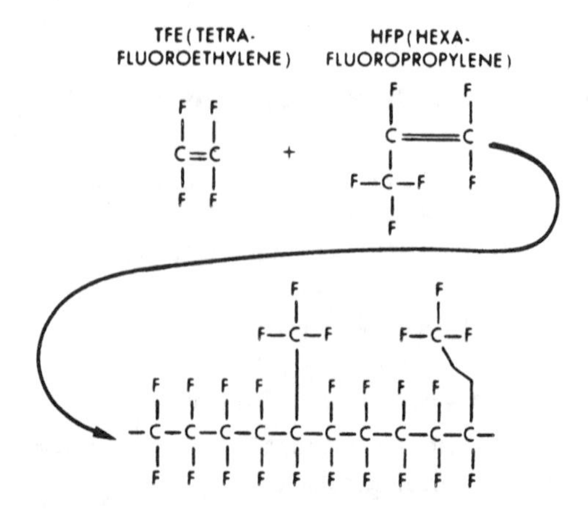

Fig. E-21. Manufacture of polyfluoroethylenepropylene.

Fig. E-22. Shrinkable PFEP antistick covers applied to rollers. (Chemplast, Inc.)

The following list shows six advantages and disadvantages of polyfluoroethylenepropylene.

Advantages of Polyfluoroethylenepropylene (PFEP)

1. Processable by normal thermoplastic methods
2. Resistant to chemicals (including oxidizing agents)
3. Excellent solvent resistance
4. Antistick characteristics
5. Nonflammable
6. Low coefficient of friction, dielectric constant, mold shrinkage, and water absorption.

Disadvantages of Polyfluoroethylenepropylene

1. Comparatively high cost
2. High density
3. Subject to creep
4. Low compressive and tensile strength
5. Low stiffness
6. Toxic upon thermal decomposition

Polychlorotrifluoroethylene (PCTFE) or (CTFE)

Polychlorotrifluoroethylene is produced in various formulations. Chlorine atoms are substituted for fluorine in the carbon chain.

$$-CF_2-\underset{\displaystyle Cl}{\underset{|}{CF}}$$

Polychlorotrifluoroethylene (PCTFE)

Monomers are obtained by fluorinating hexachloroethane, then dehalogenating (controlled removal of the halogen, chlorine) with zinc in alcohol:

$$CCl_3CCl_3 \xrightarrow[HF]{\text{Anhydrous}} CCl_2FCClF_2 \xrightarrow[\text{boiling}]{\text{Zinc}} CClF = CF_2 + Cl_2$$

Hexachloroethane ethyl alcohol

The polymerization is similar to PTFE, in that it is accomplished in an aqueous emulsion and suspension. During bulk polymerization, peroxide or Ziegler-type catalysts are used.

$$nCF_2 = CFCl \xrightarrow{\text{polymerize}} (—CF_2CFCl—)_n$$

Chlorotrifluoroethylene Polychlorotrifluoroethylene

The addition of the chlorine atoms to the carbon chain allows processing by normal thermoplastic equipment. The chlorine presence also allows selected chemicals to attack and break the partially crystalline polymer chain. PCTFE may be produced in an optically clear form depending on the degree of crystallinity. Copolymerization with vinylidene fluoride or other fluoroplastics provides varying degrees of chemical inertness, thermal stability, and other unique properties.

Polychlorotrifluoroethylene is harder, more flexible, and possesses a higher tensile strength than PTFE. It is more expensive than PTFE and has a service temperature from −400 °F to +400 °F [−240 °C to + 205 °C]. Introduction of the chlorine atom lowers their electrical insulating properties and raises the coefficient of friction. Although it is more expensive than polytetrafluoroethylene, it finds similar uses. These include insulation for wire, cable, printed circuit boards, and electronic components. The property of chemical resistance is best used in producing transparent windows for chemicals, seals, gaskets, O-rings, and pipe lining, as well as pharmaceutical and lubricant packaging (Fig. E-23). Dispersions and films may be used in coating reactors, storage tanks, valve bodies, fittings, and pipes. Films may be sealed by thermal or ultrasonic techniques. Epoxy adhesives may be used on chemically etched surfaces.

In Figure E-24, the tough plastics ethyl-enechlorotrifluoroethylene (E-CTFE) is shown. The lists below show eight advantages and five disadvantages of polychlorotrifluoroethylene.

Advantages of Polychlorotrifluoroethylene (PCTFE or CTFE)

1. Excellent solvent resistance
2. Optically clear
3. Zero moisture absorption
4. Self-extinguishing
5. Low permeability
6. Creep resistance better than PTFE or PFEP
7. Very low coefficient of friction
8. Good low temperature capability

(A) Insulation for wires and cables.

(B) Ball-valve seats.

Fig. E-23. Two uses of PCTFE. (Chemplast Inc.)

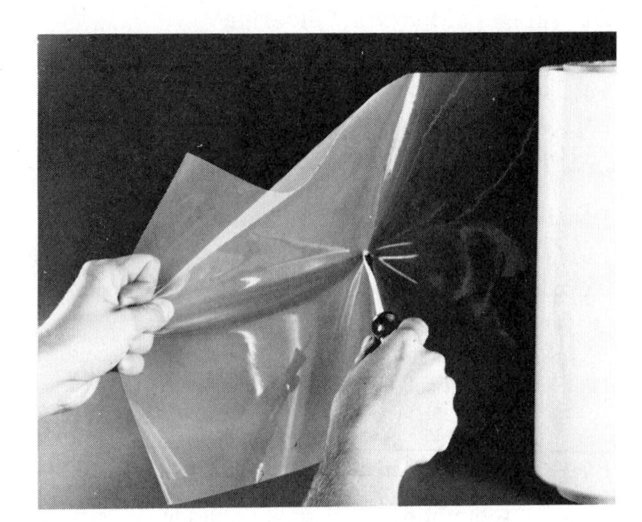

Fig. E-24. Ethylenechlorotrifluoroethylene (E-CTFE) has high-performance properties common to other fluoropolymers. It surpasses them in impermeability, tensile strength, and resistance to abrasion. (Chemplast, Inc.)

Disadvantages of Polychlorotrifluoroethylene

1. Lower electrical properties than PTFE
2. More difficult to mold than PFEP
3. Crystallizes upon slow cooling
4. Less solvent resistance than PTFE and PFEP
5. Higher coefficient of friction than PTFE and PFEP

Polyvinyl Fluoride (PVF)

The earliest preparation of vinyl fluoride gas (in 1900) was considered impossible to polymerize. In 1958, however, Du Pont announced the polymerization of vinyl fluoride (PVF). In 1933, monomer resins were prepared in Germany by the reaction of hydrogen fluoride with acetylene using selected catalysts:

$$HF + CH \equiv CH \xrightarrow[\text{catalysts}]{} CH_2 = CHF$$

Although the monomer has been known to chemists for some time, it is hard to manufacture or polymerize. Polymerization is carried out using peroxide catalysts in various aqueous solutions at high pressures.

Polyvinyl fluoride may be processed by normal thermoplastic methods. These plastics are strong, tough, flexible, and transparent. They have outstanding weather resistance. PVF has good electrical and chemical resistance, with a service temperature near 300 °F [150 °C].

Uses include protective coatings and surfaces for exterior use, finishes for plywood, sealing tapes, packaging of corrosive chemicals, and many electrical insulating applications. Coatings may be applied to automotive parts, lawn mower housings, house shutters, gutters, and metal siding. Four advantages and disadvantages are shown below. (Also see polyvinyls)

Advantages of Polyvinyl Fluoride

1. Processable by thermoplastic methods
2. Low permeability
3. Flame retardance
4. Good solvent resistance

Disadvantages of Polyvinyl Fluoride

1. Lower thermal capability than highly fluorinated polymers
2. Toxic (in thermal decomposition)
3. High dipole bond
4. Subject to attack by strong acids

Polyvinylidene Fluoride (PVDF)

Closely resembling polyvinyl fluoride is polyvinylidene fluoride (PVF_2). It became available in 1961 and was produced by the Pennwalt Chemical Corporation under the trade name Kynar. Polyvinylidene fluoride is polymerized thermally by the dehydrohalogenation (removal of hydrogen and chlorine atoms) of chlorodifluoroethane under pressure:

$$CH_3CClF_2 \xrightarrow{500-1700\ °C} CH_2 = CF_2 + HCl$$

Like polyvinyl fluoride, PVDF does not have the chemical resistance of PTFE or PCTFE. The alternating CH_2 and CF_2 groups in its backbone contribute to its tough, flexible characteristics. The presence of the hydrogen atoms reduces chemical resistance and allows solvent cementing and degradation. PVDF materials are processed by thermoplastic methods, and they are ultrasonically and thermally sealed. PVDF finds wide use in film and coating forms because of its toughness, optical properties, and resistance to abrasion, chemicals, and ultraviolet radiation. Service temperatures range from –80 °F to + 300 °F [–62 °C to + 150 °C]. A familiar use is the coating seen on aluminum siding and roofs. Molded parts may include such items as valves, impellers, chemical tubing, ducting, and electronic components (Fig. E-25). Six advantages and three disadvantages of polyvinylidene fluoride follow. (Also see polyvinyls.)

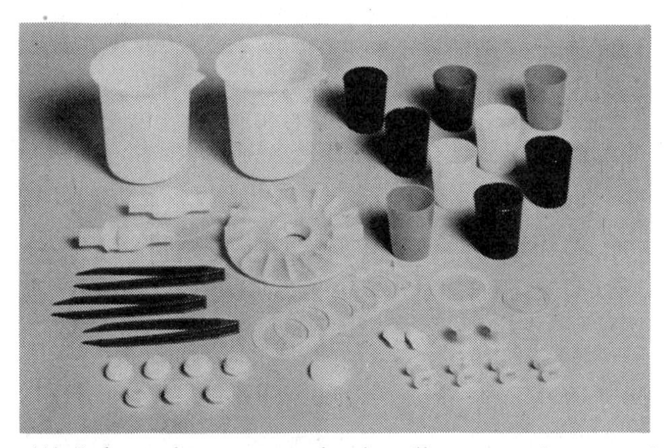

(A) Lab equipment, spools, impellers, containers, and gaskets.

(B) Film, sheet, robs, moldings, and coatings.

Fig. E-25. Polyvinylidene fluoride has many uses. (KREHA)

Advantages of Polyvinylidene Fluoride

1. Processable by thermoplastic methods
2. Low creep
3. Excellent weatherability
4. Nonflammability
5. Abrasion resistance better than PTFE
6. Good solvent resistance

Disadvantages of Polyvinylidene Fluoride

1. Lower thermal capability and chemical resistance than PTFE or PCTFE
2. Toxic in thermal decomposition
3. High dipole

Perfluoroalkoxy (PFA)

In 1972, Du Pont offered perfluoroalkoxy (PFA) under the trade name of Tefzel. It is produced from polymerization of perfluoroalkoxyethylene. PFA may be processed by conventional thermoplastics methods. This plastics has properties similar to PTFE and PFEP and is available in pellet, film, sheet, rod, and powder forms.

Uses include, dielectric and electrical insulators, coatings, and liners for valves, pipes, and pumps. The following list includes seven advantages and six disadvantages of PFA.

Advantages of Perfluoroalkoxy (PFA)

1. Higher temperature capability than PFEP
2. Excellent resistance to chemicals (including oxidizing agents)
3. Excellent solvent resistance
4. Antistick characteristics
5. Nonflammability
6. Low coefficient of friction
7. Processable by thermoplastic methods

Disadvantages of Perfluoroalkoxy

1. Comparatively high cost
2. High density
3. Subject to creep
4. Low compressive and tensile strength
5. Low stiffness
6. Toxic in thermal decomposition

Other Fluoroplastics

There are numerous other fluorine-containing polymers and copolymers. Chlorotrifluoroethylene/vinylidene fluoride is used in fabricating O-rings and gaskets:

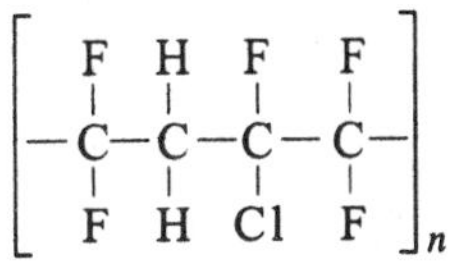

$$\left[-\underset{\textstyle F}{\overset{\textstyle F}{C}}-\underset{\textstyle H}{\overset{\textstyle H}{C}}-\underset{\textstyle Cl}{\overset{\textstyle F}{C}}-\underset{\textstyle F}{\overset{\textstyle F}{C}}- \right]_n$$

Hexafluoropropylene/vinylidene fluoride is an outstanding oil- and grease-resistant elastomer for O-rings, seals and gaskets.

The chemical structure of two fluoroacrylate elastometers is shown in Figure E-26. Polytrifluoronitrosomethane, fluorine-containing silicones and polyesters, and other fluorine-containing polymers are also being produced.

The linear chained copolymer ethylenechlorotrifluoroethylene (ECTFE) has high performance properties common to other fluoroplastics. It surpasses them in permeability, tensile strength, wear resistance, and creep. Uses include release agents, tank linings, and dielectrics.

The copolymer ethylene-tetrafluoroethylene (ETFE) has properties and applications similar to those of ECTFE.

At degradation temperatures, toxic fluorine gas is released. (See Chapter 4 Health and Safety.) The properties of the basic fluoroplastics are listed in Table E-6.

Ionomers

In 1964, Du Pont introduced a new material, known as *ionomers*, that had characteristics of both thermoplastics and thermosets. Ionic bonding is seldom found in plastics but is a characteristic of ionomers. Ionomers possess chains similar to polyethylene with ion cross-links of sodium, potassium, or similar ions (Fig. E-27). In this material both organic and inorganic compounds are linked together, and because the cross-linking is basically ionic, the weaker bonds are easily broken on heating. This material may therefore be processed as a thermoplastics. At atmospheric temperatures, the plastics has properties usually associated with linked polymers.

The basic ionomer chain is made by polymerization of ethylene and methacrylic acid. Other inexpensive polymer chains may be developed with similar cross-links.

Since they combine ionic and covalent forces in their molecular structure, ionomers can exist in a number of physical states and have a number of physical properties. They may be processed and reprocessed by any thermoplastic technique. They are more expensive

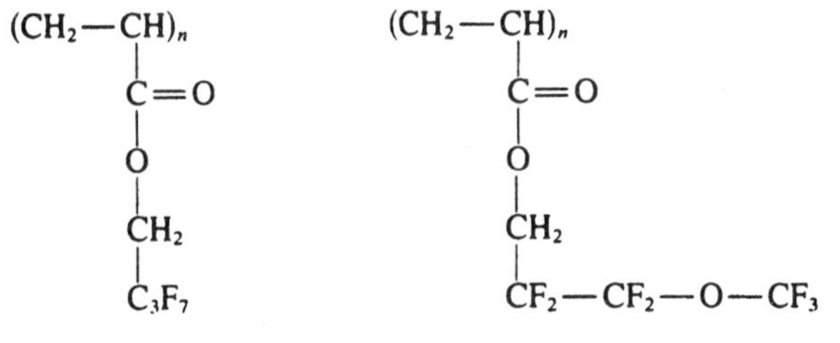

$$(CH_2-CH)_n \quad | \quad C{=}O \quad | \quad O \quad | \quad CH_2 \quad | \quad C_3F_7$$

$$(CH_2-CH)_n \quad | \quad C{=}O \quad | \quad O \quad | \quad CH_2 \quad | \quad CF_2-CF_2-O-CF_3$$

Fig. E-26. Two fluoroacrylate elastomers.

Table E-6. Properties of Fluoroplastics

Property	Polytetra-fluoroethylene (PTFE)	Polyfluro-ethylenepro-pylene (PFEP)	Polychloro-trifluoroethyl-ene (PCTFE)	Polyvinyl Fluoride (PVF)	Polyvinylidene (PVF_2)
Molding qualities	Excellent	Excellent	Excellent	Excellent	Excellent
Relative density	2.14–2.2	2.12–2.17	2.1–2.2		1.75–1.78
Tensile strength, MPa	14–35	19–21	30–40	58–124	40–50
(psi)	(2 000–5 000)	(2 700–3 100)	(4 500–6 000)		(5 500–7 400)
Compressive strength, MPa	12–14		30–50	30–50	60
(psi)	(1 700–2 000)	(4 500–7 400)	(4 500–7 400)	(8 680)	(8 680)
Impact strength, Izod, J/mm	0.40	No break	0.125–0.135	0.18–0.2	0.18–0.2
(ft lb/in)	(8)	(2.5–2.7)	(2.5–2.7)	(3.6–4.0)	(3.6–4.0)
Hardness	Shore D50–D65	Rockwell R25	Rockwell R75–R95	Shore D80	Shore D80
Thermal expansion, 10^{-4}/°C	25	20–27	11–18	70	22
Resistance to heat, °C	287	205	175–199	149	149
(°F)	(550)	(400)	(350–390)	(300)	(300)
Dielectric strength, V/mm	19 000	19 500–23 500	19 500–23 500	10 000	10 000
Dielectric constant (at 60 Hz)	2.1	2.1	2.24–2.8	8.4	8.4
Dissipation factor (at 60 Hz)	0.000 2	<0.000 3	0.001 2	0.049	0.049
Arc resistance, s	300	165+	360	50–70	50–70
Water absorption (24 h), %	0.00	0.01	0.00	0.04	0.04
Burning rate, mm/min	None	None	None	Self-extin-guishing	Self-extin-guishing
Effect of sunlight	None	None	None	Slight bleach	Slight
Effect of acids	None	None	None	Attacked by fuming sulfuric	Attacked by fuming sulfuric
Effect of alkalies	None	None	None	None	None
Effect of solvents	None	None	None	Resists most	Resists most
Machining qualities	Excellent	Excellent	Excellent	Excellent	Excellent
Optical qualities	Opaque	Transparent	Translucent to opaque	Transparent	Transparent to translucent

than polyethylene, but possess a higher moisture vapor permeability than polyethylene. Ionomers are available in transparent forms.

Uses of ionomers include safety glasses, shields, bumper guards, toys, containers, packaging films, electrical insulation, and coatings for paper, bowling pins, or other substrates (Fig. E-28). In the shoe industry they are used for inner liners and as soles and heels. Ionomers are coextruded with polyester films to produce a heat-sealable layer while improving package durability.

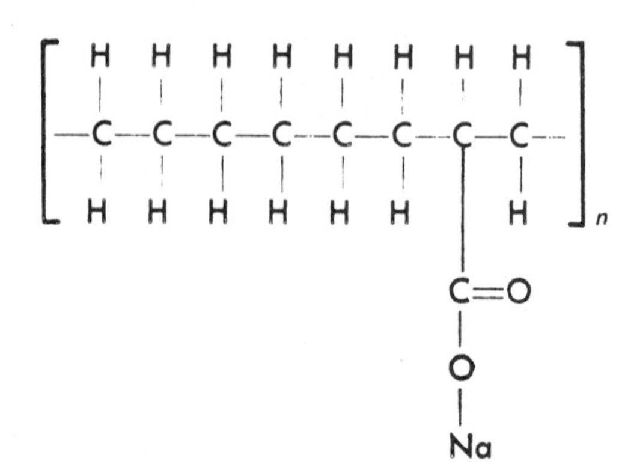

Fig. E-27. An example of ionomer structure.

Ionomers are used for a number of composite market, applications. Laminated or coextruded films are used as tear-open pouches for pharmaceutical and food packaging. Metal, foil laminates and heat-sealable skin and blister packaging continue to grow. Foamed applications include bumper guards, shoe components, wrestling mats, and ski lift seat pads. Ionomer coatings on golf balls and bowling pins extend the service lives of these products.

Table E-7 gives some properties of ionomers while the following list shows seven advantages and four disadvantages of ionomers.

Advantages of Ionomers

1. Outstanding abrasion resistance
2. Excellent shock resistance, even at low temperatures
3. Good electrical traits
4. High melt strength
5. Abrasion resistance
6. Do not dissolve in common solvents
7. Excellent transparent colors

Disadvantages of Ionomers

1. Some swelling from detergent-alcohol mixtures
2. Must be stabilized for exterior use
3. Service temperature of 161 °F [72 °C]
4. Must compete with the less expensive polyolefins

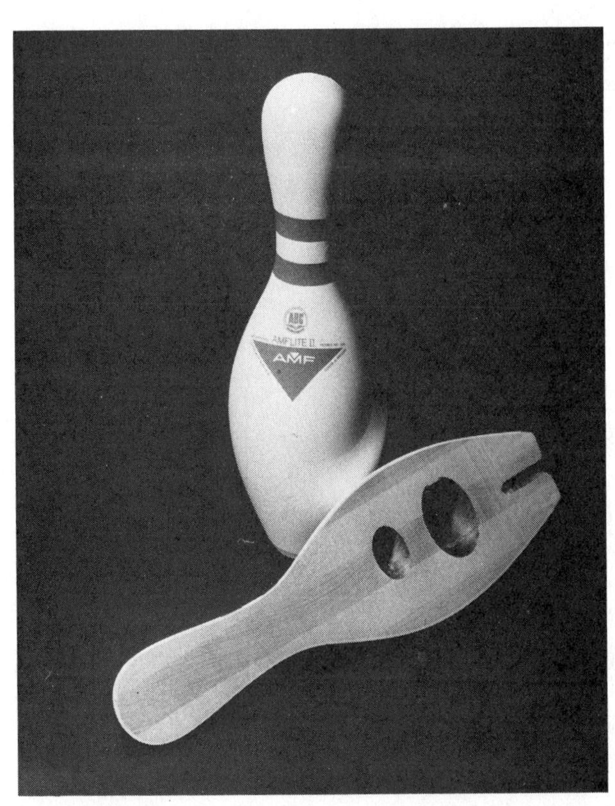

(A) Bowling pins coated with ionomer last longer than pins covered with other protective materials.

(B) Foam injection-molded bumper guards of ionomer are stronger and lighter than solid constructions.

Fig. E-28. Ionomer applications. (Du Pont Co.)

Table E-7 Properties of Ionomers

Property	Ionomer
Molding qualities	Excellent
Relative density	0.93–0.96
Tensile strength, MPa	24–35
(psi)	(3 500–5 000)
Izod, impact, J/mm	0.3–0.75
(ft lb/in.)	(6–15)
Hardness, Shore	D50–D65
Thermal expansion, 10^{-4}/°C	30
Resistance to heat, °C	70–105
(°F)	(160–220)
Dielectric strength, V/mm	35 000–40 000
Dielectric constant (at 60 Hz)	2.4–2.5
Dissipation factor (at 60 Hz)	0.001–0.003
Arc resistance, s	90
Water absorption (24 h), %	0.1–1.4
Burning rate, mm/min	Very slow
Effect of sunlight	Requires stabilizers
Effect of acids	Attacked by oxidizing acids
Effect of alkalies	Very resistant
Effect of solvents	Very resistant
Machining qualities	Fair to good
Optical qualities	Transparent

Nitrile Barrier Plastics

Formulations of copolymers with a nitrile (C≡N) functionality of over 50 percent are called nitrile polymers.

These plastics offer very low permeability and form a barrier against gases and odors. This property is the result of the high nitrile content. Their transparency and processibility make them useful as containers.

Each maker of nitrile barrier plastics varies the formulation.

Most combinations are based on acrylonitrile (AN or methacrylonitrile). Some formulations may reach 75 percent AN. All formulations are amorphous, with a slight yellow tint. Approximate monomer compositions of nitrile barrier are shown in Table E-8. The Borg-Warner product, *Cyclopac 930*, contains more than 64 percent acrylonitrile, 6 percent butadiene, and 21 percent styrene.

Although heat-sensitive, nitrile barrier plastics have been used with all thermoplastics processing techniques (Table E-9).

These plastics have found uses as beverage containers, food packages, and containers for many nonfood fluids.

The AN residual monomers of food containers cannot exceed 0. 10 p.p.m

Phenoxy

A family of resins based on bisphenol A and epichlorohydrin was introduced in 1962 by Union Carbide. These resins were called *phenoxy* but could be classed

Table E-8. Estimated Compositions of Nitrile Barrier Plastic

Composition	Borg-Warner Cycopac 930	Monsanto Lo Pac	Sohio Barex	DuPont NR-16	ICI LPT*
Acrylonitrile, %	65–75	65–75	65–75	65–75	65–75
Butadiene, %	5–10		5–8		
Methyl acrylate, %			20–25		
Methyl methacrylate, %			3–5		
Styrene, %	20–30	25–35		25–35	25–35

*LPT-Low permeability thermoplastics

Table E-9. Properties of Nitrile Barrier Plastics

Property	Nitrile Barrier (Standards)
Molding qualities	Good
Relative density	1.15
Tensile strength, MPa	62
(psi)	(9 000)
Impact strength, Izod, J/mm	0.075–0.2
(ft lb/in)	(1.5–4)
Hardness, Rockwell	M72–78
Thermal expansion, 10^{-4}/°C	16.89
Resistance to heat, °C	70–100
(°F)	(158–212)
Dielectric strength, V/mm	8 660
Dielectric constant (at 60 Hz)	4.55
Dissipation factor (at 100 Hz)	0.07
Water absorption (24 h), %	0.28
Burning rate, mm/min	—
Effect of sunlight	Slight yellowing
Effect of acids	None to attacked
Effect of alkalies	None to attacked
Effect of solvents	Dissolves in acetonitrile
Machining qualities	Good
Optical qualities	Transparent

as polyhydroxyethers. The plastics structure resembles polycarbonates, and the material has similar properties.

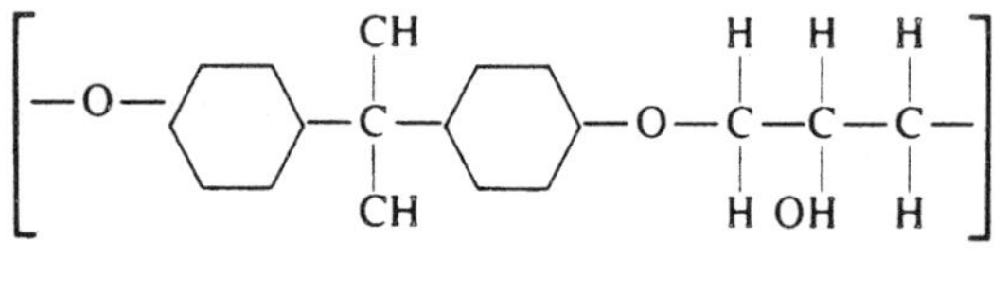

$$\left[-O-C_6H_4-C(CH)(CH)-C_6H_4-O-CH(H)-CH(OH)-CH(H)- \right]$$

Phenoxy resins are made and sold as thermoplastic epoxide resins. They may be processed on normal thermoplastic machinery, with product service temperatures in excess of 170 °F [75 °C].

Phenoxy resins, because of the reactive hydroxyl groups, may be cross-linked. Cross-linking agents may include diisocyanates, anhydrides, triazines, and melamines.

Homopolymers have good creep resistance, high elongation, low moisture absoprtion, low gas transmission, and high rigidity, tensile strength, and ductility. They find applications as clear or colored protective coatings, molded electronic parts, pipe for gas and crude oil, sports equipment, appliance housings, cosmetic cases, adhesives, and containers for food or drugs.

Table E-10 gives some of the properties of phenoxy.

Polyallomers

In 1962, a plastics distinctly different from simple copolymers of polyethylene and polypropylene was produced by Eastman. The process, called *allomerism,* is conducted by alternately polymerizing ethylene and propylene monomers. Allomerism is a variation in chemical composition without a change in crystalline form. The plastics exhibits the crystallinity usually associated with the homopolymers of ethylene and propylene. The term *polyallomer* is used to distinguish this alternately segmented plastics from homopolymers and copolymers of ethylene and propylene.

Although highly crystalline, polyallomers may have a relative density as low as 0.896. They are available in various formulations. Properties include high stiffness, impact strength, and abrasion resistance. Their flexibility has been used in the production of hinged boxes, looseleaf binders, and various other folders (Fig. E-29). Polyallomers may be used as food containers or films over a wide range of service tem-

Table E-10. Properties of Phenoxy

Property	Phenoxy
Molding qualities	Good
Relative density	1.18–1.3
Tensile strength, MPa	62–65
(psi)	(9 000–9 500)
Impact strength, Izod, J/mm	0.125
(ft lb/in)	(2.5)
Resistance to heat, °C	80
(°F)	(175)
Dielectric constant (at 60 Hz)	4.1
Dissipation factor (at 60 Hz)	0.001
Water absorption (24 h), %	0.13
Burning rate	Self-extinguishing
Effect of acids	Resistant
Effect of alkalies	Resistant
Effect of solvents	Soluble in ketones
Machining qualities	Good
Optical qualities	Translucent to opaque

Fig. E-29. The covers of these notebooks are made of polyallomer plastics.

peratures. They find limited use where other polyolefins, are marginal.

Processing may be done on all normal thermoplastic equipment. Like polyethylene, polyallomers are not cohesively bonded but may be welded.

Table E-11 lists some of the properties of polyallomers.

Polyamides (PA)

From research that began in 1928, Wallace Hume Carothers and his colleagues concluded that linear polyesters were not suitable for commercial fiber production. Carothers did succeed in producing polyesters of high molecular mass and oriented them by elongation under tension. These fibers were still inadequate and could not be successfully spun. The amino acids present in silk, a natural fiber, prompted Carothers to study synthetic polyamides. Of many formulations of amino acids, diamines, and dibasic acids, several showed promise as possible fibers. By 1938, the first commercially developed polyamide was introduced by Du Pont. It was 6,6 polyamide and was given the trade name *Nylon*. This condensation plastics was called Nylon 6,6 (also written as 66 or 6/6) because both the acid and the amine contain six carbon atoms.

Table E-11. Properties of Polyallomers

Property	Polyallomer (Homopolymer)
Molding qualities	Excellent
Relative density	0.896–0.899
Tensile strength, MPa	20–27
(psi)	(3 000–3 850)
Impact strength, Izod, J/mm	8.5–12.5
(ft lb/in)	(170–250)
Hardness, Rockwell	R50–R85
Thermal expansion, 10^{-4}/°C	21–25
Resistance to heat, °C	50–95
(°F)	(124–200)
Dielectric strength, V/mm	32 000–36 000
Dielectric constant (at 60 Hz)	2.3–2.8
Dissipation factor (at 60 Hz)	0.000 5
Water absorption (24 h). %	0.01
Burning rate	Slow
Effect of sunlight	Slight—should be protected
Effect of acids	Very resistant
Effect of alkalies	Very resistant
Effect of solvents	Very resistant
Machining qualities	Good
Optical qualities	Transparent

$$NH_2(CH_2)_6NH_2 + COOH(CH_2)_4COOH$$
Hexamethylene diamine   *Adipic acid*

$$\rightarrow n[NH_2(CH_2)_6NH{\cdot}CO(CH_2)_4COOH] \text{ —heat}$$
Nylon salt

$$\rightarrow [NH(CH_2)_4NH{\cdot}CO(CH_2)_4CO]_n\text{—} + nH_2O$$
Nylon 6,6 polymer chain

Nylon has come to mean any polyamide that can be processed into filaments, fibers, films, and molded parts.

The repeating—CONH—(amide) link is present in a series of linear, thermoplastic Nylons.

- Nylon 6—Polycaprolactam:

$$[NH(CH_2)_5CO]_x$$

- Nylon 6,6—Polyhexamethyleneadipamide:

$$[NH(CH_2)_6NHCO(CH_2)_4CO]_x$$

- Nylon 6,10—Polyhexamethylenesebacamide:

$$[NH(CH_2)_6NHCO(CH_2)_8CO]_x$$

- Nylon 11—Poly(11-aminoundecanoic acid)

$$[NH(CH_2)_{10}CO]_x$$

- Nylon 12—Poly(12-aminododecanoic acid)

$$[NH(CH_2)_{11}CO]_x$$

There are many other types of nylon currently available, including Nylon 8, 9, 46, and copolymers from more sophisticated diamines and acids. In the United States, Nylon 6,6 and Nylon 6 are by far the most used. Properties of Nylons may also be changed by introducing additives. Amino-containing polyamide resins will react with a number of materials, and cross-linked, thermosetting reactions are possible.

Although developed primarily as a fiber, polyamides find use as molding compounds, extrusions, coatings, adhesives, and casting materials. Acetals and fluorocarbons share some of the same properties and uses as polyamide. Polyamide resins are costly. They are selected when other resins will not meet service requirements. Acetal resins are superior in fatigue endurance, creep resistance, water resistance. Nylons have met increased competition from these resins.

Molding compounds were offered in 1941 and have grown in applications. They are among the toughest of plastics materials. Nylons are self-lubricating, impervious to most chemicals, and highly impermeable to oxygen. They are not attacked by fungi or bacteria. Polyamides may be used as food containers.

The largest applications of homopolymer molding compounds (Nylons 6; 6,6; 6, 10; 11; and 12) include gears, earns, bearings, valve seats, combs, furniture casters, and door catches. They are used where wear resistance, quiet operation, and low coefficients of friction are needed.

Because of their crystalline structure, polyamide products have a milky-opaque appearance (Fig. E-30). Transparent films may be obtained from Nylon 6 and 6,6 if they are cooled very rapidly. Polyamides are clear amorphous materials when melted. On cooling, they crystallize and become cloudy. This crystallinity contributes to stiffness, strength, and heat resistance. Polyamides are harder to process than other thermoplastic materials. All thermoplastic processing equipment may be used, but fairly high processing temperatures are needed. The melting point of a polyamide is abrupt or sharp; that is, they do not soften or melt over a broad range of temperatures (Fig. E-31). When sufficient energy has overcome the crystalline and molecular attractions, they suddenly become liquid and may be processed. Because all Nylons absorb water, they are dried before molding. This ensures desirable physical properties in the molded product.

Extruded and blown films are used to package oils, grease, cheese, bacon, and other products where low gas permeability is essential. The high service temperatures of Nylon film are used for boil and bake-in-the-bag food products. Although polyamides are hygroscopic (absorb water), they find many applications as electrical insulators.

Fig. E-30. Polyamide products have a milky-opaque appearance as seen in this valve and fitting. (Du Pont Co.)

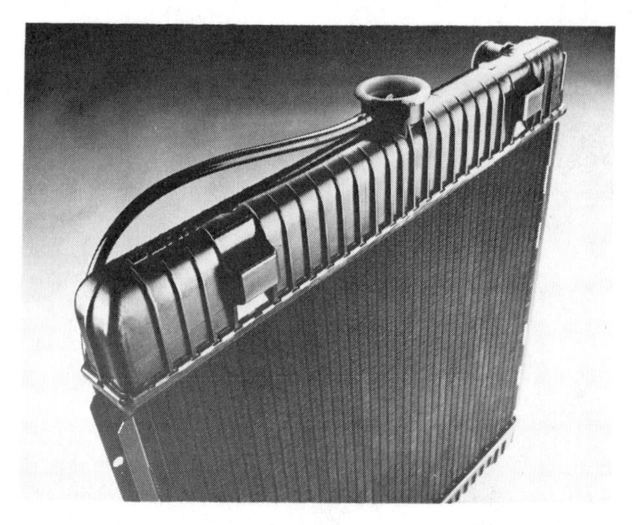

Fig. E-31. An automobile radiator can be made of polyamide because it will not melt over a wide temperature range. (BASF)

Nylon 11 may be used as a protective coating on metal substrates. Polyamides are used in a powder form by spraying or fluidized-bed processes. Typical uses include rollers, shafts, panel slides, runners, pump impellers, and bearings. Water dispersions and organic solvents of polyamide resin permit certain adhesive and coating applications on paper, wood, and fabrics.

Polyamide-based adhesives may be of the hot-melt or solution type. Hot-melt adhesives are simply heated above the melting point and applied. Aminopolyamide resins may be reacted with epoxy or phenolic resins to produce a thermosetting adhesive. These adhesives find use in bonding wood, paper laminates, and aluminum, and in adhering copper to printed circuit boards. They are used as flexible adhesives for bread wrappers, dried soup packets, cigarette packages, and bookbindings. Polyamide-epoxy combinations are used as two-part systems in casting applications, including potting and encapsulation of electrical components. When combined with pigments and other modifying agents, polyamides may be used as printing inks. The uses of polyamides in textiles and carpets are well known and need little discussion.

Clothing, light tents, shower curtains, and umbrellas are among products made from Nylon. The monofilaments, multifilaments, and staple fibers are made by melt spinning. This is followed by cold drawing to increase tensile strength and elasticity. Monofilaments are used in fishing lines, surgical sutures, tire cords, rope, sports equipment, brushes, artificial human hair, and synthetic animal fur.

Polyamides are easily machined (Fig. E-32), but drilled or reamed holes are likely to be slightly undersized, due to the resiliency of the material.

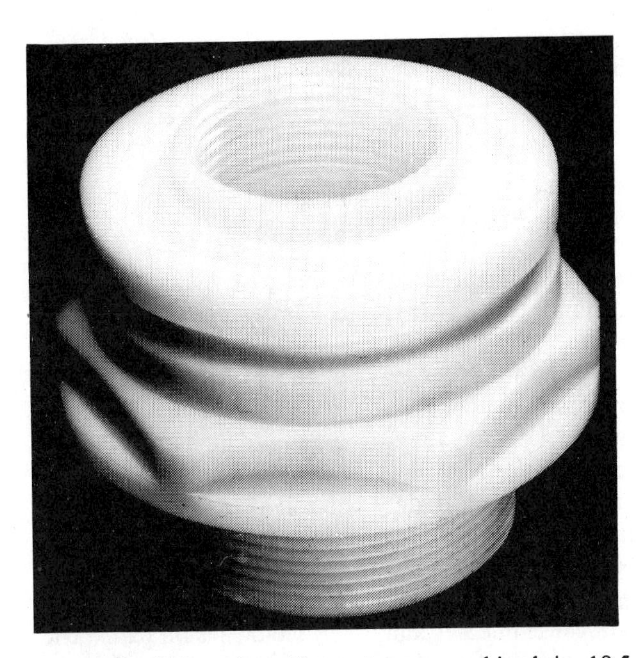

Fig. E-32. This polyamide part was machined in 18.5 seconds on an automatic lathe. The same part in light metal would take more than twice as long to machine; a free-cutting steel part, 13 times as long. (BASF)

Cementing polyamide is difficult because of its solvent resistance; however, phenols and formic acid are specific solvents that are used in cementing polyamides. Epoxy resins are also employed for this purpose.

Some of the basic traits of polyamides are shown in Table E-12. Seven advantages and five disadvantages are listed below.

Advantages of Polyamide (Nylon)

1. Tough, strong and impact resistant
2. Low coefficient of friction
3. Abrasion resistance
4. High temperature resistance
5. Processable by thermoplastic methods
6. Good solvent resistance
7. Resistant to bases

Disadvantages of Polyamide

1. High moisture absorption with related dimensional instability
2. Subject to attack by strong acids and oxidizing agents
3. Requires ultraviolet stabilization
4. High shrinkage in molded sections
5. Electrical and mechanical properties influenced by moisture content

Polycarbonates (PC)

An important material in the production of plastics is phenol. It is used in producing phenolic, polyamide, epoxy, polyphenylene oxide, and polycarbonate resins.

Phenol is a compound that has one hydroxyl group attached to an aromatic ring. It is sometimes called *monohydroxy benzene,* C_6H_5OH.

Bisphenol A (two phenol and acetone), a vital ingredient in the production of polycarbonates, may be prepared by combining acetone with phenol (Fig. E-33). Bisphenol A is sometimes referred to as diphenylol propane or bis-dimethylmethane.

Table E-12. Properties of Polyamides

Property	Nylon 6,6 (Unfilled)	Nylon 6,10 (Unfilled)	Nylon 6,10 (Glass-Filled)
Molding qualities	Excellent	Excellent	Excellent
Relative density	1.13–1.15	1.09	1.17–1.52
Tensile strength, MPa	62–82	58–60	89–240
(psi)	(9 000–12 000)	(8 500–8 600)	(13 000–35 000)
Compressive strength, MPa	46–86	46–90	90–165
(psi)	(6 700–12 500)	(6 700–13 000)	(13 000–24 000)
Impact strength, Izod, J/mm	0.05–0.1	0.06	0.06–0.3
(ft lb/in)	(1.0–2.0)	(1.2)	(1.2–6)
Hardness, Rockwell	R108–R120	R111	M94, E75
Thermal expansion, 10^{-4}/°C	20	23	3–8
Resistance to heat, °C	80–150	80–120	150–205
(°F)	(180–300)	(180–250)	(300–400)
Dielectric strength, V/mm	15 000–18 5000	13 500–19 000	16 000–20 000
Dielectric constant (at 60 Hz)	4.0–4.6	3.9	4.0–4.6
Dissipation factor (at 60 Hz)	0.014–0.040	0.04	0.001–0.025
Arc resistance, s	130–140	100–140	92–148
Water absorption, (24 h), %	1.5	0.4	0.2–2
Burning rate, mm/min	Self-extinguishing	Self-extinguishing	Self-extinguishing
Effect of sunlight	Discolors slightly	Discolors slightly	Discolors slightly
Effect of acids	Attacked	Attacked	Attacked
Effect of alkalies	Resistant	None	None
Effect of solvents	Dissolved by phenol & formic acid	Dissolved by phenols	Dissolved by phenols
Machining qualities	Excellent	Fair	Fair
Optical qualities	Translucent to opaque	Translucent to opaque	Translucent to opaque

Polycarbonates are linear, amorphous polyesters because they contain esters of carbonic acid and an aromatic bisphenol.

Another important material used in the production of polycarbonate is phosgene. Phosgene, a poisonous gas, was used in World War I.

As early as 1898, A. Einhorn prepared a polycarbonate material from the reaction of resorcinol and phosgene. Both W. H. Carothers and F. J. Natta performed research on a number of polycarbonates using ester reactions.

Research continued after World War II in Germany by Farbenfabriken Bayer and in the United States by General Electric. By 1957, both had arrived at the production of polycarbonates produced from bisphenol A. Volume production in the United States did not begin until 1959.

There are two general methods of preparing polycarbonates. The most common method is to react purified bisphenol A with phosgene under alkaline conditions (Fig. E-34). An alternate method involves the reaction of purified bisphenol A with diphenyl carbonate (meta-carbonate) in the presence of catalysts, under a vacuum (Fig. E-35).

Purity of the bisphenol A is vital if the plastics is to possess high-clarity and long linear chains with no cross-linking substances.

The phosgenation process is preferred, since it may be carried out at low temperatures using simple technology and equipment. The process requires the recovery of solvents and inorganic salts. In either method, however, the product is comparatively high in cost.

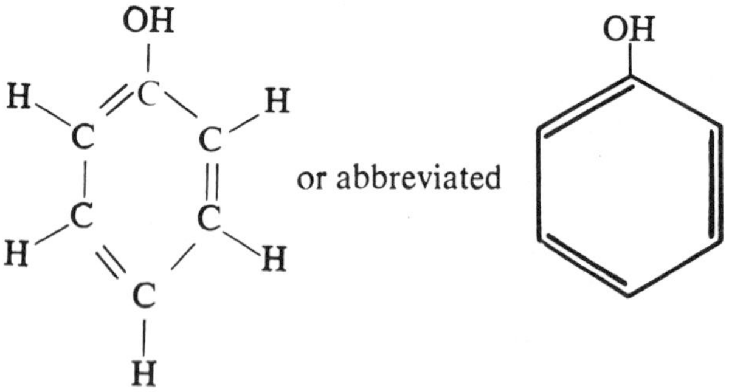

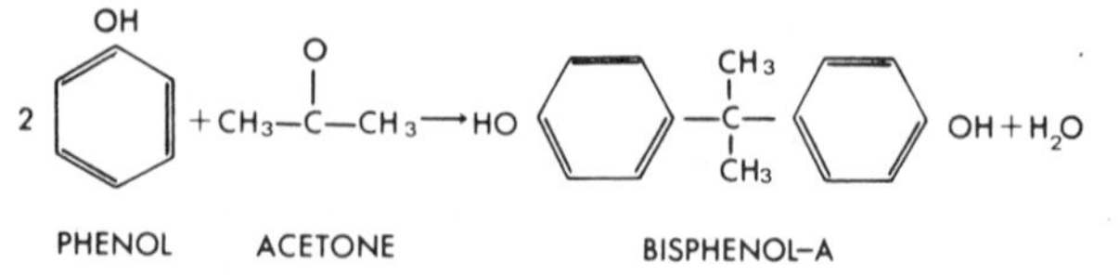

OH, H, C, or abbreviated, OH

$2\,C_6H_5OH + CH_3\text{—}CO\text{—}CH_3 \rightarrow HO\text{—}C_6H_4\text{—}C(CH_3)_2\text{—}C_6H_4\text{—}OH + H_2O$

PHENOL ACETONE BISPHENOL-A

Fig. E-33. Preparation of bisphenol-A.

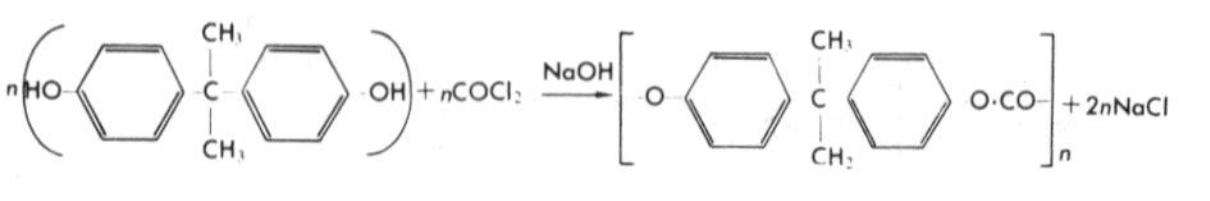

$n\,HO\text{—}C_6H_4\text{—}C(CH_3)_2\text{—}C_6H_4\text{—}OH + n\,COCl_2 \xrightarrow{NaOH} [O\text{—}C_6H_4\text{—}C(CH_3)_2\text{—}C_6H_4\text{—}O\text{·}CO]_n + 2n\,NaCl$

Fig. E-34. The first method used to prepare polycarbonates.

$n\,HO\text{—}C_6H_4\text{—}C(CH_3)_2\text{—}C_6H_4\text{—}OH + n\,C(=O)(OC_6H_5)_2 \xrightarrow{\text{SODIUM HYDROXIDE}} (\text{—}O\text{—}C_6H_4\text{—}C(CH_3)_2\text{—}C_6H_4\text{—}O\text{—}C(=O)\text{—})_n$

BISPHENOL-A DIPHENYL CARBONATE POLYCARBONATE

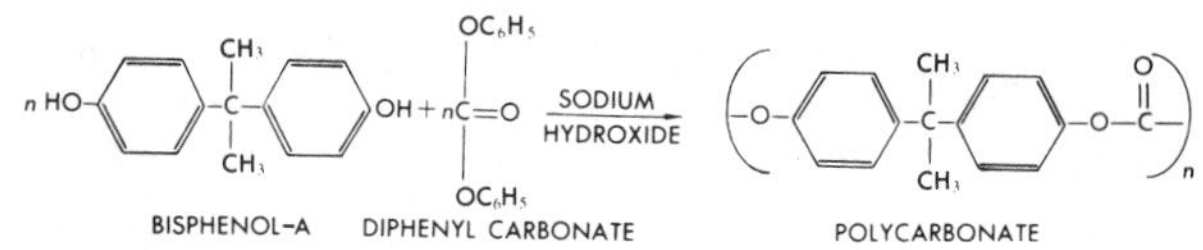

Fig. E-35. The second method used to prepare polycarbonates.

Polycarbonates may be processed by all usual themoplastic methods. Their resistance to heat and their high melt temperatures does require higher processing temperatures. The molding temperature is very critical and must be accurately controlled to produce usable products. Polycarbonates are sensitive to hydrolysis at high processing temperatures. Compounds should be dried, or vented barrel equipment used, since water will cause bubbles and other, blemishes on parts. The unique properties of polycarbonate are due to the carbonate groups and the presence of benzene rings in the long, repeating molecular chain. Polycarbonate properties include: high impact strength, transparency, excellent creep resistance, wide temperature limits, high dimensional stability, good electrical characteristics, and self extinguishing behavior. Tough transparent grades are used in lenses, films, windshields, light fixtures, containers, appliance components and tool housings (Fig. E-36). Temperature resistance is made use of in hot-dish handles, coffee pots, popcorn popper lids, hair dryers, and appliance housings. These plastics have excellent properties from –275 °F to + 270 °F [–170 °C to + 132 °C]. Polycarbonates supply the impact and flexural strength needed in pump impellers, safety helmets, beverage dispensers, small appliances, trays, signs, aircraft parts, cameras, and various packaging and film uses. Coextruded packages are used for ovenable frozen food trays and microwave pouches. Polycarbonate parts also have very good dimensional

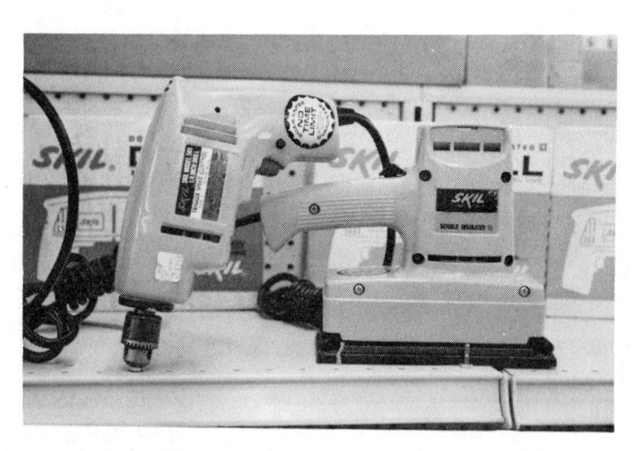

Fig. E-36. Polycarbonate used in products around the home.

stability. Glass-filled grades have improved impact, moisture, and chemical resistance (Fig. E-37).

Most aromatic solvents, esters, and ketones will attack polycarbonates. Chlorinated hydrocarbons are used as solvent cements for cohesive bonds.

There are several hundred variations on the polycarbonate structure. The structure may be changed by substituting various radicals as side groups or by separating the benzene rings by more than one carbon atom. Some possible structural combinations are shown in Figure E-38.

Some of the properties of polycarbonates are listed in Table E-13. A list of five advantages and four disadvantages follows.

Fig. E-37. High-impact polycarbonate has better mechanical properties and equivalent electrical properties when used to replace glass insulators. (H.K. Porter Co., Inc.)

Advantages of Polycarbonates

1. High impact strength
2. Excellent creep resistance
3. Available in transparent grades
4. Continuous application temperature over 120 °C (248 °F)
5. Very good dimensional stability

Disadvantages of Polycarbonates

1. High processing temperatures
2. Poor resistance to alkalies
3. Subject to solvent crazing
4. Require ultraviolet stabilization

Polyetheretherketone (PEEK)

The wholly aromatic structure of PEEK contributes to the high temperature resistance of this crystalline thermoplastic. The basic repeating unit is shown in Figure E-39. PEEK can be melt-processed in conventional thermoplastic equipment. Applications include coatings on wire and high temperature composites for aerospace and aircraft components.

Table E-13. Properties of Polycarbonates

Property	Polycarbonate (Unfilled)	Polycarbonate (10%-40% Glass-Filled)
Molding qualities	Good to excellent	Very good
Relative density	1.2	1.24–1.52
Tensile strength, Mpa	55–65	83–172
(psi)	(8 000–9 500)	(12 000–25 000)
Compressive strength, MPa	71–75	90–145
(psi)	(10 300–10 800)	(13 000–21 000)
Impact strength, Izod (ft lb/in.)	0.6–0.9	0.06–0.325
J/mm	(12–18)	(1.2–6.5)
Measurement of bar	12.7 x 3.175 mm	6.35 x 12.7 mm
Hardness, Rockwell	M73–78, R115, R125	M88–M95
Thermal expansion, 10^{-4}/°C	16.8	4.3–10
Resistance to heat, °C	120	135
(°F)	(250)	(275)
Dielectric strength, V/mm	15 500	18 000
Dielectric constant (at 60 Hz)	2.97–3.17	3.0–3.53
Dissipation factor (at 60 Hz)	0.0009	0.0009–0 001 3
Arc resistance, s	10–120	5–120
Water absorption. (24 h), %	0.15–0.18	0.07–0.20
Burning rate, mm/min	Self-extinguishing	Slow 20–30
(in/min)		(0.8–1.2)
Effect of sunlight	Slight	Slight
Effect of acids	Attacked slowly	Attacked by oxidizing acids
Effect of alkalies	Attacked	Attacked
Effect of solvents	Soluble in aromatic and chlorinated hydrocarbons	Soluble in aromatic and chlorinated hydrocarbons
Machining qualities	Excellent	Fair
Optical qualities	Transparent to opaque	Transparent to opaque

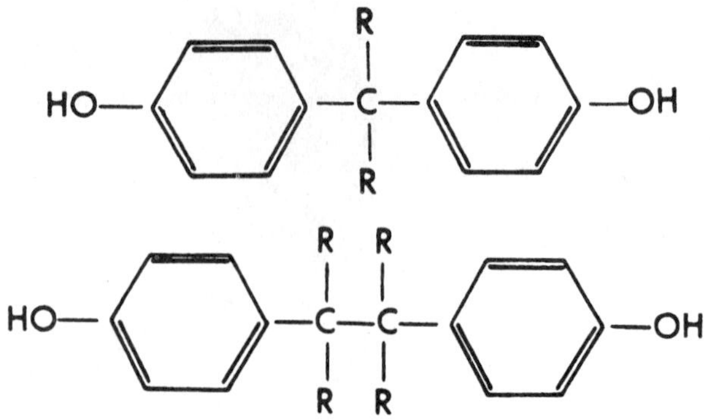

POSSIBLE RADICAL SIDE GROUPS

R	R_1
$—H$	$—H$
$—H$	$—CH_3$
$—CH_2$	$—CH_3$
$—CH_3$	$—C_2H_5$
$—C_2H_5$	$—C_2H_5$
$—CH_3$	$—CH_2—CH_2—CH_3$
$—CH_2—CH_2—CH_3$ 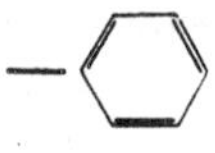	$—CH_2—CH_2—CH_3$ 

Fig. E-38. Possible combinations of polycarbonates.

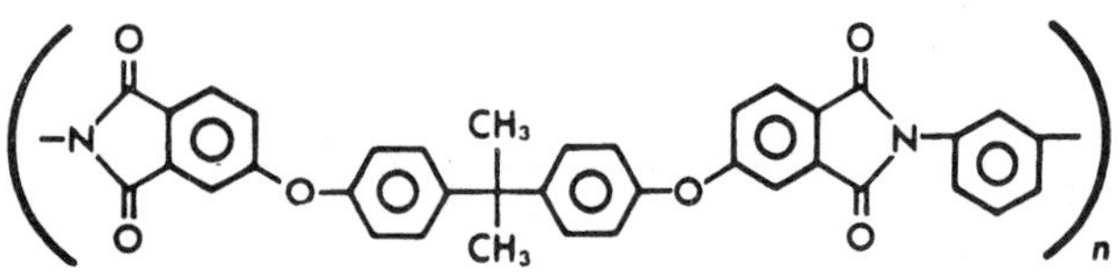

Fig. E-39. The general chemical structure of polyetherimide.

Polyetherimide (PEI)

Polyetherimide (PEI) is an amorphous thermoplastic based on repeating ether and imide units. The general chemical structure of these polymers is shown in Figure E-40. Reinforced and filled grades improve strength, temperature resistance, and creep. All grades are processed on conventional equipment. Typical applications include jet engine components, ovenable cookware, flexible circuitry, composite structures for aircraft, and food packaging.

Thermoplastic Polyesters

The thermoplastic polyester group of plastics includes saturated polyesters and aromatic polyesters. Familiar trade names of saturated polyesters are Dacron and Mylar.

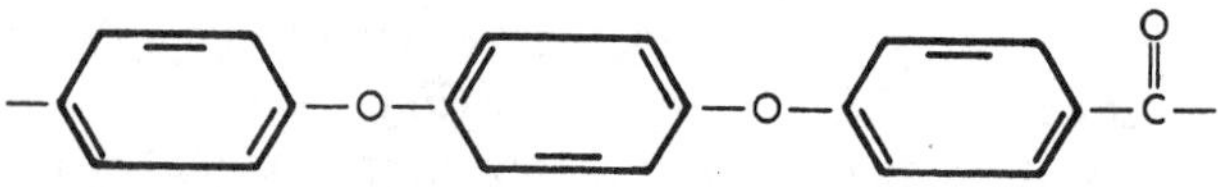

Fig. E-40. Chemical structure of PEEK.

Saturated Polyesters

Saturated polyesters are based on the reaction of terephthalic acid ($C_6H_4(COOH)_2$) and ethylene glycol ($(CH_2)_2(OH)_2$). They are linear polymers with high-molecular mass (Table E-14). Saturated polyesters are used in fiber and film production. It was with linear polyesters that W. H. Carothers did his basic research. After several years, he stopped trying to produce polyester fibers. Instead, he began investigating synthetic polyamides.

Polyethylene terephthalate (PET) may be produced by melt condensation polymerization from terephthalic acid or dimethyl terephthalate and ethylene glycol.

To reduce crystallinity, PET may be copolymerized. Copolyesters, of glycol-modified PET are called PETG. Clear shampoo and detergent bottles are familiar applications. PCTA copolyester is produced from cyclohexanedimethanol, terephthalic acid (TPA) and other dibasic acids.

This plastics has been used for food packaging, clothing fibers, carpeting, and tire cords for nearly 20 years. It has recently gained popularity in packaging carbonated beverages because of its low gas permeability.

Most applications require PET to be oriented and crystalline for optimum properties. Orientation processes are done at 212 to 248 °F [100 to 120 °C], or slightly above the glass-transition temperature (T_g).

PET is used for synthetic fibers, photographic film, videotape, dual ovenable containers, computer and magnetic tapes, and numerous beverage bottles, including distilled spirits. Reinforced and filled grades are used in gears, cowl vent grills, electrical switches, and sporting goods.

Polybutylene terephthalate (PBT) or polytetramethylene terephthalate (PTMT) were introduced in 1962. Ethylene glycol, an automobile antifreeze, is also one of the main materials in production of polyester fibers. The original development began in England by Imperial Chemical Industries (ICI). By 1953, Du Pont had purchased rights to develop Dacron fibers. Extrusion and injection grades were on the market by 1969.

Saturated (unreactive) polyesters do not undergo any cross-linking. These linear polyesters are thermoplastic. Clothing and draperies are common uses of these fibers. Industrial uses may include reinforcements for belting or tires.

Polyester films are used for recording tape, dielectric insulators, photographic film, and boil-in-the-bag food products (Fig. E-41).

Table E-14. Properties of Thermoplastic Polyesters: Saturated Polyester

Property	Polyethylene Terephthalate (PETP or PET)	Polybutylene Terephthalate (PBTP) (Unfilled)	Linear Aromatic (Decomposes at 550)	Linear Aromatic (injection grade)
Molding qualities	Good	Good	Sinters	Good
Relative density	1.34–1.39	1.31–1.38	1.45	1.39
Tensile strength, MPa	59–72	56	17	20
(psi)	(8 550–10 500)	(8 100)	(2 500)	(2 900)
Compressive strength, MPa	76–128	59–100	76–105	68
(psi)	(11 025–18 130)	(8 550–14 500)	(11 025–15 230)	(9 860)
Impact Strength, Izod (ft lb/in)	0.01–0.04	0.04–0.05		0.08
J/mm	(0.2–0.8)	(0.8–1)		(1.6)
Hardness	Rockwell M94–MI0I	Rockwell M68–M98	Shore D88	—
Thermal expansion, 10^{-4}/°C	15.2–24	155	7.1	7.36
Resistance to heat. °C	80–120	50–90		280
(°F)	(176–248)	(122–194)		(536)
Dielectric strength, V/mm	13 780–15 750	16 500		13 750
Dielectric constant (at 60 Hz)	3.65	3 29	3.22	
Dissipation factor (at 60 Hz)	0.005 5		0.004 6	
Arc resistance, s	40–120	75–192		100
Water absorption (24 h), %	0.02	0.08		0.01
Burning rate, mm/min	Slow burning	10		
Effect of sunlight	Discolors slightly	Discolors	None	
Effect of acids	Attacked by oxidizing acids	Attacked	Slight	Slight
Effect of alkalies		Attacked	Attacked	Attacked
Effect of solvents	Attacked by halogen hydrocarbons	Resistant	Resistant	Resistant
Machining qualities	Excellent	Fair	Fair	Good
Optical qualities	Transparent to opaque	Opaque	Opaque	Translucent

(A) These vegetables are packaged in films designed for "boil-in-the-bag" use.

(B) Polyester "bake-in-the-bag" film helps retain moisture and flavor, and eliminates need for cleanup.

Fig. E-41. Some uses of polyester films.

Because of their thermoplastic nature, compounds based on saturated PET and PBT may be injection or extrusion molded.

Well-known trade names include Terylene, Dacron, and Kodel fibers, and Mylar film. Other uses include gears, distributor caps, rotors, appliance housings, pulleys, switch parts, furniture, fender extensions and packaging. The following is a list of two advantages and three disadvantages of saturated polyesters.

Advantages of Saturated Polyesters

1. Tough and rigid
2. Processable by thermoplastic methods

Disadvantages of Saturated Polyesters

1. Subject to attack by acids and bases
2. Low thermal resistance
3. Poor solvent resistance

Aromatic Polyesters

Oxybenzoyl polyesters were introduced by Carborundum in 1971 and 1974 under the trade names Ekonol and Ekcel. Both materials are linear chains of *p*-oxybenzoyl units. Because Ekonol does not melt below its decomposition temeprature, it must be sintered, compression molded, or plasma-sprayed. Ekcel can be processed with injection and extrusion equipment. High-temperature stability, stiffness, and thermal conductivity are important properties.

Some formulations can be melt-processed, but might require processing temperatures between 572 to 842 °F [300 to 400 °C]. Members of this class of materials are sometimes called nematic, anisotropic, liquid crystal polymer (LCP), or self-reinforcing polymers. These terms try to describe the formation of tightly packed fibrous chains during the melt phase. It is the fibrous chain that gives the polymer its self-reinforcing qualities. Parts must be designed to accommodate the anisotropic characteristics of LCP. Applications include chemical pumps, dual ovenable cookware, engine parts, and aerospace components.

Typical uses include bearings, seals, valve seats, rotors, high performance aerospace and automotive parts, electrical insulation components, and coatings for pans.

In 1978, *polyarylate* was introduced. This light-amber colored plastics is made from iso- and terephthalic acid and bisphenol A.

Polyarylates are aromatic polyester thermoplastic materials. The term *aryl* refers to a phenyl group derived from an aromatic compound.

A number of alloy and filled grades are available.

Polyarylate must be dried before injection or extrusion molding. It has excellent ultraviolet, thermal, and heat-deflection resistance. Applications include glazing, appliance housings, electrical connectors and lighting fixtures, exterior glazing, halogen lamp lenses, and selected microwave cookware.

Typical properties are shown in Table E-15.

Thermoplastic Polyimides

Polyimides were developed by Du Pont in 1962. They are obtained from condensation polymerization of an aromatic dianhydride and an aromatic diamine (Fig. E-42). Aromatic polyimides are linear and thermoplastic, and are hard to process. They can be molded by allowing enough time for flow to occur once glass transition temperature is exceeded. Many polyimides do not melt, but must be fabricated by machining or other forming methods.

Table E-15. Properties of Polyarylate

Property	Polyarylate
Molding qualities	Good
Relative density	1.21
Tensile strength, MPa	48–75
(psi)	(6 962–10 879)
Impact strength, Izod (6 mm), J/mm	0.24
(ft lb/in)	(4)
Hardness, Rockwell	R105
Deflection temperature (at 1.82 MPa or 264 Psi), °C	280
(°F)	(536)
Water absorption (24h), %	0.01
Optical, refractive index	1.64

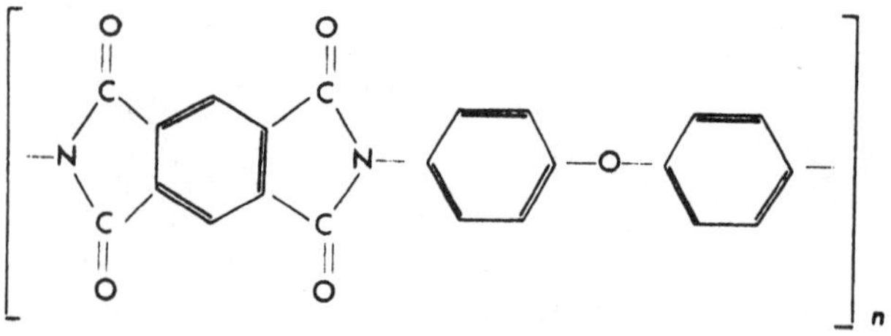

Addition polymerization provides plastics with slightly lower heat resistance than does condensation polymerization.

Polyimides compete with various fluorocarbons for applications requiring low friction, good strength, toughness, high dielectric strength, and heat resistance. They possess good resistance to radiation, but are surpassed in chemical resistance by fluoroplastics. Polyimide is attacked by strong alkaline solutions, hydrazine, nitrogen dioxide, and secondary amine compounds.

Polyimides, though costly and hard to process, are used in the making of aerospace, electronics, nuclear power, and office and industrial equipment. Other parts made include valve seats, gaskets, piston rings, thrust washers, and bushings. Films are made by a casting process (usually from prepolymer form). They are used for laminates, dielectrics, and coatings.

In Figure E-43, polyimide in a hot liquid, electrostatic spray form is being applied to electric skillets and houseware items. After curing and baking at 550 °F [290 °C] polyimide forms a hard and flexible glossy finish similar to porcelain.

Prolonged contact with this resin and its reducers may cause serious cracking or workers' skin. The solvents are no more toxic than other aromatics.

Table E-16 gives some properties of polyimides. The following list gives six advantages and disadvantages of polyimides.

Advantages of Polyimide

1. Short-exposure temperature capability of 600 °F to 700 °F [315 to 371 °C]
2. Excellent barrier
3. Very good electrical properties
4. Excellent solvent and wear resistance
5. Good adhesion capability
6. Especially suitable for composite fabrication

Fig. E-42. Basic polyimide structure.

(A) Colored polyimide finishes are used on cookware.

(B) Electric skillets are coated on an automatic spray line.

Fig. E-43. Examples of polyimide coating. (DeBeers Labs, Inc.)

Disadvantages of Polyimide

1. Difficulty of fabrication
2. Hygroscopic (moisture-absorbing)
3. Subject to attack by alkalies
4. Comparatively high cost
5. Dark color
6. Most types have volatiles or contain solvents that must be vented during cure

Table E-16. Properties of Polyimides

Property	Polyimide (Unfilled)
Molding qualities	Good
Relative density	1.43
Tensile strength, MPa	70
(psi)	(10 000)
Compressive strength, MPa	>165
(psi)	(>24 000)
Impact Strength, Izod, J/mm	0.045
(ft lb/in)	(0.9)
Hardness, Rockwell	E45–E58
Resistance to heat, °C	300
(°F)	(570)
Dielectric strength, V/mm	22 000
Dielectric constant (at 60 Hz)	3.4
Arc resistance, s	230
Water absorption (24 h), %	0.32
Burning rate	Nonburning
Effect of acids	Resistant
Effect of alkalies	Attacked
Effect of solvents	Resistant
Machining qualities	Excellent
Optical qualities	Opaque

Fig. E-44. General structural formula of polyamide-imide.

Polyamide-imide (PAI)

An amorphous member of the polyimide family is polyamide-imide. It was marketed in 1972 under the trade name Torlon by Amoco Chemicals. This material contains aromatic rings and a nitrogen linkage, as shown in Figure E-44. Polyamide-imide has striking properties (Table E-17). This material can withstand continuous temperatures of 500 °F [260 °C]. Because of its low coefficient of friction, excellent service temperature, and dimensional stability, polyamide-imide may be melt-processed into aerospace equipment, gears, valves, films, laminates, finishes, adhesives, and jet engine components. (Fig. E-45.)

Polymethylpentene

This plastics is reported to be an isotactically arranged aliphatic polyolefin of 4-methylpentene-1. Polymethylpentene was developed in the laboratory as early as 1955. It did not gain commercial value until Imperial

Table E-17. Properties of Polyamide-imide

Property	Poly(amide-Imide) (Unfilled)
Molding qualities	Excellent
Relative density	1.41
Tensile strength, MPa	185
(psi)	(26 830)
Compressive strength, MPa	275
(psi)	(39 900)
Impact strength, Izod J/mm	0.125
(ft lb/in)	(2.5)
Hardness, Rockwell	E78
Thermal expansion, 10^{-4}/°C	9.144
Resistance to heat, °C	260
(°F)	(500)
Dielectric strength, V/mm	>400
Dielectric constant (at 60 Hz)	3.5
Arc resistance, s	125
Water absorption (24 h), %	0.28
Burning rate	Nonburning
Effect of acids	Very resistant
Effect of alkalies	Very resistant
Effect of solvents	Very resistant
Machining qualities	Excellent

Fig. E-45. The piston skirt, number two ring, connecting rods, wrist pins, intake valves, valve spring retainers, tappets, timing gears, and other parts are made of (Torlon) polyamide-imide for a Polimotor Lola 2 liter engine that muscles up 318 horsepower at 9500 rpm, but weighs only 168 pounds. (Amoco Chemicals)

Chemical Industries, Ltd., announced it under the tradename of TPX in 1965.

Ziegler-type catalysts are used to polymerize 4-methylpentene-1 at atmospheric pressures (Fig. E-46). After polymerization, catalyst residues are removed by washing with methyl alcohol. The material is then compounded with stabilizers, pigments, fillers or other additives into a granular form.

Formulas for this type of plastics are shown in Figure E-47. To avoid confusion, the carbon atoms of the continuous chain must be numbered. This has been done in the formulas shown.

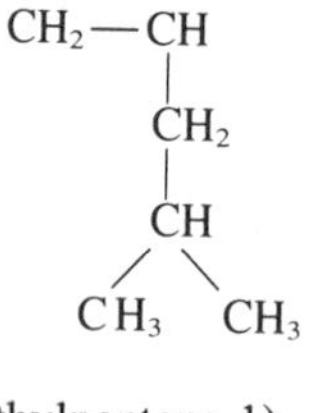

$$\begin{array}{c} CH_2-CH \\ \quad\;\; | \\ \quad\;\; CH_2 \\ \quad\;\; | \\ \quad\;\; CH \\ \quad\; / \;\; \backslash \\ CH_3 \quad CH_3 \end{array}$$

Fig. E-46. Poly(4-methylpentene-1)

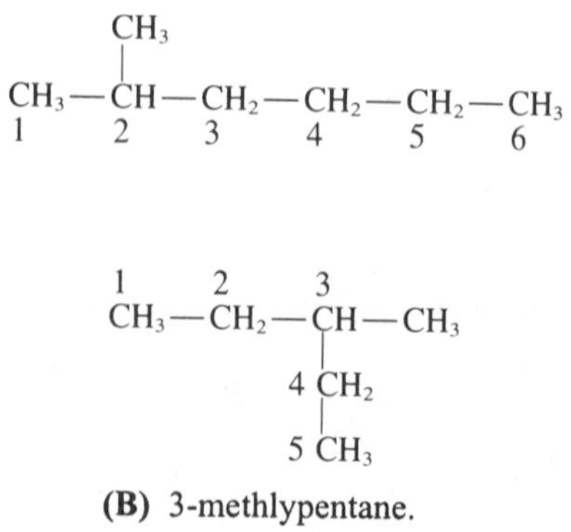

$$\begin{array}{cccccc} & CH_3 & & & & \\ & | & & & & \\ CH_3- & CH- & CH_2- & CH_2- & CH_2- & CH_3 \\ 1 & 2 & 3 & 4 & 5 & 6 \end{array}$$

$$\begin{array}{cccc} 1 & 2 & 3 & \\ CH_3- & CH_2- & CH- & CH_3 \\ & & | & \\ & & 4\ CH_2 & \\ & & | & \\ & & 5\ CH_3 & \end{array}$$

(B) 3-methlypentane.

Fig. E-47. Continuous-chain formulas with carbon atoms numbered.

Copolymerization with other olefin units (including hexene-1, octene-1, decene-1, and octadecene-1) can offer enhanced optical and mechanical properties.

Commerical poly (4-methylpentene-1) has a relatively high service temperature, which may exceed 320 °F [160 °C]. Although the plastics is nearly 50 percent crystalline, it has a light-transmission value of 90 percent. Spherulite growth may be retarded by rapid cooling of the molded mass. The open packing of the crystalline structure gives polymethylpentene a low relative density of 0.83. This is close to the theoretical minimum for thermoplastics.

Polymethylpentene may be processed on normal thermoplastic equipment, at processing temperatures that may exceed 470 °F [245 °C].

In spite of its high cost, this plastics has found uses in chemical plants, autoclavable medical equipment, lighting diffusers, encapsulation of electronic components, lenses, and metalized reflectors (Fig E-48). A well-known use is the packaging of bake-in-the-bag and boil-in-the-bag foods. These packages are used in the home and in catering services for airlines or manufacturing plants. Packaged foods may be boiled in water or cooked in normal or microwave ovens. Transparency is useful in showing materials in dispensing equipment.

Other side-branched polyolefins are possible. Three possible polymers are shown in Figure E-49. The branched side chains increase stiffness and lead to higher melting points. Polyvinyl cyclohexane melts at about 640 °F [338 °C]. Table E-18 gives some of the properties of polymethylpentene. A list of five advantages and two disadvantages of polymethylpentene follows.

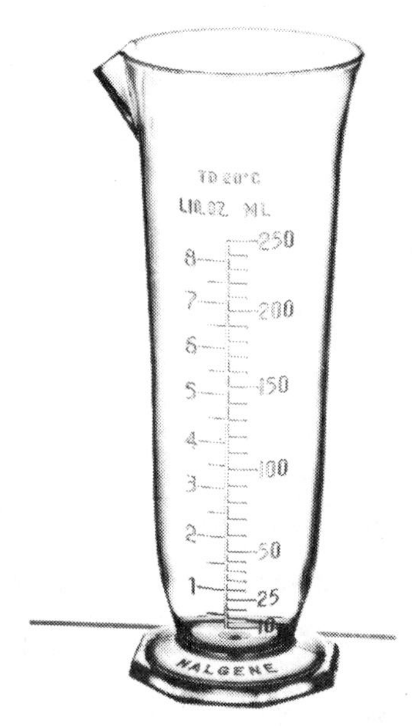

Fig. E-48. The clarity, chemical resistance, and toughness of polymethylpentene make it suitable for laboratory items. (ICI, Ltd.)

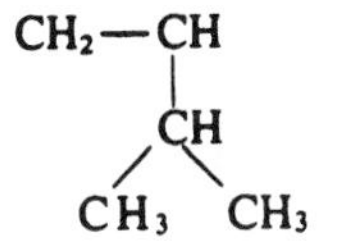

$$CH_2-CH(-CH(CH_3)-CH_3)$$

(A) Poly (3-methylbutene-1)

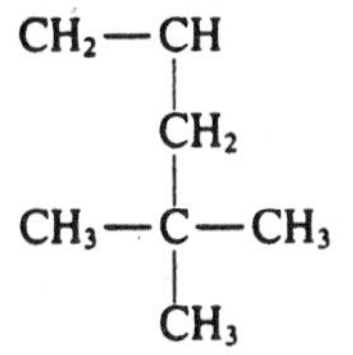

$$CH_2-CH(-CH_2-C(CH_3)_2-CH_3)$$

(B) Poly (4,4-dimethylpentene-1)

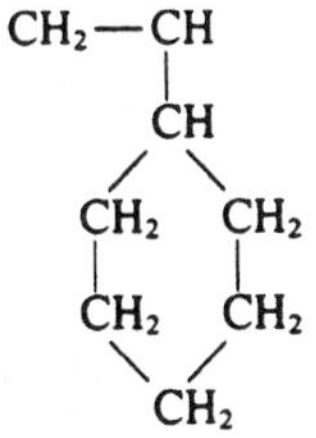

$$CH_2-CH(-C_6H_{11})$$

(C) Poly (vinylcyclohexane).

Fig. E-49. Side-branched polyolefin polymers.

Table E-18. Properties of Polymethylpentene

Property	Polymethylpentene (Unfilled)
Molding qualities	Excellent
Relative density	0.83
Tensile strength, MPa	25–28
(psi)	(3 500–4 000)
Impact strength, Izod, J/mm	0.02–0.08
(ft lb/in)	(0.4–1.6)
Hardness, Rockwell	L67–74
Thermal expansion, 10^{-4}/°C	29.7
Resistance to heat, °C	120–160
(°F)	(250–320)
Dielectric strength, V/mm	28 000
Dielectric constant (at 60 Hz)	212
Dissipation factor (at 60 Hz)	0.000 7
Water absorption (24 h), %	0.01
Burning rate, mm/min	25
(in/min)	(1.0)
Effect of sunlight	Crazes
Effect of acids	Attacked by oxidizing agents
Effect of alkalies	Resistant
Effect of solvents	Attacked by chlorinated aromatics
Machining qualities	Good
Optical qualities	Transparent to opaque

Advantages of Polymethylpentene

1. Minimum density (lower than polyethylene)
2. High light-transmission value (90%)
3. Excellent dielectric, volume resistivity, and power factor
4. Higher melting point than polyethylene
5. Good chemical resistance

Disadvantages of Polymethylpentene

1. Must be stabilized against most radiation sources
2. More costly than polyethylene

Polyolefins: Polyethylene (PE)

Ethylene gas is a member of an important group of unsaturated, aliphatic hydrocarbons called *olefins,* or *alkenes*. An *ethenic* refers to ethylene materials. The word olefin means oil forming.

It was originally given to ethylene because oil was formed when ethylene was treated with chlorine. The term olefin now applies to all hydrocarbons with linear carbon-to-carbon double bonds. Olefins are highly reactive because of this carbon-to-carbon double bond. Some of the major olefin monomers are shown in Table E-19.

In the United States, ethylene gas is readily produced by cracking higher hydrocarbons of natural gas or petroleum. The importance and relationship of ethylene to other polymers is shown in Figure E-50.

Between 1879 and 1900, several chemists experimented with linear polyethylene polymers. In 1900, E. Bamberger and F. Tschirner used the expensive material diazomethane to produce a linear polyethylene which they called "polymethylene."

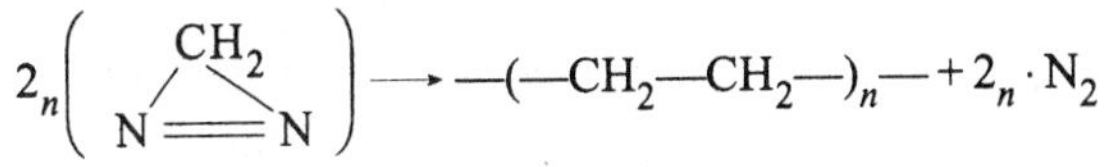

$$2_n\left(\begin{array}{c} CH_2 \\ N{=}N \end{array}\right) \longrightarrow -(-CH_2-CH_2-)_n- + 2_n \cdot N_2$$

Diazomethane *Polyethylene*

W. H. Carothers and co-workers reported producing polyethylene of low molecular mass in 1930. The commercial feasibility of polyethylene resulted from

Table E-19. The Principal Olefin Monomers

Chemical Formula	Olefin Name
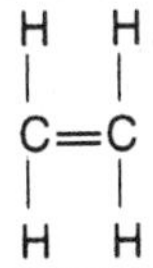 $HHC{=}CHH$	Ethylene
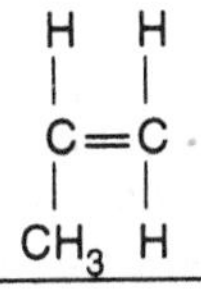 $H(CH_3)C{=}CHH$	Propylene
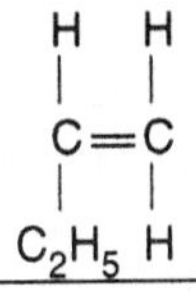 $H(C_2H_5)C{=}CHH$	Butene-1
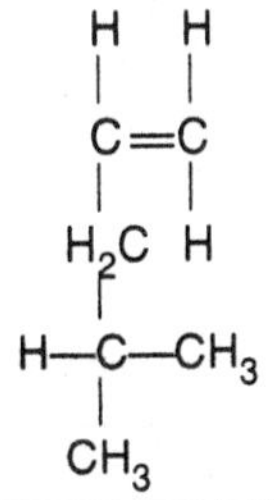 $H(H_2C-CH(CH_3)-CH_3)C{=}CHH$	4-Methylpentene

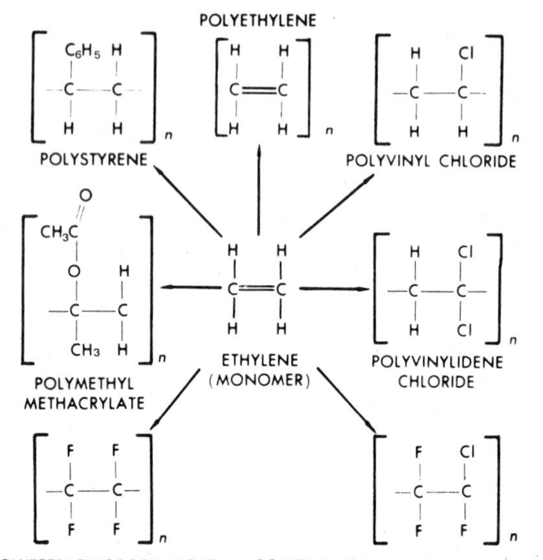

Fig. E-50. The ethylene monomer and its relationship to other monomer resins.

research by Dr. E. W. Fawcett and Dr. R. O. Gibson of the Imperial Chemical Industries (ICI) in England. Their discovery, in 1933, was a result of investigating the reaction of benzaldehyde and ethylene (obtained from coal) under high pressure and temperature. In September, 1939, ICI began commercial production of polyethylene, and the demands of World War II used all the polyethylene produced as insulation of high-frequency radar cables. By 1943, the United States was producing polyethylene by the high-pressure methods developed by ICI These early, low-density materials were highly branched, with a disorderly arrangement of molecular chains. Low-density materials are softer, more flexible, and melt at lower temperatures, thus, they may be more easily processed. By 1954, two new methods were developed for making polyethylene with higher relative densities of 0.91 to 0.97.

One process, developed by Karl Ziegler and associates in Germany, permitted polymerization of ethylene at low pressures and temperatures in the presence of aluminum triethyl and titanium tetrachloride as catalysts. At the same time, the Phillips Petroleum Company developed a polymerization process using low pressures with a chromium trioxide promoted, silica-alumina catalyst. The conversion of ethylene to polyethylene may also be done with a catalyst of molybdenum oxide on an alumina support and other promoters, a process developed by Standard Oil of Indiana. Only small quantities have been produced in the United States with this process.

The Ziegler process is used more extensively outside the United States while the Phillips Petroleum process is commonly used by U. S. firms.

Polyethylene can be produced with branched or linear chains (Fig. E-51) by either the high-pressure (ICI) or low-pressure (Ziegler, Phillips, Standard Oil)

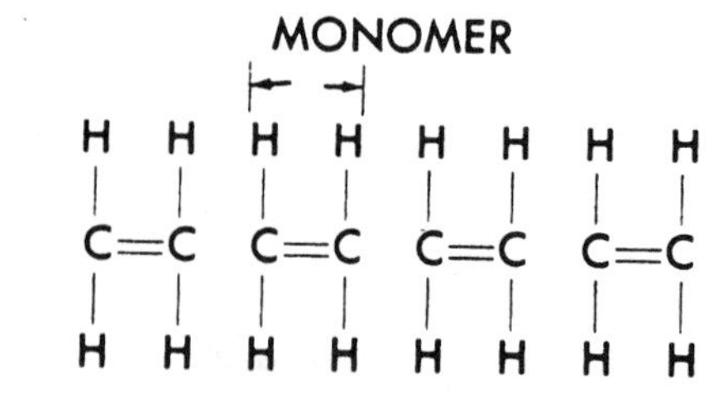

(A) Monomers of ethylene.

Mer

(B) Polymer containing many C_2H_4 mers.

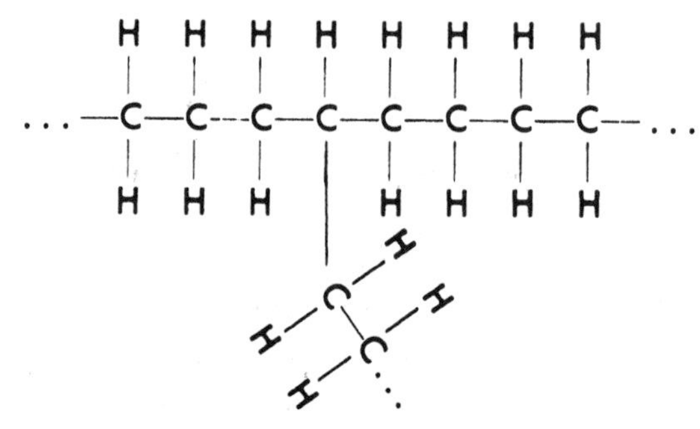

(C) Polymer with branching.

Fig. E-51. Addition polymerization of ethylene. The original double bonds of the ethylene monomer is broken, forming two bonds to connect adjacent mers.

methods. The differentiation of polymer type based on pressures used for polymerization is not employed today. The American Society for Testing and Materials (ASTM) has divided polyethylenes into four groups:

- Type 1 (Branched) 0.910–0.925 (low density)
- Type 2 0.926–0.940 (medium density)
- Type 3 0.941–0.959 (high density)
- Type 4 (Linear) 0.969 and above (high density to ultrahigh density homopolymers)

From Figure E-52, it can be seen that the physical properties of low-density (branched) and high-density (linear) polyethylenes are different. Low-density polyethylene has a crystallinity of 60 to 70 percent. Higher-density polymers may vary in crystallinity from 75 to 90 percent (Fig. E-53).

With increased density the properties of stiffness, softening point, tensile strength, crystallinity and creep resistance are increased. Increased density reduces impact strength, elongation, flexibility, and transparency.

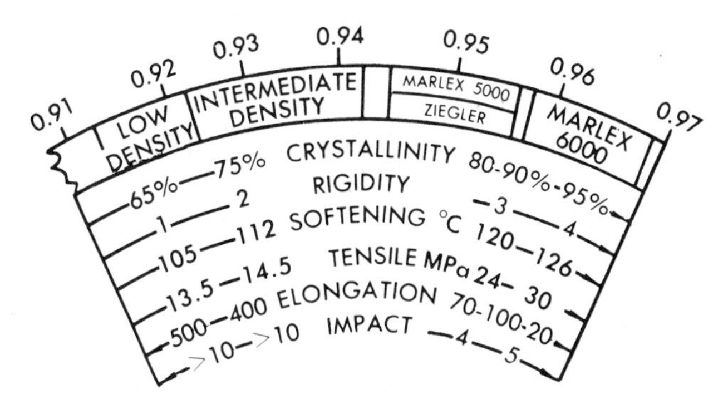

Fig. E-52. Density range of polyethylene. (Phillips Petroleum Co.)

(A) Electron micrograph shows intercrystalline links bridging radial arms of polyethylene spherulite.

(B) Small platelet crystals of polyethylene grown on intercrystalline links may be seen in this electron micrograph.

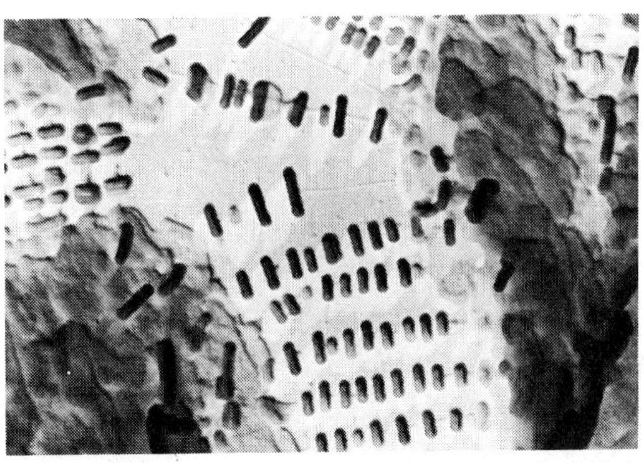

(C) Lamellar crystals of polyethylene were formed by depositing polymer from solution of links.

Fig. E-53. Closeup views of polyethylene. (Bell Telephone Laboratories)

Polyethylene properties may be controlled and identified by molecular mass and its distribution. Molecular mass and its distribution may have the effects shown in Table E-20.

Figure E-54 is a schematic representation comparing a polymer with a narrow molecular-mass distribution to a polymer with broad molecutar-mass distribution. Molecular-mass distribution is the ratio of large, medium, and small molecular chains in the resin. If the resin is composed of chains that are all of close to the average length, the molecular mass distribution is called *narrow*. Molecular chains of average length can flow past each other more easily than large ones.

A melt-index device (Fig. E-55) is used to measure the melt flow at a specified temperature and pressure.

Table E-20. Property Changes Caused by Molecular Mass and Distribution

Property	As Average Molecular Mass Increases (Melt Index Decreases)	As Molecular Mass Distribution Broadens
Melt viscosity	Increases	
Tensile strength at rupture	Increases	No significant Change
Elongation at rupture	Increases	No significant Change
Resistance to creep	Increases	Increases
Impact strength	Increases	Increases
Resistance to low temperature brittleness	Increases	
Environmental stress cracking resistance	Increases	Increases
Softening temperature		Increases

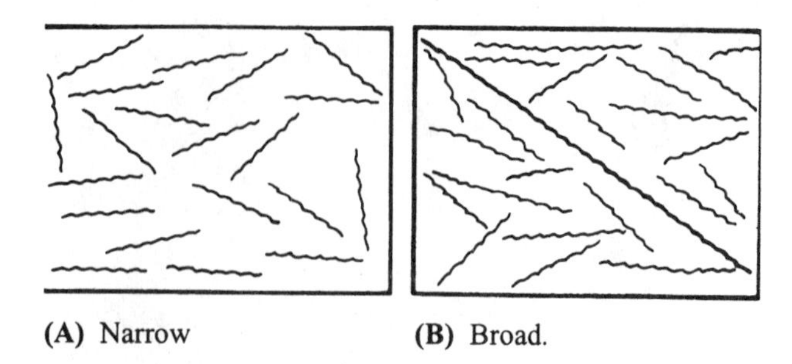

(A) Narrow (B) Broad.

Fig. E-54. Molecular mass (weight) distribution.

(A) Melt indexer with digital temperature display, optional elapsed time indicator and mass support.

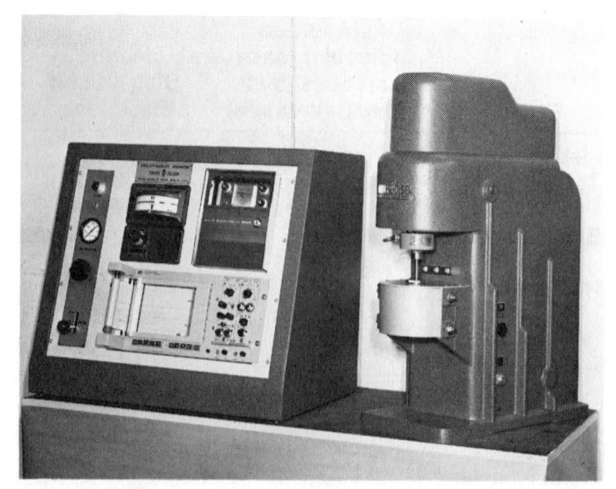

(B) Sieglaff-McKelvey rheometer.

Fig. E-55. These machines are used to test the flow of plastics materials.

This melt index depends on molecular mass and its distribution. As melt index goes down, the melt viscosity, tensile strength, elongation, and impact strength increase. For many processing methods, it is desirable to have the hot resin flow very easily indicating use of a resin with a high melt index. Polyethylenes with high molecular mass have a low melt index. By varying density, molecular mass, and molecular mass distribution, polyethylene may be produced with a wide variety of properties.

Polyethylene may be cross-linked to convert a thermoplastic material to a thermoset. Such conversion, after forming, opens up many new possibilities. This cross-linking may be accomplished by chemical agents (usually peroxides) or by irradiation. There is growing commercial use of irradiation to cause branching of polyethylene products. (See Chapter 20, Radiation Processing.) Radiation cross-linkage is rapid and leaves no objectionable residues. Irriadiated parts may be exposed to temperatures in excess of 250 °C [500 °F], as shown in Figure E-56.

Excessive radiation may reverse the cross-linking effect by breaking the main links in the molecular chain. Stabilizers or pigments of carbon black must be used to absorb or block the damaging effects of ultraviolet radiation on polyethylene.

Because of low price, processing ease, and a broad range of properties, polyethylene has become the most used plastics. As one of the lightest thermoplastics, it may be chosen where costs are based on cubic mass. Very good electrical and chemical resistance have led to wide use of polyethylene in wire coatings and dielectrics. Polyethylene is also used as containers, tanks, pipes, and coatings where chemical agents are present. At room temperature, there is no solvent for polyethylene; however, it is easily welded.

Polyethylene may be easily processed by all thermoplastic methods. Probably the largest use is in the making of containers and film consumed by the packaging industry. Blow-molded containers are seen in

Fig. E-56. Controlled radiation treatment can improve the heat-resistance of polyethylene. The center container was so treated and held its shape at 350° F (175° C).

every supermarket. These containers replace heavier ones of glass and metal (Fig. E-57). Tough plastics bags for packaging foods and films for packaging fresh fruit, frozen foods, and bakery products are but a few of the many uses for polyethylene.

Lower-density polyethylene films are made with good clarity by quick-chilling of the melt as it emerges from the die. As the hot amorphous melt is quickly chilled, wide-spread crystallization does not have time to occur. Low-density films are used to package new garments, such as shirts and sweaters, as well as sheets and blankets. At the laundry and dry cleaner, very thin films are used for packaging. Higher-density films are used where greater heat resistance is required, as for boil-in-the-bag food packs. Silage covers, reservoir linings, seedbed covers, moisture barriers, and covering for harvested crops are only a few uses in construction, agriculture, or horticulture.

Although polyethylene is a good moisture barrier, it has a high gas permeability. It should not be used under vacuum or for transporting gaseous materials. Although permeable to oxygen and carbon dioxide, films used for meats and some produce may require small vent holes to allow ventilation. Oxygen keeps the meat looking red and prevents moisture from condensing on packaged produce. Heat sealing and shrink wrapping are easily done with these films. Electronic or radio-frequency heat sealing is difficult because of the low electrical dissipation (power) factor of polyethylene.

Polyethylene is used to coat paper, cardboard, and fabrics to improve their wet strength as well as other properties. Coated materials may then be heat sealed. The packaging of milk is a well-known use. Powdered forms are used for dip coating, flame spraying, and fluidized-bed coating where a layer impervious to chemicals and moisture is required.

Injection-molded toys, small appliance housings, garbage cans, freezer containers, and artificial flowers benefit from polyethylene's toughness, chemical inertness, and low service temperatures (Fig. E-58).

Extruded polyethylene pipe and ducting is used in chemical plants and for some domestic cold water service. Corrugated drain pipe is replacing clay or concrete pipe and tiles because it is less costly, faster to install, and lighter. Monofilaments find uses for ropes and fishing nets and are woven for lawn chairs. Polyethylene is widely used as electrical wire and cable covering. Irradiated or chemically cross-linked films are used as dielectrics in winding electrical coils. They also have limited packaging applications.

Fig. E-57. Polyethylene beverage containers. (Uniloy Blow-molding Machinery Division, Hoover International.)

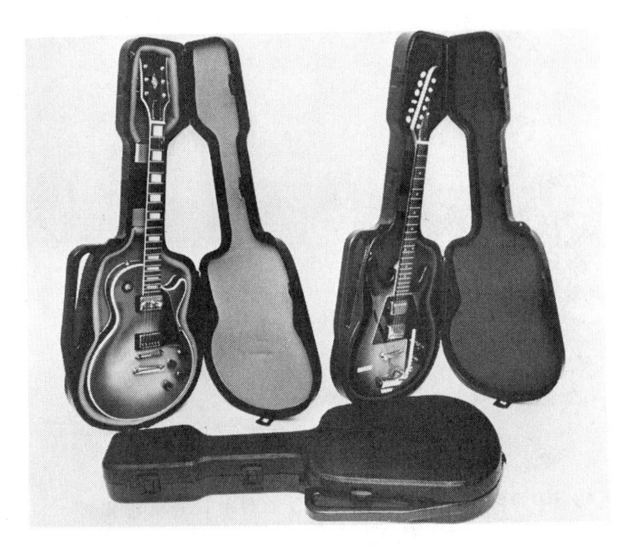

Fig. E-58. Blow-molded guitar case made of polyethylene. (Chemplex Co.)

Polyethylene may be foamed by several methods. A foaming agent that breaks down and releases a gas during the molding operation is preferred for commercial uses. A physical foaming method consists of introducing a gas, such as nitrogen, into the molten resin under pressure. While in the mold and under atmospheric pressure, the gas-filled polyethylene expands. Azodicarbonamide may be used for chemical foaming of either low- or high-density resins. Foams may be selected as gasket material and dielectric materials in coaxial cables. Cross-linked foams are suitable for cushioning, packaging, or flotation, while structural foam is used for furniture components and internal panels in automobiles. Low-density foams find use in wrestling mats, athletic padding, and flotation equipment. Figure E-59 shows some uses of foamed polyethylene.

Table E-21 gives properties of low-, medium-, and high-density polyethylene. A list of six advantages and five disadvantages of polyethylene follows.

Advantages of Polyethylene

1. Low cost (except UHMWPE)
2. Excellent dielectric properties
3. Moisture resistance
4. Very good chemical resistance
5. Available in food grades
6. Processable by all thermoplastic methods (except HMWHPE and UHMWPE)

(A) Ice cream freezer.

(B) Drinking glass

(C) Decorative wagon wheel.

Fig. E-59. Foamed polyethylene products. (Phillips Petroleum Co.)

Disadvantages of Polyethylene

1. High thermal expansion
2. Poor weathering resistance
3. Subject to stress cracking (except UHMWPE)
4. Difficulty in bonding
5. Flammable

Adding fillers, reinforcements, or other monomers may also change properties. Some of the common comonomers are shown in Table E-22.

A number of new polymerization techniques has expanded the potential application of polyethylene.

Very Low Density Polyethylene (VLDPE)

This linear, nonpolar polyethylene is produced by copolymerization of ethylene and other alpha olefins. Densities range from 0.890 to 0.915. VLDPE is easily processed into disposable gloves, shrink packages, vacuum cleaner hoses, tubing, squeeze tubes, bottles, shrink wrap, diaper film liners and other health care products.

Linear Low Density Polyethylene (LLDPE)

Production of linear low-density polyethylene is controlled through catalyst selection and regulation of reactor conditions. Densities range from 0.916 to 0.930. These plastics contain little if any chain branch-

Table E-21. Properties of Polyethylene

Property	Low-Density Polyethylene	Medium-Density Polyethylene	High Density Polyethylene
Molding qualities	Excellent	Excellent	Excellent
Relative density	0.910–0.925	0.926–0.940	0.941–0.965
Tensile strength, MPa	4–16	8.24	20–38
(psi)	(600–2 300)	(1 200–3 500)	(3 100–5 500)
Compressive strength, MPa			19–25
(psi)			(2 700–3 600)
Impact strength, Izod, J/mm	No break	0.025–0.8	0.025–1.0
(ft lb/in)		(.05–16)	(0.5–20)
Hardness, Shore	D41–D46	D50–D60	D60–D70
R10	R15		
Thermal expansion, 10^{-4}/°C	25–50	35–40	28–33
Resistance to heat, °C	80–100	105–120	
(°F)	(180–212)	(220–250)	(250)
Dielectric strength, V/mm	18 000–39 000	18 000–39 000	18 000–20 000
Dielectric constant (at 60 Hz)	2.25–2.35	2.25–2.35	2.30–2.35
Dissipation factor (at 60 Hz)	0.000 5	0.000 5	0.000 5
Arc resistance, s	135–160	200–235	
Water absorption (24 h), %	0.015	0.01	0.01
Burning rate, mm/min	Slow 26	Slow 25–26	Slow 25–26
(in./min)	(1.04)	(1–1.04)	(1–1.04)
Effect of sunlight	Crazes—must be stabilized	Crazes—must be stabilized	Crazes—must be stabilized
Effect of acids	Oxidizing acids	Oxidizing acids	Oxidizing acids
Effect of alkalies	Resistant	Resistant	Resistant
Effect of solvents	Resistant (below 60 °C)	Resistant (below 60 °C)	Resistant (below 60 °C)
Machining qualities	Good	Good	Excellent
Optical qualities	Transparent to opaque	Transparent to opaque	Transparent to opaque

Table E-22. Common Comonomers with Olefins

Formula	Name
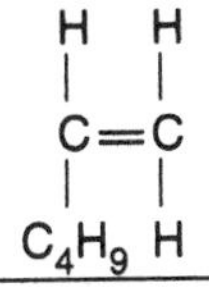 H H | | C=C | | C_4H_9 H	1-Hexene
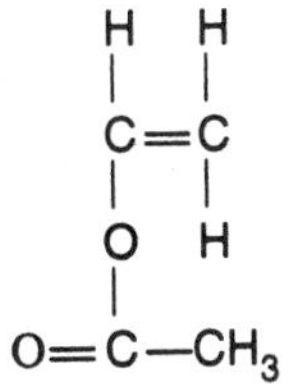 H H | | C=C | | O H | O=C—CH_3	Vinyl acetate
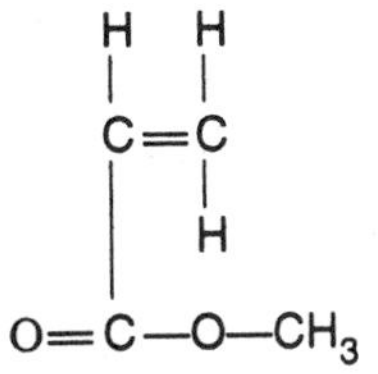 H H | | C=C | | | H | O=C—O—CH_3	Methly acrylate
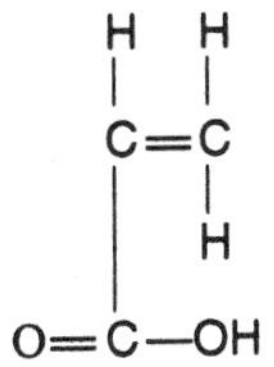 H H | | C=C | | | H | O=C—OH	Methly acrylate

ing. As a result, these plastics exhibit good flex life, low warpage, and improved stress-crack resistance. Films for ice, trash, garment, and produce bags are tough and puncture and tear resistant.

High Molecular Weight-High Density Polyethylene (HMW-HDPE)

High molecular weight-high density polyethylenes are linear polymers with molecular mass ranging from 200000 to 500000. Propylene, butene and hexene are common monomers. High molecular mass results in toughness, chemical resistance, impact strength and high abrasion resistance. High melt viscosities require special attention to equipment and mold designs. Densities are 0.941 or greater. Trash liners, grocery bags, industrial pipe, gas tanks, and shipping containers are familiar applications.

Ultrahigh Molecular Weight Polyethylene (UHMWPE)

Ultrahigh molecular weight polyethylenes have molecular weights from 3 to 6 million. This accounts for their high wear resistance, chemical inertness, and low coefficient of friction. These materials do not melt or flow like other polyethylenes. Processing is similar to methods used with polytetrafluoroethylene. (PTFE)

Sintering produces products with microporosity. Ram extrusion and compression molding are major forming methods.

Applications include chemical pump parts, seals, surgical implants, pen tips and butcher-block cutting surfaces.

Ethylene acid

A wide variety of properties similar to those of LDPE may be produced by varying the pendent carboxyl groups on the polyethylene chain. These carboxyl groups reduce polymer crystallinity, thus improving clarity, lowering the temperature required to heat seal, and improving adhesion to other substrates.

Molds should be designed to accommodate the adhesive qualities. Processing equipment should be corrosion resistant.

The FDA allows up to 25 percent acrylic acid and 20 percent methacrylic acid for ethylene copolymers in contact with food.

Most applications of ethylene acid and copolymers are for packaging of foods, coated papers, and composite foil pouches and cans.

Ethylene-ethyl acrylate (EEA)

By varying the ethyl acrylate pendent groups in the ethylene chain, properties may vary from rubbery to tough polyethylene-like polymers. The ethyl group on the PE chain lowers crystallinity.

Applications include hot melt adhesives, shrink wrap, produce bags, bag-in-box products, and wire coating.

Ethylene-methyl acrylate (EMA)

This copolymer is produced by the addition of methyl acrylate (40 percent by weight) monomer with ethylene gas.

EMA is a tough, thermally stable olefin with good elastomeric characteristics. Typical applications include disposable medical gloves, tough, heat-sealable layers, and coating for composite packaging.

EMA polymers meet the FDA and USDA requirements for use in food packaging.

Ethylene-vinyl acetate (EVA)

A wide variety of properties is available from this family of thermoplastic polymers. The vinyl acetate is copolymerized in varying amounts from 5 to 50 percent by weight onto the ethylene chain. If the vinyl acetate side groups exceed 50 percent, they are considered vinyl acetate-ethylene (VAE).

Typical EVA applications include hot melts, flexible toys, beverage and medical tubing, shrink wrap, produce bags, and numerous coatings.

Polyolefins: Polypropylene

Until 1954, most attempts to produce plastics from polyolefins had little commercial success, and only the polyethylene family was commercially important. It was in 1955 that Italian scientist F. J. Natta announced the discovery of sterospecific polypropylene. The word *sterospecific* indicates that the molecules are arranged in a definite order in space. This is in contrast to branched or random arrangements. Natta called this regular, arranged material *isotactic polypropylene.* Experimenting with Ziegler-type catalysts, he replaced the titanium tetrachloride in $Al(C_2H_5) + TiCl_4$ with the sterospecific catalyst titanium trichloride. This led to the commercial production of polypropylene.

It is not surprising that polypropylene and polyethylene have many of the same properties. They are similar in origin and manufacture. Polypropylene has become a strong competitor of polyethylene.

Polypropylene gas, $CH_3—CH═CH_2$, is cheaper than ethylene. It is obtained from high-temperature cracking of petroleum hydrocarbons and propane. The basic structural unit of polypropylene is:

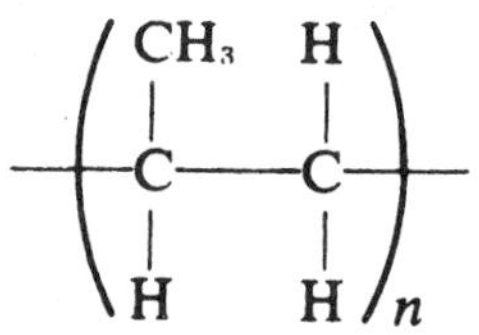

$$\left(\begin{array}{cc} CH_3 & H \\ | & | \\ —C— & —C— \\ | & | \\ H & H \end{array} \right)_n$$

Figure E-60 shows the stereostatic arrangements of polypropylene. In Figure E-60A, the molecular chains show a high degree of order with all the CH_3 groups along one side. Atactic polymers are rubbery, transparent materials of limited commercial value. Atactic and syndiotactic plastics grades are more impact resistant than isotactic grades. Both syndiotactic and atactic structures may be present in small quantities in isotactic plastics. Commercially available polypropylene is about 90 to 95 percent isotactic.

The general physical properties of polypropylene are similar to those of high-density polyethylene. However, polyethylene and polypropylene differ in four important respects:

1. Polypropylene has a relative density of 0.90; polyethylene has relative densities of 0.941 to 0.965
2. The service temperature of polypropylene is higher
3. Polypropylene is harder, more rigid, and has a higher brittle point

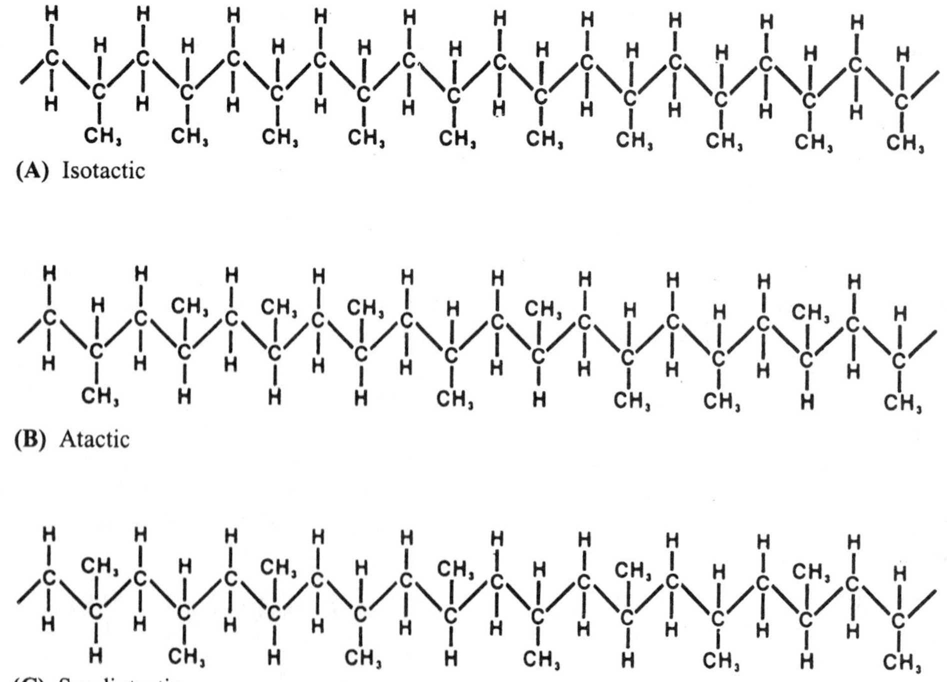

(A) Isotactic

(B) Atactic

(C) Syndiotactic

Fig. E-60. Stereotactic arrangements of polypropylene.

4. Polypropylene is more resistant to environmental stress cracking (Fig. E-61)

The electrical and chemical properties of the two materials are very similar. Polypropylene is more susceptible to oxidation and degrades at elevated temperatures.

Polypropylene may also be made with a variety of properties by adding fillers, reinforcements, or blends of special monomers (Fig. E-62). It is easily processed in all conventional thermoplastic equipment. It cannot be cemented by cohesive means, but is readily welded.

Polypropylene is competitive with polyethylene for many uses. It has the advantage of a higher service temperature (Fig. E-63). Typical uses include sterilizable hospital items, dishes, appliance parts, dishwasher components, containers, items incorporating integral hinges, automotive ducts, and trim. Extruded and cold-drawn monofilaments find use as rotproof ropes that will float on water. Some fibers are finding increasing uses for textiles and for outdoor or automotive carpeting. It may be used as tough packaging film or as electrical insulation on wire and cable. Slit film fiber, a process known as filibration, is widely used in producing ropes and fibers from polypropylene. It is coextrusion-blow-molded into numerous food containers. See Figure E-64. Also see Barrier plastics, ethylene-vinyl, and polyvinylidene chloride or polyvinyl alcohol.

Because of first rate abrasion resistance, high service temperature, and potentially lower cost, foamed polypropylene is finding growing use. Cellular polypropylene is foamed in much the same manner as polyethylene.

Fig. E-61. Chair seats and backs molded of polypropylene. (Exxon Chemical Co.)

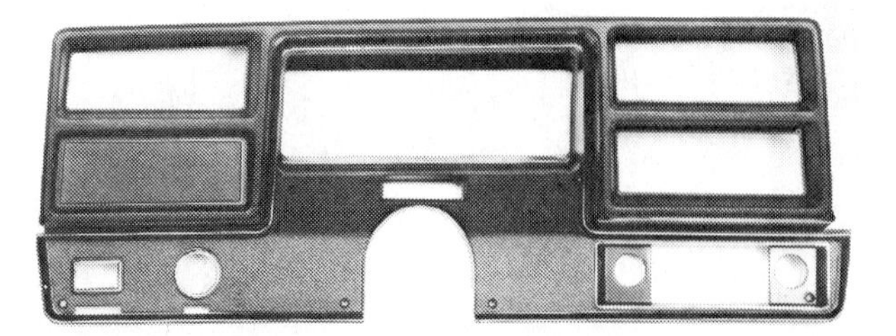

(A) Instrument panel that was injection molded of glass reinforced polypropylene. AC Spark Plug Div.)

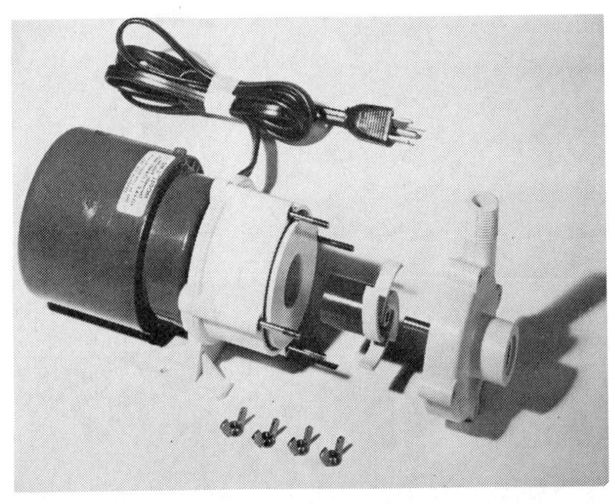

(B) Pump housing, impeller, magnetic housing, and volute are made of glass fiber reinforced polypropylene. (Fiberfill Div., Dart Industries)

Fig. E-62. Some used of polypropylene.

(A) Coffee pot.

(B) Sterilizable hospitalware.

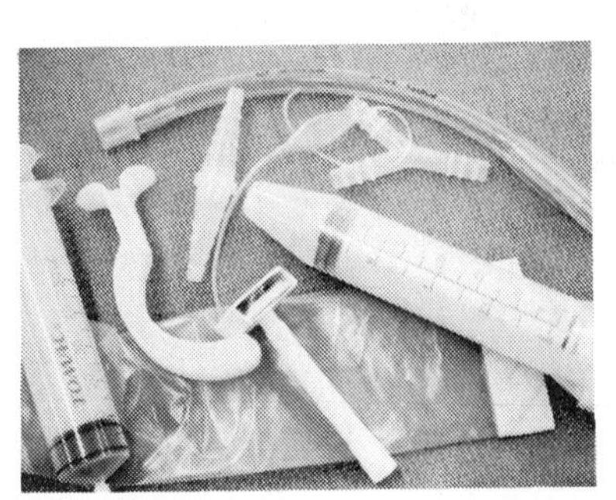

(C) Other gas- and steam-sterilizable hospital items.

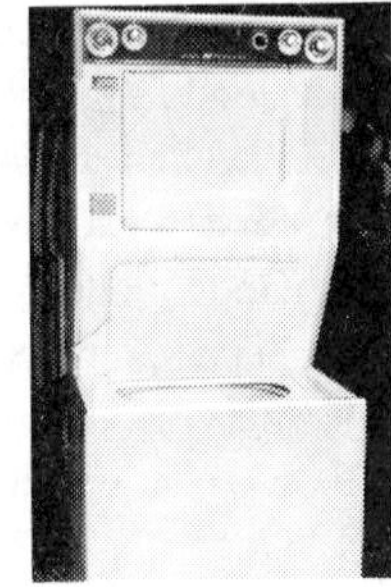

(D) Washer-dryer combination.

Fig. E-63. The high service temperature of polypropylene permits wide application.

Fig. E-64. Multilayered containers provide an effective oxygen barrier, allowing plastics to be used for foods such as tomato-based product, fruit juices, dressings, sauces, pickles, jams, and jellies. The container is made of a layer of ethylene-vinyl alcohol copolymer (EVOH) protected by inner and outer layers of polypropylene (PP). (Continental Can Co.)

Table E-23 gives some of the properties of polypropylene. Eleven advantages and six disadvantages of polypropylene are listed below.

Advantages of Polypropylene

1. Processable by all thermoplastic methods
2. Low coefficient of friction
3. Excellent electrical insulation
4. Good fatigue resistance
5. Excellent moisture resistance
6. First rate abrasion resistance
7. Good grade availability
8. Service temperature to 126 °C [260 °F]
9. Very good chemical resistance
10. Excellent flexural strength
11. Good impact strength

Disadvantages of Polypropylene

1. Broken down by ultraviolet radiation
2. Poor weatherability
3. Flammable, (flame-retarded grades are available)
4. Subject to attack by chlorinated solvents and aromatics
5. Difficult to bond
6. Oxidative breakdown accelerated by several metals

Polyolefins: Polybutylene (PB)

A polyolefin, called polybutylene (PB), was introduced by Witco Chemical Corporation in 1974. It has ethyl side groups in the linear backbone. This linear isostatic material may exist in a variety of crystalline forms. Upon cooling, the material is less than 30 percent crystalline. During aging and before complete crystalline transformation, many postforming techniques may be used. Crystallinity then varies from 50 to 55 percent after cooling. Polybutylene may be formed by conventional thermoplastic techniques.

Major uses include high performance films, tank liners, and pipes. It is used as hot-melt adhesives and coextruded as moisture barrier and heat-sealable packages. Table E-24 gives some properties of polybutylene.

Polyphenylene Oxides

This family of materials should probably be called *polyphenylene*. Several plastics have been developed by separating the benzene ring backbone of polyphenylene with other molecules, making these plastics more flexible and able to be molded by usual thermoplastic methods. Polyphenylene with no benzene-ring separation is very brittle, insoluble, and infusible.

Polyphenylene

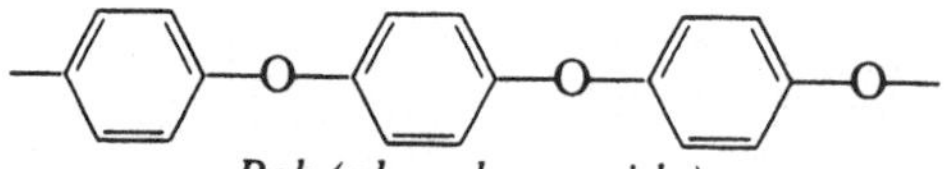

Poly(phenylene oxide)

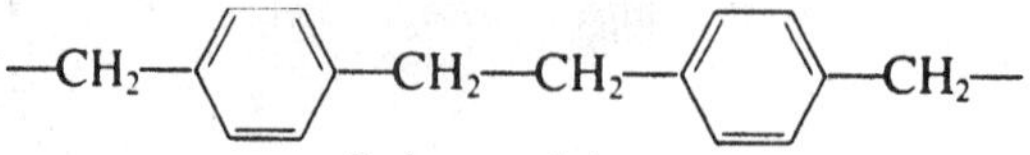

Poly-p-xylylene

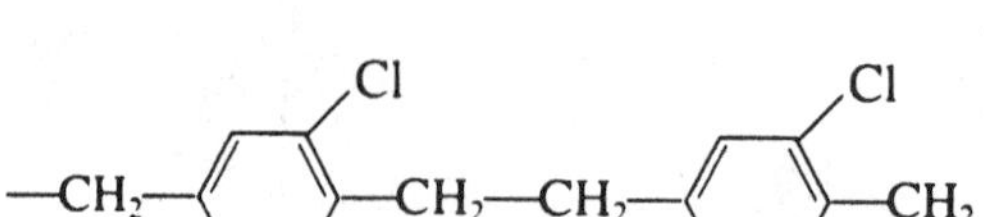

Polymonochloroparaxylylene

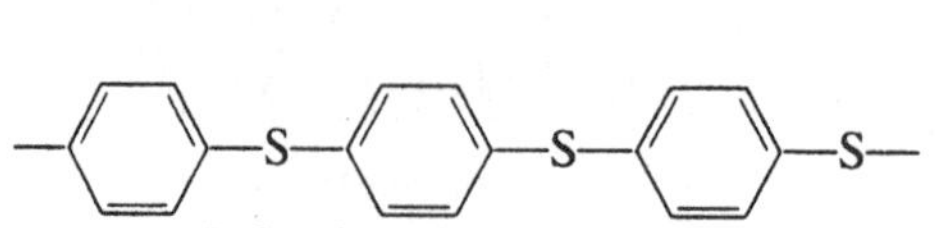

Poly(phenylene sulphide)

Table E-23. Properties of Polypropylene

Property	Polypropylene Homopolymer (Unmodified)	Polypropylene (Glass-Reinforced)
Molding qualities	Excellent	Excellent
Relative density	0.902–0.906	1.05–1.24
Tensile strength, MPa	31–38	42–62
(psi)	(4 500–5 500)	(6 000–9 000)
Compressive strength, MPa	38–55	38–48
(psi)	(5 500–8 000)	(5 500–7 000)
Impact strength, Izod, J/mm	0.025–0.1	0.05–0.25
(ft lb/in)	(0.5–2)	(1–5)
Hardness, Rockwell	R85–R110	R90
Thermal expansion, 10^{-4}/°C	14.7–25.9	7.4–13.2
Resistance to heat, °C	110–150	150–160
(°F)	(225–300)	(300–320)
Dielectric strength, V/mm	20 000–26 000	20 000–25 500
Dielectric constant (at 60 Hz)	2.2–2.6	2.37
Dissipation factor (at 60 Hz)	0.0005	0.0022
Arc resistance, s	138–185	74
Water absorption (24 h), %	0.01	0.01–0.05
Burning rate	Slow	Slow-nonburning
Effect of sunlight	Crazes—must be stabilized	Crazes—must be stabilized
Effect of acids	Oxidizing acids	Slowly attacked by oxidizing acids
Effect of alkalies	Resistant	Resistant
Effect of solvents	Resistant (below 80 °C)	Resistant (below 80 °C)
Machining qualities	Good	Fair
Optical qualities	Transparent to opaque	Opaque

Table E-24. properties of Polybutylene

Property	Polybutylene (Molding grades)
Molding qualities	Good
Relative density	0.908–0.917
Tensile strength MPa	26–30
(psi)	(3770–4350)
Impact strength, Izod, J/mm	No break
Hardness, Shore	D55–D65
Thermal expansion, 10^{-4}/°C	—
Resistance to heat, °C	<110
(°F)	(<230)
Dielectric constant (at 60 Hz)	2.55
Dissipation factor (at 60 Hz)	0.0005
Water absorption (24 h), %	<0.01–0.026
Burning rate, mm/min	45.7
(in/min)	(1.8)
Effect of sunlight	Crazes
Effect of acids	Attacked by oxidizing acids
Effect of alkalies	Very resistant
Effect of solvents	Resistant
Machining qualities	Good
Optical qualities	Translucent

Three advantages and disadvantages of polyphenylenes are listed below.

Advantages of Polyphenylene

1. Excellent solvent resistance
2. Good radiation resistance
3. High thermal and oxidative stability

Disadvantages of Polyphenylene

1. Difficult to process
2. Comparatively expensive
3. Limited availability

Polyphenylene Oxide (PPO)

Union Carbide brought out a heat-resisting plastics called polyphenylene oxide in 1964. It may be prepared by the catalytic oxidation of 2,6-dimethyl phenol (Fig. E-65).

Similar materials have been prepared making use of ethyl, isopropyl, or other alkyl groups. In 1965, the General Electric Co. introduced poly-2, 6-dimethyl-1, 4-phenylene ether as a polyphenylene oxide material. Then in 1966, General Electric announced another, similar thermoplastic with the tradename Noryl. This material is a physical blend of polyphenylene oxide and high impact polystyrene with a large diversity in formulations and properties.

Because Noryl costs less and has properties similar to polyphenylene oxides, many uses are the same

$$n\,(CH_3)_2C_6H_3{-}OH + \frac{n}{2}O_2 \longrightarrow \left[{-}(CH_3)_2C_6H_2{-}O{-} \right]_n + nH_2O$$

Fig. E-65. Preparation of polyphenylene oxide.

(Fig. E-66). This modified phenylene oxide material (Noryl) may be processed by normal thermoplastic equipment with processing temperatures from 375 °F to 575 °F [190 °C to 300 °C]. Modified phenylene oxide parts may be welded, heat-sealed, or solvent-cemented with chloroform and ethylene dichloride. Filled, reinforced, and flame-retardant grades are used as alternatives to die-cast metals, PC, PA, and polyesters. Video display terminals, pump impellers, radomes, small appliance housings, and instrument panels are typical applications. The following list shows five advantages and one disadvantage of polyphenylene oxide.

Advantages of Polyphenylene Oxide

1. Good fatigue strength and impact strength
2. Can be metal-plated
3. Thermally and oxidatively stable
4. Resistant to radiation
5. Processable by thermoplastics methods

Disadvantage of Polyphenylene Oxide

1. Comparatively high cost

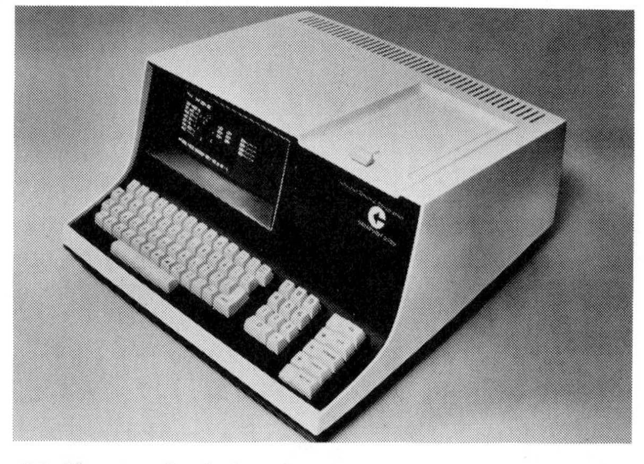

(A) Housing for desktop computer.

(B) Injection-molded housing of electric pinking shears.

Fig. E-66. Some applications of polyphenylene oxide. (General Electric Co.)

Polyphenylene Ether (PPE)

Polyphenylene polymers and alloys belong to a group of aromatic polyethers. To be useful, these polyethers are alloyed with PS to lower melt viscosity and allow conventional processing. Without this PS separation, the polymer is very brittle, insoluble, and infusible. These copolymers are used for small appliance housing and electrical components.

Parylenes

In 1965, Union Carbide introduced poly-*p*-xy-lylene under the trade name *Parylene* (Fig. E-67). Its primary market is in coating and film applications.

Parylene C (polymonochloroparaxylylene) offers improved permeability to moisture and gases. Parylenes are not formed as other thermoplastics are. They are polymerized as coatings on the surface of the product. The process is similar to vacuum metalizing. Table E-25 lists some properties of parylenes.

Polyphenylene Sulfide (PPS)

In 1968, the Phillips Petroleum Company announced a material known as polyphenylene sulfide, with the tradename Ryton. The material is available as thermoplastic or thermosetting compounds. Cross-linking is achieved by thermal or chemical means.

This rigid, crystalline polymer with benzene rings and sulfur links exhibits outstanding high temperature stability and chemical and abrasion resistance. Computer components, range components, hair dryers, submersible pump enclosures, and small appliance housings are typical uses. It is used as an adhesive, laminating resin, and as coatings for electrical parts.

Table E-26 gives some of the traits of three polyphenylene oxides and below are listed six advantages plus four disadvantages of polyphenylene sulfide.

Advantages of Polyphenylene Sulfide

1. Capable of extended usage at 450 °F [232 °C]
2. Good solvent and chemical resistance
3. Good radiation resistance
4. Excellent dimensional stability
5. Nonflammable
6. Low water absorption

Fig. E-67. Structure of poly-para-xylylene.

Table E-25. Properties of Parylene

Property	Polyparaxylyene	Polymonochloroparaxylyene
Molding qualities	Special process	Special process
Relative density	1.11	1.289
Tensile strength, MPa	44.8	68.9
(psi)	(6 500)	(9 995)
Thermal expansion, 10^{-4}/°C	17.52	8.89
Resistance to heat °C	94	116
(°F)	(201)	(240)
Water absorption (24 h), %	0.06	0.01
Effect of solvents	Insoluble in most	Insoluble in most
Optical qualities	Transparent	Transparent

Table E-26. Properties of Polyphenylene Oxide

Property	Polyphenylene oxide (Unfilled)	Noryl SE-1 SE-100	Polyphenylene Sulfides
Molding qualities	Excellent	Excellent	Excellent
Relative density	1.06–1.10	1.06–1.10	1.34
Tensile strength, MPa	54–66	54–66	75
(psi)	(7 800–9 600)	(7 800–9 600)	(10 800)
Compressive strength, MPa	110–113	110–113	
(psi)	(16 000–16 400)	(16 000–16 400)	
Impact strength, Izod, J/mm	0.25*	0.25*	0.015 at 24 °C 0.5 at 150 °C
(ft lb/in)	(5.0)*	(5.0)*	(0.3 at 75 °F) (1.0 at 300 °F)
Hardness, Rockwell	R115–R119	R115–R119	R124
Thermal expansion. 10^{-4}/°C	13.2	8.4–9.4	14
Resistance to heat °C	80–105	100–130	205–260
(°F)	(175–220)	(212–265)	(400–500)
Dielectric strength, V/mm	15 500–21 500	15 500–21 500	23 500
Dielectric constant (at 60 Hz)	2.64	2.64–2.65	3.11
Dissipation factor (at 60 Hz)	0.0004	0.000 6–0.000 7	
Arc resistance, s	75		
Water absorption (24 h), %	0.066		0.02
Burning rate	Self-ex., nondrip	Self-ex., nondrip	Nonburning
Effect of sunlight	Colors may fade	Colors may fade	
Effect of acids	None		Attacked by oxidizing acids
Effect of alkalies	None		None
Effect of solvents	Soluble in some aromatics	Soluble in some aromatics	Resistant
Machining qualities	Excellent	Excellent	Excellent
Optical qualities	Opaque	Opaque	Opaque

Disadvantages of Polyphenylene Sulfide

1. Hard to process (high melt temperature)
2. Comparatively high cost
3. Fillers needed for good impact strength
4. Subject to attack by chlorinated hydrocarbons

Polyaryl Ethers

Polyaryl ethers, polyaryl sulfone, and phenylene oxide have good physical and mechanical properties, good heat deflection temperatures, high impact strength, and good chemical resistance.

There are three different chemical groups that link the phenylene structure—isopropylidene, ether, and sulfone. The ether linkage and the carbon of the isopropylidene group impart toughness and flexibility to the plastics. See polyphenylene oxide, and polyphenylene ether.

In 1972, Uniroyal introduced a polyaryl ether plastics under the trade name of Arylon T.

Polyaryl ethers are prepared from aromatic compounds containing no sulfur links. The resulting polymer is more easily processed and service temperatures may be in excess of 170 °F [75 °C]. Uses include business machine parts, helmets, snowmobile parts, pipes, valves, and appliance components.

Table E-27 gives properties of polyaryl ether.

Polystyrene (PS)

Styrene is one of the oldest known vinyl compounds, however, industrial exploitation of this material did not begin until the late 1920s. This simple aromatic compound had been isolated as early as 1839 by the German chemist Edward Simon. Early monomer solutions were obtained from such natural resins as storax and dragon's blood (a resin from the fruit of the Malayan rattan palm). In 1851, French chemist M. Berthelot reported the production of styrene monomers by passing benzene and ethylene through a red-hot tube. This dehydrogenation of ethyl benzene is the basis of today's commercial methods.

By 1925, polystyrene (PS) was commercially available in Germany and the United States. For Germany, polystyrene became one of the most vital plastics during World War II. Germany had already embarked upon large-scale synthetic rubber production. Styrene was an essential ingredient for the production of styrene-butadiene rubber. When the natural sources of rubber were cut off in 1941, the United Stated began a crash program for the production of rubber from butadiene and styrene. This synthetic rubber became known as Government Rubber-Styrene (GR-S). There is still a large demand for styrene-butadiene synthetic rubber.

Styrene is chemically known as vinyl benzene, with the formula:

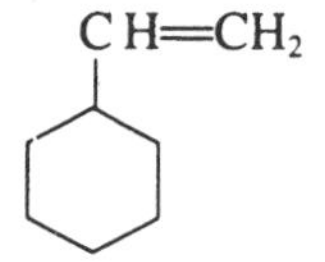

$CH{=}CH_2$

Styrene

In pure form, this aromatic vinyl compound will slowly polymerize by addition at room temperature. The monomer is obtained commercially from ethyl benzene (Fig. E-68).

Styrene may be polymerized by bulk, solvent, emulsion, or suspension polymerization. Organic peroxides are used to speed the process.

Polystyrene is an atactic, amorphous thermoplastic with the formula shown in Figure E-69. It is inexpensive, hard, rigid, transparent, easily molded and possesses good electrical and moisture resistance (Fig. E-70). Physical properties vary, depending on the molecular mass distribution, processing, and additives.

Polystyrene may be processed by all normal thermoplastic processes and may be solvent-cemented. Some common uses include wall tile, electrical parts, blister packages, lenses, bottle caps, small jars, vacuum-formed refrigerator liners, containers of all kinds, and transparent display boxes. Thin films and sheet stock have a metallic ring when struck or dropped. These forms are used in packaging foods and other items, such as some cigarette packets. Children see the use of polystyrene in model kits and toys. Adults may be aware of this material in inexpensive dishes, utensils, and glasses. Filaments are extruded and deliberately stretched or drawn to orient the molecular chains.

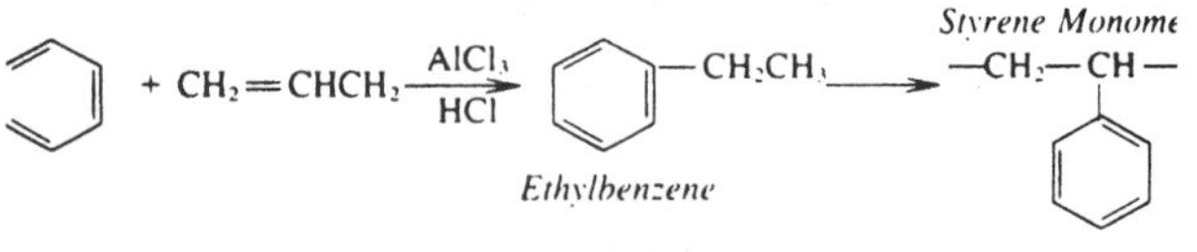

Table E-27. Properties of Polyaryl Ether

Property	Polyaryl Ether (Unfilled)
Molding qualities	Excellent
Relative density	1.14
Tensile strength, MPa	52
(psi)	(7 500)
Compressive strength, MPa	110
(psi)	(16 000)
Impact strength, Izod, J/mm	0.4
(ft lb/in)	(8)
Bar size, mm	12.7 x 7.25
(in)	1/2 x 1/4
Hardness, Rockwell	R117
Thermal expansion, 10^{-4}/°C	16.5
Resistance to heat, °C	120–130
(°F)	(250–270)
Dielectric strength, V/mm	16 930
Dielectric constant (at 60 Hz)	3.14
Dissipation factor (at 60 Hz)	0.006
Arc resistance, s	180
Water absorption (24 h), %	0.25
Burning rate	Slow
Effect of sunlight	Slight, yellows
Effect of acids	Resistant
Effect of alkalies	None
Effect of solvents	Soluble in ketones, esters, chlorinated aromatics
Machining qualities	Excellent
Optical qualities	Translucent to opaque

Fig. E-68. Production of the vinyl benzene (styrene) monomer.

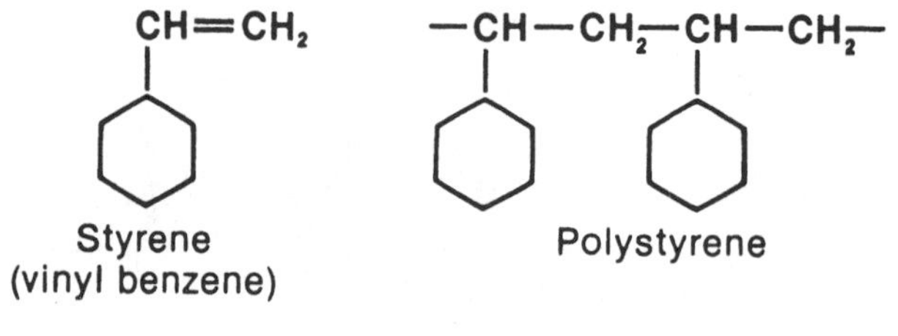

Fig. E-69. Polymerization of styrene.

(A) Drum table with marble top.

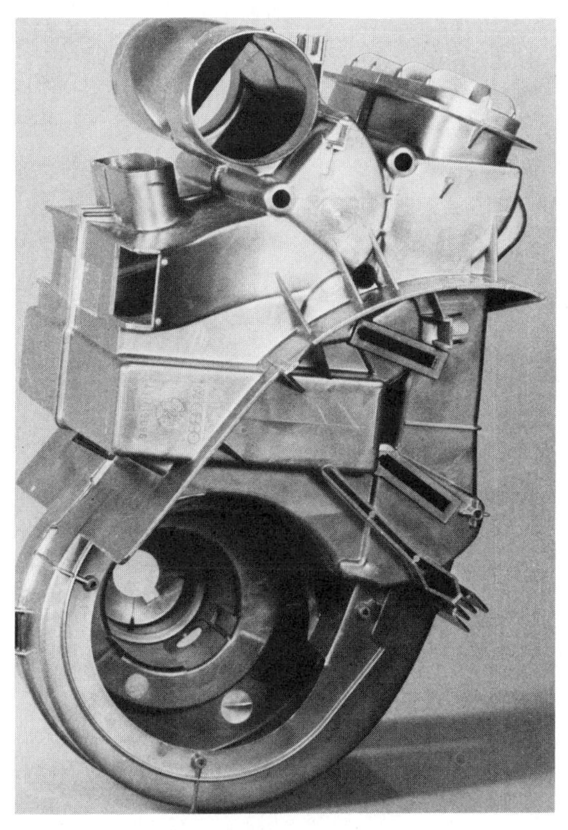

(B) Automobile heater housing. (BASF)

Fig. E-70. Some polystyrene objects.

This orientation adds tensile strength in the direction of stretching. Filaments may be used for brush bristles.

Expanded or foamed polystyrene is made by heating polystyrene containing a gas-producing or *blowing* agent. The foaming is accomplished by blending a volatile liquid such as methylene chloride, propylene, butylene, or fluorocarbons into the hot melt. As the mixture emerges from the extruder, the blowing agents release gaseous products resulting in a low density cellular material.

Expanded polystyrene (EPS) is produced from polystyrene beads containing an entrapped blowing agent. These agents may be pentane, neopentane, or petroluem ether. Upon either pre-expanding or final molding, the blowing agent volatizes causing the individual beads to expand and fuse together. Steam or other heat sources are used to cause the expansion. Both expanded and foamed forms have a closed cellular structure so that they can be used as flotation devices. Because of its low thermal conductivity, this material has found widespread use as thermal insulation (Fig. E-71) used in refrigerators, cold storage rooms, freezer display cases, and building walls. It has the added advantage of being moistureproof. There are many packaging uses because of its thermal insulation value and shock absorption characteristics. Packing in cellular polystyrene can save on shipping and breaking costs (Fig. E-72).

Foamed and expanded polystyrene sheets may be thermoformed. They are made into such familiar packaging items as egg cartons and meat or produce trays.

Fig. E-71. Expanded polystyrene beads. (Sinclair-Koppers Co.)

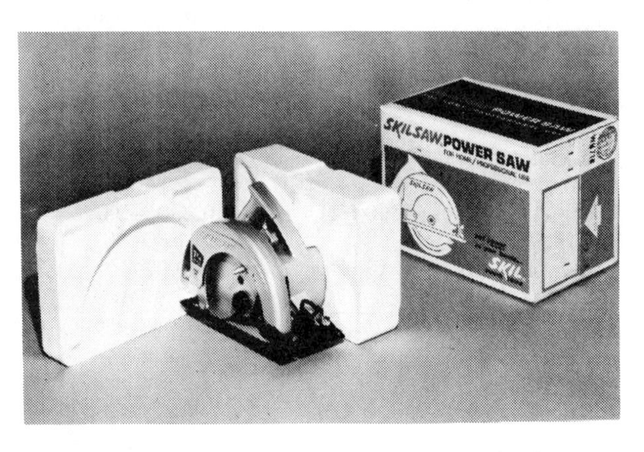

Fig. E-72. Cellular polystyrene has many packaging uses. (Sinclair-Koppers Co.)

Molded drinking cups, glasses, and *ice* chests are commonly used items.

Polystyrenes cannot withstand prolonged heat above 150 °F [65 °C] without distorting, thus they are not good exterior materials. Special grades and additives can correct this problem. Polystyrenes reinforced with glass fiber are used in automotive assemblies, business machines, and appliance housings.

The properties of polystyrene can be considerably varied by copolymerization and other modifications. Styrene-butadiene rubber has been mentioned above. Polystyrene is used in sporting goods, toys, wire and cable sheathing, shoe soles, and tires. Two of the most useful copolymers (terpolymers) are styrene-acrylonitrile and styrene-acrylonitrile-butadiene (ABS).

Table E-28 gives some of the properties of polystyrene while below are listed nine advantages and six disadvantages of polystyrene.

Advantages of Polystyrene

1. Optical clarity
2. Light mass
3. High gloss
4. Excellent electrical properties
5. Good grades available
6. Processable by all thermoplastic methods
7. Low cost
8. Good dimensional stability
9. Good rigidity

Disadvantages of Polystyrene

1. Flammable (retarded grades available)
2. Poor weatherability
3. Poor solvent resistance
4. Brittleness of homopolymers
5. Subject to stress and environmental cracking
6. Poor thermal stability

Styrene-Acrylonitrile (SAN)

Acrylonitrile ($CH_2{=}CHCHN$) is copolymerized with styrene (C_6H_6), giving products a higher resistance than polystyrene to various solvents, fats, and other compounds (Fig. E-73). These products are suitable for components requiring impact strength and chemical resistance and are used in vacuum cleaners and kitchen equipment.

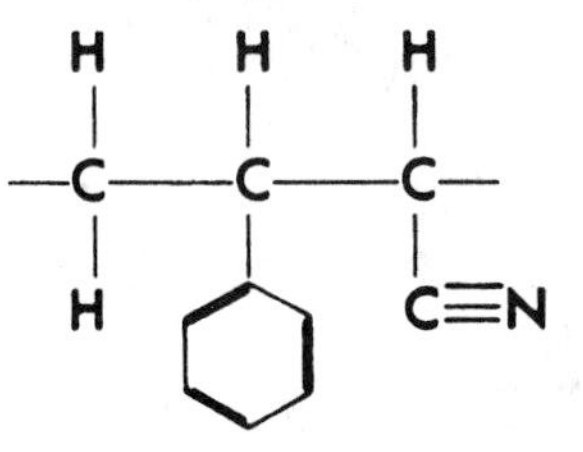

Fig. E-73. SAN mer.

Table E-28. Properties of Polystyrene

Property	Polystyrene (Unfilled)	Impact- and Heat-Resistant Polystyrene	Polystyrene (20-30% Glass-Filled)
Molding qualities	Excellent	Excellent	Excellent
Relative density	1.04–1.09	1.04–1.10	1.20–1.33
Tensile strength, MPa	35–83	10–48	62–104
(psi)	(5 000–12 000)	(1 500–7 000)	(9 000–15 000)
Compressive strength, MPa	80–110	28–62	93–124
(psi)	(11 500–16 000)	(4 000–9 000)	(13 500–18 000)
Impact strength, Izod. J/mm	0.0125–0.02	0.025–0.55	0.02–0.22
(ft lb/in)	(0.25–0.40)	(0.5–11)	(0.4–4.5)
Hardness, Rockwell	M65–M80	M20–M80, R50–RI00	M70–M95
Thermal expansion, 10^{-4}/°C	15.2–20	8.5–53	4.5–11
Resistance to heat °C	65–78	60–80	82–95
(°F)	(150–170)	(140–175)	(180–200)
Dielectric strength, V/mm	19 500–27 500	11 500–23 500	13 500–16 500
Dielectric constant (at 60 Hz)	2.45–2.65	2.45–4.75	
Dissipation factor (at 60 Hz)	0.0001–0.0003	2.45–4.75	0.004–0.014
Arc resistance, s	60–80	10–20	25–40
Water absorption (24 h), %	0.03–0.10	0.05–0.6	0.05–0.10
Burning rate	Slow	Slow	Slow-nonburning
Effect of sunlight	Yellows slightly	Yellows slightly	Yellows slightly
Effect of acids	Oxidizing acids	Oxidizing acids	Oxidizing acids
Effect of alkalies	None	None	Resistant
Effect of solvents	Soluble in aromatic and chlorinated hydrocarbons	Soluble in aromatic and chlorinated hydrocarbons	Soluble in aromatic and chlorinated hydrocarbons
Machining qualities	Good	Good	Good
Optical qualities	Transparent	Translucent to opaque	Translucent to opaque

Styrene-acrylonitrile (SAN) copolymers may have about 20 to 30 percent acrylonitrile content. By varying the proportions of each monomer, a wide range of properties and processability may be obtained. A slight yellow cast is typical of SAN, due to the copolymerization of acrylonitrile with the styrene member.

This copolymer is easily molded and processed. SAN-type materials inherently absorb more moisture than polystyrene and may result in molding defects such as silver streaking. Predrying is advised.

Methyl ethyl ketone, trichloroethylene, and methylene chloride are among the effective solvents for SAN.

This tough, heat-resistant plastics is used for telephone parts, containers, decorative panels, blender bowls, syringes, refrigerator compartments, food packages, and lenses (Fig. E-74).

Table E-29 gives some of the properties of SAN. Three advantages and disadvantages of styrene-acrylonitrile (SAN) are listed below.

Advantages of Styrene-Acrylonitrile (SAN)

1. Processable by thermoplastic methods
2. Rigid and transparent
3. Improved solvent resistance over polystyrene

Disadvantages of Styrene-Acrylonitrile

1. Higher water absorption than polystyrene
2. Low thermal capability
3. Low impact strength

Table E-29. Properties of SAN

Property	SAN (Unfilled)
Molding qualities	Good
Relative density	1.075–1.1
Tensile strength, MPa	1.075–1.1
(psi)	(9 000–12 000)
Compressive strength, MPa	97–117
(psi)	(14 000–17 000)
Impact strength, Izod, J/mm	0.01–0.02
(ft lb/in)	(0.35–0.50)
Hardness, Rockwell	M80–M90
Thermal expansion, 10^{-4}/°C	M80–M90
Resistance to heat, °C	60–96
(°F)	(140–205)
Dielectric strength, V/mm	15 750–19 685
Dielectric constant (at 60 Hz)	2.6–3.4
Dissipation factor (at 60 Hz)	0.006–0.008
Arc resistance, s	100–150
Water absorption (24 h), %	0.20–0.30
Burning rate	Slow to Self extinguishing
Effect of sunlight	Yellows
Effect of acids	None
Effect of alkalies	Attacked by oxidizing agents
Effect of solvents	Soluble in ketones and esters
Machining qualities	Good
Optical qualities	Transparent

(A) Jar for petroleum jelly.

(B) Housing for air precleaner.

Fig. E-74. Two uses of transparent SAN. (Monsanto Co.)

(Olefin-Modified) Styrene-Acrylonitrile (OSA)

A tough, heat and weather resistant polymer is produced by tailoring the molecular weight and monomer ratios of saturated olefinic elastomer with styrene and acrylonitrile. It is used almost exclusively as a coextrudant over other substrates. Topper covers, boat hulls, and decorative wood and metal construction panels are typical applications.

Styrene-Butadiene Plastics (SBP)

This armorphous copolymer consists of two blocks of styrene repeating units separated by a block of butadiene. This is in contrast to styrene butadiene (SBR), which is thermosetting.

These polymers are ideally suited for packaging applications including cups, deli containers, meat trays, jars, bottles, skin packages, and overwrap. Reinforced and filled grades are used in tool handles, office equipment housings, medical devices, and toys.

Styrene-Maleic Anhydride (SMA)

This thermoplastic is distinguished by higher heat resistance than the parent styrenic and ABS families. SMA is obtained by the copolymerization of maleic anhydride and styrene. Butadiene is sometimes terpolymerized to produce impact-modified versions. Applications include vacuum cleaner housings, mirror housings, thermoformed headliners, fan blades, heater ducts, and food service trays.

Polysulfones

In 1965, Union Carbide introduced a linear, heat-resistant thermoplastic called polysulfone. The basic repeating structure consists of benzene rings joined by a sulfone group (SO_2), an isopropylidene group (CH_3CH_3C), and also an ether linkage (O).

One basic polysulfone is made by mixing bisphenol A with chlorobenzene and dimethyl sulfoxide in a caustic soda solution. The resulting condensation polymerization is shown in Figure 7-75. The light amber color of the plastics is a result of the addition of methyl chloride, which ends polymerization. The outstanding thermal and oxidation resistance is the result of the benzene-to-sulfone linkages. Polysulfone can be processed by all normal methods. It must be dried before use and may require processing temperatures in excess of 700 °F [370 °C]. Service temperatures range from −150 to 345 °F [−100 to + 175 °C].

Polysulfone can be machined, heat-sealed, or solvent-cemented with dimethyl formamide or dimethyl acetamide.

Polysulfones are competitive with many thermosets. They may be processed in rapid-cycle thermoplastics equipment. The polysulfones have excellent mechanical, electrical, and thermal properties. They are used for hot-water pipes, alkaline battery cases, distributor caps, face shields for astronauts, electrical circuit breakers, appliance housings, dishwasher impellers, autoclavable hospital equipment, interior components for aerospace craft, shower heads, lenses, and numerous electrical insulating components (Fig. E-76). When used outdoors, polysulfones should be painted or electroplated to prevent degradation. Five

$$NaO-C_6H_4-C(CH_3)_2-C_6H_4-ONa_n + Cl-C_6H_4-SO_2-C_6H_4-Cl \xrightarrow{2NaCl}$$

$$\left[-C_6H_4-C(CH_3)_2-C_6H_4-O-C_6H_4-SO_2-C_6H_4-O-\right]$$

Fig. E-75. Basic repeating structure of polysulfone.

(A) Immersible cornpopper has molded polysulfone cover.

(B) Microwave oven cookware made of polysulfone.

(C) Polysulfone thermos pitcher liner.

Fig. E-76. Applications of polysulfone. (Union Carbide)

advantages and four disadvantages of polysulfones are found in the following list.

Advantages of Polysulfones

1. Good thermal stability
2. Excellent high-temperature creep resistance
3. Transparent
4. Tough and rigid
5. Processable by thermoplastic methods

Disadvantages of Polysulfones

1. Subject to attack by many solvents
2. Poor weatherability
3. Subject to stress cracking
4. High processing temperature

Polyarylsulfone

Polyarylsulfone is a high-temperature, amorphous thermoplastic introduced in 1983. It offers properties similar to other aromatic sulfones. Uses include circuit boards, high temperature bobbins, sight glasses, lamp housings, electrical connectors and housings, and panels of composite materials for numerous transporation components.

Polysulfones have been prepared with a variety of bisphenols with methylene, sulfide, or oxygen linkages. In polyaryl sulfone, the bisphenol groups are linked by ether and sulfone groups. There are no isopropylene (aliphatic) groups present. The term *aryl* refers to a phenyl group derived from an aromatic compound. In naming these compounds, if more than one hydrogen is substituted in the aryl group, a numbering system is normally used. Three possible disubstituted benzenes are shown in Fig. E-77. The basic properties of polysulfone are given in Table E-30.

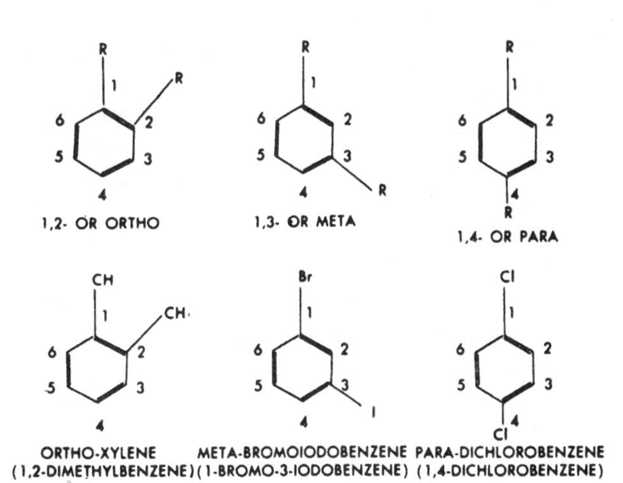

Fig. E-77. Three possible disubstituted benzenes.

Polyethersulfone (PES)

This plastics, with outstanding oxidation and thermal resistance, was introduced in 1973. Polyethersulfone has performed well under creep and stress forces, at temperatures above 390 °F [200 °C]. It is characterized by an absence of aliphatic groups, and is an amorphous structure. Polyethersulfone is very resistant to both acids and alkalies, but it is attacked by ketones, esters, and some halogenated and aromatic hydrocarbons. The basic monomer unit is shown in Figure E-78.

The distinguishing properties of PES are its high temperature performance, good mechanical strength, and low flammability.

Polyethersulfone has found uses in aerospace components, sterilizable medical components, and oven windows. Compounded grades extend their useful temperatures and improve mechanical properties. They have been used as adhesives and may be plated. Table E-31 gives properties of this plastics.

Polyphenylsulfone (PPSO)

The sulfone that best resists stress cracking is polyphenylsulfone. Introduced in 1976, polyphenylsulfone is an amorphous structure with very high impact strength that will withstand a continuous service temperature of 375 °F [190 °C]. Uses include semiconductor carriers, valves, circuit boards and aerospace components.

Polyvinyls

There is a large and varied group of addition polymers that chemists refer to as vinyls. These have the formulas:

$$CH_2{=}CH{-}R \text{ or } CH_2{=}C(R)(R)$$

Radicals (R) may be attached to this repeating vinyl group as side groups to form several polymers

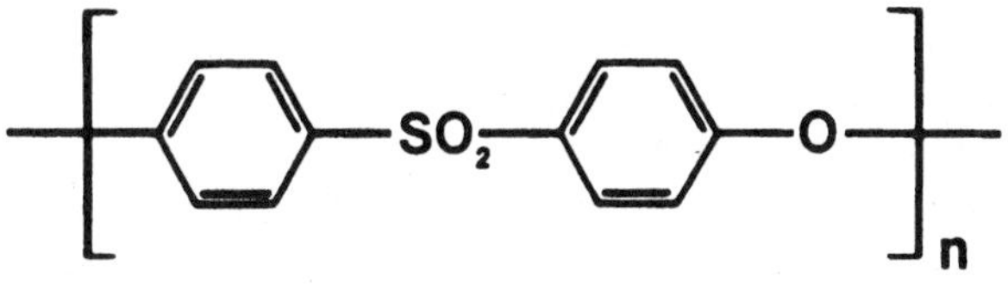

$$\left[-C_6H_4-SO_2-C_6H_4-O- \right]_n$$

Fig. E-78. Basic repeating unit of polyethersulfone.

Table E-30. Properties of Polysulfones

Property	Polysulfone (Unfilled)	Polyary Sulfone (Unfilled)
Molding qualities	Excellent	Excellent
Relative density	1.24	1.36
Tensile strength, MPa	70	90
(psi)	(10 200)	(13 000)
Compressive strength, MPa	96	123
(psi)	(13 900)	(17 900)
Impact strength, Izod, J/mm	0.06; bar 7.25 mm	0.25
(ft lb/in)	(1.3); (bar 1/4 in)	(5)
Hardness, Rockwell	M69, R120	M110
Thermal expansion, 10^{-4}/°C	13.2–14.2	11.9
Resistance to heat, °C	150–175	260
(°F)	(300–345)	(500)
Dielectric strength, V/mm	16 730	13 800
Dielectric constant (at 60 Hz)	3.14	3.94
Dissipation factor (at 60 Hz)	0.000 8	0.003
Arc resistance, s	75–122	67
Water absorption (24 h), %	0.22	1.8
Burning rate	Self-extinguishing	Self-extinguishing
Effect of sunlight	Strength loss, yellows slightly	Slight
Effect of acids	None	None
Effect of alkalies	None	None
Effect of solvents	Partly soluble in aromatic hydrocarbons	Soluble in highly polar solvents
Machining qualities	Excellent	Excellent
Optical qualities	Transparent to opaque	Opaque

Table E-31. Properties of Polyethersulfone

Property	Polyethersulfone (Unfilled)
Molding qualities	Excellent
Relative density	1.37
Tensile strength, MPa	84
(psi)	(12 180)
Impact strength, Izod, J/mm	0.08
(ft lb/in)	(1.6)
Hardness, Rockwell	M88
Thermal expansion, 10^{-4}/°C	13–97
Resistance to heat, °C	150
(°F)	(300)
Dielectric strength, V/mm	15 750
Dielectric constant (at 60 Hz)	3.5
Dissipation factor (at 60 Hz)	0.001
Arc resistance, s	65–75
Water absorption (24 h). %	0.43
Effect of sunlight	Yellows
Effect of acids	None
Effect of alkalies	None
Effect of solvents	Attacked by aromatic hydrocarbons
Machining qualities	Excellent
Optical qualities	Transparent

related to each other. Addition polymers are shown in Table E-32 with the radical side groups attached.

Through common usage, the *vinyl plastics* are those polymers with the vinyl name. Many authorities limit their discussion to include only polyvinyl chloride and polyvinyl acetate. Polyvinyl homopolymers or copolymers may include polyvinyl chloride, polyvinyl acetate, polyvinyl alcohol, polyvinyl butyral, polyvinyl acetal, and polyvinylidene chloride. Fluorinated vinyls are discussed with other fluorine-containing polymers.

The history of polyvinyls may be traced to as early as 1835. French chemist V. Regnault reported that a white residue could be synthesized from ethylene dichloride in an alcohol solution. This tough white residue was again reported in 1872 by E. Baumann. It occurred while reacting acetylene and hydrogen bromide in sunlight. In both cases, sunlight was the polymerizing catalyst that produced the white residue. In 1912, Russian chemist I. Ostromislensky reported the same sunlight polymerization of vinyl chloride and vinyl bromide. Commercial patents were granted in several countries for the manufacture of vinyl chloride by 1930.

In 1933, W. L. Semon of the B. F. Goodrich Company added a plasticizer, tritolyl phosphate, to polyvinyl chloride compounds. The resulting polymer mass could be easily molded and processed without substantial decomposition.

Germany, Great Britain, and the United States produced plasticized polyvinyl chloride (PVC) commercially during World War II. It was largely used in place of rubber.

Today, polyvinyl chloride is the leading plastics produced in Europe while in the United States it ranks second after polyethylene. The polyvinyl chloride molecule (C_2H_3Cl) is like that of polyethylene, and Figure E-79 shows this similarity.

Polyvinyl Chloride

The basic raw ingredient of polyvinyl chloride, depending on availability, is acetylene or ethylene gas.

Table E-32 Monofunctional Monomers and Their Polymers

Monomer	Polymer
$CH_2{=}CH_2$ Ethylene	$\rightarrow -CH_2-CH_2-CH_2-CH_2-CH_2-CH_2-CH_2-CH_2-$ Polyethylene
$O-COCH_3$ $CH_2{=}CH$ Vinyl acetate	$O-COCH_3$ $O-COCH_3$ $O-COCH_3$ $O-COCH_3$ $\rightarrow -CH_2-CH-CH_2-CH-CH_2-CH-CH_2-CH-\ldots$ Polyvinyl acetate
Cl $CH_2{=}CH$ Vinyl chloride	Cl Cl Cl Cl $\rightarrow -CH_2-CH-CH_2-CH-CH_2-CH-CH_2-CH-\ldots$ Polyvinyl chloride
C_6H_5 $CH_2{=}CH$ Styrene (Vinyl benzene)	C_6H_5 C_6H_5 C_6H_5 C_6H_5 $\rightarrow -CH_2-CH-CH_2-CH-CH_2-CH-CH_2-CH-\ldots$ Polystyrene
Cl $CH_2{=}C$ Cl Vinylidene chloride	Cl Cl Cl Cl $\rightarrow -CH_2-CH-CH_2-CH-CH_2-CH-CH_2-CH-\ldots$ Cl Cl Cl Cl Polyvinylidene chloride
$COOH$ $CH_2{=}CH$ Acrylic acid	$COOH$ $COOH$ $COOH$ $COOH$ $\rightarrow -CH_2-CH-CH_2-CH-CH_2-CH-CH_2-CH-\ldots$ Polyacrylic acid
$COOH$ $CH_2{=}C$ CH_3 Methacrylic acid	$COOH$ $COOH$ $COOH$ $COOH$ $\rightarrow -CH_2-CH-CH_2-CH-CH_2-CH-CH_2-CH-\ldots$ CH_3 CH_3 CH_3 CH_3 Polymethacrylic acid
CH_3 $CH_2{=}C$ CH_3 Isobutylene	CH_3 CH_3 CH_3 CH_3 $\rightarrow -CH_2-CH-CH_2-CH-CH_2-CH-CH_2-CH-\ldots$ CH_3 CH_3 CH_3 CH_3 Polyisobutylene

Ethylene is the chief source in the United States. During its manufacture, polymerization may be started by peroxides, azo compounds, persulfates, ultraviolet light, or radioactive sources. For addition polymerization, the double bonds of the monomers must be broken by the use of heat, light, pressure, or a catalyst system.

The uses of polyvinyl chloride plastics may be broadened by the addition of plasticizers, fillers, reinforcements, lubricants, and stabilizers. They may be formulated into flexible, rigid, elastomeric, or foamed compounds.

Polyvinyl chloride is most widely used in flexible film and sheet forms. These films and sheets are competitive with other films for collapsible containers, drum liners, sacks, and packages. Washable wallpapers and certain clothing such as handbags, rainwear, coats, and dresses are other uses. Sheets are made into chemical tanks and ductwork of all types. They are easily fabricated by welding, heatsealing, or solvent-cementing with mixtures of ketones or aromatic hydrocarbons.

Extruded profile shapes of both rigid and flexible polyvinyl chloride find use as architectural moldings, seals, gaskets, gutters, exterior siding, garden hose, and moldings for movable partitions (Fig. E-80). Injection-molded soles for footwear are popular in several countries.

```
    H  H  H  H  H  H  H  H
    |  |  |  |  |  |  |  |
...C—C—C—C—C—C—C—C...
    |  |  |  |  |  |  |  |
    H  H  H  H  H  H  H  H
```

(A) Polyethylene

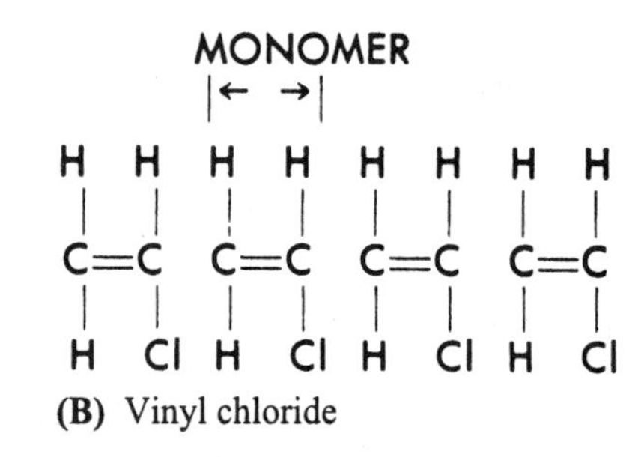

(B) Vinyl chloride

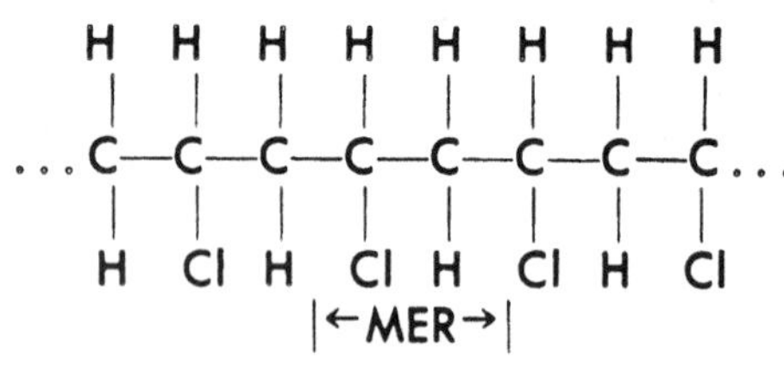

(C) Polyvinyl chloride

Fig. E-79. Similarity of polyethylene and polyvinyl chloride.

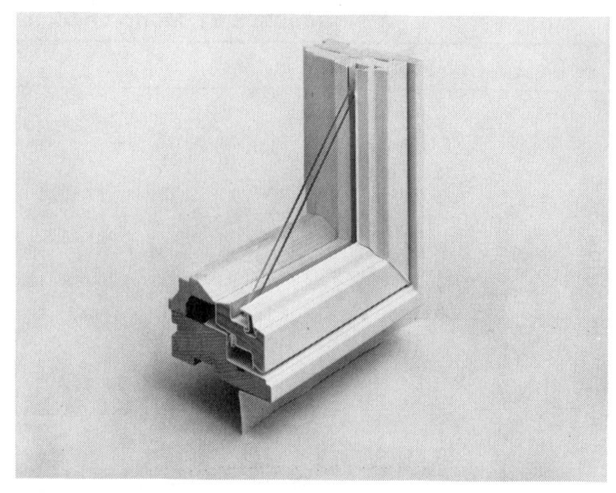

(A) Polyvinyl coating on wood for weatherproof finish and durability. (Perma-shield)

(B) Solid vinyl gutter and downspout system.(Bird & Son)

Fig. E-80. Polyvinyl coatings have many uses.

Organosols and plastisols are liquid or paste dispersions or emulsions of polyvinyl chloride. They are used for coating various substrates, including metal, wood, plastics, and fabrics. They may be applied by dipping, spraying, spreading, or slush and rotational casting. Laminates of polyvinyl film, foam, and fabric are used for upholstery materials. Dip coatings are found on tool handles, sink drainers, and other substrates as a protective layer. Slush and rotational casting of polyvinyls are used to produce hollow articles such as balls, dolls, and large containers. Heavily filled polyvinyls and their copolymers are used in the production of floor coverings and tiles. Foams have found limited application in the textile and carpeting industries. Large quantities of PVC are used as blow-molded containers and as extruded coverings for electrical wire (Fig. E-81).

Generally, materials in the vinyl family are flame, water, chemical, electrical, and abrasion resistant. They have good weatherability and may be transparent. To aid processing and provide various properties, polyvinyls are commonly plasticized. Both plasticized and unplasticized PVC compounds are available. Unplasticized grades are used in chemical plants and building industries. Plasticized grades are more flexible and soft. With increases in plasticizer, there is more bleeding or migration of the plasticizer chemical into adjacent materials. This is of prime importance in packaging of food products and medical supplies.

All thermoplastic processing techniques are employed with vinyls.

Polyvinyl chloride (PVC) is the most used and commonly thought-of vinyl. Other homopolymer and copolymer polyvinyls are finding increasing use, however. The following list shows six advantages and five disadvantages of polyvinyl chloride.

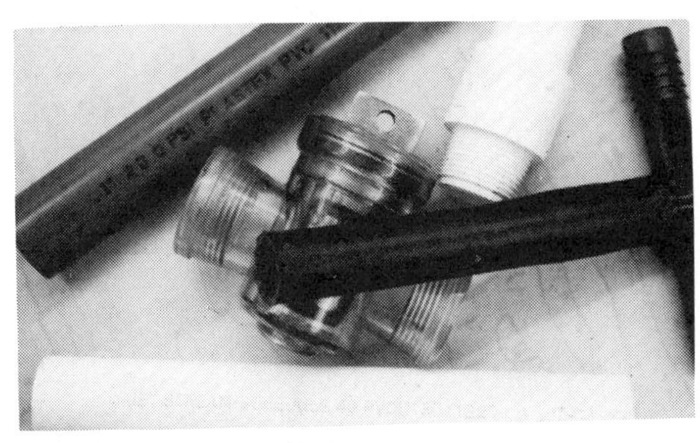

(A) Pipe and plumbing fittings

(B) Foamed flotation devices

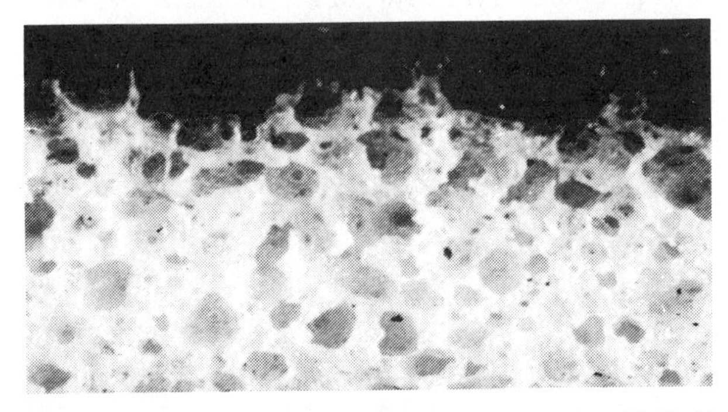

(C) Mechanically foamed carpet backing. Magnified 10 times. (Firestone Plastics Co.)

Fig. E-81. Some of the uses of polyvinyl.

Advantages of Polyvinyl Chloride (PVC)

1. Processable by thermoplastic methods
2. Wide range of flexibility (by varying levels of plasticizers)
3. Nonflammable
4. Dimensional stability
5. Comparatively low cost
6. Good resistance to weathering

Disadvantages of Polyvinyl Chloride

1. Subject to attack by several solvents
2. Limited thermal capability
3. Thermal decomposition evolves HCl
4. Stained by sulfur compounds
5. Higher density than many plastics

Polyvinyl Acetate (PVAc)

Vinyl acetate ($CH_2{=}CH{-}O{-}COCH_3$) is prepared industrially from liquid or gaseous reactions of acetic acid and acetylene. Homopolymers find only limited uses because of excessive cold flow and low softening point. They are used in paints, adhesives, and various textile finishing operations. Polyvinyl acetates are usually in an emulsion form. *White glues* are familiar polyvinyl acetate emulsions. Moisture absorption characteristics are high with chosen alcohols and ketones as solvents. Remoistenable adhesives and hot-melt formulations are other well-known uses. Polyvinyl acetates are used as binder emulsions in some paint formulations. Their resistance to degradation by sunlight makes them useful for interior or exterior coatings. Other uses may include emulsion binders in paper, cardboard, portland cements, textiles, and chewing-gum bases.

Some of the best-known commercial products are copolymers of polyvinyl chloride and polyvinyl acetate used for floor coverings and for modern phonograph records (Fig. E-82). These vinyl records have several advantages over polystyrene and the older shellac discs. Two advantages and three disadvantages of polyvinyl acetate are shown on the following page.

Advantages of Polyvinyl Acetate

1. Excellent for forming into films
2. Heat-sealable

Disadvantages of Polyvinyl Acetate

1. Low thermal stability
2. Poor solvent resistance
3. Poor chemical resistance

Polyvinyl Formal

Polyvinyl formal is generally produced from polyvinyl acetate, formaldehyde, and other additives that change

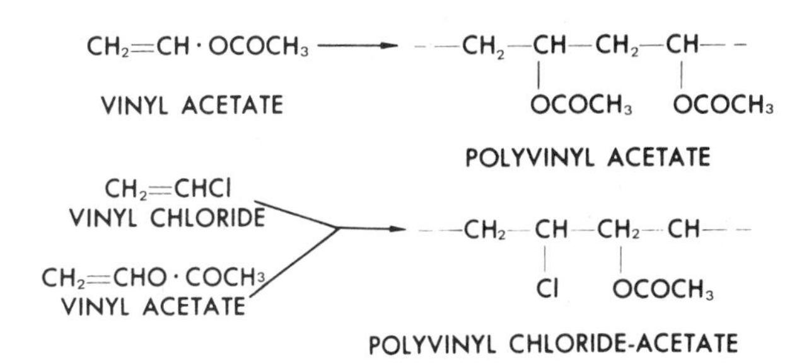

Fig. E-82. Production of polyvinyl acetate and polyvinyl chloride-acetate.

the alcohol side groups on the chain to *formal* side groups. Polyvinyl formal finds its greatest uses as coatings of metal containers and as electrical wire enamels.

Polyvinyl Alcohol (PVAl)

Polyvinyl alcohol is a useful derivative produced from the alcoholysis of polyvinyl acetate (Fig. E-83). Methyl alcohol (methanol) is used in the process. Polyvinyl alcohol (PVA) is both alcohol- and water-soluble. The properties vary depending on the concentration of polyvinyl acetate that remains in the alcohol solution. Polyvinyl alcohol may be used as a binder and adhesive for paper, ceramics, cosmetics, and textiles. It finds use as water-soluble packages for soap, bleaches, and disinfectants. It is a useful mold-releasing agent used in the manufacture of reinforced plastics products. There has been only limited use of polyvinyl alcohol for moldings and fibers.

Polyvinyl Acetate

One more useful derivative of polyvinyl acetate is polyvinyl acetal. It is produced from the treatment of polyvinyl alcohol (from polyvinyl acetate) with an acetaldehyde (Fig. E-84). Polyvinyl acetal materials find limited use as adhesives, surface coatings, films, moldings, or textile modifiers.

Polyvinyl Butyral (PVB)

Polyvinyl butyral is produced from polyvinyl alcohol (Fig. E-85). This plastics is used as interlayer film in laminated safety glass.

Polyvinylidene Chloride (PVDC)

In 1839, a substance similar to vinyl chloride was discovered but it contained one more chlorine atom (Fig. E-86). This material has become commercially important as vinylidene chloride ($H_2C{=}CCl_2$).

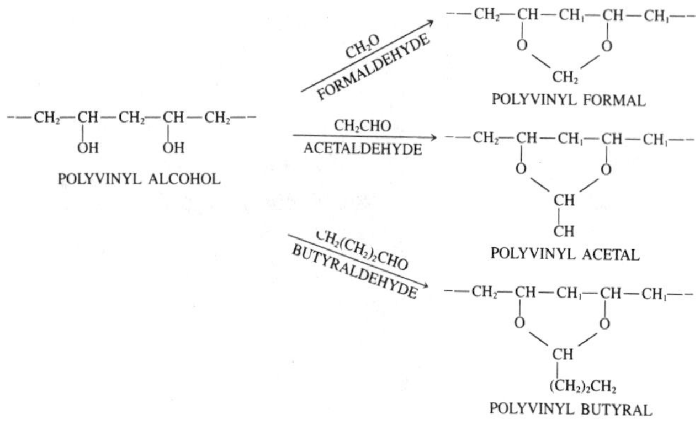

CH₂—CH—CH₂—CH—CH₂—CH—CH₂—CH—CH₂—CH (OH side groups)

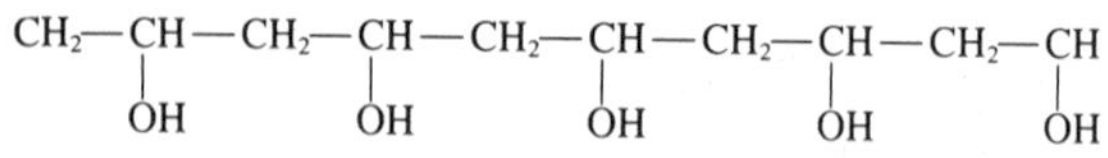

Fig. E-83. Polyvinyl alcohol (OH side groups).

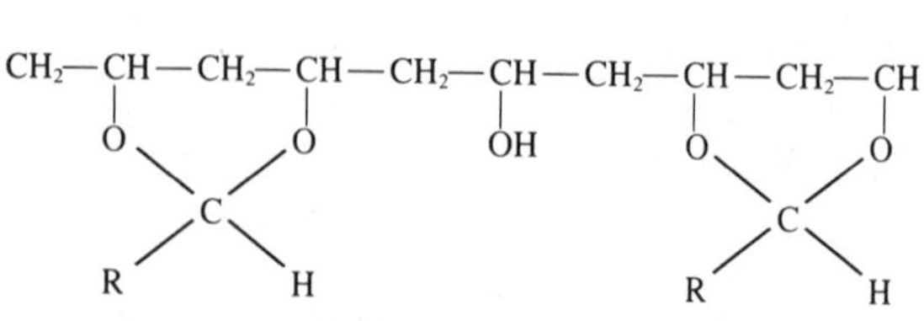

Fig. E-84. A polyvinyl acetal.

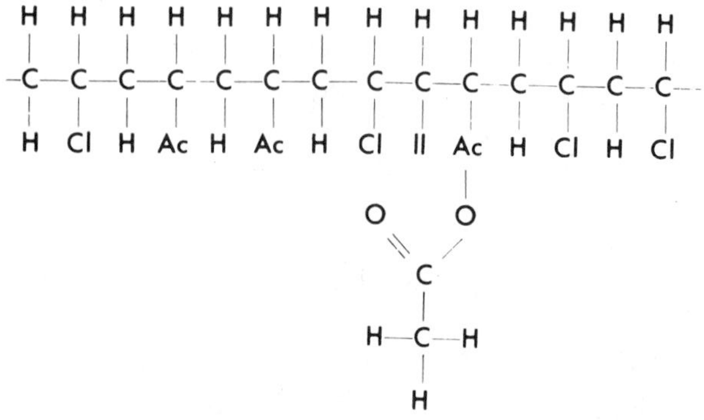

POLYVINYL ALCOHOL — CH_2O FORMALDEHYDE → POLYVINYL FORMAL; CH_3CHO ACETALDEHYDE → POLYVINYL ACETAL; $CH_3(CH_2)_2CHO$ BUTYRALDEHYDE → POLYVINYL BUTYRAL

Fig. E-85. Production of polyvinyl butyral.

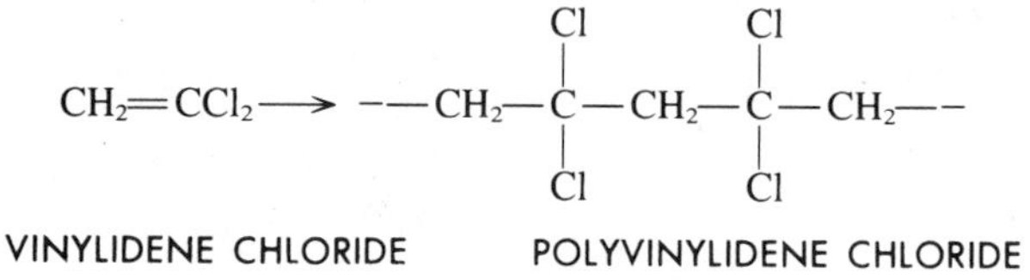

$CH_2{=}CCl_2 \longrightarrow$ —CH₂—CCl₂—CH₂—CCl₂—CH₂—

VINYLIDENE CHLORIDE POLYVINYLIDENE CHLORIDE

Fig. E-86. Polymerization of vinylidene chloride.

Polyvinylidene chloride is costly and hard to process; thus, it is normally found as a copolymer with vinyl chloride, acrylonitrile, or acrylate esters. A well-known food wrapping film, Saran, is a copolymer of vinylidene chloride and acrylonitrile. It exhibits clarity and toughness, and permits little gas or moisture transmission.

The chief use of polyvinylidene chloride copolymers (Fig. E-87) is in coating and film packaging; however, they have found some success as fibers for carpeting, automobile seat upholstery, draperies, and awning textiles. Their chemical inertness allows them to be used as pipe, pipe fittings, pipe linings, and filters.

There are many other polyvinyl polymers that warrant further research and study. Polyvinyl carbazole is used for dielectrics, polyvinyl pyrrolidone is used as a blood plasma substitute, and polyvinyl ethers

Fig. E-87. Copolymerization of vinyl chloride and vinyl acetate.

are used as adhesives. Polyvinyl ureas, polyvinyl isocyanates, and polyvinyl chloroacetate have been explored for commercial use.

All chlorinated or chlorine-containing polymers may emit toxic chlorine gas upon high temperature breakdown. Adequate venting should be provided to protect the operator during processing.

Table E-33 gives some properties of polyvinyl plastics, while a list of four advantages and two disadvantages of polyvinylidene chloride is given below.

Advantages of Polyvinylidene Chloride

1. Low water permeability
2. Approved for food wrap by FDA
3. Processable by thermoplastic methods
4. Nonflammable

Disadvantages of Polyvinylidene Chloride

1. Lower strength than PVC
2. Subject to creep

Table E-33. Properties of Polyvinyls

Property	Rigid Vinyl Chloride (PVC)	PVC-Acetate (Copolymer)	Vinylidene Chloride Compound
Molding qualities	Good	Good	Excellent
Relative density	1.30–1.45	1.16–1.18	1.65–1.72
Tensile strength, MPa	34–62	17–28	21–34
(psi)	(5 000–9 000)	(2 500–4 000)	(3 000–5 000)
Compressive strength MPA	55–90		14–19
(psi)	(8 000–13 000)		(2 000–2 700)
Impact strength, Izod, J/mm	0.02–1.0		0.06–0.05
(ft lb/in)	(0.4–20.0)		(0.3–1.0)
Hardness, Rockwell	M110–M120	R35–R40	M50–M65
Thermal expansion 10^{-4}/°C	12.7–47		48.3
Resistance to heat °C	65–80	55–60	70–90
(°F)	(150–175)	(130–140)	(160–200)
Dielectric strength, V/mm	15 750–19 700	12000–15 750	15 750–23 500
Dielectric constant (at 60 Hz)	3.2–3.6	3.5–4.5	4.5–6.0
Dissipation factor (at 60 Hz)	0.007–0.020		0.030–0.045
Arc resistance, s	60–80		
Water absorption (24 h), %	0.07–0.4	3.0+	0.1
Burning rate	Self-extinguishing	Self-extinguishing	Self-extinguishing
Effect of sunlight	Needs stabilizer	Needs stabilizer	Slight
Effect of acids	None to slight	None to slight	Resistant
Effect of alkalies	None	None	Resistant
Effect of solvents	Soluble in ketones and esters	Soluble in ketones and esters	None to slight
Machining qualities	Excellent		Good
Optical qualities	Transparent	Transparent	Transparent

Appendix F

Thermosetting Plastics

Your study of plastics resins is not complete until you become familiar with thermosetting plastics. Look for the outstanding properties and applications of each plastics described in this appendix because plastics properties affect product design, processing, economics, and service.

You should remember that thermosetting materials undergo a chemical reaction and become "set." In general, most are high molecular weight materials resulting in a hard, brittle plastics.

This appendix treats individual groups of thermosetting plastics. These are *allylics, alkyds, amino plastics (urea-formaldehyde and melamineformaldehyde), casein, epoxy, furan, phenolics, unsaturated polyesters, polyurethane and silicones.*

Alkyds

In the past, there was some confusion about ester-based alkyd resins. The term *alkyd* was once used strictly for unsaturated polyesters modified with fatty acids or vegetable oils. These resins find uses in paints and other coatings. Today, *alkyd molding compounds* refer to unsaturated polyesters modified by a nonvolatile monomer (such as diallyl phthalate) and various fillers. The compounds are formed into granular, rope, nodular, putty, and log shapes to allow continuous automatic molding (Fig. F-1).

Fig. F-1. Alkyd bulk molding compound in several different shapes. (Allied Chemical Corp.)

To obtain a resin suitable for molding compounds, the cross-linking mechanism must be modified so that rapid cure will take place in the mold. The use of initiators in the resin compound also speeds up the polymerization of the double bonds. These resins should not be confused with saturated polyester molding compounds, which are linear and thermoplastic.

R. H. Kienle coined the word *alkyd* from the "al" in *al*cohol and the "cid" from a*cid*. It is usually pronounced al-kid. Alkyd resins may be produced by reacting phthalic acid, ethylene glycol, and the fatty acids of various oils such as linseed oil, soybean oil, or tung oil (Fig. F-2). Kienle combined fatty acids with unsaturated esters in 1927, while searching for a better electrical insulating resin for General Electric. The

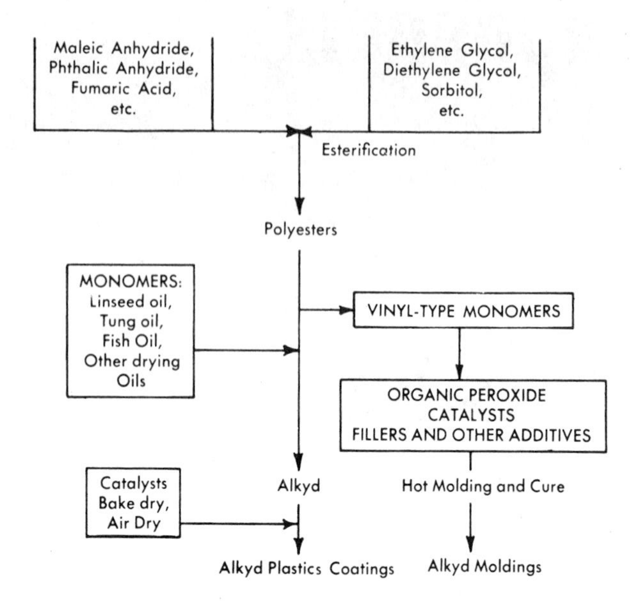

Fig. F-2. Production of alkyd coatings and moldings.

need for materials in World War II greatly speeded up the interest in alkyd finishing resins.

Alkyd resins may have a structure like the one shown in Figure F-3. When this resin is applied to a substrate, heat or oxidizing agents are used to begin cross-linking.

One expert estimates that nearly half of the surface coatings used in the United States are alkyd polyester coatings. The increasing use of silicone and acrylic latex coatings may greatly reduce that figure.

Alkyd coatings are of value because of their relatively low cost, durability, heat resistance, and adaptability. They may be changed to meet special coating needs. To increase durability and abrasion resistance, rosin may be used to modify alkyd resins. Phenolics and epoxy resins may improve hardness and resistance to chemicals and water. Styrene monomers added to the resin may extend flexibility of the finished coating and serve as cross-linking agents.

Alkyd resins are used in *oil-based* house paint, baking enamel, farm implement paint, emulsion paint, porch and deck enamel, spar varnish, and chlorinated rubber paint. Moldified alkyd resins may produce special coatings, such as the *wrinkle* and *hammer* finishes often used on equipment and machinery.

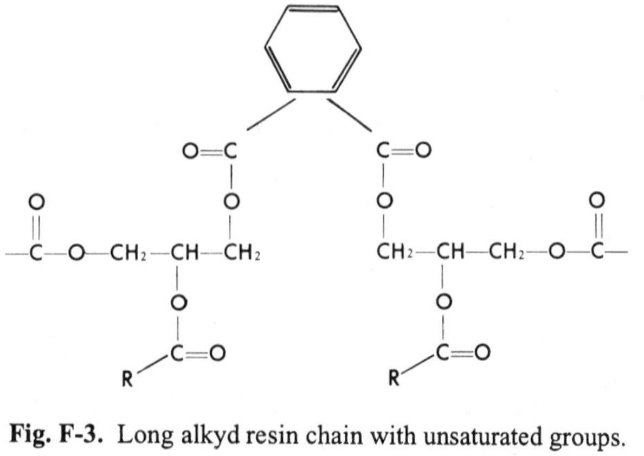

Fig. F-3. Long alkyd resin chain with unsaturated groups.

Alkyd baking finishes are referred to as *heat-convertible*. They polymerize or harden when heated. Resins that harden when exposed to air are called *air-convertible* resins.

Alkyd resins are used as plasticizers for various plastics, vehicles for printing inks, binders for abrasives and oils, and special adhesives for wood, rubber, glass, leather, and textiles.

New processing technologies and uses for thermosetting alkyd molding compounds are also of current commercial importance.

Alkyd molding compounds are processed in compression, transfer, and reciprocating-screw equipment. By using initiators such a benzoyl peroxide or tertiary butyl hydroperoxide, the unsaturated resins can be made to cross-link at molding temperatures (greater than 50 °C or 120 °F). Molding cycles may be less than twenty seconds.

Typical uses of alkyd molding compounds include appliance housings, utensil handles, billiard balls, circuit breakers, switches, motor cases, and capacitor and commutator parts (Fig. F-4). They have been used in electrical applications where less costly phenolic or amino resins are not suitable.

Alkyd compounds in putty form are used to encapsulate electronic and electrical components. Different resin formulations, fillers, and processing techniques provide a wide range of physical characteristcs.

Table F-1 gives some of the properties of alkyd plastics.

Allylics

Allylic resins usually involve the esterification of allyl alcohol and a dibasic acid (Fig. F-5). The pungent odor of allyl alcohol ($CH_2{=}CHCH_2OH$) has been known to science since 1856. The name *allyl* was coined from the Latin word *allium*, meaning garlic. These resins did not become commercially useful until 1955 although one allyl resin was used in 1941 as a low-pressure laminating resin.

Allylics and polyesters have common applications, properties, and historical background of development. For this reason, allylics are sometimes erroneously included in the study of polyesters. Allyl monomers are used as cross-linking agents in polyesters, which adds to the confusion. Allylics are a distinct family of plastics based on monohydric alcohols while the chemical basis for polyesters is polyhydric alcohols.

Allylics are unique in that they may form pre-polymers (partly polymerized resins). They may be homopolymerized or copolymerized. Monoallyl esters

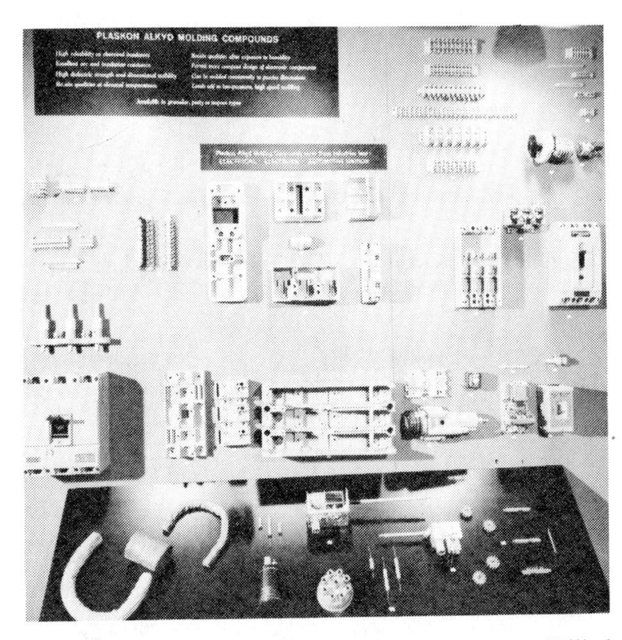

(A) Electrical, electronic, and automotive parts. (Allied Chemical Corp.)

(B) Other automotive parts.

Fig. F-4. Various alkyd products.

Table F-1. Properties of Alkyds

Property	Alkyd, (Glass-Filled)	Alkyd Molding Compound (Filled)
Molding qualities	Excellent	Excellent
Relative density	2.12–2.15	1.65–2.30
Tensile strenfth, MPa	28–64	21–62
(psi)	(4000–9 500)	(3000–9000)
Compressive strength. MPa	103–221	83–262
(psi)	(15 000–32 000)	(12000–38000)
Impact strength, Izod, J/mm	0.03–0.5	0.015–0.025
(ft lb/in)	(0.60–10)	(0.30–0.50)
Hardness, Barcol	60–70	55–80
Rockwell	E98	E95
Thermal expansion, 10^{-4}/°C	3.8–6.35	5.08–12.7
Resistance to heat, °C	230	150–230
(°F)	(450)	(300–450)
Dielectric strength, V/mm	9845–20870	13 800–17 720
Dielectric constant (at 60 Hz)	5.7	5.1–7.5
Dissipation factor (at 60 Hz)	0.010	0.009–0.06
Arc resistance, s	150–210	75–240
Water absorption (24 h), %	0.05–0.25	0.05–0.50
Burning rate	Slow to nonburning	Slow to nonburning
Effect of sunlight	None	None
Effect of acids	Fair	None
Effect of alkalies	Fair	Attacked
Effect of solvents	Fair to good	Fair to good
Machining qualities	Poor to fair	Poor to fair
Optical qualities	Opaque	Opaque

may be produced as either thermosetting or thermoplastic resins. The saturated monoallyl esters (thermoplastic) are sometimes used as copolymerizing agents in alkyd and vinyl resins. Unsaturated monoallyl esters have been used to produce simple polymers, including allyl acrylate, allyl chloroacrylate, allyl methacrylate, allyl crotonate, allyl cinnamate, allyl cinnamalacetate, allyl furoate, and allyl furfurylacrylate.

Simple polymers have been produced from diallyl esters including diallyl maleate (DAM), diallyl oxalate, diallyl succinate, diallyl sebacate, diallyl pththalate (DAP), diallylcarbonate, diethylene glycol bis-

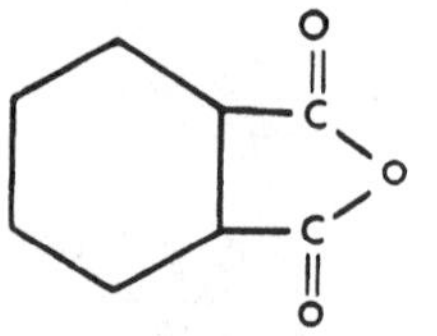

(A) Phthalic anhydride. 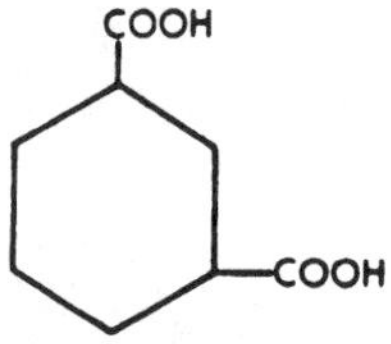 (B) Isophthalic acid.

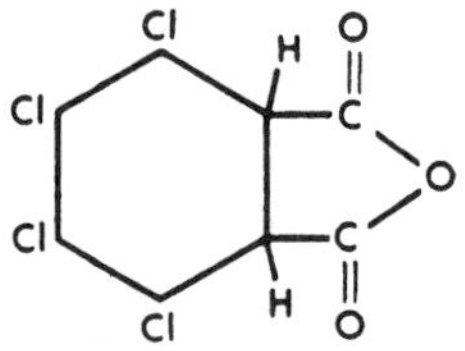

(C) Tetrachlorophthalic acid. 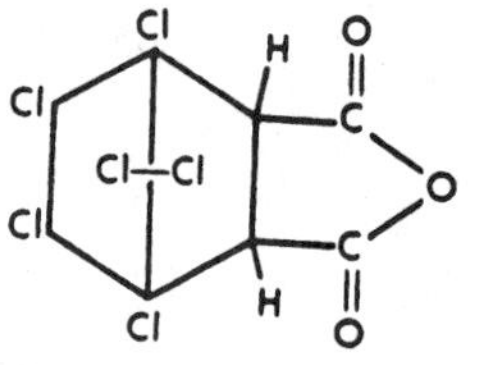 (D) Chlorendic anhydride.

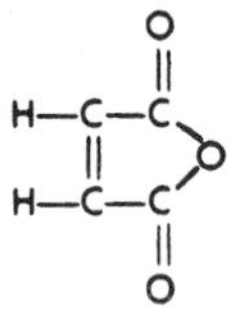

(E) Maleic anhydride.

Fig. F-5. Dibasic acid used in the manufacture of allylic monomers.

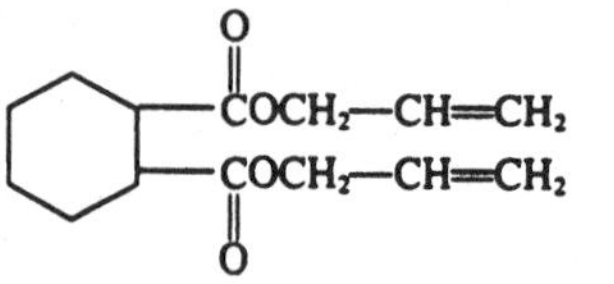

(A) Diallyl phthalate (ortho). 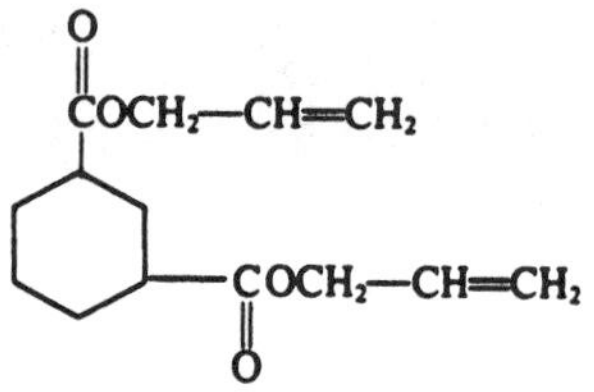 (B) Diallyl isophthalate (meta).

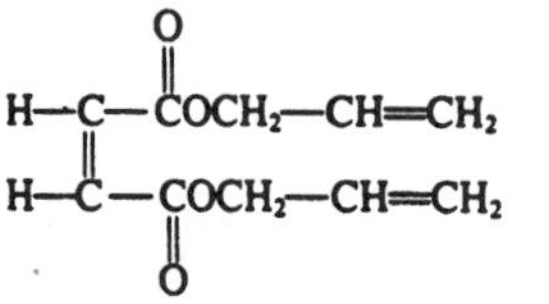

(C) Diallyl maleate. 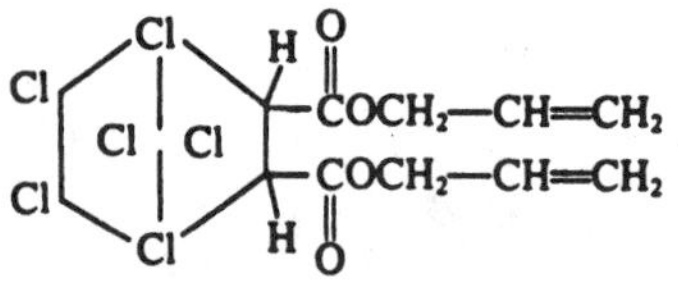 (D) Diallyl chlorendate.

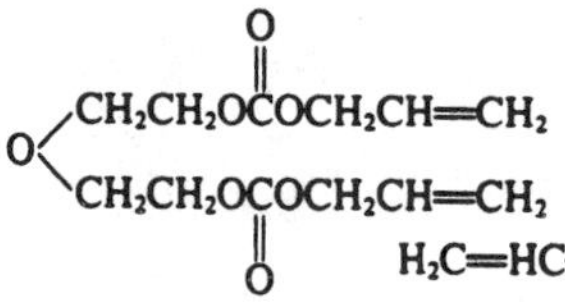

(E) Diethylene glycol bis-(allyl carbonate). 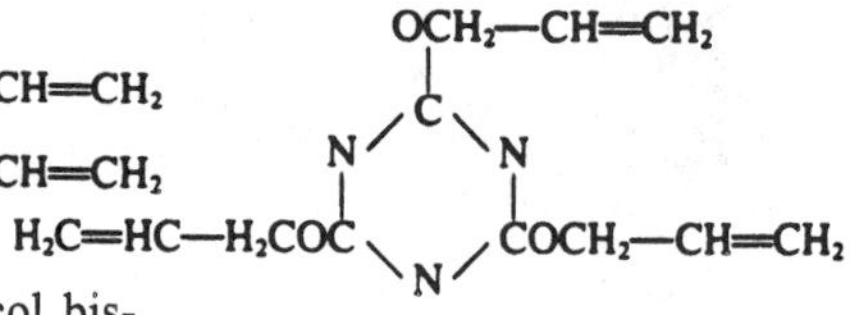 (F) Triallyl cyanurate.

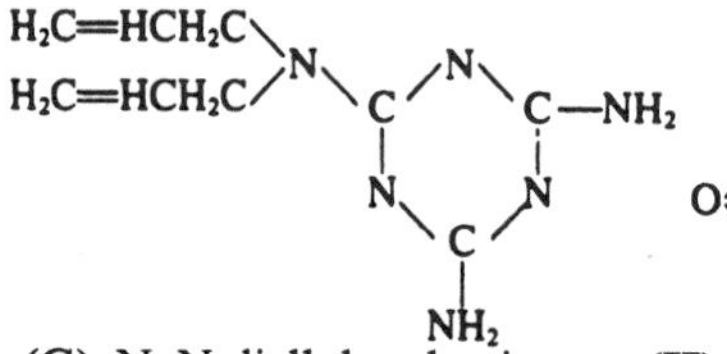

(G) N, N-diallyl melamine. 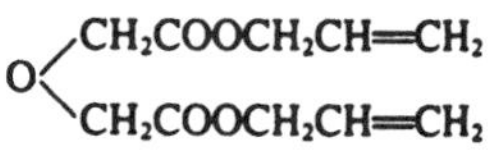 (H) Diallyl diglycollate.

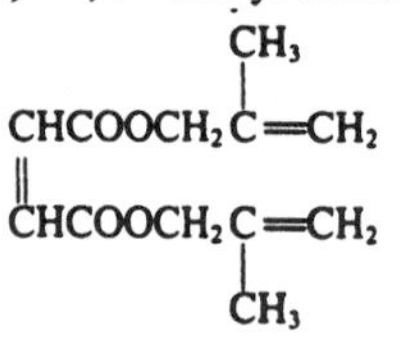

(I) Dimethallyl maleate. 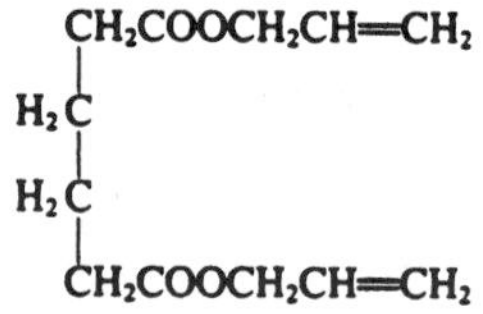 (J) Diallyl adipate.

Fig. F-6. Structural formulas of some commercial allylic monomers.

allyl carbonate, diallyl isophthalate (DAIP), and others (Fig. F-6). The most widely used commercial compounds are DAP (Fig. F-6A) and DAIP (Fig. F-6B).

Diallyl resins are usually supplied as monomers or prepolymers. Both forms are converted to the fully polymerized thermosetting plastics by the addition of selected peroxide catalysts. Benzoyl peroxide or *tert*-butyl perbenzoate are two often-used catalysts for polymerization of allylic resin compounds. These resins and compounds may be catalyzed and stored for more than a year, if kept at low ambient temperatures. When the material is subjected to normal temperatures of molds, presses, or ovens, complete cure is reached.

The diethylene glycol bis-allyl carbonate resin monomer (Fig. F-6E) used for laminating and optical castings is illustrated below. This resin could be polymerized with benzoyl peroxide at a temperature of 180 °F [82 °C].

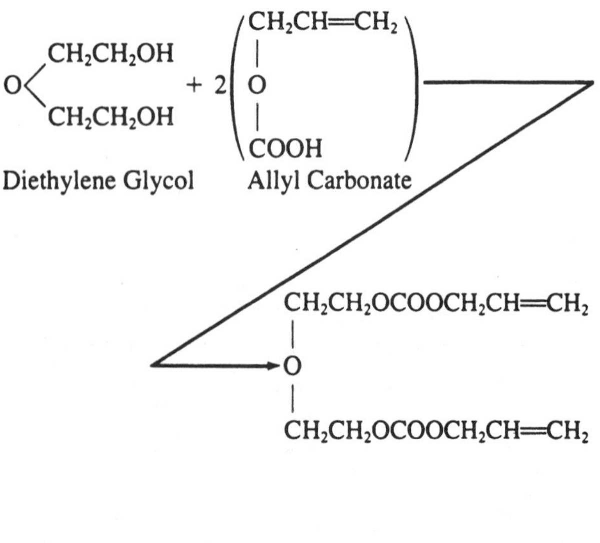

$$O(CH_2CH_2OH)_2 + 2\,(CH_2CH{=}CH_2\text{—O—COOH}) \rightarrow O(CH_2CH_2OCOOCH_2CH{=}CH_2)_2$$

Diethylene Glycol   Allyl Carbonate

Diallyl phthalate may be used as a coating and laminating material but it is probably most noted for its principal use as a molding compound. Diallyl phthalate will polymerize and cross-link because of two available double bonds. (See Fig. F-6A).

Good mechanical, chemical, thermal, and electrical properties are only a few of the attractive attributes of diallyl phthalate (Fig. F-7). It also offers long shelf life and ease of handling. Radiation and ablation resistance make these materials useful in space environments.

Nearly all allylic molding compounds include catalysts, fillers, and reinforcements and are usually blended into putty-like premixes. High cost is a factor that limits use of allylics to applications where their properties are vital. Compounds may be compression- or transfer-molded. Some meet the requirements for

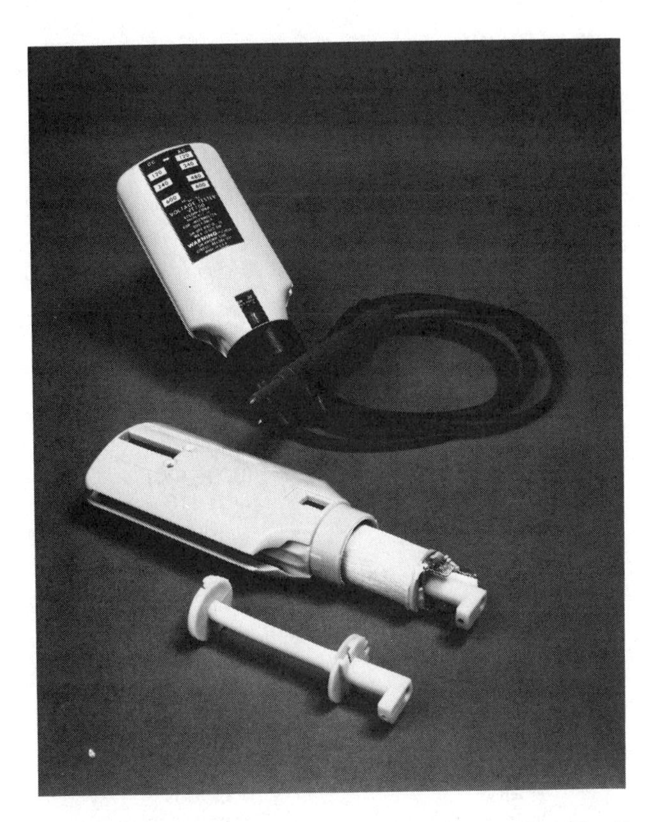

Fig. F-7. This hand-held voltage tester has a solenoid coil bobbin made of a strong diallyl phthalate molding compound.

special high-speed injection machines, low-pressure encapsulation, and extrusion.

Some allylic resin formulations are used in producing laminates. Monomer resins are used to preimpregnate wood, paper, fabrics, or other materials for lamination. Some improve various properties in paper and garments. Crease-, water-, and fade-resistance may be enhanced by the addition of diallyl phthalate monomers to selected fabrics. For furniture and paneling, preimpregnated decorative laminates or overlays are bonded to cores of less costly materials. Melamine and allylic resin-bound laminates have many of the same properties.

Allylic monomers and prepolymers also find use in prepregs (wet layup laminates). These prepregs consist of fillers, catalysts, and reinforcements. They are combined just prior to cure. Some are preshaped to ease molding. Allylic prepregs may be prepared in advance and stored until needed. Premix and prepreg molded parts have good flexural and impact strength, and their surface finishes are excellent.

Allylic monomers are used as cross-linking agents for polyesters, alkyds, polyurethane foams, and other unsaturated polymers. They are useful because the basic allylic monomer (homopolymer) does not polymerize at room temperature. This allows materials to be stored for very long periods. At temperatures of 150 °C [300 °F] and above, diallyl phthalate monomers cause polyesters to cross-link. These compounds may be molded at faster rates than those cross-linked with styrene.

Acrylics (methyl methacrylate) cross-linked with diallyl phthalate monomers have good surface hardness and elasticity.

Allylics have been used in vacuum impregnation to seal the pores of metal castings, ceramics, and other compositions. Other uses include the impregnation of reinforcing tapes that are used to wrap motor armatures. Allylics are used to coat electrical parts and to encapsulate electronic devices.

Table F-2 gives some of the basic properties of allylic (diallyl phthalate) plastics. Four advantages and three disadvantages are listed below.

Advantages of Allylic Esters (Allyls)

1. Excellent moisture resistance
2. Availability of low-burning and self-extinguishing grades
3. Service temperatures as high as 204-232 °C [400-450 °F]
4. Good chemical resistance

Disadvantages of Allylic Esters (Allyls)

1. High cost (compared to alkyds)
2. Excessive shrinkage during curing
3. Not usable with phenols and oxidizing acids

Amino Plastics

Various polymers have been produced by interaction of amines or amides with aldehydes. The two most significant and commercially useful amino plastics are produced by the condensation of urea-formaldehyde and melamine-formaldehyde.

Urea-formaldehyde polymers may have been produced as early as 1884. In Germany, Goldschmidt and his colleagues pointed their efforts toward making moldable amino plastics.

In the United States, urea-formaldehyde resins were commercially produced in 1920. Thio-ureaformaldehyde and melamine-formaldehyde resins were being produced in the period of 1934–1939.

Urea-Formaldehyde (UF)

Reaction between the white crystalline solid, urea (NH_2CONH_2), and aqueous solutions of formaldehyde (formalin) produces urea resins, which may be modified by the addition of other reagents. For complete polymerization and cross-linkage of this thermosetting resin, heat or catalysts and heat are usually needed during the molding operation (Fig. F-8).

Urea-formaldehyde polymerization is shown in Figure F-9. The excess water molecule is a result of condensation polymerization.

Table F-2. Properties of Allylic Plastic

Property	Diallyl Phthalate Compounds		
	Glass-Filled	**Mineral-Filled**	**Diallyl Isophthalate**
Molding qualities	Excellent	Excellent	Excellent
Relative density	1.61–1.78	1.65–1.68	1.264
Tensile strength, MPa	41.4–75.8	34.5–60	30
(psi)	(6 000–11 000)	(5 000–8 700)	(4 300)
Compressive strength, MPa	172–241	138–221	
(psi)	(25 000–35 000)	(20 000–32 000)	
Impact strength, Izod, J/mm	0.02–0.75	0.015–0.225	0.01–0.015
(ft lb/in)	(0.4–15)	(0.3–0.45)	(0.2–0.3)
Hardness, Rockwell	E80–E87	E61	M238
Thermal expansion, 10^{-4}/°C	2.5–9	2.5–10.9	
Resistance to heat, °C	150–205	150–205	150–205
(°F)	(300–400)	(300–400)	(300–400)
Dielectric strength, V/mm	15 550–17 717	15 550–16 535	16 615
Dielectric constant (at 60 Hz)	4.3–4.6	5.2	3.4
Dissipation factor (at 60 Hz)	0.01–0.05	0.03–0.06	0.008
Arc resistance, s	125–180	140–190	123–128
Water absorption (24 h), %	0.12–0.35	0.2–0.5	0.1
Burning rate	Self-extinguishing to nonburning	Self-extinguishing to nonburning	Self-extinguishing to nonburning
Effect of sunlight	None	None	None
Effect of acids	Slight	Slight	Slight
Effect of alkalies	Slight	Slight	Slight
Effect of solvents	None	None	None
Machining qualities	Fair	Fair	Good
Optical qualities	Opaque	Opaque	Transparent

Many products once produced with urea-formaldehyde plastics are now being made of thermoplastic materials. Faster production cycles and higher output are possible with thermoplastics. Some amino compounds are being processed in specially designed injection molding equipment. This approach increases output and makes these products able to compete with most thermoplastics.

Resins with a urea-formaldehyde base are made into molding compounds that contain several ingredients including resin, filler, pigment, catalyst, stabilizer, plasticizer, and lubricant. Such compounds speed up curing and production rates.

Molding compounds are produced with latent acid catalysts added to the resin base. Such catalysts react at molding temperatures. Many urea-formaldehyde molding compounds have prepromoters or catalysts added causing them to have a limited storage life; therefore, they should be kept in a cool place. Stabilizers may be added to help control latent catalyst reaction. Lubricants are added to improve molding quality.

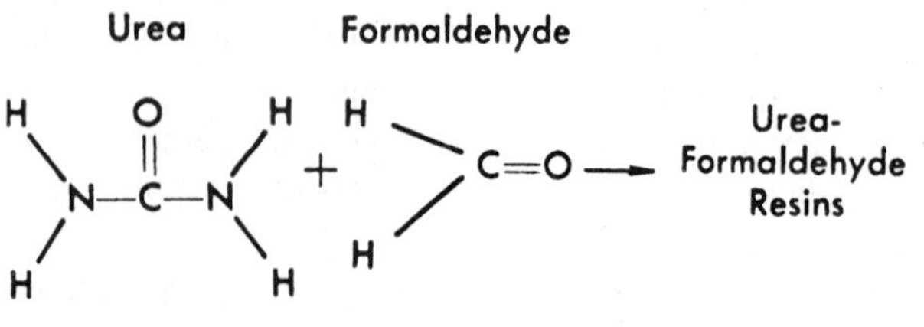

Urea + Formaldehyde → Urea-Formaldehyde Resins

Fig. F-8. Polymerization of urea and formaldehyde.

Plasticizers improve flow properties and help reduce cure shrinkage. Although the base resin is water-clear, color pigments (transparent, translucent, or opaque) may be used. If color is not needed, fillers such as alpha-cellulose (bleached cellulose) fiber, macerated

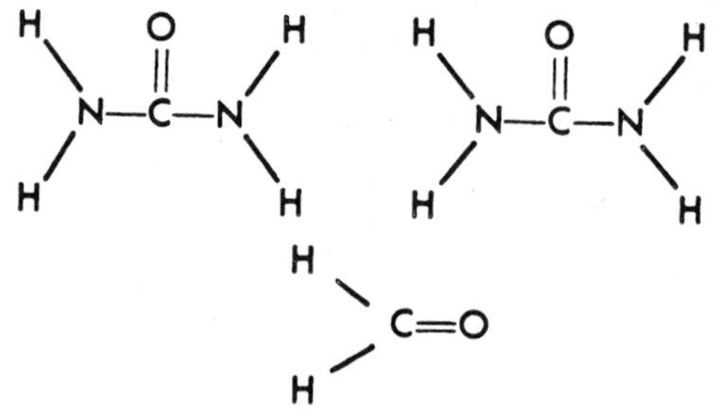

(A) Urea-formaldehyde resin.

$+H_2O$

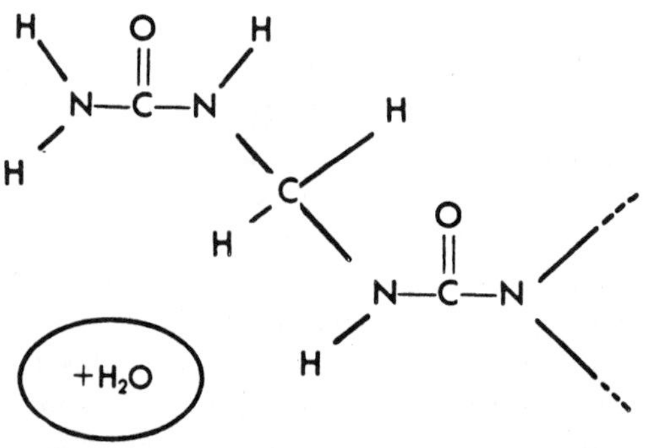

(B) Urea-formaldehyde plastics.

Fig. F-9. Formation of urea-formaldehyde resins.

fabric, or wood flour may be added to improve molding and physical characteristics and lower the cost.

Urea-formaldehyde plastics have many industrial uses because they have outstanding molding qualities and are low in cost. They are used in electrical and electronic applications where good arc and tracking resistance is needed. They provide good dielectric properties and are unaffected by common organic solvents, greases, oils, weak acids, alkalies, or other hostile chemical environments. Urea compounds do not attract dust by static electricity charges. They will not burn or soften when exposed to an open flame and they have very good dimensional stability when filled.

Urea-formaldehyde molding compounds are used to make bottle caps and electrical or thermal insulating materials.

Urea-formaldehyde products do not impart taste or odors to foods and beverages. They are preferred materials for appliance knobs, dials, handles, pushbuttons, toaster bases, and end plates. Wall plates, switch toggles, receptacles, fixtures, circuit breakers, and switch housings are only a few of many electrical insulating uses (Fig. F-10).

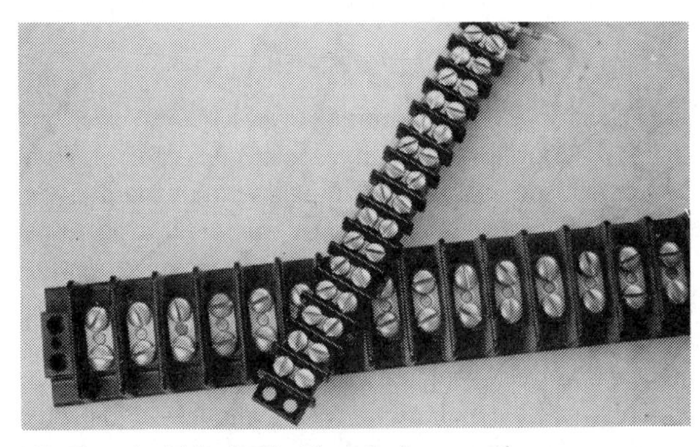

(A) Terminal blocks for electrical connections.

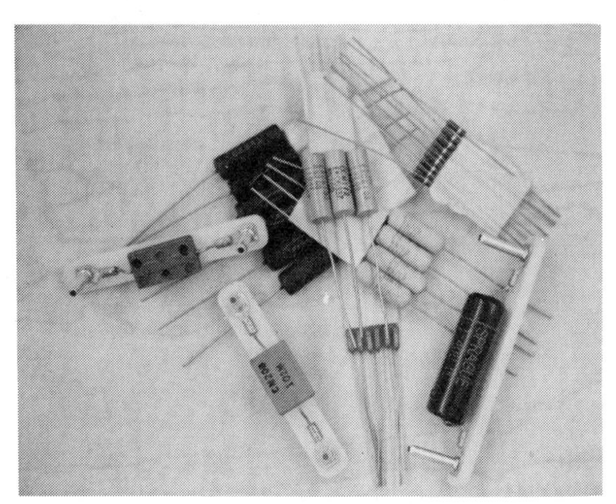

(B) Insulating blocks and packing for electronic components.

Fig. F-10. Applications of urea-formaldehyde compounds.

One of the largest current uses of urea-formaldehyde resins is in adhesives for furniture, plywood, and chipboard. Chipboard is made by combining 10 percent resin binder with wood chips which is then pressed into flat sheets. The product has no grain so it is free to expand in all directions and does not warp. Water resistance is poor, however. Plywood bonded with this resin is suitable only for interior use.

Urea-formaldehyde resins may be foamed and cured into a plastics state. Such foams are inexpensive, and various densities may be produced. They are easily produced by vigorously whipping a mixture of resin, catalyst, and a foaming detergent. Another foaming method involves introducing a chemical agent that generates a gas (usually carbon dioxide) while the resin is curing. These foams have found uses as thermal insulating materials in buildings and in refrigerators, and as low-density cores for structural sandwich construction (Fig. F-11). Undercured or partially polymerized foams have caused allergy and flu-like symptoms in some people. Degassing of molded furniture parts and wall insulation are a major source of this problem.

Urea-formaldehyde foams may be made flame-resistant, but at the expense of foam density. Large amounts of water may be absorbed, since these foams are open-celled (spongelike) structures. This ability to absorb water is made use of by florists. Stems of cut flowers are inserted into water-soaked urea foam used

(A) Flower-arranging block.

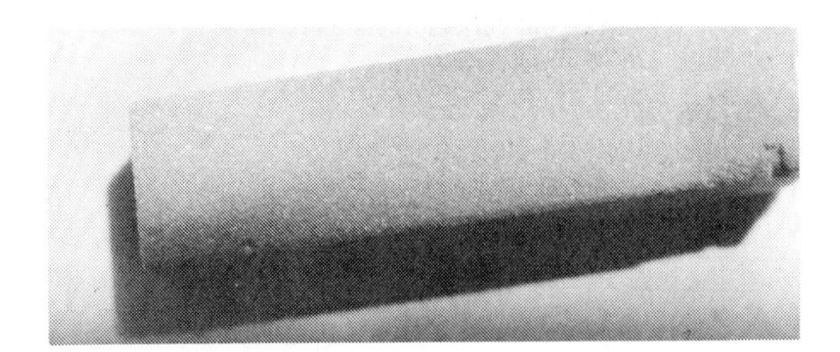

(B) Thermal insulation material.

Fig. F-11. Uses of open-celled or foamed urea-formaldehyde.

as a base for flower arrangements. Ground foam has been used as artificial snow on television and in theatrical productions. The open-celled structure may also be filled with kerosene and used as a lighting agent for fireplaces.

Urea resins find extensive use in the textile, paper, and coating industries. The existence of drip-dry fabrics is due largely to these resins.

Urea resins or powders are sometimes used as binders for foundry cores and shell molds.

The coating applications of urea-formaldehyde resins are limited. They can be applied only to substrates that can withstand curing temperatures of 100 to 350 °F [40° to 175 °C]. These resins are combined with a compatible polyester (alkyd) resin to produce enamels. From 5 to 50 percent urea resin is added to the alkyd-based resin. These surface coatings are outstanding in hardness, toughness, gloss, color stability, and outdoor durability.

These surface coatings may be seen on refrigerators, washing machines, stoves, signs, venetian blinds, metal cabinets, and many machines. At one time, they were widely applied to automobile bodies, however the curing or baking time required has made this uneconomical for mass production. The baking time depends on the temperature and the proportion of amino resin to alkyd resin.

Urea resins modified with furfuryl alcohol have been used successfully in the manufacture of coated abrasive papers. (See Furan, below.)

Urea-formaldehyde plastics are molded with ease in compression and transfer molding machines. They may also be processed by reciprocating-screw injection molding machines. Depending on the grade of resin and the filler used, allowance should be made for shrinkage after removal from the mold. Improved dimensional stability can be had by post-conditioning the product in an oven.

Table F-3 gives some properties of alpha-cellulose filled urea-formaldehyde plastics. Five advantages and four disadvantages of urea-formaldehyde are listed below.

Advantages of Urea-Formaldehyde

1. Good hardness and scratch resistance
2. Comparatively low cost
3. Wide color range
4. Self-extinguishing
5. Good solvent resistance

Disadvantages of Urea-Formaldehyde

1. Must be filled for successful molding
2. Long-term oxidation resistance is poor
3. Attacked by strong acids and bases
4. Uncured, partially polymerized plastics and foams degas

Table F-3. Properties of Urea-Formaldehyde

Property	Urea-Formaldehyde (Alpha-Cellulose Filled)
Molding qualities	Excellent
Relative density	1.47–1.52
Tensile strength, MPa	38–90
(psi)	(5 500–13 000)
Compressive strength, MPa	172–310
(psi)	(25 000–45 000)
Impact strength, Izod, J/mm	0.0125–0.02
(ft lb/in)	(0.25–0.40)
Hardness, Rockwell	M110–M120
Thermal expansion, 10^{-4}/°C	5.6–9.1
Resistance to heat, °C	80
(°F)	(170)
Dielectric strength, V/mm	11 810–15 750
Dielectric constant (at 60 Hz)	7.0–9.5
Dissipation factor (at 60 Hz)	0.035–0.043
Arc resistance, s	80–150
Water absorption (24 h), %	0.4–0.8
Burning rate	Self-extinguishing
Effect of sunlight	Grays
Effect of acids	None to decomposes
Effect of alkalies	Slight to decomposes
Effect of solvents	None to slight
Machining qualities	Fair
Optical qualities	Transparent to opaque

Melamine-Formaldehyde (MF)

Until about 1939, melamine-formaldehyde was an expensive laboratory curiosity. Melamine ($C_3H_6N_6$) is a white crystalline solid. Combining it with formaldehyde results in the formation of a compound referred to as methylol derivative (Fig. F-12). With additional formaldehyde, the formulation will react to produce tri-, tetra-, penta-, and hexamethylol-melamine. The formation of trimethylol melamine is shown in Figure F-13.

Commercial melamine resins may be obtained without acid catalysts, but both thermal energy and catalysts are used to speed polymerization and cure. Poly-

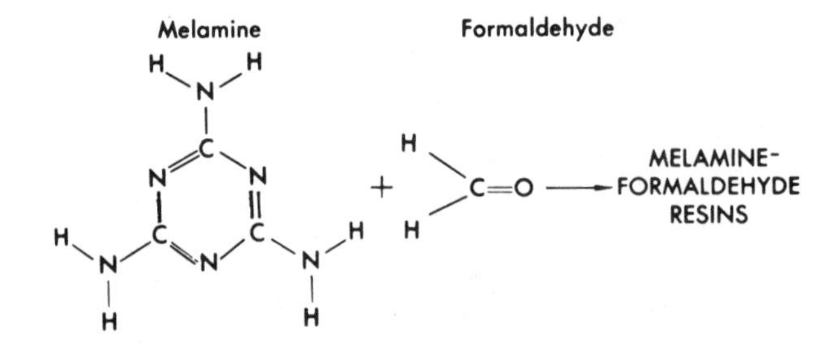

Fig. F-12. Formation of melamine-formaldehyde resins.

Fig. F-13. Formation of trimethylol melamine.

merization of urea and melamine resins is a condensation reaction, producing water, which will evaporate or escape from the molding cavity.

Resins of formaldehyde-plus benzoguanamine ($C_3H_4N_5C_6H_5$) or thiourea ($CS(NH_2)_2$) are of minor commercial importance. The reaction of formaldehyde and compounds such as aniline, dicyandiamide, ethylene urea, and sulfonamide provide more complex resins for varied applications.

In pure form, amino *resins* are colorless and soluble in warm solutions of water and methanol.

Amino resins, urea plastics, and melamine plastics are frequently grouped together as one entity. Their chemical structures, properties, and applications have much in common. But melamine-formaldehyde products are better than the urea-formaldehyde plastics in several respects.

Melamine products are harder and more water-resistant. They may be combined with a greater variety of fillers that allow the making of products with better heat, scratch, stain, water and chemical resistance. Melamine products are also much more expensive than urea products.

Probably the largest single use of melamine-formaldehyde is the manufacture of tableware (Fig. F-14). For this application, molding powders are normally filled with alpha-cellulose. Asbestos or other fillers are sometimes used for handles and utensil housings.

Melamine resin is widely used for surface coatings and decorative laminates. Paper-based laminates are sold under the trade names of Formica and Micarta. Photographic prints on fabrics or paper impregnated with melamine resin are placed on a base or core material and cured in a large press. Kraft paper impregnated with phenolic resin is commonly used for the base material since it is durable and is compatible with melamine resin. This base also costs less than multiple layers of melamine-impregnated paper. There is a broad spectrum of uses for these laminates, including surfacing wood, metal, plaster, and hardboard. A familiar use of these laminates is surfacing for kitchen counters and table tops.

Fig. F-14. Melamine plastics canister set.

A 3 percent melamine resin solution may be added during the preparation of paper pulp to improve the wet strength of paper. Paper with this resin binder has a wet strength nearly as great as its dry strength. Crush and folding resistance is also greatly increased without added brittleness.

Finishes resistant to water, chemicals, alkali, grease, and heat are formulated of amino resins. Heat or heat and catalysts are needed to cure the amino resins. These finishes are seen on stoves, washing machines, and other appliances.

Urea or melamine resins may be made compatible with other plastics to produce finishes of outstanding merit. Polyester (alkyd) resins or phenolic resins may be combined to produce finishes with the best features of both resins. These finishes are sometimes used where a hard, tough, mar-resistant finish is required.

Melamine resins are often used to make waterproof exterior plywood and marine plywood, and for other adhesive applications requiring a light-colored, nonstaining adhesive. Catalysts, heat, or high-frequency energy are used to cure melamine adhesives in plywood and panel assemblies.

Melamine-formaldehyde resins are employed commercially for textile finishes. The well-known drip-dry fabrics and fabrics with permanent glazing, rotproofing, and shrinkage control owe their existence largely to such resins. Melamine and silicone resins are used to produce waterproof fabrics.

Melamine-formaldehyde compounds are easily molded in normal compression and transfer molding machines. Special reciprocating-screw injection machines are also used.

Table F-4 gives some of the properties of melamine-formaldehyde plastics, while listed below are five advantages and three disadvantages of melamine-formaldehyde.

Advantages of Melamine-Formaldehyde

1. Good hardness and scratch resistance
2. Comparatively low cost
3. Wide color range
4. Self-extinguishing
5. Good solvent resistance

Disadvantages of Melamine-Formaldehyde

1. Must be filled for successful molding
2. Poor long-term oxidation resistance
3. Subject to attack by strong acids and bases

Casein

Casein plastics are sometimes classed as natural polymers and are referred to as *protein plastics* by many.

Table F-4. Properties of Melamine-Formaldehyde

Property	No Filler	Alpha-Cellulose Filled	Glass-Fiber Filled
Molding qualities	Good	Excellent	Good
Relative density	1.48	1.47–1.52	1.8–2.0
Tensile strength, MPa		48–90	34–69
(psi)		(7 000–13 000)	(5 000–10 000)
Compressive strength, MPa	276–310	276–310	138–241
(psi)	(40 000–45 000)	(40 000–45 000)	(20 000–35 000)
Impact strength, Izod. J/mm	0 .012–0.0175	0.03–0.9	
(ft lb/in)	(0.24–0.35)	(0.6–18.0)	
Hardness, Rockwell	Ml15–Ml25	M120	
Thermal expansion, 10^{-4}/°C	10	3.8–4.3	
Resistance to heat, °C	99	99	150–205
(°F)	(210)	(210)	(300–400)
Dielectric strength, V/mm		10 630–11 810	6690–11810
Dielectric constant (at 60 Hz)		6.2–7.6	9.7–11.1
Dissipation factor (at 60 Hz)		0.030–0.083	0.14–0.23
Arc resistance, s	100–145	110–140	180
Water absorption (24 h), %	0.3–0.5	0.1–0.6	0.09–0.21
Burning rate	Self-extinguishing	Nonburning	Self-extinguishing
Effect of sunlight	Color fades	Slight color change	Slight
Effect of acids	None to decomposes	None to decomposes	None to decomposes
Effect of alkalies		Attacked	None to slight
Effect of solvents	None	None	None
Machining qualities		Fair	Good
Optical qualities	Opalescent	Translucent	Opaque

Casein is a protein found in a number of sources including animal hair, feathers, bones, and industrial wastes. There has been little interest in these sources, and only skimmed milk is now of commercial interest in the making of casein.

The history of protein-derived plastics can probably be said to date from the work of W. Krische, a German printer, and Adolf Spitteler of Bavaria around 1895. At that time, there was a demand in Germany for what may be described as a *white blackboard.* These boards were thought to possess better optical properties than those with a black surface. In 1897, Spitteler and Krische, in attempting to develop such a product, produced a casein plastics that could be hardened with formaldehyde. Until then casein plastics had proved unsatisfactory because they were soluble in water. Galaith (milkstone), Erinoid, and Ameroid are trade names of those early protein plastics.

Casein is not coagulated by heat but must be precipitated from milk by the action of rennin enzymes or acids. This powerful coagulant causes the milk to separate into solids (curds) and a liquid (whey). After the whey is removed, the curd containing the protein is washed, dried, and made into a powder. When kneaded with water, the doughlike material may be shaped or molded. A simple drying operation then causes much shrinkage. Casein is thermoplastic while being molded. Molded products may be made water resistant by soaking in a formalin solution that creates links holding the casein molecules together. The long linear chain of casein molecules is known as the *polypeptide chain.* There are a number of other peptide groups and possible side reactions. No single formula could represent the interaction between casein and formaldehyde. A very simplified reaction is shown in Figure F-15.

It is doubtful that casein will gain popularity because it is costly to make and the raw material has value as a food. Casein plastics are seriously affected by humid conditions and cannot be used as electrical insulators. The lengthy hardening process and poor resistance to decomposition by heat make them unsuitable for modern processing rates. There are only limited commercial uses of casein plastics in the United States since they offer no property advantages over synthetic polymers, and production costs are high.

Casein plastics are used to a limited extent for buttons, buckles, knitting needles, umbrella handles, and other novelty items. They may be reinforced and filled, or obtained in transparent colors. Casein has held some of its appeal because it may be colored to imitate onyx, ivory, and imitation horn. Casein is most widely used in the stabilization of rubber latex emulsion, preparation of medical compounds and food products, preparation of paints and adhesives, and the sizing of paper and textiles. Casein finds other uses in insecticides, soaps, pottery, inks, and as modifiers in other plastics.

Films and fibers may be produced from these plastics. The woollike fibers are warm, soft, and have prop-

$$\begin{array}{c} CO \\ | \\ NH \end{array} + CH_2O + \begin{array}{c} CO \\ | \\ NH \end{array} \longrightarrow \begin{array}{c} CO \\ | \\ N \end{array}\!\!-\!CH_2\!-\!\!\begin{array}{c} CO \\ | \\ N \end{array}$$

Fig. F-15. Formation of Casein.

erties that compare favorably with natural wool. The films are generally of little use except as handy forms for the coating of paper and other materials. Casein glue is a well-known wood adhesive.

Table F-5 gives some of the properties of casein plastics. Two advantages and disadvantages of casein follow.

Advantages of Casein

1. Produced from nonpetrochemical sources
2. Excellent molding qualities and colorability

Disadvantages of Casein

1. Poor resistance to acids and alkalies, yellows in sunlight
2. High water absorption

Epoxy (EP)

Hundreds of patents have been granted for the commercial uses of epoxide resins. One of the first descriptions of polyepoxides is a German patent by I. G. Farbenindustrie in 1939.

In 1943, the Ciba Company developed an epoxide resin of commercial significance in the United States. By 1948, a variety of commercial coating and adhesive applications were discovered.

Epoxy resins are thermosetting plastics. There are several thermoplastic epoxy resins used for coatings and adhesives. Many different epoxy resin structures available today are derived from bisphenol acetate and epichlorohydrin.

Table F-5. Properties of Casein

Properties	Casein Formaldehyde (Unfilled)
Molding qualities	Excellent
Relative density	1.33–1.35
Tensile strength, MPa	48–79
(psi)	(7000–10000)
Compressive strength, MPa	186–344
(psi)	(27000–50000)
Impact strength, Izod, J/mm	0.045–0.06
(ft lb/in)	(0.9–1.20)
Hardness, Rockwell	M26–M30
Resistance to heat, °C	135–175
(°F)	(275–350)
Dielectric strength, V/mm	15500–27500
Dielectric constant (at 60 Hz)	6.1–6.8
Dissipation factor (at 60 Hz)	0.052
Arc resistance, s	Poor
Water absorption (24 h), %	7–14
Burning rate	Slow
Effect of sunlight	Yellows
Effect of acids	Decomposes
Effect of alkalies	Decomposes
Effect of solvents	Slight
Machining qualities	Good
Optical qualities	Translucent to opaque

Bisphenol A (bisphenol acetate) is made by the condensation of acetone with phenol (Fig. F-16). Epichlorohydrin-based epoxies are widely used because of availability and lower cost. The epichlorohydrin structure is obtained by chlorination of propylene:

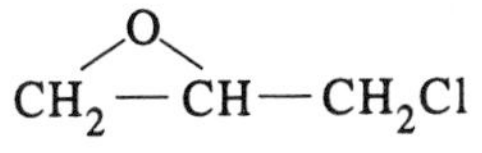

$$CH_2\overset{O}{—}CH—CH_2Cl$$

It will become evident that the epoxy group, for which the plastics family is named, has a triangular structure:

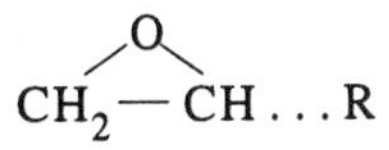

$$CH_2\overset{O}{—}CH\ldots R$$

Epoxy structures are usually terminated by this epoxide structure but many other molecular structures may terminate the long molecular chain. A linear epoxy polymer may be formed when bisphenol A and epichlorohydrin are reacted (Fig. F-17). In some literature, these polymers are called polyethers.

A typical structural formula of epoxy resin based on bisphenol A may be represented as shown in Figure F-18. Other intermediate epoxy-based resins are possible, but are too numerous to mention.

Epoxy resins are cured, as a rule, by adding catalysts or reactive hardeners. Members of the aliphatic and aromatic amine family are commonly used hardening agents. Various acid anhydrides are also used to polymerize the epoxide chain.

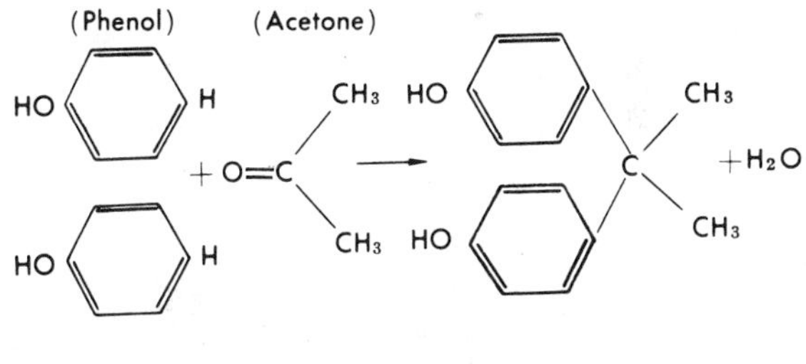

(Phenol) (Acetone)

Fig. F-16. Production of bisphenol-A.

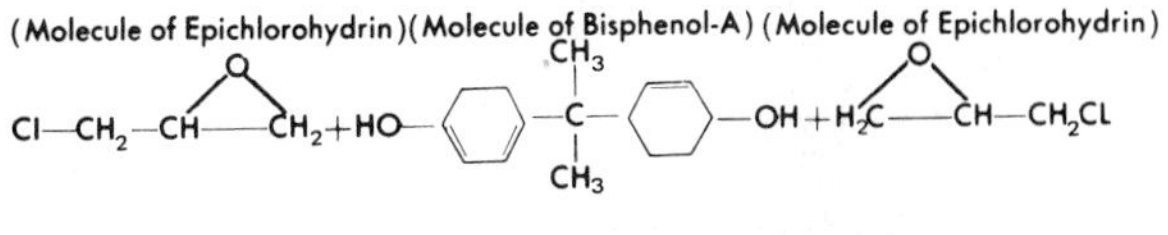

(Molecule of Epichlorohydrin)(Molecule of Bisphenol-A) (Molecule of Epichlorohydrin)

Fig. F-17. Formation of linear epoxy polymer.

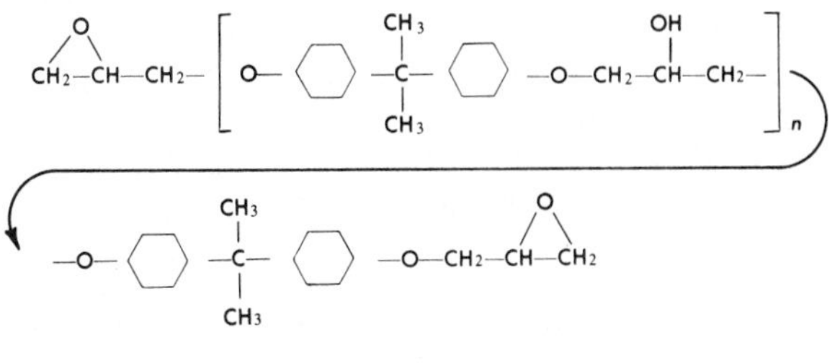

Fig. F-18. Bisphenol-A-based epoxy resin.

Epoxy resins will polymerize and cross-link as thermal energy is added. Catalysts and heat are often used to reach a desired degree of polymerization.

Single-component epoxy resins may contain latent catalysts. These react when enough heat is applied. There is a practical shelf-life expectancy for all epoxy resins.

Reinforced epoxy resins are very strong. They have good dimensional stability and service temperatures as high as 600 °F [315 °C]. Preimpregnated reinforcing materials are used to produce products by hand-layup, vacuum-bag, or filament-winding processes (Fig. F-19.) Epoxy has a very good chemical and fatigue resistance, thus epoxy resins replace the less costly unsaturated polyester resins in many applications. A saving of one-third of the resin mass is realized when epoxy resins are used in place of polyesters.

Epoxy-glass laminates find many uses because they have high strength-to-mass ratio. Superior adhesion to all materials and wide compatibility make epoxy resins desirable (Fig. F-20). Laminated circuit boards, radomes, aircraft parts, and filament-wound pipes, tanks, and containers are only a few uses for this plastics material. (Fig. F-21)

Filled epoxy resins are commonly used for special castings. These strong compounds may be used for low-cost tooling. In dies, jigs, fixtures, and molds for short production runs, epoxies are replacing other tooling materials. Faithful reproduction of details is obtained when epoxy compounds are cast against prototypes or patterns (Fig. F-22).

Many different fillers are used in caulking and patching compounds containing epoxy resins. The adhesive qualities and low shrinkage of epoxies during cure make them durable in caulking or patching applications.

Fig. F-19. Continuous fibrous glass in a matrix of epoxy is used in this composite filament wound rocket motor case. (Structural Composite Industries)

(A) Epoxy adhesives, pressure-injected into hairline cracks in concrete, restore the original load-bearing strength of the material. (Scott J. Saunders Associates)

(B) End closures and ribs of this boron slat for aircraft are bonded to the inner skin with epoxy adhesives.

Fig. F-20. High strength makes epoxy resins very useful.

Electrical potting is another casting use. Epoxies are outstanding in protecting electronic parts from moisture, heat, and corrosive chemicals (Fig. F-23). Electric motor parts, high-voltage transformers, relays, coils, and many other components may be protected from severe environments by being potted in epoxy resins.

Molding compounds of epoxy resins and fibrous reinforcements can be molded by injection, compression, or transfer processes. They are molded into small electrical items and appliance parts, and have many modular uses.

Versatility is achieved by controlling the resin manufacture, the curing agents and the rate of cure. These resins may be formulated to give results ranging from soft, flexible compounds to hard, chemical-resistant products. By incorporation of a blowing agent, low-density epoxy foams may be produced. The qualities offered by the epoxy plastics are adhesion, chemical resistance, toughness, and excellent electrical characteristics.

When the epoxy resins were first introduced in the 1950s, they were recognized as outstanding coating

(A) Graphite/epoxy composites are used on the vertical tail and the entire horizontal stabilizer of Northrop's multirole F-20 Tigershark tactical fighter. These workers are preparing the graphite/epoxy composite material, which is lighter than aluminum and stronger than steel.

(B) The F/A-18 Hornet strike fighter carries about 2300 pounds of graphite/epoxy structures. (Northrop Aircraft Division)

Fig. F-21. Graphite/epoxy materials in aircraft applications.

materials (Fig. F-24). They are more expensive than other coating materials but their adhesion qualities and chemical inertness make them competitive. Five advantages and three disadvantages of epoxy resins are shown below.

Advantages of Epoxy

1. Wide range of cure conditions, from room temperature to 350 °F [178 °C]
2. No volatiles formed during cure
3. Excellent adhesion
4. Can be cross-linked with other materials
5. Suitable for all thermosetting processing methods

Disadvantages of Epoxy

1. Poor oxidative stability; some moisture sensitivity
2. Thermal stability limited to 350–450 °F [178–232 °C]
3. Many grades are expensive

Fig. F-22. Casting mold of epoxy resin, for use in prototype or low-volume production. (Henkel Corp.)

Fig. F-23. A polyamide-epoxy resin blend was used to encapsulate this selenium rectifier. (Henkel Corp.)

(A) Epoxy spray coating is the second can from the left.

(B) Epoxy coating on metal substrate of modern farm silo storage bins. (Bolted Tank Group, Butler Mfg. Co.)

Fig. F-24. Epoxy coating.

Epoxy-based finishes are used on driveways, concrete floors, porches, metal appliances, and wooden furniture. Epoxy finishes on home appliances are a major application of this durable, abrasion-resistant finish. Epoxy coatings have replaced glass enamel finishes for tank car and other container linings that need to resist chemicals. Ship hulls and bulkheads may be coated with epoxy. More durable finishes mean fewer repairs and reduced surface tension between the ship and water. These factors reduce maintenance and fuel costs.

The flexibility of many epoxy coatings makes them popular for postforming coated metal parts. For example, sheets of metal are coated while flat. They are then formed or bent into shallow pans with no damage to the coating.

The ability of epoxy adhesives to bond to dissimilar materials has allowed them to replace soldering, welding, riveting, and other joining methods. The aircraft and automotive industries use these adhesives where heat or other bonding methods might distort the surface. Honeycomb or panel structures make use of the superb adhesive and thermal properties of epoxy.

Table F-6 gives some of the properties of various epoxies. Epoxy copolymers are made by cross-linking with phenolics, melamines, polyamide, urea, polyester, and some elastomers.

Furan

Furan resins are derivatives of furfurylaldehyde and furfuryl alcohol (Fig. F-25). Acid catalysts are used in polymerization therefore it is vital to apply a protective coating to substrates attacked by acids.

Furan plastics have excellent chemical resistance and can withstand temperatures as high as 265 °F [130 °C]. They are used primarily as additives, binders, or adhesives. Furfurylaldehyde has been co-reacted with phenolic plastics. Resins based on furfuryl alcohol are used with amino resins to improve wetting. Their wetting and adhesive capabilities make these resins ideal for impregnating agents. Reinforced and laminated products of furan plastics include tanks, pipes, ducting and construction panels. Furan resins are also used as sand binders in foundries (Table F-7). Two advantages and disadvantages of furans follow.

Advantages of Furans

1. Produced from nonpetrochemical sources
2. Excellent chemical resistance

Disadvantages of Furans

1. Hard to process, limited to fiber-reinforced plastics
2. Subject to attack by halogens

Phenolics (PF)

Phenolics (phenol-aldehyde) were among the first true synthetic resins produced. They are known chemically as phenol-formaldehyde (PF). Their history reaches back to the work of Adolph Baeyer in 1872. In 1909, the chemist Baekeland invented and patented a technique for combining phenol (C_6H_3OH, also called carbolic acid) and gaseous formaldehyde (H_2CO).

The success of the phenol-formaldehyde resins later stimulated research into urea- and melamine-formaldehyde resins.

The resin formed from the reaction of phenol with formaldehyde (an aldehyde) is known as a *phenolic.* Figure F-26 shows the reaction of a phenol with formaldehyde. This involves a condensation reaction in which water is formed as a byproduct (A-stage). The first phenol formaldehyde reactions produce a low-molecular-mass resin that is compounded with fillers and other ingredients (B-stage). During the molding

Table F-6. Properties of Epoxies

Property	Epoxy Molding Compounds: Glass-Filled	Epoxy Molding Compounds: Mineral-Filled	Epoxy Molding Compounds: Microballoon-Filled
Molding qualities	Excellent	Excellent	Good
Relative density	1.6–2.0	1.6–2.0	0.75–1.00
Tensile strength, MPa	69–207	34–103	17–28
(psi)	(10000–30000)	(5000–15000)	(2500–4000)
Compressive strength, MPa	172–276	124–276	69–103
(psi)	(25000–40000)	(18000–40000)	(10000–15000)
Impact strength, Izod, J/mm	0.5–1.5	0.015–0.02	0.008–0.013
(ft lb/in)	(10–30)	(0.3–0.4)	(0.15–0.25)
Hardness, Rockwell	M100–M110	M100–M110	
Thermal expansion, 10^{-4}/°C	2.8–8.9	5.1–12.7	
Resistance to heat, °C	150–260	150–260	
(°F)	(300–500)	(300–500)	
Dielectric strength, V/mm	11810–15750	11810–15750	14960–16535
Dielectric constant, (at 60 Hz)	3.5–5	3.5–5	
Dissipation factor (at 60 Hz)	0.01	0.01	
Arc resistance, s	120–180	150–190	120–150
Water absorption, (24 h), %	0.05–0.20	0.04	0.10–0.20
Burning rate	Self-extinguishing	Self-extinguishing	Self-extinguishing
Effect of sunlight	Slight	Slight	Slight
Effect of acids	Negligible	None	Slight
Effect of alkalies	None	Slight	Slight
Effect of solvents	None	None	Slight
Machining qualities	Good	Fair	Good
Optical qualities	Opaque	Opaque	Opaque

Furfurylaldehyde — or — Furfuryl Alcohol → Furon

Fig. F-25. Acid catalysts will cause condensation of furfurylaldehyde or furfuryl alcohol. Cross-linking occurs between furan rings.

Table Table F-7 Properties of Furan

Property	Furan (Asbestos-Filled)
Molding qualities	Good
Relative density	1.75
Tensile strength, MPa	20–31
(psi)	(2900–4500)
Compressive strength. MPa	68–72
(psi)	(9900–10450)
Hardness, Rockwell	R110
Resistance to heat, °C	130
(°F)	(266)
Water absorption (24 h), %	0.01–2.0
Burning rate	Slow
Effect of sunlight	None
Effect of acids	Attacked
Effect of alkalies	Little
Effect of solvents	Resistant
Machining qualities	Fair
Optical qualities	Opaque

process, the resin is trans-. formed into a highly cross-linked thermosetting plastics product by heat and pressure (C-stage).

Although the monomer solution of phenol is commercially used, cresols, xylenols, resorcinols, or syn-

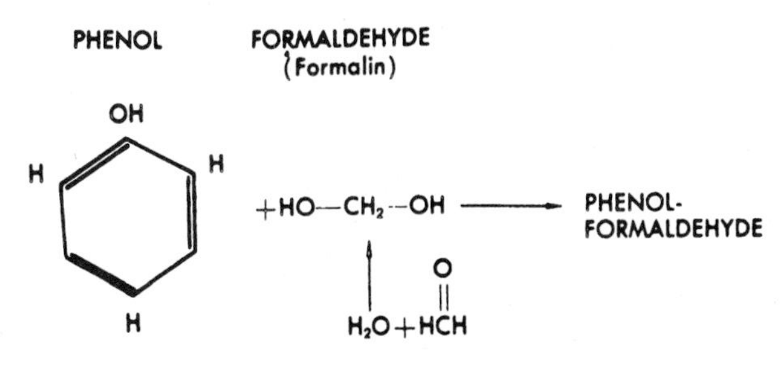

Fig. F-26. Reaction of phenol and formaldehyde.

thetically produced oil-soluble phenols may be used. Furfural may replace the formaldehyde.

In *one-stage resins*, a *resol* is produced by reacting a phenol with an excess amount of aldehyde in the presence of a catalyst (not acid). Sodium and ammonium hydroxide are common catalysts. This product is soluble and low in molecular mass. It will form large molecules without addition of a hardening agent during the molding cycle.

Two-stage resins are produced when phenol is present in excess with an acid catalyst. The low-molecular mass and soluble *novolac* resin is the result. It will remain a linear thermoplastic resin unless compounds capable of forming cross-linkage on heating are added. They are called *two-stage resins* because some agent must be added before molding (Fig. F-27).

The *A-Stage* novalac resin is a fusible and soluble thermoplastic. The *B-stage* resin is produced by thermally blending the A-stage and hexamethylenetetramine. The B-stage is usually sold in a granular or powder form. Fillers, pigments, lubricants and other additives are compounded with the resin in this stage. During molding, heat and pressure convert the B-stage resin into an insoluble, infusible, *C-stage* thermosetting plastics.

Phenolics are not used as frequently as they once were because so many new plastics have been developed (Fig. F-28). Their low cost, mold-ability, and physical properties make phenolics leaders in the thermoset field, however. These materials are widely used as molding powders, resin binders, coatings, and adhesives.

Molding powders or compounds of novolac resins are rarely used without a filler. The filler is not used simply to reduce cost. It improves physical properties, increases adaptability for processing, and reduces shrinkage. Curing time, shrinkage, and molding pressures may be reduced by preheating phenol-formaldehyde compounds. Advances in equipment and techniques have kept phenolics competitive with many thermoplastics and metals (Fig. F-29). Phenolics are used in conventional transfer and compression molding operations, and in injection and reciprocating-screw machines. Molded phenolic parts are abrasive and hard to machine. Although molded phenolics have many uses as electrical insulation, they exhibit poor tracking resistances under very humid conditions. In a few uses, they have been replaced by thermoplastics.

Phenolic resins have a major appearance drawback—they are too dark in color for use as surface layers on decorative laminates and as adhesives where glue joints may show. Phenolic-resin impregnated cotton fabric, wood, or paper are often used in the making of gear wheels, bearings, substrates for electrical circuit boards, and melamine decorative laminates

Fig. F-27. Final curing or heat-hardening should be considered a further condensation process.

(A) Billiard balls of molded phenolic.

(B) Molded parts for a humidifier. (Durez Division, Hooker Chemical Corp.)

(C) End panels for a broiler oven. (Durez Division, Hooker Chemical Corp.)

Fig. F-28. Some uses of phenolics.

(A) Automotive brake system parts of phenolic.

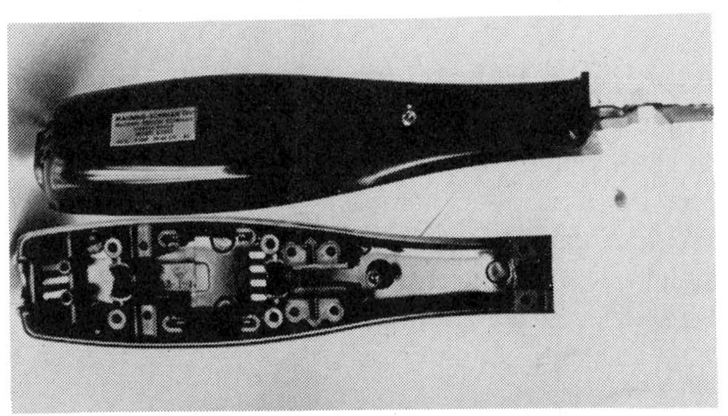

(B) High-impact phenolic used for handle of electric carving knife.

Fig. F-29. Phenolics remain competitive with thermoplastics. (Durez Division, Hooker Chemical Corp.)

(A) Paper and cloth.

(B) Electrical circuit board.

Fig. F-30. Materials impregnated with phenolic resin.

(Fig. F-30). These laminates are usually made in large presses under controlled heat and pressure. Many methods of impregnation are used, including dipping, coating, and spreading.

Phenol formaldehyde resins may be cast into many profile shapes, such as billiard balls, cutlery handles, and novelty items.

Phenolic-based resins are available in liquid, powder, flake, and film forms. The ability of these resins to impregnate and bond with wood and other materials is the reason for their success as adhesives. They improve adhesion and heat resistance, and are widely used in the making of plywood and as binder adhesives in wood-particle moldings. Wood-particle boards are used in many building applications, such as sheathing, subflooring, and core stocks.

Phenolic resins are used as binders for abrasive grinding wheels. The abrasive grit and resin are simply molded into the desired shape and cured. Resin binders are an important ingredient in the shell molds and cores used in foundries (Fig. F-31). These molds and cores produce very smooth metal castings. As heat-resistant binders, phenolic resins are used in making brake linings and clutch facings.

Because of their high resistance to water, alkalies, chemicals, heat, and abrasion, phenolics are sometimes chosen for use in finishes. They are used for coating appliances, machinery, or other devices requiring maximum heat resistance.

A high-strength, heat- and fire-resistant foam may be produced using, phenolic resins. The foam may be produced in the plant or on the site by rapidly mixing a blowing agent and catalyst with the resin. As the chemical reaction generates heat and begins the polymerization process, the blowing agent vaporizes. This causes the resin to expand into a multicellular, semipermeable structure. These foams may be used as fill for honeycomb structures in aircraft, flotation materials, acoustic and thermal insulation, and as packing materials for fragile objects.

Microballons (small hollow spheres) may be produced of phenolic plastics filled with nitrogen. These spheres vary from 0.0002 to 0.0032 in. [0.005 to 0.08 mm] in diameter. They may be mixed with other resins to produce syntactic foams. These foams find uses as insulative fillers. They serve as vapor barriers when placed on volatile liquids such as petroleum.

Table F-8 gives properties of phenolic materials, and eight advantages and four disadvantages of phenolic resins are shown below.

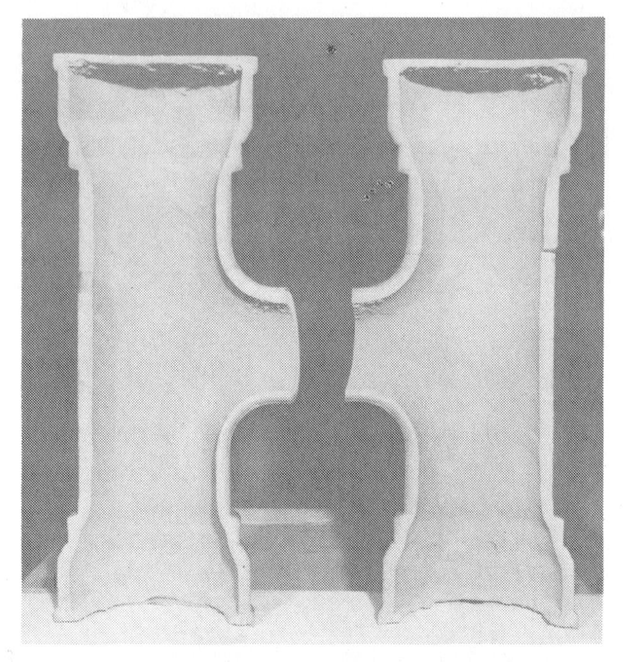

(A) Phenolic resin was used as a binder for this sand core.

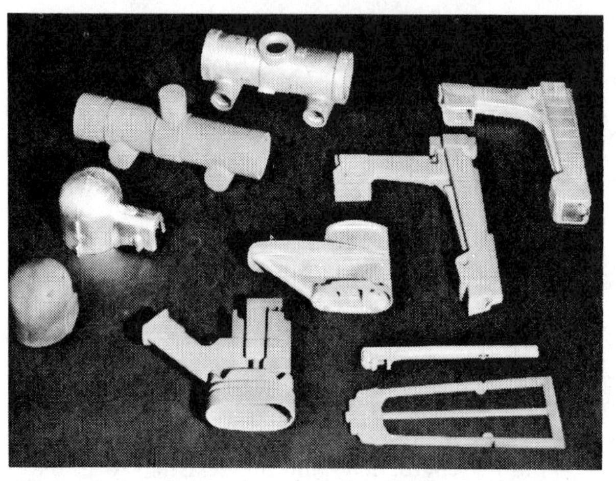

(B) Resin-bonded sand cores and resulting metal castings.

Fig. F-31. Resin binders used in foundry work. (Acme Resin Co.)

Advantages of Phenolics

1. Comparatively low cost
2. Suitable for use at temperatures to 400 °F [205 °C]
3. Excellent solvent resistance
4. Rigid
5. Good compressive strength
6. High resistivity
7. Self-extinguishing
8. Very good electrical characteristics

Disadvantages of Phenolics

1. Need fillers for moldings
2. Poor resistance to bases and oxidizers
3. Volatiles released during cure (a condensation polymer)
4. Dark color (due to oxidation discoloration)

Phenol-Aralkyl

In 1976, the Ciba-Geigy Corporation introduced a group of resins based on aralkyl ethers and phenols. Two basic prepolymer grades are available. Both are sold as 100 percent prepolymer resin. One grade cures by a condensation reaction. The other undergoes an addition reaction similar to epoxy. Condensation grades are blended with phenolic novolac resins to improve phenolic properties. Addition polymerization grades are finding uses in the fabrication of laminates. These resins are used as binders for the making of cutting wheels, printed circuit boards, bearings, appliance parts, and engine components. Because of excellent mechanical properties, processing advantages, and thermal capabilities, these prepolymers will find other uses. Table F-9 lists phenol-aralkyl properties.

Unsaturated Polyesters

The term *polyester resin* encompasses a variety of materials. It is often confused with other polyester classifications. A polyester is formed by the reaction of a polybasic acid and a polyhydric alcohol. Changes with acids, with acids and bases, and with some unsaturated reactants permit crosslinking, forming thermosetting plastics.

The term polyester resin should refer to unsaturated resins based on dibasic acids and dihydric alcohols. These resins are capable of cross-linking with unsaturated monomers (often styrene). Alkyds and polyurethanes of the polyester resin group are discussed individually.

Sometimes the term *fiberglass* has been used to indicate unsaturated polyester plastics. This term should refer only to fibrous pieces of glass. Various resins may be used with glass fiber acting as a reinforcing agent. The main use for unsaturated polyester resin is in the making of reinforced plastics. Glass fiber is the most-used reinforcement.

Credit for the first preparation of polyester resins (alkyd type) is usually attributed to the Swedish chemist Jons Jacob Berzelius in 1847 and to GayLussac and Pelouze in 1833. Further development was conducted by W. H. Carothers and by R. H. Kienle. Through the 1930s most of the work on polyesters was aimed at developing and improving paint and varnish applications. Further interest in the resin was stimulated by Carleton Ellis in 1937. He found that by adding unsaturated monomers to unsaturated polyesters, cross-linking and polymerization time was greatly reduced. Ellis has been called the father of unsaturated polyesters.

Large-scale industrial use of unsaturated polyesters developed quickly as wartime shortages spurred development of many resin uses. Reinforced polyester structures and parts were widely used during World War II.

Table F-8. Properties of Phenolics

Property	Phenol-Formaldehyde (Unfilled)	Phenol-Formaldehyde (Macerated Fabric)	Phenolic Casting Resin (Unfilled)
Molding qualities	Fair	Fair to good	
Relative density	1.25–1.30	1.36–1.43	1.236–1.320
Tensile strength, MPa	48–55	21–62	34–62
(psi)	(7000–8000)	(3000–9000)	(5000–9000)
Compressive strength, MPa	69–207	103–207	83–103
(psi)	(10000–30000)	(15000–30000)	(12000–15000)
Impact Strength, Izod, J/mm	0.01–0.018	0.038–0.4	0.012–0.02
(ft lb/in)	(0.20–0.36)	(0.75–8)	(0.24–0.40)
Hardness, Rockwell	M124–M128	E79–E82	M93–M120
Thermal expansion, 10^{-4}/°C	6.4–15.2	2.5–10	17.3
Resistance to heat, °C	120	105–120	70
(°F)	(250)	(220–250)	(160)
Dielectric strength, V/mm	11810–15750	7 875–15 750	9845–15750
Dielectric constant (at 60 Hz)	5–6.5	5.2–21	6.5–17.5
Dissipation factor (at 60 HZ)	0.06–0.10	0.08–0.64	0.10–0.15
Arc resistance, s	Tracks	Tracks	
Water absorption (24 h), %	0.1–0.2	0.40–0.75	0.2–0.4
Burning rate	Very slow	Very slow	Very slow
Effect of sunlight	Darkens	Darkens	Darkens
Effect of acids	Decomposed by oxidizing acids	Decomposed by oxidizing acids	None
Effect of alkalies	Decomposes	Attacked	Attacked
Effect of solvents	Resistant	Resistant	Resistant
Machining qualities	Fair to good	Good	Excellent
Optical qualities	Transparent to translucent	Opaque	Transparent to opaque

The word *polyester* is derived from two chemical processing terms, *poly*merization and *ester*ification. In esterification, an organic acid is combined with an alcohol to form an ester and water. A simple esterification reaction is shown in Figure F-32. (See Alkyds, above.)

The reverse of the esterification reaction is called *saponification.* In order to obtain a good yield of ester in a condensation reaction, water must be removed to prevent saponification (Fig. F-33). If a polybasic acid (such as maleic acid) is caused to react, and the water is removed as it is formed, the result will be an *unsaturated polyester. Unsaturated* means that the double-bonded carbon atoms are reactive or possess unused valance bonds. These can be attached to another atom or molecule, thus such a polyester is capable of cross-linkage. There are many other reactive or unsaturated monomers that can be used to change or tailor the resin to meet a special purpose or use. Vinyl toluene, chlorostyrene, methyl methacrylate, and diallyl phthalate are commonly used monomers. Unsaturated styrene is an ideal, low-cost monomer most often used with polyesters (Fig. F-34).

The four main functions of a monomer are as follows:

$$\underset{(ACID)}{R{-}C{-}OH + HO{-}\overset{\overset{O}{\|}}{C}{-}R} \xrightarrow{ESTERIFICATION} \underset{(ESTER)}{R{-}C{-}O{-}\overset{\overset{O}{\|}}{C}{-}R} + \underset{(WATER)}{HOH}$$

Fig. F-32. Examples of esterification reaction.

Table Table F-9. Properties of Phenol-Aralkyl

Property	Phenol-Arakyl (Glass Filled)
Molding qualities	Good
Relative density	1.70–1.80
Tensile strength, MPa	48–62
(psi)	(6900–9000)
Compressive strength, MPa	206–241
(psi)	(3000–3500)
Impact strength, Izod, J/mm	0.02–0.03
(ft lb/in)	(0.4–0.6)
Hardness, Rockwell	
Resistance to heat, °C	250
(°F)	(480)
Dielectric strength. V/mm	
Dielectric constant (at 1 MHz)	2.5–4.0
Dissipation factor (at 1 MHz)	0.02–0.03
Water absorption (24 h), %	0.05
Effect of acids	None to slight
Effect of alkalies	Attacked
Effect of solvents	Resistant
Machining qualities	Fair
Optical qualities	Opaque

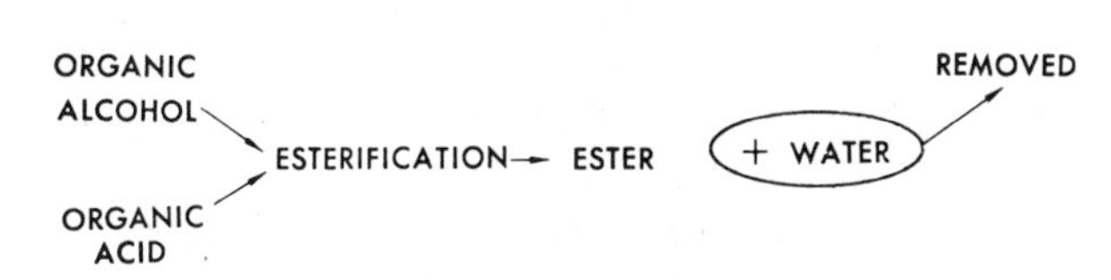

Fig. F-33. To prevent saponification, water should be removed in an esterification reaction.

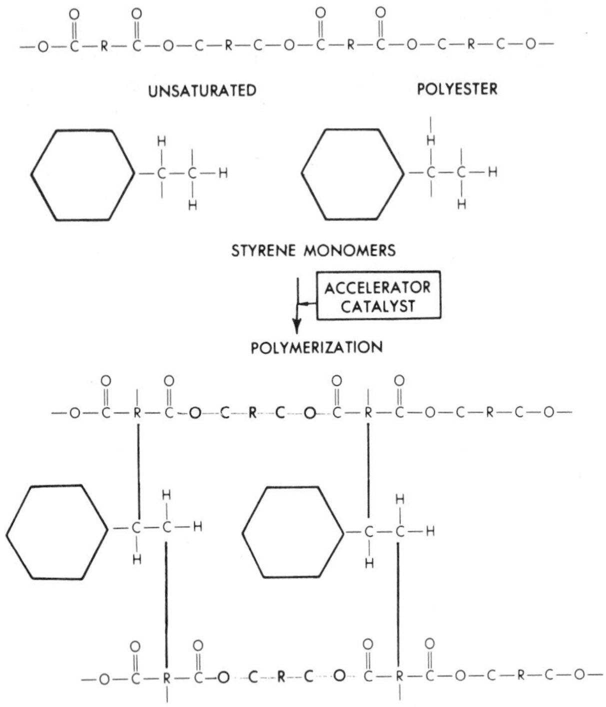

Fig. F-34. Polymerization reaction with unsaturated polyester and styrene monomers.

1. To act as a solvent carrier for the unsaturated polyester
2. To lower viscosity (thin)
3. To enhance selected properties for specific uses
4. To provide a rapid means of reacting (cross-linking) with the unsaturated linkages in the polyester

As the molecules randomly collide and occasional bonds are completed, a very slow polymerization (cross-linking) process will occur. This process may take days or weeks in simple mixtures of polyesters and monomers.

To speed up polymerization at room temperature, accelerators (promoters) and catalysts (initiators) are added. The accelerators commonly used are cobalt naphthenate, diethyl aniline, and dimethyl aniline. Polyester resins will usually have the accelerator added by the manufacturer unless otherwise specified. Resins that contain an accelerator require only a catalyst to provide rapid polymerization at room temperatures. With the addition of an accelerator the shelf life of the resin is appreciably shortened. Inhibitors such as hydroquinone may be added to stabilize or retard premature polymerization. These additives do not interfere with the final polymerization to any great extent. The speed of cure can be influenced by temperature, light, and the amounts of additives.

Polyester resins may be formulated without accelerators. All resins should be kept in a cool, dark storage area until used.

CAUTION: If the accelerator and catalyst are supplied separately, never mix them together directly. A violent explosion may result.

Methyl ethyl ketone peroxide, benzoyl peroxide, and cumene hydroperoxide are the three common organic peroxides used to catalyze polyester resins. These catalysts break down, releasing free radicals, when they come in contact with accelerators in the resin. The free radicals are attracted to the reactive unsaturated molecules, thus beginning the polymerization reaction.

By the strictest definition, the term *catalyst* is incorrectly used when referring to the polymerization mechanism of polyester resins. By the strict definition, a catalyst is a substance that by its mere presence aids a chemical action, without itself being permanently changed. In polyester resins, however, the catalyst breaks down and becomes a part of the polymer structure. Since these materials are consumed in initiating the polymerization, the term *initiator* would be more accurate. A true catalyst is recoverable at the end of a chemical process.

Exposure to radiation, ultraviolet light, and heat have also been used to begin the polymerization of double-bonded molecules. If catalysts are used, the resin mix becomes correspondingly more sensitive to heat and light. On a hot day or in the sunlight, less catalyst is required for polymerization. On a cold day, more catalyst would be needed. The resin and catalyst also could be warmed to produce a rapid cure.

The final curing reaction is called *addition polymerization* because no byproducts are present as a result of the reaction. In phenol-formaldehyde reactions, the curing reaction is called *condensation polymerization* because a byproduct, water, is present. (Fig. F-35) See polyallyl esters and allylics.

Polyester may be specially modified for a wide variety of uses by altering the chemical structure or by using additives. With higher percentages of unsaturated acid, more cross-linkage is possible. A stiffer, harder product results. The adding of saturated acids will increase toughness and flexibility. Thixotropic fillers, pigments, and lubricants also may be added to the resin.

Polyester resins that contain no wax are susceptible to *air inhibition*. When exposed directly to air, such resins remain undercured, soft, and tacky for some time after setting. This is desirable when multiple layers are to be built up. Resins purchased from the manufacturer without wax are referred to as *air-inhibited* resins. The absence of wax permits better bonds between multiple layers in hand-layup operations.

In some cases a tackfree cure of air-exposed surfaces is wanted. For one-step castings, moldings, or surface coats, such a cure is obtained with a *non-air-inhibited* polyester. Non-air-inhibited resins contain a wax that floats to the surface during the curing opera-

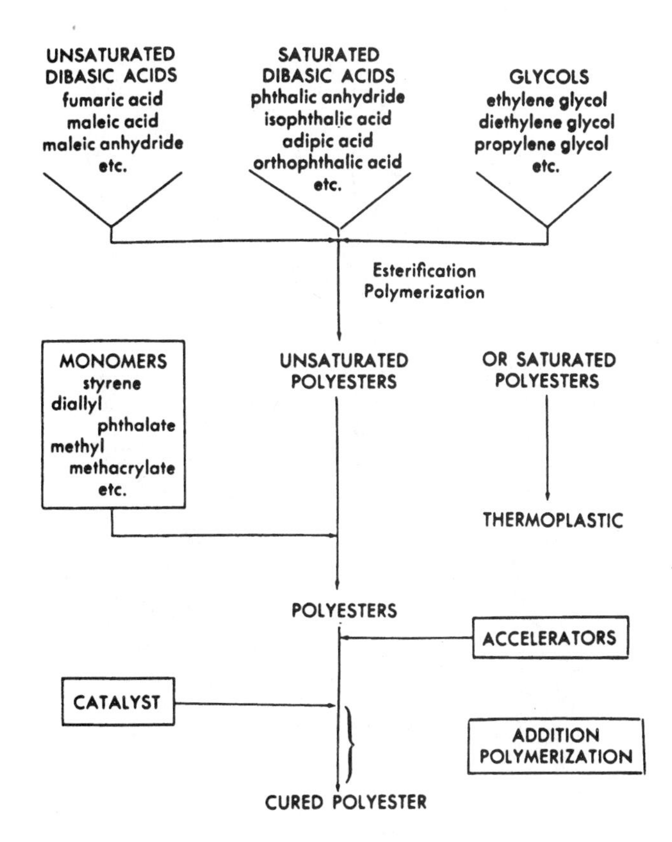

Fig. F-35. Production scheme of cured polyester.

tion blocking out the air and allowing the surface to cure tackfree. Many waxes may be used in such resins including household paraffin wax, carnauba wax, beeswax, stearic acid, and others. The use of waxes adversely affects adhesion; therefore, if more layers are to be added, all wax must be removed from the surface by sanding.

By altering the basic combination of raw materials, fillers, reinforcements, cure time, and treatment technique, a wide range of properties is possible.

Polyester finds its main use in making composite products. The primary value of reinforcement is to gain high strength-to-mass ratio (Fig. F-36). Glass fiber is the most common reinforcing agent. Asbestos, sisal, many plastics fibers, and whisker filaments are also used. The type of reinforcement chosen depends on the end use and method of fabrication. Reinforced polyesters are among the strongest materials known. They have been used in automobile bodies and in boat hulls and because of their high strength-to-mass ratio, they have aircraft and aerospace uses as well. (Fig. F-37)

Other applications include radar domes, ducts, storage tanks, sports equipment, trays, furniture, luggage, sinks, and many kinds of ornaments. (Fig. F-38)

Unreinforced casting grades of polyesters are used for embedding, potting, casting, and sealing. Resins filled with wood flour may be cast in silicone molds to produce precise copies of wood carvings and trim. Special resins may be emulsified with water, further reduc-

(A) The roof of this pavilion consists of 640 panels of fibrous glass-reinforced plastics. (Hooker Chemical Corp.)

(B) Glass-reinforced polyester is used for this welding helmet.

Fig. F-36. Products of fibrous-glass-reinforced polyester have a high strength-to-mass ratio.

ing costs. These resins, referred to as *water-extended resins*, may contain up to 70 percent water. The castings undergo some shrinkage due to water loss.

Fabrication methods for reinforced polyesters include hand layup, sprayup, matched molding, premix molding, pressure-bag molding, vacuum-bag. molding, casting, and continuous laminating. Other molding modifications are sometimes used. Compression molding equipment is sometimes used with doughlike premixes containing all ingredients. Table F-10 gives some of the properties of many polyesters. The following list contains six advantages and two disadvantages of polyesters.

Advantages of Unsaturated Polyesters

1. Wide curing latitude
2. May be used for medical devices (artificial limbs)
3. Accept high filler levels
4. Thermosetting materials
5. Inexpensive tooling
6. Non-burning halogenated grades available

Disadvantages of Unsaturated Polyesters

1. Upper service temperature limited to 93 °C (200 °F)
2. Poor resistance to solvents

Fig. F-37. Classic composite 1953 Chevrolet Corvette was first composite production body. It was composed of glass fibers in a matrix of polyester.

There are a number of thermoset interpenetrating polymer networks (IPN) involving a cross-linked unsaturated polyester, vinyl ester, or polyester-urethane copolymer in a urethane network. You will recall that an IPN is a configuration of two or more polymers, each existing in a network. There is a synergistic effect when one of the polymers is synthesized in the presence of the other. Urethane/polyester network resins have good wet-out and may be used in pultrusion, filament winding, RIM, RTM, sprayup, and other composite reinforcing methods. Thermoplastic IPNs do not

Fig. F-38. This vessel is made of a reinforced corrosion resistant polyester resin. It is used to remove sulfuric acid mist from other gasses. (ICI Americas, Incorporated)

Table F-10. Properties of Thermosetting Polyesters

Property	Thermosetting Polyester (Cast)	Thermosetting Polyester (Glass Cloth)
Molding qualities	Excellent	Excellent
Relative density	1.10–1.46	1.50–2.10
Tensile strength, MPa	41–90	207–345
(psi)	(6000–13000)	(30000–50000)
Compressive strength, MPa	90–252	172–345
(psi)	(13000–36500)	(25000–50000)
Impact strength, Izod, J/mm	0.01–0.02	0.25–1.5
(ft lb/in)	(0.2–0.4)	(5.0–30.0)
Hardness, Rockwell	M70–M115	M80–M120
Thermal expansion, 10^{-4}/C	14–25.4	3.8–7.6
Resistance to heat, °C	120	150–180
(°F)	(250)	(300–350)
Dielectric strength, V/mm	14960–19685	13780–19690
Dielectric constant, (at 60 Hz)	3.0–4.36	4.1–5.5
Dissipation factor (at 60 Hz)	0.003–0.028	0.01–0.04
Arc resistance, s	125	60–120
Water absorption (24 h), %	0.15–0.60	0.05–0.50
Burning rate	Burns to self-extinguishing	Burns to self-extinguishing
Effect of sunlight	Yellows slightly	Slight
Effect of acids	Attacked by oxidizing acids	Attacked by oxidizing acids
Effect of alkalies	Attacked	Attacked
Effect of solvents	Attacked by some	Attacked by some
Machining qualities	Good	Good
Optical qualities	Transparent to opaque	Transparent to opaque

form chemical crosslinks as do thermoset IPNs. There is a physical entanglement that interferes with polymer mobility. Thermoplastic IPN has been produced using PA, PBT, POM and PP with silicone as the IPN.

Thermosetting Polyimide

Polyimides may exist as either thermoplastic or thermosetting materials. Addition polyimides are available as thermosets. Condensation polyimides, thermally decompose before they reach their melting point during processing.

Thermosetting polyimides are molded by injection, transfer, extrusion, and compression methods.

Thermosetting polyimides find uses in aircraft engine parts, automobile wheels, electrical dielectrics, and coatings.

Table F-11 gives properties of thermosetting polymides.

Both thermosetting and thermoplastic polyimides are considered high temperature polymers.

Some improvements in resin systems in recent years have resulted in a high temperature polymer that is less brittle and more easily processed. Some systems begin with a thermoplastic polymer based on an aromatic tetracarboxylic dianhydride and an aromatic diamine. A completely imidized powder is the result. Unlike addition polyimides, this product can be processed from many common organic solvents (e.g., cyclohexanone). Although the polyimide is completely imidized, additional cross-linking occurs during processing

Bismakimides (BMI). These are a class of polyimides with a general structure containing reactive double bonds on each end of the molecule.

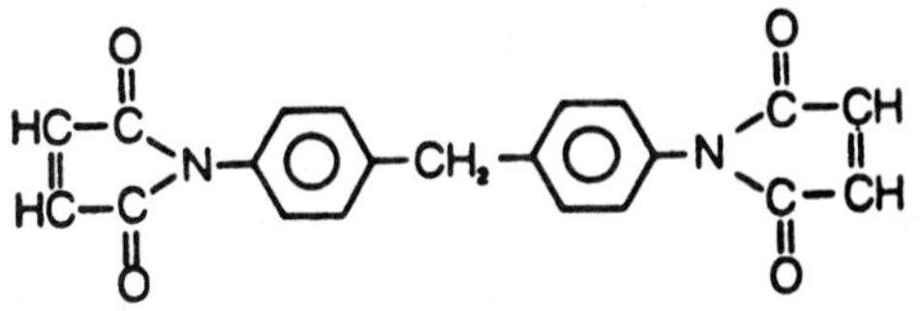

Various functional groups, such as vinyls, allyls, or amines, are used as co-curing agents with bismaleimides to improve the properties of the homopolymer.

In one bismaleimide system, component A (4,4'-bismaleimidodiphenylmethane) and component B (o-,o'-diallyl-bisphenol A) are reacted together to provide a BMI molecule with a tougher, flexible backbone.

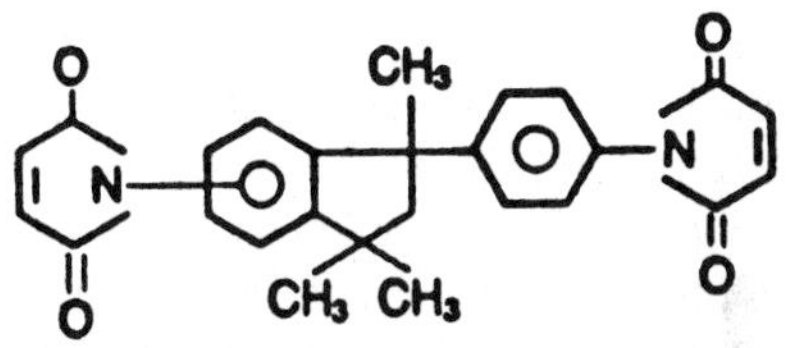

Tests have shown improved strength and toughness with formulations using a higher ratio (1.0:0.87) of BMI to diallylbisphenol. A. Typical properties of two bismaleimide systems are shown in Table F-11.

Polyurethane (PU)

The term *polyurethane* refers to the reaction of polyisocyanates (-NCO-) and polyhydroxyl (-OH-) groups.

Table F-11. Properties of Thermosetting Polyimide and Bismaleimide

Properties	Thermosetting Polyimide (Unfilled)	Bismaleimide 1:1	Bismaleimide 10:0.87
Molding qualities	Good	good	good
Relative density	1.43	—	—
Tensile strength, MPa	86	81	92
(psi)	(12500)	(11900)	(13600)
Compressive strength, MPa	275	201	207
(psi)	(39900)	(29 900)	(30500)
Impact strength, Izod, J/mm	0.075	—	—
(ft lb/in)	(1.5)	—	—
Hardness, Rockwell	E50		
Thermal expansion, 10^{-4}/°C	13.71		
Resistance to heat, °C	350	272	285
(°F)	(660)	(523)	(545)
Dielectric strength, V/mm	22050		
Dielectric constant (at 60 Hz)	3.6		
Dissipation factor (at 60 Hz)	0.001 8		
Water absorption (24 h), %	0.24		
Effect of acids	Slowly attacked		
Effect of alkalies	Attacked		
Effect of solvents	Very resistant		
Matching qualities	Good		
Optical qualities	Opaque		

A simple reaction of isocyanate and an alcohol is shown below. The reaction product is urethane, not a polyurethane.

$$R \cdot NCO + HOR_1 — R \cdot NH \cdot COOR_1$$

PolyIsocyanate PolyHydroxyl PolyUrethane

The German chemists Wurtz, in 1848, and Hentschel, in 1884, produced the first isocyanates.

These later led to the development of polyurethanes. It was Otto Bayer and his coworkers who actually made possible the commercial development of polyurethanes in 1937. Since that time, polyurethanes have developed into many commercially available forms that include coatings, elastomers, adhesives, molding compounds, foams, and fibers.

The isocyanates and di-isocyanates are highly reactive with compounds containing reactive hydrogen atoms. For this reason, polyurethane polymers may be reproduced. The recurring link of the polyurethane chain is NHCOO or NHCO.

More complex polyurethanes have been developed based on toluene di-isocyanates (TDI) and polyester, diamine, castor oil, or polyether chains. Other isocyanates used are diphenylmethane diisocyanate (MDI), and polymethylene polyphenyl isocyanate (PAPI).

The first polyurethanes were made in Germany to compete with other polymers produced at that time. Linear aliphatic polyurethanes were used to make fibers. Linear polyurethanes are thermoplastic. They may be processed by all normal thermoplastic techniques, including injection and extrusion. Because of cost, they find limited use as fibers or filaments.

Polyurethane coatings are noted for their high abrasion resistance, unusual toughness, hardness, good flexibility, chemical resistance, and weatherability (Fig. F-39). ASTM has defined five distinct types of polyurethane coatings, as shown in Table F-12.

Polyurethane resins are used as clear or pigmented finishes for home, industrial, or marine use. They improve the chemical and ozone resistance of rubber and other polymers. These coatings and finishes may be simple solutions of linear polyurethanes or complex systems of polyisocyanate and such OH groups as polyesters, polyethers, and castor oil.

Many polyurethane elastomers (PUR) (rubbers), may be prepared from di-isocyanates, linear polyesters or polyether resin, and curing agents (Fig. F-40). If formulated into a linear thermoplastic urethane, they may be processed by normal thermoplastic processing equipment. They find uses as shock absorbers, bumpers, gears, cable covers, hose jacketing, elastic thread (Spandex), and diaphragms. Common uses of cross-linked thermosetting elastometers include industrial tires, shoe heels, gaskets, seals, O-rings, pump impellers, and tread stock for tires. Polyurethane elastomers have extreme resistance to abrasion, ozone aging, and hydrocarbon fluids. These elastometers cost more than conventional rubbers but are tough, elastic, and show a wide range of flexibility at temperature extremes.

Polyurethane foams are widely used and well known. They are available in flexible, semirigid, and rigid forms in a number of different densities. Various

(A) Foams, insulation, sponges, belts andgaskets of polyurethane.

(B) This self-locking polyurethane container is shockproof, fire retardant and moistureproof. (Poly-Con Industries, Inc.)

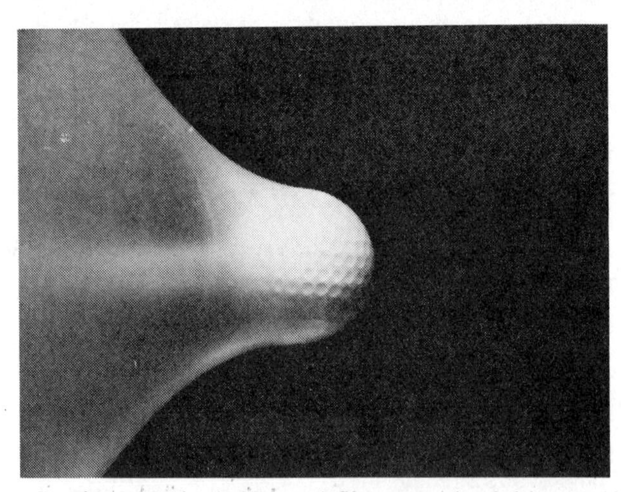

(C) This 0.003-in. (0.076-mm) film stops a hard-driven golf ball, demonstrating the puncture resistance and tensile strength of polyurethane. (B.F. Goodrich Chemical Co.)

Fig. F-39. Applications of polyurethane.

Table F-12. ASTM Designations for Polyurethane Coatings

ASTM Type	Components	Pot Life	Cure	Clear or Pigmented Uses
(I) Oil-modified	One	Unlimited	Air	Interior or exterior wood and marine. Industrial enamels
(II) Prepolymer	One	Extended	Moisture	Interior or exterior. Wood, rubber and leather coatings
(III) Blocked	One	Unlimited	Heat	Wire coatings and baked finishes
(IV) Prepolymer + Catalyst	Two	Limited	Amine/ catalyst Air	Industrial finishes and leather, rubber products
(V) Polyisocyanate + Polyol	Two	Limited	NCO/OH reaction	Industrial finishes and leather, rubber products

$$\underset{\text{Polyester}}{n\text{HOR}-(\text{OR}-)_x-\text{OH}} + \underset{\text{Di-isocyanate}}{n\text{OCNR}_1\text{NCO}} \rightarrow \underset{\text{Polyurethane}}{(-\text{OR}-(\text{OR})_x\text{OCONHR}_1\text{NHCO}-)_n}$$

$$\underset{\text{Polyether}}{n\text{HOR}(-\text{OCOR}_2\text{CO}\cdot\text{OR}-)_x\text{OH}} + \underset{\text{Di-isocyanate}}{n\text{OCNR}_1\text{NCO}} \rightarrow \underset{\text{Polyurethane}}{-\text{OR}(-\text{OCOR}_2\text{CO}\cdot\text{OR})_x\text{OCONHR}_1\text{NHCO}-_n}$$

Fig. F-40. Production of polyurethane elastomers.

flexible foams are used as cushioning for furniture, automobile seating, and mattresses. They are produced by reacting toluene di-isocyanate (TDI) with polyester and water in the presence of catalysts. At higher densities, they are cast or molded into drawer fronts, doors, moldings, and complete pieces of furniture. Flexible foams are open-celled structures that may be used as artificial sponges. These foams are used by the garment and textile industries for backing and insulation (Fig. F-41).

Semirigid foams find use as energy-absorbing materials in crash pads, arm rests, and sun visors.

The three largest uses of rigid polyurethane foam are in the making of furniture, automotive and construction moldings, and various thermal insulation uses. Replicas of wood carvings, decorative parts, and moldings are produced from high-density, self-skinning foams (Fig. F-42). The insulation value of these foams makes them an ideal choice for insulating refrigerators and refrigerated trucks and railroad cars. They may be foamed in place for many architectural uses. They may be placed on vertical surfaces by spraying the reaction mixture through a nozzle. They have found use as flotation devices, packing, and structural reinforcement.

Fig. F-41. A polyurethane foam backing is used for this carpet.

Fig. F-42. This china cabinet of polyurethane, produced by Jasper Stylemasters Plastics, has the appearance, mass and feel of wood. (The Upjohn Company)

Rigid polyurethane is a closed cellular material produced by the reaction of TDI (prepolymer form) with polyethers and reactive blowing agents such as monofluorotrichloromethane (fluorocarbon). Diphenylmethane di-isocyanate (MDI) and polymethylene polyphenyl isocyanate (PAPI) are also used in some rigid foams. MDI foams have better dimensional stability, while PAPI foams have high temperature resistance.

Polyurethane-based caulks and sealants are inexpensive polyisocyanate materials used for encapsulation, and for construction and manufacturing uses. Various polyisocyanates are also useful adhesives. They produce strong bonds between flexible fabrics, rubbers, foams, or other materials.

Many blowing or foaming agents are explosive and toxic. When mixing or processing polyurethane foams, make certain that proper ventilation is provided.

Table F-13 gives some properties of urethane plastics. Six advantages and four disadvantages of polyurethane plastics are listed below.

Advantages of Polyurethane

1. High abrasion resistance
2. Good low-temperature capability
3. Wide variability in molecular structure
4. Possibility of ambient curing
5. Comparatively low cost
6. Prepolymers foam readily

Disadvantages of Polyurethane

1. Poor thermal capability
2. Toxic (isocyanates are used)
3. Poor weatherability
4. Subject to attack by solvents

Silicones (SI)

In organic chemistry, carbon is studied because of its capability of forming molecular structures with many other elements. Carbon is considered to be a reactive element. Carbon is capable of entering into more molecular combinations than is any other element. Life on earth is based on the element carbon.

The second most abundant element on earth is silicon. It has the same number of available bonding sites as carbon. Some scientists have speculated that life on other planets may be based on silicon. Others find this possibility hard to accept because silicon is an inorganic solid with a metallic appearance. Most of the earth's crust is composed of SiO_2 (silicon dioxide) in the form of sand, quartz, and flint.

The tetravalent capacity of silicon interested chemists as early as 1863. Friedrich Wohler, C. M. Crafts, Charles Friedel, F. S. Kipping, W. H. Carothers, and many others did work that led to the development of silicone polymers.

By 1943, Dow Corning Corporation was producing the first commercial silicone polymers in the United States. There are thousands of uses for these materials. The word *silicone* should be applied only to polymers containing silicon-oxygen-silicon bonding, however it is often used to denote any polymer containing silicon atoms.

In many carbon-hydrogen compounds, silicon may replace the element carbon. Methane (CH_4) may be changed to silane or silicomethane (SiH_4). Many structures similar to the aliphatic series of saturated hydrocarbons may be formed.

Table F-13. Properties of Polyurethane

Property	Cast Urethane	Urethane Elastomer
Molding qualities	Good	Good to excellent
Relative density	1.10–1.50	1.11–1.25
Tensile strength, MPa	1–69	31–58
(psi)	(175–10000)	(4500–8400)
Compressive strength, MPa	14	14
(psi)	(2000)	(2000)
Impact strength, Izod, J/mm	0.25 to flexible	Does not break
(ft lb/in)	(5)	
Hardness, Shore	10A–90D	30A–70D
Rockwell	M28, R60	
Thermal expansion, 10^{-4}/°C	25.4–50.8	25–50
Resistance to heat, °C	90–120	90
(°F)	(190–250)	(190)
Dielectric strength, V/mm	15750–19690	12990–35435
Dielectric constant (at 60 Hz)	4–7.5	5.4–7.6
Dissipation factor (at 60 Hz)	0.015–0.017	0.015–0.048
Arc resistance, s	0.1–0.6	0.22
Water absorption (24 h), %	0.02–1.5	0.7–0.9
Burning rate	Slow to self-extinguishing	Slow to self-extinguishing
Effect of sunlight	None to yellows	None to yellows
Effect of acids	Attacked	Dissolves
Effect of alkalies	Slight to attacked	Dissolves
Effect of solvents	None to slight	Resistant
Machining qualities	Excellent	Fair to excellent
Optical qualities	Transparent to opaque	Transparent to opaque

The following general types of bonds may be of value in understanding the formation of silicone polymers:

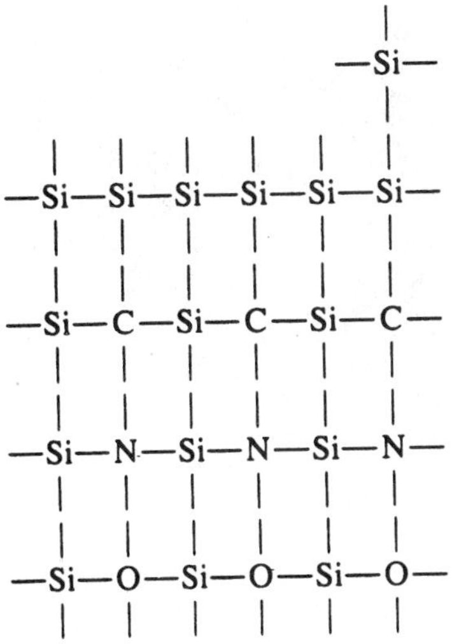

```
       |
     —Si—     Tetravalent silicone
       |

  |   |   |   |   |   |
—Si—Si—Si—Si—Si—Si—   Silanes
  |   |   |   |   |   |

  |   |   |   |   |   |
—Si—C—Si—C—Si—C—      Silcarbanes
  |   |   |   |   |   |

  |   |   |   |   |   |
—Si—N—Si—N—Si—N—      Sianzanes
  |   |   |   |   |   |

  |   |   |   |   |   |
—Si—O—Si—O—Si—O—      Siloxanes
  |   |   |   |   |   |
```

Compounds with only silicon and hydrogen atoms present are called *silanes,* When the silicon atoms are separated by carbon atoms, the structure is called a *silcarbane* (sil-CARB-ane). A *polysiloxane* is produced when more than one oxygen atom separates the silicon atoms in the chain.

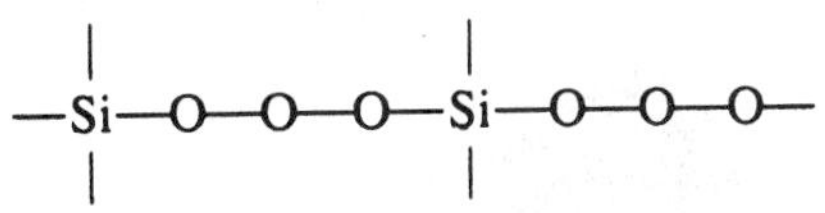

```
  |           |
—Si—O—O—O—Si—O—O—O—
  |           |
```

A polymerized silicone molecular chain could be based on the structure shown in Figure F-43 modified by radicals (R).

Many silicone polymers are based on chains, rings, or networks of alternating silicon and oxygen atoms. Common ones contain methyl, phenyl, or vinyl groups on the siloxane chain (Fig. F-44). A number of polymers are formed by varying the organic radical groups on the silicon chain. Many copolymers are also available.

The amount of energy needed to produce silicone plastics makes the price high. Silicone plastics may still be economical if one considers longer product life, higher service temperatures, and flexibility at temperature extremes.

Silicones are produced in five commercially available categories: fluids, compounds, lubricants, resins, and elastomers (rubber).

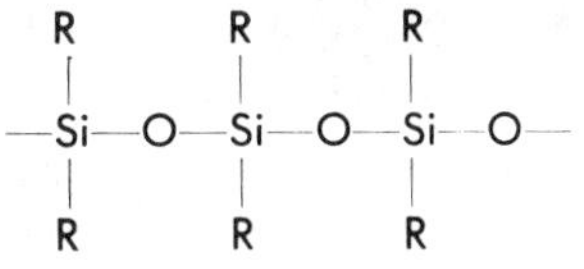

```
   R      R      R
   |      |      |
—Si—O—Si—O—Si—O—
   |      |      |
   R      R      R
```

Fig. F-43. An example of a polymerized silicone molecular chain.

```
  CH3    CH3    CH3
   |      |      |
—Si—O—Si—O—Si—O—
   |      |      |
  CH3    CH3    CH3
```

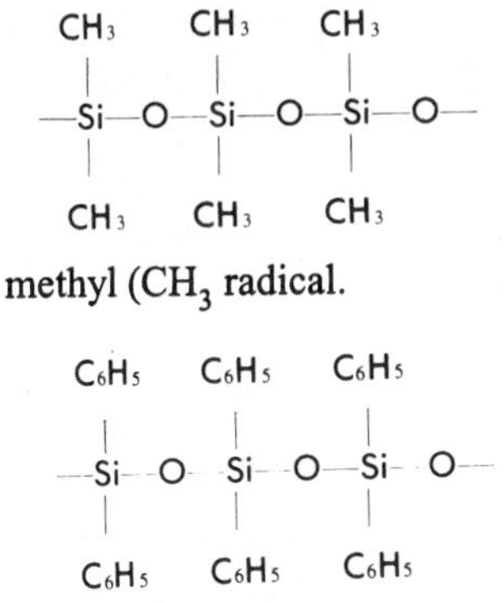

(A) Based on methyl (CH_3 radical.

```
  C6H5   C6H5   C6H5
   |      |      |
—Si—O—Si—O—Si—O—
   |      |      |
  C6H5   C6H5   C6H5
```

(B) Based on phenyl (C_6H_5) radical.

Fig. F-44. Two siloxane polymers.

Probably the best-known silicon plastics are associated with oils and ingredients for polishes. Examples are lens-cleaning tissues or waterrepellent fabrics treated with a thin-film coating of silicone.

Silicone fluids are added to some liquids to prevent foaming (antifoaming), prevent transmission of vibrations (damping), and improve electrical and thermal limits of various liquids. Fluid silicones are used as additives in paints, oils and inks, as mold-release agents, as finishes for glass and fabrics, and for paper coating.

Silicone compounds are usually granular or fibrous filled materials. Because of their outstanding electrical and thermal properties, mineral- and glass-filled silicone compounds are used for encapsulation of electronic components. Figure F-45 shows such uses of silicone coupounds.

As adhesives and sealants, silicone plastics are limited by their high cost. Their high service temperature and elastic properties make them useful for sealing, gasketing, caulking, and encapsulating, and for repairing all types of materials (Fig. F-46).

The chemical inertness of foamed silicone is useful in breast and facial implants in plastic surgery. Its main uses include electrical and thermal insulation of electrical wires and electrical components.

As lubricants, silicones are prized because they do not deteriorate at extreme service temperatures. Silicones are used for lubricating rubber, plastics, ball bearings, valves and vacuum pumps.

Silicone resins have many uses such as releasing agents for baking dishes. Silicone resins are also found in flexible, tough coatings used as high-temperature paints for engine manifolds and mufflers. Their waterproofing capability makes them useful in treating masonry and concrete walls.

Excellent waterproofing, thermal, and electrical properties make these resins valuable for electrical insulation in motors and generators.

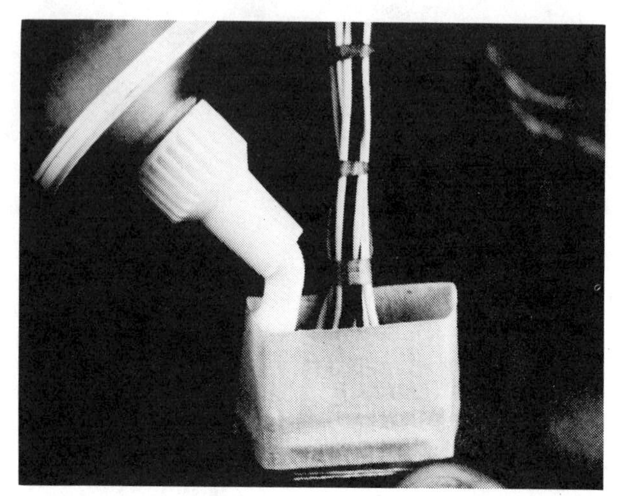

(A) Potting a small electrical component with silicone compound.

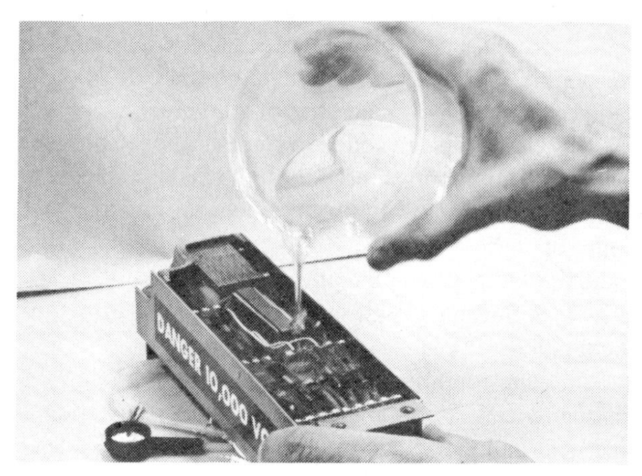

(B) Use of silicone casting resin for potting electronic components.

Fig. F-45. Uses of silicone compounds. (Dow-Corning Corp.)

Laminates reinforced with glass cloth find uses for structural parts, ducts, radomes and electronic panel boards. These silicone laminates are characterized by their excellent dielectric and thermal properties and strength-to-mass ratio.

Diatomaceous earth, glass fiber, or asbestos may be used as fillers in preparing a premix or putty for the molding of small parts from silicone resins.

Some of the best known silicones are in the form of elastomers. Few industrial rubbers or elastomers can withstand long exposure to ozone (O_3) or hot mineral oils. Silicone "rubbers" are stable at elevated temperatures and remain flexible when exposed to ozone or oils.

Silicone elastomers find use as artificial organs, O-rings, gaskets, and diaphragms. They are also used as flexible molds forecasting of plastics and low-melting-point metals.

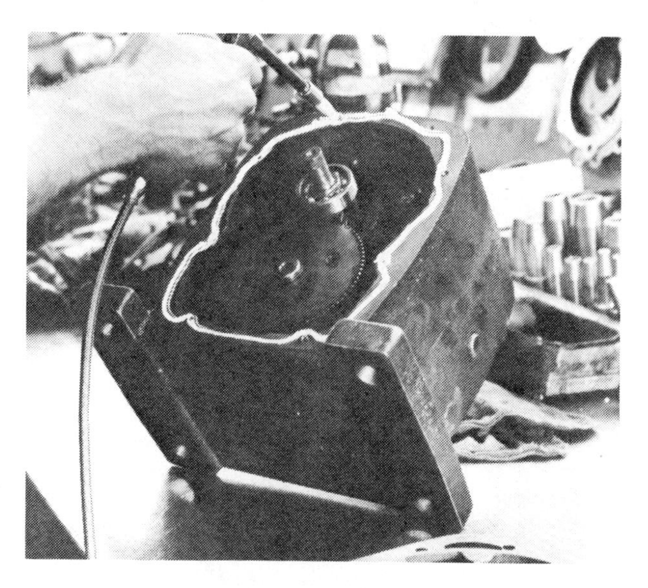

Fig. F-46. Silicone sealant is used in the assembly of this gear box. (Dow-Corning Corp.)

Room-temperature vulcanizing (RTV) elastorners are used to copy intricate molded parts, seal joints, and adhere parts (Fig. F-47).

Silly Putty and *Crazy Clay* are novelty silicone products (Fig. F-48). This bouncing putty is a silicone elastomer that is also used for damping noise and as a

(A) Intricate details, such as this picture-frame decoration, may be reproduced with silicone RTV.

(B) Silicone reproduces fine detail and allows severe undercuts.

Fig. F-47. Use of room-temperature-vulcanizing (RTV) silicone in molding. (Dow-Corning Corp.)

Fig. F-48. A novelty toy made of a silicone compound.

sealing and filling compound. A hard bouncing putty will rebound to 80 percent of the height from which it is dropped. Other super-rebounding novelty items are produced from this compound.

Silicone molding compounds may be processed in the same way as other thermosetting organic plastics. Silicones are often used in resin form as castings, coatings, adhesives, or laminating compounds.

Research and development is being carried out on various other elements with covalent bonding capacity. Boron, aluminum, titanium, tin, lead, nitrogen, phosphorus, arsenic, sulphur, and selenium may be considered for inorganic or semiorganic plastics. The formulas in Figure F-49 show a few of many possible chemical structures of inorganic or semi-organic plastics.

Table F-14 gives properties of silicone plastics. Seven advantages and three disadvantages of silicone are shown below.

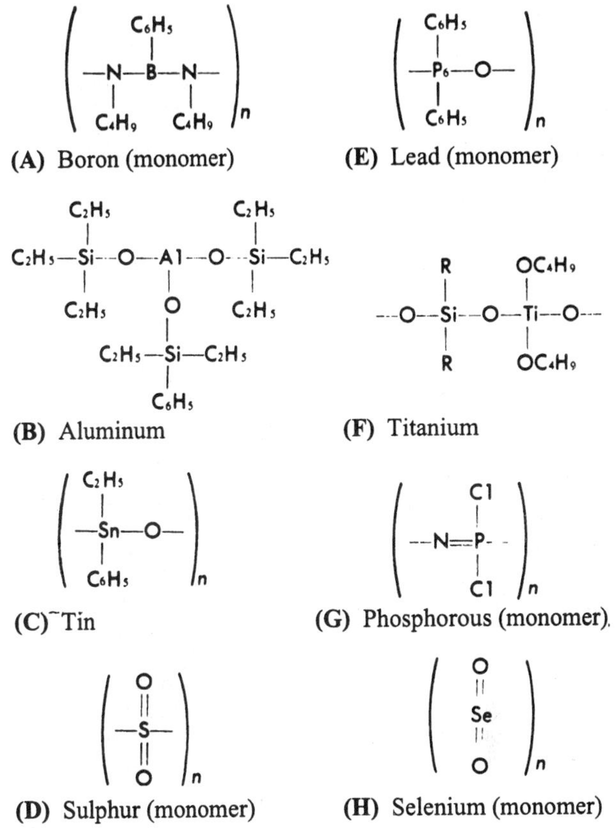

(A) Boron (monomer) — $(-N(C_4H_9)-B(C_6H_5)-N(C_4H_9)-)_n$

(E) Lead (monomer) — $(-P_b(C_6H_5)_2-O-)_n$

(B) Aluminum — $C_2H_5-Si(C_2H_5)_2-O-Al(O-Si(C_2H_5)_2(C_6H_5))-O-Si(C_2H_5)_2-C_2H_5$

(F) Titanium — $-O-SiR_2-O-Ti(OC_4H_9)_2-O-$

(C)$^-$Tin — $(-Sn(C_2H_5)(C_6H_5)-O-)_n$

(G) Phosphorous (monomer) — $(-N{=}PCl_2-)_n$

(D) Sulphur (monomer) — $(-S(=O)_2-)_n$

(H) Selenium (monomer) — $(Se(=O)_2)_n$

Fig. F-49. Possible chemical structures of inorganic or semi-inorganic plastics.

Advantages of Silicones

1. Wide range of thermal capability, from −100 °F to 600 °F [−73 °C to 315 °C]
2. Good electrical traits
3. Wide variation in molecular structure (flexible or rigid forms)
4. Available in transparent grades
5. Low water absorption
6. Available in flame-retardant grades
7. Good chemical resistance

Disadvantages of Silicones

1. Low strength
2. Subject to attack by halogenated solvents
3. Comparatively high cost

Table F-14. Properties of Silicones

Property	Cast Resin (Including RTV)	Molding Compounds (Mineral-Filled)	Molding Compounds (Glass-Filled)
Molding qualities	Excellent	Excellent	Good
Relative density	0.99–1.50	1.7–2	1.68–2
Tensile strength, MPa	2–7	28–41	28–45
(psi)	(350–1 000)	(4 000–6 000)	(4 000–6 500)
Compressive strength. MPa	0.7	90–124	69–103
(psi)	(100)	(13 000–18 000)	(10 000–15 000)
Impact strength, Izod, J/mm		0.013–0.018	0.15–0.75
(ft lb/in)		(0.26–0.36)	(3–15)
Hardness	Shore A15–A65	Rockwell M71–M95	Rockwell M84
Thermal expansion, 10^{-4}/°C	20–79	5–10	0.61–0.76
Resistance to heat, °C	260	315	315
(°F)	(500)	(600)	(600)
Dielectric strength, V/mm	21 665	7 875–15 750	7 875–15 750
Dielectric constant (at 60 Hz)	2.75–4.20	3.5–3.6	3.3–5.2
Dissipation factor (at 60 Hz)	0.001–0.025	0.004–0.005	0.004–0.030
Arc resistance, s	115–130	250–420	150–205
Water absorption, %	0.12 (7 days)	0.08–0.13 (24 h)	0.1–0.2 (24 h)
Burning rate	Self-extinguishing	None to slow	None to slow
Effect of sunlight	None	None to slight	None to slight
Effect of acids	Slight to severe	Slight	Slight
Effect of alkalies	Moderate to severe	Slight to marked	Slight to marked
Effect of solvents	Swells in some	Attacked by some	Attacked by some
Machining qualities	None	Fair	Fair
Optical qualities	Transparent to opaque	Opaque	Opaque

G

Useful Tables

ENGLISH METRIC CONVERSION

	If You Know	You Can Get	If You Multiply By*
LENGTH	Inches	Millimetres (mm)	25.4
	Millimetres	Inches	0.04
	Inches	Centimetres (cm)	2.54
	Centimetres	Inches	0.4
	Inches	Metres (m)	0.0254
	Metres	Inches	39.37
	Feet	Centimetres	30.5
	Centimetres	Feet	4.8
	Feet	Metres	0.305
	Metres	Feet	3.28
	Miles	Kilometres (km)	1.61
	Kilometre	Miles	0.62
AREA	Inches^2	Millimetres^2 (mm)	645.2
	Millimetres^2	Inches^2	0.0016
	Inches^2	Centimetres^2 (cm^2)	6.45
	Centimetres^2	Inches^2	0.16
	Foot^2	Metres^2 (m^2)	0.093
	Metres^2	Foot^2	10.76
CAPACITY-VOLUME	Ounces	Millilitres (ml)	30
	Millilitres	Ounces	0.034
	Pints	Litres (l)	0.47
	Litres	Pints	2.1
	Quarts	Litres	0.95
	Litres	Quarts	1.06
	Gallons	Litres	3.8
	Litres	Gallons	0.26
	Cubic Inches	Litres	0.0164
	Litres	Cubic Inches	61.03
	Cubic Inches	Cubic Centimetres (cc)	16.39
	Cubic Centimetres	Cubic Inches	0.061
WEIGHT (MASS)	Ounces	Grams	28.4
	Grams	Ounces	0.035
	Pounds	Kilograms	0.45
	Kilograms	Pounds	2.2
FORCE	Ounce	Newtons (N)	0.278
	Newtons	Ounces	35.98
	Pound	Newtons	4.448
	Newtons	Pound	0.225
	Newtons	Kilograms (kg)	0.102
	Kilograms	Newtons	9.807
ACCELERATION	Inch/Sec^2	Metre/Sec^2	.0254
	Metre/Sec^2	Inch/Sec^2	39.37
	Foot/Sec^2	Metre/Sec^2 (m/s^2)	0.3048
	Metre/Sec^2	Foot/Sec^2	3.280
TORQUE	Pound-inch (Inch-Pound)	Newton-Metres (N-M)	0.113
	Newton-Metres	Pound-Inch	8.857
	Pound-Foot (Foot-Pound)	Newton-Metres	1.356
	Newton-Metres	Pound-Foot	.737
PRESSURE	Pound/sq. in. (PSI)	Kilopascals (kPa)	6.895
	Kilopascals	Pound/sq. in.	0.145
	Inches of Mercury (Hg)	Kilopascals	3.377
	Kilopascals	Inches of Mercury (Hg)	0.296
FUEL PERFORMANCE	Miles/gal	Kilometres/litre (km/l)	0.425
	Kilometres/litre	Miles/gal	2.352
VELOCITY	Miles/hour	Kilometres/hr (km/h)	1.609
	Kilometres/hour	Miles/hour	0.621
TEMPERATURE	Fahrenheit Degrees	Celsius Degrees	5/9 (F° -32)
	Celsius Degrees	Fahrenheit Degrees	9/5 (C° +32) = F

*Approximate Conversion Factors to be used where precision calculations are *not* necessary

CONVERT CENTIGRADE TEMPERATURE TO FAHRENHEIT AND VICE VERSA

Centigrade C° = 5/9 (F° – 32) TEMPERATURE CONVERSION TABLES To Fahrenheit F° = (9/5XC°) + 32

C.		F.	C.		F.	C.		F.	C.		F.
– 17.8	**0**	32	8.89	**48**	118.4	35.6	**96**	204.8	271	**520**	968
– 17.2	**1**	33.8	9.44	**49**	120.2	36.1	**97**	206.6	277	**530**	986
– 16.7	**2**	35.6	10.0	**50**	122.0	36.7	**98**	208.4	282	**540**	1004
– 16.1	**3**	37.4	10.6	**51**	123.8	37.2	**99**	210.2	288	**550**	1022
– 15.6	**4**	39.2	11.1	**52**	125.6	37.8	**100**	212.0	293	**560**	1040
– 15.0	**5**	41.0	11.7	**53**	127.4	38	**100**	212	299	**570**	1058
– 14.4	**6**	42.8	12.2	**54**	129.2	43	**110**	230	304	**580**	1076
– 13.9	**7**	44.6	12.8	**55**	131.0	49	**120**	248	310	**590**	1094
– 13.3	**8**	46.4	13.3	**56**	132.8	54	**130**	266	316	**600**	1112
– 12.8	**9**	48.2	13.9	**57**	134.6	60	**140**	284	321	**610**	1130
– 12.2	**10**	50.0	14.4	**58**	136.4	66	**150**	302	327	**620**	1148
– 11.7	**11**	51.8	15.0	**59**	138.2	71	**160**	320	332	**630**	1166
– 11.1	**12**	53.6	15.6	**60**	140.0	77	**170**	338	338	**640**	1184
– 10.6	**13**	55.4	16.1	**61**	141.8	82	**180**	356	343	**650**	1202
– 10.0	**14**	57.2	16.7	**62**	143.6	88	**190**	374	349	**660**	1220
– 9.44	**15**	59.0	17.2	**63**	145.4	93	**200**	392	354	**670**	1238
– 8.89	**16**	60.8	17.8	**64**	147.2	99	**210**	410	360	**680**	1256
– 8.33	**17**	62.6	18.3	**65**	149.0	100	**212**	413	366	**690**	1274
– 7.78	**18**	64.4	18.9	**66**	150.8	104	**220**	428	371	**700**	1292
– 7.22	**19**	66.2	19.4	**67**	152.6	110	**230**	446	377	**710**	1310
– 6.67	**20**	68.0	20.0	**68**	154.4	116	**240**	464	382	**720**	1328
– 6.11	**21**	69.8	20.6	**69**	156.2	121	**250**	482	388	**730**	1346
– 5.56	**22**	71.6	21.1	**70**	158.0	127	**260**	500	393	**740**	1364
– 5.00	**23**	73.4	21.7	**71**	159.8	132	**270**	518	399	**750**	1382
– 4.44	**24**	75.2	22.2	**72**	161.6	138	**280**	536	404	**760**	1400
– 3.89	**25**	77.0	22.8	**73**	163.4	143	**290**	554	410	**770**	1418
– 3.33	**26**	78.8	23.3	**74**	165.2	149	**300**	572	416	**780**	1436
– 2.78	**27**	80.6	23.9	**75**	167.0	154	**310**	590	421	**790**	1454
– 2.22	**28**	82.4	24.4	**76**	168.8	160	**320**	608	427	**800**	1472
– 1.67	**29**	84.2	25.0	**77**	170.6	166	**330**	626	432	**810**	1490
– 1.11	**30**	86.0	25.6	**78**	172.4	171	**340**	644	438	**820**	1508
– 0.56	**31**	87.8	26.1	**79**	174.2	177	**350**	662	443	**830**	1526
– 0	**32**	89.6	26.7	**80**	176.0	182	**360**	680	449	**840**	1544
0.56	**33**	91.4	27.2	**81**	177.8	188	**370**	698	454	**850**	1562
1.11	**34**	93.2	27.8	**82**	179.6	193	**380**	716	460	**860**	1580
1.67	**35**	95.0	28.3	**83**	181.4	199	**390**	734	466	**870**	1598
2.22	**36**	96.8	28.9	**84**	183.2	204	**400**	752	471	**880**	1616
2.78	**37**	98.6	29.4	**85**	185.0	210	**410**	770	477	**890**	1634
3.33	**38**	100.4	30.0	**86**	186.8	216	**420**	788	482	**900**	1652
3.89	**39**	102.2	30.6	**87**	188.6	221	**430**	806	488	**910**	1670
4.44	**40**	104.0	31.1	**88**	190.4	227	**440**	824	493	**920**	1688
5.00	**41**	105.8	31.7	**89**	192.2	232	**450**	842	499	**930**	1706
5.56	**42**	107.6	32.2	**90**	194.0	238	**460**	860	504	**940**	1724
6.11	**43**	109.4	32.8	**91**	195.8	243	**470**	878	510	**950**	1742
6.67	**44**	111.2	33.3	**92**	196.7	249	**480**	896	516	**960**	1760
7.22	**45**	113.0	33.9	**93**	199.4	254	**490**	914	521	**970**	1778
7.78	**46**	114.8	34.4	**94**	201.2	260	**500**	932	527	**980**	1796
8.33	**47**	116.6	35.0	**95**	203.0	266	**510**	950	532	**990**	1814

DECIMAL EQUIVALENTS OF FRACTIONS OF ONE INCH

Fraction	Decimal
1/64	.015625
1/32	.031250
3/64	.046875
1/16	.062500
5/64	.078125
3/32	.093750
7/64	.109375
1/8	.125000
9/64	.140625
5/32	.156250
11/64	.171875
3/16	.187500
13/64	.203125
7/32	.218750
15/64	.234375
1/4	.250000
17/64	.265625
9/32	.281250
19/64	.296875
5/16	.312500
21/64	.328125
11/32	.343750
23/64	.359375
3/8	.375000
25/64	.390625
13/32	.406250
27/64	.421875
7/16	.437500
29/64	.453125
15/32	.468750
31/64	.484375
1/2	.500000
33/64	.515625
17/32	.531250
35/64	.546875
9/16	.562500
37/64	.578125
19/32	.593750
39/64	.609375
5/8	.625000
41/64	.640625
21/32	.656250
43/64	.671875
11/16	.687500
45/64	.703125
23/32	.718750
47/64	.734375
3/4	.750000
49/64	.765625
25/32	.781250
51/64	.796875
13/16	.812500
53/64	.828125
27/32	.843750
55/64	.859375
7/8	.875000
57/64	.890625
29/32	.906250
59/64	.890625
15/16	.937500
61/64	.953125
31/32	.968750
63/64	.984375
1	1.000000

STANDARD DRAFT ANGLES

Depth	¼°	½°	1°	1½°	2°	2½°	3°	5°	7°	8°	10°	12°	15°	Depth
1/32	.0001	.0003	.0005	.0008	.0011	.0014	.0016	.0027	.0038	.0044	.0055	.0066	.0084	1/32
1/16	.0003	.0006	.0011	.0016	.0022	.0027	.0033	.0055	.0077	.0088	.0110	.0133	.0168	1/16
3/32	.0004	.0008	.0016	.0025	.0033	.0041	.0049	.0082	.0115	.0132	.0165	.0199	.0251	3/32
1/8	.0005	.0010	.0022	.0033	.0044	.0055	.0066	.0109	.0153	.0176	.0220	.0266	.0335	1/8
3/16	.0008	.0016	.0033	.0049	.0065	.0082	.0098	.0164	.0230	.0263	.0331	.0399	.0502	3/16
1/4	.0011	.0022	.0044	.0066	.0087	.0109	.0131	.0219	.0307	.0351	.0441	.0531	.0670	1/4
5/16	.0014	.0027	.0055	.0082	.0109	.0137	.0164	.0273	.0384	.0439	.0551	.0664	.0837	5/16
3/8	.0016	.0033	.0065	.0098	.0131	.0164	.0197	.0328	.0460	.0527	.0661	.0797	.1005	3/8
7/16	.0019	.0038	.0076	.0115	.0153	.0191	.0229	.0383	.0537	.0615	.0771	.0930	.1172	7/16
1/2	.0022	.0044	.0087	.0131	.0175	.0218	.0262	.0438	.0614	.0703	.0882	.1063	.1340	1/2
5/8	.0027	.0054	.0109	.0164	.0218	.0273	.0328	.0547	.0767	.0878	.1102	.1329	.1675	5/8
3/4	.0033	.0065	.0131	.0196	.0262	.0328	.0393	.0656	.0921	.1054	.1322	.1595	.2010	3/4
7/8	.0038	.0076	.0153	.0229	.0306	.0382	.0459	.0766	.1074	.1230	.1543	.1860	.2345	7/8
1	.0044	.0087	.0175	.0262	.0349	.0437	.0524	.0875	.1228	.1405	.1763	.2126	.2680	1
1 1/4	.0055	.0109	.0218	.0327	.0437	.0546	.0655	.1094	.1535	.1756	.2204	.2657	.3349	1 1/4
1 1/2	.0064	.0131	.0262	.0393	.0524	.0655	.0786	.1312	.1842	.2108	.2645	.3188	.4019	1 1/2
1 3/4	.0076	.0153	.0305	.0458	.0611	.0764	.0917	.1531	.2149	.2460	.3085	.3720	.4689	1 3/4
2	.0087	.0175	.0349	.0524	.0698	.0873	.1048	.1750	.2456	.2810	.3527	.4251	.5359	2
Depth	¼°	½°	1°	1½°	2°	2½°	3°	5°	7°	8°	10°	12°	15°	Depth

CONVERSION OF SPECIFIC GRAVITY TO GRAMS PER CUBIC INCH

16.39 x Specific Gravity = Grams/In.³

Specific Gravity	Grams/In.³	Specific Gravity	Grams/In.³
1.20	19.7	1.82	29.8
1.22	20.0	1.84	30.2
1.24	20.3	1.86	30.5
1.26	20.7	1.88	30.8
1.28	21.0	1.90	31.1
1.30	21.3	1.92	31.5
1.32	21.6	1.94	31.8
1.34	22.0	1.96	32.1
1.36	22.3	1.98	32.5
1.38	22.6	2.00	32.8
1.40	22.9	2.02	33.1
1.42	23.3	2.04	33.4
1.44	23.6	2.06	33.8
1.46	23.9	2.08	34.1
1.48	24.3	2.10	34.4
1.50	24.6	2.12	34.7
1.52	24.9	2.14	35.1
1.54	25.2	2.16	35.4
1.56	25.6	2.18	35.7
1.58	25.9	2.20	36.1
1.60	26.2	2.22	36.4
1.62	26.6	2.24	36.7
1.64	26.9	2.26	37.0
1.66	27.2	2.28	37.4
1.68	27.5	2.30	37.7
1.70	27.9	2.32	38.0
1.72	28.2	2.34	38.4
1.74	28.5	2.36	38.7
1.76	28.8	2.38	39.0
1.78	29.2	2.40	39.3
1.80	29.5		

To Determine the Cost/Cu./In.:
Price/Lb. x Sp. Gravity x .03163.
$1.32 x 1.76 x .03163 = $0.09/Cu./In.

DIAMETERS AND AREAS OF CIRCLES

Diam.	Area	Diam.	Area	Diam.	Area	Diam.	Area
1/64"	.00019	7/8"	2.7612	11/16"	17.257	7/8"	61.862
1/32	.00077	15/16	2.9483	3/4	17.721		
3/64	.00173			13/16	18.190	9-"	63.617
1/16	.00307	2-"	3.1416	7/8	18.665	1/8	65.397
3/32	.00690	1/16	3.3410	15/16	19.147	1/4	67.201
1/8	.01227	1/8	3.5466			3/8	69.029
5/32	.01917	3/16	3.7583	5-"	19.635	1/2	70.882
3/16	.02761	1/4	3.9761	1/16	20.129	5/8	72.760
7/32	.03758	5/16	4.2000	1/8	20.629	3/4	74.662
1/4	.04909	3/8	4.4301	3/16	21.125	7/8	76.589
9/32	.06213	7/16	4.6664	1/4	21.648	10-"	78.540
5/16	.07670	1/2	4.9087	5/16	22.166	1/8	80.516
11/32	.09281	9/16	5.1572	3/8	22.691	1/4	82.516
3/8	.11045	5/8	5.4119	7/16	23.211	3/8	84.541
13/32	.12962	11/16	5.6727	1/2	23.758	1/2	86.590
7/16	.15033	3/4	5.9396	9/16	24.301	5/8	88.664
15/32	.17257	13/16	6.2126	5/8	24.850	3/4	90.763
1/2	.19635	7/8	6.4918	11/16	25.406	7/8	92.886
17/32	.22165	15/16	6.7771	3/4	25.967		
9/16	.24850			13/16	26.535	11-"	95.033
19/32	.27688	3-"	7.0686	7/8	27.109	1/2	103.87
5/8	.30680	1/16	7.3662	15/16	27.688		
21/32	.33824	1/8	7.6699			12-"	113.10
11/16	.37122	3/16	7.9798	6-"	28.274	1/2	122.72
23/32	.40574	1/4	8.2958	1/8	29.465		
3/4	.44179	5/16	8.6179	1/4	30.680	13-"	132.73
25/32	.47937	3/8	8.9462	3/8	31.919	1/2	143.14
13/16	.51849	7/16	9.2806	1/2	33.183		
27/32	.55914	1/2	9.6211	5/8	34.472	14-"	153.94
7/8	.60132	9/16	9.9678	3/4	35.785	1/2	165.13
29/32	.64504	5/8	10.321	7/8	37.122		
15/16	.69029	11/16	10.680			15-"	176.71
31/32	.73708	3/4	11.045	7-"	38.485	1/2	188.69
		13/16	11.416	1/8	39.871		
1-"	.7854	7/8	11.793	1/4	41.282	16-"	201.06
1/16	.8866	15/16	12.177	3/8	42.718	1/2	213.82
1/8	.9940			1/2	44.179		
3/16	1.1075	4-"	12.566	5/8	45.664	17-"	226.98
1/4	1.2272	1/16	12.962	3/4	47.173	1/2	240.53
5/16	1.3530	1/8	13.364	7/8	48.707		
3/8	1.4849	3/16	13.772			18-"	254.47
7/16	1.6230	1/4	14.186	8-"	50.265	1/2	268.80
1/2	1.7671	5/16	14.607	1/8	51.849		
9/16	1.9175	3/8	15.033	1/4	53.456	19-"	283.53
5/8	2.0739	7/16	15.466	3/8	55.088	1/2	298.65
11/16	2.2465	1/2	15.904	1/2	56.745		
3/4	2.4053	9/16	16.349	5/8	58.426	20-"	314.16
13/16	2.5802	5/8	16.800	3/4	60.132	1/2	330.06

STEAM TEMPERATURE VERSUS GAUGE PRESSURE

Gauge Pressure Lbs.	Temp. Deg. F
50	297.5
55	302.4
60	307.1
65	311.5
70	315.8
75	319.8
80	323.6
85	327.4
90	331.1
95	334.3
100	337.7
105	341.0
110	344.0
115	347.0
120	350.0
125	353.0
130	356.0
135	358.0
140	361.0
145	363.0
150	365.6
155	368.0
160	370.3
165	372.7
170	374.9
175	377.2
180	379.3
185	381.4
190	383.5
195	385.7
200	387.5

WEIGHT OF 1000 PIECES IN POUNDS BASED ON WEIGHT OF ONE PIECE IN GRAMS

Weight Per Piece in Grams	Weight Per 1000 Pieces in Pounds	Weight Per Piece in Grams	Weight Per 1000 Pieces in Pounds
1	2.2	51	112.3
2	4.4	52	114.5
3	6.6	53	116.7
4	8.8	54	118.9
5	11.0	55	121.1
6	13.2	56	123.3
7	15.4	57	125.5
8	17.6	58	127.7
9	19.8	59	129.9
10	22.0	60	132.1
11	24.2	61	134.3
12	26.4	62	136.5
13	28.6	63	138.7
14	30.8	64	140.9
15	33.0	65	143.1
16	35.2	66	145.3
17	37.4	67	147.5
18	39.6	68	149.7
19	41.8	69	151.9
20	44.0	70	154.1
21	46.2	71	156.3
22	48.4	72	158.5
23	50.6	73	160.7
24	52.8	74	162.9
25	55.0	75	165.1
26	57.2	76	167.4
27	59.4	77	169.6
28	61.6	78	171.8
29	63.8	79	174.0
30	66.0	80	176.2
31	68.2	81	178.4
32	70.4	82	180.6
33	72.6	83	182.8
34	74.8	84	185.0
35	77.0	85	187.2
36	79.2	86	189.4
37	81.4	87	191.6
38	83.7	88	193.8
39	85.9	89	196.0
40	88.1	90	198.2
41	90.3	91	200.4
42	92.5	92	202.6
43	94.7	93	204.8
44	96.9	94	207.0
45	99.1	95	209.2
46	101.3	96	211.4
47	103.5	97	213.6
48	105.7	98	215.8
49	107.9	99	218.0
50	110.1	100	220.2

EQUIVALENT WEIGHTS
1 Gram = .0353 Oz.
.0625 Pounds = 1 Ounce = 28.3 Grams
454 Grams = 1 Pound

LENGTH EQUIVALENTS

Millimeters to Inches

Millimeters	Inches	Millimeters	Inches	Millimeters	Inches
1	.03937	34	1.33860	67	2.63779
2	.07874	35	1.37795	68	2.67716
3	.11811	36	1.41732	69	2.71653
4	.15748	37	1.45669	70	2.75590
5	.19685	38	1.49606	71	2.79527
6	.23622	39	1.53543	72	2.83464
7	.27559	40	1.57480	73	2.87401
8	.31496	41	1.61417	74	2.91338
9	.35433	42	1.65354	75	2.95275
10	.39370	43	1.69291	76	2.99212
11	.43307	44	1.73228	77	3.03149
12	.47244	45	1.77165	78	3.07086
13	.51181	46	1.81102	79	3.11023
14	.55118	47	1.85039	80	3.14960
15	.59055	48	1.88976	81	3.18897
16	.62992	49	1.92913	82	3.22834
17	.66929	50	1.96850	83	3.26771
18	.70866	51	2.00787	84	3.30708
19	.74803	52	2.04724	85	3.34645
20	.78740	53	2.08661	86	3.38582
21	.82677	54	2.12598	87	3.42519
22	.86614	55	2.16535	88	3.46456
23	.90551	56	2.20472	89	3.50393
24	.94488	57	2.24409	90	3.54330
25	.98425	58	2.28346	91	3.58267
26	1.02362	59	2.32283	92	3.62204
27	1.06299	60	2.36220	93	3.66141
28	1.10236	61	2.40157	94	3.70078
29	1.14173	62	2.44094	95	3.74015
30	1.18110	63	2.48031	96	3.77952
31	1.22047	64	2.51968	97	3.81889
32	1.25984	65	2.55905	98	3.85826
33	1.29921	66	2.59842	99	3.89763
				100	3.93700

VOLUME EQUIVALENTS

1 c.c. = .061 cu. in.
1 cu. in. = 16.387 c.c.

Appendix H

Sources of Help and Bibliography

Sources of Help

The following alphabetical list of service organizations, standards and specifications groups, trade associations, professional societies, reference, and U.S. governmental agencies may serve as sources for further information:

American Chemical Society
1155 16th Street, NW
Washington, DC 20036
(202) 872-4600

American Conference of Governmental Industrial Hygienists (ACGIH)
6500 Glenway Avenue
Cincinnati, OH 45201
(513) 661-7881

American Industrial Hygiene Association (AIHA)
66 S Miller Road
Akron, OH 44130
(216) 762-7294

American Insurance Association (AIA)
85 John Street
New York, NY 10038
(212) 669-0400

American Medical Association (AMA)
535 N Dearborn Street
Chicago, IL 60610
(312) 645-5003

American National Standards Institute (ANSI)
1430 Broadway
New York, NY 10018
(212) 354-3300

American Petroleum Institute
1801 K Street, NW
Washington, DC 20006
(202) 682-8000

(The) American Society for Testing and Materials (ASTM)
1916 Race Street
Philadelphia, PA 19103
(215) 299-5400

(The) American Society of Mechanical Engineers (ASME)
United Engineering Center
345 E 47th Street
New York, NY 10017
(212) 705-7722

American Society of Safety Engineers
850 Busse Highway
Park Ridge, IL 60068
(312) 692-4121

Center for Plastics Recycling Research (CPRR)
PO Box 189
Kennett Square, PA 19348
(215) 444-0659

Chemical Manufacturers Association
2501 M Street, NW
Washington, DC 20037
(202) 887-1100

Defense Standardization Program Office (DSPO)
5203 Leesburg Pike, Suite 1403
Falls Church, VA 22041-3466

Department of Defense (DOD)
Office for Research and Engineering
Washington, DC 20301
(202) 545-6700

Department of Transportation (DOT)
Hazardous Materials Transportation
400 7th Street, SW
Washington, DC 20590
(202) 426-4000

Environmental Protection Agency (EPA)
401 M Street, SW
Washington, DC 20460
(202) 829-3535

Factory Mutual Engineering Corporation
1151 Providence Highway
Norwood, MA 02062
(617) 762-4300

Federal Emergency Management Agency
PO Box 8181
Washington, DC 20024
(202) 646-2500

Federal Register
U.S. Government Printing Office
Superintendent of Documents
Washington, DC 20402
(202) 783-3238

Food and Drug Administration (FDA)
200 Independence Avenue
Washington, DC 20204
(202) 245-6296

General Services Administration (GSA)
Federal Supply Service
18th and F Streets
Washington, DC 20406
(202) 566-1212

Global Engineering Documentation Services, Inc.
3301 W MacArthur Boulevard
Santa Ana, CA 92704
(714) 540-9870

Industrial Health Foundation Inc. (IHF)
34 Penn Circle
Pittsburgh, PA 15232
(412) 363-6600

Instrument Society of America
400 Stanwix Street
Pittsburgh, PA 15222
(412) 261-4300

International Organization for Standardization (ISO)
1 rue de Varembe,
CH 1211
Geneve 20 Switzerland/Suisse

Leidner, Jacob. *Plastics Waste Recovery of Economic Value.* New York: Marcel Dekker, Inc., 1981.

Manufacturing Chemists Association, Inc.
1825 Connecticut Avenue, NW
Washington, DC 20009
(202) 887-1100

National Association of Manufacturers
1776 F Street, NW
Washington, DC 20006
(202) 737-8551

National Bureau of Standards (NBS)
Standards Information & Analysis Section
Standards Information Service (SIS)
Building 225, Room B 162
Washington, DC 20234
(301) 921-1000

National Conference on Weights and Measures
c/o National Bureau of Standards
Washington, DC 20234
(301) 921-1000

National Fire Protection Association (NFPA)
470 Atlantic Avenue
Boston, MA 02210
(617) 770-3000

National Institute for Occupational Safety and Health (NIOSH)
U.S. Department of Health, Education, and Welfare
Parklawn Building
5600 Fishers Lane
Rockville, MD 20852
(301) 472-7134

National Safety Council
444 N Michigan Avenue
Chicago, IL, 60611
(312) 527-4800

Navy Publications and Printing Service Office
700 Robbins Avenue
Philadelphia, PA 19111
(215) 697-2000

Occupational Safety and Health Administration (OSHA)
U.S. Department of Labor
Department of Labor Building
Connecticut Avenue, NW
Washington, DC 20210
(202) 523-9361

Office of the Federal Register
1100 "L" Street NW, Rm 8401
Washington, DC 20408
(202) 523-5240

Plastics Education Foundation
Society of Plastics Engineers, Inc.
14 Fairfield Drive
Brookfield Center, CT 06805
(203) 775-0471

Safety Standards
U.S. Department of Labor
Government Printing Office (GPO)
Washington, DC 20402
(202) 783-3238

Society of Plastics Engineers, Inc.
Plastics Education Foundation
14 Fairfield Drive
Brookfield Center, CT 06805
(203) 775-0471

(The) Society of the Plastics Industry, Inc. (SPI)
1025 Connecticut Avenue, NW
Ste 409
Washington, DC 20036
(202) 822-6700

Underwriters Laboratories (UL)
333 Pfingston Road
Northbrook, IL 60062
(312) 272-8800

U.S. Government Printing Office
Superintendent of Documents
Washington, DC 20402
(202) 783-3238

Bibliography

The following bibliography list may be useful for further study and more detailed discussion of selected topics presented:

Advanced Composites: Conference Proceedings, American Society for Metals, December 2–4, 1985.

Allegri, Theodore. *Handling and Management of Hazardous Materials and Waste.* New York: Chapman and Hall, 1986.

Bernhardt, Ernest. *CAE Computer Aided Engineering for Injection Molding.* New York: Hanser Publishers, 1983.

Billmeyer, Fred W. *Textbook of Polymer Science.* 3rd ed. New York: Wiley, 1984.

Brooke, Lindsay. "Cars of 2000: Tomorrow Rides Again!" *Automotive Industries,* May 1986, pp 50–67.

Broutman, L., and R. Krock. *Composite Materials.* 6 vols. New York: Academic Press, 1985.

Budinski, Kenneth. *Engineering Materials: Properties and Selection.* 2nd ed. Reston: Reston Publishing Company, Inc., 1983.

Carraher, Charles E., Jr., and James Moore. *Modification of Polymers.* New York: Plenum Press, 1983.

"Chemical Emergency Preparedness Program Interim Guidance," Revision 1, #9223.01A. Washington, DC: United States Environmental Protection Agency, 1985.

Composite Materials Technology, Society of Automotive Engineers, 1986.

"Defense Standardization Manual: Defense Standardization and Specification Program Policies, Procedures and Instruction," DOD 4120. 3-M, August 1978.

Dreger, Donald. "Design Guidelines of Joining Advanced Composites," *Machine Design,* May 8, 1980, pp 89-93.

Dym, Joseph. *Product Design with Plastics: A Practical Manual.* New York: Industrial Press, 1983.

Ehrenstein, G., and G. Erhard. *Designing with Plastics: A Report on the State of the Art.* New York: Hanser Publishers, 1984.

English, Lawrence. "Liquid-Crystal Polymers: In a Class of Their Own," *Manufacturing Engineering,* March 1986, pp 36-41.

English, Lawrence. "The Expanding World of Composites," *Manufacturing Engineering,* April 1986, pp 27-31.

Fitts, Bruce. "Fiber Orientation of Glass Fiber-Reinforced Phenolics," *Materials Engineering,* November 1984, pp. 18-22.

Grayson, Martin. *Encyclopedia of Composite Materials and Components.* New York: John Wiley and Sons Inc., 1984.

Johnson, Wayne, and R. Schwed. "ComputerAided Design and Drafting," *Engineered Systems,* March/April 1986, pp 48-51.

Kliger, Howard. "Customizing Carbon-Fiber Composites: For Strong, Rigid, Lightweight Structures," *Machine Design,* December 6, 1979, pp 150-157.

Levy, Sidney, and J. Harry Dubois. *Plastics Product Design Engineering Handbook.* 2nd ed. New York: Chapman and Hall, 1984.

Lubin, George. *Handbook of Composites. New* York: Van Nostrand Reinhold Company, Inc., 1982.

Modern Plastics Encyclopedia. Vol 63 (10A), October 1986.

Mohr, G., and others. *SPI Handbook of Technology and Engineering of Reinforced Plastics/ Composites.* 2nd ed. Malabar: Robert Krieger Publishing Company, 1984.

Moore, G. R., and D. E. Kline. *Properties and Processing of Polymers for Engineers.* Englewood Cliffs: Prentice-Hall, Inc., 1984.

Naik, Saurabh, and others. "Evaluating Coupling Agents for Mica/Glass Reinforcement of Engineering Thermoplastics," *Modern Plastics, June* 1985, pp 1979–1980.

Plunkett, E. R. *Handbook of Industrial Toxicology.* New York: Chemical Publishing Company, 1987.

Powell, Peter C. *Engineering with Polymers. New* York: Chapman and Hall, 1983.

Richardson, Terry. *Composites: A Design Guide.* New York: Industrial Press, 1987.

Schwartz, Mel - *Fabrication of Composite Materials: Source Book,* American Society for Metals, 1985.

Schwartz, M. M. *Composite Materials Handbook.* New York: McGraw-Hill Book Company, 1984.

Seymour, Ramold B., and Charles Carraher, Poly*mer Chemistry.* New York: Marcel Dekker, Inc., 1981.

Shook, Gerald. *Reinforced Plastics for Commercial Composites: Source Book,* American Society for Metals, 1986.

"Standardization Case Studies: Defense Standardization and Specification Program," Department of Defense, March 17, 1986.

Stepek, J. and H. Daoust. *Additives for Plastics.* New York: Springer Verlag, 1983, p 260.

Von Hassell, Agostino. "Computer Integrated Manufacturing: Here's How to Plan for It," *Plastics Technology Productivity Series, No.* 1, 1986.

Wigotsky, Victor. "Plastics are Making Dream Cars Come True," *Plastics Engineering, May* 1986, pp 19–27.

Wigotsky, Victor. "U.S. Moldmakers Battle Foreign Prices for Survival," *Plastics Engineering,* November 1985, pp 22–23.

Wood, Stuart. "Patience: Key to Big Volume in Advanced Composites," *Modern Plastics,* March 1986, pp 44–48.

Index